BRYOZOAN STUDIES 2001

PROCEEDINGS OF THE TWELFTH INTERNATIONAL BRYOZOOLOGY ASSOCIATION CONFERENCE
DUBLIN / IRELAND / 16-21 JULY 2001

Bryozoan Studies 2001

Edited by

Patrick N. Wyse Jackson
Department of Geology, Trinity College, Dublin, Ireland

Caroline J. Buttler
Department of Geology, National Museums and Galleries of Wales, Cardiff, Wales, UK

Mary E. Spencer Jones
Department of Zoology, The Natural History Museum, London, UK

A.A. BALKEMA PUBLISHERS LISSE / ABINGDON / EXTON (PA) / TOKYO

Cover illustration: *Cribrilina punctata* (Hassall, 1841) a cheilostome bryozoan species first described from Dublin Bay. [Drawing courtesy of Peter Hayward]

Published by: A.A. Balkema Publishers, a member of Swets & Zeitlinger Publishers
www.balkema.nl and www.szp.swets.nl

ISBN 90 5809 388 3

Printed in The Netherlands

Bryozoan Studies 2001, Wyse Jackson, Buttler & Spencer Jones (eds)
© 2002 Swets & Zeitlinger, Lisse, ISBN 90 5809 388 3

Table of Contents

Bryozoan Studies 2001, Wyse Jackson, Buttler & Spencer Jones (eds)
© 2002 Swets & Zeitlinger, Lisse, ISBN 90 5809 388 3

Dedication

This volume is a Festschrift in honour of Richard Stanton Boardman, a founding member of the International Bryozoology Association and pre-eminent scholar of stenolaemate bryozoans. Rich's work forms a highly influential core of research on stenolaemate bryozoans, spanning their entire temporal distribution from Ordovician to Recent.

Rich's entire career has been spent at the Smithsonian's Natural History Building, where he first was employed as a palaeontologist by the U.S. Geological Survey and later was recruited as a Curator in the Department of Paleobiology of the National Museum of Natural History (NMNH). He is currently Curator Emeritus, living in Sarasota, Florida, with his wife Phyllis.

A beneficiary of the American G.I. Bill following World War II, Rich developed a passion for stenolaemate bryozoans as he worked on a Ph.D. dissertation on Devonian trepostomatous bryozoans. The dissertation was published by the U.S. Geological Survey (1960. *Professional Paper* 340). Rich's career-long interest in the processes, patterns, and taxonomic implications of growth in stenolaemates was established during his Ph.D. studies, leading to his first publication (1954. Morphologic variation and mode of growth of Devonian trepostomatous Bryozoa: *Science* 120: 322-323). This was only the first of many papers on aspects of growth in stenolaemate bryozoans, ranging from skeletal microstructure to colony-wide growth patterns. Rich's papers on stenolaemate growth continue and include his most recent publication prior to this volume (2001. *Journal of Paleontology* 75: 225-240) as well as the paper included herein.

Rich's focus on growth has been in part to understand better which characters are most important in taxonomy and phylogeny, and papers on these topics have spanned his entire career, from the U.S. Geological Survey monograph on Devonian trepostomatous bryozoans of New York State, to a large paper on morphology, anatomy, evolution, and classification of stenolaemates (1998. *Smithsonian Contributions to Paleobiology* 86), and continuing in his work on revision of trepostomatous bryozoans for the *Treatise on invertebrate paleontology*.

Overcoming the limited ways in which stenolaemate bryozoans were studied has also gotten Rich's attention, and he has introduced important improvements in techniques of study, in collaboration with students and technicians at the NMNH. His collaboration with Donald Dean was especially productive. Rich introduced the use of impressions, on thick acetate sheets, of highly polished and etched surfaces of Palaeozoic stenolaemates as a more rapid supplement to or replacement for thin sections. For the study of living stenolaemates, which formerly were studied either as stained soft tissues after dissolution of the skeletal material or as cleaned skeletons after soft tissues were bleached away, he and Don Dean demonstrated beautifully that epoxy could be driven into the tissues of modern bryozoans and then thin sections prepared with the soft tissues attached to and suspended across and within the skeleton in their life position. For the first time, extensive direct observation of the skeletal-soft tissue relationship could be made rather than inferred.

Also deriving ultimately from his interest in growth of zooids and entire colonies, Rich and his NMNH colleague Alan Cheetham, along with Pat Cook (all three of whom were founding members of the IBA), explored the meaning, range, and implications of coloniality, not just in bryozoans but in all colonial invertebrates. This collaboration, largely in the late 1960s and early 1970s, resulted in several seminal papers and a book (1973. *Animal colonies*, edited by R.S. Boardman, A.H. Cheetham, and W.A. Oliver, Jr) that played a large role in the renaissance in the study of colonies, colonial integration, and the function of invertebrate animal colonies.

There is much to celebrate in the focus, quality, and timeliness of Rich's research that has been briefly outlined above, as well as his service to palaeontology as co-editor of additional books aimed at a broad audience of palaeontologists and students. It is with pleasure that we dedicate this volume to his past, present, and future research accomplishments.

Bryozoan Studies 2001, Wyse Jackson, Buttler & Spencer Jones (eds)
© 2002 Swets & Zeitlinger, Lisse, ISBN 90 5809 388 3

Foreword

This volume contains forty nine papers presented at the 12th International Conference of the International Bryozoology Association held in the Department of Geology, Trinity College, Dublin from Monday 16th to Friday 20th July 2001. This meeting was attended by 108 delegates together with 35 accompanying members from 27 countries, making it the largest IBA meeting by attendance to date. A submeeting, co-sponsored by the Society for the History of Natural History, was convened on Saturday 21st July and was devoted the history of bryozoology. These papers will be published separately by the Association in 2002.

The scientific programme of the conference was spread over four days. The Conference was presided over by Professor Ken McKinney (Applachian State University, North Carolina) and the conference host was Dr Patrick Wyse Jackson (Trinity College, Dublin). 85 oral presentations and 16 poster presentations were made, which covered all aspects of bryozoological research. The Gilbert Larwood Award for the best first time contribution to an IBA conference was made to Africa Gómez of the University of Hull. In her paper on *Celleporella* and contribution of sibling speciation to marine diversity, she demonstrated genetic differentiation using nuclear and mitochrondial genetic markers.

On the evening of the first day, delegates and accompanying members attended a Civic Reception hosted by the Deputy Lord Mayor of Dublin. In the beautiful surroundings of the 18th century Mansion House, the residence of the Lord Mayor, the IBA was officially welcomed to Dublin. Traditionally the delegates enjoy a mid-Conference excursion. On Wednesday 18th July after a visit to the Old Library in Trinity College, Dublin and a viewing of the celebrated Book of Kells, the group headed south into the delightful surroundings of County Wicklow and stopped at the monastic site at Glendalough. After an enjoyable lunch at the Glendalough House Hotel, delegates visited the remains of the 10th century monastic complex complete with a fine example of an Irish Round Tower. Others treked through the glaciated valley to visit the remains of an 18th century mining site, where the base metal minerals galena and sphalerite were collected from abandoned spoil heaps. Bryozoological research for some was not abandoned during this day of leisure: some sampling of the lakes for freshwater bryozoans was undertaken.

The conference was concluded with the Business Meeting during which Dr Dennis Gordon was elected President for the next three year cycle and the venue for the 13th IBA Conference was confirmed. We look forward to meeting next in Concepción, Chile. The Conference Dinner took place at Johnnie Fox's Pub in the Dublin Mouintains where after a good dinner all were treated to a hooley of traditional Irish music and dance, and to a rendition of Waltzing Matilda from Robin Wass. It was a measure of how charged the atmosphere in the pub had become by the end of the evening, that most delegates had abandoned their seats and had taken to the aisles which served as makeshift, if somewhat narrow, dancefloors.

Unfortunately both the pre-Conference geological and post-Conference zoological field excursions had to be cancelled due to the outbreak of foot and mouth disease in the United Kingdom. As a result of this problem movement by groups and scientific students throughout the Irish countryside was severely restricted by the Irish Government.

The IBA conference was a resounding success and as host Patrick Wyse Jackson warmly thanks all those who helped in its organisation. Caroline Buttler and Mary Spencer Jones were responsible for the distribution of submitted papers to reviewers, and with Patrick Wyse Jackson have undertaken the editing of this volume. We thank all those who reviewed papers and in particular the contributing authors for their speed in returning manuscripts to comply with the short deadline. We are pleased with the relationship with A.A. Balkema of Rotterdam, the publishers of this volume. Accommodation was provided in Trinity College and in Mercer's Court and we thank the staff of their respective Accommodation Offices, and in particular Sarah Campbell for their assistance. The Technical Staff and Post-Graduate Students of the Department of Geology provided expert audio-visual facilities and refreshments. For this we thank Neil Kearney, Francis Hendron, Declan Burke, Peter Keogh, John Murray, Claire Mulhall, Rebecca Gould, Mags Duncan and Paul Grehan.

The International Bryozoology Association thanks the sponsors of the 12th conference whose subventions were most valuable: the Department of Geology, Trinity College, Dublin, the Geological Survey of Ireland, and Marine Institute - Foras na Mara, Dublin. Considerable assistance was also forthcoming from the Convention Bureau of Ireland and Bord Failte.

Patrick Wyse Jackson
Caroline Buttler
Mary Spencer Jones

18th December 2001

Bryozoan Studies 2001, Wyse Jackson, Buttler & Spencer Jones (eds)
© 2002 Swets & Zeitlinger, Lisse, ISBN 90 5809 388 3

Bryozoan communities in the Weddell Sea, Antarctica: a first overview

Beate Bader
Institut für Geowissenschaften der Universität Kiel, Olshausenstraße 40, D-24098 Kiel, Germany

ABSTRACT: During *Polarstern* cruise ANTXVII/3 bryozoans were sampled in the eastern Weddell Sea and the Antarctic Peninsula. For the Weddell Sea 144 species were previously known and 12 species could be added to the taxonomic list. Reproductive stages were observed in 20 species, and it was shown that autumn is a reproductive season for bryozoans in the Weddell Sea. Within the growth forms the erect skeletal morphology dominates. In situ photographs demonstrate that iceberg grounding causes considerable damage to the benthic communities. Sessile organisms are eradicated and pioneer species begin to grow in high abundances on the devastated substratum. A preliminary analysis shows that *Cellarinella* spp. and *Melicerita obliqua* (Thornely, 1924) are the first species which recolonise the highly disturbed areas.

1 INTRODUCTION

The Weddell Sea is one of the most "polar" parts of the Southern Ocean (Hempel 1992), and its southern part represents one of the largest shelf areas around the Antarctic continent. Most of the Weddell Sea is covered by sea ice all year round. Along the eastern coast, the sea ice breaks up frequently, resulting in large polynyas. In summer the coastal polynya houses an important phytoplankton bloom. Overall primary production in the Weddell Sea is low and mainly limited to the very short summer period (Hempel 1992).

Faunistically, the benthos and fish fauna of the Weddell Sea are well known (Voß 1988, Barthel & Gutt 1992, Gutt et al. 1994). A detailed community analysis has been published by Voß (1988), who separated three different communities. The eastern shelf community is rich in suspension feeders, mainly sponges and bryozoans. Bryozoans are the principal component of the southern shelf community, whilst holothurians dominate the southern trench.

Bryozoans are a major component of the Antarctic benthos in many of the areas that have been studied (Bullivant 1961, 1967, Winston 1983, Voß 1988, Winston & Heimberg 1988, Galeron et. al. 1992, Barnes1995a, b, Barnes & Clarke 1998, Moyano & Ristedt 2000). The last few decades have seen a dramatic explosion of taxonomic and ecological work in southern polar waters. Many new Antarctic cheilostomatid bryozoan species and genera have been described, whilst the taxonomic status of many others has been elucidated (see Hayward 1995 and references therein). In contrast with other areas of Antarctica the bryozoan fauna of the shelf of the Weddell Sea is not well known. A first description of the bryozoan fauna was given by Ristedt (1995) and a first taxonomic list by Zabala et al. (1997).

The aim of the present paper is to present a first overview for the bryozoans of the Weddell Sea. These include the (1) taxonomic work, (2) growth forms, (3) reproduction season and growth, and (4) recolonisation of disturbed areas.

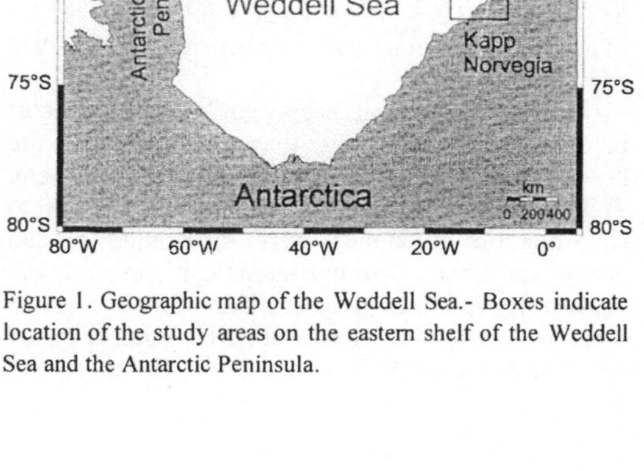

Figure 1. Geographic map of the Weddell Sea.- Boxes indicate location of the study areas on the eastern shelf of the Weddell Sea and the Antarctic Peninsula.

2 MATERIAL

During the *Polarstern* cruise ANTXVII/3, from 17th March to 11th May 2000, bryozoans were sampled in the eastern Weddell Sea and the Antarctic Peninsula in 100 – 800 m water depth (Fig. 1). Bryozoans were mainly collected, together with other benthic organisms in general, by means of TV grab, giant box corer (GKG), bottom trawl (GSN) and Agassiz trawl (AGT). Only the vertical grabs (GKG, TV grab) sampled entire colonies; in the GSN all colonies were broken. Special attention was paid to the study of colours, which can help in the discrimination of species but are quickly lost after fixation, so each species was photographed with a digital camera directly on board. Reproductive status was studied under the microscope by noting the presence of embryos.

For identification specimens were observed with optical and scanning electron microscopy (CAM-SCAN-Serie-2-CS-44). SEM observations were made by cleaning with sodium hypochlorite and coating in gold-platinum.

To study different stages of colonization an underwater photograph transect taken off Kapp Norvegia was analysed. The photographs were done by Julian Gutt from the Alfred-Wegerner Institut, Bremerhaven, Germany during *Polarstern* cruise 1996 and show a total of 80 images in a length of 300 m 15 images with an interval of approximately twenty metres were selected and for each picture bryozoans were identified; if possible as taxa otherwise as genus. Then the size of single colonies or clusters were measured in percent. In Figure 3 the total amount of bryozoan cover for each photograph is represented.

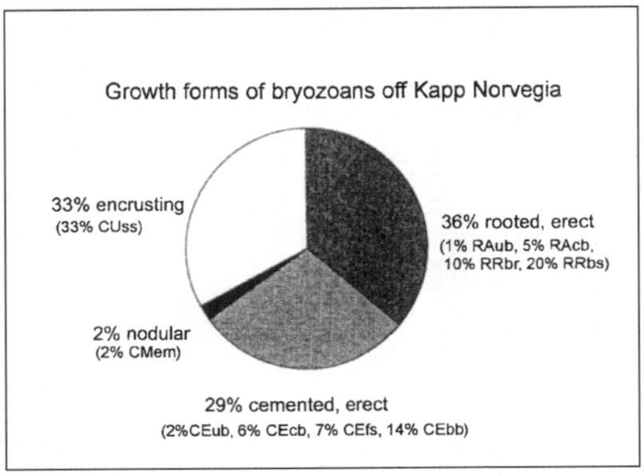

Figure 2. Growth forms of bryozoan species in the Weddell Sea (after Hageman et al. 1997). Abbreviations: Rooted, articulated, unilaminar branches (RAub), Rooted, rigid, bilaminar sheet (RRbs), Rooted, articulated, cylindrical branches (RAcb), Rooted, rigid, branches (RRbr), Cemented, erect, bilaminar branches (CEbb), Cemented, unilaminar, solid substrate (CUss), Cemented, erect, cylindrical branches (CEcb), Cemented, erect, unilaminar sheet (CEub), Cemented, multilaminar, encrusting massive (CMem), Cemented, erect, fenestrate sheet (CEfs).

3 TAXONOMY

In Hayward's *Synopsis* (1995) about ten species were described, which also occur in the Weddell Sea. First taxonomic studies for the Weddell Sea were done by Zabala et al. (1997) and Gontar (pers. comm.) who worked with bryozoan material from *Polarstern* cruise ANTXIII/3. Zabala et al. found 122 species and Gontar another 22 species for the Weddell Sea, which agree with the described species of Hayward (1995). Taxonomic work in this study of bryozoan material of *Polarstern* cruise ANTXV/3 adds 12 species.

These first taxonomic works on bryozoan species in the Weddell Sea show that nearly half of the known Antarctic cheilostome bryozoans occur there. This verifies the assumption of Hayward (1995) that all Antarctic bryozoans were distributed around Antarctica. Four bryozoogeographical zones are recognised in Antarctic waters: East Antarctic seas, West Antarctic seas, the Antarctic Peninsula und Scotia Arc (Barnes & De Grave 2000). Based on statistical analysis of all known cheilostome species for the southern polar region or the area south of 47° the species revealed two homogenous site groupings, corresponding to an Antarctic and a Magellanic group, and a third much more heterogeneous Subantarctic Island group. Ordination of just the Antarctic grouping shows that the Weddell Sea occupies an extremely isolated position (Barnes & De Grave 2000).

4 GROWTH FORMS

Within the bryozoan taxa of the Weddell Sea the diversity of skeletal morphology is large. All growth forms are present: encrusting sheets (e.g. *Escharella watersi* (Waters, 1904)) to erect flexible (e.g. *Cellaria incula* (Hayward & Ryland, 1993)), erect rigid (e.g. *Cellarinella nodulosa* (Waters, 1904)), erect unilaminar (e.g. *Smittoidea malleata* (Rogick, 1956)) and erect bilaminar (e.g. *Isoschizoporella secunda* (Hayward & Thorpe, 1984)) colonies. The erect cheilostomate taxa (65 %; 36 % of them being species with roots) predominate within the bryozoan fauna (Fig. 2). Also Winston & Heimberg (1988) found the same pattern at Low Island (South Shetland Islands) and Barnes (1995a) at Signy Island (North Weddell Sea).

The surface sediments on the Weddell Sea shelf consists of soft bottom with varying proportions of mud, sand, gravel and boulders; distribution patterns can be regarded as considerably heterogeneous

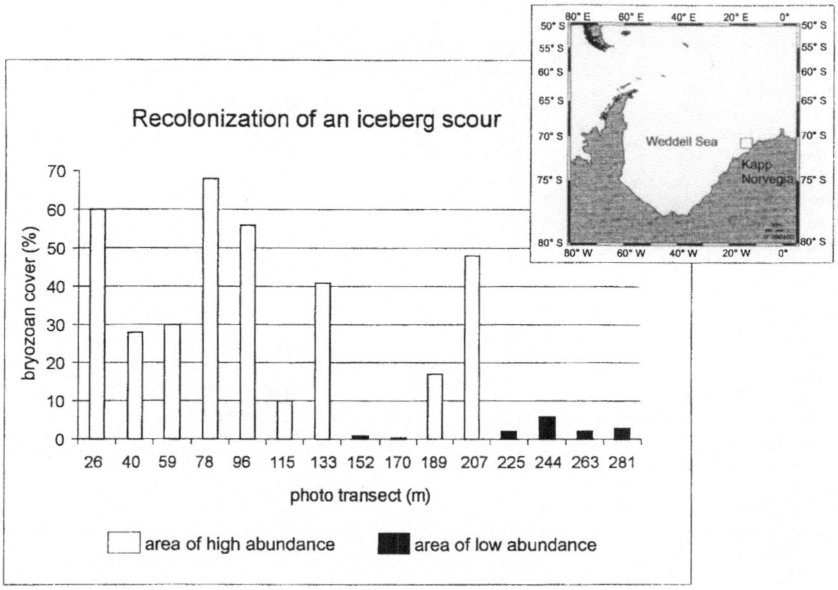

Figure 3. Distribution of bryozoan cover in the underwater photograph transect 006 P, ANTXVIII/3 in 230 m water depth.

(Gerdes et al. 1992). The appearance of coarser sediment particles like sandy grains allow larger, weak calcified bryozoan species with roots to settle. Large colonies with rooted, rigid growth forms dominate in areas with fine sediment and pebbles, which were almost completely covered by sediment. If stones are only partially covered by sediment, small, rigid colonies with a laminar growth form and small, encrusting species can settle. In pebbled areas large, rigid colonies dominate while rooted species are seldom.

5 REPRODUCTION AND GROWTH

Almost all species showed reproductive structures (ovicells or gonozooids) and many were observed to be fertile. Not all species were easy to check for the presence of growing embryos because, like *Cellarinella* species, the ovicells were not visible externally being immersed in the frontal wall and heavily calcified. Observations show that 20 species are reproductive in autumn, 33 fertile species were found in summer (Zabala et al. 1997) (Table 1). Also Barnes & Clarke (1998) found sexually reproductive zooids in *Isoseculiflustra rubefacta* (Kluge, 1914), *Nematoflustra flagellata* (Waters, 1904), *Himantozoum antarcticum* (Calvet, 1905) (north of the Weddell Sea) which were formed in late summer and the larvae were brooded for ±10 months before being re leased in January/February (*Nematoflustra flagellata*) or February/March (*Himantozoum antarcticum*). Hence, contrary to the suggestion of Zabala et al. (1997) autumn is also a reproductive season for several species in the Weddell Sea.

An interesting aspect of propagation is the asexual and sexual reproduction of bryozoan species. Some *Cellarinella* species are able to originate from broken colony fragments and asexual propagation in these species is more significant in maintaining the population than sexual propagation (Winston 1983). For sponges an asexual propagation is supposed in iceberg disturbed areas (Arntz pers. comm.) but there is no evidence of asexual propagation for bryozoans in such intense disturbed areas. Observations of underwater videos and samples from TV-grabs show that *Cellarinella* species are one of the first bryozoans to colonise iceberg disturbed areas. Their distribution is assumed to originate from sexual production when larvae settle. All colonies of the same species were of nearly the same height and appearance. If specimens originated from broken colony fragments, it is likely that the colonies would have different heights, and ages ranging from large old colonies to small young ones.

Skeletal growth check lines occur in different bryozoan species of the Weddell Sea. *Cellarinella* spp. and *Melicerita obliqua* which are one of the conspicious organisms of the Antarctic shelf show a distinctive nodular appearance. Stable isotope studies an broken segments of *Melicerita obliqua* indicate segment formation during periods of high primary production (austral summer), and node formation during periods of winter (Brey et al. 1998). This annual formation of growth bands has been also veri-

Table 1. Reproductive status and color of embryos.

Cheilostomata	summer*	autumn**	colour of embryos
Klugeflustra antarctica (Hastings)	x		
Isosecuriflustra angusta (Kluge)		x	white
Isosecuriflustra tenuis (Kluge)	x	x	light beige
Isosecuriflustra thysanica (Moyano)		x	white
Austroflustra vulgaris (Kluge)		x	white
Amphiblestrum inermis (Gray)	x		
Amphiblestrum irossi Hayward and Thorpe	x		
Camptoplites latus (Kluge)	x		
Cornuplites pectogemma (Goldstein)	x		
Cornuplites polymorpha (Kluge)	x		
Klugella echinata (Kluge)	x		
Notoplites drygalskii (Kluge)	x		
Cellaria diversa Livingstone		x	red
Cellaria moniliorata (Rogick)	x		
Swanomia membranacea (Thornley)		x	orange
Melicerita flabellifera Hayward and Winston		x	light beige
Melicerita obliqua (Thornley)		x	light beige
Arachnopusia latiavicularis (Moyano)	x		
Arachnopusia avicularia Hayward and Thorpe	x		
Astochoporella cassidula Hayward and Thorpe	x	x	light beige
Escharoides praestita (Waters)	x		
Exochella hymanae (Rogick)		x	orange-red
Escherella watersi (Hayward and Thorpe)	x		
Lageneschara lyrulata (Calvet)	x	x	white
Cellarinella watersi Calvet	x		
Cellarinella njegovanae Rogick		x	white
Kymella polaris (Waters)		x	white
Buffonelloides antarctica Hayward	x		
Lacerna watersi Hayward and Thorpe	x		
Smittina alticollaria Rogick	x		
Smittina anecdota Hayward and Thorpe	x	x	orange-red
Smittina antarctica (Waters)	x		
Smittina rogickae (Hayward and Taylor)	x		
Smittina ornatipectoralis (Rogick)			
Smittoidea albula Hayward and Taylor	x		
Smittoidea malleata Hayward and Thorpe	x	x	red
Thypticocirrus rogickae Hayward and Thorpe		x	red
Thypticocirrus phylactelloides (Calvet)	x		
Pemmatoporella marginata (Calvet)	x		
Reteporella gelida (Waters)	x		
Reteporella frigida (Waters)	x		
Reteporella antarctica (Waters)	x		
Reteporella lepralioides (Waters)	x		
Reteporella hippocrepis (Waters)	x	x	white
Orthoporida compacta (Waters)		x	red
Bostrychopora dentata (Waters)	x	x	white

* Zabala et al. 1997 **own data

fied for *Cellarinella watersi* at Signy Island (Barnes, 1995b).

6 ICEBERG SCOURS

Apparent irregularities in the benthic structure were identified as iceberg scour marks if an abrupt change in the epibenthic structure could not be explained by other physical or biological processes. First analysis of an underwater photo transect west of Kapp Norvegia in 230 m water depth showed a changing gradient of colonization: (1) areas with a poor cover, and (2) areas with a high cover of epibenthic species (Figure 3). In these communities either ascidians or bryozoans and sponges dominate. A first stage of recolonisation was defined by having only a poor fauna composed of different taxa, or by being dominated by one or two highly abundant pioneer species. In the poor covered areas, ascidians dominate the sessile benthos community with a high abundance (Figure 4). Within the bryozoans only a few single species occur; *Cellarinella* spp. and *Melicerita obliqua* being characteristic. Areas with high bryozoan cover represent a later stage of succession (Figure 5). At this later stage erect, flexible and erect,

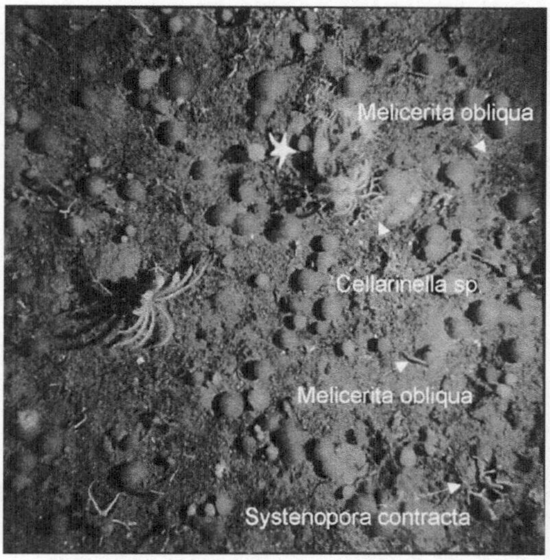

Figure 4. Area with a high colonization of bryozoans.

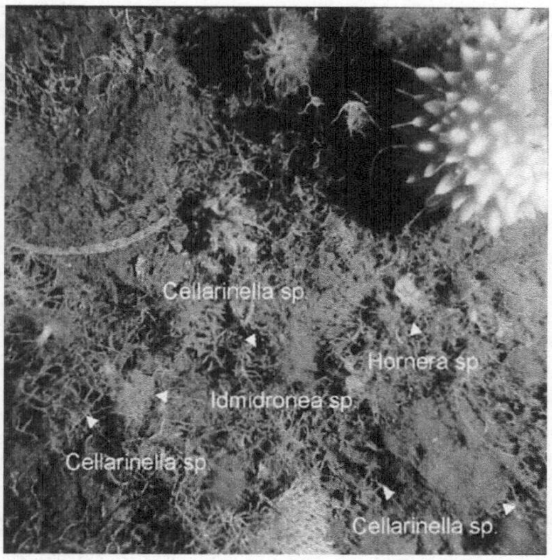

Figure 5. Area with a poor cover of bryozoans. Ascidians dominate the epifauna.

rigid colonies are common; mainly forms with roots such as *Cellarinella* spp., *Melicerita obliqua* and *Cellaria* spp. In shelf areas disturbed by iceberg grounding, *Cellaria* is a typical pioneer species (Gutt et al. 1996). However, areas with a high cover of bryozoans do not portray a mature and diverse community, which consists mostly of sessile suspension feeders like gorgonarians, ascidians and ophiuriods (Starmans 1997).

7 CONCLUSION

The *Polarstern* cruise ANTXVII/3 allowed the study of bryozoan community of the Weddell Sea. The taxonomic work shows that the number of species increase with each new sample and also means that around the one-half of the bryozoan fauna as yet known for the whole Antarctic are present in the Weddell Sea.

All growth forms are represented but the erect species are the dominant part in the benthic communities. Furthermore the distribution of the species with their growth forms is addicted to the substrate. One aspect of future work is to study the structural potential of bryozoan communities in forming different habitats, and the special settlement strategies of bryozoans in bryozoan dominated communities. By means of these results the structure and history of these communities should be evaluated.

The colonies of *Melicerita obliqua* and *Cellarinella* spp. appear nodular in outline. Stable isotope studies on selected species with entire colonies should allow the interpretation of their life cycle and age.

First analysis of iceberg disturbed areas show different stages of recolonisation. In a first stage with a poor colonization of benthic organisms *Cellarinella* spp. and *Melicerita obliqua* are the first species which are able to occupy spatial niches produced by iceberg grounding on the Antarctic shelf. In a second stage the abundance of bryozoan species increase and erect, flexible growth forms are common. The study of the change between destruction and recolonisation allows testimonies of the development of the benthic communities and also the time span of development.

ACKNOWLEDGEMENTS

This research was supported by the Deutsche Forschungsgemeinschaft project Scha 355/18-2 "Ecology and phylogeny actuopaleontology of bryozoans of Antarctica and Arctic". I thank J. Gutt who make the underwater pictures available and N. Teixido for discussion. I would also like to thank D.K.A. Barnes and an anonymous reviewer for revision of early draft.

REFERENCES

Bader, B. 1998. Temperierte Schelfkarbonate des Holozäns der Nordbretagne: Sedimentologischen und ökologische Prozesse zur Bildung von Bryozoe/Mollusken-Sedimenten (BRYOMOL-Fazies). *Berichte-Reports, Geologisch-Paläontologisches Institut der Universität Kiel* 89: 1-204.

Barnes, D.K.A. 1995a. Sublittoral epifaunal communities at Signy Island, Antarctica. II. Below the ice-foot zone. *Marine Biology* 121: 565-572.

Barnes, D.K.A. 1995b. Seasonal and annual growth in erect species of Antarctic bryozoans. *Journal of Experimental Marine Biology and Ecology* 188: 181-198.

Barnes & Clarke 1998. Seasonality of polypide recycling and sexual reproduction in some erect Antarctic bryozoans. *Marine biology* 131: 647-658.

Barnes, D.K.A. & De Grave, S. 2000: Biogeography of Southern Polar Bryozoans. *Vie et Milieu* 50(4): 261-273.

Bullivant, J.S. 1961: Photographs of Antarctic bottom fauna. *Polar Research* 10: 505-508.

Bullivant, J.S. 1967: Ecology of the Ross Sea. *New Zealand Department of Scientific and Industrial Research* 176: 49-75.

Barthel, D. & Gutt, J. 1992. Sponge associations in the eastern Weddell Sea. *Antarctic Science* 4: 137-150.

Brey, T., Gutt, J., Mackensen, A. & Starmans, A. 1998: Growth and productivity of the high Antarctic bryozoan *Melicerata obliqua*.- *Marine Biology* 132: 327-333.

Eggelston, D. 1972. Patterns of reproduction in the marine Ectoprocta of the Isle of Man. *Journal of Natural History* 6: 31-38.

Galeron, J., Herman, R.L., Arnoud, P.M., Arntz, W.E., Hain, S. & Klages, M. 1992: Macrofaunal communities on th continental shelf and slope of the southeastern Wddell Sea, Antarctica. *Polar Biology* 12: 283-290.

Gerdes, D., Klages, M., Arntz, W.E., Herman, R.L., Galeron, J. & Hain, S. 1992. Quantitative investigations on macrobenthos communities of the southeastern Weddell Sea shelf based on multibox corer samples. *Polar Biology* 12: 291-301.

Gutt, J., Ekau, W. & Gorny, M. 1994. New results on the fish and shrimp fauna of the Weddell Sea and Lazarev Sea (Antarctic). *Proceedings of NIPR Symposium of Polar Biology* 7: 91-102.

Gutt, J., Starmans, A. & Dieckmann, G. 1996: Impact of iceberg scouring on polar benthic habitats. *Marine Ecology Progress Series* 137:311-316.

Hageman, S.J., Bone, Y., McGowran, B. & James, N.P. 1997. Bryozoan colonial growth-forms as paleoenvironmental indicators: evaluation of methodology. *Palaios* 12: 405-419.

Hayward, P.J. 1995. *Antarctic Cheilostomatous Bryozoa*. Oxford, Oxford Universtiy Press.

Moyano, H.I. & Ristedt, H. 2000: The Sclerodomidae (Bryozoa) of the eastern Weddell Sea collected during the Antarctic I, II, V expeditions of the R/V Polarstern. Gayana 64(1): 83-107.

Ristedt, H. 1995. Die Bryozoen-Fauna der östlichen Weddel-See. *Berichte zur Polarforschung* 155: 75-78.

Hempel, G. 1992. The Weddell Sea, A high polar ecosystem. In K. Sherman, L.M. Alexander & B.D. Gold (eds), Large marine ecosystems: 5-18. Washingto D.C., American Association for the Advancement of Science.

Starmans, A. 1997. Vergleichende Untersuchungen zur Ökologie und Biodiversität des Mega-Epibenthos der Arktis und Antarktis. *Beriche zur Polarforschung* 250: 1-150.

Voß, J. 1988. Zoogeographie und Gemeinschaftsanalyse des Makrozoobenthos des Weddelmeeres (Antarktis). *Berichte zur Polarforschung* 45: 1-145.

Winston, J. 1983. Patterns of growth, reproduction and mortality in bryozoans from the Ross Sea, Antarctica. *Bulletin of Marine Science* 33(3): 688-702.

Winston, J.E. & Heimberg, B.F. 1988. The role of bryozoans in the benthic community at Low Island, Antarctica. *Antarctic Journal of U.S* 21: 188-189.

Zabala, M., Orejas, C. & Alva, V. 1997. Bryozoans of the Weddell Sea. *Berichte zur Polarforschung* 249: 55-71.

Bryozoan Studies 2001, Wyse Jackson, Buttler & Spencer Jones (eds)
© 2002 Swets & Zeitlinger, Lisse, ISBN 90 5809 388 3

Modelling multivariate determinants of growth in Antarctic bryozoans

David K. A. Barnes
British Antarctic Survey, High Cross, Madingley Road, Cambridge, UK

Sammy De Grave
Oxford University Museum of Natural History, Parks Road, Oxford, UK

ABSTRACT: Bryozoans seem fairly typical amongst Antarctic benthos, in both the timing and magnitude of growth. Growth varies interspecifically from nearly comparable to much slower than warm water equivalent species. Factors including species identity, age, year, competition, substratum stability and locality have been suggested as influential. We analysed 5000 colonies of three of the most ubiquitous cheilostomatids in shallow Antarctic waters, *Celleporella bougainvillei, Fenestrulina rugula* and *Inversiula nutrix*. Substratum surface area and depth were treated as separate factors, but both act as proxies for substratum stability. Intra- and inter-specific competition were also treated as separate factors. Growth of each species was scored as the number of zooids per year, due to high seasonal variability. The relative importance of factors was ranked Age > Inter-specific competition > Intra-specific competition/ latitude > species identity/ depth/ rock surface area. So for each age-class competition for space was the most dominant force shaping their growth. Even in their first year, most colonies of the three study species experienced interference competition for space. Inter-specific competition typically reduced colony size as the study species were poor competitors. Intra-specific competition slows expansion rate by restricting the growing margins of colonies. Latitude probably has no direct effect but may have many indirect influences such as altering the light dependent primary productivity, the frequency of ice-mediated disturbance, the suite of competitors and frequency of competition. We suggest that the high number of model terms significant and the high level of overall model significance not only validates this multivariate approach but indicates it may be a widely appropriate method to increase power and resolution in ecological investigations.

1 INTRODUCTION

Growth rate in animals varies with a bewildering array of factors. Some factors or combinations of factors are, however, so over-bearing that virtually all animals, irrespective of lifestyle or taxon, are similarly influenced. The polar nearshore marine environment is such a place and marine invertebrates which live there nearly all grow slowly (Everson 1977, Stockton 1984, Berkman 1990, Arntz et al. 1994, Brey et al. 1995, Peck et al. 1997). Growth rates are quite variable, however, both within and between taxa. Some grow very slowly compared to ecologically similar species in warm waters whilst others are only a little slower or comparable (Arntz et al. 1994, Barnes 2000a). A few (some sponges and an ascidian) are capable of fast growth (Dayton et al. 1974, Rauschert 1991). Polar animals are not small though; some of the representative species attain larger sizes than elsewhere in the worlds water masses, largely due to increased oxygen availability (Chapelle & Peck 1999). Increased maximum size may not be limited to solitary forms, some colonial animals such as the cheilostomatid *Arachnopusia inchoata* reach >1 m in diameter, making them the largest living bryozoans (pers. obs., cf. Cocito et al. 2000). As with almost anywhere in shallow water, there are, in polar waters large differences between the size of species within the same taxon. The growth rates of such species, and the module size of colonies, may or may not differ but the largest or oldest survivors differ by several orders of magnitude. In Antarctica, colonies of the bryozoan *A. inchoata* grow to 10s of cm, whilst *Celleporella bougainvillei* rarely exceeds a few cm (see Barnes 2000a). Growth of southern polar bryozoans has now been quantified in a number of quite different species (see Winston 1983, Brey et al. 1999, Barnes 2000a, Barnes & Arnold 2001) and our knowledge is approaching that for other parts of the globe. Little remains known on causative or influential agents involved in growth.

Although growth within polar waters seems to be slow, it is not uniformly so, there are likely to be patterns and processes influencing rate in the Southern Ocean (or arctic seas). Bryozoans are a particularly appropriate taxon to analyse determinant of growth with for a number of reasons. Firstly they are highly abundant and ubiquitous in shallow water on a local, regional and global scale. Secondly their sessile and skeletal (in the case of cheilostomes) nature facilitates collection and historical investigation. The history of each colony is recorded in bryozoans (from seasonal environments) as growth check lines. Furthermore deformation of zooids on experiencing competition is recorded in their growth patterns. Perhaps surprisingly, growth tempo of two species of bryozoans, *Celleporella bougainvillei* and *Inversiula nutrix* was recently found to increase towards the pole (Barnes & Arnold 2001). Differential interference competition was suggested as a causative agent. When many factors or parameters are clearly influencing or have the potential to influence an animal's performance, meaningful analysis needs to be inclusive. Log-linear modelling has often been used for quantitative, but categorised, data. This can simultaneously assess the influence of all possible combinations of variables measured and is particularly useful in highlighting less obvious factors. Still more resolution can be gained by collecting non-categorised quantitative data.

Multivariate approaches are now widely used in the analyses of communities and habitat attributes. Although initially these analytical methods were used in a descriptive sense to delineate communities and species assemblages, they have proven powerful tools to analyse larger data sets in which several variables are simultaneously operating. However, these methods have not generally been applied to help explain more specific influences in growth studies, where normally only one influential variables is studied. Nevertheless, it has been recognised that growth is regulated by a suite of variables, making the use of multivariate methods necessary. Here we construct a multiple regression model, following transformation of the variables by PCA. The objective of this technique is to assess the relative statistical importance of the variables and their interactions, rather than using the derived model in a predictive sense. Such multivariate approaches have been pioneered in terrestrial and marine botanical studies (e.g. Santos 1993). We investigate three species which have similar distributions, ecological ranges, growth format and are all fairly poor spatial competitors. We hypothesis that 1) there is no effect of species identity, 2) there is no significant difference between biological and abiotic factors within each species, 3) there is no significant difference between biological and abiotic factors with age, and, finally 4) interspecific competition has more negative impact on growth than intra-specific encounters as within-species meetings generally result in perseverance (ties or stand-offs, see Barnes 2000a).

2 METHODS

2.1 *Study sites*

Rocks from cobble to boulder size on the Wentworth scale were collected from rocky shores at different latitudes. The major localities included South Georgia (54°S), South Orkney Islands (60°S), and islands of the Antarctic Peninsula (63-68°S) which are shown in Figure 1 (labelled 1, 2 and 3 respectively). Collection from 55-58°S is difficult as the only land is the South Sandwich Islands which are very exposed and difficult to access. Similarly there is little land from 61-62°S. All study localities were enclosed within the Polar Frontal Zone or Antarctic Convergence (marking the 'boundary' of the Southern Ocean) and experience scour from sea ice and negative sea temperatures. Although sea temperatures and salinities are similar between these regions the extent and frequency of disturbance mediated by floating ice differs (Barnes 2000a).

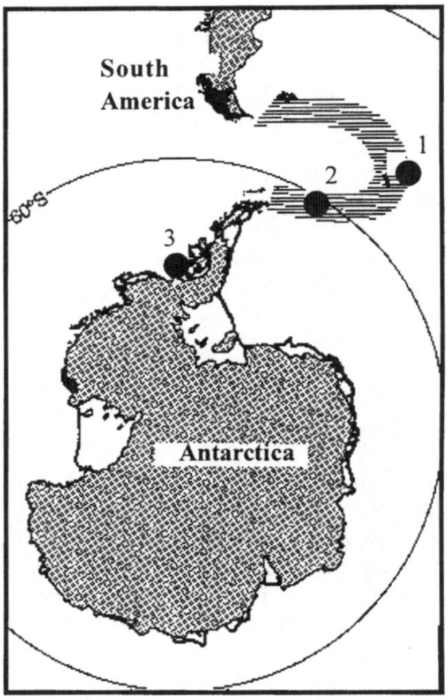

Figure 1. The position of the three study regions in the Southern Ocean relative to the continents of Antarctica, South America and the adjoining mountain chain: the Scotia Arc (shown line-shaded).

2.2 *Species*

The three study species are ubiquitous in shallow Antarctic seas (Hayward 1995). *Celleporella bougainvillei* (d'Orbigny), *Fenestrulina rugula* Hayward & Ryland and *Inversiula nutrix* Jullien were both common and abundant on substrata at most sample localities. All three species are easily identifiable, both alive or dead and do not resemble each other. There are, however, numerous species of *Celleporella* and *Fenestrulina* in Antarctic waters. All are lithophyllic species, predominantly occurring on the lower surfaces of rocks (see Barnes 2000a). The age of each species can usually be determined by faint growth check lines which are concentric and visible in the colony exterior. These check lines have been shown to be annual in periodicity (Barnes & Arnold 2001). The number of zooids in each colony was counted under a binocular microscope and the proportional area of the colony margin occluded by competitors was measured. Competitive interactions were recorded when there was physical contact between two or more colonies and encounters were classed as either intra-specific, inter-specific or both.

2.3 *Statistical techniques*

Species specific data sets were constructed, incorporating a measure of growth (number of modular units or zooids created per year) and the associated variables: latitude, rock surface, rock area, rock depth, age and a measure of intra- and inter-specific competition. All variables, with the exception of the growth term were transformed by means of a varimax rotated Principal Component Analysis (PCA). The main reason for transforming a set of variables into principal components, is that PCA substitutes orthogonal linear combinations of variables for the original variables, essentially removing colinearity, whilst maintaining the original multivariate data structure. Furthermore, the uncorrelated nature of the resulting variables makes it easier to assess the relative importance of predictors entered into the final multiple regression model, as the order of entering does not affect the regression model. PCA biplots were constructed as an aid in data interpretation, although the final discussion was solely based on the multiple regression model.

Growth itself was arcsin transformed before entering in the multiple regression model. In the multiple regression model, each original variable was replaced by the PCA score with which that variable was highly correlated. Because of the varimax rotation, only one original variable was strongly correlated with a given PCA score variable. In the multiple regression model, all variables were simultaneously entered, and their partial correlation coefficients extracted, as well as overall goodness-of-fit measures of the model. Partial correlation coefficients take account of the variability induced by the other variables and effectively partial this out. Therefore, a comparison between partial correlation coefficients allows a direct comparison of the effect of any variable in the model with any other, without having to allow for additional variability.

The proportion of colony surfaces occluded by competitors was regressed against colony size for each species. The advantage of simple linear regression was the separation of some potential causes for relationships. Whilst the correlation coefficients from the multiple regression gave the relative significance of different parameters, they do not show the full picture as information on the intensity or causes of influences is limited. For example, intra-specific competition might have a higher significance (than inter-specific competition) due to a greater frequency of occurrence rather than any intrinsic effect. The slope and intercepts of regression lines were compared using General Linear Model, ANalysis Of VAriance (GLM ANOVA).

3 RESULTS

3.1 *Proportional importance of factors*

An example of a varimax rotated PCA biplot is given in Fig. 2, demonstrating the considerable interplay between the variables on bryozoan growth. Tables 1-4 show the multiple regression model parameters (A), their partial correlation coefficients (B) and the variance explained by the overall model and its significance. The results for four models are shown in each table: ages combined, age 1, age 2 and ages 3+ (3-5 combined due to paucity of data). All four models are significant in the species *I. nutrix*, with most variance being explained by the inclusion of age as a parameter (Table 1). Although coefficient for age was larger than for any other parameter (so is the most important explanatory factor) all other parameters are significant. This clearly illustrates the usefulness and power of the multivariate approach (see also Fig. 2). The environmental factors depth (positive) and rock surface area (negative) were the next most important parameters. However in the models

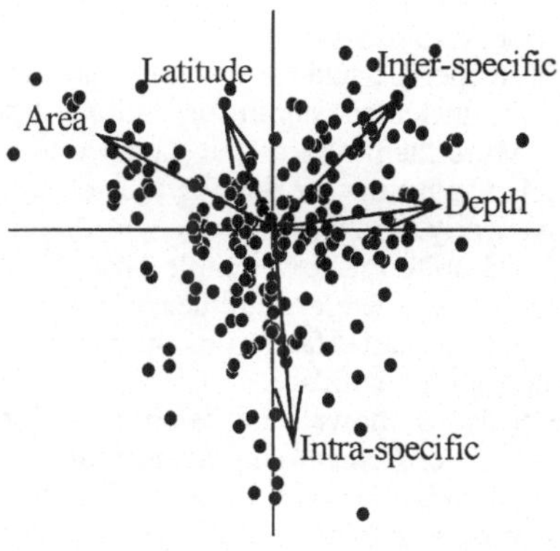

Figure 2. Principal Component Analysis of growth with potentially influential factors of *Inversiula nutrix* (arrows). All year classes.

Table 1. Model parameters, partial correlation coefficients and their significances to growth of *Inversiula nutrix*. Coefficients in bold are significant (P<0.05, Bonferonni corrected). The results of each overall model are shown as ANOVA F ratios, their significance (** = P<0.001, * = P<0.05, NS = Not significant) and variance explained (adjusted r^2 value). Significant partial correlation coefficients (P<0.05) are shown in bold. Abbreviated terms are rock surface area (S. Area) and inter-specific competition (Inter-S) and intra-specific competition (Intra-S).

(A)	(B) All ages	Year1	Year2	Yrs 3+
Age	**0.816**			
S.Area	**-0.299**	**-0.442**	**-0.514**	**-0.357**
Depth	**0.473**	**0.589**	**0.629**	**0.499**
Inter-S	**-0.176**	**-0.456**	**-0.278**	**-0.414**
Intra-S	**-0.206**	**-0.204**	**-0.463**	**-0.410**
Latitude	**0.191**	**0.415**	**0.375**	**0.284**
df	6, 402	5, 231	5, 110	5, 50
F *ratio*	167.2**	59.4**	33.7**	9.7**
Adj r^2	0.71	0.55	0.59	0.44

Table 2. Model results and correlation coefficients for species *Celleporella bougainvillei* as for Table 1.

(A)	(B) All ages	Year1	Year2	Yrs 3+
Age	**0.949**			
S.Area	-0.049	-0.052	**-0.349**	0.054
Depth	**0.583**	**0.345**	**0.433**	0.113
Inter-S	**-0.304**	**-0.488**	**-0.601**	**-0.374**
Intra-S	0.083	**-0.156**	**-0.284**	-0.013
Latitude	-0.023	**0.127**	**0.329**	**0.227**
df	6, 329	5, 222	5, 71	5, 25
F *ratio*	533.9**	29.6**	16.2**	NS
Adj r^2	0.91	0.39	0.5	0.15

for each year-class there is negligible difference between parameters in terms of importance. Increased rock surface area or spatial competition (of either inter or intra-specific nature) increased growth whilst conversely increased depth or latitude increased growth. More than 50% of the variance in the data was explained by the models for year-classes 1 and 2, whilst nearly 50% was explained by the model for years 3-5 despite the low sample size. The growth of *I. nutrix* was, therefore, influenced by a mesh of positive and negative factors of similar importance and changing little with age.

The analyses showed the determinant f mactors to the growth of *Celleporella bougainvillei* (Table 2) had many similarities with those of *I. nutrix* (Table 1). Age was, again the most important parameter. Rock area and competition were typically negative influences and depth and latitude positive influences.

Table 3. Model results and correlation coefficients for species *F. rugula* as for Table 1.

(A)	(B) All ages	Year1	Year2	Yrs 3+
Age	**0.822**			
S.Area	-0.035	-0.136	-0.085	0.067
Depth	**0.308**	**0.357**	0.130	-0.034
Inter-S	**-0.357**	**-0.584**	**-0.423**	**-0.573**
Intra-S	**-0.218**	**-0.437**	**-0.282**	**-0.321**
Latitude	**0.323**	**0.293**	**0.375**	0.214
df	6, 511	5, 355	5, 122	5, 23
F *ratio*	213.7**	71.9**	12.0**	4.3*
Adj r^2	0.71	0.5	0.3	0.37

Table 4. All species combined. Species included as a model paraterer. Explanation as Table 1.

(A)	(B) All ages	Year1	Year2	Yrs 3+
Age	**-0.194**			
S.Area	**-0.130**	**-0.171**	**-0.259**	**-0.236**
Depth	**-0.090**	**0.362**	**0.347**	**0.251**
Inter-S	**0.764**	**-0.378**	**-0.250**	**-0.352**
Intra-S	-0.060	**-0.167**	**-0.195**	**-0.313**
Latitude	**0.250**	**0.362**	**0.426**	**0.389**
Species	**0.105**	**0.327**	0.09	-0.176
df	71,255	6,819	6,314	6,109
F *ratio*	292.6**	88.4**	28.6**	10.7**
Adj r^2	0.618	0.389	0.341	0.37

The analyses also showed some major differences between *C. bougainvillei* and *I. nutrix*. Only three parameters were significant in the all-ages model and they explained 92% of data variance. Variance explained by each year-class model were lower and that for years 3-5 (3+) was non-significant. Inter-specific competition was the only parameter which was significant in all four models and was the most important in each of the specific year-class models.

Growth of the third species, *F. rugula*, showed similar and, to some extent, intermediate trends in comparison with the two previous species. All four models were significant, but only at a reduced level in the case of that for years 3-5 (Table 3). As with *I. nutrix*, most parameters were significant in the all ages model which similarly explained about 70% of data variance. In common with *C. bougainvillei* were the lower levels of variance explained by the specific year-class models, the non-significance of some parameters and the dominant influence of inter-specific competition in the specific year-class models. Uniquely in *F. rugula* was the importance of intra-specific competition in specific year-class models and the decrease in number of significant parameters with age. When specific age classes were considered the growth of *F. rugula*, and to a lesser extent *C. bougainvillei*, was dominated by spatial competition.

A fourth set of models were executed with species included as a parameter to assess the relative importance of identity with respect to other terms (Table 4). The first model, in Table 4 (all ages) is thus the only model to incorporate all factors for relative comparison. It is also the least generally applicable, as the species included are only a subset of all competitors and thus inclusion of other species could alter the results of the model substantially. The all species models show that the species term, whilst significant did not emerge as a dominant parameter.

3.2 *The influence of spatial competition on growth*

As an important (but equal) factor in the growth of *I. nutrix*, and the dominant influence for *C. bougainvillei* and *F. rugula*, the effect of spatial competition was considered in more detail. Two dimensional linear regression was used to explore the negative relationships of inter- and intra-specific interference competition with growth shown by the models. The regressions were age separated due to the overwhelming influence of age on growth in models.

The proportion of colony margins in contact with competitors (degree of spatial competition) ranged from 90% to just 2% (Figs 3-5). The intercepts of the relationships give the extreme sizes of colonies under spatial competition and the slopes show the magnitude of influence. The high variability of the data around the lines of best fit were in part due to the lack of inclusion of other significant explanatory parameters but also because, even with all terms included, more than 40% of the variance was unexplained (Tables 1-3). Although there was a high degree of overlap the influence of intra-specific competition was generally more severe to *I. nutrix* than inter-specific competition (Fig. 3). Typically, a colony of this species at age 1 only exceeded 100 zooids if not occluded on more than 30% of its colony margin. The slopes (magnitude of influence) did not, however, change with age (GLM ANOVA, P>0.05).

As in *I. nutrix* there was no clear separation of the influences of intra- and inter-specific competition on

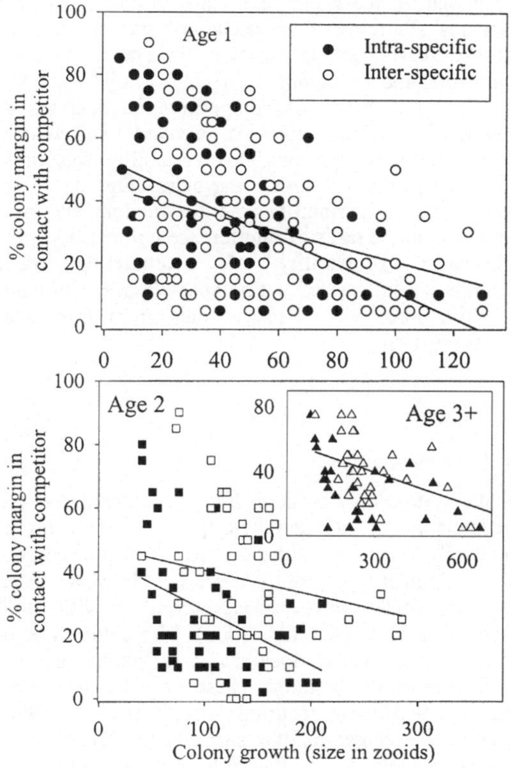

Figure 3. Proportion of colony margin in spatial competition vs colony growth for the species *Inversiula nutrix*. Ages are shown. Key for intra-specific and interspecific competition is in the legend (top right) of all three species. All regressions are significant

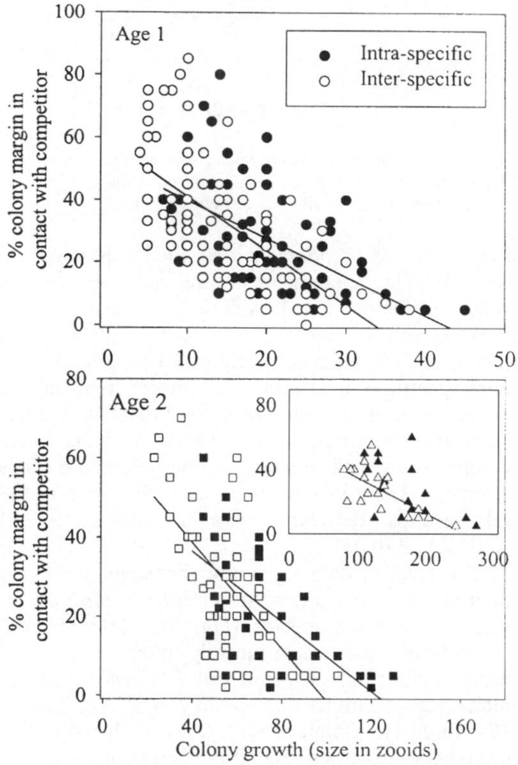

Figure 4. Proportion of colony margin in spatial competition vs colony growth for species *Celleporella bougainvillei*. Age, competition and regressions as in Fig. 3.

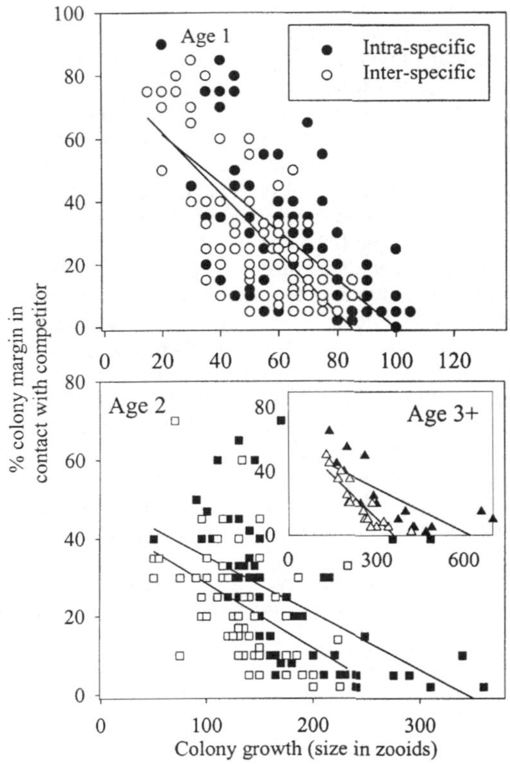

Figure 5. Proportion of colony margin in spatial competition vs colony growth for species *Fenestrulina rugula*. Age, competition and regressions as in Figures 3 and 4.

the growth of *C. bougainvillei* when superimposed (Fig. 4). The relationships were, however, significantly different (GLM ANOVA, F *ratio* = 108, $P<0.01$) and, in contrast to *I. nutrix*, inter-specific competition had a more severe effect on growth than intra-specific competition. The greater frequency of higher levels of occlusion of colony margins in inter-specific competition causes distortion at the lower colony sizes and results in the regression lines crossing. There was no significant difference in the relationships bewteen years 1 and 2 (GLM ANOVA, $P>0.05$).

The level of data separation between intra- and inter-specific effects was greater in *F. rugula* than either of the other species (Fig. 5). Inter-specific competition had a significantly more severe and negative influence on growth of *F. rugula* than did intra-specific competition (GLM ANOVA, F *ratio* = 171, $P<0.01$). Colonies with 20% of their margins occluded were 40% smaller if the competitor was not conspecific.

The multiple regression models identified the proportional importance of competition to growth of the study species (Tables 1-4). The linear regressions (Figs 3-5) quantified the intensity of influence of spatial competition on growth and showed inter-specific and intra-specific differences to be small but significant. No overall conclusions can still be formulated, however, without quantification of the frequency of competition (and of each type of competition) occurring in each of the three species.

3.3 *Prevalence of competition experienced by study species*

Likelihood of experiencing spatial competition increased with age in all species, from 50-70% in year 1, to 90-100% in year 4 (Fig. 6). The incidence of intra-specific competition was not greatly lower than that of inter-specific encounters despite the fact that there is a clearly greater potential for the latter given the number of species present on Antarctic rocks. In all species the proportion of inter-specific encounters (and simultaneous intra- and inter-specific encounters) increased with age. The chance of any colony experiencing both intra- and inter-specific competition was low and only approached 20% by year 4. In the youngest year-class inter-specific encounters of *I. nutrix* formed about twice the proportion than in the same year class of the other two species. Patterns of prevalence of spatial competition were very similar between the species *C. bougainvillei* and *F. rugula*.

The probability of competitive encounters, although inevitable if colonies survived long enough, varied between species. This variability was most obvious in the youngest year-class (Fig. 7). Furthermore, the variability was non-random and predictable on the basis of interference competition performance. Probability of encountering other competitors increased within year-classes with competitive rank (wins / losses, data from Barnes & Rothery 1996).

4 DISCUSSION

4.1 *Growth in the freezer*

Polar waters may be cold, but sea-temperatures are reasonably constant and there may be little or no direct effect of temperature on growth of polar organisms. Annual temperature range has, however, been demonstrated to have a strong correlation with zooid size (O'Dea & Okamura 2000). The cold temperature of polar waters does have a substantial indirect influence on the benthos. These vary from ice production (leading to scouring – see Gutt et al. (1996) raising the oxygen holding capacity (leading to a degree of gigantism – see Chapelle & Peck (1999). Although Chapelle & Peck (1999) only compared amphipod sizes, gigantism is obvious in other animals such as isopods, pycnogonans and some non-arthropod taxa (Arntz et al. 1994). Oxygen effects size rather than growth and although ice scour might

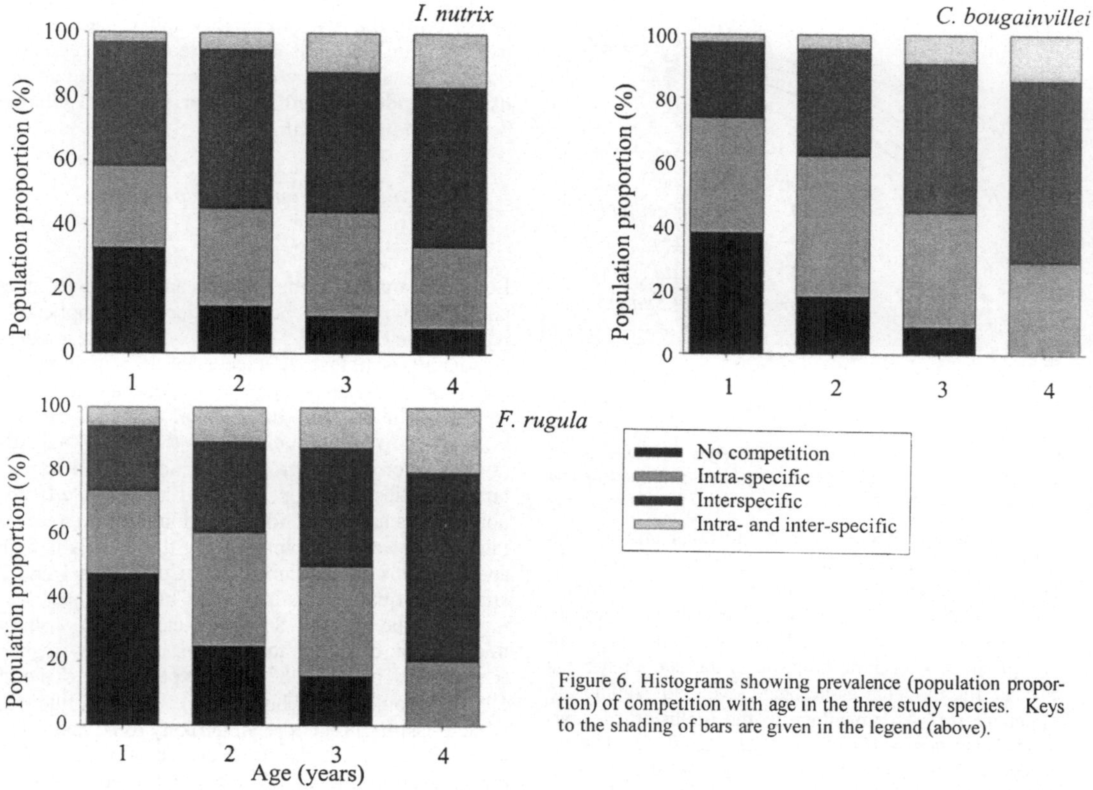

Figure 6. Histograms showing prevalence (population proportion) of competition with age in the three study species. Keys to the shading of bars are given in the legend (above).

typically influence survival of solitary (non-colonial) species it would be unlikely to alter growth. Colonial animals, however, may endure sub-lethal mortality – that is some or many zooids might be scraped off but the colony survives (see Thomsen & Håkansson 1995). Thus an animal may appear to have negligible or even negative growth over time as parts are removed (Stanwell-Smith & Barnes 1997). In erect species, like those of the genera *Cellarinella* or *Melicerita*, each of these fragments may fall onto the sediment, develop rootlets and grow as new colonies (Winston 1983, Brey et al. 1999). In both cool and warm waters predators have been reported as doing likewise to sponges (Dayton 1989 and Chanas & Pawlik 1996 respectively). In polar waters, though, the abundance and rate of consumption of direct predators probably have little overall effect on populations (Barnes 2000a). These factors, besides being difficult to quantify, did not warrant inclusion in the model. Growth output of any animal is to some extent geared to input so can we explain growth on the basis of food limitation (Clarke 1988). Certainly when food is plentiful (during the polar summer) many organisms grow quickly, or at least as quickly as similar animals in warm water (Barnes 2000a). But food is superabundant for only short periods and otherwise rare (Clarke & Leakey 1996) and is, therefore, similarly available to rock encrusters, especially those which all colonise the same (under)surface.

Age, as with all animals, was likely to be a major factor and this variable dominated the explanation of variance of our models. Young animals are generally small but grow proportionally quicker – this is not restricted to any group of organisms it is universal. What is gained from the models, is just how important age is with respect to other variables and how this alters with species identity. Within the limits of food availability, and after age, what factors may be expected to effect growth? Perhaps one of the most intuitively reasonable variables might be species identity. The three species examined here only showed slight differences in zooid production (growth) despite having different sized zooids with differential calcification (see Hayward 1995). Thus the models confirmed species identity as important, but - not that important. This result may be due to the specific species chosen, which are all relatively thin-calcified r-stategists (Barnes 2000a). It seems likely that had K species with very high levels of skeletal investment, such as *Arachnopusia inchoata* been included a species effect would have been more evident. The three species selected for inclusion into the models on the basis that they were not only the

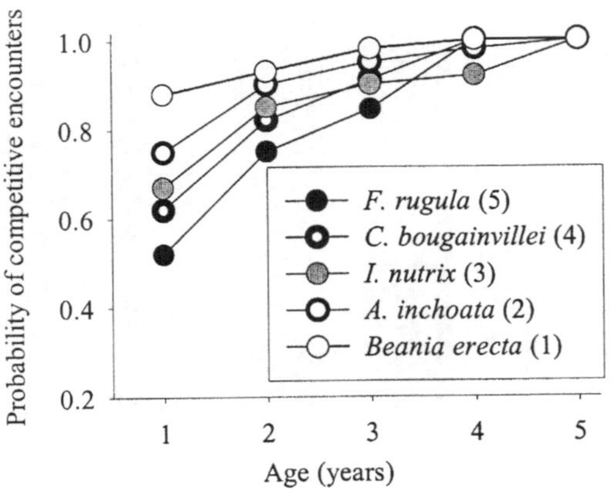

Figure 7. Probability of competitors meeting with species and age. Competitive rankings, *Arachnopusia inchoata* and *Beania erecta* data from Barnes (2000a).

only common species across latitudes and depths within polar waters, they were virtually the only species common to all sample points. With such a situation of lack of specific commonality, there are two options, 1) only include species common to all sample points as done here, or 2) include all species but run the model at assemblage level and exclude the species term. An advantage to the inclusion of cosmopolitan genera (such as *Celleporella* and *Fenestrulina* – see Hayward & Ryland 1999) is facilitation of future and larger (global) growth rate comparisons (within taxa). Previous comparisons of populations of the same three species at different latitudes have found that, in contrary to the general polar invertebrate trend, growth rates increased with increasing latitude in Antarctic waters (Barnes & Arnold 2001). Whilst the models bore out such a relationship, this was weak in comparison to other factors. Depth and rock surface area, not factors which have previously been considered to be linked with growth performance at any latitude, were similarly significant but weak. For a given age the most important overall factor to emerge from the models, either in isolation or with a species term included, was spatial competition. Barnes & Arnold (2001) suggested that changes in competition might explain observed patterns of differential growth with latitude as competition also varies along this gradient (Barnes 2000b). More specifically interspecific competition was particularly important, despite intraspecific encounters being of similar frequency (Fig. 6). Finally little consistent variability in band width of Antarctic bryozoans has been reported with year (Winston 1983, Brey et al. 1999, Barnes 2000a) so time and food variability is likely to be low on a ranking of factors. As an overview we can state the models indicate a ranking Age > Inter-specific competition > Intra-specific competition/latitude > species identity/depth/rock surface area. We can add time and predation to the lowest category and visualise environmental influences if intrinsic factors are removed. So inter-specific competition > intra-specific competition/ latitude > all others is our suggested relative determination of growth.

4.2 *How satisfactory and applicable are the models?*

Before the output of the models can be considered as universal hypotheses both accuracy and applicability must be examined. We can consider the models reasonably satisfactory (accurate) if a number of conditions are met: 1) there are good biological reasons for believing that fundamental terms are not absent, 2) growth measurement of the study animals (averaged across year groups) has not masked important subtleties, 3) the species used are discrete (taxonomically sound) and identification has been accurate, 4) the variance explained by the models is high and the terms highly significant. Conditions 1 and 2 are hard to quantify, as important but unrealised factors could be present. Subtleties can include colony microhabitat on substrata such as local flow regime (Cancino & Hughes 1987), genotype (Bayer & Todd 1996) or fouling (Stebbing 1971). Such influences while possibly present seem unlikely to be major because of the relatively small degree of variance left unexplained by the model. Only in the year 3+ year group can condition 4 be doubted to any degree. Fortunately the species appear to be both morphologically and taxonomically distinct meeting condition 3. Although all suggested conditions for model accuracy seem to hold, new findings could degrade the current evaluations of 1 –3 so negating the results of 4.

If the models are considered biologically satisfactory can they be applied in a wider sphere or are the findings restricted to the current species, time and space. That species was not a highly significant term suggests that the ranking of factors should be applicable to at least other encrusting cheilostomatid species in Antarctic waters. The findings will be of virtually no use in interpreting the determinants of erect bryozoan growth, as spatial competition (dominant in the models presented) is likely to have little role in animals with so little basal attachment to substratum. It is unlikely that the findings would be generalisable to cyclostomatid bryozoans, despite the similar encrusting nature of many, because of their very different growth format (see McKinney 1992). It is possible that other colonial and clonal encrusting organisms, such as ascidians and cnidarians, might have similar determinants of growth and that they may be of similar relative importance for the same reasons.

Although time has been considered on an annual (Recent) basis, we must consider wider scale temporal processes. Trends of growth traits, such as zooid formation (Lidgard & Jackson 1989) and growth habits (McKinney & Jaklin 1993) show significant change from Palaeozoic to present. Whilst such trends and direct consideration of macro-scale time are not appropriate to our models they are important to their overall interpretation. Ultimately, of course, taphonomic evidence could be included to generate models that would span Recent and non-Recent periods. The main problem in doing so would be the short perseverance of individual species in geological time. Likewise full comprehension of our, and future, models requires appreciation of a wide spatial scale. Whether these polar models would hold outside the margins of the Southern Ocean (the Polar Frontal Zone) may prove interesting and should not be hard to test. One of the most appropriate places to start might be to investigate similar communities at similar latitudes from the other pole. Very little is known about growth of encrusting Arctic bryozoans although there has been preliminary investigation of erect forms (Schäfer 1994). Intense competition for space has been described on sub-arctic rocks (Barnes & Dick 2000) with assemblages forming similar hierarchies to those known from the Antarctic.

4.3 *Why is competition so important?*

Despite the fact that bryozoans, together with tubicolous polychaetes, are pioneer taxa on newly available substrata a degree of spatial competition is inevitable. That inter-specific competition followed by intra-specific competition emerge as the most important extrinsic determinants of growth may have two principal causes. Firstly competition, unlike other extrinsic factors, is direct and frequent in its effects and secondly it's effects are actual loss of energy and space. Competition is always going to generate negative effects on organisms as even those with the most superior abilities will expend (extra) energy on overcoming a rival no matter how trivial. When bryozoans meet a spatial competitor there are a number of possible outcomes. The net result for a given competitor is that it, 1) loses energy but not space by overgrowing the rival, 2) loses energy and space (to grow into) by being restricted along the margin in contact with the competitor, 3) loses energy and space (to grow into) but fuses with it's competitor (Knight-Jones & Moyse 1961) and 4) loses energy and space by being overgrown by its rival. All of these outcomes involve loss of energy which could have been used for production of new zooids (growth). Most potential outcomes (especially for the three study species as all are poor competitors) involve restriction by reduction of the growing margin of the colony. The last potential outcome, which would occur in the event of any of the three study species meeting most other types of Antarctic bryozoan involves not only loss of energy and restriction of growing edge but also reduction of colony size.

4.4 *Why are species differentially influenced by intra- vs interspecific competition?*

The frequency of the four possible net results of spatial competition (discussed above) is not evenly distributed between inter-specific and intra-specific competition. A tied outcome (options 3 and 4) is much more likely and fusion restricted to intra-specific encounters (Knight-Jones and Moyse 1961, Craig 1994, Barnes 2000a, b). In contrast, inter-specific interactions generally involve competitors less evenly matched, thus one competitor is effected much more seriously at the expense of the other. This can be viewed as sessile equivalent strategies of Maynard-Smith's (1976) original game theory. Intra-specific encounters are typically of 'dove' nature, whilst inter-specific meetings are 'hawk'-like involving destruction of all or part of the losing competitor. As all our study species are poor competitors inter-specific encounters are likely to involve a reduction in size (negative growth) at any level from 1-100%. Clearly such a mechanism explains the asymmetric influence of competition within and between species on growth. The effects of competition may be even more asymmetrical when more distantly related competitors are considered (Diamond 1987). Over the entire benthic assemblage in our study area the mean influence on growth may differ little between intra- and inter-specific competition. Overall space gains by winning competitors should approximately equal those lost by losing competitors. This balance is, therefore, similar to most species being involved in tied interactions and neither gaining or losing area (except potential area to expand into). The variance of growth as a result of inter-specific encounters over the entire benthic assemblage in our study area should, however, be much larger than the variability associated with intra-specific encounters. In this study we are not concerned with the overall community, merely three representatives of it. As these representatives are all poor competitors they are chiefly losers in inter-specific competition and as such this factor is most important as a determinant of growth.

5 CONCLUSIONS

Many ecological studies have treated associations between variables has effectively showing that one influences the other. Some famous examples of anthropogenic processes, such as the correlation between increasing fridge ownership and crime level, amongst others have helped dispel quick conclusions. This is why we have used multiple regression in a non-predictive sense (i.e. non causal). Monsteller & Tukey (1977) spelled out that consistency, responsiveness and, most importantly, a mechanism are required to evidence causality. Our models generate evidence of consistency and responsiveness but are there convincing mechanisms to support influences on growth of the factors investigated by this study? Of greater relevance, as our emphasis is the multivariate level of approach, is whether any proposed mechanism(s) can explain the differential influence of the factors as suggested by the model. We suggest that there are mechanisms to explain that the influence of the factors (investigated) on growth and that they do so in an uneven manner. Inter-specific competition generally results in reduction of colony size as the three species are poor competitors – this is the most severe influence. Intra-specific competition restricts the growing margins of colonies slowing their rate of expansion. Latitude probably has no direct effect but many indirect influences such as altering the light dependent primary productivity, the frequency of ice-mediated disturbance, the suite of competitors and frequency of competition (Barnes 2000b). These are the next most important effects. There is little evidence to separate the specific factors in final (least important) category.

REFERENCES

Arntz, W.E., Brey, T. & Gallardo, V.A. 1994. Antarctic zoobenthos. *Oceanography and marine Biology an Annual Review* 32: 251-303

Barnes, D.K.A. 2000a. Life patterns of encrusting and erect cheilostomatida. In A. Herrera Cubilla & J.B.C. Jackson (eds) *Proceedings of 11th International Bryozoology Conference*: 145-153. Bilboa, Panama, Smithsonian Tropical Research Institute.

Barnes, D.K.A. 2000b. Diversity, recruitment and competition on island shores at polar locations compared with lower latitudes; encrusting community examples. *Hydrobiologia* 440: 37-44

Barnes, D.K.A. & Arnold, R.J. 2001. A growth cline in encrusting benthos along a latitudinal gradient within Antarctic waters. *Marine Ecology Progress Series* 210: 85-91

Barnes, D.K.A. & Dick, M.A. 2000. Overgrowth competition in encrusting bryozoan assemblages of the intertidal and infralittoral zones of Alaska. *Marine Biology* 136: 813-822

Bayer, M.M. & Todd, C.D. 1996. The effect of polypide regression and other parameters on colony growth in the cheilostomate *Electra pilosa* (L.) In D.P. Gordon, A.M. Smith & J.A. Grant-Mackie (eds) *Bryozoans in space and time.* Proceedings of the International Bryozoology Association *10th conference:* 29-38. New Zealand: N.I.W.A., Wellington.

Berkman, P.A. 1990. The population biology of the Antarctic scallop *Adamussium colbecki* (Smith 1902) at New Harbour, Ross Sea. In: K.R. Kerry & G. Hempel (eds) *Antarctic ecosystems: ecological change and conservation,* Springer-Verlag, Berlin, p. 281-288

Brey, T., Peck, L.S., Gutt, J., Hain, S. & Arntz, W.E. 1995. Population dynamics of *Magellania fragilis*, a brachiopod dominating a mixed bottom macro-benthic assemblage on the Antarctic shelf. *Journal of Experimental Marine Biology and Ecology* 95: 857-869.

Brey, T., Gerdes, D., Gutt, J., Mackensen, A. & Starmans, A. 1999. Growth and age of the Antarctic bryozoan *Cellaria incula* on the Weddell Sea shelf. *Antarctic Science* 11: 408-414

Cancino, J.M. & Hughes, R.N. 1987. The effect of water flow on growth and reproduction of *Celleporella hyalina* (L.) (Bryozoa: Cheilostomata). *Journal of Experimental Marine Biology and Ecology* 112: 109-130

Chanas, B. & Pawlik, J.R. 1996. Does the skeleton of a sponge provide a defense against predatory fish. *Oecologia* 107: 225-231

Chapelle, G. & Peck, L.S. 1999. Polar gigantism dictated by oxygen availability. *Nature* 399: 114-115

Clarke, A. 1988. Seasonality in the Antarctic marine ecosystem. *Comparative Biochemistry and Physiology*, 90: 461-473.

Clarke, A. & Leakey, R.J.G. 1996. The seasonal cycle of phytoplankton, macronutrients and microbial community in a nearshore Antarctic marine ecosystem. *Limnology and Oceanography*, 41: 1281-1294

Cocito, S. Bianchi, C.N., Morri, C. & Peirano, A. 2000. First survey of sessile communities on subtidal rocks in an area with hydrothermal vents: Milos Island, Aegean Sea. *Hydrobiologia* 426: 113-121.

Craig, S. 1994. Intraspecific fusion in the encrusting bryozoan *Fenestrulina* sp. In P.J. Hayward, J.S. Ryland & P.D. Taylor (eds) *Biology and Palaeobiology of Bryozoans*: 51-55. Olsen & Olsen, Fredensborg.

Dayton, P.K. 1989. Interdecadal variation in an Antarctic sponge and it's predators from oceanographic climate shifts. *Science*, New York, 245: 1484-1486

Dayton, P.K., Robilliard, G.A., Paine, R.T. & Dayton, L.B. 1974. Biological accommodation in the benthic community at McMurdo Sound, Antarctica. *Ecological Monographs* 44: 105-128.

Dell, R.K. 1972. Antarctic benthos. *Advances in Marine Biology* 10: 1-216.

Diamond, J.M. 1987. Competition among different taxa *Nature* 326: 241.

Everson, I. 1977. Antarctic marine secondary production and the phenomenon of cold adaptation. *Philosophical Transactions of the Royal Society* 279: 55-66

Gutt, J., Starmans, A. & Dieckmann, G. 1996. Impact of iceberg scouring on polar benthic habitats. *Marine Ecology Progress Series* 137: 311-316

Hayward, P.J. 1995. *Antarctic Cheilostomatous Bryozoa.* Oxford University Press, 355 pp

Hayward, P.J. & Ryland, J.S. 1990. *The marine fauna of the British Isles and north-west Europe.* Clarendon, Oxford.

Knight-Jones, E.W. & Moyse, J. 1961. Intra-specific competition in sedentary marine animals. *Symposium of the Society for Experimental Biology* 15: 72-95

Lidgard, S. & Jackson, J.B.C. 1989. Growth in encrusting cheilostome bryozoans I Evolutionary trends. *Palaeobiology* 15: 255-282

Maynard-Smith, B. 1976. A comment on the red queen. *American Naturalist* 110: 325-330

McKinney, F.K. 1992. Competitive interactions between related clades: evolutionary implications of overgrowth interactions between encrusting cyclostome and cheilostome bryozoans. *Marine Biology* 114: 645-652

McKinney, F.K. & Jaklin, A. 1993. Living populations of free-lying bryozoans: implications for post-Palaeozoic decline of the growth habit. *Lethaia* 26: 171-179

Monsteller, F. & Tukey, J.W. 1977. data analysis and regression. A second course in statistics. Addison-Wesley, Reading, MA.

O'Dea, A. & Okamura, B. 2000. Intracolony variation in zooid size in cheilostome bryozoans as a new technique for investigating palaeoseasonality. *Palaeogeography, Palaeoclimatology and Palaeoecology* 162: 319-332

Peck, L.S., Brockington, S. & Brey, T. 1997. Growth and metabolism in the Antarctic brachiopod *Liothyrella uva*. *Phillosophical Transactions of the Royal Society* 352: 851-858

Rauschert, M. 1991. Ergebnisse der faunistischen Arbeiten im Benthal von King George Island (Südshetlandinseln, Antarktis). *Berichte Polarforschung* 76: 1-75.

Santos, R. 1993. A multivariate study of biotic and abiotic relationships in a subtidal algal stand. *Marine Ecology Progress Series* 94: 181-190

Schäfer, P. 1994. Growth strategies of Arctic Bryozoa in the Nordic Seas. In P.J. Hayward, J.S. Ryland & P.D. Taylor (eds) *Biology and Palaeobiology of Bryozoans*: 173-176. Fredensborg: Olsen & Olsen.

Stanwell-Smith, D. & Barnes, D.K.A. 1997. Benthic community development in Antarctica: recruitment and growth on settlement panels at Signy Island *Journal of Experimental Marine Biology and Ecology* 212: 61-79.

Stebbing, A.R.D. 1971. Growth of *Flustra foliacea* (Bryozoa). *Marine Biology* 9: 267-273.

Stockton, W.L. 1984. The biology and ecology of the epifaunal Scallop *Adamussium colbecki* on the west side of McMurdo Sound, Antarctica. *Marine Biology* 78: 171-178

Thomsen, E. & Håkansson, E. 1995. Sexual versus asexual dispersal in clonal animals: examples from cheilostome bryozoans. *Paleobiology* 21: 496-508

Winston, J.E. 1983. Patterns of growth, reproduction and mortality in bryozoans from the Ross Sea, Antarctica. *Bulletin of marine Science* 33: 688-702.

Bryozoan Studies 2001, Wyse Jackson, Buttler & Spencer Jones (eds)
© 2002 Swets & Zeitlinger, Lisse, ISBN 90 5809 388 3

Seasonality and inter-annual variability in recruitment patterns of temperate encrusting fauna

David K.A. Barnes
British Antarctic Survey, High Cross, Madingley Road, Cambridge, UK

Ben C. Maughan
Maritime Environment Information Centre, UK Hydrographic Office, Admiralty Way, Taunton, UK

ABSTRACT: Sets of triplicate slate panels were immersed at two sites in SW Ireland and at three depths and analysed monthly between 1997 and 2001. The mean number of recruits varied on both seasonal and inter-annual time-scales, peaking during summer periods. Relative sizes of recruitment peaks varied both between sites and depths. Recruitment on intertidal panels was an order of magnitude lower than on those in the sub-tidal, even when allowing for immersion period. The mean number of species followed a predictable seasonal pattern, peaking in summer, and varied little on an inter-annual basis. Per cent cover varied less predictably with season and year. Typically just two of three colonist species were responsible for 90% of space occupation, but this also varied with time on a variety of scales. Recruitment of individual taxa showed a variety of patterns both in terms of the timing of initiation or cessation and duration. These included predictable bursts of settlement of similar seasonal timing (e.g. *Aetea* sp.) and predictable long duration bouts (e.g. *Chorizopora brongniartii*) to less predictable bursts (e.g. the sponge *Scypha ciliatum*) and low level fluctuations (e.g. *Scrupocellaria* sp.). The variability of the study assemblages in both time and space illustrate the importance of within- and between-site replication and the use of long-term data sets.

1 INTRODUCTION

Observations of biota colonising artificial substrata, natural substrata and cleared natural substrata have enabled coastal biologists to investigate and experiment with a wide range of fundamental ecological concepts. Panel studies have elucidated major recruitment variability with exposure (Todd & Turner 1986), latitude (Barnes 1996), light and flow rate (Maughan & Barnes 2000) and sediment load (Maughan 2001). Despite the artificial nature of panels they have generally proved to be the most popular tool for a number of reasons: 1. Homogeneity. The area of surfaces are uniform and can be accurately estimated and recruits area can be easily measured. 2. Replication. Unlimited near-identical panels can be made and used. 3. Utilisation. Panels are easily stored, manipulated and set up into replicate arrays. Bryozoans too, have been used widely as a taxon to interpret ecological pattern and process. Seasonal trends in bryozoan recruitment have been documented from environments as different as temperate estuaries (Maturo 1959) and polar boulder fields (Stanwell-Smith & Barnes 1997).

It was realised by early ecologists that much of the processes and factors that impacted on sessile adult marine organisms were, to a large extent, determined by events prior to settlement (see Todd & Turner and references therein). Many decades later after intense studies on competition and predation on adults, supply-side ecology again became in-vogue (Underwood & Fairweather 1987). Clearly the timing of larval release, free-living duration, timing of settlement, specificity of settlement coupled with gregarious or conspecific attraction or inhibition will have a major influence on where, when and if an adult colonist occurs. There is a considerable stochastic element to these processes, from local to global anomolies such as the El-Niño-Southern-Oscillation.

Lough Hyne (Ine) Marine Nature Reserve (LHMNR), Europe's oldest marine reserve, has been the site of extensive marine ecological research since the 1920s. Initial work concentrated on the role, behaviour and predators of a key (echinoid) species (Kitching 1987). Lilly et al. (1953) revealed a diverse epilithic community to be present on the boulders in the fast flowing rapids but only recently has attention focused on the rich and substantial benthic communities (Maughan & Barnes 2000a,b, Bell & Barnes 2001, Maughan 2001). Recently panels have been deployed which revealed 91 species in less than two years. Despite the wealth of work carried out at Lough Hyne over many decades and the new found appreciation for long-term data sets, few studies have extended over a two year period (partly because this is the typical length of field data sets of PhDs). The few medium to long term studies (e.g. sea temperature, echinoid population size and algal abundance have established superannual patterns (Kitching 1987). Here we report patterns and analyses of bryozoan recruitment over the years 1997-2001. Although in itself this data set is the longest continuously sampled one of its kind, it is intended as part of a larger, longer running programme. In this paper we look at the variability of recruitment with species identity, depth, site and time (at monthly, seasonal and annual scales).

2 METHODS

2.1 *Study sites*

Lough Hyne is a small sea lough in south-west Ireland, western Europe. The prevalent environmental conditions, such as flow rates, temperature, salinity and oxygen content, have been described in detail by numerous authors and are summarised in Kitching (1987). During the study period the sea temperature varied from localised freezing (and temporary formation of sea-ice) to 18°C. The study was carried out at three depths (0, 6 and 12 m) at each of two sites (Whirlpool Cliff and Labhra Cliff). The peak flow rates of the two sites were 100 cms^{-1} and 5 cms^{-1} at Whirlpool Cliff and Labhra Cliff respectively, whilst flow minima were 0 at both sites. Other than flow rate the two sites experienced similar wave action and were composed of similar vertical solid rock faces (see Maughan & Barnes 2000a). The apparatus was assembled on natural horizontal shelves in each of the cliff faces.

2.2 *Protocol*

The apparatus, preparation, positioning and analysis of the artificial substrata used was as for Todd & Turner (1986) and Stanwell-Smith & Barnes (1997). Machined slate panels of 15 cm x 15 cm x 1 cm size were attached in triplicate to weighted rods which were sited (horizontally) to rest 1 cm above the substratum. Only the middle 10 cm x 10 cm of the (down-facing) under-surfaces of the panels were analysed. The triplicate panel sets were serially replaced (using SCUBA) at all depths and sites to give monthly resolution of recruitment between October 1997 and May 2001. All recruits were identified to the lowest taxonomic level (mostly species) and counted. A number of cyclostomes and some cheilostome ancestrulae were not identified beyond genus. Data was standardised to 30 days. Each month panels were individually photographed on retrieval and colonist surface area recorded. The transparencies were then projected onto a grid of points and the total area colonised estimated by the proportion of points overlain by fauna. The number of interactions between species were recorded using a binocular microscope. Interactions were only counted if a) both colonies were alive at time of retrieval, b) at least 5% of the colony periphery was in contact for one of the competitors and c) the encounter angle was frontal (between 30°-330° see Maughan & Barnes 2000a). Colony recruitment onto an existing colony (fouling) was not considered an interaction in this context.

2.3 *Statistical techniques*

The significance of variability in taxa recruitment with station and month was interpreted using a two factor General Linear Model (GLM ANOVA). Community-time analysis was performed using Detrended Correspondence Analysis (DCA). This is a similar ordination to Correspondence Analysis and is useful when the underlying structure of data is more of a gradient than divisively clustered. Ordinations organise sites in multidimensional space such that proximity of points reflects similarity. DCA has advantages over CA in reducing distortion along and linkage between axes. DCA biplots were constructed to interpret underlying community structure.

3 RESULTS

3.1 *Gross recruitment with time, depth and site*

The mean total numbers of recruits generally peaked in the latter part of each year and the peaks represented levels which were several orders of magnitude higher than background levels (Fig. 1). The timing and relative magnitude of peaks were consistent across sites and depths but the absolute values were depth linked. Whilst a maximum of only 200 or less recruits per 30 days occurred in the intertidal, up to 700 occurred at 12 m and up to 2400 recruited on panels at 6 m. Inter-annual variability was apparent along with monthly and seasonal changes, for example the peak of recruitment in 2000 (Fig. 1) was longer in duration whilst that of 1998 was bimodal. Inter-annual variability also differed between the two sites. Biomodal peaks were apparent at the Labhra subtidal sites in year 2000 but were unimodal at Whirpool subtidal sites.

The proportion of recruit individuals formed by bryozoans was small, rarely exceeding 20%, and negligible at both intertidal sites.

3.2 *Species richness with time, depth and site*

Patterns of species richness had some features in common with those of recruit numbers, such as a fluctuating amplitude with summer peaks and depauperate intertidal levels, but were otherwise quite different. Peaks were only one order of magnitude higher than background levels (Fig. 2). The

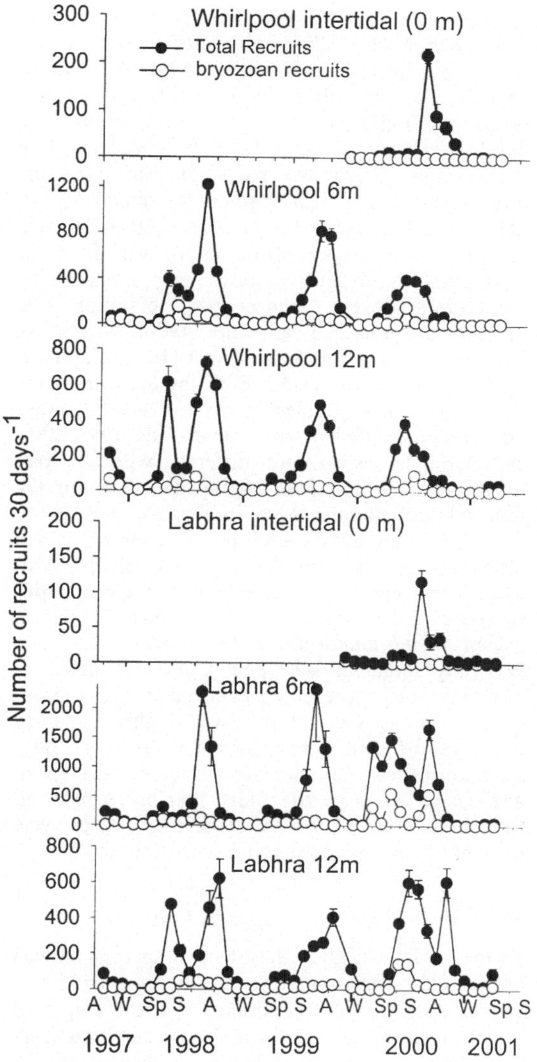

Figure 1. Total and bryozoan recruit numbers per 30 day period on panels at Lough Hyne Marine Nature Reserve with year. Data are presented as mean and standard error of 3 panel replicates for each depth (at each site). No data was collected in the intertidal zones until late 1999.

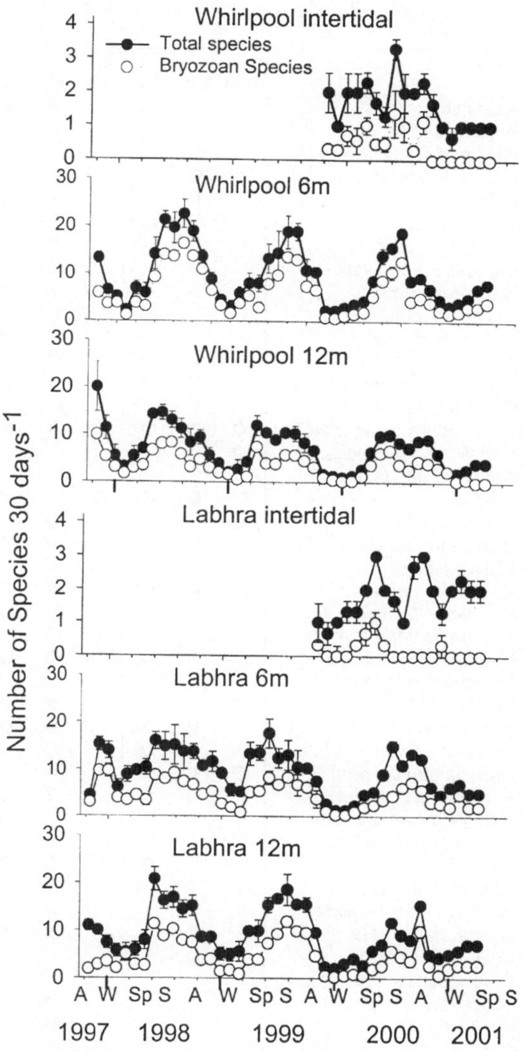

Figure 2. Species numbers per 30 day period on panels at Lough Hyne Marine Nature Reserve with year. Data are presented as mean and standard error of 3 panel replicates for each depth (at each site).

timing and size of peaks were similar across sites and subtidal depths. A maximum of about 20 species recruited per 30 days but the fluctuation from winter to summer values was gradual (rather than the dramatic changes in recruit numbers).

No data was collected in the intertidal zones until late 1999. Inter-annual and monthly variability were apparent but minor in comparison with seasonal changes, for example the peak in 2000 was longer in duration whilst that of 1998 was bimodal. Differences between panel replicates were generally small as for recruit numbers and indicated by the error bars. In addition the extent of inter-annual variability differed between the two sites. No biomodality was obvious in species number peaks in contrast to recruit numbers. Perhaps the most significant difference in the pattern between species and recruit numbers was, however, the proportion constituted by the bryozoans (mainly cheilostomes). More than 50% of species were generally bryozoans and (as a consequence) they were largely responsible for the overall trend in species richness (as opposed to be inconsequential to the overall recruit pattern).

Table 1. Bryozoan species prevalence on panels at LHMNR with depth and site. Key to symbols; present for 1-2 months (○), 3-5 months (**○**), 6-9 months (◉), and 10+months (●).

Species	L Inter	L 6m	L 12m	W Inter	W 6m	W 12m
Aetea sp.		◉	●	○	●	●
Alcyonidium sp.			○	○	○	
Bicelleriella ciliata		○				○
Bugula flabellata		**○**				
Bugula sp.		●	●		●	◉
Caberea boryii		○	○		○	○
Callopora aurita	○				○	
Callopora craticulata					○	
Callopora dumerlii		●	●		●	●
Callopora lineata	○	○	○		◉	○
Callopora rylandii			○		◉	○
Cellepora pumicosa		◉	**○**		◉	○
Celleporella hyalina		**○**	◉	○	●	●
Celleporina hassalii						○
Chorizopora brongniartii	**○**	◉	**○**	○	●	●
Cribrilina cryptooecium	**○**					
Cribrilina punctata		○				
Crisia sp.	**○**	◉	○	**○**	●	●
Diastoporidae	○		**○**		**○**	**○**
Disporella hispida		**○**	●		●	**○**
Electra pilosa	○	**○**			○	**○**
Escharella immersa	○	●	◉		●	○
Escharella labiosa			**○**		◉	○
Escharella variolasa					**○**	
Escharella ventricosa			○		**○**	○
Escharina vulgaris			○		**○**	
Escharoides coccinea	**○**	○	**○**		●	◉
Fenestrulina malusii		●	●		●	◉
Figularia figularis			**○**		○	
Filicrisia geniculata	○	●	●		●	●
Haplopoma graniferum		**○**			**○**	◉
Haplopoma sciaphilum						○
Hincksina flustroides					○	
Hippothoa sp.		○	◉		●	○
Lagenoporina sp.						○
Membraniporella nitida					●	**○**
Microporella ciliata		●	●	○	●	●
Nolella dilatata		**○**	**○**			
Omalasecosa ramulosa			○			○
Phaeostachys spinifera					◉	○
Puellina gattyae			○		◉	
Schizomavella auriculata	○	○	◉	○	●	◉
Schizomavella discoidea		○				
Schizomavella linearis		**○**	**○**		◉	**○**
Schizoporella dunkeri					○	
Schizoporella unicornis			**○**		●	**○**
Schizotheca fissa			**○**		**○**	○
Scruparia ambigua						○
Scruparia chelata		○	○		○	◉
Scrupocellaria reptans		●	**○**	○	●	◉
Scrupocellaria scrupea		●	●		●	◉
Scrupocellaria scruposa	○	●	●	○	●	●
Smittoidea amplissima		○				
Tubulipora sp.	**○**	●	●	○	●	●
Turbicellepora sp.		**○**			○	○
Umbonula ovicellata	**○**	●	◉		**○**	○

3.3 *Bryozoan species prevalence with depth and site*

Although 56 different bryozoans were recorded from panels during the study period, only 47 were identified to specific species. Of these 56 types, however, 55 were placable within the 39 genera found. The bryozoans found and their prevalence by site and depth are shown in Table 1. Five bryozoans, two cyclostomes (*Crisia* sp. and *Tubulipora* sp.) and three cheilostomes (*Chorzipora brongniartii* (Audoin), *Schizomavella auriculata* (Hassall) and *Scrupocellaria scruposa* (Linnaeus)), were found at all depths at both sites. Conversely 11 bryozoans (cheilostomes) were found at just one station. The most species found at any one station was 43 at Whirlpool Cliff 6 m and the fewest (10) in the intertidal zone at the same site. Both the total bryozoan species and their prevalence was low at both intertidal stations. The bryozoan fauna of the two different intertidal sites was quite different, with only half their species in common. The species lists for the four subtidal stations were, in contrast, similar but did differ in the duration of species occurrence. All bryozoan species with the highest level of prevalence (occurring for 10+ months over the study period) occurred at both sites and more than one depth, except *Membraniporella nitida* (Johnston) which was only found at Whirlpool Cliff. Ctenostomes were less well represented than either cyclostomes or cheilostomes. Cyclostomes were highly numerous (see Figs. 3-6) but were unspeciated by this study, so were difficult to critically compare. *Callopora* and *Escharella* were the most speciose genera, although few representative species of either were present at many stations or for many months.

3.4 *Temporal patterns in specific taxa*

At the Labhra Cliff 6 m station most species showed both a clear annual pattern and a degree of interannual variability of recruitment. The strength of both of these along with the absolute recruit numbers did, however, vary considerably with species. A number of the selected taxa shown in Figure 3 have one major recruitment event in each year, at about the same time in each year (e.g. *Tubulipora* sp., *Scrupocellaria scruposa* and *Microporella ciliata* (Pallas)). The variability between the patterns in these taxa is in the magnitude, duration and timing of the peaks.

Bugula sp. and *Fenestrulina malusii* (Audouin) had almost continuous but low levels of recruitment during the entire study period. Some species such as *Callopora dumerlii* (Audouin) and *Umbonula ovicellata* Hastings showed non-regular patterns of recruitment. The major recruitment event of the former species was in 1998 but had high levels through the 1998/1999 winter but showed no recruitment in the winters of 1997, 1999 or 2000. *U. ovicellata* re-

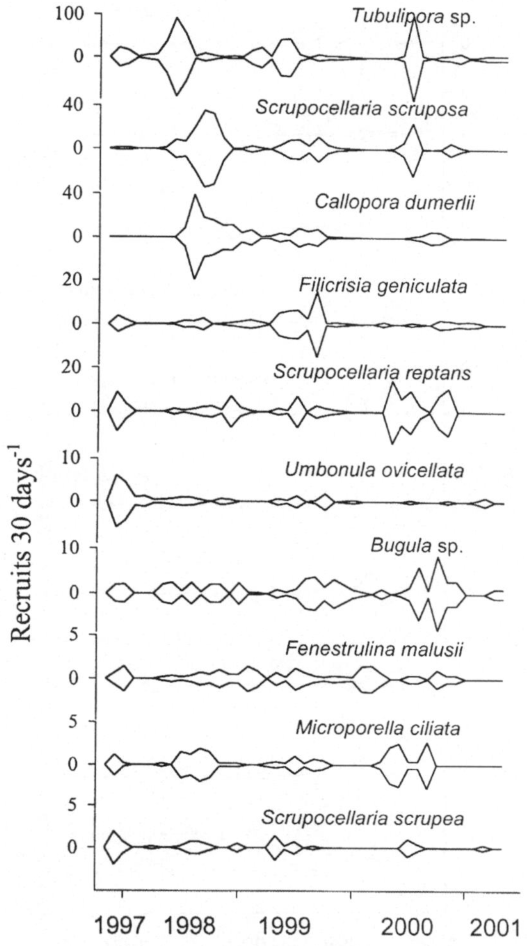

Figure 3. Recruit numbers of selected bryozoan species per 30 day period on panels at Labhra Cliff 6 m (Lough Hyne Marine Nature Reserve) with time. Data are means (from three panel replicates) in 'kite' format for pattern enhancement.

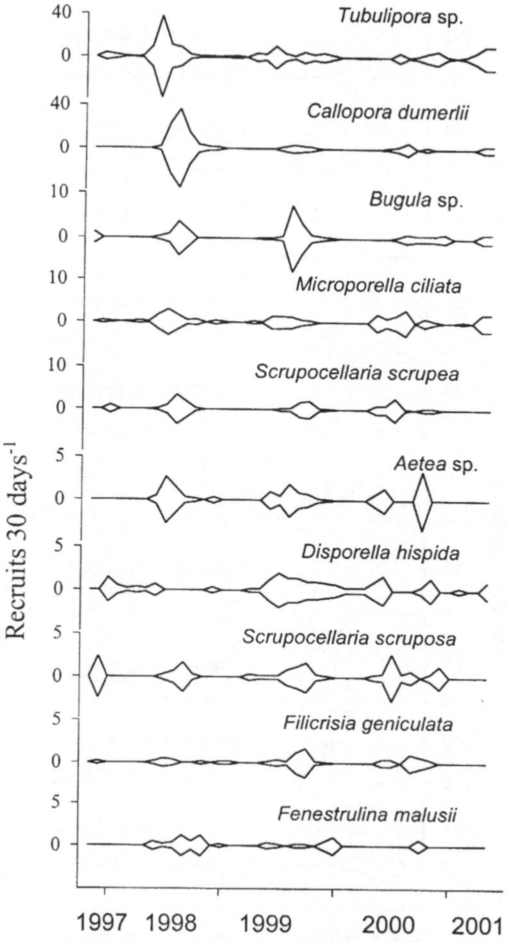

Figure 4. Recruit numbers of selected bryozoan species per 30 day period on panels at Labhra Cliff 12 m (Lough Hyne Marine Nature Reserve) with time. Data are presented as in Figure 3.

cruitment patterns showed even greater annual variation as the only real event (during the study period) occurred in the winter of 1997.

At the 12 m station of the same site (Labhra Cliff) the suite of major recruiting species, the levels and patterns of recruitment differed from those at 6 m (Fig. 4). At Labhra Cliff 12 m (in comparison with 6 m) recruitment patterns were generally more regular with distinct summer peaks, there was less recruitment in 1997 and more in 1998. The peaks of the different taxa were closely aligned in timing at 12 m in 1998, 1999 and 2000 (Fig. 4), only in 1997 is this apparent at 6 m (Fig. 3). Whilst some species, e.g. *Fenestrulina malusii* and *Microporella ciliata* showed virtually identical patterns at both depths, others e.g. *Bugula* sp. were very different. Most species, however, showed subtle differences, such as recruitment in winter 1998 at 6 m but not 12 m in *Callopora dumerlii.*

At the 6 m Whirlpool Cliff station the major colonist species differed from those at either subtidal depth at Labhra Cliff as did subtleties of patterns. At this station, more than any other, strong seasonality and temporally coincident multi-species recruitment was evident. In common with the 12 m Labhra (and 12 m Whirlpool Cliff) station there was little recruitment in late 1997. There were, however, also depressed values at this point in the year - every year, unlike at other stations. *Chorizopora brongniartii*, unlike any other bryozoan species in this study, recruited almost throughout the entire 4 year study period. Some taxa, e.g. *Aetea* sp. and

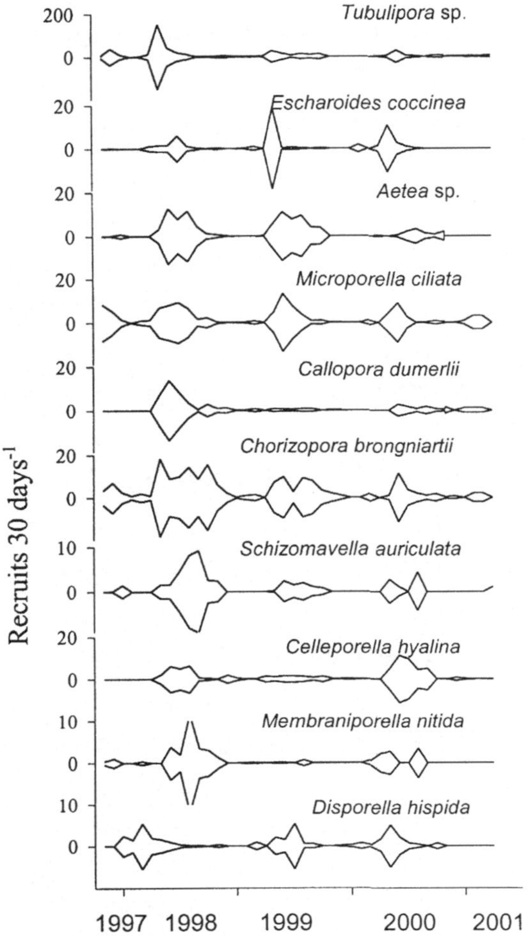

Figure 5. Recruit numbers of selected bryozoan species per 30 day period on panels at Whirpool Cliff 6 m (Lough Hyne Marine Nature Reserve) with time. Data are presented as in Figure 3.

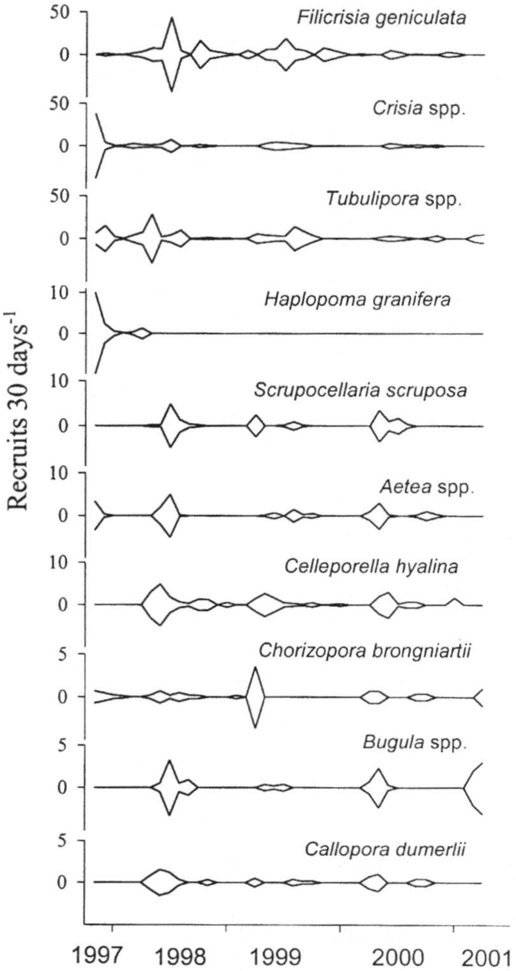

Figure 6. Recruit numbers of selected bryozoan species per 30 day period on panels at Whirlpool Cliff 12 m (Lough Hyne Marine Nature Reserve) with time. Data are presented as in Figure 3.

Disporella sp. showed very consistent patterns in both duration, timing and magnitude of recruitment. Other taxa (*Tubulipora* sp. and *Callopora dumerlii*), whilst showing markedly different year to year recruit numbers, were highly consistent in pattern between depths and sites. Despite the consistency of pattern (timing and duration) the actual numbers of recruits of these two taxa in peaks at each station were quite different.

The seasonal signal at Whirlpool Cliff 12 m was, similar to 6 m at the same site, strong but less synchronised between species, particularly in 1999, *Tubulipora* sp. was the major recruiting bryozoan in the subtidal except at Whirlpool Cliff 12 m. However, as at other sites, the taxa with most numerous recruits was still a cyclostome (*Filicrisia geniculata*). The recruitment characteristics of *F. geniculata* at Whirlpool Cliff 12 m were, as at Labhra 6 m, throughout the year. In contrast to the shallower site the main peak was in summer 1998 rather than summer 1999. The highest peak of *Bugula* sp. was in 1998 in comparison with 1999 at Labhra 12 m and 2000 at Labhra 6 m. *Haplopoma granifera* (Johnston) only had high recruitment in 1997 (as did *Unbonula ovicellata* at Labhra 6 m),. The complete absence of *H. granifera* recruits after early 1998 made this species the most variable on an inter-annual basis.

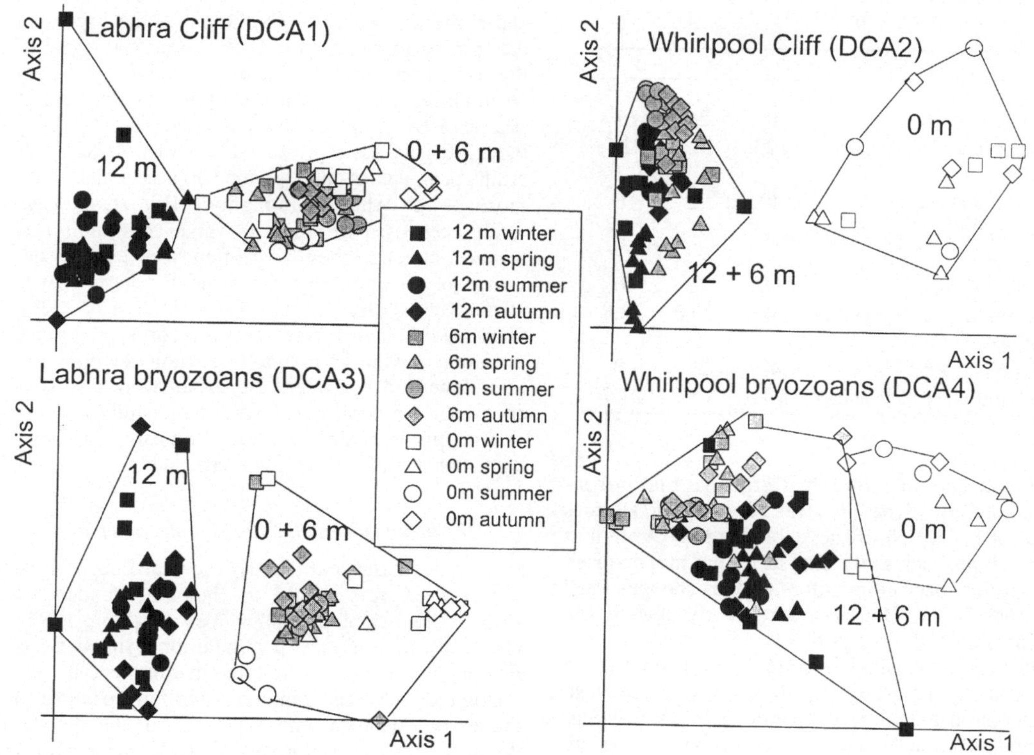

Figure 7. Detrended Correspondence Analysis (DCA) of community mean monthly abundance data with site, depth and season. DCA1 and 2 were performed on the whole community and DCA3 and 4 were performed on just the bryozoan community. Associated Eigenvalues and cumulative percentage variance explained by axes are illustrated in Table 2.

3.5 *Underlying community structure similarity*

An initial ordination (DCA) was performed on the entire data set which showed the main factor responsible for community separation to be depth. No clear gradient emerged with respect to depth, site, season or year as the structure at the sites distorted the ordination in different directions. As a result the data set was split on the basis of site and separate ordinations were performed on Labhra and Whirlpool. As with the overall analysis depth was the major structuring parameter in both Labhra and Whirpool communities. Although the pattern, a cline from 12 m to the intertidal zone, was consistent in both communities, there was a significant difference (in pattern) between Labhra and Whirlpool. At both sites there were essentially two distinct clusters (Fig. 7). At Labhra these were 12 m and a combined 6 m /intertidal grouping, whereas at Whirlpool clusters were subtidal and intertidal (DCA 1,2 in Fig. 7). In both cases there was a degree of separation within the mixed depth clusters but it was not complete.

Likewise there was some separation of seasonal data but the pattern was not consistent (winter data had high axis 2 values but other seasons were high and low at Labhra; summer data had high axis 2 values but other seasons were high and low at Whirlpool). On reanalysis of the data using just the bryozoan proportion of the community (the most speciose taxon), the gross patterns at both sites remained unchanged. Ordination of the bryozoan component, therefore, was adequate to investigate whole community structure irrespective of site. At both sites the DCAs of bryozoan community structure increased the separation within the mixed depth cluster but seasonal patterns still remained relatively obscured.

4 DISCUSSION

4.1 *Benthic community structure at Lough Hyne*

On a mega-benthic scale, the population dynamics of a single species, the sea-urchin *Paracentrotus lividus* (Lamark) may have been very important at structuring the lithophylic epifauna at Lough Hyne (Kitching 1987). The wide-scale nature of sea-urchin barrens suggests that this is not a feature restricted to this site, country or even region (Elner & Vadas 1990). The most recent times of prolific sea-urchin

Table 2. Eigenvalues and cumulative percent variance explained by axes of the DCAs (Fig. 7).

Axes	1	2	3
DCA1			
Eigenvalues	.46	.23	.17
Cumulative % variance	12.2	18.4	22.9
DCA2			
Eigenvalues	.49	.22	.16
Cumulative % variance	13.7	20.0	24.6
DCA3			
Eigenvalues	.80	.28	.20
Cumulative % variance	14.8	20.0	23.8
DCA4			
Eigenvalues	.43	.23	.20
Cumulative % variance	10.8	16.5	21.7

abundance and, therefore their principal role in benthic community dynamics, was 20 years ago (Barnes et al. 2001). Much of the stable hard substratum at Lough Hyne, such as the cliff faces, are monopolised by superior space competitors such as sponges (Bell & Barnes 2001). Boulders, particularly those in the high flow areas do have a rich and varied community but like cliffs one which is quite different from those observed on the panels in this study. The taxa which dominated the recruit numbers, such as barnacles and polychaetes, are common on boulders as are the taxon which dominated the species numbers; the bryozoans (Lilly et al. 1953, Maughan & Barnes 2000a, b). Like the sponges, bryozoans are clearly important to the benthic biodiversity of Lough Hyne as at many localities around the globe (Barnes 1996). The results presented indicate that a large number of individuals (particularly cyclostomes) and species (particularly cheilostomes) settle and recruit at sites of very different environmental character at Lough Hyne. The current study has also shown mixed contribution and inferences that can be drawn with respect to community structure. In recruit numbers the role of bryozoans was minor but in species numbers it was major. Overall and specific taxa patterns in the time-abundance plots (Figs 2-6) suggested season was the key factor but ordinations showed depth and site to be more important.

4.2 *Using bryozoans to analyse community structure*

As bryozoans composed the major taxon in terms of species and showed distinct patterns against the factors quantified here, can we confidently use bryozoans to dissect underlying community structure? The ordinations performed (DCA) would strongly suggest they are highly appropriate as bryozoan-only patterns mirrored overall community patterns at each site with respect to both the depth gradient and clustering (Fig. 7). Furthermore, the bryozoan data preserved the faint but mixed seasonal signal of the overall community. Such a seasonal signal may be amplified during shorter duration or polar studies (Barnes 1996, Stanwell-Smith & Barnes 1997). Even though bryozoans represented less than 10% of recruits, peak values corresponded approximately with those of the whole community. Bryozoans have rarely been used to measure whole community attributes, although they have often been used to develop key community assessment criteria (Buss 1979, Rubin 1982, Barnes 1996, Maughan 2001). Such studies incorporate a variety of spatial scale and span a wide latitudinal spectrum but little has been done in deep water. Bryozoan work also covers a huge temporal range, with some studies focussing on pre-Mesozoic events. Ironically this highlights the lack of Recent 'long term data'.

4.3 *The importance of measuring time at wide scale*

Sea-urchin studies, particularly those at Lough Hyne have proved very useful to the understanding of lithophyllic biota dynamics (see Kitching 1987). The study of echinoderm population dynamics has also helped ecologists studying recruitment by exposing the problems of inference without recourse to studies on adequate time scales. Both the debates about the periodic outbreaks of *Acanthaster plancii* (crown-of-thorns starfish) and various species of sea-urchins, and their causes, effects and remedy in the 1980s and 1990s highlighted the importance and lack of long term data collection. Lough Hyne has one of the longest term sea temperature and echinoid population censi data sets anywhere (Barnes et al. 2001). The cryptic and lithophyllic benthos has been studied intermittently over the last fifty years which has at least established some of the changes in species abundance (Little et al. 1992). Recording the 'end-effect' established community may tell us what has happened but it will not tell us why, how or even when (as colonists of some species grow very slowly). Spatial and temporal data of larval abundance, their settlement and recruitment, however, can supply many of these answers. Supply-side ecology has in the last couple of decades again become in vogue with ecologists and rightly so (Underwood & Fairweather 1987). Although the plankton at Lough Hyne have been one of the more intense areas of study it has either been of particular taxa (Holmes 1985, Ballard & Myers 1997) or all taxa at higher taxonomic level (Greenwood et al. 2001). Only recently has settlement or recruitment been investigated (Maughan & Barnes 2000a, b, Maughan 2001), so by default this has been on short time scale.

5 CONCLUSIONS

This medium-term data set clearly illustrates the effect and value of analysis of recruitment at the inter-annual level as well as more typical lower-scale time studies. Whilst distinct changes were visable at monthly, seasonal and inter-annual in recruit numbers, only seasonal patterns were evident in species numbers. Seasonal patterns were most obvious in individual taxa. We suggest, however, that short term studies overestimate the strength of the seasonal signal through lack of medium or long term data available. Using multiple seasonal data from differing years seems to be responsible for a reduction in the emergence of season as a structuring parameter in community analysis. Both time (at month, season and annual scales) and site were inferior to depth in structuring the overall community. The ubiquitous, abundant, speciose and fast recruiting nature of bryozoans makes them a good candidate for modelling systems on various time and spatial scales. The summits, agreements and new laws have shown we can 'talk the talk' with respect to environmental impact, biodiversity and global change in the marine environment, bryozoans may help us to 'walk the walk' with the action to follow the words.

ACKNOWLEDGEMENTS

We would like to thank Declan O'Donnell for permits to work at Lough Hyne MNR, James Bell and Amy Greenwood for dive buddying, Bob MacNamara for manufacture of some apparatus and Simon Harrison for aid with the ordinations (DCA).

REFERENCES

Ballard, L. & Myers, A. 1997. Vertical distribution, morphology and diet of *Proboscidactyla stellata* (Cnidaria: Limnomedusae) in Lough Hyne Marine Nature Reserve, Co. Cork, Ireland. *Journal of the Marine Biological Association of the UK* 77: 999-1009

Barnes, D.K.A. 1996. Low levels of colonisation in Antarctica: the role of bryozoans in early community development. In Gordon, D.P., Smith, A.M. & Grant-Mackie, J.A. (eds) *Bryozoans in space and time*. National Institute of Water and Atmospheric Research, Wellington.

Barnes, D.K.A., Crook, A.C., O'Mahoney, M., Steele, S. & Maguire, D. 2001. Sea temperature variability and *Paracentrotus lividus* (Echinoidea) population fluctuations. *Journal of the Marine Biological Association of the UK* 81: 359-360

Bell, J.J. & Barnes, D.K.A. 2001. A sponge diversity centre within a marine island. *Hydrobiologia* 440: 55-64

Buss, L.W. 1979. Bryozoan overgrowth interactions – the interdependence of competition for space and food. *Nature* 281: 2745-2747

Elner, R.W. & Vadas, R.L. 1990. Inference in ecology: the Sea urchin phenomenon in the northwestern Atlantic. *American Naturalist*, 136: 108-125

Greenwood, A., O'Riordan, R. & Barnes, D.K.A. 2001. Seasonality of zooplankton in a temperate sea lough. *Journal of the Marine Biological Association of the UK* 81: in press

Holmes, J.M.C. 1985. *Anchistrotos lucipetus* sp. nov. (Copepoda, Taeniacanthidae), a parasitic copepod from Lough Ine, south west Ireland. *Crustaceana* 48: 18-25

Kitching, J.A. 1987. Ecological studies at Lough Hyne. *Advances in Ecological Research* 17: 115-186

Lilly, S.J., Sloane, J.F., Bassindale, R., Ebling, F.J. & Kitching, J.A. 1953. The ecology of Lough Ine rapids with special reference to water currents. IV. The sedentary fauna of sublittoral boulders. *Journal of Animal Ecology* 22: 87-122

Little, C.D., Morritt, D. & Stirling, P. 1992. Changes in the fauna and flora of Lough Hyne. *Irish Naturalists' Journal* 24: 87-95.

Maturo, F.J.S. 1959. Seasonal distribution and settling rates of estuarine Bryozoa. *Ecology* 40: 116-127.

Maughan, B. 2001. The effects of sedimentation and light on recruitment and development of a temperate subtidal epifaunal community. *Journal of Experimental Marine Biology and Ecology* 256: 59-71

Maughan, B. & Barnes, D.K.A. 2000a. Seasonality of competition in early development of subtidal encrusting communities. *Marine Ecology* 21: 205-220

Maughan, B. & Barnes, D.K.A. 2000b. Epilithic boulder communities of Lough Hyne, Ireland: the influences of water movement and sedimentation. *Journal of the Marine Biological Association of the UK* 80: 767-776

Rubin, J.A. 1982. The degree of intransitivity and its measurement in an assemblage of encrusting cheilostome Bryozoa. *Journal of Experimental Marine Biology and Ecology* 60: 119-128

Stanwell-Smith, D. & Barnes, D.K.A. 1997. Benthic community development in Antarctica: recruitment and growth on settlement panels at Signy Island. *Journal of Experimental Marine Biology and Ecology* 212: 61-79.

Todd, C.D. & Turner, S. 1986. Ecology of intertidal and sublittoral cryptic epifaunal assemblages. I. Experimental rationale and the analysis of larval settlement. *Journal of Experimental Marine Biology and Ecology* 99: 199-231

Underwood, A.J. & Fairweather, P.G. 1987. Supply-side ecology and benthic marine assemblages. *Trends in Ecology and Evolution* 4: 16-20.

Bryozoan Studies 2001, Wyse Jackson, Buttler & Spencer Jones (eds)
© 2002 Swets & Zeitlinger, Lisse, ISBN 90 5809 388 3

Use of radioactive labelled food to assess the role of the funicular system in the transport of metabolites in the cheilostome bryozoan *Membranipora membranacea* (L.)

M.A. Best & J.P. Thorpe
School of Biological Sciences, University of Liverpool, Port Erin Marine Laboratory, Port Erin, Isle of Man, IM9 6JA, UK

ABSTRACT: Colonies of the large cheilostome *Membranipora membranacea* (L.) were fed radioactive labelled algal cells in a follow up to the earlier study of Best & Thorpe (1985). It was intended to try to establish the path taken by labelled metabolites when these were transported across the colony. The earlier work showed that transport occurs, but the levels of activity used were too great to allow resolution of detail in autoradiographs. Here we used a range of lower levels of radioactivity in an attempt to increase resolution. At higher magnifications the autoradiographs are somewhat 'fuzzy', but the results support the assumption that the funiculus transports metabolites between zooids in a colony.

1 INTRODUCTION

Best & Thorpe (1985) used radioactive isotope tracers to provide the first direct evidence of the interzooidal transport of metabolites in cheilostome bryozoans. They fed *Membranipora membranacea* on phytoplankton containing ^{14}C. However, as early as the 1870s, anatomical and morphological studies were considered to show structures suggestive of a food transport system (Nitsche 1871 a,b). This view was strengthened by the work of Davenport (1891) and Brien & Huysmans (1937). Hyman (1959), noting that interconnecting pores in the Gymnolaemata were plugged, suggested that the communication between zooids probably consisted of a slow percolation of fluid through the cells at the pores. She also speculated that "this process is further aided by mesenchymal cords, also called funicular cords."

Much detailed anatomical work has centred on the development of the interconnecting pores (pore plates or rosette plates) (see Silen 1944, Bobin 1958 a, b, 1962, 1964, 1965, 1971, Banta 1969, Lutaud 1979). However, some attention also has been paid to the funicular system. Lutaud (1961) observed that the funicular cells of *Membranipora membranacea* contained mainly lipids and non glycogenic periodic acid-Schiff (PAS) positive granules (the periodic acid-Schiff reaction is a test for carbohydrate material), but the structure and contents of the cells varied even within a single zooid. Moreover, Lutaud (1961) also observed a peripheral canal, which unites all the ends of the funicular system and the communication plates in each cystid. She concluded that the canal carries PAS-positive material and fat globules. Bobin (1977) subsequently observed similar canalicular patterns in another cheilostome, *Electra pilosa*. She considered that the funicular pathways were not constant, although longitudinal trunks always abut onto the transverse rosettes, and numerous branches are stretched between these trunks, the digestive loop of the polypide, and the lateral rosettes. She suggested that these patterns are transitory and that this "is the periodic reflection of the functional state that I would especially attribute to colonial nutrition." She further states that "concrete information on such physiological problems is still lacking for the cheilostomes."

It has also been suggested (Bobin 1977) that the polarity of the intercommunicating rosette plates (as evidenced from developmental studies) indicates the direction of metabolite circulation, with only the rosettes within the distal half of a zooid (termed "donor" by Lutaud 1961) properly belonging to that zooid (see Fig. 1).

Previous autoradiographic studies (Best & Thorpe 1985) illustrated the colonial transport of metabolites. This work was followed up by a technically more advanced study using double labelling (^{14}C and ^{35}S) aimed primarily at quantifying colonial metabolism and rates of transport (Miles et al. 1995). This confirmed and extended the main findings of the earlier work, but neither study indicated the mechanisms by which transport occurred. Best & Thorpe (1985) proposed the funicular system as a likely candidate for the transport system. Here we attempt to furnish further evidence to support this hypothesis.

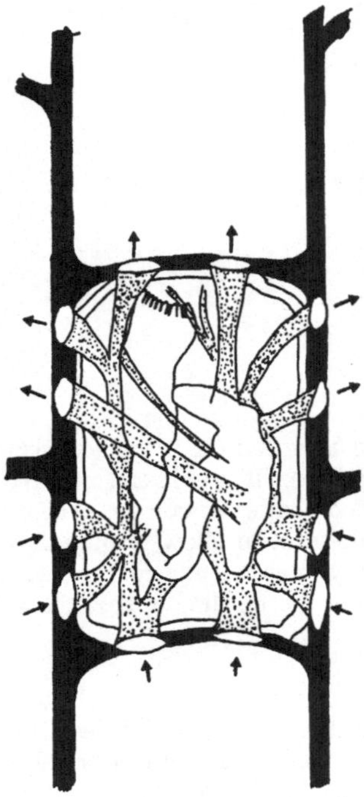
Figure 1. Diagram to show the general features of a cheilostome funicular system (after Ryland 1967, Bobin 1977 and others).

Figure 2 a) and b) illustrate likely areas of ^{14}C accumulation (heavy stippling) shown in Plate 1A.

2 MATERIALS AND METHODS

The methods used followed those described previously by Best (1985) and Best & Thorpe (1985). *Membranipora membranacea* was again chosen as the experimental species. Healthy young actively growing colonies (area 50 - 500 cm^{2}) were placed in a tank of 2 μm filtered, autoclaved seawater, and aerated with dust free air for a period of 24 hours. The colonies were then fed with a culture of the green unicellular flagellate alga *Tetraselmis suecica*, which had been radioactively labelled by growing for 24 hours in a culture medium containing 4 ml of 10 μCi bicarbonate solution (Sodium C^{14} bicarbonate; Amersham). Colonies were encouraged to feed by adding three drops of a non-radioactive *Tetraselmis suecica* culture from a 25 μm pipette.

Each colony was mapped and the precise locations of a number of zooids within the colony were noted. To each of these regions was added 1, 2, 3, 4 or 5 drops of the labelled algal culture from a 5 μm pipette. This procedure was used in an attempt to improve the resolution of the final autoradiograph. It had previously been found that if a polypide ingested a large amount of radioactive algae the image produced an autoradiograph which was too dark for detail to be observed in that or adjacent zooids. Alternatively, if the selected polypides ingested a few cells, only a faint grey area would be visible on the autoradiographs (see Best 1985, also Best & Thorpe 1985).

After feeding, colonies were killed at time intervals of 0, 5, 10, 20, 40 and 80 minutes and 3, 12, 24 and 48 hours. The colonies were thoroughly washed in particle free sea water, blotted dry, and then placed on a card, surrounded by blotting paper. This bundle was then placed in a plant press which was then put in an oven at 48°C for 18 hours (this corresponds to the "Method B" of Best & Thorpe 1985). After this treatment the colonies were removed and placed in contact with Kodak direct exposure film, sealed in envelopes and again placed in a plant press. This was then sealed in a number of black, plastic bags and placed in a freezer at -20 °C for 6 weeks.

The films were developed for 4.5 minutes in Kodak dental x-ray developer diluted 1 + 3, placed briefly in a water stop bath and fixed for at least five minutes in Kodafix solution (diluted 1 + 7). The films were then washed and dried (for further details see Best 1985, pp. 264-268 and Best & Thorpe 1985). The films were examined by eye, by transmitted light microscopy, and by using a microfiche reader (in a darkened room to reduce reflected light on the screen).

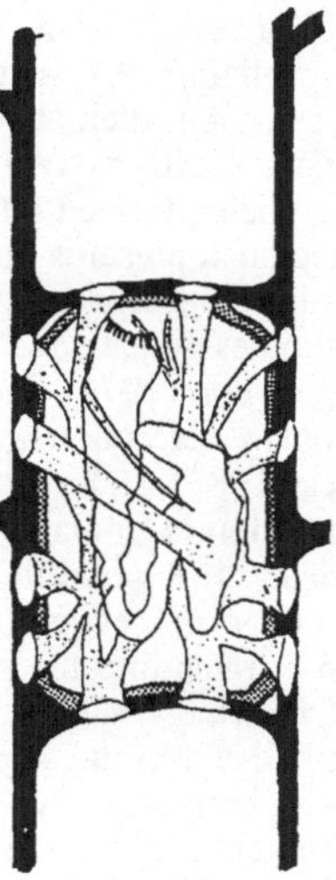

Figure 3. Illustrates likely areas of ^{14}C accumulation (heavy stippling) shown in Plate 1B.

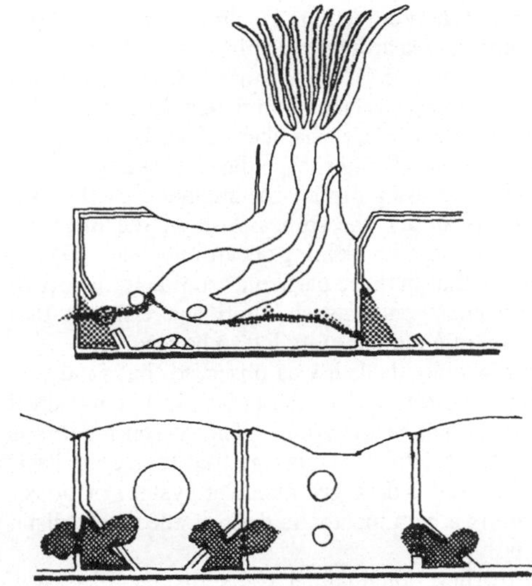

Figure 4a) and b) illustrate likely areas of ^{14}C accumulation, particularly around pore plates (heavy stippling) shown in Plate 1C.

3 RESULTS AND DISCUSSION

The attempt to improve resolution of areas of radioactivity within the colonies was only moderately successful. Areas of high grain density (i.e. very dark areas on the film) were found, but did not occur as would have been predicted from the number of drops of radioactive suspension fed to the

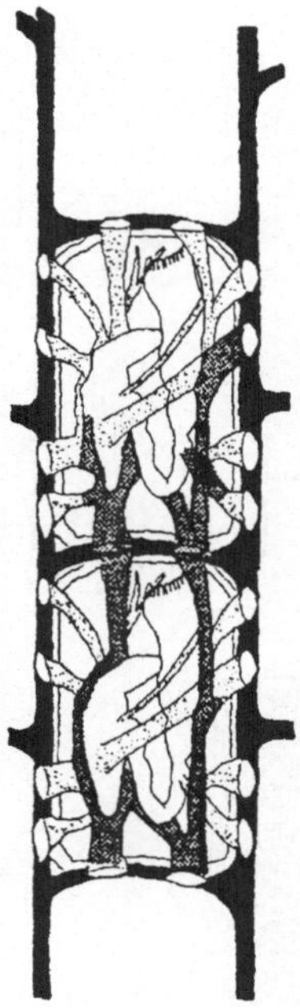

Figure 5. Illustrates likely areas of ^{14}C accumulation (heavy stippling) shown in Plate 1D.

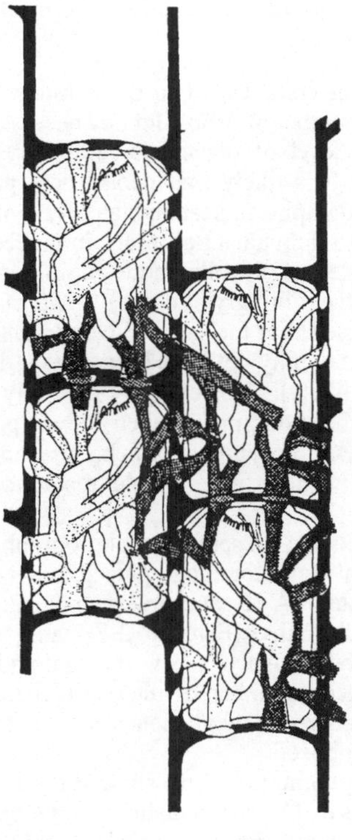

Figure 6. Illustrates likely areas of ^{14}C accumulation (heavy stippling) shown in Plate 1I.

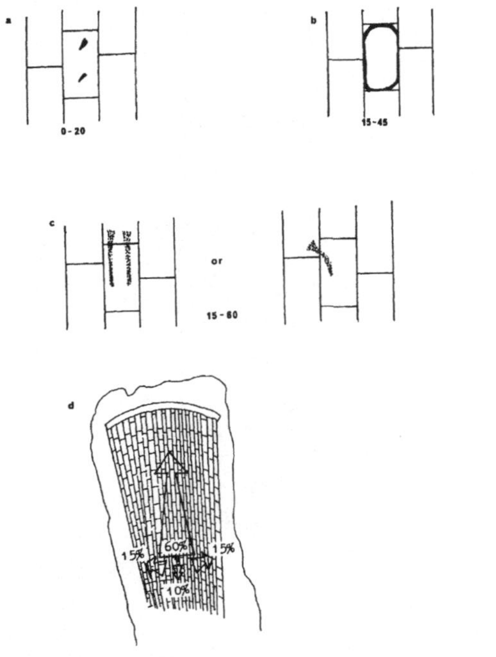

Figure 7. Hypothetical scheme for metabolite transport within a bryozoan colony. a) ingestion; b) distribution through zooid; c) transfer to adjacent zooid; d) colonial distribution. Times are in minutes.

polypides; the colonies fed on the cultures containing the highest number of drops did not necessarily show the greatest level of detectable radioactivity within the colony. It is likely that dense, dark patches on the autoradiographs, indicating high levels of activity, correspond to individual polypides in the feeding area which ingested algal cells more rapidly than their neighbours (i.e. "hungry" zooids). Alternatively, it could be argued that the labelled carbon taken up by the algal culture may not have been evenly distributed (this is less likely). However, by carefully searching through all the radiographs taken it was possible to find individual polypides which, presumably by chance, had a clear and suitable resolution on the autoradiographs.

The results are presented as photographs of various areas on the autoradiographic plates (autoradiographs) (Plate 1 A-K) and are accompanied by explanatory diagrams (Figures 2-7) and text (see below). Careful analysis of the autoradiographic plates revealed a number of interzooidal interactions and allows a hypothetical metabolite transport system to be proposed.

Within five minutes of ingestion of the labelled algae two areas of C^{14} accumulation can be observed in each feeding zooid (Plate 1A). There are two possible interpretations of this pattern. The more likely is that this represents two areas of concentrated food accumulation - the upper area being the pharynx and oesophagus, the lower region being food accumulated in the stomach, probably the area of the lower caecum (see Fig. 2a). An alternative explanation is that the upper region of activity represents food collected in the rectum being formed into a faecal pellet while the lower region represents food remaining in the stomach; again in the caecum (Fig. 2b). This is a less likely explanation as gut passage time can take at least 20 minutes. Menon (1974) quotes a minimum time from ingestion to defecation as 29 minutes at 22°C (see also Best 1985). The C^{14} pattern of the distal food concentration is similar in shape to a retracted lophophore and it is possible that food particles may be trapped on or between the tentacles as the tentacle crown retracts into the polypide. However, closer study indicates that the area of activity (Plate 1A) is too small and in the wrong position and orientation to be a lophophore.

In the autoradiographs of zooids killed 10 - 20 minutes after feeding the label is dispersed within the zooid and interzooidal transport is beginning. C^{14} concentration appears most dense towards the perimeter of the zooid (Plate 1B) and this may reflect the accumulation of carbon metabolites in the peripheral canal of the zooid (Fig. 3). In the same time period a limited area around the spine imprints indicates an accumulation of C^{14} in this area (Plate 1C). This may be explained by metabolites entering the lateral and/or transverse pore chambers (Fig. 4) and is the first indication of interpolypidal transport of metabolites. Occasionally an accumulation of the isotope occurs around the spines without the intervening stages of zooid darkening occurring (Plate 1D). It is possible that in these particular zooids transport to the pore chambers was so rapid and efficient that there was little radioactivity left to be recorded.

In a few individuals it was observed that food was not dispersed throughout the polypide but remained in the distal half of the zooid. The reason for this is difficult to ascertain. It may be that the zooid itself is aberrant with a damaged funicular system or possibly there is a high metabolic drain immediately distal to the polypide.

The earliest longitudinal transport recorded was 15 minutes after ingestion (see Best 1985, Fig. 4.2.c; Best & Thorpe 1985). However in most such transfers this was not observed until between 30 minutes and two hours after feeding. In a number of cases it was possible to distinguish two distinct channels of activity (Plate 1E). These may well correspond with the longitudinally running caecal funicular trunk and the tentacular funicular trunk (Fig. 5). It is interesting to note that after transfer the pathways become less distinct (Plate 1F), suggesting diffusion and dilution of the metabolite.

Over 70% of all clearly identified interzooidal translocations were parallel to the longitudinal axis of

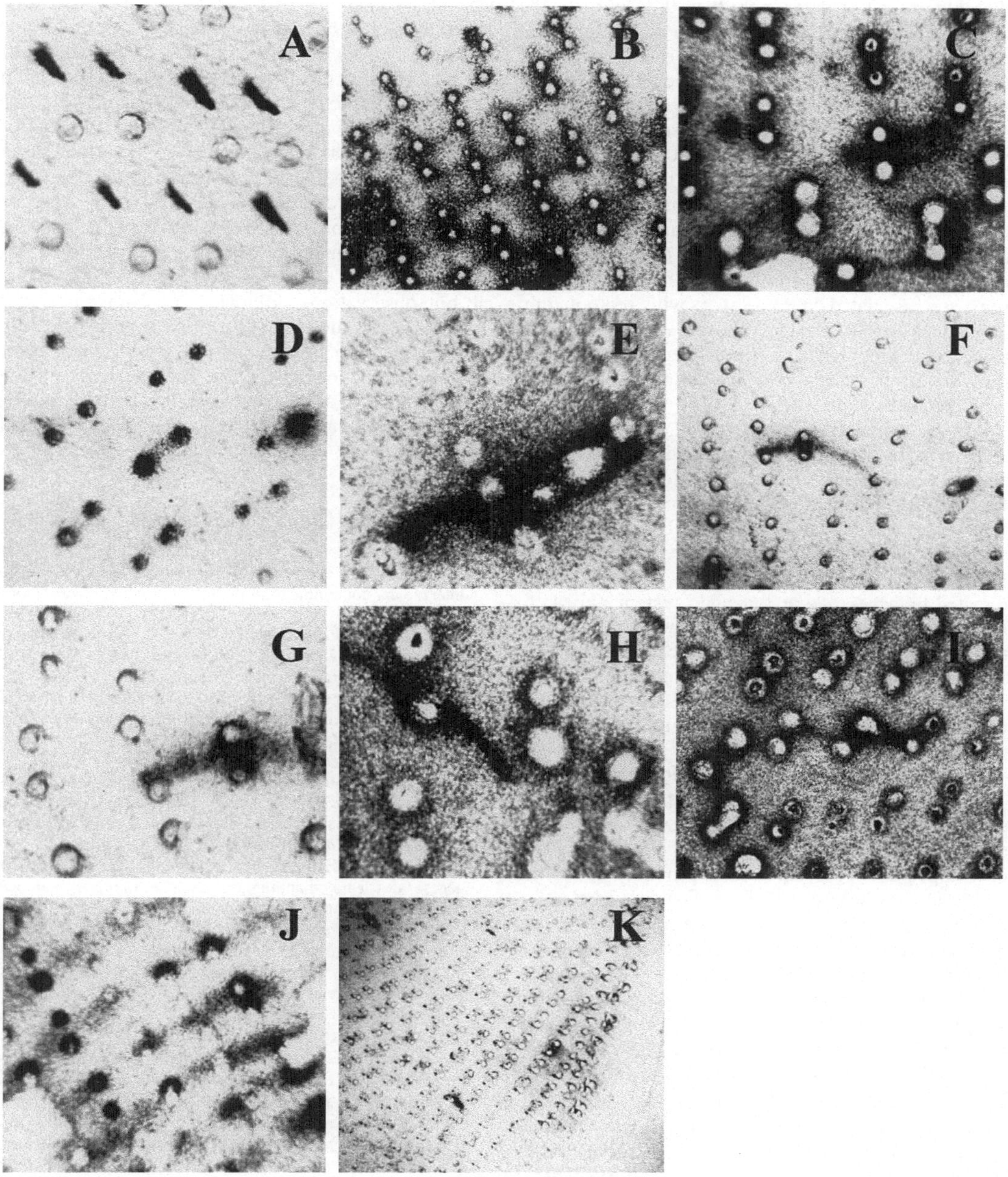

Plate 1. A. Autoradiograph of several zooids 5 minutes after ingestion of labelled food. For interpretation see text; also Figure 2; B. Autoradiograph of several zooids 10-20 minutes after ingestion of labelled food. For interpretation see text; also Figure 3; C. Autoradiograph of several zooids 10-20 minutes after ingestion of labelled food. For interpretation see text; also Figure 4; D. Autoradiograph showing longitudinal transfer after ingestion of labelled food. For interpretation see text; also Figure 5; E. Autoradiograph showing accumulation of label around spines of zooids. For interpretation see text; F. Autoradiograph illustrating diminution of label with distance. For interpretation see text; G. Autoradiograph illustrating label entering a tower zooid from an autozooid. For interpretation see text; H. Autoradiograph illustrating oblique transfer of label. For interpretation see text; I. Autoradiograph illustrating complex cross-transfer of label. For interpretation see text; also Figure 6; J. Autoradiograph illustrating parallel transport of label between zooids along the longitudinal axis of the colony towards the growing edge. For interpretation see text; K. Autoradiograph illustrating accumulation of label in intermediate zooids as well as at the growing edge.

the colony, and of these more than 95% were in the distal direction (i.e. metabolites were being transported towards the growing edge). The majority of these longitudinal translocations travelling proximally were observed to enter regions where there was an accumulation of tower cells, and in one case an observation was made of the C^{14} stream entering a "tower zooid" (Plate 1G). Best & Thorpe (1985) previously observed a significant accumulation of C^{14} in the tower zooids. This accumulation suggests that such zooids are metabolically active possibly due to the upward growth of the frontal membrane, or possibly they are associated in some way with reproduction (see Hincks 1880, Best & Thorpe 1985). In either case, since they have no feeding apparatus, whilst they are growing the tower zooids must constitute a local metabolic sink.

Of the lateral transfers observed, fewer that 1% were at right angles to the longitudinal axis of the zooids. Many apparent right angle transfers were found to be in fact accumulations of C^{14} along the transverse walls of the zooids. However, lateral transfers did occur as an oblique (i.e. diagonal) transfer (Plate 1H). This is quite feasible as the laterally directed branches of the funicular system appear to intercept the septulum of the lateral wall of the zooid at an oblique angle (see Ryland 1967, 1976, Bobin 1977). Two or more simultaneous transfers from one zooid can produce a complex transfer such as the cross-over illustrated (Plate 1I and Fig. 6), which appears to involve four zooids.

During the first few hours of food transference (i.e. 1 - 3 hours after ingestion) adjacent polypides seem to transfer metabolites in close sequence with each other (Plate 1J). After three hours this sequencing is not observed. This is probably due to the dispersion and dilution of the radioactive label causing the metabolite transference pathway to be less distinct. It is possible that differential retention of the label could also cause this effect.

After C^{14} has passed through 3 - 4 zooids it is no longer recognisable as a discrete compact zone of activity and is indicated by various diffuse shades of grey on the autoradiograph. After between 12 and 48 hours C^{14} is detectable at the growing edge of the colony. It appears mainly as a diffuse dark grey area, although certain zooids show a greater accumulation of radioactivity than others (see Best & Thorpe 1985). Although the C^{14} has accumulated at the growing edge it is noticeable that the polypides between the site of ingestion and the site of accumulation appear to have retained a small amount of radioactive label, particularly near the spines (Plate 1K).

From our observations it is possible to suggest how the transport of metabolites through the bryozoan colony utilising the funicular system may occur. This is illustrated diagrammatically in Fig. 7 a-d.

It is possible to propose some other mechanism which would explain the results obtained; by suggesting, for example, that metabolites may be transferred through a system involving the coelomic fluids and pores in the septum. However the indications are that at least most pores are plugged (see Bobin 1977). (This is to be expected since unplugged pores would severely reduce the efficiency of the hydrostatic skeleton which is required to effect lophophore extension, although such problems may be overcome by various strategies such as membrane sacs, as found in cyclostomes). Simple diffusion is also unlikely as the transport rates are very high; over 1.5 mm/hr. (see Best 1985, Best & Thorpe 1985). Overall, our observations, following on from our earlier work (Best & Thorpe 1985), provide an insight into how metabolites are transported across colonies of cheilostome Bryozoa, but for further detail of the precise mode of funicular transport between zooids autoradiography at the electron microscope level will be required.

REFERENCES

Banta, W.C. 1969. The body wall of cheilostome Bryozoa. II. Interzooidal communication organs. *Journal of Morphology* 129: 149-170.

Best, M.A. 1985. *Some aspects of the ecology and physiology of feeding marine Bryozoa.* Unpublished thesis. University of Liverpool. Port Erin, Isle of Man.

Best, M.A. & Thorpe, J.P. 1985. An autoradiographic study of feeding and the colonial transport of metabolites in the marine bryozoan *Membranipora membranacea. Marine Biology* 84: 295-300.

Bobin, G. 1958a. Structure et genese des diaphragmes autozoeciaux chez *Bowerbankia imbricata* (Adams) (Bryozoaire Ctenostome, Vesicularine). *Archives de Zoologie Experimentale et Generale* 96: 59-99.

Bobin, G. 1958b. Histologie des bourgeons autozoeciaux et genese de leurs diaphragmes chez *Vesicularia spinosa* (Linne). (Bryozoaire Ctenostome). *Bulletin de la Societe Zoologique de France* 83: 132-144.

Bobin, G. 1962. Histogenese des diaphragmes septaux stoloniaux et valeur des rosettes chez les Vesicularines (Bryozoaires Ctenostomes). *Archives de Zoologie Experimentale et Generale* 101: 14-42.

Bobin, G. 1964. Cytologie des rosettes de *Bowerbankia imbricata* (Adams). (Bryozoaire Ctenostome, Vesicularine). Hypothese sur leur fonctionnement. *Archives de Zoologie Experimentale et Generale* 104: 1-44.

Bobin, G. 1965. Introduction a l'histoire des *Bowerbankia imbricata* (Adams). (Bryozoaire Ctenostome Vesicularine). Les Ribonucleines. *Archives de Zoologie Experimentale et Generale* 105: 1-25.

Bobin, G. 1971. Histophysiologie du systeme rosettes-funicules de *Bowerbankia imbricata* (Adams). (Bryozoaire Ctenostome). Les lipides. *Archives de Zoologie Experimentale et Generale* 112: 771-792.

Bobin, G. 1977. Interzooecial communications and the funicular system. In: Woollacott, R.M. & Zimmer, R.L. (eds.) *Biology of bryozoans*: 307-333. New York: Academic Press.

Brien, P. & Huysmanns, G. 1937. La croissance et la bourgeonnment du stolon chez les Stolonifera (*Bowerbankia* Farre) (Evolution du stolon et de la zoecie chez les Bryo-

zoaires). *Annales de la Societe Royale Zoologique de la Belgique* 68: 13-40.
Davenport, C.B. 1891. Observations on budding in *Paludicella* and some other Bryozoa. *Bulletin of the Museum of Comparative Zoology, Harvard* 22: 1-114.
Hincks, T. 1880. *A history of the British marine Polyzoa.* London: Van Voorst.
Hyman, L.H. 1959. *The invertebrates. Vol. 5. Smaller coelomate groups.* New York: McGraw-Hill..
Lutaud, G. 1961. Contribution a l'e'tude du bourgeonnement et de la croissance des colonies chez *Membranipora membranacea* (Linne), Bryozoaire Chilostome. *Annales de la Societe royale zoologique de la Belgique* 91: 157-300.
Lutaud, G. 1969. Le plexus parietal de Hiller et la coloration du systeme nerveux par le bleu de methylene chez quelques bryozoaires chilostomes. *Zeitschrift fur Zellforschung und Mikroscopische Anatomie, Berlin* 302: 303-314.
Menon, N.R. 1974. Clearance rates of food suspension and food passage rates as a function of temperature in two North Sea bryozoans. *Marine Biology* 24: 65-67.
Miles, J.S., Harvell, C.D., Griggs, C.M. & Eisner, S. 1995. Resource translocation in a marine bryozoan: quantification and visualization of ^{14}C and ^{35}S. *Marine Biology* 122: 439-445.
Nitsche, H. 1871a. Beitrage zur Kenntnis der Bryozoen. IV. Uber die Morphologie der Bryozoen. *Zeitschrift fur Wissenschaftliche Zoologie, Leipzig* 21: 92-119.
Nitsche, H. 1871b. Beitrage zur Kenntnis der Bryozoen. III. Uber die Anatomie und Entwicklungsgeschichte von *Flustra membrananacea* L. *Zeitschrift fur Wissenschaftliche Zoologie, Leipzig* 21: 416-498.
Ryland, J.S. 1967. Polyzoa. *Oceanography and Marine Biology an Annual Review* 5: 343-369.
Ryland, J.S. 1976. Physiology and ecology of marine bryzoans. *Advances in Marine Biology* 14: 285-433.
Silen, L. 1944. On the division and movements of the alimentary canal of the Bryozoa. *Arkiv for Zoologi* 35A: 1-40.

Bryozoan Studies 2001, Wyse Jackson, Buttler & Spencer Jones (eds)
© 2002 Swets & Zeitlinger, Lisse, ISBN 90 5809 388 3

Use of radioactive labelled silt to show depletion of food supply by upstream colonies of *Flustrellidra hispida* (Fabricius)

M.A. Best & J.P. Thorpe
School of Biological Sciences, University of Liverpool, Port Erin Marine Laboratory, Port Erin, Isle of Man, IM9 6JA, UK

ABSTRACT: Colonies of the large ctenostome bryozoan *Flustrellidra hispida* (Fabricius) were placed in a chamber with a slow flow through it and were offered a mixture of food. This consisted of a suspension of the single celled alga *Tetraselmis suecica* mixed with fine particles of clay silt, which had been allowed to acquire a bacterial coating. The bacteria had been previously labelled with radio isotope so that if they were eaten off the ingested silt particles, the colonies would show labelling. The colonies did take up the label and in all cases the upstream colonies showed higher, often much higher, levels of radioactivity. If bacterial inhibitor was used during labelling, significantly reduced levels of label were found in the Bryozoa. It is concluded that the bryozoans are removing the bacteria from the silt and that the upstream colonies depleted the food supply to the downstream colonies.

1 INTRODUCTION

Over many years ecological competition in bryozoans has received considerable attention. Earlier work concentrated on spatial competition (e.g. Stebbing 1973 a, b, Jackson & Buss 1975, Jackson 1977, 1979 a, b), but competition for food was largely ignored or considered unimportant. This followed from the assumption that localised depletion of food by filter feeding organisms is rarely a limiting factor in the marine environment; indeed it had long been considered that for sessile marine invertebrates on hard substrata only space was limiting (see e.g. Knight-Jones & Moyse 1961). However Buss (1979, 1981) and Buss & Jackson (1981) provided evidence that food depletion can occur and may be severe enough to limit the size of a bryozoan population. Various subsequent studies have also suggested that competition for food may be of direct or indirect ecological significance (e.g. Best & Thorpe 1986 a, b, Thorpe et al. 1986, Okamura 1987, 1988, Sanderson & Thorpe 1996, Gruenbaum 1997, Lisbjerg & Petersen, 2000).

Mechanisms and rates of feeding have been quite extensively studied over many years (see e.g. Atkins 1932). A number of older studies provided estimates of feeding rates or feeding efficiency in bryozoans (Bullivant 1968, Strathmann 1973, 1982, Winston 1977, Best & Thorpe, 1983) and these have been extended by various experimental and theoretical approaches (e.g. Best & Thorpe, 1986 a, 1994, 1996; Hunter & Hughes, 1993; Sanderson et al., 1994; Riisgard & Goldson, 1997; Eckman & Okamura, 1998). One approach highly relevant to feeding competition has been to examine the effect of the feeding of upstream colonies or polypides on the food supply to other bryozoans further downstream, where food particles are being carried in a unidirectional flow. From several studies it appears that in such situations significant depletion may occur (Okamura, 1984, 1985, 1988, 1990).

The aim of our work was to provide data on the extraction efficiency of colonies serially placed in a food bearing water flow (as might occur in, for example, a water current passing around the fronds of adjacent *Fucus serratus* L. plants covered with encrusting bryozoan colonies). Radioactively labelled silt was used as the nutritional substance; these experiments being, in effect, a "spin off" from an earlier investigation into the role of resuspended sedimentary organic matter and its associated microorganisms in the nutrition of *Flustrellidra hispida* (Fabricius) (Best 1985).

2 MATERIALS AND METHODS

Colonies of the ctenostome bryozoan *Flustrellidra hispida* (Fabricius), which is an epiphyte of fucoid algae, were collected from the lower intertidal zone at Port St. Mary ledges (Isle of Man, northern Irish Sea) by pulling off intact plants of *Fucus serratus*. Only dense and healthy bryozoan colonies were used. For each experiment two plants with colonies of *Flustrellidra hispida* were carefully washed in 2

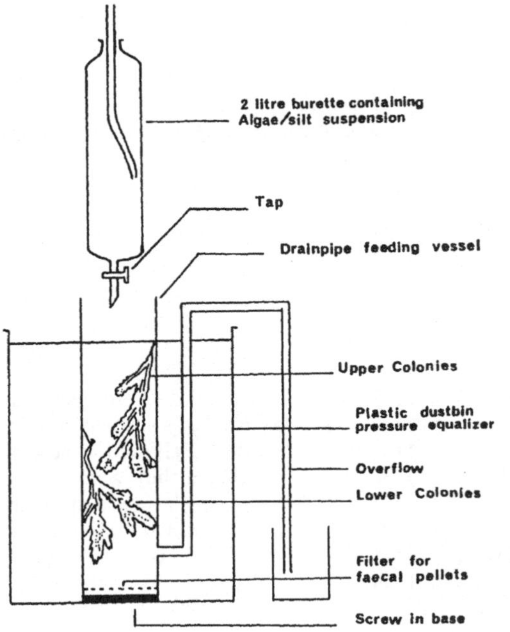

Figure 1. Diagram of the apparatus used to feed labelled silt to colonies of *Flustrellidra hispida* on two plants of *Fucus serratus*. The tap of the burette was adjusted to provide a continuous slow drip of labelled silt. For further details see Best (1985).

µm filtered autoclaved seawater and were placed one above the other in a "feeding chamber" (see Fig. 1).

Labelled silt was produced by the standard method (see Best 1985). ^{14}C label was incubated with natural silt particles both with and without bacterial inhibitor (30% NaCl). In the absence of the inhibitor the label was taken up by both the organic layer on the surface of the inorganic silt particle and by the bacteria present on the surface of the particle. With the inhibitor the label is absorbed only by the organic layer.

The bryozoans were then drip fed a mixture of radioactive sediment and a non-radioactive culture of the alga *Tetraselmis suecica*, the alga being supplied at a feeding chamber concentration of 50 cells per µl. The alga was used because earlier work showed that this stimulated feeding by the bryozoan (Best & Thorpe 1983, Best 1985). In each experiment feeding was continued for six hours (each zooid processed approximately 3.5 ml of water per hour). Faecal pellets from the bryozoans and uneaten silt were collected for subsequent analysis of ^{14}C content.

After each experiment the two plants were removed from the flow through system, thoroughly washed and dried. All the bryozoan colonies from each plant were cut into pieces with a new scalpel blade, pooled, and pieces randomly selected and placed into 5 ml scintillation tubes. The contents of each tube were dissolved and prepared for scintillation counting (for details of methods used see Best 1985).

3 RESULTS AND DISCUSSION

The results, presented as radioactivity levels measured in disintegrations per second, are given in Table 1, where the fourth column presents these data as percentage of activity found in the upper (proximal) colonies when compared with the lower (distal) colonies. Table 2 indicates the overall percentage distribution of the radioactivity assuming that all the ^{14}C in the system is accounted for by : (a) the bryozoan colonies; (b) the faecal pellets; and (c) the remaining uneaten silt (totals for (b) and (c) :20,257 DPM and 52,874 DPM respectively).

It can be clearly seen that the bryozoan colonies attached to the alga at the proximal end of the flow through system have more ^{14}C in their tissues than those positioned more distally to the incoming nutrients. In some experiments this was up to five times more in the proximal colonies. There was a large variation between experimental colonies and in one case (Experiment 6) the difference between proximal and distal colonies was minimal. The differences between the experiments can probably be explained by variation between colonies in levels of silt ingestion, variation in the exact positioning of the colonies in the feeding chamber, and possible variations in the flow. However, the overall results from the autoradiography confirm the earlier work of Okamura (e.g. 1984, 1985, 1990), who used different methodology, indicating that in flow regimes food supply can be depleted by the presence of other colonies feeding upstream.

An unexpected feature of our results is that over the course of each of the two series (with and without bacterial inhibitor) of three experiments (replicates) levels of radioactive label incorporated into the colonies declined (Table 1). The likely reason for this is that the replicates were not true replicates and despite efforts to maintain the silt in an even suspension, more entered the feeding chamber in the first experiment of each of the two sets, and subsequently the density of the silt suspension declined. This probably occurred because, although very fine, the silt was denser than the sea water medium, and thus presumably tended to sink towards the bottom of the burette. Thus when the burette tap was opened the greatest density of silt would be delivered and this would subsequently decline.

All colonies contained over 2 % of the available ^{14}C in the system and in one exceptional case nearly 30%. Such levels indicate that sediment associated carbon may be nutritionally important to bryozoans.

Table 1. Levels of radioactivity (measured as disintegrations per minute, DPM) for proximal (upper) and distal (lower) colonies of *F. hispida.* Note that the levels are always higher in the proximal colonies, indicating depletion of the food supply before it reaches the distal colonies. The fourth column indicates the difference as a percentage. Experiments 1-3 were conducted without any microbial inhibitor in the mixture when the silt was radioactively labelled, whereas inhibitor was present in experiments 4-6.

Experiment No.	Proximal	Distal	Difference (%)
1	95,666	16,731	572
2	27,454	15,310	179
3	22,533	7,737	291
4	21,350	10,222	209
5	11,943	9,999	119
6	6,725	6,531	103

Table 2. Relative distributions of radioactivity for the system as a whole (each measured as percentages of the total) for proximal (upper) and distal (lower) colonies of *F. hispida.* Experiments 1-3 were conducted without any microbial inhibitor in the mixture when the silt was radioactively labelled, whereas inhibitor was present in experiments 4-6. Other levels measured were in faecal pellets (6.2%) and uneaten silt (16.3%).

Experiment No.	Proximal	Distal
1	29.4	5.1
2	8.4	4.7
3	6.9	2.4
4	6.6	3.1
5	3.7	3.1
6	2.1	2.0

A large proportion of the available carbon was present in the uneaten silt (16.25%) probably indicating that bryozoans are not primarily adapted for feeding in resuspended sediment. The filtering efficiency of the feeding apparatus generally is lower than for the capture of algal cells (see Best & Thorpe 1983, Best 1985). It is, however, possible that the uneaten silt collected was mainly sediment that did not pass through the bryozoan lophophores and was not available to the colony. The overall percentage of ^{14}C in the faecal pellets (>6%) is quite high, but the contribution from any individual colony would probably be very low, suggesting an efficient extraction of sediment associated carbon once sediment particles have been ingested.

REFERENCES

Atkins, D. 1932. The ciliary feeding mechanisms of the entoproct Polyzoa and a comparison with that of the ectoproct Polyzoa. *Quarterly Journal of Microscopical Science* 75: 393-423.

Banta, W.C., McKinney, F.K. & Zimmer, R.L. 1974. Bryozoan monticules; excurrent water outlets? *Science* 185: 783-784.

Best, M.A. 1985. *Some aspects of the ecology and physiology of feeding marine Bryozoa.* Unpublished thesis. Port Erin, Isle of Man: University of Liverpool.

Best, M.A. & Thorpe, J.P. 1983. Effects of particle concentration on clearance rate and feeding current velocity in the marine bryozoan *Flustrellidra hispida. Marine Biology* 77: 85-92.

Best, M.A. & Thorpe, J.P. 1986a. Effects of food particle concentration on feeding current velocity in six species of marine Bryozoa. *Marine Biology* 93: 255-262.

Best, M.A. & Thorpe, J.P., 1986 b. Feeding current interactions and competition for food among the bryozoan epiphytes of *Fucus serratus. Marine Biology* 93: 371-375.

Best, M.A. & Thorpe, J.P. 1994. Particle size, clearance rate and feeding efficiency in marine Bryozoa. In: Hayward, P.J., Ryland, J.S. & Taylor, P.D. (Eds.). *Biology and palaeobiology of bryozoans*: 9-14. Fredensborg, Denmark: Olsen & Olsen.

Best, M. A. & Thorpe, J. P. 1996. The effect of suspended particulate matter (silt) on the feeding activity of the intertidal ctenostome bryozoan *Flustrellidra hispida* (Fabricius). In: Gordon, D.P., Smith, A.M. & Grant-Mackie, J.A. (Eds.). *Bryozoans in space and time*: 39 - 46. Wellington, New Zealand: National Institute of Water and Atmospheric Research.

Bullivant, J.S. 1968. The rate of feeding of the bryozoan *Zoobotryon verticillatum. New Zealand Journal of Marine and Freshwater Research* 2: 134.

Buss. L.W. 1979. Bryozoan overgrowth interactions - the interdependence of competition for food and space. *Nature* 281: 475-477.

Buss. L.W. 1981. Mechanisms of competition between *Onychocella alula* (Hastings) and *Antropora tincta* (Hastings) on an eastern Pacific rocky shoreline. In: Larwood, G.P. & Nielsen, C. (Eds.) *Recent and fossil Bryozoa*: 39-49. Fredensborg, Denmark: Olsen & Olsen.

Buss, L.W. & Jackson, J.B.C. 1981. Planktonic food availability and suspension-feeder abundance : Evidence of *in situ* depletion. *Journal of Experimental Marine Biology and Ecology* 49: 151-161.

Eckman, J.E. & Okamura, B. 1998. A model of particle capture by bryozoans in turbulent flow: significance of colony form. *American Naturalist* 152: 861-880.

Gruenbaum, D. 1997. Hydromechanical mechanisms of colony organisation and cost of defense in an encrusting bryozoan, *Membranipora membranacea. Limnology and Oceanography* 42: 741-752.

Hunter, E. & Hughes, R.N. 1993. The effect of cell concentration on colony growth and feeding in the bryozoan *Celleporella hyalina. Journal of the Marine Biological Association of the United Kingdom* 73: 321-331.

Jackson, J.C.B. 1977. Competition on marine hard substrata: the adaptive significance of solitary and colonial strategies. *American Naturalist* 111: 743-767.

Jackson, J.B.C. 1979a. Morphological strategies of sessile animals. In: Larwood, G.P. & Rosen, B.R. (Eds.) *Biology and systematics of colonial organisms*: 499-555. London: Academic Press.

Jackson, J.B.C. 1979b. Overgrowth competition between encrusting cheilostome ectoprocts in a Jamaican cryptic reef community. *Journal of Animal Ecology* 48: 805-824.

Jackson, J.B.C. & Buss, L.W. 1975. Allelopathy and spatial competition among coral reef invertebrates. *Proceedings of the National Academy of Sciences* 72: 5160-5163.

Knight-Jones, E.W. & Moyse, J. 1961. Interspecific competition in sedentary marine animals. *Symposia of the Society for Experimental Biology* 15: 72-95.

Lisbjerg, D. & Petersen, J.K. 2000. Clearance capacity of *Electra bellula* (Bryozoa) in seagrass meadows of Western Australia. *Journal of Experimental Marine Biology and Ecology* 244: 285-296.

Okamura, B. 1984. The effects of ambient flow, colony size, and upstream colonies on the feeding success of Bryozoa I *Bugula stolonifera* (Ryland), an arborescent species. *Journal of Experimental Marine Biology and Ecology* 83: 179-193.

Okamura, B. 1985. The effects of ambient flow, colony size, and upstream colonies on the feeding success of Bryozoa II. *Conopeum reticulum*, an encrusting species. *Journal of Experimental Marine Biology and Ecology* 85: 60-80.

Okamura, B. 1987. Particle size and flow velocity induce an inferred switch in bryozoan suspension feeding behaviour. *Biological Bulletin* 173: 222-229.

Okamura, B. 1988. The influence of neighbors on the feeding of an epifaunal bryozoan. *Journal of Experimental Marine Biology and Ecology* 120: 105-123.

Okamura, B. 1990. Particle size, flow velocity and suspension feeding by erect bryozoans *Bugula neritina* and *Bugula stolonifera. Marine Biology* 105: 33-35.

Riisgard, H.U. & Goldson, A. 1997. Minimal scaling of the lophophore filter pump in ectoprocts (Bryozoa) excludes physiological regulation of filtration rate to nutritional needs. Test of hypothesis. Marine Ecology Progress Series 156: 109-120.

Sanderson, W.G., Thorpe, J.P. & Clarke, A. 1994. A preliminary study of feeding rates in the Antarctic cheilostome bryozoan *Himantozoum antarcticum.* In Hayward, P.J., Ryland, J.S. & Taylor, P.D. (eds). *Biology and palaeobiology of bryozoans*: 167-172. Fredensborg, Denmark: Olsen & Olsen.

Sanderson, W.G. & Thorpe, J. P. 1996. Effects of temperature on the feeding activity of some temperate intertidal Bryozoa. In Gordon, D.P., Smith, A.M. & Grant-Mackie, J.A. (eds). *Bryozoans in space and time*: 271-282. Wellington, New Zealand: National Institute of Water and Atmospheric Research.

Stebbing, A.R.D. 1973a. Competition for space between the epiphytes of *Fucus serratus* L. *Journal of the Marine Biological Association of the United Kingdom* 53: 247-261.

Stebbing, A.R.D. 1973b. Observations on colony overgrowth and spatial competition. In: Larwood, G.P. (ed.) *Living and fossil Bryozoa*: 247-261. London: Academic Press.

Strathmann, R.R. 1973. Function of lateral cilia in suspension feeding lophophorates (Brachiopoda, Phoronida, Ectoprocta). *Marine Biology* 23: 129-136.

Strathmann, R.R. 1982. Cinefilms of particle capture by an induced local change of beat of lateral cilia of a bryozoan. *Journal of Experimental Marine Biology and Ecology* 62: 225-236.

Thorpe, J.P., Clarke, D.R.K. & Best, M.A. 1985. Natural variation in tentacle number in marine bryozoans and the possible effects of interspecific competition for food. In: Nielsen, C. & Larwood, G.P. (eds) *Bryozoa: Ordovician to Recent*: 319-328. Fredensborg, Denmark: Olsen & Olsen.

Winston, J.E. 1977. Feeding in marine bryozoans. In: Woollacott, R.M. & Zimmer, R.L. (eds) *Biology of bryozoans*: 233-272. New York: Academic Press.

Bryozoan Studies 2001, Wyse Jackson, Buttler & Spencer Jones (eds)
 ISBN 90 5809 388 3

Colony life then and now: Lower Paleozoic trepostomes (500-350 mya) and living cyclostomes: a review

Richard S. Boardman

Curator emeritus, Department of Paleobiology, National Museum of Natural History, Smithsonian Institution, Washington D.C. 20560, USA; present address 3612 East Forest Lake Drive, Sarasota, Florida 34232, USA

ABSTRACT: Nearly all zooidal budding in stenolaemates of all ages is done by interior walls of free-walled zooids under outermost cuticular membranes of colonies. Interior walls apparently grow continuously and walls of neighboring zooids grow outwardly at equal rates to establish basic growth habits of colonies. The controlling difference in skeletal growth between trepostomes and living cyclostomes is the presence of interzooidal communication pores in the interior walls of living colonies. Lack of pores in walls of trepostomes required that skeletons above basal walls of colonies were entirely interior, nutrients were distributed through outermost confluent colony-wide body cavities so that growth occurred only on outermost skeletal surfaces, and orificial walls of autozooids were free-walled. As a result, trepostome colonies developed few growth habits but could continue outward growth to form colonies that were typically robust and "stony". In contrast, interzooidal communication pores in interior walls of living cyclostomes distribute nutrients throughout colonies permitting skeletal growth on all interior skeletal surfaces and inner surfaces of cuticular membranes to form exterior skeletal walls. Exterior skeletons of living colonies make possible numerous growth habits and commonly halt outward growth of interior walls early so that living colonies typically are relatively delicate. Polypide anatomy of living colonies is unexpectedly varied, complex in function, and seems often to be uncorrelated with skeletal characteristics. Polypides range from relatively short to extremely long in different taxa, grow faster than their skeletal walls, undergo relatively few cycles, and start feeding early and steadily during regeneration so that inward growth from stationary attachment organs requires inward shifting of retractor muscle connections during a cycle of constant feeding. Undersized regenerating polypides in Lower Paleozoic trepostomes also must have started feeding early in each cycle, but in contrast, grew away from basal diaphragms at roughly the same rate as their living chamber walls to a uniformly short size, and underwent many cycles indicated by ontogenetic sequences of basal diaphragms in long autozooids. In some trepostomes, closely spaced skeletal diaphragms in many autozooids were structural diaphragms that were too close to colony surfaces or to each other to have accommodated feeding polypides. Temporarily at least, they were nonfeeding and out-of-phase with actively feeding neighbors, just as many autozooids in living cyclostomes stop feeding when they develop skeletal terminal diaphragms.

1 INTRODUCTION

The modern study of the order Trepostomata Ulrich, 1882, began with H. Alynne Nicholson in the 1870s. He first demonstrated the necessity of internal studies of colonies using thin sections, followed by Edward O. Ulrich in the 1880s. Both sectioned Paleozoic trepostomes for the first time and realized that few morphological characteristics were revealed externally in contrast to the great number that were available internally in sections. Ulrich's accurate discussion of the structures of Paleozoic Bryozoa (1890, p. 287-322) is amazingly comprehensive by today's standards. A listing of rediscoveries of Ulrich's 1890s discoveries through the 20th century by following authors would be most revealing. Cumings and Galloway (1915) made the first major attempt to consider the biologic interpretations of trepostomes but their work, as was Ulrich's, was limited at the time largely by the lack of biological information on living stenolaemates and bryozoans in general. The seminal work of Folke Borg (1926, 1944) provided the basic modes of growth and biology of living cyclostomes that made possible many of the comparisons of living and trepostome colonies presented here. Tavener-Smith and Williams (1972) produced a major paper on skeletal secretion and structure of living and fossil Bryozoa. The first thorough explanation and illustration of the fixed-walled morphology and mode of growth of frontal exterior autozooidal walls after Borg was by Hinds (1975). J.-G. Harmelin (1976), produced a monograph on living fixed-walled cyclostomes of the Mediterranean that was one of the first to illustrate sections of polypides. A beautifully illustrated monograph that emphasized the brood chambers of stenolaemates was published by Priska Schafer in 1991.

The skeletal morphologies of the Lower Paleozoic trepostomes and living cyclostomes demonstrate two different modes of colony growth and architecture within the class Stenolaemata, both designed to accommodate the special problems of repeated degeneration-regeneration cycles of polypides. Despite these differences, shared functional and growth characteristics common to the class are available in detail in living cyclostomes and provide biologic interpretations that otherwise could not be realized in the extinct order. Many of the basic growth and functional relationships of living cyclostomes have yet to be discovered however, so that continued study of both cyclostomes and trepostomes is equally urgent and rewarding.

2 LIVING COLONY ARCHITECTURE

Colony growth habits in living cyclostomes are many and varied, largely because *interior* skeletal walls (present in nearly all fossil and living stenolaemates) have interzooidal communication pores that permit distribution of nutrients and resulting growth throughout colony interiors. Therefore, portions of cuticular membranes are calcified on their inner surfaces to form *exterior skeletons* that include frontal zooidal walls, fascicular zooidal walls, peristomes, terminal diaphragms, and some brood chamber covers. Living colonies have membranous orificial walls of polypides either *free-walled*, *fixed-walled*, or a *combination.*

Free-walled colonies with communication pores. Nearly all zooidal budding has been done by *interior* zooidal walls in stenolaemates of all ages. *Interior skeletons* partition existing body cavities of colonies. Body cavities of colonies are enclosed by outermost cuticular membranes that loosely cover colony skeletons (Fig. 1). Outermost cuticular membranes are held in place over colonies primarily by a combination of exterior colony basal walls and connections to orificial membranes of functioning polypides. Nutrient transfer among zooids and extrazooidal skeleton apparently occurs through the confluent and colony-wide cavities between the outermost cuticular membranes and colony skeletons and certainly occurs through the interzooidal communication pores of *interior* skeletal walls (Fig. 3). Therefore, nutrients are available throughout colony interiors so theoretically, tissues and skeleton can grow throughout colonies.

Stenolaemate colonies of all ages with separate colony wide outermost cuticular membranes are termed *free-walled* because orificial walls of their polypides are not skeletally attached but are attached to and open through the cuticular membranes to obtain outside nutrients (Fig. 1). Living horneroid cyclostomes are entirely free-walled with communication pores but so far have not been found with *exterior* skeleton above colony bases. Their *interior* skeletons are grown on their broad outermost surfaces as indicated by growth laminae, so they apparently calcify using nutrients largely from autozooids and confluent outer cavities of their colonies (Fig. 2). Rare *interior* diaphragm membranes are calcified from both sides indicating some use of communication pores in horneroids.

Fixed-walled colonies have membranous orificial walls of polypides *fixed* to skeletal apertures of frontal *exterior* zooidal walls or their extended *exterior* peristomes (Fig. 3). *Frontal walls* are *exterior* walls of zooids attached to outer ends of vertical *interior* walls (Figs 3-5). *Vertical zooidal walls* are the primary skeletal walls of stenolaemates and typically give length and breadth to zooids, establish the basic architecture of colonies, provide the skeletal foundation for fastening muscles and any attachment organs of polypides, and house retracted polypides. In fixed-walled colonies, vertical walls of many autozooidal living chambers are *interior* walls only (Figs 3, 5). In others they are a combination of *interior* walls internally and *exterior* frontal walls externally (Fig. 4; Boardman, 1998, Figs 14, 42, 43). Frontal *exterior* walls of zooids commonly cover skeletal surfaces of colonies so that confluent outer body cavities of colonies are closed off below distal budding zones and transfer of nutrients is entirely through communication pores of *interior* walls. Polypides typically retract inwardly through any peristomes present to their vertical walls.

In fixed-walled colonies, new zooids bud as *interior* walls under unattached cuticular membranes of distal-most budding zones (Figs 5, 27). *Exterior frontal* walls of autozooids calcify on inner surfaces of cuticular colony membranes just under budding zones and fasten to outer edges of *interior* walls to become outer parts of individual autozooidal walls and end *interior* wall growth permanently, typically producing relatively delicate colonies.

Tubuliporine cyclostomes have been defined largely by the presence of primary *exterior* frontal walls. The taxonomic distinction between genera that are free-walled with interior vertical walls and fixed-walled with exterior frontal walls begun by Borg (1926, 1944) at the suborder level was certainly reasonable considering the relatively small number of characters considered and the few species he had before him. The family Cinctiporidae Boardman, McKinney, and Taylor, 1992, however, has both free and fixed-walled genera in the same family, based on several shared skeletal characteristics apparently of phylogenetic significance and on similar polypide anatomy.

Fasciculate colonies typically have long parallel autozooids in clusters called *fascicles* (Fig. 6). They are either free-walled throughout (Boardman, 1998, Figs 52-61) or have *combined* orifical walls. In colonies with combined orificial walls the outermost ring of autozooids of a fascicle has *exterior* vertical walls, the *fascicle walls*, which coalesce to isolate a fascicle. The other vertical walls of the outermost autozooids and those wholly within a fascicle are *interior* so the colonies are a combination of mature autozooids arranged in an outer ring that is *fixed-free* and all inner autozooids that are *free-walled* (Fig. 6).

3 TREPOSTOME COLONY ARCHITECTURE

Colony growth habits of trepostomes were few (commonly encrusting, massive, dendroid, frondescent, and bifoliate) largely because all colonies lacked interzooidal communication pores internally so were restricted to *interior* skeletal growth. It is evident from the growth laminae of zooidal walls that skeletons grew largely on their outermost surfaces and wall edges that were directly available to the nutrients of feeding polypides and outer confluent body cavities (Fig. 7). Skeletal diaphragms of zooids were impervious so the last formed diaphragms kept outer nutrients from inert colony interiors. Trepostomes had protective outermost cuticular membranes colony wide and were free-walled (Fig. 8). Continued growth in outer regions of trepostomes typically produced large robust "stony" colonies (Boardman, 1960, pl. 3 fig. 2b; pl. 7, fig. 1a; pl. 9, fig. 1a).

4 EQUAL GROWTH OF INTERIOR ZOOIDAL WALLS

In stenolaemates of all ages, nearly all autozooids start as *interior* walls in budding zones. All *interior* autozooidal walls increase in length outwardly by addition to their outer edges and surfaces, so necessarily grow under *free* cuticular colony membranes (Figs 1, 2, 7, 8). Growth rates of *interior* walls of neighboring autozooids are essentially equal as indicated by local colony surfaces that are typically level. Growth rates are equal largely because *interior* zooidal walls are *compound*, that is, they are grown cooperatively with adjacent autozooids so that outward growth stays equal. The equal growth rate is independent of polypide cycles, either in-phase (Figs 2, 5), or out of-phase (Fig. 9). In trepostomes, polymorphs generally parallel autozooids and grow outwardly with them. In living cyclostomes, polymorphs are more varied in configuration and only the *interior* walls of autozooids all grow outwardly at the same rates. Many living free-walled colonies therefore develop level colony surfaces locally. *Interior* walls of others, however, develop uneven colony surfaces. Their autozooids grow outwardly isochronously with each other beyond intervening polymorphs to develop *interior* peristomes (Boardman, 1998, figs 36, 37).

As a result: the *first growth constant* of interior zooidal walls of all stenolaemates is simply that skeletal units (for example, a diaphragm and its outward extension into the autozooid wall, (Figs 10, 18, 22)) of adjacent zooids that abut or merge on shared boundaries of compound *interior* walls were calcified at the same time (Fig. 10). The *second growth constant* states that *interior* zooidal walls within colonies were grown isochronously parallel to their local colony surfaces, indicated by in-phase arrangements of skeletal diaphragms in trepostomes (Fig. 10) or differences in wall thicknesses or other discernable wall morphology in living cyclostomes.

5 CONTINUOUS GROWTH OF INTERIOR ZOOIDAL WALLS

Compound *interior* skeletal walls of autozooids and some polymorphs are not only equal in growth rate within local regions of stenolaemate colonies of all ages, but also apparently grow continuously without internal interruption in growing parts of colonies. Continuous *interior* wall growth is suggested by equal outward growth of some polymorphs with neighboring autozooids and equal growth rates of walls during different stages of polypide cycles of out-of-phase autozooids. Neighboring autozooids are out-of-phase where polypides are of different sizes and degenerated in living colonies (Fig. 9) and living chambers were of different depths in trepostomes. Apparently, *interior* wall growth rates have been independent of different stages of polypide cycles within stenolaemate colonies of all ages (the *cystid* concept).

6 LIVING SECONDARY EXTERIOR SKELETONS

Secondary exterior skeletons in living cyclostomes are typically scattered or clustered rather than colony-wide and calcify after vertical zooidal walls and autozooidal functions are established in a colony. Secondary *zooidal* skeletal structures of living species include scattered thin *exterior* peristomes that grow from apertures of established vertical autozooidal walls and all *exterior terminal diaphragms* calcified apparently after living functions of autozooids terminate. Secondary *colony* skeleton seems rare and apparently originates from the calcification of inner sides of limited portions of outermost cuticular colony membranes of free-walled colonies. Outer walls of brood chambers, either exterior or in-

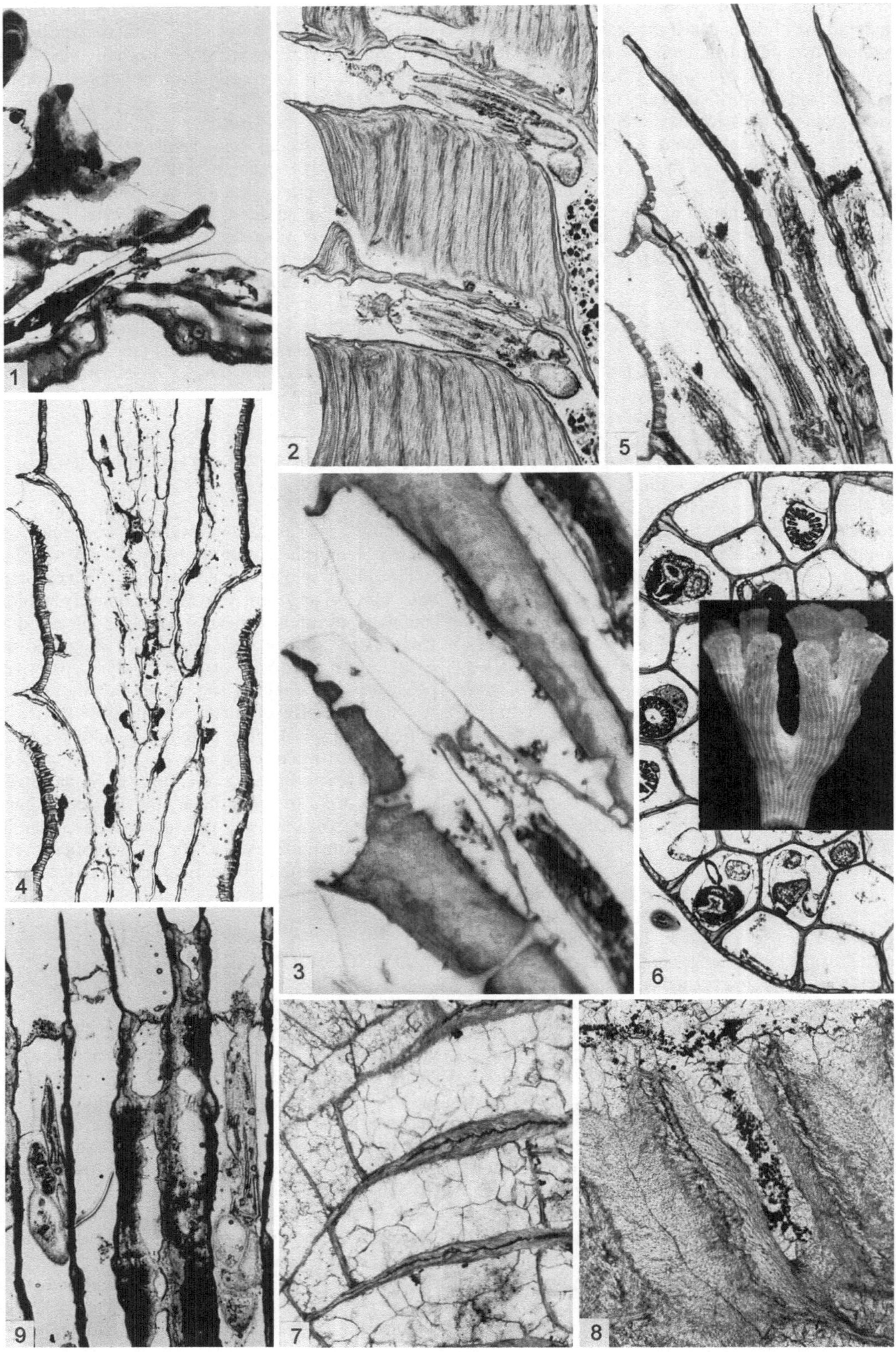
1
2
5
4
3
6
9
7
8

Figures 1-6, 9, living cyclostomes. 1, outermost membranous cuticular wall of freewalled *Densipora corrugata* MacGillivray, Phillip Island, Australia, x75. 2, growth laminae added to outer surfaces of extrazooidal skeleton between free-walled autozooids of horneroid, Arctic, x75. 3, cone-shaped attachment organ fastened to interior wall of living chamber, cylindrical membranous vestibular wall and flaring orificial wall attached to skeletal aperture of exterior frontal wall of fixed-walled idmidroneid, communication pore through interior wall, pseudopores in frontal wall, New Zealand, x225. 4, vertical walls of autozooids including inner interior walls and outer exterior frontal walls of fixed-walled cyclostome, Poor Knights Island, New Zealand, x38. 5, budding zone of interior walls at growing tip, beginning growth of exterior frontal walls proximally of idmidroneid, New Zealand, x75. 6, colony x2.3 and transverse section x38 of *Fasciculipora ramose* d'Orbigny with outer ring of combined fixed-free autozooids with exterior fascicular walls and interior free walls within fascicular walls of branch, McMurdo Sound, Anarctica. 9, out-of-phase polypide cycles of different sized regenerating polypides and one completely degenerated capped by attachment organ, smaller regenerating polypide was feeding with fecal pellet, in *Neofungella* sp., Palmer Penninsula, Antarctica, x75. Figures 7, 8, Lower Paleozoic trepostomes. 7, skeletal laminae indicate growth on outer edges of autozooids, terminal diaphragms on lower two, Upper Ordovician *Amplexopora septosa* (Ulrich), Kentucky, x75. 8, indication of free-walled polypide and outermost cuticular colony membrane in Upper Ordovician *Homotrypa nodulosa* Bassler, Ohio, x75.

terior, also are secondary skeletal structures and are not considered here (see Schäfer, 1991).

Secondary exterior peristomes (the emergent peristomes of Boardman et al, 1992) are scattered on colonies of some free-walled and fewer fixed-walled taxa where they calcify directly to apertures of established vertical autozooidal walls, either free (Figs 11, 12) or fixed (Boardman et al, 1992, Fig. 43a). Polypides are attached to secondary *exterior* peristomes only by their fixed orificial walls to form scattered fixed-walled autozooids in otherwise free-walled colonies. In free-walled colonies, access of individual autozooids to confluent outer body cavities by secondary peristomes is lost and in some colonies outward growth of *interior* walls of autozooids is terminated (Fig. 12). Secondary peristomes calcify on the inner surfaces of *exterior* cuticular membranes of autozooids as extensions of living chambers so apparently use nutrients from their individual autozooids during calcification. Theoretically then secondaw peristomes calcify while autozooids are feeding before terminal diaphragms are formed. Secondaw, exterior peristomes are typical of some cinctiporids (Fig. 11) and heteroporoids (Fig. 12).

Exterior terminal diaphragms are common in living cyclostomes and occur just inside skeletal apertures of autozooids that are either *interior* and free (Fig. 13), or *exterior* and fixed (Hinds, 1973, Fig. 1). Interzooidal communication pores supply nutrients from within so that, diaphragms calcify the inner sides of existing cuticular membranous diaphragms by edgewise calcification (Fig. 14) from surrounding autozooidal walls towards diaphragm centers (Fig. 15). Subtriangular pseudopores typically point in the direction of calcification in developing *exterior* skeleton (Fig. 15; Weedon and Taylor, 1997, 119-121). Terminal diaphragms end autozooidal feeding that generally is continued by more distal autozooids in living colonies.

Outermost exterior colony skeleton was first reported by Borg (1926, p. 197). Apparently it covered most of a colony in a single species but was not figured. *Exterior* colony skeleton (Fig. 16) occurs in patches on a few free-walled cinctiporid colonies as observed externally. This secondary skeleton apparently is formed by calcification of the inner surfaces of portions of cuticular, colony-wide membranes that cover established vertical walls of one or more zooids. They are not frontal walls of fixed-walled autozooids restricted to single autozooids. They are apparently patches of outermost colony skeleton that preserve the position of colony membranes during life. As a result, a patch of *exterior* skeleton is significantly inflated outwardly (Boardman, 1998, Fig. 113b) and can cover the *interior* walls of several neighboring autozooids (Fig. 16) up to their skeletal apertures. Initially, the thin *exterior* skeleton was interpreted to be primary *exterior* frontal walls of autozooids (Boardman et al, 1992, Fig. 16) placing both free and fixed autozooids in the same colony (Boardman et al, 1992, p. 29). This interpretation has been reversed, favoring the secondary origin of a calcified outer cuticular membrane of the colony covering *interior* autozooidal walls (Boardman, 1998, p. 43). An internal study is needed to confirm this interpretation. This is an excellent test of just how much can be safely determined by observation of colony exteriors alone.

7 TREPOSTOME SECONDARY ENCRUSTING OVERGROWTHS

Secondary *exterior* skeleton is ruled out in trepostomes by the lack of interzooidal communication pores in *interior* walls. However, secondary encrusting overgrowths arose locally from continued intracolony growth of the *interior* walls of clusters of autozooids of established colonies (Figs 17, 18). The secondary encrusting growth proceeded laterally in all directions from the originating clusters onto colony surfaces, presumably with no genetic change (Fig.18; Boardman, 1960, p. 39). Outermost cuticular membranes and confluent body cavities of the colonies covered both originating clusters of autozooids and the encrusted overgrowths without interruption. The basal colony walls of the encrusting overgrowths were necessarily *exterior*, probably

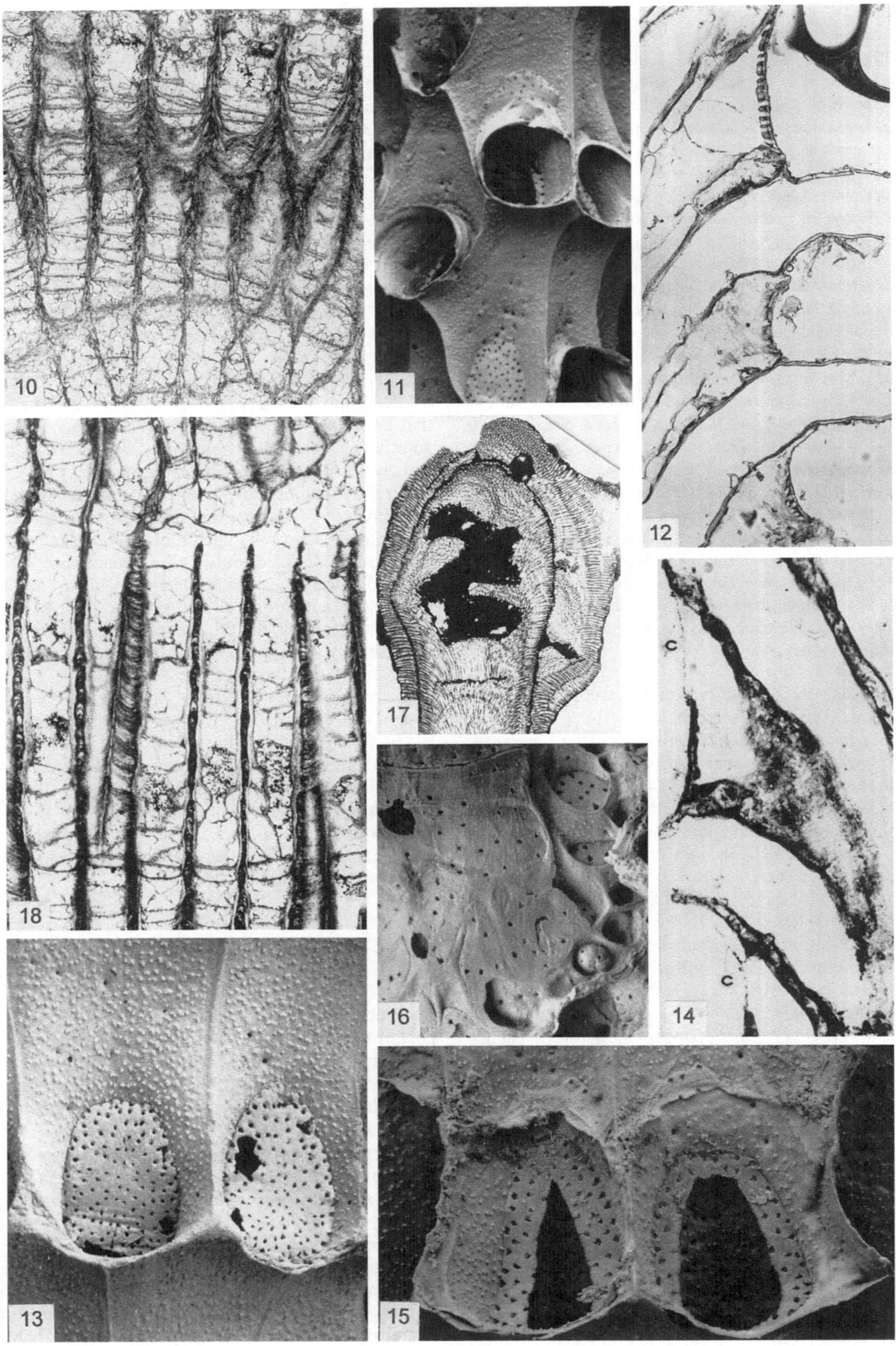
10
11
12
18
17
c
c
14
16
13
15

Figures 10, 17, 18, Lower Paleozoic trepostomes. 10, isochronous development of closely spaced structural diaphragms followed by living chambers in transverse section of *Rhombotrypa quadrata* (Rominger), 1866, Upper Ordovician, Indiana, x38. 17, secondary intracolony encrusting overgrowth supporting dendroid branch to right in Middle Devonian *Leptotrypella multitecta* Boardman, 1960, New York State, x2.3.18, continuing growth of autozooids to left, intracolony encrusting growth to right, approximate isochronous alteration of polypides and structural diaphragms in *Heterotrypa* sp., Upper Ordovician, Ohio, x38. Figures 11-16, living cyclostomes. 11, 13, 15, 16, *Cinctipora elegans*, Hutton, 1873, New Zealand, 11, secondary (emergent) peristomes in 5 of 6 autozooids, x26. 13, secondary exterior terminal diaphragms capping interior walls of free-walled autozooids, x40. 15, secondary exterior terminal diaphragms calcifying towards centers, direction indicated by pointing pseudopores, x52. 16, outer cuticular membranous wall of colony calcified over small polymorphs from left to right, x55. 12, 14, both from same colony of *Heteropora* sp. Middleton Island, Alaska. 12, secondary terminal diaphragm in polymorph above, secondary exterior peristomes attached to edges of vertical interior walls of autozooids, x75. 14, terminal exterior diaphragms forming on cuticular membranes 6) by edgewise growth, x92.

capturing the outer cuticular walls of the free-walled zooids being overgrown. Colonies that encrusted themselves were characteristic of Lower Paleozoic trepostomes, and in some species were the rule. The concentration of growth on outer regions of colonies and the lack of *exterior* skeletal walls that could halt outward growth of *interior* walls that are present in living colonies certainly contributed to the prevalence of encrusting overgrowths in the trepostomes.

8 TREPOSTOME POLYPIDES

Indications of retracted polypides of Lower Paleozoic trepostomes have been preserved in full length living chambers generally bounded by basal and terminal diaphragms (Figs 8, 18-22). The retracted polypides and living chambers of Lower Paleozoic trepostomes were consistently shorter but were comparable in cross sectional size to those of living autozooids (Boardman and McKinney, 1976, Figs 1, 2). Preserved polypides are rare and include indications of tentacles, orificial and vestibular membranous walls, and dark granular masses that represent digestive tracts (Boardman, 1983, p. 97; 1999, p. 804). Attachment organs have not been recognized. Apparently, ancient polypides were roughly comparable in anatomy and function to living polypides.

Presumably, regenerating undersized polypides in trepostomes began feeding early in polypide cycles, as they do in living colonies (see below). If so, polypides must have established membranous orificial walls open to outside nutrients early in regeneration. Also, beginning colonies and distal regions of older colonies surely required nourishment from their autozooids, even though their fully regenerated living chambers and polypides were undersized relative to those of later ontogenetic growth in proximal regions of colonies. Theoretically then, both in-phase and out-of-phase autozooids with feeding polypides of varying sizes could have actively contributed to colony growth.

Progressive polypide degeneration-regeneration cycles are indicated in autozooids of Lower Paleozoic trepostomes by outward successions of basal diaphragms that floored ontogenetic sequences of living chambers and their single polypides. *Progressive polypide cycles* occurred where polypides of a new cycle developed outwardly beyond the polypides of the last cycle. All evidence indicates only one polypide to a living chamber begun by a basal diaphragm (Boardman, 1971, Pl. 7, Fig. 4). Therefore, single regenerating polypides grew outwardly from their fixed basal diaphragms equally and apparently continuously with their new skeletal living chamber walls.

Full sized living chambers bounded by basal diaphragms and capping terminal diaphragms were relatively few (Figs 18-22). The great majority of living chambers are in ontogenetic sequences that lack terminal diaphragms. The spacing of basal diaphragms and their living chambers in continuous sequences was generally less than the length of full sized living chambers because the living chambers overlapped with each other (Fig. 22). The overlapping living chamber walls in succession combined to form the walls of the autozooids. Assuming that the full living chamber length remained the same during successive growth cycles within an autozooid, the spacing of the last two basal diaphragms and presumably the advancement of the new polypide equaled the outward growth of the previous living chamber during its cycle. If true, tentacle protrusion distances were set automatically from cycle to cycle.

Short retracted polypides were positioned and restricted in length by the basal diaphragms of the typically shallow living chambers. During a polypide growth cycle, living chambers and their polypides increased in length to full size from their basal diaphragms that were fixed in position in the autozooidal walls. Assuming continuous wall growth, the next growth cycle started "immediately" by forming new basal diaphragms within the just abandoned living chambers. Broad exozones of many trepostomes have 20 or 30 diaphragms in sequence that are widely spaced enough to have been basal diaphragms. Apparently, these diaphragms indicate 20 or 30 polypide cycles with each cycle growing a new basal diaphragm, living chamber, and polypide (Boardman, 1971, Pl. 7, Fig. 4). Measurements of full sized living chambers of trepostomes for species descriptions therefore, need to consider stages of polypide growth cycles and ontogenetic position in colonies. Repeated polypide growth cycles with a continuing outside source of nutrients developed

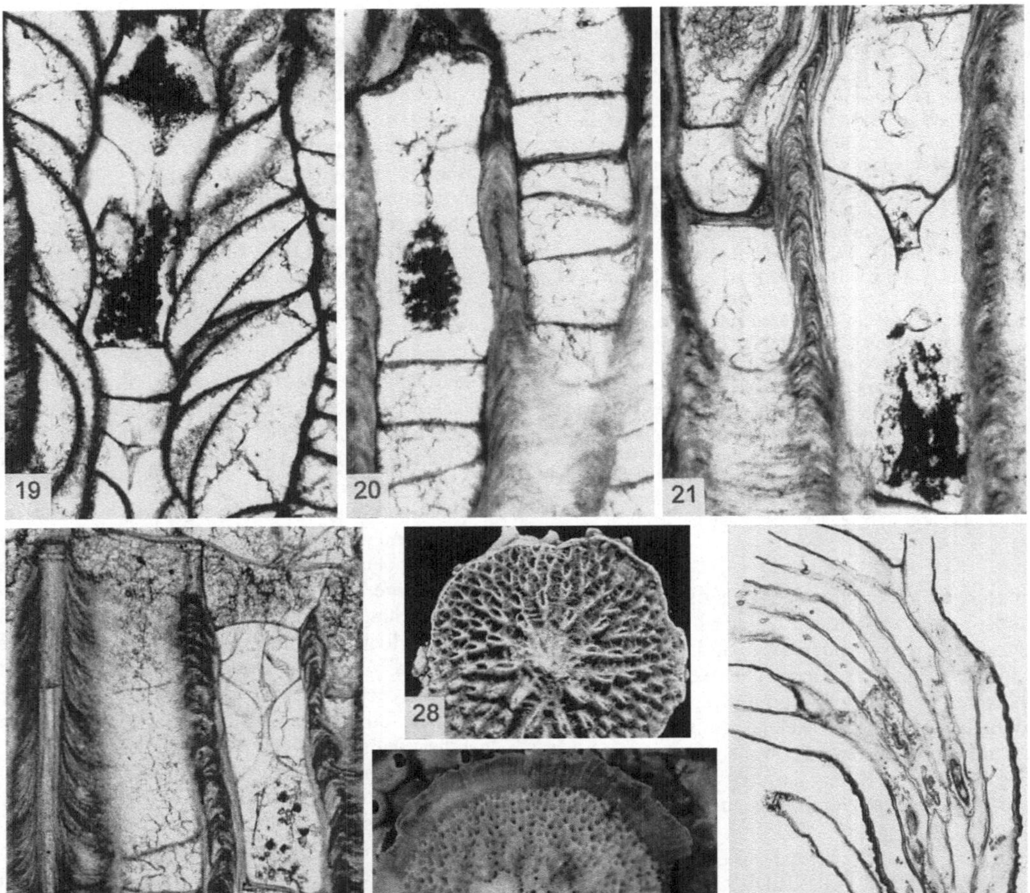
19
20
21
22
28
29
27
23
24
25
26

Figures 19-22, Lower Paleozoic trepostomes. 19, preserved polypide with dark granular gut, narrow tubular vestibular wall and flaring orificial wall, fragment of vestibular and orificial wall below separated from polypide above by terminal and basal diaphragms in *Prasopora grayae* Nicholson and Etheridge, 1877, Middle Ordovician, Scotland, x75. 20, feeding autozooid to left, out-of -phase with non-feeding autozooid to right that developed series of structural diaphragms while feeding autozooid grew a living chamber and polypide across compound skeletal walls in *Atactotoechus fruticosus* (Hall), 1877, Middle Devonian, New York State, x75. 21, feeding autozooid to right capped by interior terminal diaphragm, non-feeding out-of-phase autozooid to left with structural diaphragms in *Leptotrypella furcata* (Hall) 1877, Middle Devonian, New York State, x75. 22, in-phase feeding autozooids with living chamber on right capped by an interior terminal diaphragm, following series of basal diaphragms closely spaced by overlapping living chambers of ontogenetic series, large style to left in *Heterotrypa* sp, Upper Ordovician, Kentucky, x75. Figures 23-27, 29, living cyclostomes. Figs 23-26, 29, disporelloid sps. 23, simple, thin attachment organ fixed by ligaments to living chamber wall, Greenland, x225. 24, thick attachment organ attached perimetrically to chamber wall, large deeply stained atrial sphincter muscle, Alaska, x225. 25, pleated attachment organ attached by ligaments, New Zealand, x150. 26, unique bilateral (in plane at right angles) attachment organ attached to mural spines, New Zealand, x225. 29, typical encrusting disporelloid growth habit with extended apron of basal colony wall, brood chamber with secondary exterior cover, same species as fig. 24, Alaska, x7.5. 27, free-walled budding zone above, exterior frontal walls with peristomes just below, polypides without attachment organs retracted to bottoms of living chambers in *Mecynoecia delicatula* Harmelin, 1976, Mediterranean, x38. Figure 28, Lower Paleozoic disporelloid growth habit developed in order Cystoporata, Middle Silurian, Tennessee, x3.8.

long autozooids of many basal diaphragms, wide exozones, and robust "stony" colonies.

9 TREPOSTOME NON-FEEDING AUTOZOOIDS

The currently held assumption that all skeletal diaphragms of all autozooids were basal to actively feeding polypides in living chambers is not true for some trepostomes, just as not all autozooids were actively feeding in living cyclostomes because of terminal diaphragms (Figs 13-15). Applying the *first growth concept*, in some trepostomes closely spaced *skeletal diaphragms* of many autozooids were *structural* diaphragms that were calcified in sequences during growth of a neighboring living chamber (Figs 20, 21). In some exozones, *structural* diaphragms were grown so close to each other in sequences that they never had the space to function as basal diaphragms in support of living chambers with polypides (Fig. 10; Boardman, 2001, Fig. 2.5). Autozooids with last formed *structural* diaphragms were temporarily, at least, non-feeding and out-of-phase with polypide cycles of actively feeding neighbors. *Structural* diaphragms either alternated irregularly with basal diaphragms and their living chambers (Fig. 10, Boardman, 2001, Figs 2.6, 4.7) or occupied entire exozones, indicating long non-feeding intervals by their autozooids.

10 LIVING POLYPIDES

Polypides of living cyclostomes can be divided into two groups by the presence or absence of attachment organs. *Attachment organs* of retracted polypides are collar or cone-shaped membranes that fix polypides to their vertical living chamber walls within feeding distances of skeletal apertures during a polypide cycle (Figs 1-3, 5, 9, 23-26). During outward growth of vertical walls, polypides with attachment organs undergo *progressive polypide cycles* in which attachment organs and polypides of new cycles develop outwardly beyond the polypides of the last cycles. The feeding distance between attachment organs and skeletal apertures automatically remains the same from cycle to cycle when the advancement of the polypides of each new cycle equals the outward growth of the vertical skeletal walls of the last cycle.

In colonies with attachment organs, progressive polypide cycles become stationary without changing positions when vertical walls of living chambers stop outward growth. In stationary cycles of cinctiporids, attachment organs are retained from cycle to cycle leaving scars of their attachment ligaments embedded in the thickening living chamber walls (Boardman et al, 1992, Figs, 3, 9b,c).

Undersized polypides begin to feed early in regeneration as indicated by fecal pellets in alimentary canals (Fig. 9; Boardman et al, 1992, Figs 66, 79-81). In contrast to Lower Paleozoic trepostomes, living polypides grow in length faster than enclosing skeletal walls in order to regain full size in each of their several cycles. During the regeneration stage of a cycle, vestibular and orificial membranous walls probably grow outwardly from their fixed attachment organs equal to the accompanying growth of their *interior* skeletal walls. Polypides necessarily regenerate most of their length inwardly from their fixed attachment organs. Assuming that polypides can function continuously during regeneration, skeletal retractor muscle connections to living chamber walls necessarily shift inwardly without losing contact with the walls (Boardman et al, 1992, p. 35). In contrast to trepostomes, polypide cycles of living autozooids seem to be relatively few so that colonies characteristically are more fragile.

Polypides lacking attachment organs apparently have no method of migrating outwardly with each growth cycle (with one know exception) so they retract permanently at or near bottoms of living chambers throughout autozooidal life (Fig. 27). These polypide cycles are *stationary* because they lack attachment organs to hold polypides in progressively changing outer positions. Tentacle protrusion distances to skeletal apertures increase as living chambers of vertical walls grow in length. Where skeletal living chambers far exceed retracted polypides in length, full protrusion of tentacles apparently is accomplished by pleated and movable membranous sacs that carry tentacle crowns to distant skeletal apertures (Boardman, 1998, Figs 20, 21).

In modern colonies, relatively long living chambers can contain polypides that range from relatively short to long polypides that nearly fill their long living chambers. The extremes occur in the fasciculates with their characteristically long living chambers. Illogically, a species with the shortest polypides lacks attachment organs and retracts to the bottoms of living chambers, requiring long distances to skeletal apertures (Boardman, 1998, Figs 20, 21). Two species with long polypides that nearly fill the long living chambers during the last stages of their regeneration cycles have attachment organs that seemingly are needed only when undersized during earlier regeneration stages (Boardman, 1998, Figs 58-68).

In a small sampling of sections of 40 living cyclostomes species the anatomy of polypides is both unexpectedly diverse and confusing in distribution. The differences are most apparent in their attachment organs. Polypides of disporelloids have at least five different attachment organs (Figs 23-26), horneroids have four, tubuliporines have five, and fasciculates have two (Boardman, 1998, fig. 129). To add to the confusion, in addition to species with attachment organs in these 40 cyclostomes, one horneroid, two tubuliporines, three fasciculates, and one heteroporoid lack them entirely and retract polypides to bottoms of living chambers (Boardman, 1998, fig. 129). The mode of growth and morphology of colony skeletons has determined the informal classification of living cyclostomes used here. Skeletal morphology seems much less varied than polypide anatomy. How do such different polypides grow comparable skeletons within any one of these informal taxonomic groupings?

The more detail discovered about the stenolaemates themselves, the more inherently detailed the phylogenies and classifications of stenolaemates become and the less adequate current interpretations seem. The lack of correlation between the characteristics of living colony skeletons and their polypide anatomy raises basic phylogenetic and taxonomic questions that require the internal study of a host of unstudied stenolaemates, both fossil and living, to produce realistic interpretations.

For example, the ubiquitous living disporelloids are identified on their simple encrusting button-shaped growth habit (Figs 23-26, 29). The same growth habit is scattered in different orders and serves as the founding astogeny of many larger colonies throughout stenolaemate history, beginning in the Paleozoic (Fig. 28). The best guess lacking adequate distributional information suggests that the growth habit has evolved many times in different orders and that the five polypides indicate different and unknown origins.

11 CONCLUSIONS

Improved knowledge of the biology of both living and fossil stenolaemates, their morphologic and anatomical details and their inter-related functions and modes of growth, are required to produce modern classifications having at least some phylogenetic content, as in most other classes of the animal kingdom.

ACKNOWLEDGEMENTS

Again, special thanks to D.A. Dean of the Department of Paleobiology, the Natural Museum of Natural History of the Smithsonian Institution, for his uniquely superior sections and photomicrographs that make these details visible and interpretations possible. Thanks also to Paul Taylor for the SEMs illustrated here and a most useful technical review of the manuscript.

REFERENCES

Boardman, R.S., 1960. Trepostomatous Bryozoa of the Hamilton Group of New York State. *U.S. Geological Survey Professional Paper* 340: 1-87.

Boardman, R.S. 1971. Mode of growth and functional morphology of autozooids in some Recent and Paleozoic Tubular Bryozoa. *Smithsonian Contributions to Paleontology* 8: 1-55.

Boardman, R.S. 1983. General features of the Class Stenolaemata. In R.A. Robison (ed.), *Treatise on invertebrate paleontology. Part G. Revised. Volume 1*: 49-137. Boulder and Lawrence: Geological Society of America and University of Kansas Press.

Boardman, R.S. 1998. Reflections on the morphology, anatomy, evolution, and classification of the Class Stenolaemata (Bryozoa). *Smithsonian Contnbutions to Paleobiology* 86: 1-59.

Boardman, R.S. 1999. Indications of polypides in feeding zooids and polymorphs in Lower Paleozoic Trepostomata (Bryozoa). *Joumal of Paleontology* 73(5): 803-815.

Boardman, R.S. 2001. The growth and function of skeletal diaphragms in the colony life of Lower Paleozoic Trepostomata (Bryozoa). *Joumal of Paleontology* 75(2): 225-240.

Boardman, R.S. & McKinney, F.K. 1976. Skeletal architecture and preserved organs of foursided zooid in convergent genera of Paleozoic Trepostomata (Bryozoa). *Journal of Paleontology* 50: 25-78.

Boardman, R.S., McKinney, F.K.& Taylor, P.D. 1992. Morphology, anatomy, and systematics of the Cinctiporidae, new family (Bryozoa: Stenolaemata). *Smithsonian Contributions to Paleontology* 70: 1-81.

Borg, F. 1926. Studies on Recent cyclostomatous Bryozoa. *Zool. Bidr. Uppsala* 10: 181-507.

Borg, F. 1944. The stenolamatous Bryozoa. In S. Bock (ed.), *Further zoological results of the Swedish Antarctic Expedition 1901-1903* 3(5): 1-276

Cumings, E.R. & Galloway, J.J. 1915. Studies of the morphology and histology of the Trepostomata or monticuliporids. *Geological Society of America Bulletin* 26: 349-374.

Harmelin, J.-G. 1976. Le sous-ordre de Tubuliporina (Bryozoaires Cyclostomata) en Méditerranée, écologie et systématique. *Mem. Inst Oceanogr. Monaco* 10: 1-326.

Hinds, R.W. 1973. Intrazooidal structures in some tubuliporinid cyclostome Bryozoa. In G.P. Larwood (ed.), *Living and Fossil Bryozoa*: 299-306. London: Academic Press.

Hinds, R.W. 1975. Growth mode and homeomorphism in cyclostome Bryozoa. *Joumal of Paleontology* 49: 875-910.

Nicholson, H.A. 1881. *On the structure and affinities of the genus* Monticulipora *and its subgenera, with critical descriptions of illustrative species.* Edinburgh: William Blackwood and Sons.

Schäfer, P. 1991. Brutkammern der Stenolaemata (Bryozoa): konstruktionsmorphologie und phylogenetische bedeutung. *Courier Forschungsinstitut Senckenburg* 136: 1-147.

Tavener-Smith, R. & Williams, A. 1972. The secretion and structure of the skeleton of living and fossil Bryozoa. *Philosophical Transactions of the Royal Society* 264(859): 97-159.

Ulrich, E.O. 1890. Paleozoic Bryozoa. *Illinois Geological Survey* 8: 283-688.

Weedon, M.J. & Taylor, P.D. 1997. Skeletal ultrastructure in some tubuliporine cyclostome bryozoans. *Acta Zoologica* 78: 107-122.

Bryozoan Studies 2001, Wyse Jackson, Buttler & Spencer Jones (eds)
© 2002 Swets & Zeitlinger, Lisse, ISBN 90 5809 388 3

Bryozoans from temperate Pleistocene deep-water mounds, Great Australian Bight, Australia

Y. Bone
Dept. of Geology and Geophysics, Adelaide University, South Australia 5005, Australia

N.P. James
Dept. of Geological Sciences, Queen's University, Kingston, Ontario K7l 3N6 Canada

ABSTRACT: The Great Australian Bight, an enormous latitude-parallel shelf, has been the site of heterozoan carbonate deposition since Eocene time. Facing the Southern Ocean, the shelf is a high-energy, storm-dominated, mid-latitude carbonate system, with wave abrasion depth of ~70 meters and swell wave- base of ~120mwd. The facies transition upslope is marked by numerous bryozoan-rich build-ups. These structures, visible on seismic images, were drilled during ODP Leg 182. The mounds, at ~200 to ~ 350mwd, are muddy and characterised by prolific bryozoan growth. Sea-floor relief is up to 40 m and hundreds of metres laterally. No mounds are presently growing. The youngest mounds probably grew during glacial low-stands when upwelling provided the nutrients needed for high productivity. Bryozoans are diverse, with 74 genera identified, representing all growth forms in both cheilostomes and cyclostomes. Sand/mud fractions comprise the same genera as coarser fractions, but with delicate branching, articulated branching and articulated zoöidal forms more numerous.

1 INTRODUCTION

The continental shelf of southern Australia, an east-west, high-energy, mid-latitude, cool-water environment, has been the production, deposition and accumulation site of cool-water carbonates throughout the Cenozoic (James et al. 1992, 1997, 1999, 2001; Hageman et al. 1995). The sediments, formed predominantly of skeletal-fragments, have been studied extensively by many researchers, with the study by Wass and Connelly (1982) one of the first to appreciate the importance of bryozoans. Later, more-widespread studies have broadened the scope of investigations and presented detailed analyses of the diverse biogenic components and sedimentological processes across the southern Australian continental margin are now published in a series of papers by James et al. (1992, 1997, 1999, 2001).

One of the most important findings has been the visual confirmation during cruise FR98/05 aboard *RV Franklin,* using real-time underwater video, that patchiness is an important aspect of sea-floor cover on this temperate, open-ocean, high-energy continental shelf. Such patchiness, a well-known attribute of many marine benthic communities, is in this case dominated by consolidated/hard substrates.

This is the 2-D observation that is revealed by seismic of 3-D structures (mounds) within the packages of sediments that have accumulated in this area at times during the Pleistocene (James et al. 2001).

One of the aims of Ocean Drilling Project (ODP) 182 in 1999 was to drill these structures. The drilling was successful, and Hole 1131, located in the central part of the Great Australian Bight, is the topic of this paper (Fig. 1).

1.1 *The Great Australian Bight*

The current sea-floor of the Great Australian Bight (GAB) is an excellent example of patchiness, comprising large areas of carbonate sand dotted with isolated aggregations of divers sessile invertebrate communities growing on consolidated/hard substrates. The areal extent of these benthic communities ranges from the micro (a few cm) to the macro (a m or so) to the mega (nautical miles). individual patches are usually consist of abundant upright sponges, with prolific numbers of attached, encrusting and rooted bryozoans, and varying numbers of tunicates, molluscs, foraminifers and hydroids. The biota-covered patches are the areas of sediment deposition for as long as the patch is maintained.

2 BRYOZOAN MOUNDS

Muddy bryozoan-rich reef-mounds are a conspicuous part of many ancient carbonate ramp systems (Cuffey 1985, Monty et al. 1995) e.g. the Ordovician Mayesville bioherm, the Mississippian Mule

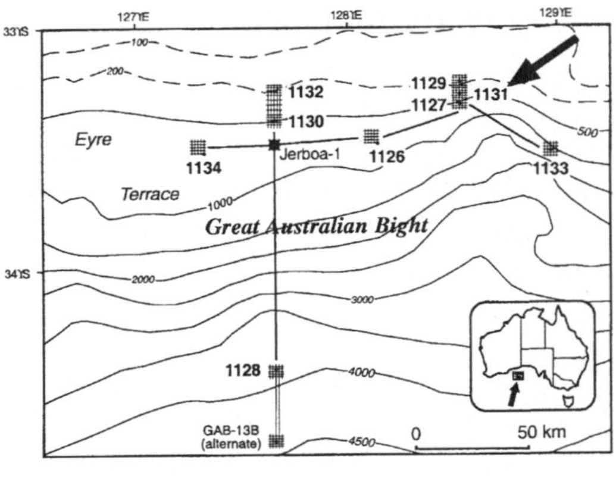

Figure 1. Location of ODP Leg 182 Drillhole 1131 in the Great Australian Bight, southern Australia. The hole was sited in 330m water depth.

shoe, New Mexico, the complex Waulsortian mounds of Europe and the Paleocene Danian mounds. These structures have features in common with the GAB mounds (e.g. association with brachiopods and echinoderms), but also have major differences (e.g. do not have extensive sea-floor diagenetic cavity infills). There were no examples in the literature of Quaternary bryozoan reef mounds prior to Leg 182.

The structures were first recognised on seismic images (Feary and James 1995). A cruise aboard CSIRO RV*Franklin* in 1995 located these mound-like features in water depths of ~300m. Extensive bottom profiling determined that they have a surface expression of <10m. Repeated attempts to sample the mounds produced sea-floor sediments that did not differ from those obtained from off-mound sites because the mounds are buried by up to 10m of Holocene slope carbonate (Fig. 2). Radiometric age dates confirm that the mounds at the site of 1131B range in age from ~20Ka to ~60Ka. Seismic profiles suggested the mounds are stacked and prograded upwards slightly during growth.

Drilling revealed that bryozoans dominated the fauna present on the sea-floor and that the mounds formed from the deposition and accumulation of predominantly bryozoan skeletal grains. Sponge spicules and foraminifers can be important, with echinoids, molluscs, ascidian spicules, serpulids, brachiopods, ostracods and peloids occasionally present as trace components. The grains are mainly unabraded and uncemented, except in cases where hardgrounds are present, when abrasion and/or cementation are visible (5 – 10%).

The mounds are represented in the cores obtained from the ODP Leg182 drilling by packages of rudstones, grainstones and floatstones. These can either be individual mounds or stacked mounds, with thin interlayers of packstones and wackestones (Fig. 2).

3 SAMPLES AND METHODS

The cores from Hole 1131B were logged, enabling the division of the core into describable rock units. These were sampled so that every representative facies was present, along with samples to represent a time-span component within the sample-suite.

Individual samples were split, with one split allocated to the bryozoan study. These samples were sieved into coarse and sand/mud fractions. All samples were individually examined under the binocular microscope to determine the bryozoan growth forms present, the genera within each growth form, and the relative abundance of each. The quantification of each genus in a sample was taken as a visual estimate of fragments + volume in relation to the total bryozoans (fragments + volume) within that sample.

Quantification was further based on calculating the % present of each genus in the sample, and allocating the following presence criteria:

dominant	>25% of all	bryozoans	in	sample	
common	>10 - 24%	"	"	"	"
present	>5 – 9%	"	"	"	"
trace	<5%	"	"	"	"

These data were then allocated a numerical equivalent (10 for dominant, 6 for common, 3 for present, and 1 for trace) so that a final ranking could be allotted to each genus that occurs in the samples studied in each of the four different environments, i.e. youngest mounds, thin mounds, off-mounds and oldest mounds. This method is far from statistically faultless, but it is impossible to devise such a package involving colonial animals that fragment into different sizes upon the death of the entire colony. It is further exacerbated by the manner and size of such fragments is dependent on a number of intrinsic (e.g. growth form) and extrinsic (e.g. hydrodynamic regime) parameters. The method does provide a semi-quantitative picture of the relative importance of each genus in the mounds.

4 BRYOZOAN DIVERSITY AND DENSITY IN THE MOUNDS

4.1 *Bryozoan diversity*

The diversity of the bryozoans in the mounds in drill-hole 1131B is remarkable. It has a total of 72 genera, with 56 in the coarse fraction and 35 in the fine fraction (Table 1).

Large bryozoans are mostly fenestrate, flat robust branching, encrusting and delicate branching growth forms. Fenestrates are not diverse and the 4 important genera evenly distributed. Flat robust branching forms are more diverse with *Adeonellopsis, Bracebridgia, Caleschara, Labioporella* and *Porina* the most abundant. Delicate branching forms are like

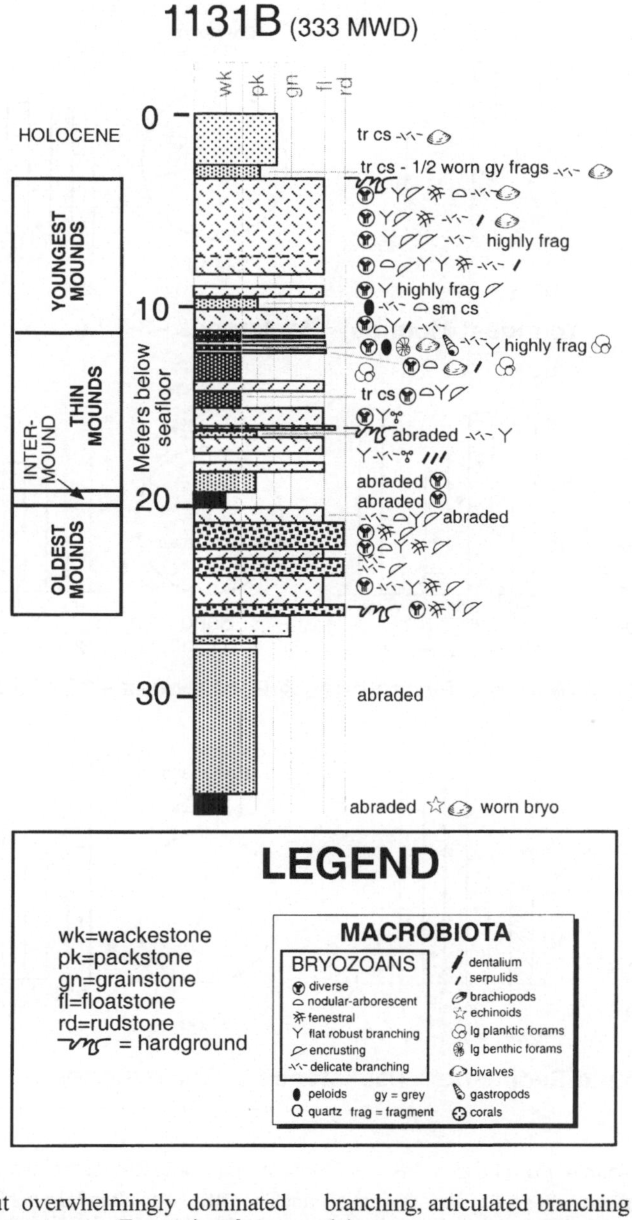

wise fairly diverse, but overwhelmingly dominated by *Idmidronea* and *Nevianipora*. Encrusting forms are the most diverse, with 28 genera, but no single genus dominant. Nodular-arborescent types, although conspicuous because of their size, are not diverse and dominated by *Celleporaria*. Sand and finer-size bryozoan-fragments within the mounds comprise the same genera as the coarser fraction. The main difference in the finer fraction is that delicate branching, articulated branching and articulated zooidal genera are more numerous.

4.2 *Bryozoan density*

Analysis of the sediments shows that the density (or distribution) of the bryozoans within the mounds is high relative to off-mound samples (Table 1), with other phyla only playing a minor role. Sponges,

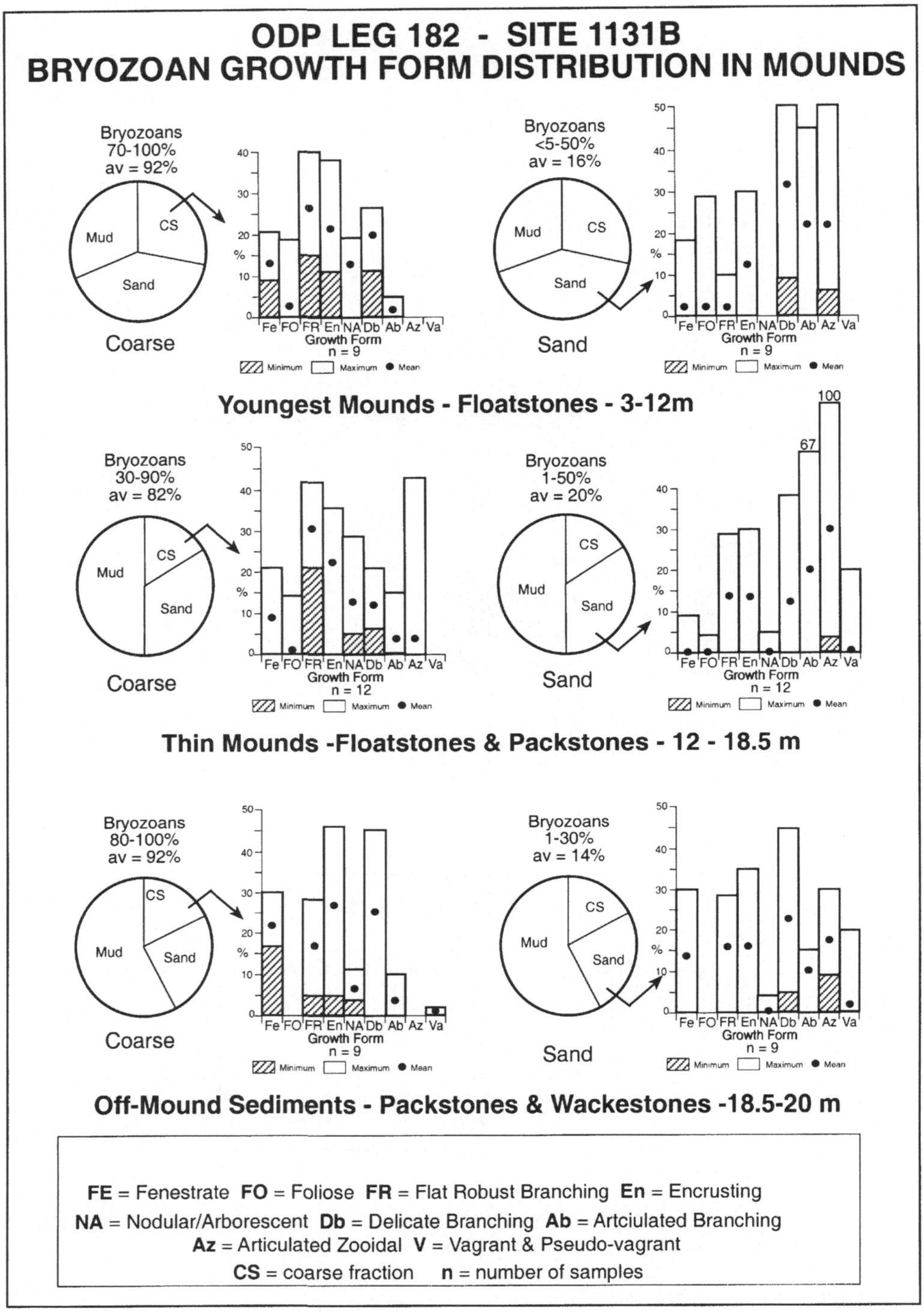

Figure 3. Bryozoan growth form distribution within the individual mounds in the different facies. The hatched area represents the % range of bryozoans with this specific growth form. and the • is the average for all samples (= n) analysed for that facies and/or mound. The mud fraction does not contain any bryozoans. CS is the coarse fraction.

Table 1. Genera distribution in sediment fractions.

GROWTH	Yng. Mds		Thin Mds		Off Mds		Old. Mds	
FORM	coarse	fine	coarse	fine	coarse	fine	coarse	fine
FENESTRATE								
Adeona							tr	
Iodictyum	tr				com		dom	pres
Reteporellina	pres	tr	pres	tr	tr		pres	
Sertella	pres		com	tr	pres		dom	dom
Triphyllozoon	tr	tr	tr	tr			tr	
FOLIOSE								
Caleschara	tr	tr						
Sinupetraliella			tr	tr				
FLAT ROBUST BRANCHING								
Adeonella	tr		tr	tr			tr	tr
Adeonellopsis	pres	tr	dom	dom	dom	tr	dom	com
Bracebridgia	pres	tr	tr	tr			tr	tr
Caleschara	pres		tr	com				
Cucullipora	tr		tr					
Labioporella	tr		tr		tr	tr	tr	
Melicerita		tr	tr		tr	tr	tr	
Porina	tr		tr		com		tr	tr
Reteporella	tr		tr	tr			tr	
Schizoporella							pres	tr
Steginoporella								tr
DELICATE BRANCHING								
Crisia	tr	pres	tr	pres			tr	tr
Entiapora		tr	tr				tr	tr
Diaporecea	tr	tr	tr	pres	tr			
Entalophora	tr						tr	
Galeopsis	tr	tr	tr	pres		tr	tr	
Hornera	pres	tr	pres	tr	tr	tr	pres	pres
Idmidronea	com	com	com	dom	tr	tr	dom	dom
Nellia							tr	tr
Nevianipora	tr	pres	tr	tr	tr		dom	pres
Tubiopora	tr	dom		tr			tr	tr
ARTICULATED BRANCHING								
Caberea	tr	tr	pres	tr	tr		pres	tr
Canda	tr	pres	tr	tr			tr	tr
Caleschara	com	dom	dom	com	pres	tr	tr	com
Cellaria	tr	dom	tr	dom	tr	tr	tr	dom
Margaretta		tr						tr
ARTICULATED ZOOIDAL								
Bugula		dom						tr
Calpidium		pres				pres		
Cornuticella		tr				tr		
Orthoscuticella		pres		pres		com		pres
Paracribricellina		tr		tr		dom		pres
Rhabdozoum								trt
Vittaticella		dom		dom		dom		dom
VAGRANT (PSEUDOVAGRANT)								
Celleporaria	tr	tr			tr			tr
Conescharellina	tr		tr		pres		tr	
Lunulites							tr	
Selenaria			tr					
NODULAR-ARBORESCENT ENCRUSTERS								
Celleporaria	dom	tr	com		com		pres	tr
Celleporina			tr		pres		tr	tr
Osthimosia	tr							
Sphaeropora			pres		pres			
ENCRUSTING								
Amphiblestrum			tr	tr			tr	
Arachnopusia	tr		tr	tr			tr	tr
Arthropoma	tr	tr	tr	tr	tr		tr	
Calloporina	tr	tr	tr	pres			tr	tr
Chaperia			tr				tr	tr
Crassimarginata	tr		tr	tr	tr		tr	tr
*Cribrilina see**								
Cucullipora							tr	
Cryptosula	tr						tr	
Exochella	tr		tr				tr	
**Figularia*		tr			pres		tr	tr
Gigantopora			tr				tr	
Hiantopora	tr	tr	tr	tr			tr	tr
Hippomenella	tr	tr	tr	tr	tr		tr	
Hippomonavella			tr					tr
Hippoporella		pres	tr					
Lichenopora			tr				tr	
Membranipora	tr	tr	tr				tr	
Micropora			tr	tr	tr		tr	
Microporella	tr	tr	tr		tr		tr	tr
Mucropetraliella								tr
Monoporella			tr				tr	
Plagioecia	tr		tr	tr	tr			
Parellisina			tr				tr	
Parasmittina	tr						tr	tr
Pyripora							tr	
Schizobrachiella	tr		tr				tr	
Schizomavella	tr		tr				tr	tr
Schizoporella	tr	pres	com	pres	pres		pres	tr
Serripetraliella							tr	tr
Sinupetraliella	pres			tr				pres
Smittina	tr	tr	tr	tr			tr	tr
Smittoidea	pres	tr	tr		tr		tr	tr
Spiralaria			tr					
Stylopoma			tr		tr		tr	tr
Thalamoporella								

Bryozoan genera present in ODP Leg 182 drill-hole 1131B, listed under their growth forms.

D = >25% of bryozoans present in the sample are this genus;

C = >10% of bryozoans present in the sample are this genus;

P = >5% of bryozoans present in the sample are this genus

T = 1-<5% of bryozoans present in the sample are this genus.

which probably were in even higher abundance than the bryozoans whilst the mounds were active, based on our studies of the modern-day equivalents in the same location, are now represented by considerable volumes of silica spicules in the fine fraction of the sediments, particularly in the youngest mound in the core. The significance of each of the bryozoan forms in the mounds can be seen in Fig. 3. The consistent prevalence of the delicate branching cyclosomes, *Idmidronea* and *Nevianopora,* was unexpected, based on our previous work in the GAB. Whether this is an artifact of our sampling and sieving technique aboard RV *Franklin* is uncertain.

5 DISCUSSION

The accumulation of the bryozoan-rich sediments produced in the faunally-rich patches on the GAB is controlled mainly by oceanographic and climatic parameters, i.e. bottom-water temperature and sea-level. Where the sea-floor is swept daily by high-energy currents, i.e. at depths 0 - ~100 m, the sediment fragments are continually tumbled and/or abraded, except for some small physically-protected areas, e.g. in the lee of sea-floor highs, etc. Overall, however, the sea floor is an exceptionally high-energy area. The relative position of this surface is largely controlled by sea-level, which itself is climatically controlled globally and tectonically controlled locally. Consequently, in such areas the net accumulation of sediment is minimal and a starved shelf results.

It is only below storm-wave base that sediment accumulates at a measurable rate. So, not surprisingly, the mounds are preserved in water depths >200m. Thus, the unabraded nature of the bulk of the bryozoan fragments is explained. Additionally, this lack of grain abrasion supports the notion of their autochthonous production, i.e. there has been minimal downslope movement.

The bryozoans present are typical of a cool-water assemblage (Hageman et al. 1995). However, the proportion of cool-water, low-Mg calcite cyclstomes was unexpectedly high, and was closer in composition (and in species richness and volume of bryozoans present in sediment sample) to modern sea-floor samples from Antarctica rather than to the GAB (Winzar 1999). This supports the notion that the mounds grew at times of upwelling and enhanced trophic resources (James et al 2001) similar to nutrient-level found in cold Antarctic waters today.

The bulk of the volume of the bryozoan fragments were derived from forms that use calcite (either low-Mg calcite or intermediate-Mg calcite) for their skeletal elements (Bone and James 1993). Those fragments that have a powdery coating of cement (<5%) are from bryozoans that use aragonite for their skeletal elements, and probably represents minor sea-floor diagenesis.

6 CONCLUSIONS

The bryozoan-sponge mounds of the GAB continental margin are not active to-day. The youngest true mounds are Pleistocene in age. These have a rich assemblage of bryozoans, with delicate erect cyclostomes numerically the most common forms present. Periods of mound growth correspond with times of glaciation, when upwelling of nutrient-rich cold waters would have promoted high productivity. These mounds have a geometry (James et al. 2001) that may well present itself as major fossil structures, if and when the sediments become lithified.

To date diagenesis has been minimal. The climate operating at the time of their formation meant that the bryozoans present in the cold waters were mainly dominated by geochemically-stable low-Mg calcite forms (Bone and James 1993, 1997), particularly cyclostomes, and thus they do not alter. The presence of aragonitic forms such as the flat, robust branching bryozoans, will eventually lead to minor lithification.

These deep-water structures lack large aragonitic bryozoans such as the shallower water *Adeona* spp. (Bock and Cook 1998) and vagrant free-living forms. Thus they do not have a high diagenetic potential for lithification when subject to percolating meteoric and shallow-burial fluids, similarly to that seen in the adjacent onshore Oligo-Miocene Abrakurrie Limestone (James and Bone 1991). Since they also lack early sea-floor cements, they would be preserved as large mounds in bedded sediments, all of which are subject to extensive early mechanical and chemical compaction. The matrix will be dominated by the ubiquitous low-Mg calcite cyclostomes.

ACKNOWLEDGEMENTS

We would like to thank the Captains and crews of CSIRO RV *Franklin,* the Schlumberger officers, crew and drillers, the ODP laboratory personnel on Leg 182 and our numerous colleagues who assisted with sample acquisition at sea. The paper benefited by S. Hageman's assistance and suggestions. YB was supported by CSIRO and ARC. NPJ was supported by NSERC and ODP.

REFERENCES

Bock, P.E. & Cook, P.L. 2000. Early astogeny of *Adeona* colonies. In A. Herrera Cubilla and J.B.C. Jackson (eds), *Proceedings of the 11th International Bryozoology Association Conference*: 162-167. Panama: Smithsonian Tropical Research Institute.

Bone, Y. & James, N.P. 1993. Bryozoa as sediment producers, Lacepede Shelf, Southern Australia. *Sedimentary Geology* 86: 247-271.

Bone, Y. & James, N.P. 1997. Bryozoan stable isotope survey from the cool-water Lacepede Shelf, southern Australia. In N.P. James & J.A.D. Clarke (eds), *Cool Water Carbonates*: 94-105. SEPM Special Publications Series 56.

Cuffey, R.J., 1977. Bryozoan contributions to reefs and bioherms through geologic time. *Studies in geology* 4: 181-194.

Feary, D.A. & James, N.P. 1995. Cenozoic biogenic mounds and buried Miocene(?) barrier reef on a predominately cool-water carbonate continental margin - Eucla Basin, western Great Australian Bight. *Geology* 23: 427-431.

Feary, D.A. & James, N.P., 1998. Seismic stratigraphy and geological evolution of the Cenozoic, cool-water Eucla Platform, Great Australian Bight: *American Association of Petroleum Geologists Bulletin* 82: 792-816.

Hageman, S., Bone, Y., McGowran, B. & James, N.P. 1995. Modern bryozoan assemblages and distribution on the cool-water Lacepede Shelf, southern Australian margin. *Australian Journal of Earth Sciences* 42: 571-580.

James, N.P. & Bone Y. 1991. Origin of a cool water, Oligo-Miocene deep shelf carbonate, Eucla Platform, Southern Australia. *Sedimentology* 38: 323-341.

James, N.P., Bone, Y., von der Borch, C.C. & Gostin, V.A. 1992. Modern carbonate and terrigenous clastic sediments on a cool-water, high-energy, mid-latitude shelf; Lacepede Shelf, Southern Australia. *Sedimentology* 39: 877-904.

James N.P., Bone, Y., Hageman, S.J., Feary, D.A. & Gostin, V.A. 1997. Palimpsest temperate carbonate sedimentation and oceanography; Lincoln Shelf, southern Australia. In N.P. James & J.A.D. Clarke (eds), *Cool Water Carbonates*: 73-93. SEPM Special Publications Series 56.

James, N.P., Collins, L.B., Bone, Y. & Hallock, P. 1999. Sub-tropical carbonates in a temperate realm: Modern sediments on the southwest Australian shelf. *Journal of Sedimentary Research* 69: 1297-1321.

James, N.P., Bone, Y., Collins, L.B. & Kyser, T.K., 2001. Surficial sediments of the Great Australian Bight: facies dynamics and oceanography on a vast cool-water carbonate shelf. *Journal of Sedimentary Research* 71: 549-567.

James, N.P., Feary, D.A., Surlyk, F., Simo, J.A., Betzler, C., Holbourn, A., Li, Q., Matsuda, H., Machiyama, H., Brooks, G.R., Andres, M.S., Hine, A.C. & Malone, M.J., 2001. Quaternary bryozoan reef mounds in cool-water, upper slope environments; Great Australian Bight. *Geology*.

Monty, C.L.V., Bosence, D.W.J., Bridges, P.H. & Pratt, B.R. 1995. Carbonate mud-mounds: their origin and evolution. *International Association of Sedimentologists Special Publication* 23.

Wass, R.E., Connolly, J.R. & McIntyre, J. 1970 Bryozoan carbonate sand continuous along southern Australia. *Marine Geology* 9: 63 – 73.

Winzar, G. 1997. Sir Douglas Mawson's bryozoan samples, collected from the Antarctic continental shelf over 60 years ago. Unpublished B.Sc. (Hons) thesis, University of Adelaide.

Bryozoan Studies 2001, Wyse Jackson, Buttler & Spencer Jones (eds)
© 2002 Swets & Zeitlinger, Lisse, ISBN 90 5809 388 3

Observations on ecological adaptations of *Lanceopora smeatoni* (MacGillivray), West Island, South Australia

K.M. Brown, R. Schmidt & Y. Bone
Department of Geology & Geophysics, Adelaide University, South Australia. 5005.

ABSTRACT: This paper reports on observations of the ecological adaptations of *Lanceopora smeatoni* (MacGillivray) occurring at West Island, South Australia. At this location, *L. smeatoni* is found in 5 m water depth, ranges in size from 0.5 to 5 cm in width, and has a density of up to 240 specimens per/m^2. An extrazooidal attachment anchors the colony to the straw-like sheaths of the seagrass *Posidonia sinuosa*. Observations over a 2.5 year period show that *L. smeatoni* has an annual life cycle and a maximum growth rate of 0.3 mm/day. It is proposed that the annual life cycle is an adaptation to the short life-span of *P. sinuosa* and rapid colonisation of other epizoans. The ecological benefits of the distinct discoid morphology and the extrazooidal attachment are also addressed.

1 INTRODUCTION

The understanding of the ecological parameters governing the occurrence of modern Bryozoa can be used to identify environments in the rock record. Understanding the evolutionary benefits of particular architectural shapes to certain environments and knowledge of the life cycle of Bryozoa, are examples of the issues that should be addressed.

L. smeatoni is a particularly good species to examine because of its characteristic morphology and specific habitat. *L. smeatoni* has a bifoliate discoid morphology, a stem-like extrazooidal attachment and is bright orange in colour.

This paper examines previously recorded habitat information for *Lanceopora* spp. and incorporates new habitat information from West Island to derive additional ecological information for this genus.

1.1 *West Island, South Australia*

West Island, South Australia, is a pristine area designated as a marine reserve. The island is subjected to an average 2m swell generating from the Southern Ocean during most of the year. Some protection is afforded on the north-west side of the island and this is where the *L. smeatoni* colonies were observed (Fig.1). They were first investigated in February 1999 and the site was revisited every 3 months for the following 28 months.

1.2 *Taxonomy of* Lanceopora smeatoni

Family Lanceoporidae Harmer, 1957
Lanceopora d'Orbigny
Lanceopora d'Orbigny, 1851
Parmularia MacGillivray, 1887

Lanceopora smeatoni (MacGillivray)
Fig. 2
Schizoporella smeatoni MacGillivray, 1890
Calyptotheca smeatoni Harmer, 1957

Description: Colony bilaminate, irregularly subcircular to reniform, orange color; soft but turgid extrazooidal rhizoid originating from the ancestral area attaches to substrate with finer fibrils; substrate is frequently the sheaths of *Posidonia*, sometimes shells. Zooids elongated rectangular, margins defined by often sinuous narrow ridge; frontal wall tuberculate and perforated by pores (sometimes several in depressions between large tubercules), except for smooth shallower area proximal to orifice, usually one prominent rounded conical umbo placed centrally proximal to this area; orifice terminal, semi-circular with broad rounded sinus, proximal margin slightly raised, very small chondyles in proximolateral corners. Zooids in early astogenetic areas lack the umbo and are less tuberculate (Fig. 2a). Ovicell initially forms as large smooth depression with endooecium immersed in proximal half of distal zooid and partly in distolateral zooids (Fig. 2b); frontal walls of lateral and distal zooids gradually grow over at same rate as endooecium (Fig. 2c),

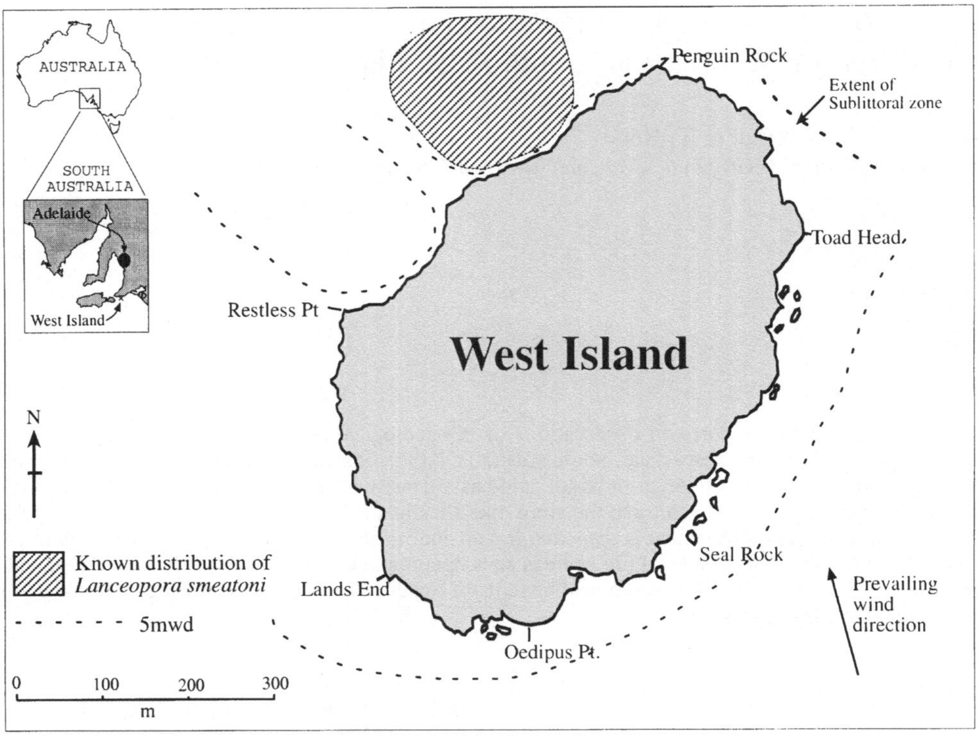

Figure 1. Location Map *Lanceopora smeatoni* at West Island, South Australia.

finally covering almost all of distal and most of the lateral zooids (except the orifice) with a convex globular ectooecium, which is also perforated and tuberculate with occasional larger tubercules; generally Y-shaped ridge at suture between the three (or four) components from the neighboring zooids (Fig. 2d); ooecial orifice slightly larger and wider than autozooecial but similar shape; ovicells only develop after approximately 12 generations and are from there on quite common. No avicularia were observed.

2 ECOLOGY OF *LANCEOPORA SMEATONI*

2.1 *Habitat and distribution*

The general properties of *L. smeatoni* found at West Island are significantly different from previous descriptions (Bock 1982, Bone & James 1997, Chimonides & Cook 1993, Cook & Chimonides 1985, James *et al* 1997, Shepherd 1983). A summary of these differences is shown in Table 1.

The most significant difference between this study and previous studies is the substrate type. All previous records of the genus *Lanceopora* describe colonies living in a sandy substratum and attached to sand grains. There were no colonies living attached to sand grains at West Island, even in relatively large sand patches. Over 90% of the *L. smeatoni* specimens were found attached to the 'straw-like' sheaths of *Posidonia sinuosa* (Fig. 3). The others were found attached to the rhizomes of *Amphibolis antarctica*, or attached to sponges, shell debris and rhodoliths. In contrast Shepherd (1983) reported the presence of *Lanceopora obliqua* amongst active sand ripples, with not a single specimen found in nearby beds of *Posidonia australis*.

It is known that larvae of *Lanceopora* spp. actively seek out a site for settlement (Cook & Chimonides 1985). It therefore appears that at West Island larvae of *Lanceopora* spp. are positively selecting the seagrass substratum for attachment. This poses various questions regarding this selectivity: Does their food source occur in association with the seagrass? Is there some benefit in the slowing of current speed that occurs in seagrass beds? Or is the fibrous nature of the sheaths an easy attachment site? These questions certainly demand further investigation, but currently remain unanswered.

2.2 *Hydrological considerations*

The distinctive colony form of *Lanceopora* spp. suggests an adaptation to extrinsic ecological factors. Its similarity to a weather-vane may lead to the suggestion that the colony rotates about its attachment process and aligns itself parallel with the current direction as a strategy to avoid the foliose sec-

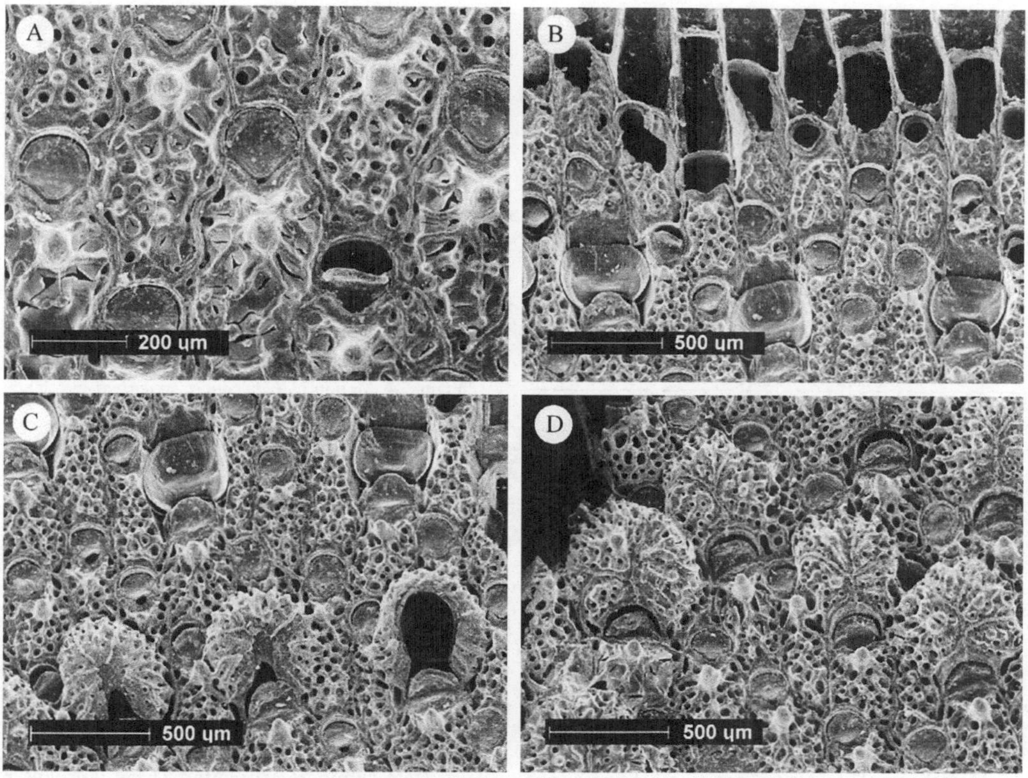

Figure 2. SEM's of *Lanceopora smeatoni* from West Island, South Australia. a) mature zooids, b) growing edge of colony, c) fully developed ovicells, d) ovicells at various stages of growth.

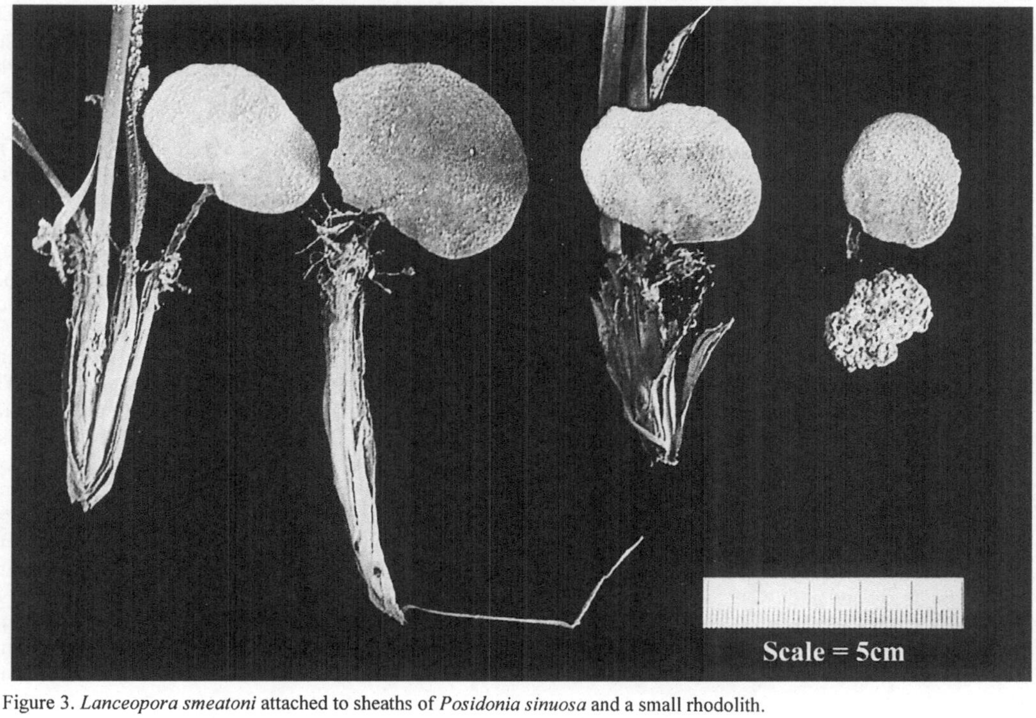

Figure 3. *Lanceopora smeatoni* attached to sheaths of *Posidonia sinuosa* and a small rhodolith.

Table 1. Summary of differences between previous findings and current findings of *Lancepora* spp. (*Bock, 2000 **Bone & James, 1997, [$]James *et al*, 1997, [#]Cook & Chimonides, 1985, [$$]Shepherd, 1983, [x]S. Hageman, pers. Comm., 1997)

	Previous Reports	This Report
Location	S. Australian margin, Queensland Australia; China and Indo-Pacific region	West Island, South Australia.
Water depth	*Common >20mwd. **45.5mwd in G.A.B. [$]126mwd, Lincoln Shelf [$$,x]15-20mwd Spencer Gulf [#]15-20mwd Townsville	Abundant 5mwd
Substrate type & energy	Soft and sandy (all reports). [$,$$]Rippled sand & high energy.	Predominantly sheaths of seagrass *Posidonia sinuosa*. Also shells, rhodoliths & pebbles. Not found attaching to soft sandy substrate. Moderate energy
Abundance	[$]<10 m^{-2} Lincoln Shelf	<240 m^{-2}
Size	*<30mm diameter	<53mm diameter

tion of the colony from breaking during high-energy hydrodynamic events.

The observations do not agree with this weather-vane-hypothesis. The "stems" are flexible and allow movement in any direction. The twisting (side to side) movement, however, is restricted and the colony rapidly returns to the original position (Fig. 4a). Backward and forward motion is more flexible and allows the colonies to flatten against the seabed in a strong current. They always return to the original vertical position (Fig. 4b). At West Island the majority of individual colonies are aligned in one direction. This orientation is perpendicular to the predominant current in the area. This orientation allows maximum water flow over both sides of the colonies (Fig. 4b).

The resultant benefits obtained from this specific erect, planar/foliose morphology, can be therefore described as twofold:

1. The "stem" holds the colony up into the water column. Here there is an increase in current flow, which enhances the interception of food particles.
2. The flexibility of the "stem" allows the colony to move with the swell, resulting in both maximum water flow over the entire colony, as well as protection in strong or changing currents by allowing it to bend and twist.

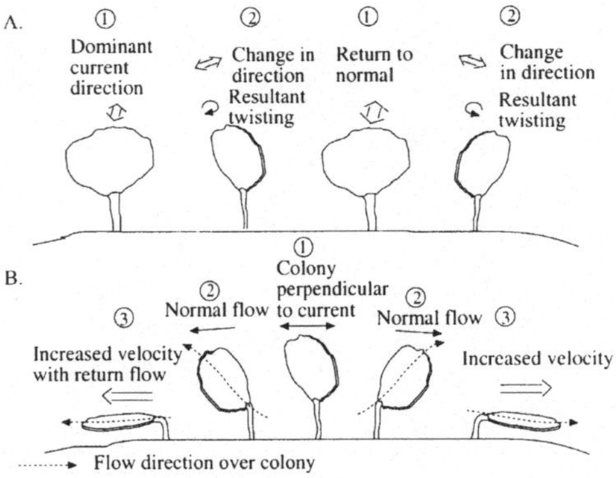

Figure 4. Motion of *Lanceopora smeatoni* in relation to current flow.

2.3 *Annual Growth*

A significant seasonal variation in both the density and size of colonies of *L. smeatoni* was observed during the study.

Growth was first visible in June/July (Australian winter) of each year. The colonies were small at this stage (<10 mm). By October, the colonies were still small, but their density had increased. The colonies reached their maximum size and density four months later, in January/February (summer). By March/April (late summer/early autumn) the colonies had begun to die. This was characterised by browning of the pigmentation, followed by bleaching. By June of the second year only a few older colonies were left and growth of new colonies had already started. The older colonies from the previous year were usually encrusted by epizoans such as serpulid worms and coralline algae.

This annual life-cycle has also been recognized in the upper Spencer Gulf, South Australia (Shepherd 1983). Here they grow in amongst migrating sand waves, which often bury the colonies, hence the need for rapid growth. Observations from this environment recorded the age of colonies as 8 months, reaching a size of 25 mm. This equates with rate of approximately 0.1 mm/day (Shepherd 1983).

Observations from the West Island location suggest that colonies can live for up to a maximum of 1 year. If a growth rate occurred a steady pace, then a growth rate of 0.15 mm/day is achieved by those colonies with a maximum size of 53 mm. However, this study also indicates that a constant growth rate does not occur, but that there is a significant change in growth rate, with maximum growth occurring between October and February. If the colonies are 10 mm in length by October, then the growth rate is 0.3 mm/day. This is significantly higher than the upper Spencer Gulf specimens, and more in line with the rate shown by *Conopeum aciculata* colonies in the Coorong Lagoon, South Australia (Sprigg & Bone 1993).

The resultant benefits of the rapid growth rate of *L. smeatoni* is, again, twofold:

1. Allows colonisation in ephemeral environments, e.g. the migrating sands in Spencer Gulf, and short lived (1-2 years) seagrass sheaths and roots.
2. Acts as a defense mechanism against epipzoan encrustation.

3 CONCLUSIONS

Lanceopora smeatoni is conventionally known as a sand fauna species, but at West Island it is found predominantly attached to the sheaths of *Posidonia sinuosa* and not found attached to sand particles. The reason for this preference is conjectural, although the main cause is likely to be substrate selectivity. The observed specimens grow up to 53 mm across (20 mm greater then previously recorded), and have a population density of $\leq$240 specimens/m^2 in a water depth of 5 m. Colonies are aligned perpendicular to the dominant current direction, which allows the greatest volume of water to flow over all of the zooids and not just those at the periphery of the colony. The flexible "stem" allows the colony to move in response to the current, which reduces the likelihood of damage. *L. smeatoni* has a 1 year life-span, initiating in June/ July. An average growth rate of 0.15 mm per day is calculated for a constant growth rate, but a peak rate of 0.3mm per day occurs over a 6 month period between October and February. Thereby, suggesting that growth rate should be reported as ranging between 0.15-0.3 mm/day, depending on colony age and season. This rapid growth rate is beneficial in allowing colonization on relatively temporal substrates and acts a defense mechanism against epizoan encrustation.

ACKNOWLEDGMENTS

We would like to acknowledge all the dive buddies for allowing the field work to happen; Anthony Cheshire for financial support; Phil Bock for assistance with the taxonomy; and Miles Davies for his support and suggestions. This paper is dedicated to Rex.

REFERENCES

Bock, P.E. 1982. Bryozoans (Phylum Bryozoa). In S.A. Shepherd & I.M Thomas (eds), *Marine invertebrates of South Australia. Pt. 1. South Australian Government handbook of the Flora and Fauna of South Australia.* Government printer: South Australia. 319-394.

Bone, Y. & James, N.P. 1993: Bryozoa as sediment producers, Lacepede Shelf, Southern Australia. *Sedimentary Geology.* 86: 247-271.

Bone, Y. & James, N.P. 1997. Bryozoan stable isotope survey from the cool-water Lacepede shelf, southern Australia. In: N.P. James & A.D. Clarke (eds), *Cool-water Carbonates.* SEPM (Society for Sedimentary Geology) Special Publication 56: 93-105.

Chimonides, P.J. & Cook, P.L. 1993. Notes on *Parmularia* MacGillivray (Bryozoa: Cheilostomida) from Australia. *Memoirs of the Queensland Museum.* 33 (1): 41-48.

Cook, P.L. & Chimonides, P.J. 1985. Larval settlement and early astogeny of *Parmularia* (Cheilostomata) In C. Nielsen & G.P. Larwood (eds.), *Bryozoa: Ordovician to recent.* Olsen & Olsen, Denmark. 71-78.

Harmer, S.E. 1957. Polyzoa of the Siboga Expedition. Pt IV Cheilostomata Ascophora II (Ascophora, except Reteporidae, with additions to part II, Anasca) *Siboga-Expedition reports 28D.* 641-1147.

James, N.P. 1997. The cool-water carbonate depositional realm. In N.P. James & A.D. Clarke (eds), *Cool-water Carbonates.* SEPM (Society for Sedimentary Geology): Special Publication 56: 1-22.

James, N.P., Bone, Y., Hageman, S.J., Feary, D.A & Gostin, V.A. 1997. Cool-water carbonate sedimentation during the terminal Quaternary sea-level cycle: Lincoln Shelf, southern Australia. In N.P. James & A.D Clarke (eds), *Cool-water Carbonates.* SEPM (Society for Sedimentary Geology): Special Publication 56: 53-75.

MacGillivray, P.H. 1890. An additional list of South Australian Polyzoa. *Transactions and Proceedings and Report of the Royal Society of South Australia.* 13(1): 1-7.

Shepherd, S.A. 1983. The epifauna of megaripples: species' adaptations and population responses to disturbance. *Australian Journal of Ecology.* 8: 3-8.

Sprigg, M. & Bone, Y., 1993. Bryozoa in Coorong-type Lagoons in southern Australia. *Transactions of Royal Society of South Australia.* 117: 87-95.

Wass, R.E. & Yoo, J.J. 1983. Cheilostome Bryozoa from the southern Australian continental shelf. *Australian Journal of Marine and Freshwater Research.* 34: 303-354.

Bryozoan Studies 2001, Wyse Jackson, Buttler & Spencer Jones (eds)
© 2002 Swets & Zeitlinger, Lisse, ISBN 90 5809 388 3

Larval release pattern in Antarctic bryozoans

J. M. Cancino & F. Torres
Facultad de Ciencias Universidad Católica de la Santísima Concepción, Casilla 297, Concepción, Chile

H.I. Moyano
Departamento de Zoología, Facultad de Ciencias Naturales y Oceanográficas, Universidad de Concepción, Casilla 160-C, Concepción, Chile

ABSTRACT: Most marine bryozoans in temperate shallow waters incubate their larvae and release them following a circadian rhythm regulated by daylight. Antarctic bryozoans have been reported to sexually reproduce mainly in the summer months, which prompts the question whether larval release in such bryozoans follows a similar circadian pattern to those from lower latitudes. The present study aims at answering such a query. In January 1999, a total of 35 species of bryozoans were collected by scuba diving between 15 and 36 m depth at Doumer Island, (Base Yelcho, Instituto Antártico Chileno, at 64°52'24"S, 63°36'00"W) and kept in running seawater. The cycle of larval release was studied for 2-3 days after collection using larval release chambers. The dominant species, *Nematoflustra flagellata* (Waters 1904) and *Himantozoum antarcticum* (Calvet 1909), and their associated species, released larvae almost exclusively in daytime when photon flux density was above ~0.5 µE m-2 s-1. Larval release was almost absent from 23 h at night until 05 h in the morning. Therefore, larval release in shallow waters Antarctic bryozoans follows a similar circadian pattern to bryozoans in temperate and tropical waters. Since shallow water Antarctic bryozoans restrict their larval release to summer and autumn, seasons at which sunlight is present, it can be concluded that the process of larval release in shallow water bryozoans is controlled world-wide by the daylight cycle. The adaptive selective advantage of releasing larvae in response to light is discussed. The results of the present study also provide evidence that fecundity is low in Antarctic bryozoans.

1 INTRODUCTION

Bryozoans provide excellent material to study reproductive-ecology strategies related to environment. Most bryozoans are colonial and have brood chambers in which their larvae are incubated (Ryland 1960, 1974). Larvae are mostly lecithotrophic, and because they have a short planktonic life they settle soon after release into the water (Ryland 1960, Ramírez & Cancino 1991, Cancino & Gallardo 2000), closely linking larval release with larval settlement patterns. In all the species studied in detail so far, but one, larval release takes place only in daytime (Cancino et al. 1991b, 1992, 1994). Consequently, in the natural environment, larval release follows a typical circadian pattern with most larvae being released soon after sunrise (Cancino et al. 1991b, 1992, 1994). An experimental study by Cancino et al. (1991b) showed that a long exposure to light inhibits larval release in *Celleporella hyalina*. The former fact made us wonder what would be the larval release pattern, if any, in Antarctic bryozoans subjected to the long summer day photoperiod.

Bryozoans are dominant components of the rocky benthic communities in the Antarctic subtidal and they have been subjected to numerous taxonomic studies since the beginning of the 20th century (for a review see Hayward 1995). Distribution patterns and their causes have begun to be studied only recently (Winston & Heimberg 1988, Barnes 1995; Barnes & Rothery 1996). Due to logistic problems, the functional biology and sexual reproductive patterns of Antarctic bryozoans remain mostly unexplored, even for shallow subtidal species, which are relatively easy to collect by SCUBA diving. Only recently have the first studies related to this topic been published (Barnes & Clarke 1994, 1998, Barnes 1995). Based on monthly sampling over 14 months, Barnes & Clarke (1998) showed, that at Signy Island (60°43'24"S; 45°36'W), *Isoseculiflustra rubefacta* Moyano 1995[(*)], *Nematoflustra flagellata* (Waters 1904) and *Himantozoum antarcticum* (Calvet 1909) have a seasonal pattern of biological activity, feeding and reproduction being restricted to spring and summer. Therefore the aim of the present study was to determine if larval release in Antarctic bryozoans takes place following a circadian rhythm regulated

[(*)] See footnote to Table 1

by daylight, similar to the circadian pattern known for bryozoans from temperate waters.

2 MATERIALS AND METHODS

Bryozoans were collected by SCUBA diving on the rocky subtidal zone at Base Yelcho of the Instituto Antártico Chileno, in Doumer Island (64°52'24"S, 63°36'00"W). Samples were kept in running seawater systems and larval release patterns were studied using the chambers devised by Cancino et al. (1991b). Two species-systems were studied. The first one involved *Himantozoum antarcticum* (Calvet 1909), which was collected from 15-20 m on January 15th, 1999 and that included 13 other species, which were using the colonies of *H. antarcticum* as a substratum (Table 1). The second system involved *Nematoflustra flagellata* (Waters 1904) which was collected on January 19th, 1999 mostly from 20-36 m, and included 34 other species that used *N. flagellata* colonies as a substratum (Table 1).

At around 20.00 h in the collecting day, 2 subsamples of bryozoans colonies, each approximately 0.7 kg wet-weight, were placed into larval release chambers (Cancino et al. 1991b, 1992, 1994). Each chamber had a 20 litre capacity and was provided with a seawater input on the base and an outlet on the top. The translucent, convex upper lid of the chamber allowed the colonies to receive natural light. While the seawater circulating through the chamber removed the released larvae, which were caught by an 80 µm mesh plankton-net sieve receiving the outlet water (see Cancino et al. 1991b for more details regarding the larval release chambers). The larvae collected in the filter were counted under a binocular microscope at 1 h interval for at least 41 hours after collection. Temperature was controlled by immersing the chambers into a 200 litre tank filled with running seawater, which was pumped directly from the sea. The seawater flowing through the larval release chambers was pumped also directly from the sea, providing natural food to the colonies while in the chambers.

For the *H. antarcticum* species-system sampling for 41 hours was carried out without any experimental manipulation of photon flux densities. The sampling of the *N. flagellata* species-system lasted for 76 hours. Experimental manipulations, aimed at understanding the role of light intensity as a factor regulating larval release pattern, were carried out on this latter species-system. One chamber was kept with the normal light intensity until 11.00 on the first day of the experiment, followed by 2 ½ days of exposure to a reduced photon flux density, a condition created by covering the chamber with black netting. Such a light-reduced condition was meant to mimic the light environment of this species complex, which lives at greater depth than *H. antarcticum*. Another chamber with *N. flagellata* colonies was kept for 1½ days in darkness (simulated by covering the chamber with several layers of black plastic bags) followed, as above, by exposure to a reduced photon flux density condition.

Table 1. Bryozoan species associated with *Himantozoum antarcticum (H. a.)* and *Namatoflustra flagellata (N. f.)* in Doumer Island, Antarctic Peninsula (Determined following Moyano 1978).

Associated species	Species acting as a substratum *H. a.*	*N. f.*
Order CHEILOSTOMATIDA		
Amphiblestrum inermis (Kluge, 1914)		X
Arachnopusia monoceros (Busk, 1854)		X
Beania erecta Waters, 1904	X	X
Beania inermis (Busk, 1852)		X
Bugula longissima Busk, 1884		X
Camptoplites bicornis Busk, 1884		X
Camptoplites sp.		X
Celleporella antarctica Moyano & Gordon, 1980	X	X
Celleporella bougainvillei (d'Orbigny, 1842)	X	X
Chaperiopsis sp.		X
Ellisina antarctica Hastings, 1945		X
Escharoides tridens (Calvet, 1909)		X
Fenestrulina cervicornis Hayward & Ryland, 1990.	X	X
Filaguria spatulata (Calvet, 1909)	X	X
Harpecia spinosissima (Calvet, 1904)	X	X
Himantozoum antarcticum (Calvet, 1909)		X
Hippadenella inerma (Calvet, 1909)	X	X
Inversiula nutrix Jullien 1888	X	X
Isoseculiflustra rubefacta Moyano 1995 (**)		X
Kymella polaris (Waters, 1904)		X
Micropora brevissima Waters, 1904		X
Nematoflustra flagellata (Waters, 1904)		X
Notoplites tenuis (Kluge, 1914)		X
Osthimosia milleporoides (Calvet, 1909)	X	X
Smittina rogickae Hayward & Taylor 1984		X
Smittina sp.		X
Orden CTENOSTOMATIDA		
Alcyonidium eightsi Winston & Hayward, 1994		X
Bowerbankia antarctica Winston & Hayward, 1994		X
Orden CYCLOSTOMATIDA		
Crisia sp.	X	X
Crisidia delicatissima Moyano 2000	X	X
Fasciculipora sp.		X
Idmidronea sp.	X	X
Lichenopora canaliculata (Busk, 1876)	X	X
Tubulipora tubigera (Busk, 1886)		X
Tubulipora sp.		X

(**) *Isoseculiflustra rubefacta* was described by Moyano 1995 and not by Kluge as cited by Barnes and Clarke (1998). The species described by Kluge corresponds to *Isoseculiflustra tenuis*, which compared to *I. rubefacta*, grows in deeper waters, and has a pale-yellow colour and slender rami.

The data on the number of larvae obtained per hour from the two replicate chambers with *H. antarcticum* colonies were averaged for 2-hour intervals and plotted as a function of time. In the case of *N. flabellata* we did not have a replicate for each chamber, and the difficulties of carrying out experimental studies in the Antarctic prevented us from replicating this experiment at a different time. In this case the data are presented as the average number of larvae released per hour, in a 2-hour interval, using the data from the single chamber available for each experimental situation.

Throughout the experiment the photon flux density was measured every 30 minutes, at the water surface beside the larval release chamber, with a LICOR light meter provided with a spherical light sensor. In order to know the photon flux density reaching the environments where the bryozoans live, measurements were carried out at different depths in the water column at the location from which bryozoans were collected.

3 RESULTS

Larval release in the *Himantozoum antarcticum* species-system took place only while light was detected and measurable (Fig. 1). In general, larvae were not released from 23.00 until 04.00 in the morning, period at which the photon flux densities was below 6.75 μE m^{-2} s^{-1} and above 0.02 μE m^{-2} s^{-1}. However, photons flux densities as low as 0.5 μE m-2 s-1, were enough to induce larval release at dawn, while at sunset larval release stopped at photon flux densities of ~2-7 μE m^{-2} s^{-1}. The second day data set (January 17th) showed two additionally interesting phenomena: (a) the daily larval release period can last for up to 18 hours and (b) high photon flux densities seem to inhibit larval release (Fig. 1). Note also that the total number of larvae obtained in a given day decreases from the first (n= 98) to the third day (n= 15) studied.

Larval release in the *Nematoflustra flagellata* species-system also occurred mainly while light was detected and measurable (Fig. 2). Larvae were not released from 21.00 h until 04.00 in the morning, period at which the photon flux densities ranged between 5.57 and < 0.01 μE m^{-2} s^{-1}. The first day data set (January 20th) showed that in this species-system the daily larval release period also last for up to 18 hours (Fig. 2). There was also evidence that larval release decrease following exposure to high values of photon flux densities, such an effect was even present after shading the experimental chamber from January 20th at 11.00 h onwards (Fig. 2). As in the previous species-system, the total number of larvae obtained in a given day decreased from the first (n=28) to the third day (n= 19) of the experiment.

Larval release in the *Nematoflustra flagellata* species-system also took place in continuous darkness, but release was delayed until later in the day (compare the first 2 days of experiment in Fig. 3 with the 3rd day in this same figure, and also see Fig. 2). Note also that the total number of larvae obtained on the day the colonies were put back into light (January 21st) was the largest obtained for this species-system in a given day (Figs. 2, 3).

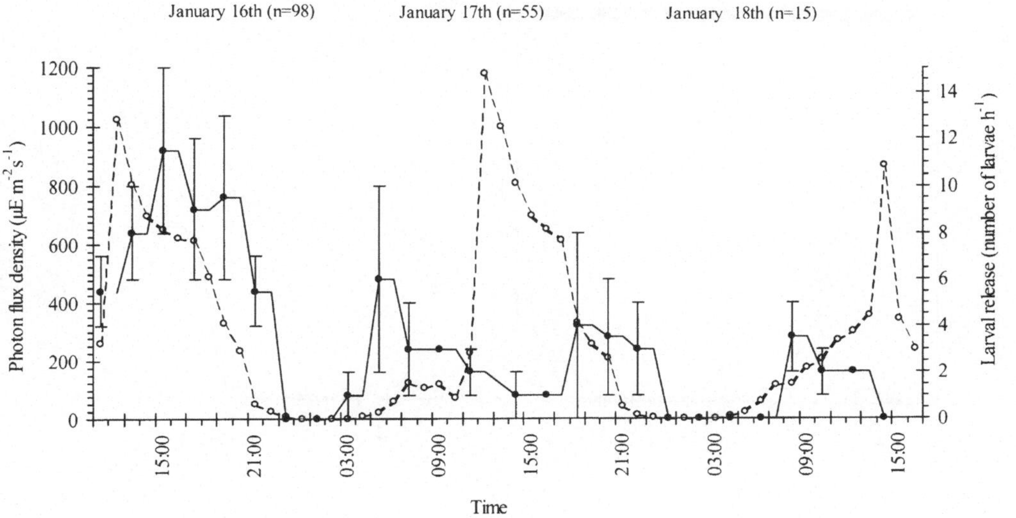

Figure 1. Photon flux density at Doumer Island, Antarctic Peninsula (open circles, dashed line) and average number of larvae released (black dots, continuous line) by the *Himantozoum antarcticum* species-system between 11:00, January 16th to 16:00, January 18th, 1999. The values given represent the average number of larvae ($\pm$ 1 s. e.) obtained from two chambers and calculated at 2 hours interval (n=4).

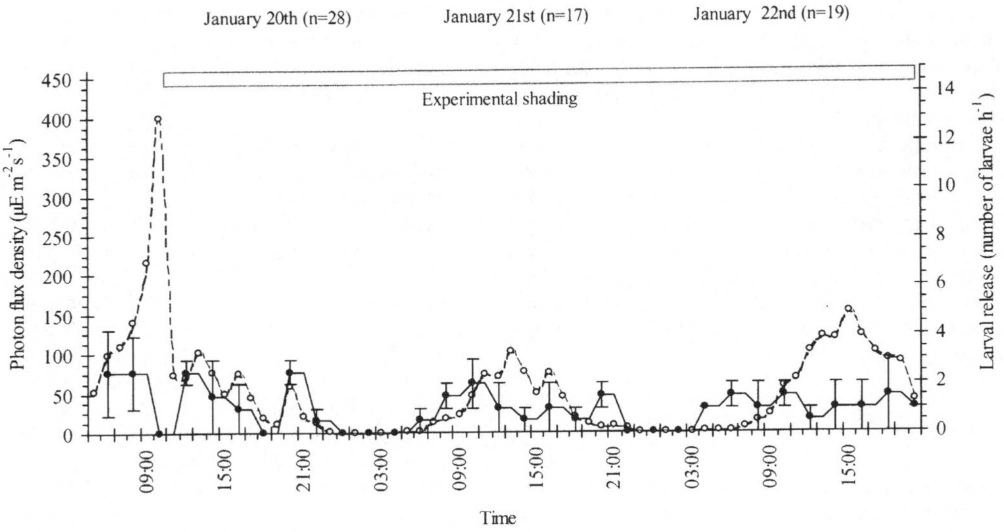

Figure 2. Photon flux density at Doumer Island, Antarctic Peninsula (open circles, dashed line) and average number of larvae released (dark circles, continuous line) by the *Nemato-flustra flagellata* species-system from 05:00, January 20th to 20:00, January 22nd, 1999. The values given are the average number of larvae obtained per hour in one chamber, calculated at two hours interval (± 1 SD, n=2). From 11.00 in the first day of experiment the chamber was covered with black netting for experimentally reducing the photon flux density.

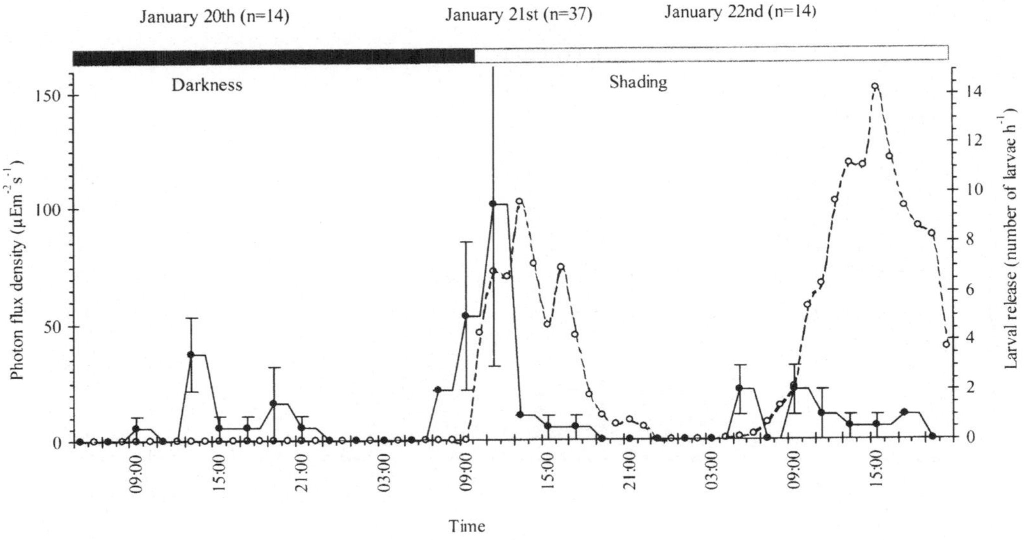

Figure 3. Photon flux density at Doumer Island, Antarctic Peninsula (open circles, dashed line) and average number of larvae obtained (dark circles, continuous line) from 05:00, January 20th to 20:00, January 22nd, 1999, with the *Nemato-flustra flagellata* species-system. Note that photon flux density was experimentally manipulated, the colonies being subjected to darkness during day one, until 09:00 of day 2. Data as in Figure 2.

Light in the natural environment, measured at 10.00 on January 25th reduced from 1077 $\mu E\ m^{-2}\ s^{-1}$ at the surface to 35 $\mu E\ m^{-2}\ s^{-1}$ at 30 m depth. The data available allowed prediction of the magnitude of incident light as a function of depth and time of the day. In such a day the minimum photon flux density required to start inducing larval release (~0.5 $\mu E\ m^{-2}\ s^{-1}$) would be available even at 30m as early as 05 h in the morning. While the lower flux densities at which larval release stops at sunset (~2 $\mu E\ m^{-2}\ s^{-1}$) will not be available below 15m at 22.00, suggesting that in the natural environment larval release should last for a shorter time than in our experiments at surface level.

4 DISCUSION

The present study has shown that, in summer, larval release in shallow waters Antarctic bryozoans follows a similar circadian pattern to bryozoans in temperate waters (Cancino et al. 1991b, 1992, 1994, Cancino & Gallardo 2000). Moreover, Barnes & Clark (1998), reported that at Signy Island the bryozoans *Isoseculiflustra rubefacta* Moyano, *Nematoflustra flagellata* (Waters) and *Himantozoum antarcticum* (Calvet) restrict feeding to spring, summer and early autumn, while larval release is likely to be restricted exclusively to summer. It can be concluded therefore that larval release in shallow water bryozoans worldwide is under the control of the daylight cycle. Such widespread bryozoan behaviour must have a selective advantage, if not a random pattern regarding timing of larval release would be expected. What, therefore, is the selective advantage of such widespread bryozoan behaviour? We can provide three possible answers:

Firstly, it is commonly thought that releasing larvae in response to light is selectively advantageous because larvae require the presence of light for orientation towards the right place of settlement (Ryland 1960, Cancino & Gallardo 2000). The need for settlement soon after release has been documented clearly for the lecithotrophic larvae of bryozoans, since the ability to settle is lost as the free-living larval time increases (Woollacott et al. 1989, Orellana & Cancino 1991, Orellana et al. 1996). The idea that larval release, induced by light, correlates with choosing the right place for settlement is supported by the data Ryland (1960, 1976) provided for the bryozoan *Celleporella hyalina*, and by evidence from tunicates (Olson 1983). This idea is also partially supported by the data of Cancino & Gallardo (2000) dealing with larval settlement on encrusting bryozoans from Chile. If settlement in Antarctic bryozoan larvae takes place soon after release, light could well be the orienting factor for selecting the right place for settlement.

Secondly, larval release in response to light is likely to be related to a maturation process requiring light for the larvae to be ready to settle and undergo the complex process of metamorphosis. Ruiz et al. (1996) showed that the length of the exposure to light significantly affected larval settlement, both in *Celleporella hyalina* and *Chaperia acanthina*. However, exposure to high photon flux densities was disadvantageous for the latter species. Such a negative affect of light offers an explanation for the inhibition of larval release at midday, reported for both *Celleporella hyalina* and *Chaperia acanthina* from the Chilean coast (Cancino et al. 1992), and reported also in the present study for Antarctic bryozoans (Figs. 1 and 2).

Thirdly, larval release in response to light might be regarded as a mechanism synchronising larvae in order to generate groups of conspecific colonies of the same age (Ramírez & Cancino 1991; Cancino et al. 1991a, 1994, Cancino & Gallardo 2000). There is increasing evidence that the presence of conspecific neighbours is required for cross-fertilisation in some bryozoans (Cancino et al. 1991a, Manríquez et al. 2001, Hughes et al. 2002).

The above three answers seem not to be of general application to all bryozoans that release their larvae in response to light. Therefore, we need more information in order to understand the selective advantage of the widespread behaviour of larval release in response to light. Likewise, information is required regarding both the larval release pattern and the stimuli inducing larval release in places where no light reaches the colonies. Many bryozoans live in the deep sea and in places in which darkness is permanent, nothing is known regarding the stimuli that induce larval release in such species.

Finally, another result of the present study that deserves attention is the small number of larvae obtained daily. In temperate waters a similar bryozoans sample size would have released up to several thousands of larvae (for a comparison see Cancino et al. 1991b). The present results give support to the ideas of Winston (1983) who has suggested that fecundity is very low in Antarctic Bryozoa. This is also supported by the data provided by Barnes & Clarke (1998), who have documented that for some Antarctic bryozoans the larvae are incubated for up to 10 months before being released into the water. Certainly reproduction, as other biological processes, seem to take place at a very low rate in Antarctic marine bryozoans.

5 ACKNOWLEDGEMENTS

Instituto Antártico Chileno (INACH) financed the present study through grant 05/97. We are grateful to Antonio Larrea (Santiago, Chile) and José Darías (from Universidad La Laguna, Canarias) for their help in collecting bryozoans, and to Dr. Aurelio San Martin (Universidad de Chile, Santiago) for his help during the field work. Suggestions from an anonymous referee, which greatly improved the final manuscript, are also acknowledged.

6 REFERENCES

Barnes, D.K.A. 1995. Seasonal and annual growth in erect species of Antarctic bryozoans. *Journal of Experimental Marine Biology and Ecology* 188: 181-198.

Barnes, D.K.A. & Clarke, A. 1994. Seasonal variation in the feeding activity of four species of Antarctic bryozoans in relation to environmental factors. *Journal of Experimental Marine Biology and Ecology* 181: 117-133.

Barnes D.K.A. & Clarke, A. 1998. Seasonality of polypide recycling and sexual reproduction in some erect Antarctic bryozoans. *Marine Biology* 131: 647-658.

Barnes, D.K.A. & Rothery, P. 1996. Competition in encrusting Antarctic bryozoan assemblages: outcomes, influences and implications. *Journal of Experimental Marine Biology and Ecology* 196: 267-284.

Cancino, J.M., Castañeda, B. & Orellana, M.C. 1991a. Reproductive strategies in bryozoans: experimental test of the effects of conspecific neighbours. In F.P. Bigey (ed.) *Bryozoaires actuels et fossiles: Bryozoa living and fossil; Bulletin de la Société des Sciences Naturelles de l'Ouest de la France, Mém. HS1*: 81-88. Nantes, France.

Cancino, J.M. & Gallardo, J. 2000. Effects of light on bryozoan larval settlement. In A. Herrera & J.B.C. Jackson (eds.), *Proceedings of the 11th Bryozoology Association Conference*: 168-175. Balboa, República de Panamá: Smithsonian Tropical Research Institute.

Cancino, J.M., Hughes, R.N. & Orellana, M.C. 1994. A comparative study of larval release in bryozoans. In P.J. Hayward, J.S. Ryland & P.D. Taylor (eds.), *Biology and paleobiology of bryozoans*: 41-46. Fredensborg, Denmark: Olsen & Olsen.

Cancino, J.M., Hughes, R.N. & Ramírez, C. 1991b. Environmental cues and the phasing of larval release in the bryozoan *Celleporella hyalina* (L.). *Proceedings of the Royal Society, London B* 246: 39-45.

Cancino, J.M., Orellana, M.C., Muñoz. M.R & Hughes, R.N. 1992. Ciclo diario de liberación larval en dos especies de briozoos. *Revista de Biología Marina, Valparaíso* 27 (2): 213-223.

Hayward, P.J. 1995. *Antarctic Cheilostomatous Bryozoa*. Oxford: University Press.

Hughes, R.N., Wright, P.J., Manríquez, P.H. & Bishop, J.D.D. 2002. Predominance of obligate outbreeding in the simultaneous hermaphrodite *Celleporella hyalina* sensu lato. This volume.

Manríquez, P.H., Hughes, R.N., & Bishop, J.D.D. 2001. Age-dependent loss of fertility in water-born sperms of the bryozoan *Celleporella hyalina* (L.). *Marine Ecology Progress Series* in press.

Moyano, H. I. 1978. Bryozoa de bahias antárticas: algunos aspectos ecológicos. *Serie Científica Instituto Antártico Chileno* 24: 35-60.

Olson, R.R. 1983. Ascidian-*Prochloron* symbiosis: the role of larval photoadaptations in midday larval release and settlement. *Biological Bulletin of the Marine Biological Laboratory, Woods Hole* 165: 221-240.

Orellana, M.C & Cancino, J.M. 1991. Effects of delaying larval settlement on metamorphosis and early colonial growth in *Celleporella hyalina* (Bryozoa: Cheilostomata). In F.P. Bigey (ed.) *Bryozoaires actuels et fossiles: Bryozoa living and fossil; Bulletin de la Société des Sciences Naturelles de l'Ouest de la France, Mém. HS1*: 309-316. Nantes, France.

Orellana, M.C., Cancino, J.M. & Hughes, R.N. 1996. Is settlement in lecithotrophic bryozoan larvae constrained by energy reserves?. In D.P. Gordon, A.M. Smith, & J.A. Grant-Mackie (eds.), *Bryozoans in space and time*: 221-226. Wellington, New Zealand: National Institute of water & Atmospheric Research Ltd.

Ramírez, C. & Cancino J.M. 1991. Respuestas a la luz y conducta de asentamiento de larvas de *Celleporella hyalina* (L.) (Bryozoa). *Revista Chilena de Historia Natural* 64(1): 29-35.

Ruiz, C., Díaz, H. & Cancino, J.M. 1996. Efectos de la densidad de flujo fotónico en el asentamiento larval de dos especies de briozoos. *Gayana Oceanología* 4: 69-75.

Ryland, J.S. 1960. Experiments on the influence of light on the behaviour of polyzoan larvae. *Journal of Experimental Biology* 37: 783-800.

Ryland, J.S. 1974. Behaviour, settlement and metamorphosis of bryozoan larvae: a review. *Thalassia Jugoslavica* 10: 239-262.

Winston, J.E. 1983. Patterns of growth, reproduction and mortality in bryozoans from the Ross Sea, Antarctica. *Bulletin of Marine Science*. 33: 688-702.

Winston, J.E. & Heimberg, B.F. 1988. The role of bryozoans in the benthic community at Low Island, Antarctica. *Antarctic Journal of the U.S.* 21:188-189.

Woollacott, R.M., Pechenik, J.A. & Imbalzano, K.M. 1989. Effects of duration of larval swimming period on early colony development in *Bugula stolonifera* (Bryozoa: Cheilostomata). *Marine Biology* 102: 57-63.

Bryozoan Studies 2001, Wyse Jackson, Buttler & Spencer Jones (eds)
© 2002 Swets & Zeitlinger, Lisse, ISBN 90 5809 388 3

Asexual propagation in the cheilostomes *Metrarabdotos* and *Coscinopleura*: effects on genetic variation and larval productivity

A.H. Cheetham
Department of Paleobiology, National Museum of Natural History, Smithsonian Institution, Washington, DC 20560-0121, USA

ABSTRACT: Although separated by 30 million years, *Metrarabdotos* and *Coscinopleura* share an arborescent growth habit potentially capable of propagating asexually by fragmentation and regeneration, and a relative paucity of (ovicelled) zooids modified for production of larvae. This combination of characters suggests a tradeoff between vegetative growth and sexual productivity; yet the potential for asexual propagation is abundantly realized in *Coscinopleura* and only rarely so in *Metrarabdotos*. The evolutionary significance of such a tradeoff is the extent to which a switch to asexual propagation depresses genetic variability. Estimated from the ratio between among-colonies and within-colonies variance in zooid morphology, however, genetic variation does not differ significantly between species of *Metrarabdotos* showing asexual propagation and those lacking it, or between *Metrarabdotos* and *Coscinopleura*. Moreover, total numbers of ovicelled zooids are at least as great in *Coscinopleura* as in coeval cheilostomes in which the percentage of such zooids is much higher. Thus, vegetative proliferation appears to be an added capability, perhaps related to growth in areas of nutrient enrichment, and not a tradeoff diverting resources from the maintenance of genetic diversity.

If the purpose of sex is to promote the genetic diversity of offspring and thus maximize evolutionary rates, a switch to asexual propagation should constrain the pool of genetic variation from which new species can be drawn and, at the same time, attenuate the longevities of individual genotypes. Thus, rates of both speciation and extinction in asexually propagating lineages could be expected to decelerate relative to the rates in exclusively sexually reproducing lineages (Jackson & Coates 1986).

Results from rearing experiments with cheilostome bryozoans including *Stylopoma* (Jackson & Cheetham 1990, 1994, Cheetham et al. 1993) and *Electra pilosa* (Hageman et al. 1999) strongly suggest that the ratio between among-colonies and within-colonies variance in zooid morphology provides a useful proxy for estimating genetic variation. The amount of deviation between experimentally determined heritability and among-colonies variance varies considerably from trait to trait (Cheetham et al. 1993); but if averaged across traits, among-colonies variance components can estimate genetic diversity closely, because genotype-by-environment interaction is generally quite small even though rarely entirely absent (Cheetham et. al. 1995, Hageman et al. 1999).

The cheilostome genera *Metrarabdotos* (Oligocene-Recent) and *Coscinopleura* (Late Cretaceous-Paleocene) are especially suitable for exploring the consequences of sexual vs. asexual propagation for several reasons. First, output of sexually produced larvae is unambiguously linked to distinctive zooidal modifications associated with brooding in both genera (Figures 1a, g). Secondly, the rigidly erect, arborescent growth habits predominant in both genera are prone to fragmentation and thus, at least potentially, for asexual propagation by regeneration. However, this potential is unequally realized. Evidence of regenerative growth occurs only rarely in *Metrarabdotos* amid abundant evidence of larvally produced colonies (Cheetham et al. 2001), whereas the reverse is true in *Coscinopleura* (Thomsen & Håkansson 1995, Håkansson & Thomsen 2001). Thirdly, zooids in both genera are complex in skeletal morphology, providing enough characters for reliable estimation of average among-colonies variance. And finally, the great abundance of both genera in fossil assemblages minimizes sampling biases in all of these properties.

This paper presents new data for *Metrarabdotos* from the Pliocene of Rhodes, Greece, and for *Coscinopleura* from the Paleocene of New Jersey, USA, and a new analysis including data from previous work. The results suggest modification of previous conclusions regarding the tradeoff between sexual and asexual propagation in these taxa.

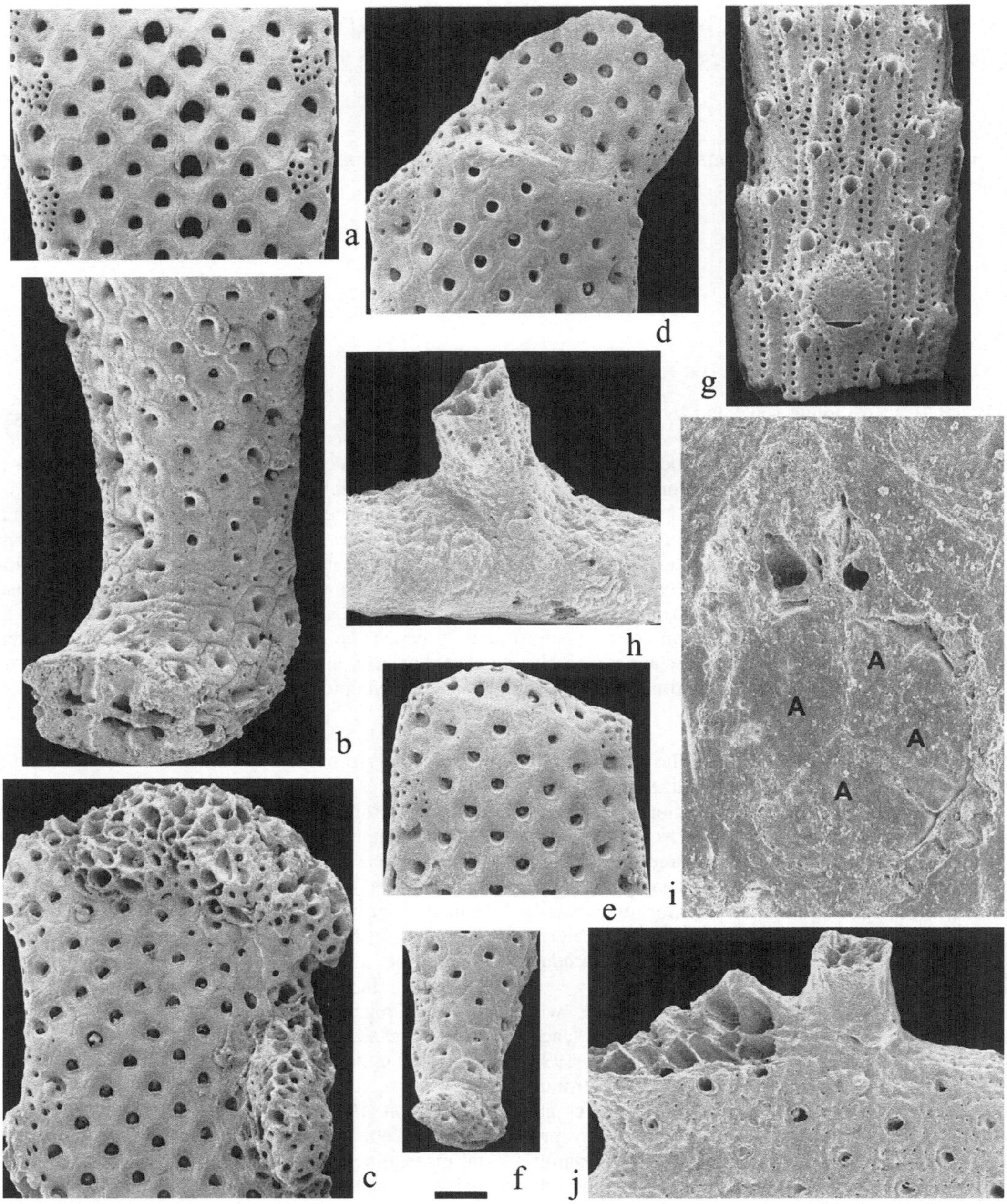

Figure 1. a-f, *Coscinopleura digitata* (Morton), Paleocene, Vincentown Formation, Vincentown, NJ, USA. a, Portion of bilaminar branch showing six ovicelled zooids, many non-ovicelled zooids, and several marginal polymorphs (coscinozooids). b, Proximal portion of new colony regenerated perpendicularly from surface of bilaminar branch fragment. c, Branch fragment with two groups of kenozooids initiating bases for new colony growth. d, New growth with initiating kenozooids emanating from proximal end of branch fragment. e, Kenozooidal growth emanating from proximal end of branch fragment. f, Proximal portion of a young, probably sexually produced colony with kenozooids reinforcing colony base. g-j, *Metrarabdotos* new species, Pliocene, Rhodes, Greece. g, Portion of bilaminar branch with one ovicelled among non-ovicelled zooids. h, Proximal portion of new colony regenerated perpendicularly from branch fragment in which zooids have become occluded by extrazooidal skeleton. i, Basal surface of sexually produced colony showing asymmetrical tetrad of ancestrular zooids (A) followed by two erect zooids distally and surrounded by extrazooidal skeleton. j, Proximal portion of regenerated colony similar to that in h. Scale bar is 500 μm in all but i, in which it is 100 μm.

1 MATERIAL AND METHODS

Data for *Metrarabdotos* are from Cheetham et al. (2001) for 12 species from the Neogene of tropical America, and from a new species from the Pliocene of Rhodes, Greece (Figures 1g-j). The Rhodes specimens were collected by Erik Thomsen from the shallow-water, bryozoan-rich beds underlying the marls from which a deeper-water bryozoan fauna was described by Moisette & Spjeldnæs (1995). Those illustrated in Figures 1g-j are catalogued in the collections of the National Museum of Natural History, Smithsonian Institution, under USNM numbers 517474-517477, respectively. More than 500 additional specimens were used for counts of zooids, ovicells, and colony bases, and for measurements of zooidal morphology (Table 1).

Data for *Coscinopleura angusta* Berthelsen are from Thomsen & Håkansson (1995) for the Danian of Denmark, and from additional specimens from the Danian of Limhamn, Sweden (sample number $IIN_1 17$ of Cheetham 1971). The specimens of *C. digitata* (Morton) illustrated in Figures 1a-f (catalogued under USNM numbers 517468-517473, respectively) are among more than 6700 on which counts and measurements were based (Table 1). All are from the samples listed for the type section of the Vincentown Formation, Vincentown, New Jersey, USA, in Cheetham (unpubl.). The illustrated specimens are from the samples numbered CN-5-1-67-14 (Figures 1a, b), CN-5-1-67-11 (Figure 1c), CC-10-23-76-2 (Figure 1d), and CN-5-1-67-10 (Figures 1e, f).

Table 1. Numbers of colony fragments in which modes of attachment, ovicell frequency, and zooidal morphology were counted or measured.

Species	Ances. bases	Regen. bases	Ovicell counts	Zooid morph.
M. n. sp.	2	8	526	5
C. angusta	2*	82*	40	5
C. digitata	2(?)	200+	6526	35
E. torta	?	100+	556	5
C. intermedia	5	0	8	5
E. septentrionalis	5	0	8	5
T. incrassata	5	0	7	5

*Data from Thomsen & Håkansson (1995). Ances. = ancestrular. Regen. = regenerated.

Comparative data for other co-occurring taxa were obtained either from Cheetham et al. (2001) and Thomsen & Håkansson (1995), or from measurements and counts in the Vincentown samples (*Euritina torta* (Gabb & Horn), *Crassimarginatella intermedia* Canu & Bassler, *Exochella septentrionalis* Canu & Bassler, *Tricephalopora incrassata* Canu & Bassler) (Table 1).

Counts and measurements in the new material were made with the methods described in detail in Cheetham et al. (2001). Except in the encrusting species, in which the numbers of both colony fragments and zooids are small enough to be tractable, the numbers of zooids were estimated from up to seven fragments in each of three size classes (those retained on sieves of 2 mm, 1 mm and 0.5 mm mesh). As in the species studied by Cheetham et al. (2001), fragments in which zooids are occluded by extrazooidal calcification were excluded from the counts. For the new species of *Metrarabdotos* from Rhodes, this method yielded 49 ovicells among approximately 30,500 zooids. For *Coscinopleura digitata*, the estimated numbers totaled 1160 ovicells and more than 1 million zooids.

To estimate genetic variation, five of the best-preserved colony fragments of each species were chosen, and four to 15 morphological characters were measured or counted, depending on the species. In *Coscinopleura digitata*, five such specimens were used from each of seven stratigraphic levels in the Vincentown Formation. Within- and among-colonies variance components were calculated by a single-classification analysis of variance on each character, and the results averaged over the four to 15 characters, as described in Cheetham et al. (2001). The use of a standard number of fragments and zooids per species or stratigraphic level serves to equalize degrees of freedom. Applied to five patches of five zooids each in a single large encrusting colony of the modern species *Metrarabdotos unguiculatum* Canu & Bassler as a control, this procedure produced an estimate of average genetic variation among patches that was close to the expected value of zero (Cheetham et al. 2001, table 2).

2 SEXUAL REPRODUCTIVE MORPHOLOGY

Despite their paucity, ovicells are morphologically distinctive (Figures 1a, g) and occur in all species of both *Metrarabdotos* and *Coscinopleura*. In *Metrarabdotos*, the percentage of zooids that are ovicelled ranges from a low of 0.01-0.03% in two of the tropical American Neogene species (*Metrarabdotos* new species 3 and 4) to a high of only 7.6% in the living *M. tenue* (Busk) (Cheetham et al. 2001). The new species from Rhodes is near the lower end of the range at 0.16%. Comparative figures for the Neogene tropical American species *Cigclisula porosa* (Canu & Bassler) and *Gemelliporella punctata* Canu & Bassler are both well above the *Metrarabdotos* range, at 8.2% and 21.7%, respectively (Cheetham et al. 2001). Both of these latter species are like most *Metrarabdotos* species in having erect, arborescent colonies.

In *Coscinopleura angusta* 0.2-1.2% of zooids are ovicelled (Thomsen & Håkansson 1995), and in *C. digitata* the range is 0.02-0.20%, all at or near the lower end of the *Metrarabdotos* range. In species of other erect Danian genera, the proportion of ovicelled zooids averages slightly higher, at 1.2-3.4%, whereas it can be much higher in species of encrusting genera, 8.0-25.3% in Danish species and up to 49.2% in those from New Jersey.

In addition to this evidence for larval production, 12 tropical American species of *Metrarabdotos*, plus the species from Rhodes, preserve features in colony bases showing their larval origin. The basal walls of the four zooids simultaneously produced by metamorphosis of a larva (ancestrular tetrad) are discernible on the encrusting base of erect colonies, in the same pattern as described by Cook (1973) in the living encrusting species *Metrarabdotos cookae* Cheetham. In the specimen illustrated in Figure 1i, the tetrad is more asymmetrical than those observed in other species (e.g. figures 4-2, 4-6 in Cheetham et al. 2001); the zooid on the left is larger than that on the right, crowding the distal zooid out of alignment with the plane of symmetry. Except in the earliest stages of colony development, the ancestrular tetrad is surrounded by extrazooidal skeleton that enlarges the colony's footprint and occludes the frontal surfaces of zooids from external view.

Larvally produced colony bases are much rarer in *Coscinopleura*. Thomsen & Håkansson (1995) found only two among more than 80 bases formed by regeneration from previously existing fragments. The sexually produced bases in *C. angusta* are readily recognizable by their incorporation of the foreign substrates on which the larvae settled. No such substrates are preserved in either of the two probably larvally produced colony bases found in *C. digitata*, however. In the specimen shown in Figure 1f, the small zooids and narrow proximal end beginning with kenozooids strongly resemble the morphology of the specimen illustrated by Thomsen & Håkansson (1995, figure 2b).

3 ASEXUAL VS. SEXUAL PROPAGATION

In both *Metrarabdotos* and *Coscinopleura*, colony bases produced by regeneration from previously existing fragments mimic those produced sexually in their general morphology. The points of similarity include narrowness in comparison to more distal branches and thickening with skeletal material of either extrazooidal (*Metrarabdotos*) or kenozooidal (*Coscinopleura*) origin. However, the polarity of the new growth distinctly contrasts with that in the older branch. It may take place at a fracture (Figure 1d), originating with the budding of kenozooids (Figure 1e), completely reversing the direction of original growth; or it may grow at a high angle to the frontal surface (Figures 1b, h) or lateral margin of the original fragment (Figure 1j), again accompanied by kenozooidal (Figure 1c) or extrazooidal (Figures 1h, j) modification that converts the previously existing fragment into a substitute for the encrusting base.

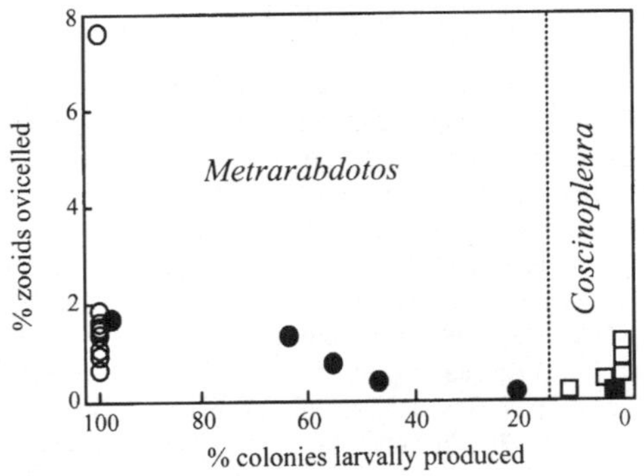

Figure 2. Percentages of ovicelled zooids and larvally produced colonies in *Metrarabdotos* (solid circles, five species with regenerated colonies; open circles, eight species without) and *Coscinopleura* (solid square, *C. digitata* ; open squares, five samples of *C. angusta*).

The relationship between the relative frequencies of asexual and sexual propagation in the two genera is shown in Figure 2. Among the five species of *Metrarabdotos* in which colony bases produced by regeneration from fragments were found, ovicell frequency decreases with increasing incidence of regeneration. The trend is significant with $R^2 = 0.9$, $P < 0.02$. However, the biological significance of this correlation is weakened by the fact that ovicell frequencies in these species are almost completely overlapped by those of species in which all colonies were apparently produced from larvae. Without the new species from Rhodes, in which 80% of the colony bases are regenerated, the statistical significance of this relationship also disappears. Moreover, the relationship is reversed in *Coscinopleura angusta*, five samples of which show an increasing frequency of ovicelled zooids with decreasing percentage of larvally produced colonies (Thomsen & Håkansson 1995, tables 1b, 2).

4 EFFECTS ON GENETIC VARIATION

The major difference in the incidence of asexual propagation in *Metrarabdotos* and *Coscinopleura* makes them ideal for a comparison of effects on genetic variation in zooid morphology. However, the two genera also differ in the number of morphological characters that can be quantified from zooid to zooid. For *Metrarabdotos*, 15 such characters are available (Cheetham et al. 2001), whereas only eight (zooid length and width, opesia length and width, length and width of coscinozooid, number of frontal pores on coscinozooid, and width of ovicell opening) can be consistently quantified in *Coscinopleura*. Therefore, species of other Neogene and Paleocene genera were included as controls in the following comparison of variance components.

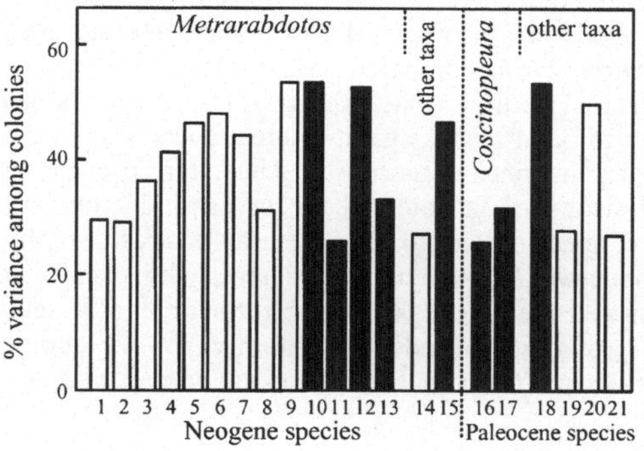

Figure 3. Among-colonies components of variance in characters of zooid morphology in *Metrarabdotos*, *Coscinopleura*, and some co-occurring species of other genera (values in Table 2 and Cheetham et al. 2001). Species with asexual propagation are indicated by solid bars, those with sexual propagation only by open bars. 1-12, Tropical American *Metrarabdotos*; 13, *Metrarabdotos* new species from Rhodes; 14, *Cigclisula porosa*; 15, *Gemelliporella punctata*; 16, *Coscinopleura angusta*, Limhamn, Sweden; 17, *C. digitata*; 18, *Euritina torta*; 19, *Crassimarginatella intermedia*; 20, *Exochella septentrionalis*; 21, *Tricephalopora incrassata*.

Table 2. Among-colonies variance components from one-way analysis of variance on characters (traits) of zooid morphology.

Trait	Sp. 13	Sp. 16	Sp. 17	Sp. 18	Sp. 19	Sp. 20	Sp. 21
LZ	0.7088	0.5946	0.5165	0.6932	-0.0158	0.3133	0.4234
WZ	-0.0264	0.4213	0.3356	0.2049	0.2962	0.5103	-0.0594
HO	0.8719	0.1145	0.3976	0.7234			
LO	0.6329	0.1341	0.2488	0.4939			
LD	0.8532						
LAVS	0.0396						
LAVL	0.2680						
P1S	0.0984						
O1S	0.0323						
O2S	-0.0484						
PL	-0.1765						
O1L	0.3182						
O2L	0.1071						
NA	0.4828						
ND	Invar.						
LC			0.1972				
WC			0.1673				
Pores			0.3358				
LAV				0.8669			
WAV				0.5131			
LOV				0.6186	0.6977	0.7211	0.4131
WOV				0.1458	0.1275	0.4406	0.2974
WOVO			0.3358	0.8669			
Mean	0.2973	0.2558	0.3157	0.5325	0.2764	0.4963	0.2686

Trait abbreviations for *Metrarabdotos* explained in Cheetham et al. (2001); for other species, LC, WC, length and width of coscinozooid; LAV, WAV, length and width of avicularium; LOV, WOV, length and width of ovicell; WOVO, width of ovicell opening. Species numbered as in Figure 3. Invar., invariant (but varies in other species).

As shown in Figure 3 and Table 2, genetic variation, estimated by the mean among-colonies components of the available morphologic traits, is similar in magnitude in *Metrarabdotos*, *Coscinopleura*, and co-occurring species of other genera. The mean values lie within a narrow range of 26-50% of total variance, irrespective of whether species exhibit erect growth (labelled 1-18 in Figure 3) or are encrusting (labelled 19-21), or whether colonies are exclusively larval in origin (open bars in Figure 3) or include those formed by regeneration (solid bars). The grand means of among-colonies variance for species of *Metrarabdotos* with and without regeneration are 40% and 38%, respectively, a difference that is clearly non-significant (t = 0.015, P > 0.9). Among the Paleocene species, the two species of *Coscinopleura* and the erect *Euritina torta* (labelled 16-18 in Figure 3) have a mean among-colonies variance of 37% compared to 35% for the three encrusting Paleocene species (labeled 19-21), also a non-significant difference (t = 0.015, P > 0.9).

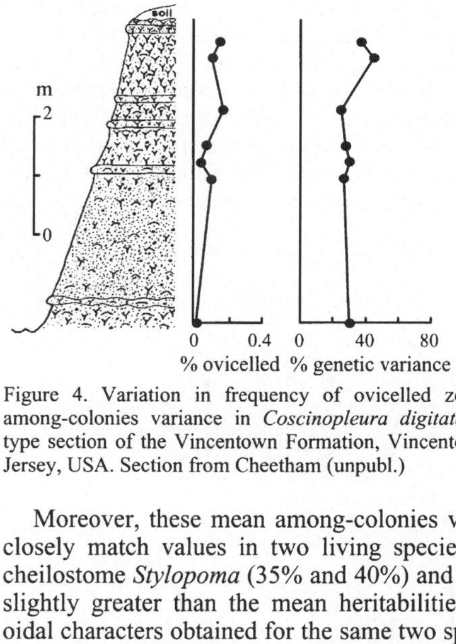

Figure 4. Variation in frequency of ovicelled zooids and among-colonies variance in *Coscinopleura digitata* through type section of the Vincentown Formation, Vincentown, New Jersey, USA. Section from Cheetham (unpubl.)

Moreover, these mean among-colonies variances closely match values in two living species of the cheilostome *Stylopoma* (35% and 40%) and are only slightly greater than the mean heritabilities of zooidal characters obtained for the same two species in rearing experiments, 34% in both cases (Cheetham et al. 1993). This estimate of genetic diversity appears to be stable regardless of whether as few as four or 10 or more characters are involved in calculating a mean (Table 2). However, the considerable variation from trait to trait suggests that smaller numbers of characters can significantly bias the estimation of mean among-colonies variance.

Further, genetic variation appears to be stable through time, at least in *Coscinopleura digitata*, even if temporal variation in its ovicell frequency is significant (Figure 4). The percentage of ovicelled zooids in this species increases significantly through the Paleocene Vincentown Formation (correlation with stratigraphic level, $r = 0.76$, $P < 0.02$). However, change in among-colonies variance is non-significant ($r = 0.5$, $P > 0.1$), and there is virtually no association between ovicell frequency and among-colonies variance ($r = 0.05$, $P >> 0.1$).

All of these results suggest that genetic variation in these bryozoans is virtually unaffected either by variation in ovicell frequency or by increasing incorporation of asexual propagation into life histories.

5 RELATION TO LARVAL OUTPUT

The maintenance of apparently normal levels of genetic diversity in species of *Metrarabdotos* and *Coscinopleura* with increasing levels of asexual propagation could be a consequence of the fact that, from the viewpoint of population genetics, even "a little recombination is as good as a lot" (Burt 2000). However, there may be another way of looking at the relation between ovicell frequency and sexual productivity. Data on ovicell frequency in *Coscinopleura digitata* and the four other species from the Vincentown section were obtained from bulk samples. Thus, the total numbers of ovicelled and non-ovicelled zooids recorded for these species should be approximately proportional to the relative abundance of the species in the section.

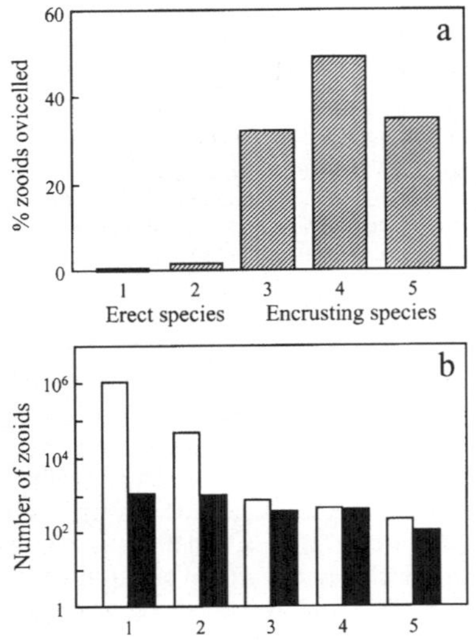

Figure 5. Percentages (a) and numbers (b) of ovicelled zooids in species from the type section of the Vincentown Formation. Solid bars, ovicelled zooids; open bars, non-ovicelled zooids. 1, *Coscinopleura digitata*; 2, *Euritina torta*; 3, *Crassimarginatella intermedia*; 4, *Exochella septentrionalis*; 5, *Tricephalopora incrassata*.

The percentages and numbers of ovicelled zooids are compared in Figure 5. Even though the percentage of ovicelled zooids is vanishingly small in *Coscinopleura digitata* and the other erect species, *Euritina torta*, compared to the encrusting species (Figure 5a), the actual numbers of such zooids appear to be at least as great in erect species as in the encrusting ones (Figure 5b). Thus, far from being deficient in their potential for larval production, the erect species seem to have been at least as sexually productive as the encrusting ones. The excess of non-ovicelled zooids (Figure 5b) then could be seen as providing an added capability for vegetative growth and the potential for asexual propagation.

6 DISCUSSION

The potential for asexual propagation in erect, arborescent cheilostomes such as *Metrarabdotos* and *Coscinopleura* may well be linked to an ability to produce a large surplus of zooids not committed to sexual reproduction, i.e. non-ovicelled, and thus a burgeoning of vegetative growth (Thomsen & Håkansson 1995). However, the low ratio of ovicelled to non-ovicelled zooids holds for all erect species of *Metrarabdotos*, whether or not the potential for asexual propagation is realized, and for encrusting species of this genus, with little or no such potential, as well (Cheetham et al. 2001).

The relationships shown in Figure 5b suggest that a large surplus of non-ovicelled zooids need not be balanced against a decline in sexual productivity. This suggestion is strongly supported by the fact that genetic diversity, at least as estimated by among-colonies variance in traits of zooid morphology, is maintained in these genera irrespective of the relative frequency of ovicelled zooids or the relative incidence of asexual colony propagation.

If asexual propagation indeed does not imply a diversion of resources away from sexual reproduction, then it probably does imply an increase in resources that are available for both processes. One such possibility is growth in an area of strong upwelling with nutrient enrichment and increased primary production, as has been suggested for the three asexually propagating *Metrarabdotos* species in the Neogene of Venezuela (Cheetham et al. 2001).

It is thus possible that the occurrence of asexual propagation in these genera is entirely an ecophenotypic response, and that all species are capable of such response given the necessary environmental conditions. So far, support for such an interpretation is weak. *Metrarabdotos* new species 10 from the Neogene of the Dominican Republic is the only species in which both exclusively sexually reproducing populations and one including both sexually and asexually produced colonies appear to have existed. Although seven samples of this species distributed over an interval of 1.5 million years have yielded more than 100 colony bases, only a single asexually produced base was found (Cheetham et al. 2001). The four species of *Metrarabdotos* in which the bulk of the asexually produced bases have been found (new species 11-13 of Cheetham et al. 2001, plus the new species from Rhodes) each occurs at only a single locality, thus giving no clue concerning the possible environmental effect on propagation mode.

Indeed, there is ample evidence that the potential for asexual propagation, like the ratio between ovicelled and non-ovicelled zooids, is at least partly taxon-specific in these cheilostomes. The major difference in the relative incidence of asexual propagation in *Coscinopleura* and *Metrarabdotos* (Figure 2) is a case in point. In this sense, these properties could be considered to be under "phylogenetic constraint," even though the species of *Metrarabdotos* in which asexual propagation is known appear not to be closely related to each other (Cheetham et al. 2001). Moreover, *Metrarabdotos* new species 10, with its single instance of asexual propagation, must have experienced a range of environments similar to that experienced by the eight other *Metrarabdotos* species, which apparently relied entirely on sexual reproduction, co-occurring in the Neogene of the Dominican Republic (Cheetham et al. 2001). Erect species of other genera in the same sections in the Dominican Republic show similar differences (*Cigclisula porosa*, all sexually produced; *Gemelliporella punctata*, 25% asexually; Cheetham et al. 2001).

Despite the evidence for taxon-specific differences among these cheilostomes in the potential for propagation by fragmentation and regeneration, the morphological basis for this property is more elusive than in cheilostomes with a free-living growth form (Thomsen & Håkansson 1995, Håkansson & Thomsen 2001). The fragility of species of *Metrarabdotos* compared to many other cheilostomes with a similar erect growth form seemingly should make them more prone to propagation by fragmentation and regeneration. The rate at which branches thicken with colony growth, a major determinant of the ability of arborescent cheilostomes to resist breakage (Cheetham & Thomsen 1981), is less than 2% of branch length for a variety of Paleogene and Neogene *Metrarabdotos* species, well below the values for species of other genera (Cheetham 1986). However, branch-thickening rates of species with asexually produced colonies range from slightly below (1.3%) to well above (2.6%) the mean for the genus (1.4% of branch length, Cheetham et al. 2001; in the new species from Rhodes the rate is 1.8%, also well above the mean for the genus). Judged by its rate of branch thickening (mean value of 2.1%, Cheetham et al. 1981), *Coscinopleura* should be at least as resistant to fragmentation as the strongest species of *Metrarabdotos*, and yet shows a much higher incidence of asexual propagation.

Thus, on present evidence, it seems likely that the incidence of asexual propagation in *Metrarabdotos*, *Coscinopleura*, and co-occurring cheilostomes with erect arborescent growth forms is controlled by a combination of taxon-specific and environmental factors. One of the least understandable results of this analysis is why *Metrarabdotos* and *Coscinopleura* were able to take advantage of environments favoring vegetative growth while coeval cheilostomes apparently were unable to do so.

7 ACKNOWLEDGMENTS

I thank Erik Thomsen for the material from Rhodes, JoAnn Sanner for specimen measurements, counts, scanning electron micrographs, and much other assistance, and Steve Hageman for a helpful review of the manuscript. Research support from the Marie Bohrn Abbott Fund of the National Museum of Natural History, Smithsonian Institution, is gratefully acknowledged.

REFERENCES

Burt, A. 2000. Perspective: Sex, recombination, and the efficacy of selection—Was Weismann right? *Evolution* 54: 337-351.

Cheetham, A.H. 1971. Functional morphology and biofacies distribution of cheilostome Bryozoa in the Danian Stage (Paleocene) of southern Scandinavia. *Smithsonian Contributions to Paleobiology* Number 6.

Cheetham, A.H. unpublished. Notes on the Vincentown Bryozoa, International Bryozoology Association Cenozoic field trip, 18-20 September 1977.

Cheetham, A.H. 1986. Branching, biomechanics and bryozoan evolution. *Proceedings of the Royal Society, London B* 228: 151-171.

Cheetham, A.H., Hayek, L.C. & Thomsen, E. 1981. Growth models in fossil arborescent bryozoans. *Paleobiology* 7: 68-86.

Cheetham, A.H., Jackson, J.B.C. & Hayek, L.C. 1993. Quantitative genetics of bryozoan phenotypic evolution, I: Rate tests for random change versus selection in differentiation of living species. *Evolution* 47: 1526-1538.

Cheetham, A.H., Jackson, J.B.C. & Hayek, L.C. 1995. Quantitative genetics of bryozoan phenotypic evolution, III: Phenotypic plasticity and the maintenance of genetic variation. *Evolution* 49: 290-296.

Cheetham, A.H., Jackson, J.B.C. & Sanner, J. 2001. Evolutionary significance of sexual and asexual modes of propagation in Neogene species of the bryozoan *Metrarabdotos* in tropical America. *Journal of Paleontology* 75: 564-577.

Cheetham, A.H. & Thomsen, E. 1986. Functional morphology of arborescent animals: strength and design of cheilostome bryozoan skeletons. *Paleobiology* 7: 355-383.

Cook, P.L. 1973. Settlement and early colony development in some Cheilostomata. In G.P. Larwood (ed.), *Living and fossil Bryozoa: Recent advances in research*: 65-71. London & New York: Academic Press.

Hageman, S.J., Bayer, M.M. & Todd, C.D. 1999. Partitioning phenotypic variation: Genotypic, environmental and residual components from bryozoan skeletal morphology. *Journal of Natural History* 33: 1713-1735.

Håkansson, E. & Thomsen, E. 2001. Asexual propagation in cheilostome Bryozoa, evolutionary trends in a major group of colonial animals. In J.B.C. Jackson, S. Lidgard & F.K. McKinney (eds), *Evolutionary patterns, growth form and tempo in the fossil record*: 326-347. Chicago & London: University of Chicago Press.
Jackson, J.B.C. & Cheetham, A.H. 1990. Evolutionary significance of morphospecies: A test with cheilostome Bryozoa. *Science* 248: 579-583.
Jackson, J.B.C. & Cheetham, A.H. 1994. Phylogeny reconstruction and the tempo of speciation in cheilostome Bryozoa. *Paleobiology* 20: 407-423.
Jackson, J.B.C. & Coates, A.G.1986. Life cycles and evolution of clonal (modular) animals. In J.L. Harper, B.R. Rosen & J. White (eds), *The growth and form of modular organisms*: 7-22. London: The Royal Society.
Moissette, P. & Spjeldnæs, N. 1995. Plio-Pleistocene deepwater bryozoans from Rhodes, Greece. *Palaeontology* 38: 771-779.
Thomsen, E. & Håkansson, E. 1995. Sexual versus asexual dispersal in clonal animals: Examples from cheilostome bryozoans. *Paleobiology* 21: 496-508.

Bryozoan Studies 2001, Wyse Jackson, Buttler & Spencer Jones (eds)
© 2002 Swets & Zeitlinger, Lisse, ISBN 90 5809 388 3

Origin and early development of the International Bryozoology Association

A.H. Cheetham
Department of Paleobiology, National Museum of Natural History, Smithsonian Institution, Washington, D.C., USA

ABSTRACT: The International Bryozoology Association (IBA) was founded in May 1965 at the conclusion of a conference on the post-Paleozoic Bryozoa of the North Sea region, held in Sweden at the Geological Institute of Stockholm University. Sixteen of the participants in that conference, E. Annoscia, O. Berthelsen, R.S. Boardman, G. Braga, A.H. Cheetham, P.L. Cook, Y.-V. Guatier, M. Labracherie, R. Lagaaij, G.P. Larwood, A.W. Medd, J. Prud'homme, S. Schager, L. Silén, N. Spjeldnaes, and E. Voigt, became the Association's first members. Motivated by a spirit of internationalism that pervaded the proliferation of new scholarly societies in the two decades following World War II, IBA's founders sought to establish an organization that would promote the exchange of information, ideas, and techniques among bryozoologists of all nations. The first duty of the newly chosen chairman (Cheetham) and secretary (Cook) was to initiate plans for the first IBA conference, targeted for 1968 at a venue that we judged likely to maximize participation in terms of both attendees and countries represented. One consideration in deciding the venue was the likelihood that proximity (in time and place) to an International Geological Congress (IGC) or International Congress of Zoology (ICZ) might increase the chances that either paleobiologists or biologists would obtain travel funding. This also influenced our planning a triennial cycle for the IBA conferences, some of which would fall in IGC years and others in ICZ years. The IBA's first conference, at the laboratories of the petroleum company AGIP in San Donato Milanese, Italy, was held in August 1968, just before the IGC meeting in Prague. Although 46 bryozoologists from 14 countries attended, participation by 17 others from three countries, including all of those from the Soviet Union, was cut short as the military action ending the "Prague Spring" got underway. Proximity to the Prague IGC, however, did seem to result in a representation of paleobiological papers unequaled at any of the IBA's next ten conferences (through 1998). Efforts by the succeeding IBA officers (Spjeldnaes and Cook) to gain formal international standing for the IBA, through affiliation with the International Union of Biological Sciences and the International Geological Union, were rewarded with the attendance of three of our colleagues from the Soviet Union at the IBA's second conference, at the University of Durham, England, in September 1971. However, by the time of the third conference (September 1974 at the Université Claude Bernard in Villeurbanne, France), participation of our colleagues from the Soviet Union was again canceled, perhaps because of renewed international tensions. As a further step toward formalizing the organization of the IBA and its relationship to other scholarly societies, participants in the 1974 conference adopted a constitution, thus ending nearly a decade of efforts to keep the IBA as informal as possible. Despite yielding to the necessity for this degree of formality, the Association avoided other proposals, such as publishing a taxon-specific journal, changing its name to accommodate taxonomic usage, or obtaining funding to bring particular groups of bryozoologists to the conferences, any of which could have constrained the inclusiveness and internationalism of the organization. Although the IBA's purpose was simply to bring bryozoologists into dialogue rather than to affect the direction of their research, anecdotal evidence suggests that since its founding, bryozoology has undergone a synergism resulting from a new level of understanding between biologists and paleobiologists, thus "putting a human face" on researchers who had previously worked in isolation. A more analytical approach to this question, however, remains to be undertaken.

The full text of the paper from which this account was abstracted will be published as: Cheetham, A.H. 2002. The founding and early history of the International Bryozoology Association, 1965-1974. In P.N. Wyse Jackson & M. Spencer Jones (eds), *Annals of bryozoology: aspects of the history of bryozoological research.* Dublin: International Bryozoology Association.

Bryozoan Studies 2001, Wyse Jackson, Buttler & Spencer Jones (eds)
© 2002 Swets & Zeitlinger, Lisse, ISBN 90 5809 388 3

Influence of colony morphology on associated biota diversity in four Bryozoa

S. Cocito & F. Ferdeghini
ENEA, Marine Environment Research Centre, P.O. Box 224, 19100 La Spezia, Italy

S. Pisaroni & D. Bedulli
Dipartimento di Biologia Evolutiva e Funzionale, Università di Parma, v.le delle Scienze, 43100 Parma, Italy

ABSTRACT: The potential role of four erect, calcified (Cheilostomatida, Ascophorina) bryozoans in providing habitat for other organisms has been analysed. Samples of *Myriapora truncata*, *Smittina cervicornis*, *Reteporella couchii*, *Pentapora fascialis*, collected in the Ligurian Sea (NW Mediterranean), were assigned to four morphological types according to their increasing morphological complexity: 'tree', 'bush', 'convolute laminae', 'anastomosed, convolute laminae'. 22 species, belonging to 7 higher taxonomic levels, were found associated with the 'tree' morphotype, 27 species (7 higher taxonomic levels) were found in 'bush' morphotype, 37 species (9 higher taxonomic levels) were identified among 'convolute laminae', whereas 'anastomosed, convolute laminae', provided habitat for 84 species (13 higher taxonomic levels). Species richness of associate biota increased with increasing morphological diversity. The position of associated biota inside the colonies was correlated to organism growth form and feeding behaviour. In particular, a relationship was found between species richness, abundance of associated bryozoan species and morphological diversity. All the growth habits displayed by associated bryozoan species found suitable habitat in the most complex morphotype.

1 INTRODUCTION

Habitat structural complexity is ecologically important because it has influence on the diversity and abundance of animal communities. It provides habitat structure for juvenile and adult animals, plays a role in regulating foraging patterns, alters boundary-layer flow, thus affecting larval settlement and subsequent population performance and influences flow-mediated erosion, transport and deposition of sediment, meiofauna and macrofauna (Thompson et al. 1996, Commito & Rusignuolo 2000).

Many marine benthic organisms, by building a carbonate erect framework, increase morphological diversity and thus habitat complexity. That biological diversity is induced by habitat complexity has already been demonstrated for some species among calcareous algae, vermetids, *Mytilus* beds, polychaete reefs and bryozoans (Hong 1982, Ben-Eliahu & Safriel 1982, Günther 1996, Bianchi & Morri 1996, Bradstock & Gordon 1983, Ferdeghini & Cocito 1999).

Bryozoans display a wide range of colony morphologies, which can reflect adaptive evolutionary or phenotypic responses to selected environmental pressures under which the organisms live (Stach 1936, McKinney & Jackson 1989). It has been shown that diversity of microenvironments across different portions of the bryozoan colony provisions for associated biota with different feeding strategies (Maluquer 1985).

In this paper we describe the potential role of four bryozoan morphotypes in providing habitat for other organisms, thus influencing biological diversity. Samples of four erect, calcified (Cheilostomatida, Ascophorina) bryozoan species displaying different morphology, from slender, branching structure to anastomosed laminae, were collected in the Ligurian Sea (northwestern Mediterranean): *Myriapora truncata* (Pallas 1766), *Smittina cervicornis* (Pallas 1766), *Reteporella couchii* (Hincks 1878), *Pentapora fascialis* (Pallas 1766). Species richness of associated biota has been analysed in relation to the four different bryozoan morphotypes. In particular, abundance, growth habit and preferred habitat of associated bryozoan species have been examined in the different portions of colony.

2 MATERIALS AND METHODS

From 1998 to 1999, 31 colonies of *Myriapora truncata*, 30 colonies of *Smittina cervicornis*, 27 colonies of *Reteporella couchii* and 32 colonies of *Pentapora fascialis* were collected in plastic bags by SCUBA diving on rocky bottoms in the Ligurian Sea ranging from 11 to 24 m deep. Colony morphologies were described and assigned to four morphological types according to their increasing morphological complexity: *M. truncata* to 'tree' type, *S. cervicornis* to 'bush', *R. couchii* to 'convolute laminae', *P. fascialis* to 'anastomosed, convolute laminae'. Colonies were rinsed with water to remove sediment and associated epibiota (both sessile and motile > 4 mm) were separated using a binocular with tweezers and identified to species or genus.

Among taxa identified, particular attention has been paid to bryozoans living as epibionts. Their abundance was expressed with a semi-quantitative index, ranging from scarce to very abundant. Position of bryozoans at the base of the colony, inside cavities or on branch surfaces was also reported. Their growth habit, as defined by Hageman et al. (1998), was classified according to orientation relative to substrate (encrusting, erect, nodular, creeping) and construction (weakly calcified, calcified, rigid, flexible). Information from literature on preferred habitat (vegetable, hard bottom, cryptic, etc.) were also taken into account (Gautier 1962, Harmelin 1976, Prenant & Bobin 1956, Prenant & Bobin 1966, Zabala 1986).

3 RESULTS

22 species belonging to 7 higher taxonomic levels were found associated with the 'tree' morphotype of *Myriapora truncata* colonies, whereas 27 species (7 higher taxonomic levels) were associated with *Smittina cervicornis* displaying 'bush' morphotype (Table 1). For both species bryozoans, which will be analysed in detail, were the most represented group, followed by annelids (*Serpula vermicularis, Hydroides pseudouncinata pseudouncinata, Vermiliopsis labiata*).

37 species belonging to 9 higher taxonomic levels were identified among colonies of *Reteporella couchii* with 'convolute laminae'; bivalves (*Modiolarca subpicta, Striarca lactea*) and gastropods (*Calliostoma conulum, Pollia scabra, Bittium reticulatum*), followed by crustaceans, mainly represented by motile fauna, annelids and bryozoans were the most abundant organisms.

Pentapora fascialis, which displayed 'anastomosed, convolute laminae', provided habitat for 84 species (13 higher taxonomic levels): with 27 species bryozoans were the most represented group. Bivalves (*Anomia ephippium, Chlamis varia*), gastropods (*Bittium reticulatum*) and cnidarians, mainly hydroids, were also abundant. Protozoa (*Miniacina miniacea*), ascidians (*Phallusia mammillata*) and fishes (*Chromis chromis, Gobius* sp., *Parablennius gattorugine, P. rouxi*) were found exclusively associated with *Pentapora fascialis* colonies (Table 1).

For three out of four bryozoan species (and relative morphotype), bryozoans were the most represented group (Table 2). They were mainly small-sized species and in some cases living epibiotically on hydroids or algae.

Myriapora truncata had from scarce to abundant bryozoans (11 species or genera) both at the base and on branches. Apart from representatives of the genus *Tubulipora*, which can develop both encrusting and erect growth forms, the majority displayed an erect-rigid growth habit. Two species were calcified encrusting and two creeping (Fig 1). Almost all were species common on hard substrata, some preferring a sciaphilic environment (*Beania magellanica, Smittina cervicornis*).

A total of 7 species and 5 genera were found as epibionts on *Smittina cervicornis*, the majority living

Table 1. Biota associated with *Myriapora truncata, Smittina cervicornis, Reteporella couchii, Pentapora fascialis* colonies. The number of associated species and relative higher taxonomic levels are indicated.

	M. truncata	*S. cervicornis*	*R. couchii*	*P. fascialis*
morphotype	tree	bush	convolute laminae	anastomosed, convolute laminae
Rhodophyta	1	-	-	2
Chromophyta	-	1	1	1
Chlorophyta	-	-	2	2
Foraminifera	1	-	-	1
Porifera	1	1	1	2
Cnidaria	-	2	1	11
Bryozoa	11	12	5	27
Annelida	2	8	5	8
Mollusca	4	2	14	19
Crustacea	-	1	6	4
Echinodermata	2	-	2	2
Ascidiacea	-	-	-	1
Fishes	-	-	-	4
associated species	22	27	37	84
taxonomic levels	7	7	9	13

Table 2. List of the bryozoan species found associated with *Myriapora truncata*, *Smittina cervicornis*, *Reteporella couchii*, *Pentapora fascialis* colonies (+ = scarce; ++ = abundant; +++ = very abundant). Growth habit (WE = weakly calcified encrusting; CE = calcified encrusting; ER = erect-rigid both rigid contiguous and articulated; EF = erect-flexible; N = nodular; S = creeping) and most used habitat (V = vegetable; H = hard bottom; Cr = cryptic; bw = brackish waters and harbours) are indicated. * develops both encrusting and erect growth form.

Myriapora truncata	base	branches	growth habit	most used habitat
Aetea truncata		+	S	V, H
Beania magellanica	+		WE	H, V
Crisia sp.		+	ER	-
Nolella stipata		++	S	V, H
Pentapora ottomulleriana	++		CE	H
Reteporella couchii		++	ER	H
Reteporella sp.	++	+	ER	-
Scrupocellaria delilii	++		EF	H
Smittina cervicornis		++	ER	H
Spiralaria gregaria	+	++	CE	H
Tubulipora sp.	+		*	-

Smittina cervicornis	base	cavities	branches	growth habit	most used habitat
Annectocyma sp.	+			CE	-
Annectocyma tubulosa			+	CE	H
Cellaria salicornioides			++	ER	H
Celleporina lucida	+			N	H
Crisia denticulata			+	ER	V, H
Crisia sp.		+	+	ER	-
Idmidronea sp.			+	ER	-
Reteporella sp.	+		++	ER	H
Scrupocellaria sp.	+		+	EF	V, H
Smittoidea reticulata	+		++	CE	H
Spiralaria gregaria	++		+	CE	H
Valkeria uva			+	S	V, H

Reteporella couchii	base	cavities	branches	growth habit	most used habitat
Bantariella verticillata			+	S	V, H
Caberea boryi	+	+++		EF	H, V
Cellaria salicornioides			+++	ER	H, V
Scrupocellaria sp.	+			EF	V, H
Scrupocellaria delilii	++			EF	H

Pentapora fascialis	cavities	branches	growth habit	most used habitat
Aetea sp.		+	S	V, H
Amathia lendigera		+	EF	V, H
Amathia pruvoti		+	EF	V, H
Bantariella verticillata		++	S	V, H
Beania hirtissima cylindrica	+		EF	H
Beania hirtissima hirtissima		+	WE	H
Beania magellanica		+	WE	H, V
Beania mirabilis		+	S	H
Bugula fulva	++		EF	H, bw
Buskea nitida		+	ER	Cr, H
Caberea boryi	++		EF	H, V
Celleporina globulosa	+	+++	N	H, Cr
Cellaria fistulosa	++		ER	H
Chartella papyrea	+		EF	H
Crisia sp.	+		ER	-
Diplosolen obelia		+++	CE	H
Hornera sp.	+		ER	H
Nolella sp.		+	S	V, H
Pherusella tubulosa	+		EF	V, H
Reteporella septentrionalis	+		ER	H
Savignyella lafontii	++		EF	H, V
Scrupocellaria reptans	+++	++	EF	H, V
Scrupocellaria scrupea	++		EF	H
Scrupocellaria scruposa	++		EF	H
Smittina cervicornis	++		ER	H
Tubulipora sp.	++		ER	-
Valkeria uva		++	S	V

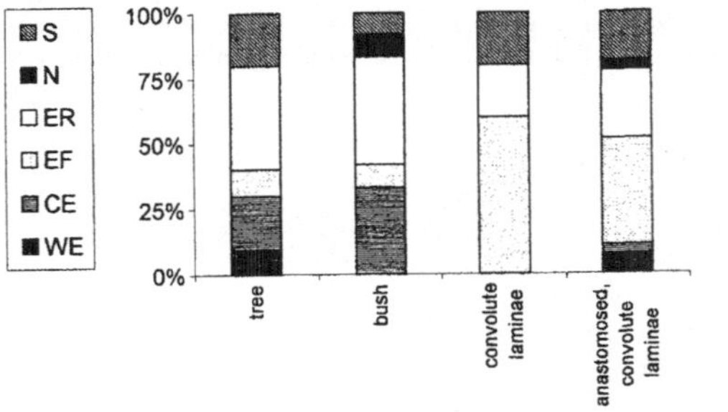

Figure 1. Percent distribution of growth habits of associated bryozoans on the four morphotypes (see Table 2 for legend explanation).

on branches, several at the base of colonies and one (*Crisia* sp.) developed among the branching structure. Erect-rigid and calcified encrusting forms were the most common growth habits, the former spreading on branches, the latter attached at the base of the colony. Their most used habitat are hard substrata, with some preferences for a sciaphilic environment (*Cellaria salicornioides, Smittoidea reticulata*).

Only 5 bryozoan species colonised *Reteporella couchii* and among these two were very abundant. *Caberea boryi* filled with erect-flexible colonies, the space formed among the convolute laminae. *Cellaria salicornioides* grew attached to the inner surface of the laminae. Both species are typically sciaphilic found attached to other bryozoans, on algae and coarse gravel.

27 bryozoan species lived associated with anastomosed, convolute laminae and inhabited cavities among laminae of *Pentapora fascialis*. The most abundant were the nodular *Celleporina globulosa,* the calcified, thin crusts of *Diplosolen obelia* which covered large portions of laminae, and the erect, branching *Scrupocellaria reptans* which filled space available among laminae. All growth habits were present: about 65% of bryozoans showed an erect growth habit, the majority being flexible, 10% were encrusting, 18% creeping (Fig. 1). Some bryozoans belonged to ubiquitous species living on hard substrata such as rock, other bryozoans, shells or on algae (*Caberea boryi, Scrupocellaria reptans, Savignyella lafontii*), some were characteristics of sciaphilic biotopes, typically attached to shells or organic substrata (*Beania magellanica, Bugula fulva, Diplosolen obelia, Scrupocellaria scrupea, S. scruposa, Reteporella septentrionalis, Smittina cervicornis*).

4 DISCUSSION

Three-dimensional growth increases the surface area of a sessile animal. It has been demonstrated that an erect sheet, at least up to 4 cm height, displays an increase of surface area greater than that performed by branching of an erect colony, and that increase of surface area in a branched, erect colony depends on the frequency of branch division and on branch diameter (McKinney & Jackson 1989). As a consequence, the surface area suitable for colonisation increases with the increasing morphological complexity performed by the four bryozoan morphological types, i.e. from 'tree' to 'bush' to 'convolute laminae' to 'anastomosed, convolute laminae'.

Biotic diversity inhabiting the four bryozoan species showed a clear relationship with morphological diversity and, consequently, habitat complexity: species richness of associate biota increased with increasing morphological diversity. If compared with the morphotypes 'convolute laminae' of *Reteporella couchii* and the 'anastomosed, convolute laminae' of *Pentapora fascialis*, the 'tree' (*Myriapora truncata*) and the 'bush' (*Smittina cervicornis*) morphotypes only occasionally hosted motile fauna, mainly represented by crustaceans, vertebrates and echinoderms. Algae and other organisms generally spreading in the space available among branching structure were rather scarce as well.

On the contrary, particularly in the 'anastomosed, convolute laminae' morphotype, the associated biota mainly inhabited cavities among branching structures, where the spatial complexity was higher. Algae, erect bryozoans, serpulids and hydroids developed from a narrow base attached to branches and, due to their growth form, tended to occupy and fill the cavities. The presence of these species, being also suspension-feeders, together with hindrance effect generated by the complex morphology of hosting bryozoan (Winston 1979) can enhance water flow deceleration process, helping sediment and dead organisms to settle in the basal portion of cavity. Consequently, many gastropods, bivalves and errant polychaetes, generally deposit feeders, filter feeders or carnivores, thrived in the deeper portion of cavities. Also motile organisms, generally deposit feeders or carnivore-scavengers, were found or seen inside cavities among branches. Associated with the more spatially complex morphotype were all feeding strategies, and consequently all food sources: suspended microorganisms entering with sea water flowing through; algae and/or microorganisms attached to the laminae; deposited silt and organic matter; animal prey and animal debris washed in. These are utilised respectively by filter-feeders, grazers, deposit feeders, and carnivore-scavengers (Crisp & Boniface 1989).

With the exception of the 'convolute laminae' morphotype, the other morphotypes, with increasing morphological complexity, provided habitat for an increasing number of bryozoan species. All the six growth habits displayed by associated bryozoan species were in the most complex morphotype, which provided different microenvironments across different portions of the colony. Different relative positions of growth habit were also noticed: erect, mostly flexible, growth habit occupied cavities and the distal part of laminae surfaces, whereas encrusting and nodular species spread exclusively in the proximal part. Differences in local hydrodynamic conditions on the laminae surface, and the relationship between growth form and feeding behaviour could explain such zonation (Bishop 1989, McKinney & Jackson 1989, McKinney 1990).

Associated bryozoans were prevalent on the inner surface of both 'convolute laminae' and 'anastomosed, convolute laminae'. These portions of colony are important source of available secondary substratum, providing shelter, protection from predation and, to some extent, favouring sciaphilic species. Furthermore, these morphotypes with large, twisted laminae, closeness of framework and scarce porosity offer in the internal portion of colony relatively less interaction with current impact than morphotypes with thin and widely spaced branches. It has been recognised that both external water flow velocity and growth form, including colony porosity, branching pattern and relative branch size, can affect the rate of flow through colony (Chamberlain & Graus 1975). In this sense, 'tree' and 'bush' colonies constitute habitat relatively less sheltered from current impact. Diversity of microenvironments across different portions of these morphotypes resulted with erect-rigid (both rigid contiguous and articulated) growth habits spreading preferably on branches and calcified encrusting attached at the base of the colony.

Functional analyses of different morphotypes across a range of environments are required in order to confirm and further characterise the direct relationship between flow, morphology and feeding performance.

5 CONCLUSIONS

The study illustrates the potential role of four erect, calcified (Cheilostomatida, Ascophorina) bryozoans in providing habitat for other organisms, thus influencing biological diversity. Four morphological types were analysed according to their increasing morphological complexity: 'tree', 'bush', 'convolute laminae', 'anastomosed, convolute laminae'. Species richness of associated biota increased with increasing morphological diversity. The position of associated biota inside the colonies was correlated to organism growth form and feeding behaviour. In particular, a relationship was found between species richness, and abundance of associated bryozoan species and morphological diversity. The most complex morphotype provided suitable habitat for all the growth habits displayed by associated bryozoan species. Functional analyses of the different morphotypes across a range of environments should be addressed to confirm and further characterise the relationship between morphological diversity and associated biota diversity.

ACKNOWLEDGEMENTS

A. Balduzzi (Genoa) helped with bryozoans identifications and provided bibliography. Research falls within the scope of the project 'Marine Biodiversity' (MURST) of the Marine Environment Research Centre of ENEA.

REFERENCES

Ben-Eliahu, M.N. & Safriel, U.N. 1982. A comparison between species diversities of polychaetes from tropical and temperate structurally similar rocky intertidal habitats. *Journal of Biogeography* 9: 371-390.

Bianchi, C.N. & Morri, C. 1996. Ficopomatus 'reefs' in the Po River Delta (Northern Adriatic): their constructional dynamics, biology, and influence of brackish water biota. *Marine Ecology Pubblicazioni della Stazione Zoologica di Napoli* I 17 (1-3): 51-66.

Bishop, J.D.D. 1989. Colony form and the exploitation of spatial refuges by encrusting Bryozoa. *Biological Review* 64: 197-218.

Bradstock, M. & Gordon, D. 1983. Coral-like bryozoan growths in Tasman Bay, and their protection to conserve commercial fish stocks. *New Zealand Journal of Marine Freshwater Research* 17: 159-163.

Chamberlain, J.A. & Graus, R.R. 1975. Water flow and hydromechanical adaptations of branched reef corals. *Bulletin of Marine Science* 25: 112-125.

Commito, J.A. & Rusignuolo, B.R. 2000. Structural complexity in mussel beds: the fractal geometry of surface topography. *Journal of Experimental Marine Biology and Ecology* 255: 133-152.

Crisp, D.J. & Boniface, M. 1989. Diversity in intertidal communities with special reference to the *Corallina officinalis* community. *Scientia Marina* 53 (2-3): 365-372.

Ferdeghini, F. & Cocito, S. 1999. Biologically generated diversity in two bryozoan buildups. *Biologia Marina Mediterranea* 6 (1): 191-197.

Gautier, Y.V. 1962. Recherches écologiques sur les bryozoaires chilostomes en Méditerranée Occidentale, *Recueil des Travaux de la Station Marine d'Endoume*, 38 (24), 163-168.

Günther, C. 1996. Development of small *Mytilus* beds and its effects on resident intertidal macrofauna. *Marine Ecology Pubblicazioni della Stazione Zoologica di Napoli I* 17 (1-3): 117-130.

Hageman, S.J., Bock, P.E., Bone, Y. & McGowran, B. 1998. Bryozoan growth habits: classification and analysis. *Journal of Paleontology* 72 (3): 418-436.

Harmelin, J.G. 1976. Le sous–ordre des Tubuliporina (Bryozoaires cyclostomes) en Méditerraneé Ecologie et systématique, *Mémoires de l'Institut Océanographique, Monaco* 10: 19-22.

Hong, J.-S. 1982. Contribution à l'étude des peuplements d'un fond de Concrétionnement Coralligène dans la région marseillaise en Méditerranée Nord-occidentale. *Bulletin of Kordi* 4: 963-984.

Maluquer, P. 1985. Algunas consideraciones sobre la fauna asociada a las colonias de *Schizoporella errata* (Waters,

1878) del puerto de Mahón (Menorca, Baleares). *Publicaciones Departamento Zoologia de Barcelona*, 11: 23-28.

McKinney, F.K. 1990. Feeding and associated colonial morphology in marina bryozoans. *Aquatic Sciences* 2(2): 255-280.

McKinney, F.K. & Jackson J.B.C. 1989. *Bryozoan Evolution*. Chicago: Chicago University Press.

Prenant, M. & Bobin, G. 1956. Bryozoares, 1. Entoproctes, Phylactolèmes, Cténostomes, *Faune de France*, 60. Fédération française des Societés de Sciences Naturelles, Lechevalier, Paris.

Prenant, M. & Bobin, G. 1966. Bryozoares, 2. Chilostomes Anasca, *Faune de France*, 68. Fédération française des Societés de Sciences Naturelles, Lechevalier, Paris.

Stach, L.W. 1936. Correlation of zoarial growth form with habit. *Journal of Geology* 44: 60-65.

Thompson, R.C., Wilson, B.J., Tobin, M.L., Hill, A.S. & Hawkins, S.J. 1996. Biologically generated habitat provision and diversity of rocky shore organisms at a hierarchy of spatial scale. *Journal of Experimental Marine Biology and Ecology* 202: 73-84.

Winston, J.E. 1979. Current-related morphology and behaviour in some Pacific Coast bryozoans. In G.P. Larwood & M.B. Abbott (eds), *Advances in byozoology*: 247-268. London: Academic Press.

Zabala, M. 1986. Fauna dels briozous dels Països Catalans, Institut des Estudis Catalans, *Arxius de la secció de ciències* 84: 1-833.

Bryozoan Studies 2001, Wyse Jackson, Buttler & Spencer Jones (eds)
© 2002 Swets & Zeitlinger, Lisse, ISBN 90 5809 388 3

Identification key for North American Ordovician trepostome families

R.J. Cuffey
Department of Geosciences (412 Deike Bldg.), Pennsylvania State University, University Park, PA 16802, USA

ABSTRACT: Many paleontologic studies depend critically upon species-level data, and bryozoan paleoecologic, reefal, biostratigraphic and evolutionary investigations are no exception. A consistent and useable taxonomic framework is essential for understanding the similarities and relationships among those species. Experiences with North American Ordovician bryozoans suggest that the lowest (genus down) and highest (suborder up) taxa as recently modified are adequate, but that – for trepostomes -- intermediate-level ones (families) in the present literature are difficult to diagnose, recognize, and apply. However, several morphologic characters can be used to construct a pragmatically useable identification key for 15 such trepostome families. Colony size and form, branch thickness, exozonal wall thickness, wall microstructure, zooecial diametre, mesozooecia abundance and appearance, acanthostyle abundance and size, cystiphragms, curved diaphragms, straight diaphragms, and hemiphragms all contribute to this taxonomic improvement.

1 INTRODUCTION

Species-level data are essential for fully understanding the paleoecology, biostratigraphy, and evolution of bryozoans, as well as of any phylum.

Ordovician bryozoans, especially trepostomes, in eastern and central North America are very diverse. To aid ongoing investigations of these fossils, my students and I have extracted the available published species illustrations into a large catalogue like those compiled fifty years ago for the important microfossil groups. Approximately 2000 species descriptions are included, although the high frequency of probable synonyms makes the number of actual species much lower, perhaps on the order of several hundred for these fossils.

This catalogue has proven extremely useful in identifying the trepostome species composition of a number of deposits analyzed by us (for example Butler & Cuffey 1996, Cuffey 1998, Cuffey & Cuffey 1995, Werts, Cuffey & Cuffey 2001, Bolton & Cuffey, in press, Cuffey et al. in press). Species and genera are treated satisfactorily (as by Utgaard & Perry 1964, and McKinney 1971, for instance). High-rank higher taxa, suborder and above, are also presently adequate for such investigations (Cuffey 1973, Cuffey & Blake 1991, Boardman et al. 1983).

However, intermediate-level taxa (i.e. families) in these trepostomes are extraordinarily difficult to recognize, diagnose, and apply in a practical and helpful fashion. Treatises have struggled with them (Bassler 1953, Astrova 1978, Ross & Osen 1980). Current workers are beginning to cluster genera together into families (Key, 1990, 1991) whose composition is evident mostly from listings (e.g. Ross & Ross 1996, Anstey & Pachut 1995). The great variation in content and number of trepostome families recognized by previous authors makes comparative analyses beyond the scope of this paper, but rather indicates the need for a key like that presented herein, which contains 15 families.

To assist in our identification process, especially to cope with the very large numbers of taxa to examine, we have developed an identification key for recognizing family-level clusters of genera containing the Ordovician American trepostome species most appropriate to compare with our unknown specimens. We offer it here for the use of other paleobryozoologists concerned with these fossils, in the hope of making species-level identifications more practical, efficient, and consistent, and thereby helping to advance knowledge of these important fossils.

The relationship between identification keys and systematic classifications (whether phenetic, cladistic, or evolutionary) has been discussed elsewhere (Blackwelder 1967, Mayr 1969, Wallace & Dallwitz 1981, Wiley 1981, Hull 1988, Winston 1999). Traditionally and ideally, a classification considers all characters (the total morphologic pattern) across all subgroups simultaneously, whereas a key applies selected characters sequentially only to certain sub-

Table 1. Characters (and character states) used in Ordovician trepostome identification key, in morphologic order.

Colony size and form:
(medium-sized to large and branching to massive, very thin branching).
Branch thickness:
(very thin, moderately thin, moderately thick, thick).
Maximum thickness of zooecial walls in exozone:
(thin, moderate but slightly thinner, moderate but slightly thicker, very thick).
Wall microstructure:
(granular, laminated, recrystallized).
Zooecial diameters:
(very large, moderate-sized).
Mesozooecia abundance:
(none, few, some, many).
Mesozooecia appearance or content in tangential section:
(open [empty], closed [filled with skeletal tissue]).
Acanthostyle abundance and sizes:
(none or tiny, many medium-sized to large).
Cystiphragms:
(present, absent).
Curved diaphragms:
(present, absent).
Straight complete diaphragms:
(present, absent).
Straight incomplete diaphragms (= hemiphragms):
(present, absent).

groups in turn. When as little is known about a group's evolutionary history as is the case with early trepostome bryozoans, practical keys like that presented herein may help shed light on possibilities for more formal taxonomic relationships, as more information becomes available through studies using them, and culminating in full-fledged revision of the entire order (R. Boardman, pers. comm.).

2 KEY CONSTRUCTION

As the many trepostome species illustrated from the North American Ordovician are repeatedly examined while identifying our specimens, one finds one's eye drawn, in sequential order, to particular combinations of morphologic character states. One is thus led more efficiently through the mass of visual data, and that is the essence of this classificatory key.

The morphologic characters (and their corresponding states) found useful in constructing the trepostome family key are given in Table 1 in an anatomically logical order, from looking at the overall colony, next at tangential sections, and finally at longitudinal sections. Then, those observations can be compared with the key's choices, certain characters being utilized in more than one place. The exact phrasing of the character states later in the key itself attempts to incorporate some of the observed variability in expression or distribution of the characters. Most of the alternative choices are dichotomous, but one is trichotomous where that is more concise but still as clear.

Ordovician time saw the origin, diversification, and initial flourishing of the bryozoans, particularly trepostomes. Central and eastern North America, notably the type-Cincinnatian, was an especially prolific province late in that period. Thus, these fossils naturally form a sizeable but coherent taxocene for analysis, not yet complicated by various events later in the Paleozoic, and therefore a key like ours has a greater chance of working reliably in the present state of knowledge.

3 CONCLUSION: THE TREPOSTOME KEY

The following lines present the identification key for discriminating among the North American Ordovician trepostome families, in the form of 13 choices of morphologic alternatives.

1a. Colony very thin branches	Batostomellidae
1b. Colony medium-size to large, branching to massive	2
2b. Some zooecial walls in exozone up to moderate thickness	6
2c. Some zooecial walls in exozone up to very thick	Stenoporidae
3a. (2a) Wall microstructure granular, zooecial diameters very large	Dianulitidae
3b. Wall microstructure laminated, zooecial diameters moderate sized	4
4a. (3b) No mesozooecia	5
4b. Few or some mesozooecia (and no or minute acanthostyles)	Diplotrypidae
5a. (4a) No or tiny acanthostyles	Monotrypidae
5b. Many medium – sized to large acanthostyles	Orbiporidae
6a. (2b) At least some zooecia with cystiphragms	Monticuliporidae
6b. No cystiphragms	7
7a. (6b) At least some zooecia with curved diaphragms	Mesotrypidae
7b. No curved diaphragms	8
8a. (7b) Diaphragms all straight and complete, though in a few forms are very rare	9
8b. Some diaphragms straight and incomplete (hemiphragms)	Esthonioporidae
9a. (8a) No mesozooecia	10
9b. Few to many mesozooecia	12
10a. (9a) Branches moderately thin	Atactotoechidae
10b. Branches moderately thick	11
11a. (10b) Maximum thickness of zooecial walls in exozone moderate but slightly thinner	Amplexoporidae
11b. Maximum thickness of zooecial walls in exozone moderate but slightly thicker	Heterotrypidae.
12a. (9b) All or most mesozooecia in tangential section open	Halloporidae
12b. Some to many mesozooecia in tangential section closed	13
13a. (12b) Wall microstructure laminated	Trematoporidae
13b. Wall microstructure recrystallized	Nicholsonellidae

The trepostome species in the relevant North American Ordovician literature have been referred to

60 genera, some of which may well be synonymous. These are distributed among the 15 families recognized as shown in Table 2.

Table 2. Genera included in the Ordovician trepostome families herein; arrangement alphabetical.

Amplexoporidae:
Amplexopora, Anaphragma, Rhombotrypa.
Atactotoechidae:
Champlainopora (= *Atactotoechus* in part).
Batostomellidae:
Batostomella, Bythopora, Eridotrypa.
Dianulitidae:
Dianulites.
Diplotrypidae:
Diplotrypa.
Esthonioporidae:
Esthoniopora, Esthonioporella.
Halloporidae:
Bimuropora, Calloporella, Hallopora (= *Calopora* = "*Callopora*"), *Halloporina, Panderopora, Parvohallopora, Revalotrypa, Sonninopora, Tarphophragma.*
Heterotrypidae:
Atactopora, Dekayella, Dekayia, Eridotrypella, Heterotrypa, Leptotrypa, Leptotrypella, Petigopora, Stigmatella.
Mesotrypidae:
Mesotrypa, Orbignyella.
Monotrypidae:
Cyphotrypa, Discotrypa, Monotrypa, Monotrypella.
Monticuliporidae:
Atactoporella, Gortanipora, Homotrypa, Homotrypella, Monticulipora, Monticuliporella, Peronopora, Peronoporella, Prasopora, Prasoporella, Prasoporina.
Nicholsonellidae:
Nicholsonella.
Orbiporidae:
Orbipora.
Stenoporidae:
Lamottopora.
Trematoporidae:
Aspidopora, Balticopora, Balticoporella, Batostoma, Dittopora, Hemiphragma, Phragmopora, Trematopora.

ACKNOWLEDGEMENTS

P.D. Taylor, C.A. Cuffey, and P.D. Zell constructively criticized preliminary versions of the identification key, and M.M. Key the entire manuscript. S.E. Lauver prepared this paper for publication.

REFERENCES

Anstey, R.L. & Pachut, J.F. 1995. Phylogeny, diversity history, and speciation in Paleozoic bryozoans. In D.H. Erwin & R.L. Anstey (eds), *New approaches to speciation in the fossil record*: 239-284. New York: Columbia University Press.

Astrova, G.G. 1978. Istoriya razvitiya, sistema i filogeniya mshanok: Otryad Trepostomata. *Academiya Nauk SSR, Paleontologicheskogo Instituta, Trudi* 169: 1-240. (English translation by Brown, D.A. 1985. Historical development, systematics and phylogeny of Bryozoa: Order Trepostomata. *[typescript]*. Canberra: Australian National University.)

Bassler, R.S. 1953. Bryozoa. In R.C. Moore (ed.), *Treatise on invertebrate paleontology*. New York and Lawrence: Geological Society of America and University of Kansas Press.

Blackwelder, R.E. 1967. *Taxonomy: A text and reference book*. New York: Wiley.

Boardman, R.S., Cheetham, A.H., Blake, D.B., Utgaard, J., Karklins, O.L., Cook, P.L., Sandberg, P.A., Lutaud, G. & Wood, T.S. 1983. Bryozoa (revised). In R.A. Robison (ed.). *Treatise on invertebrate paleontology. Part G. Revised. Volume 1*: 1-625. Boulder and Lawerence: Geological Society of America and University of Kansas Press.

Bolton, T.E. & Cuffey, R.J. In press. Bryozoa of the Lower (Romaine Formation) and Middle (Mingan Formation) Ordovician of the Mingan Islands, Quebec. *Geological Survey of Canada Bulletin*.

Butler, K.L. & Cuffey, R.J. 1996. Reduced bryozoan diversity and paleoenvironmental stress in the Saluda Dolomite (uppermost Ordovician, southeastern Indiana). In D.P. Gordon, A.M. Smith & J.A. Grant-Mackie (eds), *Bryozoans in space and time*: 55-61. Wellington: National Institute of Water & Atmospheric Research.

Cuffey, R.J. 1973. An improved classification, based upon numerical-taxonomic analyses, for the higher taxa of entoproct and ectoproct bryozoans. In G.P. Larwood (ed.), *Living and fossil Bryozoa*: 549-564. London: Academic Press.

Cuffey, R.J. 1998. The Maysville bryozoan reef mounds in the Grant Lake Limestone (Upper Ordovician) of north-central Kentucky. *Ohio Geological Survey Guidebook* 13: 38-44.

Cuffey, R.J. & Blake, D.B. 1991. Cladistic analysis of the phylum Bryozoa. In F.P. Bigey (ed.) *Bryozoa: living and fossil*: 97-108. Nantes: Société des Sciences Naturelles de l'Ouest de la France, Mémoire, hors série 1.

Cuffey, R.J., Cawley, J.C., Lane, J.A., Bernarsky-Remington, S.M., Ansari, S.L., McClain, M.D., Ross-Phillips, T.L., & Savill, A.C. In press. Bryozoan reefs and bryozoan-rich limestones in the Ordovician of Virginia. *Proceedings of the 9th International Coral Reef Symposium (Bali)*.

Cuffey, C.A. & Cuffey, R.J. 1995. The Chickasaw bryozoan reef in the Middle Ordovician of south-central Oklahoma. *Proceedings of the 7th International Symposium on the Ordovician System (Las Vegas)* 77: 435-438.

Hull, D.L. 1998. *Science as a process*. Chicago: University of Chicago Press.

Key, M.M. 1990. A new family of trepostome bryozoans from the Ordovician Simpson Group of Oklahoma. *Journal of Paleontology* 64: 700-724.

Key, M.M. 1991. The halloporid trepostome bryozoans from the Ordovician Simpson Group of Oklahoma: *Journal of Paleontology* 65: 200-212

McKinney, F.K. 1971. Trepostomatous Ectoprocta (Bryozoa) from the lower Chickamauga Group (Middle Ordovician), Wills Valley, Alabama. *Bulletins of American Paleontology* 60(267): 1-337.

Mayr, E. 1969. *Principles of systematic zoology*. New York: McGraw-Hill.

Ross, J.R.P. & Osen, B. 1980. Review of Istoriya Razvitiya, Sistema i Filogeniya Mshanok. *Journal of Paleontology* 54: 491-492.

Ross, J.R.P. & Ross, C.A. 1996. Bryozoan evolution and dispersal and Paleozoic sea-level fluctuations. In D.P. Gordon, A.M. Smith & J.A. Grant-Mackie (eds), *Bryozoans in space and time*: 243-258. Wellington: National Institute of Water and Atmospheric Research.

Utgaard, J. & Perry, T.G. 1964. Trepostomatous bryozoan fauna of the upper part of the Whitewater Formation (Cincinnatian) of eastern Indiana and western Ohio. *Indiana Geological Survey Bulletin* 33: 1-111.

Wallace, C.C. & Dallwitz, M.J. 1981. Writing coral identification keys that work. *Proceedings of the 4th International Coral-Reef Symposium (Manila)* 2: 187-190.

Werts, S.P., Cuffey, R.J. & Cuffey, C.A. 2001. Bryozoan species in the Chickasaw bryozoan reef (Ordovician, Oklahoma). *Geological Society of America Abstracts with Programs* 33(5): A-54-55.

Wiley, E.O. 1981. *Phylogenetics: The theory and practice of phylogenetic systematics.* New York: Wiley.

Winston, J.E. 1999. *Describing species: Practical taxonomic procedure for biologists.* New York: Columbia University Press

Bryozoan Studies 2001, Wyse Jackson, Buttler & Spencer Jones (eds)
© 2002 Swets & Zeitlinger, Lisse, ISBN 90 5809 388 3

Deconstructing bryozoans: origin and consequences of a unique body plan

Ruth Ann Dewel
Department of Biology, Appalachian State University, Boone, NC 28608, USA
Judith E. Winston
Virginia Museum of Natural History, 1001 Douglas Ave., Martinsville VA 24112, USA
Frank K. McKinney
Department of Geology, Appalachian State University, Boone, NC 28608, USA

ABSTRACT: The Bryozoa (or Ectoprocta) is an enigmatic phylum of colonial organisms whose phylogenetic position has long been problematic. Although traditionally they have been considered to form a separate "lophophorate" clade with phoronids and brachiopods, genome analyses indicate that they are lophotrochozoans. One of the greatest mysteries regarding bryozoans is the delayed appearance of their body plan in the fossil record. From their earliest record in the Ordovician they comprise five discrete stenolaemate orders. An explanation for their late emergence may be linked to their unusual development. Bryozoans appear to lack tissues derived from endoderm and to possess a gut derived from ectoderm. During metamorphosis or budding the polypide arises from undifferentiated cells of the body wall of the preancestrula or cystid. This unusual mode of development and the absence of endoderm suggest that the bryozoan stem lineage underwent a unique transformation involving loss of the original gut and gain of a new ectodermally derived gut.

1 THE PROBLEM

The Bryozoa (or Ectoprocta) is an enigmatic phylum of sessile aquatic organisms that exists as budded colonies of individuals or zooids. Many bryozoans exhibit polymorphism with zooids specialized for reproduction, cleaning, defense, or other functions. Some of these zooids lack feeding or other structures, but even those that are unspecialized seem to be simplified when compared to other triploblasts. All adult bryozoans appear to lack tissues derived from endoderm and assemble structures normally derived from this germ layer, such as the gut, *de novo* from ectoderm. Traditionally, the unusual characteristics of bryozoans have been attributed to their miniaturization, coloniality and sessile life style. Although these are certainly important considerations, focusing on them alone can lead us to ignore other important characteristics of bryozoans. Characteristics, which although reported in the literature by respected authorities, have been skewed by an interpretation that attempts to make bryozoans something that they are not. Our standard interpretation of bryozoan embryology and morphology is a construct resulting from over 100 years of attempts to synthesize a single framework for all invertebrates. This construct, however, may be inadequate for bryozoans because it is not supported by the available evidence. The purpose of this paper then, is to offer a new hypothesis for the origin of the bryozoan body plan based on their fossil record, morphology, and embryology.

2 THE EVIDENCE

2.1 *Fossil record*

Bryozoans are the only skeletalized marine invertebrates that apparently have no Cambrian fossil record (e.g. Boardman et al. 1987). The earliest described representatives of all of the orders of the Stenolaemata – Cyclostomata, Cystoporata, Trepostomata, Cryptostomata, and Fenestrata – appear essentially simultaneously, in the Arenigian Stage of the Early Ordovician (e.g. Taylor 1993), although a fauna of Tremadocian (earliest Early Ordovician) trepostomes has been reported (Hu & Spjeldnaes 1991). Over 20 of the Arenigian genera are multiserial and free-walled, i.e. covered by a continuously free cuticle, separated by fluid and epidermis from the skeleton that separates zooidal chambers (Taylor & Larwood 1990), although the earliest cyclostome (*Wolinella,* Dzik 1981) is erect, uniserial, and fixed-walled, i.e. with the outer skeleton in direct, continuous contact with the surficial cuticle.

The oldest bryozoan fossils, as a result, are seemingly uninformative about the phylogenetic relations of the phylum. All were colonial and most were multiserial, generally monomorphic, and covered by a continuous, external cuticle that floated above the skeleton, although uniserial, fixed-walled forms were also present. Either condition could be the ancestral or derived state. Perhaps the most informative aspect of the oldest bryozoan fossils is that they post-date all other skeletalized phyla, plus most of the macroscopic non-skeletalized invertebrate phyla, which are represented in Burgess Shale type faunas. This suggests that they are highly derived organisms, and there is a broad field of potential precursors.

2.2 *Phylogenetic position*

The phylogenetic position of bryozoans has long been problematic. On the basis of a shared feeding structure, a set of ciliated, coelom-cored tentacles ringing the mouth, they were traditionally considered to form a monophyletic "lophophorate" clade together with phoronids and brachiopods. However, since embryological characteristics in some members seem deuterostome-like, yet other characters (the presence of chitin, larval similarities) seem protostome-like, they often have been placed in a separate branch between the two main bilaterian clades. Genome analysis has confirmed the protostome relationship of these three phyla (see also Fig. 1a). They are now classed along with annelids and mollusks, in the lophotrochozoan clade of the protostomes on the basis of 18S ribosomal DNA gene sequences (Halanych et al. 1995, Mackey et al. 1996, Winnepenninckx et al. 1996, Aguinaldo et al. 1997, Littlewood et al. 1998). Additional molecular work has shown brachiopods and phoronids to form a separate clade from the bryozoans (a finding in agreement with earlier work on embryology and morphology (see also Nielsen 1996, 2001)), and they have been united as Brachiozoa (Cavalier-Smith 1998), Phoronozoa (Zrzavy et al. 1998), or Phoroniformea (Cohen 2000) basal in Lophotrochoza (Giribet et al. 2000). Several analyses of relationships have failed to support the monophyly of the bryozoans (Mackey et al. 1996; Halanych 1998; Giribet et al. 2000). Many of these studies, however did not look at a significant number of taxa, and some, because of the minute size of individual bryozoans, may have had been compromised by contamination.

Bryozoans also share several morphological apomorphies with lophotrochozoans (Fig. 1a). Gizzard teeth in the cardia of ctenostomes and chaetae cheilostomes (Gordon 1975, cf. Fig. 5) are similar to or cheatae or chaetae-like structures in brachiopods, annelids, and mollusks. They appear, in fact, to be more similar ultrastructurally to chaetae-like Kolliker's organs in mollusk juveniles (Brocco et al. 1974; cf. Fig. 8) than to chaetae in annelids and brachiopods (Gustus & Cloney 1972). The calcified zooecium (ectocyst) of bryozoans is covered by a cuticle that is functionally and structurally similar to the periostracum of brachiopods and mollusks, and yet it is secreted in a manner that is more similar to the synthesis of calcified shells in mollusks than brachiopods (Tavener-Smith & Williams 1972, Weedon & Taylor 1995).

Although these characters do not firmly resolve the relationships of bryozoans within the Lophotrochozoa, the presence of multiciliated epithelial cells, which are found in bryozoans as well as most spiralian taxa but generally are absent in phoronids and brachiopods, suggests that bryozoans are more closely allied with the former (Nielsen 2001). Moreover, the pattern of cilia on adult tentacles and the larval ciliary ridge is remarkably similar to that on bivalve gills (Nielsen, 1987) despite the fact that lateral cilia in the larvae may have a very different function (Strathmann & McEdward 1986).

Bryozoan development, in addition, appears to be more similar to development in spiralians than brachiopods and phoronids. The pattern of cell lineage during cleavage is spiral-like in the origin of the prototroch and meso- and endoderm (Nielsen 2001). Their larvae are morphologically closest to the trochophore, and coronate larvae may be broadly interpreted as a type of pericalymma larva (Salvini-Plawen 1980). Bryozoan larvae have a foot-like structure that secretes adhesive material. It is comparable to foot-like structures in entoproct (and perhaps sipunculan) larvae (Scheltema 1993, 1996), the foot in veliger and adult mollusks, and the ventral sole inferred to have been present in fossil taxa such as *Halkieria* and *Wiwaxia* (Conway Morris 1995). Bryozoan larvae, moreover, have a preoral structure similar to the preoral "foot" in sipunculans. Finally, the infolding and degeneration of the corona at metamorphosis also has been described for the prototroch of entoprocts and mollusks (Nielsen 2001).

Although Ectoprocta and Entoprocta have been postulated to be sister taxa in a superphylum, Bryozoa (Nielsen 2000, 2001), arguments supporting this relationship require that each evolved body plans different from those of their last common ancestor for several reasons. 1) Ectoprocts and entoprocts have different body plans (i.e. ectoprocts have coelomic cavities while entoprocts lack them; ectoprocts have an anus outside the ring of tentacles of the lophophore while entoprocts have it inside; ectoproct have "upstream" ciliary feeding while entoprocts have a "downstream" collecting system). 2) All possible outgroups to these taxa within the protostomes have different body plans. 3) Metamorphosis in ectoprocts and entoprocts is different. 4) Given these differences it would seem to be as difficult, if not more difficult, to convert from either body plan to the other as it would be to change from an organization that was different from both. Thus, even if the two groups are phylogenetically close, they may have evolved from an ancestor having a different body organization.

2.3 *Development*

Although bryozoans are hypothesized to have evolved coloniality through asexual budding and miniaturization from a macroscopic lophophorate or "worm-like" ancestor, neither miniaturization nor a sessile lifestyle can account for the uniqueness of bryozoan development. For example, miniaturization has been given as an explanation for the simplification of bryozoan body structure relative to that of larger bodied metazoans. However, when organisms in the same size range of bryozoan zooids are examined, this argument is not supported. Virtually all bilaterians that are not parasitic or do not contain large numbers of nutrient-providing endosymbionts have a gut. In fact the digestive system, along with the nervous system, is present in cases where enclosed circulatory and excretory systems do not appear to be necessary (Table 1). Moreover, the gut, when present, is derived directly from embryonic endoderm or an endodermally derived larval gut.

The coelomic cavities of bryozoans, in addition, do not appear to be homologous with those of other metazoans. They are absent in gymnolaemate larvae and the early mesoderm lined cavities of the stenolaemates and phylactolaemates seem to be unrelated to either schizocoely or archenteric outpocketing (Zimmer 1997). Some mesoderm appears to be of epidermal origin. In addition, the mouth and anus of sexually produced ancestrulae are formed far from the site of invagination during gastrulation, which contrasts with their placement in other adult metazoans (Zimmer 1997). The mouth in the few larvae having functional (or non-functional rudimentary) guts forms near the site of blastopore closure, and the larval mouth (where present) is unrelated to the mouth of the ancestrular zooid. In fact, the digestive organs of the ancestrula do not originate from the internalized cells of the embryo or larval gut but from parts of the larval epidermis (or perhaps mesoderm in stenolaemates) (Zimmer 1997).

Table 1. Body structures in bryozoan-sized organisms from other phyla (mostly meiofaunal).

Group	Digestive	Circulatory	Nervous	Excretory
Gnathustomulids 0.5-0.8 mm	+	-	+	+
Turbellarians	+	-	+	-/+
Polychaetes 0.3-2 mm	+	-	+	+
Tardigrades	+	-	+	-/+
Entoprocts	+	-	+	+
Tunicates	+	In some	+	+
Rotifers Most <1 mm	+	-	+	+
Gastrotrichs Most <1 mm	+	-	+	-/+
Kinorynchs <1 mm	+	-	+	+
Nematodes most 1-3 mm	+	-	+	+
Copepods 0.2-2.5 mm	+	+	+	+

Thus, as Silén (1944) pointed out, the bryozoan alimentary canal is not homologous to those of triploblastic metazoans. In bilaterians, the alimentary canal contains three regions, defined on the basis of the body layers from which they have developed. The stomodaeum and proctodaeum, the first and last parts of the alimentary canal, originate from ectodermal cells, whereas the middle region or mesenteron, originates from larval endoderm. Nearly all larval tissues degenerate in the preancestrula. Even in the few larvae that have a functional gut with an endodermal mid-region, mesenteron, the gut is entirely lost during metamorphosis. "...*[There] can be made only a superficial comparison with other groups of animals but no comparison containing a real homologization of corresponding parts of the canal.*" (Silén 1944, p. 3).

Post-metamorphic bryozoans also lack endodermally derived tissues. The new polypide, including the "gut", forms entirely from the cystid, that is, from blastema or a thickening of the cystid wall derived from epidermis and peritoneum (ectoderm and mesoderm) that invaginates to form a sac. The sac then becomes bi-lobed. From the superficial lobe the tentacle sheath, lophophore, mouth, and pharynx develop, while the rest of the digestive tract is differentiated from the deeper lobe (Zimmer 1997). Thus the gut and lophophore form from invaginated epidermis and peritoneum, just as they do in formation of the ancestrula (Zimmer 1997) (Fig. 1b).

2.4 *Other unique features?*

The apparent construction of bryozoan zooids from only ectoderm and mesoderm makes them qualitatively different from adults of all triploblastic metazoans. The most immediate and far-reaching results of the loss of endoderm are limitations on digestive processes and the consequent inability to form permanent individuals. Bryozoans are efficient at capturing food particles and have evolved an elaborate array of feeding behaviors and current patterning by individuals and colonies; one would expect their digestive systems to be effective, also. However, this may not be the case. From a physiological perspective two types of digestion can be distinguished. Intracellular digestion, the only type found in protistans, is the most primitive. It takes place in vesicular organelles within cells called lysomes and has its own suite of evolutionarily conserved enzymes. Extracellular digestion takes place in the alimentary canal. It utilizes a different suite of enzymes, plus mucus, secreted into the digestive tube by intracellular glands, hepatopancreas, etc. The extracellular digestive process is secretory and absorptive; it provides for storage of glycogen and fat, and it allows efficient elimination of unprocessed or indigestible material (Prosser 1973).

Intracellular digestion would appear to be less efficient. Over time, indigestible material, if it is not eliminated through exocytosis, accumulates in cells until they are no longer capable of digestive activity. This may be the case for bryozoans. The alimentary tract of bryozoans consists of a single layer of epidermal cells, some ciliated, some non-ciliated, and some with luminal microvilli depending on their location in the tract. Histologically they are composed of highly vacuolated columnar epithelia (or myoepthelia in gymnolaemates) with vesicles containing lysosomal enyzmes and inclusion bodies, the by products of intracellular digestion (Gordon 1975, Mukai et al. 1997). Epithelial cells of the stomach and caecum can phagocytize fat droplets and may be able to ingest whole diatoms (Marcus 1941). When cheilostomes were fed polychaete oocytes stained with methylene blue, these cells took up the stained material (see Mukai et al. 1997). Thus, evidence of intracellular digestion appears to be strongly supported. Evidence of extracellular digestion, however, is less well substantiated. Hyman (1959) stated, "Ries was unable to demonstrate the presence of formed enzymatic secretions, but the cardiac and caecal epithelia are generally believed to be glandular." On this basis, she added, "one may suppose that digestion in gymnolaemates is partly extracellular and partly intracellular." The breakdown of particles that takes place in the food whirls or "kneading" processes in the central stomach may be mostly mechanical. Studies on digestive enzymes, summarized by Mukai et al. (1997), indicate the occurrence of acid and alkaline phosphatases, amylase, esterases, and leucine aminopeptidase, but in the few species tested so far a lack of proteinase and lipase activity.

Gordon (1975), in addition, noticed no mitoses in stomach cells, indicating no replacement of these cells (as occurs in gut cells of most other organisms), making their usefulness even more limited. The polypides, perhaps because of this limitation, degenerate into brown bodies once they are "used up" for absorbing food and storing waste material. In some species brown bodies are eliminated as the first fecal pellet of a newly formed polypide; in other species they accumulate until the entire zooid is no longer capable of regeneration (Palumbi & Jackson 1983, Mukai et al. 1997). Energy acquired, rather than being used to maintain a permanent individual, is deployed in cycles of regeneration, which continue for the life of the colony.

From the earliest fossil record of bryozoans to the present, zooid size has remained remarkably constant (McKinney & Jackson 1989, Jackson & McKinney 1990, McKinney 1993). This is based on distance between centers of lophophore protrusion of nearest-neighbor zooids, a consistently comparable attribute of both stenolaemates and gymnolaemates that has a high correlation with lophophore size. Ordovician through Permian stenolaemates had a median zooid spacing of 0.283 mm, which is bracketed by slightly smaller spacing in modern stenolaemates (0.209 mm) and slightly greater spacing in modern cheilostome gymnolaemates (0.350 mm). Early Cenozoic cheilostomes have average zooid size identical to modern cheilostomes, and a very few post-Paleozoic stenolaemates and cheilostomes achieve nearest-neighbor spacing of approximately 1 mm, which is slightly larger than the largest Paleozoic stenolaemates. This constancy of zooid size apparently reflects a strong phylogenetic constraint. There have been three suggested bases for the constraint: 1) limitation of oxygen absorption by the small, non-proliferated surface area (Mangum & Schopf 1967; but see Massaro & Fatt 1967), 2) lack of nephridia and the resultant accumulation of toxins (reviewed in Gordon 1977), and 3) hydrodynamic constraints on the ciliary feeding process. Given the above evidence, it seems reasonable also to consider functional constraints imposed by the lack of a true gut.

3 A PLAUSIBLE EXPLANATION

As a result of their unusual embryology and metamorphosis, Zimmer (1977, 1997) proposed that metamorphosis in bryozoans has incorporated certain features of colonial growth by adultation (the occurrence of adult characters in the larva (Jägersten 1972)). This includes in particular the ability to generate a new individual (the ancestrula) from aggregates of undifferentiated cells (blastema). While this hypothesis is attractive because it can provide an explanation for the proximal cause of polypide duality during metamorphosis, budding, and polypide replacement, it does not examine the possibility that there is a phylogenetic reason for this peculiar combination of characters. For example, why is not the entire ancestula derived from blastema within the larva? If budding, or formation of a new individual from undifferentiated blastema, moved by adultation into the larva, would not the entire adult form from blastema as it does in some entoprocts? The cystid, or adult body wall, forms directly from part of the body wall (ecto- and mesoderm) of the larva, and one would expect that at least some of the adult digestive tract would form from larval gut. Even in echinoderms, some of which undergo what is perhaps the most dramatic metamorphoses in all of Metazoa, a portion of the larval gut is retained and reorganized to form the adult digestive system, but in bryozoans

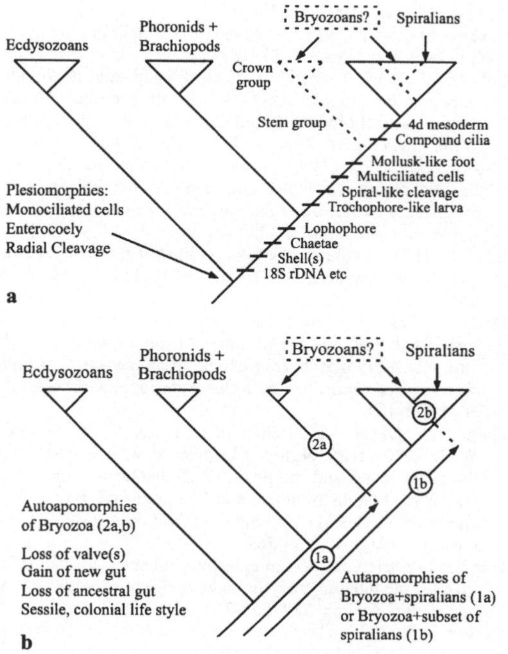

Figure 1. Development of apomorphies in lophotrochozoans (a) and bryozoans (b).

it is not. Thus events in bryozoan evolution in addition to the acquisition of coloniality could have profoundly impacted the development and final organization of the larva and adult. In fact, it is the latter aspects of their biology that should be examined closely for clues about these events. The relatively late appearance of bryozoans as skeletalized taxa in the Paleozoic, the unusual mode of development, the apparent absence of endoderm, and the duality of the polypide and cystid have important implications. These attributes suggest that prior to the appearance of the crown group (all living taxa), body plans within the stem lineage (only fossil taxa) underwent a unique transformation (Fig. 1a).

The cyclostome and cyphonautes-type larvae appear to reveal several things about the evolutionary history of bryozoans. For example, the cuticular valves (in cyclostomes there is a single central cone with concentric rings (Nielsen 1970)) on these larvae are likely to have evolved originally in an adult ancestor (Fig. 2a). If this is true, their development must have been accelerated into the larva by adultation. Because the paired valves of the cyphonautes larvae are lost at metamorphosis, they are probably not homologous with the zooecial wall (ectocyst) found in the bryozoan crown group. Their apparent formation on the episphere contrasts with the development of shells on the hyposphere in veliger larvae. Their position in bryozoans, however, may be secondary if originally there was only one shell, or if they developed first on the hyposphere but were forced later onto the episphere by invagination of the hyposphere into the animal half of the embryo. A shift in their development may have been driven by evolutionary trends occurring in the adult. Scenarios in which the trunk and associated structures retracted into the mantle cavity of the shell(s) or in which the shell(s) overgrew the posterior (metasomal) part of the body are consistent with the observed pattern of development in bryozoan larvae. Regardless of the original position of bryozoan valves, however, both their presence in larvae and their fate during metamorphosis suggest that adults in the bryozoan stem group (or in earlier ancestors) possessed a valve or valves that became lost or greatly altered before the origin of the bryozoan crown group (Fig. 1b).

Development of the polypide may also reflect events in the evolution of bryozoans. During metamorphosis the polypide develops from the ectodermal blastema of the apical organ and/or infracoronal cells and associated mesoderm. During polypide replacement and asexual budding the polypide forms from the epidermis and peritoneum of the cystid (Reed 1991, Zimmer 1997). It seems possible because of this that the lophophore or polypide evolved entirely from ancestral body wall (epidermis and underlying mesoderm) (Fig. 2b). Thus, the polypide may have arisen solely from structures associated with the body wall.

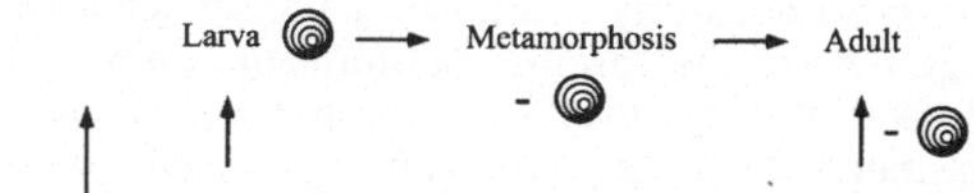

Figure 2. Acquisition of shell or shell-like sclerite (a) and polypide (b) in bryozoans.

Such structures may have been preadaped for feeding. For example, they may have been analogous (or even homologous) to structures that have ciliary tracts functioning in suspension feeding such as the tentacles, palps, or gills of mollusks. Because none of the larval gut in bryozoans contributes to the adult digestive tract, it appears that the entire ancestral gut was lost, and that the present adult gut evolved *de novo* in the bryozoan stem group through co-option of ectodermally and associated mesodermally derived tissue. The "gut" would be merely body wall turned inside, and may have formed from structures that were already folded into the body, such as those residing in a mantle cavity. The potential nutritional role of dissolved organic material for soft-bodied marine inverts via integumental nutrient transport is well established. The extent to which various organisms make use of this food source remains to be worked out (Petersen et al. 1998, Goome 2001). In mollusks for example, Stewart and Dean (1978) have demonstrated rapid amino acid uptake by the ctenidia and mantle of *Bankia gouldi*.

Finally, the orientation of the adult, as can be deduced from the migration of the polypide rudiment during metamorphosis and the location of the anus, appears to be "backwards" from that of the larvae (Zimmer & Woollacott 1977). If the distal-proximal axis of the adult, which is expected to be equivalent to the anterior-posterior axis of the larva, is reversed from the ancestral condition, the polypide may have been co-opted from posteriorly directed gill(s) rather than more anteriorly positioned palps or tentacles. Regardless of the specific transformation, however, events in the bryozoan stem group appear to have constrainded both the development and evolution of the clade.

ACKNOWLEDGEMENTS

We thank the Departments of Biology and Geology, Schools of Arts and Sciences and Graduate Studies and Research, and International Studies at Appalachian State University, and the Virginia Museum of Natural History for financial assistance.

REFERENCES

Aguinaldo, A.M.A., Turbeville, J.M., Linford, L.S., Rivera, M.C., Garey, J.R., Raff, R.A. & Lake, J.A. 1997. Evidence for a clade of nematodes, arthropods and other moulting animals. *Nature* 387: 489-493.

Boardman, R.S., Cheetham, A.H. & Rowell, A.J. (eds). 1987. *Fossil invertebrates*. Palo Alto: Blackwell Scientific Publications.

Brocco, S.L, O'Claire, R.M. & Cloney RA. 1974. Cephalopod integument: the ultrastructure of Kolliker's organs and their relationship to setae. *Cell and Tissue Research* 151: 293-308.

Brusca, R.C. & Brusca,G. 1990. *Invertebrates*. Sunderland, MA: Sinauer Associates, Inc.

Cavalier-Smith, T. 1998. A revised six-kingdom system of life. *Biological Reviews* 73: 203-266.

Cohen, B.L. 2000. Monophyly of brachiopods and phoronids: reconciliation of molecular evidence with Linnaean classification (the subphylum Phoroniformea nov.). *Proceedings of the Royal Society London, B* 267: 225-231.

Conway Morris, S. 1995. Articulated halkieriids from the Lower Cambrian of North Greenlan and their role in early protostome evolution. *Philosophical Transactions of the Royal Society London B* 347: 305-358.

Dzik, J. 1981. Evolutionary relationships of the early Palaeozoic 'cyclostomatous' Bryozoa. *Palaeontology* 24: 827-861.

Dzik, J. 1991. Possible solitary bryozoan ancestors from the early Palaeozoic and the affinities of the Tentaculita. Early Ordovician bryozoans from China. *Bulletin de la Société des Sciences Naturelles de l'Ouest de la France Mémoire HS* 1:121-131.

Giribet, G., Distel D.L., Polzm M., Sterrer, W. & Wheeler, W.C. 2000. Triploblastic relationships with emphasis on the acoelomates and the position of Gnathostomulida, Cycliophora, Platyhelminthes, and Chaetognatha: a combined approach of 18S rDNA sequences and morphology. *Systematic Biology* 49: 539-562.

Goome, J. 2001, Transport of exogenous organic substances by invertebrate integuments: the field revisited. *Journal of Experimental Zoology* 289: 254-265.

Gordon, D.P. 1973. The occurrence of a gizzard in a bryozoan of the order Cheilostomata. *Acta Zoologica* 36: 279-282.

Gordon, D. P. 1975. Ultrastructure and function of the gut of a marine bryozoan. *Cahiers de Biologie Marine* 16: 367-382.

Gordon, D.P. 1977. The aging process in bryozoans. In R.M. Woollacott & R.L. Zimmer (eds), *Biology of Bryozoans*: 335-376. Academic Press: New York.

Gustus, R.M. & Cloney, R.A. 1972. Ultrastructural similarities between setae of brachiopods and polychaetes. *Acta Zoologica* 53: 229-233.

Halanych, K.M. 1998. Considerations for reconstructing metazoan history: signal, resolution, and hypothesis testing. *American Zoologist* 38: 929-941.

Halanych, K.M., Bacheller, J.D., Aguinaldo, A.M.A., Liva, S.M., Hillis, D.M. & Lake, J.A. 1995. Evidence from 18S ribosomal DNA that the lophophorates are protostome animals. *Science* 267: 1641-1643.

Higgins, R P. & Thiel, H. (eds). 1988. *Introduction to the Study of Meiofauna*. Washington: Smithsonian Institution Press.

Hu, Z.-X. & Spjeldnaes, N. 1991. Early Ordovician bryozoans from China. *Bulletin de la Société des Sciences Naturelles de l'Ouest de la France Mémoire HS* 1: 179-185.

Hyman, L.H. 1959. *The Invertebrates, vol. V. Smaller Coelomate Groups*. New York: McGraw-Hill.

Jackson, J.B.C. & McKinney, F.K. 1990. Ecological processes and progressive macroevolution of marine clonal benthos. In R.M. Ross & W.D. Allmon (eds), *Causes of Evolution*: 173-209. Chicago: University of Chicago Press.

Jägersten, G. 1972. *Evolution of the Metazoan Life Cycle: A comprehensive theory*. London: Academic Press.

Littlewood, D.T.J., Telford, M.J., Clough, K.A. & Rohde K. 1998. Gnathostomulida - an enigmatic metazoan phylum from both morphological and molecular perspectives. *Molecular Phylogenetics and Evolution* 9: 72-79.

MacBride, E.W. 1914. *Text-book of Embryology. Vol. I. Invertebrata*. London: MacMillan and Co., Ltd.

Mackey, L.Y., Winnepenninckx, B., De Wachter, R., Backljau,T., Emschermann, P. & Garey, J.R. 1996. 18S

rRNA suggests that Entoprocta are protostomes, unrelated to Ectoprocta. *Journal of Molecular Evolution* 42: 552-559.
Mangum, C.P. & Schopf, T.J.M. 1967. Is an Ectoprocta possible? *Nature* 213: 264-266.
Massaro, T.A. & Fatt, I. 1967. Oxygen consumption of ectoprocts. *Nature* 216: 59-60.
McKinney, F.K. 1993. A faster-paced world?: contrasts in biovolume and life-process rates in cyclostome (Class Stenolaemata) and cheilostome (Class Gymnolaemata) bryozoans. *Paleobiology* 19: 335-351.
McKinney, F.K. & Jackson, J.B.C. 1989. *Bryozoan Evolution.* Boston: Unwin & Hyman.
Mukai, H., Terakado, K. & Reed, C.G. 1997. Bryozoa. In F. W. Harrison & R. M. Woollacott (eds). *Microscopic Anatomy of Invertebrates, vol. 13:* 45-206. *Lophorphorates, Entoprocta, and Cycliophora.* New York: Wiley-Liss.
Nielsen, C. 1987. Structure and function of metazoan ciliary bands and their phylogenetic significance. *Acta Zoologica* 68: 205-262.
Nielsen, C. 2000. The phylogenetic position of Entoprocta and Ectoprocta. *Proceedings of the 11th International Bryozoology Association Conference* 2000: 66-73.
Nielsen, C. 2001. *Animal Evolution. Interrelationships of the Living Phyla.* Oxford: Oxford University Press.
Nielsen, C, Scharff, N. & Eibye-Jacobsen, D. 1996. Cladistic analysis of the animal kingdom. *Biological Journal of the Linnean Society* 57: 385-410.
Palumbi, S. R. & J. B.C. Jackson. 1983. Aging in modular organisms: ecology of zooid senescence in *Steginoporella* ps. (Bryozoa: Cheilostomata). *Biological Bulletin* 164: 267-278.
Petersen, S., Arlt, G., Faubel, A. & Carman, K.R. 1998. On the nutritive significance of dissolved free amino acid uptake for the cosmopolitan oligochaete *Nais elinguis* Müller (Naididae). *Estuarine, Coastal and Shelf Science* 46: 85-91.
Prosser, C.L. 1973. *Comparative Animal Physiology.* 3rd ed. Philadelphia: W. B. Saunders.
Reed, C.G. 1991. Bryozoa. In A.C. Giese, J.S. Pearse & V.B. Pearse (eds.). *Reproduction of Marine Invertebrates*: 85-245 Pacific Grove, CA: Boxwood Press.
Salvini-Plawen, von L.V. 1980. Phylogenetical position and significance of the mesenchymatous Bilateria. *Zoologische Jahrbücher Abteilung fur Anatomie und Ontogenie der Tiere* 103: 354-373.
Scheltema, A.H. 1993. Aplacophora as progenetic aculiferans and the coelomate origin of mollusks as the sister taxon of Sipuncula. *Biological Bulletin* 184: 57-78.
Scheltema, A.H. 1996. Phylogenetic position of Sipuncula, Mollusca and the progenetic Aplacophora. In J.D. Taylor (ed). *Origin and Evolutionary Radiation of the Mollusca*: 53-58. Oxford: Oxford University Press.
Silén, L. 1944. On the division and movements of the alimentary canal of the Bryozoa. *Arkiv för Zoologi* (Stockholm) 35A: 1-40.
Steward, M.G. & Dean, R.C. 1978. Amino acid absorption and assimilation in the shipworm *Bankia fouldi. American Zoologist* 18: 575.
Strathmann, R.R. & Eernisse, D.J. 1994. What molecular phylogenies tell us about the evolution of larval forms. *American Zoologist* 34: 502-512.
Strathmann, R.R & McEdward L.R. 1986. Cyphonautes' ciliary sieve breaks a biological rule of inference. *Biological Bulletin* 171: 754-760.
Tavener-Smith, R. & Williams, A. 1972. The secretion and structure of the skeleton of living and fossil Bryozoa. *Philosophical Transactions of the Royal Society London, Series B* 264: 97-159.
Taylor, P.D. 1993. Bryozoa. In M.J. Benton (ed.), *The Fossil Record 2*: 465-488. London: Chapman & Hall.
Taylor, P.D. & Larwood G.P. 1990. Major evolutionary radiations in the Bryozoa. In P.D. Taylor & G.P. Larwood (eds), *Major Evolutionary Radiations.* Systematics Association Special Volume, 42: 209-233
Turbeville, J.M., Linford, L.S., Rivera, M.C., Garey, J.R., Raff, R.A. & Lake, JA. 1997. Evidence for a clade of nematodes, arthropods and other moulting animals. *Nature* 387: 489-493.
Weedon, M.J. & Taylor, P.D. 1995. Calcitic nacreous ultrastructures in bryozoans: implications for comparative biomineralization of lophophores and molluscs. *Biological Bulletin* 188: 281-292.
Winnepenninckx, B., Backeljau, T. & De Wachter, R. 1996. Investigation of molluscan phylogeny on the basis of 18S rRNA sequences. *Molecular Biology and Evolution* 13: 1306-1317.
Zimmer, R.L. & Woollacott, R.M. 1977. Metamorphosis, ancestrulae, and coloniality in bryozoan life cycles. In R.M. Woollacott & R.L. Zimmer (eds). *Biology of Bryozoa*: 91-142. New York: Academic Press.
Zimmer, R.L. 1997. Phoronids, brachiopods, and bryozoans, the lophophorates. In Gilbert S.F. & A.M Raunio (eds). *Embryology: Constructing the Organism*: 279-305. Sunderland, MA: Sinauer Associates, Inc.
Zrzavy, J., Mihulka, S., Kepka, P., Bezdek, A. & Tietz, D. 1998. Phylogeny of the Metazoa based on morphological and 18S ribosomal DNA evidence. *Cladistics* 14: 249-285.

Bryozoan Studies 2001, Wyse Jackson, Buttler & Spencer Jones (eds)
© 2002 Swets & Zeitlinger, Lisse, ISBN 90 5809 388 3

Colony morphologies and missed opportunities during the Cincinnatian (Late Ordovician) bryozoan radiation: examples from *Heterotrypa frondosa* and *Monticulipora mammulata*

J.M. Erickson
Department of Geology, St. Lawrence University, Canton, NY, 13617, USA

D.A. Waugh
Department of Geology, Kent State University, Kent, OH 44242, USA

ABSTRACT: Renewed interest in Paleozoic Bryozoa is focusing upon zoarial architecture and colony morphologies. A reassembled zoarium of *Heterotrypa frondosa* (d'Orbigny) measuring 13 cm high and 26 cm wide has provided basis for study of functional morphology, habitat tolerances and growth strategies. Morphologies of *H. frondosa* and sympatric *Monticulipora mammulata* d'Orbigny in the same Upper Ordovician (Cincinnatian: Maysvillian) strata of the lower Bellevue Formation provide evidence that fair-weather wave resistance was achieved by multiple trepostomate species during this depositional sequence. A growth sequence from the reassembled *H. frondosa* includes erection of a triad of vertical coalescing blades forming archways and windows, then coalescence of adjoining blades to produce chambers which extended into fronds as the cycle repeated. Such heavily buttressed, wave-resistant colonies were numerous for a brief time interval in shallow waters above storm wave base. Their presence suggests that creation of biogenic structures analogous to modern cnidarian reefs lay briefly within the ecological "grasp" of trepostomes. Only a self-replicating community structure, or seral succession, leading to deposition of an interlocking bioclastic framework failed to evolve. That this opportunity was not seized appears, regionally, to have been the result of sea level or tectonic instability, shelf-ramp morphology, climatic instability, sediment influx or a combination of these extrinsic factors. Thus, environmental instability, not biological inability, prohibited evolution of a stable seral succession of a local fringing-reef type. The terminal Ordovician extinction ended this "experiment" before its global application.

1 INTRODUCTION

In Paleozoic limestone, the contribution of skeletal bioclasts represented by bryozoan fragments was sometimes prodigious. One spectacular example is that of the richly fossiliferous, shallow-marine rudstones (Cuffey 1985) of the Late Ordovician (Maysvillian) of the Cincinnati Arch of Ohio, North America. Bryozoa contribute an average 26% of the volume (Martin 1975), the most striking of which by the sizes of their clasts are fragments of trepostome taxa. To the paleonaturalist these rocks and their fossils have raised significant ecological, morphological and functional questions, not the least of which have been, "What must a living colony of species x or y have looked like, how did it function and what did an array of this bryozoan-dominated community look like as a seascape?"

Given the bryozoan taxonomic diversity and population richness of many of these limestone tempestites (Tobin 1982), it is remarkable that no quarry, road cut or stream section has yet revealed a glimpse of this community fully preserved in its life habit to demonstrate organismal interactions, space competitions and structural tiering that surely were factors of the Maysvillian shallow marine habitat. Was there never such an interactive bryozoan community structure? Is the density and diversity one finds due only to the winnowing and gathering effects of occasional great storms? No Cincinnatian diorama has yet convinced us that an adequate synthesis of these communities has been produced.

2 SOURCES OF SPECIMENS

Field trips to the Cincinnati, Ohio, region over the past 25 years have provided numerous opportunities to note morphologies of trepostome colony fragments and a few opportunities to reconstruct more complete zoaria. A pair of specimens are worthy of description because their morphologies suggest that they grew in settings of high wave energy. Waugh and Erickson (this volume) have detailed the form and function of such a specimen of *Heterotrypa frondosa* (D'Orbigny) (St. Lawrence University #FR525). It was collected from a shale bed several metres above a prominent bioclastic limestone bed

a
b
window
chamber
c
d
f
e

Figure 1a-e. *Heterotrypa frondosa* (d"Orbigny); Bellevue Formation, Sharonville, Ohio, USA; Late Ordovician (Maysvillian: Cincinnatian); St. Lawrence University #FR525. 1a. Top view of re-assembled zoarium, revealing eroded crown and chamber morphology. Arrows indicate chamber exits. 1b. Basal view of SLU #FR526, a cast of the specimen in Fig. 1a, with box around area in Fig. 1c. 1c. Enlargement of exterior of well-preserved colony wall illustrating shape of the wave-resistant buttress form and windows in basal, exterior view. 1d. Frond of *Heterotrypa frondosa* (SLU #FR531) from same locality representing distal frondose growth form. Note windows and (broken) chambers at pointers consistent with those on the re-assembled zoarium. 1e. Lateral view of cast of partially reconstructed *H. frondosa* demonstrating upward (astogenetic) changes in morphology described in text. Scale bar is 5 cm.

Figure 1f. *Monticulipora mammulata* d'Orbigny; Bellevue Formation, Covington, Kentucky USA; Late Ordovician (Maysvillian: Cincinnatian). Oblique view of SLU #FR530, a massive, wave-resistant zoarium on which wide crenulate blades have been eroded and polished by wave action or transport. Scale bar is 10 cm.

in the Bellevue Formation at Sharonville, Ohio. At this site (loc. OH-HA-0030, Davis et al. 1998), limestone hardgrounds occur with increasing frequency upwards at the expense of shale, a record of shallowing conditions of the C2 sequence of Holland and Patzkowsky (1998). Multiple hardgrounds, many now preserved with erosional upper surfaces, resulted. Encrusting Bryozoa, including many that formed the bases of erect trepostome colonies, are found attached to these surfaces. Conditions suggest that the *H. frondosa* colony was founded on one such hardground.

A specimen of *Monticulipora mammulata* (SLU #FR530) collected as float from the lower part of the Bellevue Formation at Reidlin (=Mason) Road, Covington, Kentucky, (KY-KE-0001, Davis et al. 1998) and identified by Cuffey (pers.com. 1999) is also pertinent to this research. At this locality the Bellevue Formation is very thin in the portion of outcrop where the specimen originated. We interpret our *H. frondosa* and *M. mammulata* specimens to have been from similar stratigraphic positions though they are not proven correlatives. Prevalence of recurring bioclastic limestones in this portion of the section implies upward shallowing depositional conditions as at the Sharonville Site.

Many studies have addressed the mechanics (Cheetham 1986; McKinney & Jackson 1989; Brown 1952) of bryozoan growth and the relationships of genetic versus phenotypic expression in astogeny (Boardman & Cheetham, 1969; Anstey 1987), but many paleoecological relationships that should convey important phyletic information by understanding colony ethology of morphology have not been established for most taxa due to lack of complete colonies as fossils. Modern analogues do not exist, but perhaps one may infer answers to some of these questions by comparison of trepostome colony growth forms with modern scleractinian corals analogues. Our objectives are to describe and interpret some aspects of these specimens and thereafter to apply those ideas to the much broader interpretation of Cincinnatian paleoenvironmental adaptation of trepostomates in the province.

3 MAYSVILLIAN TREPOSTOME MORPHOLOGIES

3.1 *Heterotrypa frondosa colony form*

Waugh and Erickson (this volume) describe reconstruction of a zoarium of *Heterotrypa frondosa* that measured 13 cm high and 26 cm maximum width (Fig. 1a). The morphology of that colony gave insights about current energy and feeding. Astogeny of the specimen also suggests that both phylogenetic and ecotypic developments may be evaluated from it, at least in first approximation. Initiation of growth was in one of two forms, either as rejuvenation from a broken frond which became embedded in a firm rubble substrate, or was begun *de novo* by larval settlement on a hardground surface. The basal colony attachment was not recovered, therefore we may only infer the initiation event. Large size and mass of this colony implies a very firm attachment giving credence to the idea that it grew from settlement of the ancestrula attached to a hardground.

Figure 1b displays a basal view of the zoarium which reveals the fractured erect blades that formed the foundation of the colony. Three centres of vertical growth arose from the encrusting base, grew upward for several centimetres before they merged to form a triad composed of one coalesced blade that marked the longest axis of the colony and a second blade almost perpendicular to it. Subsequently, this triad bifurcated several times by initiating new growth centres, grew upward and then expanded these secondary blades distally resulting in coalescence of the axial elements once more. As seen in Figures 1c and 1e, the reconstructed form is one of buttressed arches and windows that generated strength to support a skeletal mass above and a system of passages for transmitting water flow while offering little resistance to current at the base.

Erect growth of three or four bifoliate blades is often seen from basal encrustations on hardground surfaces (Wilson & Palmer 1992, Fig. 26) indicating the morphology was a standard phylogenetic growth response of *H. frondosa*. We suggest that the orientations of the initial blades were very likely influenced, even "controlled," by prevailing flow directions of local currents across the encrusting base. Resulting arches and windows were thus oriented to best suit needs of the colony, a phenotypic response. Flume experiments reported by Waugh and Erickson (this volume) demonstrate that current flow through the colony base may have been a vital component of water exchange during feeding, assuring adequate distribution of food-bearing flow.

After creating archways, a new episode of zoarial expansion took place. Generation of a second set of vertical rami arose above the archways, but the colony quickly redirected maximum growth rate of these fronds into a horizontal, or lateral vector. Result of this effect was to coalesce blades into chambers of sub-cylindrical shape with somewhat constricted openings above and below. Bases of the chambers opened to the archways in the centre of the zoarium. This pattern was repeated with some minor changes in the subsequent level. Mechanical breakage and abrasion removed most of the skeleton from our specimen above this level as described below with the taphonomy. Complete development of the zoarium is not known but additional specimens from the same site suggest probable forms.

The complete marginal portion of the colony is of considerable importance (Figs. 1b and 1c). There the outer wall demonstrates the smooth, wave-resistant buttress that formed the margin of the zoarium. We interpret this morphology to have been capable of withstanding normal littoral wave action on a wide gently sloping carbonate platform comparable to such modern reef corals as *Acropora palmata.* Each of the buttresses displays a contact spot indicating that a neighboring structure, presumably biologic, was present (Fig. 1c). At the opposite margin of the zoarium a blade extending from the primary basal element, seen in Figures 1a and 1b, grew laterally outward and over another bed roughness, again assumed to have been biological and therefore a space competitor. These lateral growths accentuated colony expansion, perhaps over neighboring Bryozoa in a thicket-type habit, as indicated by the varying angle of colony expansion mentioned by Waugh and Erickson (this volume).

We agree with Brown (1952) who stated that rigid bilaminate branched, adoeniform colonies follow regular and precise, species-specific rules of growth, however in our example the interpreted high energy of its habitat superimposed somewhat more fluid colony response when regularity of branching and branch angles are considered. Initial triad, arches, chambers, and marginal buttresses and the sequence of their development in our specimen of *Heterotrypa frondosa* agree with Brown's (1952) growth sequence concepts. They represent genotypic morphology, whereas features such as height of basal archways, angle and rate of lateral expansion, total height of zoarium and the primary orientation of the colony were produced in response to environmental factors within the local habitat.

Frond thickness may also be environmentally determined (Brown 1952). Relationships between blade thickness and strength have been well discussed (Key 1991). From the same horizon and nearby, but not demonstrably related to the zoarium described above, we collected a 20 cm - long portion of a single large frond of *H. frondosa* of 5 mm thickness. As seen in Figure 1d, this incomplete specimen reiterates the pattern of anastomosing growth that produced archways and windows that lessened the surface area impacted by current. On both sides of this structure chambers also were produced. These were generated in the direction of elongation of the entire structure (primary growth axis) on both sides of the blade to which they conveyed strength as well as feeding efficiency by temporarily slowing what we interpret to have been fairly rapid, high-energy water flow. We believe that fronds such as this represent distal portions of *Heterotrypa frondosa* zoaria that in total exceeded a metre from colony base to margin and therefore the complete colony may have exceeded 1.5 metres in diameter at the crown. On a stable platform, a thicket of such colonies should have possessed wave-resistant characteristics similar to those one might imagine from a modern scleractinian morphology hybridized between those of *Acropora palmata* and *A. cervicornis* also found in Caribbean fringing reefs.

3.2 *Monticulipora mammulata colony form*

The zoarium of *Monticulipora mammulata* mentioned above is a third specimen, which suggests to us that erect trepostomes in this interval of the Cincinnatian were colonizing near shore intertidal shelf settings. This specimen is an oblate spheriod of approximately 22 cm length, 17 cm width and 11 cm height (Fig. 1f) that represents a massive *M. mammulata* colony. As preserved, it is an abraded mass showing remnants of fronds or blades that once grew erect but have been broken off and worn nearly away by wave action. Some tumbling of the mass may have occurred although this is not demonstrated by abrasion. A limonite weathering rind indicates that the specimen was likely overturned and rested at an oblique angle in a shale bed before discovery by a student. Because neither of us was present as the specimen was taken from the outcrop, its orientation and context can't be certainly established. In appearance, the specimen is reminiscent of a modern meandrenoid-type coral.

When living, this massive structure served as the base of a colony that supported thick (1.5 cm) crenulate blades of perhaps 4 centimetres height. Presently 2 cm of this height are preserved around the colony margins but only a centimeter is preserved over the crown of the colony due to breakage or abrasion. We believe that the specimen illustrated in Figure 1e represents an individual colony established in the littoral zone near the level of mean low tide. It was a wave-resistant structure under conditions of normal wave energy for the region and able to flush coarse wave-tossed sediment when required (Waugh and Erickson this volume).

4 MAYSVILLIAN BRYOZOAN HABITATS AND COMMUNITIES

Prodigious study of type Cincinnatian paleontology, sedimentology and stratigraphy has been accomplished over the past 150 years, nevertheless significant new interpretations of regional geology and paleoecology are continuing to appear (Drummond and Sheets 2001; Kolata et al 2001; Ross and Ross, this volume). We believe that our interpretations of the functional morphologies of *Heterotrypa frondosa* and *Monticulipora mammulata* support these reinterpretations both of species habitat width and local paleogeomorphology and warrant further examination.

Presently, the setting and general conditions that seem understood and that support our belief that a reef-building bryozoan community of framework organization was evolving during the Maysvillian include the following:

(1) Late Ordovician seas were "calcite seas" assumed to favor secretion of low magnesium calcite skeletons (Stanley & Hardie 1999).

(2) The Sebree Trough conducted cool, nutrient-rich bottom waters to the immediate region of the carbonate platform on which Maysvillian bryozoan communities grew (Kolata et al 2001)

(3) Epicontinental seas covering Laurentia fluctuated frequently, perhaps in part responding to glacial cycles, demonstrated by sequence stratigraphic interpretations (Holland 1999).

(4) The site lay south of the Cincinnatian paleoequator in a region of sub-tropical latitude and of temperate marine waters (Holland & Patzkowsky 1996, 1997, & 1998).

(5) Tempestites resulting from large storms have significantly influenced Maysvillian stratigraphy and the depositional/preservational record on this very gently sloping carbonate shelf (Tobin 1982; Diekmeyer 1999) although frequency of storm events is unknown as yet. The subtropical convergence zone lay across the region (Kolata et al 2001).

(6) The bryozoan fauna of the Cincinnatian Province was, "characterized by a cohort of endemic North American genera and species." (Anstey 1986, p. 3, Anstey & Chase 1974).

(7) "Bryozoan reefs represent ecological extremes for their phylum and exotic rarities among bioherms in general" (Cuffey 1999, p. 38). Two small reef mounds of *Stigmatella personata* from Maysvillian strata of Kentucky, are the only known occurrences from the type-Cincinnatian (Cuffey 1999) where they developed roughly seven metres above the Bellevue Member (of the Grant Lake Fm.). They are quite reminiscent of the earliest *Stigmatella* reef mounds of the Chazyan on Garden Island, New York (Pitcher 1974).

(8) Reef systems in general were in decline in the Late Ordovician around Laurentia (Copper 2001) perhaps in response to clastic influx, perhaps related to other factors.

(9) The Ordovician radiation collapsed at the close of the Cincinnatian.

5 MISSED OPPORTUNITES

It appears to us that some elements of the unique or endemic trepostome fauna of which *Heterotrypa* was a member (Anstey 1986) were morphologically suited to flourish under wave energy conditions in the low intertidal and high subtidal zones. The tidal range of the Maysvillian epicontinental sea is not known, but the authors have not seen evidence of a significant one in these rocks. Given a setting of broad shelves or a gentle ramp with low tidal range, wave energy would be dissipated over a wide swath of shelf surrounding the shallowest of carbonate islands without encouraging framework-type reef growth. Bryozoan thickets carpeted this terrain (Cuffey 1985).

Reefs are both lithologic features (when fossil) and biologic features constructed in response to energy and geomorphology with their component species each filling a niche in the system. We believe that thickets which developed on beach rock or hardground shore faces had wave resistant trepostome species that interacted spatially and biologically as elements of a zoned community. This community had shallower and deeper water members. *Stigmatella personata* may have been a former whereas various halloporids belonged to the latter.

The nearly unique diversity of the type-Cincinnatian bryozoan fauna (Anstey & Chase 1974) strongly supports the presence of such an integrated, tiered community. Observations of bryozoan faunal changes under shallowing conditions during the Maysvillian Dillsboro deposition, well documented by Brown and Daly (1987), were marked by an influx of thicker, more robust taxa including *Heterotrypa* species and *Monticulipora mammulata*. Those authors interpreted the data as a faunal shift in response to shallowing water. We would agree and might as readily recognize this as a biofacies change wherein a broad benthic community shift occurred over a significant area of the shelf. The role of the Sebree Trough in conducting nutrient-rich, cool water to the carbonate flats and islands represented by eroded hardgrounds of the the Bellevue Formation may have been an essential ingredient supporting the growth of large and massive trepostome colonies such as those we describe here. As sea levels dropped, the surface area of the high-energy peritidal zone expanded over the paltform and the thicket of wave-resistant bryozoans should have done likewise. When storms wracked that thicket, shattered colonies were spread both landward and seaward as blankets of thick-bladed trepostome rudstone.

Where then was the "missed opportunity"? We believe these massive trepostomes were able to withstand normal wave zone conditions in Maysvillian seas. A complex community of massive erect forms, of massive encrusters, of "back reef" and "fore reef" ramose species, and a diverse array of non-trepostomate encrusting, boring and cryptic species were present at this juncture, yet they seem not to have matured to their geologic potential by building massive skeletal reefs.

Our supposition is that at the geologic instant when trepostomes were biologically prepared to produce fringing, dare we even suggest barrier, reefs there seem to have been no place to put them. Proximity to a subsiding basin, or presence of a significant stable landmass in the absence of clastic influx on the Laurentian continental coast may have provided such a setting. A gently sloping shelf-ramp bathymetry across which great storms frequently swept gave no long-term stable depositional site for growth of framework reefs. The unique Cincinnatian trepostome experiment in Laurentia did not achieve reef-building status for want of reef-creating geologic conditions coupled with very unstable storm-generating climatic conditions, which were frequent enough to inhibit long-term community seral development and structure. The Taconic Orogeny was likely an added impediment that created sources, some volcanic, of smothering clay influxes. End-Ordovician extinctions prohibited replay of the same experiment later in the Paleozoic.

ACKNOWLEDGMENTS

We wish to thank those who have assisted with and encouraged this research. Thomas Hudson has been an influential field companion during the search for complete colonies. R.L. Trammel generously granted access to the Sharonville Industrial Park for collecting on many occasions. R.J. Cuffey kindly identified the principal specimens referred to in this paper, and he also provided authoritative discussions regarding bryozoan reefal development. Conversations with J.R.P. and Charles Ross have also been very helpful. Discussions of stratigraphic relationships at Sharonville with D. Meyer were helpful for understanding positions of our fossils in that fast-changing location. We thank F.D. Holland, Jr. for the use of his research library and B. Enslow for processing the manuscript. It is also a pleasure to note the support of St. Lawrence University for every aspect of this research. Editorial and scientific suggestions from P.N Wyse Jackson and anonymous reviewers have been most helpful. If errors of fact or reasoning are found in the text, they belong to the authors.

REFERENCES

Anstey, R.L. 1986. Bryozoan provinces and patterns of generic evolution and extinction in the Late Ordovician of North America. *Lethaia* 19(1): 3-51.

Anstey, R.L. 1987. Astogeny and phylogeny: evolutionary heterochrony in Paleozoic bryozoans. *Paleobiology* 13(1): 20-43.

Anstey, R.L. & Chase, T.L. 1974. Geographic diversity of Late Ordovician corals and bryozoans in North America. *Journal of Paleontology* 48(6): 1141-1148.

Boardman, R.S. & Cheetham, A.H. 1969. Skeletal growth, intracolony variation and evolution in Bryozoa: a review. *Journal of Paleontology* 43: 205-243.

Brown, D.A. 1952. The Tertiary cheilostomatous Polyzoa of New Zealand. *British Museum Natural History Monographs*, 405p.

Brown, G.D. Jr., & Daly. E.J. 1987. Trepostome bryozoan distributions in the Dillsboro Formation (Cincinnatian Series), Indiana: Water depth controls? In Ross, J.R.P. (ed.), *Bryozoa: present and past*: 41-47. Bellingham: Western Washington University.

Cheetham A.H. 1986. Branching biomechanics and bryozoan, evolution. *Proceedings of the Royal Society, London*, Series B 288: 151-171.

Copper, P. 2001. Reefs during the multiple crises towards the Ordovician – Silurian boundary: Anticosti Island, eastern Canada, and worldwide. *Canadian Journal of Earth Sciences* 38: 153-171.

Cuffey, R.J. 1985. Expanded reef-rock textural classification and the geologic history of bryozoan reefs. *Geology* 13: 307-310.

Cuffey, R.J. 1998[1999]. The Maysville bryozoan reef mounds in the Grant Lake limestone (Upper Ordovician) of North-Central Kentucky: 10-29. In R.A. Davis and R.J. Cuffey, (eds), *Sampling the Layer Cake that isn't: the stratigraphy and paleontology of the type-Cincinnatian*, Ohio Geological Survey Guidebook No. 13.

Davis, R.A., Deikmeyer, S.C.St.L., Goldman, L.I., Datillo, B.F., Holland, S.M. & Cuffey, R.J. 1998[1999]. Appendix A – Type Cincinnatian localities: 153-166. In R.A. Davis & R.J. Cuffey (eds*), Sampling the Layer Cake that isn't: the stratigraphy and paleontology of the type-Cincinnatian*, Ohio Geological Survey Guidebook No. 13:

Diekmeyer, S.C.St.L. 1998[1999]. Kope to Bellevue Formations: The Reidlin Road/Mason Road Site (Upper Ordovician, Cincinnati, Ohio, region: 10-29. In R.A. Davis & R.J. Cuffey (eds), *Sampling the Layer Cake that isn't: the stratigraphy and paleontology of the type-Cincinnatian*, Ohio Geological Survey Guidebook No. 13.

Drummond, C. & Sheets, H. 2001. Taphonomic reworking and stratal organization of tempestite deposition: Ordovician Kope Formation, Northern Kentucky, USA. *Journal of Sedimentary Research* 71(4): 621-627.

Holland, S.M. 1998 [1999]. Sequence stratigraphy of the Cincinnatian Series (Upper Ordovician, Cincinnati, Ohio, region:135-151. In R.A. Davis & R.J. Cuffey (eds), *Sampling the Layer Cake that isn't: the stratigraphy and paleontology of the type-Cincinnatian,* Ohio Geological Survey Guidebook No. 13.

Holland, S.M. & Patzkowsky, M.E. 1996. Sequence stratigraphy and long-term lithologic change in the Middle & Upper Ordovician of the eastern US: 119-130. In B.J. Witzke, G.A. Ludvigsen, & J.E. Day (eds), Paleozoic Sequence Stratigraphy: Views from the North American Craton: *Geological Society Special Paper* 306.

Holland, S.M. & Patzkowsky, M.E. 1997. Distal orogenic effects on peripheral bulge sedimentation: Middle and Upper

Ordovician of the Nashville Dome. *Journal of Sedimentary Research* 67(2): 250-263.

Holland, S.M. & Patzkowsky, M.E. 1998. Sequence stratigraphy and relative sea-level history of the Middle and Upper Ordovician of the Nashville Dome, Tennessee. *Journal of Sedimentary Research* 63(4): 684-699.

Key, M.M., Jr. 1991. How to build a ramose trepostome: 201-208, In F.P Bigey, *Bryozoa living and fossil.* Nantes: Société des Sciences Naturelle de l'Ouest de la France, Mémoire hors série 1.

Kolata, D.R., Huff, W.D. & Bergstrom, S.M. 2001. The Ordovician Sebree Trough: An oceanic passage to the Midcontinent United States. *Geological Society of America Bulletin* 113(8): 1067-1078.

Martin, W.D.1975. The petrology of a composite vertical section of Cincinnatian Series limestones (Upper Ordovician) of southwestern Ohio, southeastern Indiana, and northern Kentucky. *Journal of Sedimentary Petrology* 45: 907-925.

McKinney, F.K. & Jackson, J.B.C. 1989. *Bryozoan Evolution.* Unwin Hyman, Boston. 238p.

Pitcher, M. 1974. Evolution of Chazyan (Ordovician) reefs of eastern United States and Canada. *Canadian Petroleum Geologists Bulletin* 12: 632-791.

Stanley, S.M. & Hardie, L.A. 1999. Hypercalcification: Paleontology links to plate tectonics and geochemistry to sedimentology. *GSA Today* 9(2): 2-7.

Tobin, R.C. 1982. *A model for cyclic deposition in the Cincinnatian Series of southwestern Ohio, northern Kentucky, and southeastern Indiana.* Unpublished Ph.D. dissertation, University of Cincinnati, 483 p.

Waugh, D.A., & Erickson, J.M. 2002. Functional morphology of the anastomosing frondose growth form reported in *Heterotrypa frondosa* (D'Orbigny) (Bryozoa:Trepostomata) from the Cincinnatian (Late Ordovician) of Ohio. (this volume).

Wilson, M.A. & Palmer, T.J. 1992. Hardgrounds and hardground faunas. *University of Wales, Aberystwyth, Institute of Earth Studies Publication* 9: 131p.

Bryozoan Studies 2001, Wyse Jackson, Buttler & Spencer Jones (eds)
© 2002 Swets & Zeitlinger, Lisse, ISBN 90 5809 388 3

Systematics and biogeography of the Permian bryozoans in Europe

Andrej Ernst
Institut für Geowissenschaften der Christian-Albrechts-Universität zu Kiel, Germany

ABSTRACT: The systematics, biogeography and palaeoecology of bryozoans from marine sediments of the Zechstein in Germany and of the marine Permian in South Europe (NW-Tethys) were studied. This study points on the significance of the inner autozooecial morphology for the fenestrate taxonomy. Lower Permian bryozoan faunas of NW-Tethys display distinct relations to those of the Arctic and Russian Platform. Upper Permian bryozoan faunas of the NW-Tethys show similarities to the bryozoan faunas of Australia and the Far East. Bryozoan biogeography points on closing of a seaway between NW-Tethys and Uralian Sea apparently by the End of the Early Artinskian. No connection between Upper Permian Zechstein Sea and NW-Tethys existed. The age of the Zechstein is supposed to be Kazanian.

1 INTRODUCTION

The NW-Tethys is regarded here as a part of the Tethys Ocean openly connected with Central and East Tethys during almost the entire Permian. Different biotopes existed here. The bryozoan fauna of the NW-Tethys consists of taxa belonging to the orders Cystoporida, Trepostomida, Rhabdomesida, Cryptostomida, and Fenestrida. The bryozoan distribution was investigated in detail in the Permian of the Carnic Alps (Ernst 2000).

Bryozoans in the Upper *Pseudoschwagerina* Formation and Trogkofel Formation (Sakmarian - Artinskian) in the Carnic Alps are more diverse than those of the Lower *Pseudoschwagerina* Formation in the Upper Carboniferous. The Upper Permian *Bellerophon* Formation (Tatarian) in the Carnic Alps, stratigraphically correlating to the Zechstein deposits, contains no bryozoans.

Contrary to the NW-Tethys the Zechstein Basin was a nearly enclosed epicontinental sea connected only by a narrow gateway between Greenland and Norway via Norwegian-Greenland Sea to the open Arctic Sea (Ziegler 1990). The Zechstein Basin covered most parts of Central and Northern Europe. The deposits of the lower part of the Zechstein sequence are well known in Germany, Northeast England, Denmark and Poland (Fig. 1).

Bryozoan faunas are restricted to reefs and bedded sediments from off-shore regions of the Zechstein Sea (Ernst 2001). They are restricted to the carbonates of the first sedimentational cycle Z1 (Ca1). The Bryozoa of the Zechstein belong mainly to the orders Fenestrida and Trepostomida. Type material as well as material collected by the author were studied. The investigated material is housed in Senckenberg Museum, Frankfurt a. m. Main, Germany.

Fenestrate Bryozoa are the most distributed group in the Permian basins and allow exact biogeographical comparisons. The aim of this paper is to concern some problems of fenestrate systematics and to

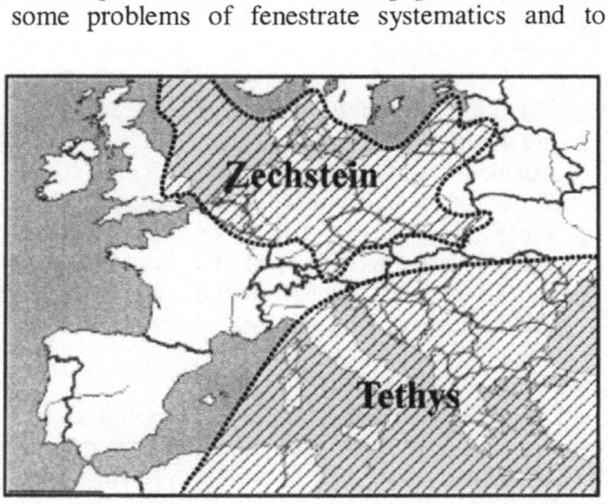

Figure 1. Permian basins in the European region.

overview the bryozoan paleobiogeography of Permian in European region.

2 FENESTRATE SYSTEMATICS

The existing systematics of fenestrate bryozoans does not consider enough the variability of their colony forms in different environments. The number of rows on the stems can be variable and should be carefully used. Many genera are defined on the basis of the colony form and the number of the zooecial rows on their stems. Some of these genera are clearly polyphylethic because their species reveal significant differences in their internal morphology.

The variability of bryozoan colony forms has been discussed in recent years (e.g. Snyder 1991, Morozova 1974, 1989). The most important morphological features in fenestrates are considered to be the shape of the autozooecial chamber as well as the presence and type of heteromorphs. The Zechstein bryozoan fauna reveals a diversification pattern that excellently confirms the significance of the iternal morphology in fenestrate bryozoans.

The transgression of the Zechstein Sea was apparently very fast and allowed an introduction of a restricted marine fauna from the Arctic Sea. The connection to the open shelf of the Arctic Sea was at times closed. Therefore, Zechstein Sea represents something like a natural "laboratory" for investigating the colony form variability. Some of the Zechstein bryozoans, regarded earlier as belonging to different groups within Fenestrida, are very closely related. Regarding their colony form and the number of the autozooecial rows they were placed in the "acanthocladiid" genera *Acanthocladia, Kalvariella* and *Penniretepora,* "septoporid" *Synocladia* and "polyporid" *Thamniscus.* The genera *Acanthocladia, Synocladia, Thamniscus,* and *Kalvariella* were originally described from the Zechstein. In addition to some nomenclature problems they were defined on the basis of outer morphological features, with an exception of the genus *Kalvariella* Morozova 1970.

Investigation of the inernal morphology of these bryozoans showed, that taxa with different colony forms possessed autozooecia of the same shape and the same kind of polymorphs. The polymorphs are ovicells, representing spherical chambers on proximal part of the autozooecial vestibule (Fig. 2). So far this kind of ovicells is known exclusively in the Zechstein bryozoans (Southwood 1985, Bancroft 1986).

The use of the colony form for description of fenestrate bryozoans has resulted in the reporting of many Zechstein species from everywhere in the world. They differ, however, radically from the Zechstein fauna in their internal morphology. A typical example is the genus *Synocladia.* This genus was characterised by the colony formed by branches joined by dissepiments carrying autozooecia. Some species from the Permian of Russia, Australia and China were placed in this genus (Crockford 1957, Trizna & Claucen 1961, Yang & Lu 1962, Morozova 1970). Detailed investigations has shown radical differences in the shape of the autozooecia and in the kind of polymorphs. Zechstein *Synocladia* species possess relatively long autozooecia with distinct vestibule and ovicells. *Synocladia* species from out-

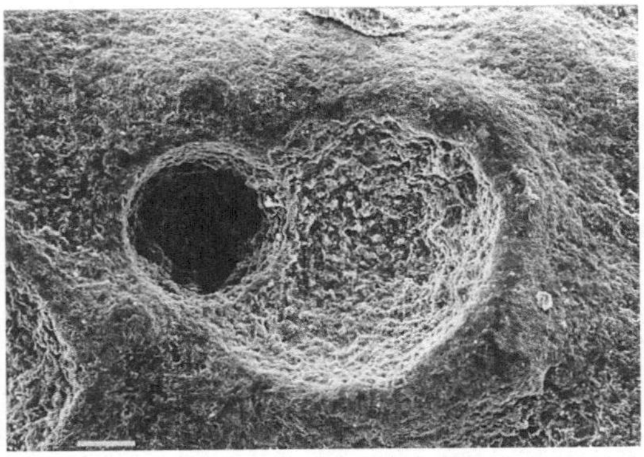

Figure 2. Ovicell of *Acanthocladia anceps* (Schlotheim 1820). Beeckerwert bei Duisburg, Germany. Zechstein, Lower Werrakarbonat. X11887, Federal Institute for Geosciences and Natural Resources, Berlin. Scale bar 30 µm.

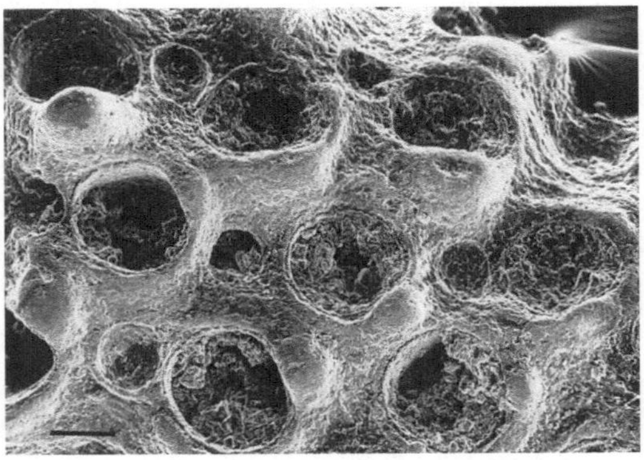

Figure 3. Obverse colony surface of an undescribed genus ("*Thamniscus*" sp.) having branched colony form from the Zechstein. Larger openings are the autozooecial apertures, smaller - cyclozooecia. Beeckerwerth near Duisburg, Germany. Zechstein, Upper Werrakarbonat. X11895, Federal Institute for Geosciences and Natural Resources, Berlin. Scale bar - 100 um.

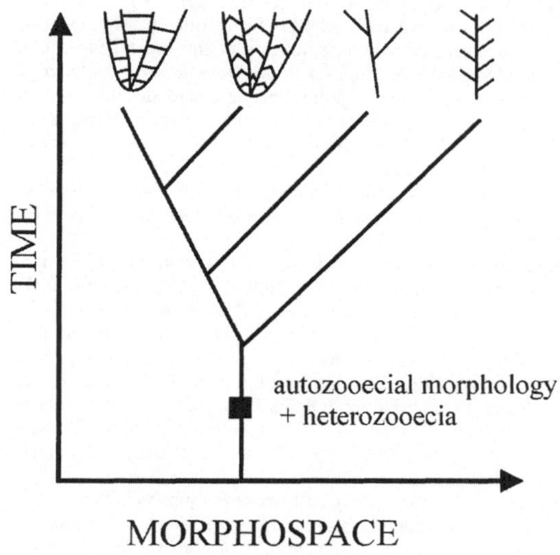

Figure 4. Proposed diversification pattern of fenestrates.

side of the Zechstein possess shorter autozooecia with weekly developed vestibule and cyclozoocea instead of ovicells. Herewith they should be excluded from the genus *Synocladia* and added to a separate genus within the family Septoporidae (Fig. 3). The cyclozooecia appear also in other genera having different colony form.

In a summary, a diversification pattern of fenestrate bryozoans can be suggested. Significant changes concerning autozooecial morphology and polymorph type allow an occupancy of the new adaptation zone (Fig. 4). Further diversification within this adaptation zone takes place within a morphospace formed by available colony forms including such a feature as the number of autozooecial rows.

3 PERMIAN PALEOBIOGEOGRAPHY

Refined and consistent taxonomy is necessary for paleobiogeographical reconstructions. The following interpretation is possible on the basis of the taxonomic composition of Permian bryozoan faunas in the European region.

The Lower Permian faunas from the Carnic Alps, Sicily and Tunisia display a mixed character including taxa from various basins. They show a close similarity to the Lower Permian faunas of the Urals, Pamir and Australia as well as to the Upper Carboniferous and Lower Permian faunas of the Arctic. Following taxa were found here: *Fistulipora regularis* Sakagami 1961, *Fistulamina lata* Crockford 1957, *Prismopora digitata* Crockford 1957, *Streblascopora germana* Bassler 1929, *Primorella serena* Gorjunova 1975, *Rectifenestella tuberculifera* (Schulga-Nesterenko 1952), *Penniretepora trapezoida* Nikiforova 1938, *Alternifenestella tenuiseptata* (Schulga-Nesterenko 1941), *Alternifenestella* aff. *tenuiseptata* (Schulga-Nesterenko 1941), *Alternifenestella nalivkini* (Elias & Condra 1957). Thus, a free faunal exchange through these regions can be suggested.

The Upper Permian faunas of Sicily and Tunisia, also investigated in the course of the present study, show no similarity neither to the Uralian Sea nor to the Zechstein. However, their generic composition, including among others such genera as *Fistulipora*, *Goniocladia*, *Stenopora*, and *Hayasakopora*, is very close to the faunas from Indonesia and Australia. That points to a closing of the connection between the Tethys and the Uralian Sea by the Early Artinskian. This is also supported by fusulinid and conodont studies (Ross 1995, Swift & Aldridge 1985).

In the Upper Permian (probably Kazanian) the Zechstein Sea was established as a result of repeated transgressions from the Norwegian-Greenland Sea. As a cause of these transgressions rifting processes has been discussed (Ziegler 1990). The bryozoans migrated with initial transgressions of the fist sedimentation cycle Z1 from the north in the Zechstein Basin. This fauna diverged in this extreme biotope and a few new taxa evolved. The connection to the open shelf of the Arctic Sea was at times closed. The environmental conditions got worse and finally caused the extinction of the entire fauna of the Zechstein Sea. Transgressions in subsequent producing by sedimentation cycles Z2-Z7 were not accompanied by new invasions of bryozoans.

4 AGE OF ZECHSTEIN

The Zechstein fauna is strongly endemic. Therefore, it is very difficult to correlate Zechstein sediments with other Permian formations. The last compilation of Menning (1995) suggests the Upper Tatarian age of the Zechstein on the base of the paleomagnetic data. The analyses of the bryozoan fauna give another picture.

Previous reports of Zechstein bryozoans from the Tethys are herein considered to be wrong (Waagen & Wentzel 1887, Bassler 1929, Termier & Termier 1971). However, two species occur in the Permian of the Russian Platform - *Rectifenestella retiformis* (Schlotheim 1820) and *Dyscritella angularis* Morozova 1970. *Rf. retiformis* occurs also in the Permian of the Arctic (Morozova & Krutchinina 1986, Nakrem 1991). A third species *Kingopora ehrenbergi* (Geinitz 1861) has been reported from the Kazanian of Nevada (Gilmour et al. 1997). The genus *Kingopora* Morozova 1970 is relatively widely distributed in the Sakmarian and Artinskian of the Arctic and in Kazanian of the Russian Platform (Lisitsyn, pers. comm.). Thus, the bryozoans indicate at least a

Kazanian age for the lowermost part of the Zechstein.

5 CONCLUSIONS

Taxonomic studies on Permian bryozoans of the European region allow the following conclusions:

- Some fenestrate bryozoans possessing different colony forms and regarded therefore as belonging to different groups, reveal the same autozooecial morphology and the same heteromorph type. Close phylogenetic relations between these forms are suggested.
- no connection between NW-Tethys and Zechstein by the end of the Permian existed;
- Seaway between Uralian Sea and Tethys has closed by the end of Early Artinskian;
- The age of the lowest Zechstein is considered to be Kazanian.

6 ACKNOWLEDGEMENTS

The research was supported by the German Research Foundation (DFG), Project Scha 355/13-1 u. 2. Figured material was kindly placed to disposal by the Federal Institute for Geosciences and Natural Resources, Berlin. I thank to Hans Arne Nakrem, Oslo for critical reading of the manuscript, and to Andrew N. Ostrovski, St. Petersburg for some helpful comments.

REFERENCES

Bancroft, A. J. 1986. Ovicells in the palaeozoic Bryozoan order Fenestrata. *Palaeobiolgy* 29(1): 155-164.

Bassler, R. S. 1929. The Permian Bryozoa of Timor. *Paläontologie von Timor* 16(28): 37-90.

Crockford, J. 1957. Permian Bryozoa from the Fitzroy Bazin, Western Australia. *Bulletin of the Bureau of Mineralogical Resources, Geology and Geophysics* 34: 9-131.

Ernst, A. 2000. Permian bryozoans of the NW-Tethys. *FACIES* 43: 79-102.

Ernst, A. 2001. Bryozoa of the Upper Permian Zechstein Formation of Germany. *Senckenbergiana lethaea* 81(1): ... . (in press)

Gilmour, E. H., McColloch, M. E. & Wardlaw, B. R. 1997. Bryozoa of the Murdock Mountain Formation (Wordian, Permian), Leach Mountains, Northeastern Nevada. *Journal of Paleontology* 71(2): 214-236.

Menning, M.: A numerical time scale for the Permian and Triassic periods: An Integrative Time Analysis. In Scholle, P. A., Peryt, T. M. & Ulmer-Scholle, D. S. (eds.), *The Permian of Northern Pangea. 1 (Paleogeography, Paleoclimates, Stratigraphy)* 77-97, Berlin-Heidelberg: Springer-Verlag.

Morozova, I. P. 1970. Upper Permian Bryozoa. *Trudy Paleontologicheskogo Instituta Akademii Nauk SSSR* 122: 1-347, Moscow: Nauka. [In Rusian]

Morozova, I. P. 1974. Revision of the genus Fenestella. *Paleontological Journal* 2: 54-67 [In Russian]

Morozova, I. P. 1989. Morphogenez, sistema i kolonial'naya integratzia mshanok otryda Fenestrida [Morphogenese, system and colonial organization of the order Fenestrida]. In *Morphogenez i puti razvitia kolonialnosti mshanok i kishechnopolostnykh [*Morphogenese and development of bryozoans and coelenterates]. Trudy Paleontologicheskogo Instituta Akademii Nauk SSSR 222: 70-88, Moscow: Nauka. [In Rusian]

Morozova, I. P. & Krutschinina, O. N. 1986. *Permian Bryozoans of the Arctic (Western Sector).* Moscow: Nauka. [In Russian]

Nakrem, H. A. 1991. Distribution of Bryozoan in the Permian Succession of Svalbard (preliminary data). In Bigey, F. P. (ed), *Bryozoa living and fossil. Bulletin de la Société des Sciences Naturelles de l'Quest de la France, Hors serie 1.* 291-298, 3 Figs., Nantes.

Ross, C. A. 1995. Permian Fusulinceans. In Scholle, P. A., Peryt, T. M. & Ulmer-Scholle, D. S. (eds.), *The Permian of Northern Pangea. 1 (Paleogeography, Paleoclimates, Stratigraphy).* 167-185, Berlin-Heidelberg: Springer-Verlag.

Snyder, E. M. 1991. Revised taxonomic approach to acanthocladiid Bryozoa. In Bigey, F. P. (ed.), *Bryozoaires actuels et fossil: Bryozoa living and fossil; Bulletin de la Société des Sciences Naturelles de l'Quest de la France, Hors serie 1.* 431-445, Nantes.

Southwood, D. A. 1985. Ovicells in some fenestrata from the Permian of N. E. England. In Nielsen, C. & Larwood, G. P. (eds), *Bryozoa: Ordovician to Recent.* 301-310; Fredensborg: Olsen & Olsen.

Swift, A. & Aldridge, R.J (1985). Conodonts of the Permian System from Great Britain. In Higgins, A. C. & Austin, R. L. (eds.), *A stratigraphical index of conodonts. British Micropalaeontological Society series* 7: 228-236, West Sussex: Ellis Horwood Ltd.

Termier, H. & Termier, G. 1971. Bryozoaires du Paleozoique superieur de Làfganistan. *Documents des Laboratoires de Geologie de la Faculte des Sciences de Lyon.* 47: 1-52 Lyon.

Trizna, V. B. & Claucen, R. A. 1961. Bryozoans of the Artinsk-Stage of the Ufimian-Plate and their role in the stratigraphy of this Stage in Urals. *Trudy VNIGRI* 179th ed., *Microfauna of the USSR* 13: 331-453 [In Russian]

Waagen, W. & Wentzel, I. 1887. Salt Range Fossils. 1. Productus Limestone Fossills. *Memoirs of Geological Survey of India, Paleontologica Indica* 13: 925-988.

Yang, J. & Lu, L. 1962. Paleozoic Bryozoans of Qilianshan. In *Geology of Qilianshan* 4: 1-114, 24 pls., Beijing: Science Press. [In Chinese]

Ziegler, P. A. 1990. *Geological atlas of Western and Central Europe.* Bath: Geological Society Publishing House.

Bryozoan Studies 2001, Wyse Jackson, Buttler & Spencer Jones (eds)
© 2002 Swets & Zeitlinger, Lisse, ISBN 90 5809 388 3

Late Cretaceous-Paleocene "porinids" – mixed frontal shields and evidence of polyphyly

Dennis P. Gordon
National Institute of Water & Atmospheric Research (NIWA), P.O. Box 14-901 Kilbirnie, Wellington, New Zealand

ABSTRACT: Putative porinids (Cheilostomata) from the Late Cretaceous and Paleocene of northwestern Europe and the USA exhibit a varied range of external frontal-shield morphologies. Some frontal shields are relatively featureless or merely dimpled; others have numerous foramina of various diameters and these are surrounded by raised ridges in some species. The complex surface topography of some frontal shields defies interpretation when viewed externally but exceptional preservation of many taxa allows fracture of zooids and SEM observation of frontal-shield interiors. A number of species have umbonuloid shields whereas others have frontally concealed spinocysts of well-developed costae. External foramina in many species correspond to frontally opening areolar septular-pore canals, spiramina, and loci of avicularia. All the species examined are unrelated to Recent *Porina gracilis*, the type species of the family, with a pseudoporous frontal shield and lepralioid ascus sac. Among the taxa now excluded from the Porinidae, *Pachythecella* is restricted to its type species, the balance of species attributed to it being transferred to *Frurionella* (Lepraliellidae). Spiraminate, umbonuloid-shielded *Pachythecella, Beisselina*, and *Pachydermopora* gen. nov. are included in the Tessaradomidae. Spiraminate *Columnotheca, Diplobeisselina*, and *Scoriopora* gen. nov. have a primary spinocystal shield of large costae. For these genera a new family, Scorioporidae, and superfamily, Scorioporoidea, are proposed.

1 INTRODUCTION

The family Porinidae d'Orbigny, 1852 was established to include erect and encrusting cheilostomes with a "*pore spécial*", proximal to the orifice, that connects the zooidal interior with the exterior, i.e., an ascopore or spiramen. Although d'Orbigny mistook avicularian chambers for such pores in some taxa, the type genus, *Porina* d'Orbigny, 1852, does have a well-developed spiramen. D'Orbigny included eleven species in *Porina*, from which Lang (1917) selected Recent *Eschara gracilis* Lamarck, 1816 as type species. The morphological heterogeneity of the genus *Porina* as d'Orbigny conceived it may be appreciated by considering the other species he included – Recent, encrusting *Porina africana* d'Orbigny, 1852 was identified by Waters (1905) as *Microporella*; erect, Campanian *P. filiformis* d'Orbigny, 1852 became type species, by original designation, of *Pachythecella* Bassler, 1934; and two other species were previously attributed by d'Orbigny (1850, 1851) to the cyclostome genus *Bidiastopora*. D'Orbigny (1852) included four other new genera in the family: *Sparsiporina*, with Miocene type species *Retepora elegans* Reuss, 1848, is a phidoloporid; *Semiporina*, with Eocene type species *S. elegans* d'Orbigny, 1852 (unillustrated and apparently lost) is unknown; *Reptoporina*, with Coniacian type species *Escharina micropora* d'Orbigny, 1850, may be a malacostegine genus related to electrids or to *Fissuricella* Voigt, 1959; and *Multiporina*, with Oligocene type species *M. ostracites* d'Orbigny, 1852, is a synonym of *Schizoporella* Jullien, 1883 (see Buge 1975).

Bassler (1935) included six genera in the Porinidae in the *Fossilium Catalogus*. Apart from *Porina*, these were: *Beisselina* Canu, 1913, *Columnotheca* Marsson, 1887, *Gastropella* Canu & Bassler, 1917, *Rotiporina* Brydone, 1930, and *Pachythecella*. Bassler (1953) added *Beisselinopsis* Voigt, 1951 in the Bryozoa *Treatise*. In the years since d'Or-bigny's monumental study of French Cretaceous Bryozoa, several other fossil species have been added to the genus *Porina* (e.g. Marsson 1887, Canu 1913, Voigt 1924, 1987, Brydone 1930).

The exquisite preservation of many Late Cretaceous and Paleocene bryozoans from northwestern Europe and the USA has meant that SEM examination of these forms has been able to yield much helpful information. It has become clear from these studies that a range of frontal-shield morphologies obtain for the family as presently constituted, which is therefore polyphyletic even if less heterogeneous than in d'Orbigny's day. Taxa must be based as far as possible, however, on monophyletic clades. Can we circumscribe a monophyletic Porinidae and where does this leave excluded genera? It is the goal of this paper to describe the range of frontal shields encountered and discuss the implications for classification.

2 MORPHOLOGICAL EXAMINATION OF SELECTED SPECIES

2.1 *Porina gracilis* (Lamarck, 1816)

Porina gracilis (Recent, southern Australia) forms erect, lobate, bilaminar colonies *c.* 20 mm high on other bryozoans or hydroids, to which they attach by means of a small encrusting base (Bock 1982, fig. 9.19). The autozooidal frontal shield is evenly and densely perforated by pseudopores that are visible internally in cleaned specimens, as well as by numerous marginal areolar-septular pore canals of the same diameter. Externally, the fully formed secondary or peristomial orifice is transversely oval or subhemispherical (with a slightly arcuate proximal rim at a lower level within the peristome) and surrounded by *c.* five small oval avicularia with complete crossbars (Fig. 1). The primary orifice, seen at growing tips of colonies, is longer than wide and narrows proximally towards a shallow U-shaped sinus flanked by a pair of condylar shoulders. The operculum, while narrower proximally, lacks a distinct poster, as illustrated precisely by Levinsen (1909, pl. 16, fig. 1b; as *Haswellia coronata*); the proximal part does not correspond to the sinus of the primary orifice, resting under the condylar shoulders. Frontally, immediately proximal to the suboral avicularia and larger than either of them, there is a longitudinally narrow, almost slit-like, pore. This opens *above* the operculum at the bottom of the peristome (Figs 2, 3), instead of directly into the compensation sac. It is thus a peristomial spiramen, not an ascopore. Owing to the depth of the peristome and secondary calcification, ovicells (Figs 1, 3) are concealed in mature zooids and nearly flush with the frontal surface of the colony. Technically, however, they are hyperstomial, i.e. situated above the orifice and opening into the peristome. Lateral interzooidal communications are via tiny uniporous septular pores.

Thus, based on *P. gracilis*, typical morphological characters of the family would include: pseudoporous lepralioid frontal shield; concealed primary orifice; tubular spiraminal canal that opens, at its inner end, into the peristome above the operculum; hyperstomial ovicells not closed by the zooidal operculum; adventitious avicularia; areolar septular-pore canals; and tiny uniporous mural septula. Other genera that share all or most of these characters include:

Cyphonella Koschinsky, 1885 (Lutetian): usually included in the Mucronellidae, i.e. Smittinidae, but had an erect growth habit and a well-developed spiramen; ovicells have not yet been found.

Haswelliporina Gordon & d'Hondt, 1997 (Aquitanian-Recent): erect, peristomes and spiramen tubes typically project from the zooid surface; pseudopore size and density less than in *Porina*.

Mosaicoporina Gordon & d'Hondt 1997 (Recent): erect to suberect, spiramen tube fully incorporated into peristome, the external opening in its proximal rim.

Semihaswellia Canu & Bassler, 1917 (?Tertiary-Recent): erect, branches with frontal and abfrontal surfaces, projecting tubiform spiramen, sparse pseudopores, and prominent ovicell.

Tremotoichus Canu & Bassler, 1917 (Lutetian-Bartonian): erect, frontal and abfrontal surfaces, peristome somewhat projecting, spiramen asymmetrical, opening at zooid surface.

We now discuss some Cretaceous-Paleocene genera and species included in the Porinidae by Bassler (1953) and other authors, but herein excluded from the family.

2.2 *Beisselina striata* (Goldfuss, 1826)

Beisselina Canu, 1913 includes the type species *B. striata* (Maastrichtian, NW Europe) and eight others, all thoroughly described by Wiesemann (1963), both externally and internally. Understanding spatial relationships among the zooidal chamber, peristome, peristomial ovicells, spiramen, adventitious avicularia, and areolar septular-pore canals was aided by artificial casts. Briefly, *B. striata* formed erect, branching, bilamellar colonies from a small encrusting base. In frontal view, the colony surface appears perforated by holes of various sizes (Fig. 4). The largest, quincuncially arranged, are the circular peristomial orifices. Zooidal outlines are not clear but, around each orifice, there is a shallow diamond-shaped (or quadrangular) furrow. At each of the four corners defined by these furrows is a spiraminal pore; other pores in the furrows are areolar septular pores, some of which are loci for small adventitious avicularia.

An interior view of the frontal shield, which is at some depth, shows an umbonuloid morphology, with a clear ring scar and the internal opening of the spiramen near the anterior end of the shield (Fig. 5). Internal openings of some areolar septular-pore canals and/or avicularia are seen at the narrowed proximal end of the zooidal interior.

Although Wiesemann (1963) included *Beisselina* in the Porinidae, he nevertheless remarked that the genus "is to be derived from membranimorph forms". The presence of an umbonuloid shield, with a membranous frontal wall in life, precludes inclusion of *Beisselina* in the Porinidae.

2.3 *Columnotheca cribrosa* Marsson, 1887

Columnotheca cribrosa (Maastrichtian, NW Europe) had erect colonies with cylindrical branches and zooids in whorls. Depending on the age of a branch and consequent degree of secondary calcification, the frontal pores and foramina that are evident can appear small and mostly of uniform size, or nearly as varied as in *Beisselina* (Fig. 22). Using natural casts, Voigt (1968) determined the spatial relationships and identities of the various foramina, ascertaining that *C. cribrosa* has paired spiramina. SEM examination of well-preserved branch tips with less secondary calcification also helps to elucidate morphology. Zooidal secondary orifices are obvious, being larger, whorled, somewhat projecting, and more or less circular. Proximal to each orifice at a about one-third zooidal length is a pair of foramina, initially relatively large but becoming reduced in diameter by calcification – these are the spiramina (Fig. 23). [In contrast, *C. monopora* Voigt, 1968 has only a single spiramen.] Most other external foramina, regardless of size, seem to represent areolar septular openings, and a few are loci of avicularia.

SEM examination of zooidal interiors shows that the spiraminal canals are separate for their entire length and open near the base of the peristomial shaft above the primary orifice. In the interior of the frontal shield, the most conspicuous feature is a longitudinally narrow, parallel-sided area resembling a costal field. Although preservation of the examined specimen is exceptional, there is no distinction in the skeletal ultrastructural fabric inside and outside the field. There is, however, a definite trend in the orientation of the mineral grains within the field to indicate that much of its surface resembles fused costal elements. Whether this fusion is ontogenetic or diagenetic is not certain. There are no indications, outside of the field, of entrances to costal lumina, nevertheless gaps along the inner margin of the field are here taken to represent, or be homologous with, intercostal lacunae. Their number and disposition suggest there may have been *c.* 14 costal elements making up this area. There is no indication, either from SEM examination of fractured material or by internal casts (see Voigt 1968) that the lacunae are expressed frontally as canals or openings.

2.4 *Gastropella ventricosa* (Canu & Bassler, 1917)

Gastropella, whose two species have branching cylindrical stems, resembles *Beisselina* in zooidal characters. The type species, *G. ventricosa* (Paleocene, Arkansas), has projecting tubular peristomes, comprising most of the visible zooidal surface, and fewer areolar septular-pore openings than in *Beisselina*, mainly confined to the margins. Avicularia are rarer in *Gastropella*, sometimes incorporated into one side of a peristome, or borne on its frontal side. A single spiraminal pore occurs midfrontally in each peristome (Fig. 6), opening towards its base above the primary orifice. As in *Beisselina*, ovicells open into the peristome.

The frontal shield is umbonuloid, as evidenced by SEM examination of the zoecial interior. Proximal to the umbonuloid area and ring scar is a longer, narrow area of interior wall (Fig. 7).

2.5 *Pachythecella filiformis* (d'Orbigny, 1852)

Porina filiformis d'Orbigny, 1852 (?Campanian, Royan, France), the type species of *Pachythecella* Bassler, 1934, is little known, although specimens attributed to it have been found in Denmark, Rügen, and Aachen (see Voigt 1930; Berthelsen 1962). D'Orbigny's original illustrations are stylised, especially his figure 14 which exaggerates the peristomes. These are not so well developed in the actual specimen, illustrated here (Figs 8, 9).

In putative conspecific material from other localities, Voigt (1930) interpreted the "*pore spécial*" on the front of each zooid to be an avicularium, not an ascopore or spiramen. This appears not to be the case in d'Orbigny's type specimen, however. In the cleaned specimen the pores representing spiramina are simple, circular, and deep and do not resemble avicularia. Although it is not possible to fracture the specimen to view the interior of the frontal shield, I do not believe the species is congeneric with those included by Voigt 1930 and Berthelsen 1962 in *Pachythecella*. Danish and Rügen species attributed by Voigt (1930) and Berthelsen (1962) to *P. filiformis* strictly differ from the type specimen. Not only is the midfrontal spiramen missing in these northern forms, details of the peristomial orifice differ as well. This and some other species have peristomial avicularia and at least one has vicarious avicularia. Species with both peristomial and vicarious avicularia closely resemble *Frurionella* Canu & Bassler, 1925 (see Gordon 1993). Although the Maastrichtian type species of *Frurionella*, *F. parvipora* Canu & Bassler, 1925, has a more robust colony form than the gracile European species attributed to *Pachythecella*, they nevertheless could be included in the scope of the genus definition of *Frurionella* (otherwise a new genus would be required). Danish and Rügen specimens previously referred to as *P. filiformis* may be called *Frurionella dubiosa* (Levinsen, 1925) n. comb., included in the synonymy of *P. filiformis* by Voigt (1930) and Berthelesen (1962). The peristomial orifice of *F. dubiosa* (Stevns Klint material) is very variable. It can be "schizoporelloid" with a median sinus near (but not confined to) branch tips, circular, or transversely reniform to oval and wider than long (Fig. 10). This is in contrast to the uniformly circular orifices in *Pachythecella filiformis*. Interzooidal boundaries are not in evidence except at branch tips but tiny, sparse areolar-septular pores can be discerned in interzooidal positions.

Within or below the proximal rim of many peristomes of *F. dubiosa* is a small oval avicularium (rarely two) with a complete, median crossbar. Some peristomial orifices show the tips of one or a pair of spines laterally (Fig. 11). Perhaps it is an artifact of preservation but it appears that spines can also be short or lacking, for some orifices in the same branch give no evidence of them. A fractured section of one peristome shows that spines are hollow throughout their length and curve distally from the distal rim of the primary orifice (Fig. 12). The frontal shield is umbonuloid, with a distinct ring scar, occupying about half the length of the zooidal interior. There is no spiramen.

Frurionella lundgreni (Pergens & Meunier, 1886) n. comb. (Danian) is also umbonuloid (Fig. 13). Peristomial orifices are consistently reniform and the avicularium is placed suborally. Oral spines are seen only at the bottom of the peristomial shaft (Fig. 14). Other species examined lack oral spines but may be related to *Frurionella* on the basis of other characters. These are: *Entalophora variegata* d'Orbigny, 1851; *Porina angustata* d'Orbigny, 1852; and *Porina cylindrica* Voigt, 1924. In *F. variegata* n. comb. the peristomial avicularium is concealed from frontal view, occurring below the proximal rim on its inner side. This is the only gracile species of *Frurionella* in which ovicells have been observed; they occur at the bottom of the peristomial shaft above the primary orifice (Fig. 17). *Frurionella parvipora* has partly concealed peristomial ovicells (Gordon 1993).

2.6 *Diplobeisselina nobilis* (Levinsen, 1925)

Voigt (1987) established the genus *Diplobeisselina* for *Monoporella nobilis* Levinsen, 1925 and *Mucronella sutneri* Koschinsky, 1885. In the erect, bilamellar type species, *D. nobilis*, zooids are regularly quincuncially arranged, each having somewhat of a diamond shape with distal, proximal, and two lateral corners (Fig. 25). Within this arrangement is a rounded, transversely D-shaped or subcircular secondary orifice, generally with a pair of avicularia proximally, adjacent to which is a pair of small spiramen pores. The four corners of the diamond are marked by relatively large areolar-septular pores, any of which can be occupied by an avicularium (Fig. 26). All such avicularia are oval with a complete cross-bar.

SEM examination of shield interiors shows a well-developed costal field (Fig. 27). In exceptionally well-preserved material from Maastricht, costal surfaces clearly show radiating fans of planar-spherulitic microstructure (Fig. 28). Sutures between adjacent costae indicate three pairs of relatively large costae – proximal, central, and distal – with five smaller structures – one midproximal, two lateral, and two subdistal – between them. Small entrance pores to the costae are evident outside the perimeter of the costal field in some zooids, but there are no intercostal lacunae. A pair of foramina adjacent to the large distal pair of costae corresponds to the inner openings of the spiramen. The anterior tips of the distalmost pair of costae are reflexed against the bottom of the peristomial shaft against its proximal side.

Porina salebrosa Marsson, 1887 may be included in *Diplobeisselina*. Externally, the arrangement of orifices, heterozooids, and pores is similar to that in *D. nobilis* but often three avicularia are developed near the orifice and the spiramen may appear single at its entrance while internally divided, just below the entrance. Internally, the frontal shield is costate as in *D. nobilis* (Fig. 30) with seven large costae (including a fused oral pair that curves into the peristome) and what appear to be small ones between the bases of the larger. One fractured specimen shows an ovicell at the base of the peristome (Fig. 31).

2.7 *Porina pachyderma* Marsson, 1887

This Maastrichtian species formed erect, dichotomously branching bilamellar colonies remarkable for their extraordinarily thick secondary calcification, densely dimpled zooidal surface, and irregular, even nodular, relief in older parts of colonies. Peristomial orifices are regularly arranged in quincunx such that they occur in straight oblique rows across the colony surface, an arrangement that is apparent even in thickly calcified specimens. Adjacent to each circular peristome is a pair of oval avicularia, each with a complete crossbar. Their disposition in relation to the orifice varies from peristome to peristome, however, while the small round spiraminal pore is consistent in its position proximal to the orifice (Fig. 18). Zooidal surfaces are densely dimpled, with fine granularity in each dimple; at high magnification the continuous network of ridges around the dimples imparts a reticulated appearance to the zooids. Tiny scattered pores in the zooids are evidently the openings of areolar-septular canals. Very long bispatulate avicularia occur sporadically on the faces and edges of the branches; each has a pyriform opesia, a long rostral palate with a median groove, and no crossbar (Fig. 19). Very few such avicularia lack some investing secondary calcification and, indeed, the entire colony surface gives indication that adding of calcification was a continuous feature in life, possibly as a consequence of the activity of commensal organisms. Where peristomes and avicularia are invested and concealed by calcification the colony surface appears nodular (Fig. 20).

Fractured colony fragments show the thickness of frontal calcification that can build up, as well as zooidal interiors. The frontal shield is umbonuloid, and ring scars are obvious. The spiramen opens near

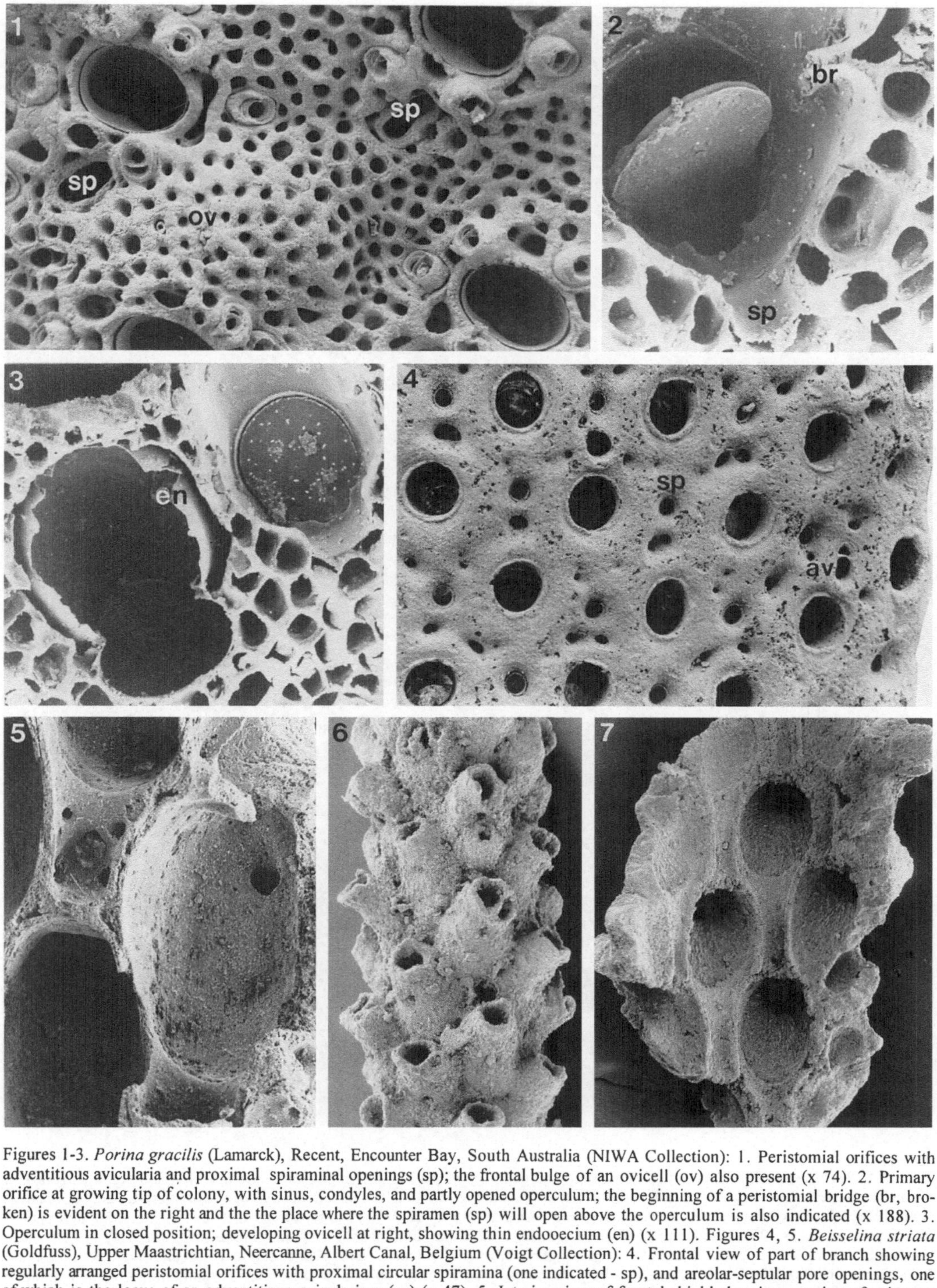

Figures 1-3. *Porina gracilis* (Lamarck), Recent, Encounter Bay, South Australia (NIWA Collection): 1. Peristomial orifices with adventitious avicularia and proximal spiraminal openings (sp); the frontal bulge of an ovicell (ov) also present (x 74). 2. Primary orifice at growing tip of colony, with sinus, condyles, and partly opened operculum; the beginning of a peristomial bridge (br, broken) is evident on the right and the the place where the spiramen (sp) will open above the operculum is also indicated (x 188). 3. Operculum in closed position; developing ovicell at right, showing thin endooecium (en) (x 111). Figures 4, 5. *Beisselina striata* (Goldfuss), Upper Maastrichtian, Neercanne, Albert Canal, Belgium (Voigt Collection): 4. Frontal view of part of branch showing regularly arranged peristomial orifices with proximal circular spiramina (one indicated - sp), and areolar-septular pore openings, one of which is the locus of an adventitious avicularium (av) (x 47). 5. Interior view of frontal shield showing opening of spiramen tube (x 79). Figures 6, 7. *Gastropella ventricosa* Canu & Bassler, Paleocene, Lawrenceville, Henry Co., Alabama (Voigt Collection): 6. Part of branch; spiraminal pores visible on frontal faces of peristomes (x 32). 7. Interior view of frontal shields (x 45).

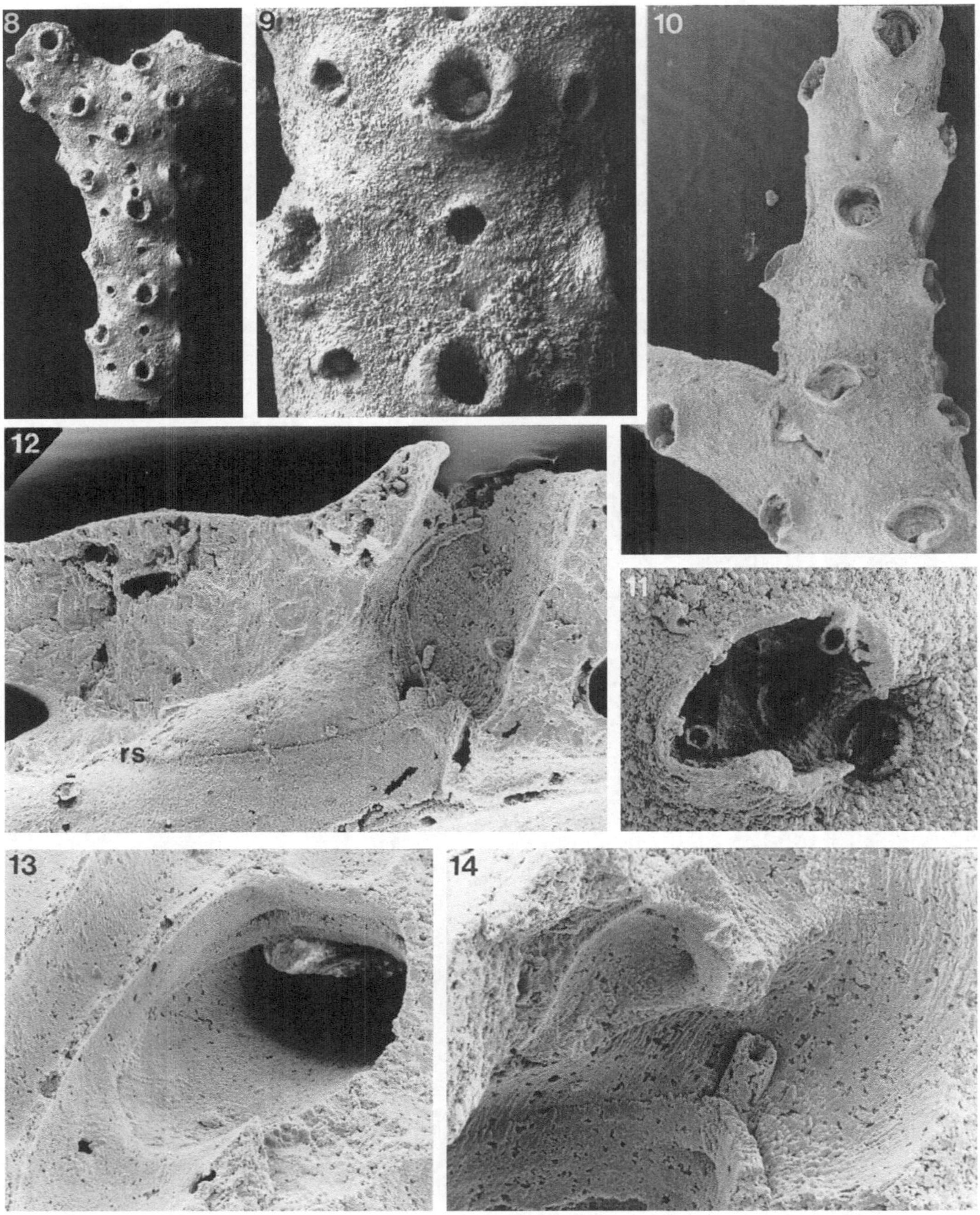

Figures 8, 9. *Pachythecella filiformis* (d'Orbigny), Senonian, Royan, France (d'Orbigny Collection, MNHN holotype R61706): 1. Cleaned branch fragment (x 26). 9. Close-up of same (x 75). Figures 10-12. *Frurionella dubiosa* (Levinsen) n. comb., Danian, Stevns Klint (Voigt Collection): 10. Branch fragment near growing tip (x 38). 11. Peristomial orifice showing spine tips and intraoral avicularium (x 139). 12. Longitudinal section through distal part of zooid showing umbonuloid ring scar (rs) and a hollow, curving oral spine in the peristomial shaft (x 139). Figures 13, 14. *Frurionella lundgreni* (Pergens & Meunier) n. comb., Danian, from glacial drift boulder, Daerstorf near Hamburg, Germany (Voigt Collection): 13. Interior view of umbonuloid frontal shield (x 165). 14. Fractured zooid showing distal end with part of avicularium chamber above and short spine at bottom of curved peristomial shaft (x 242).

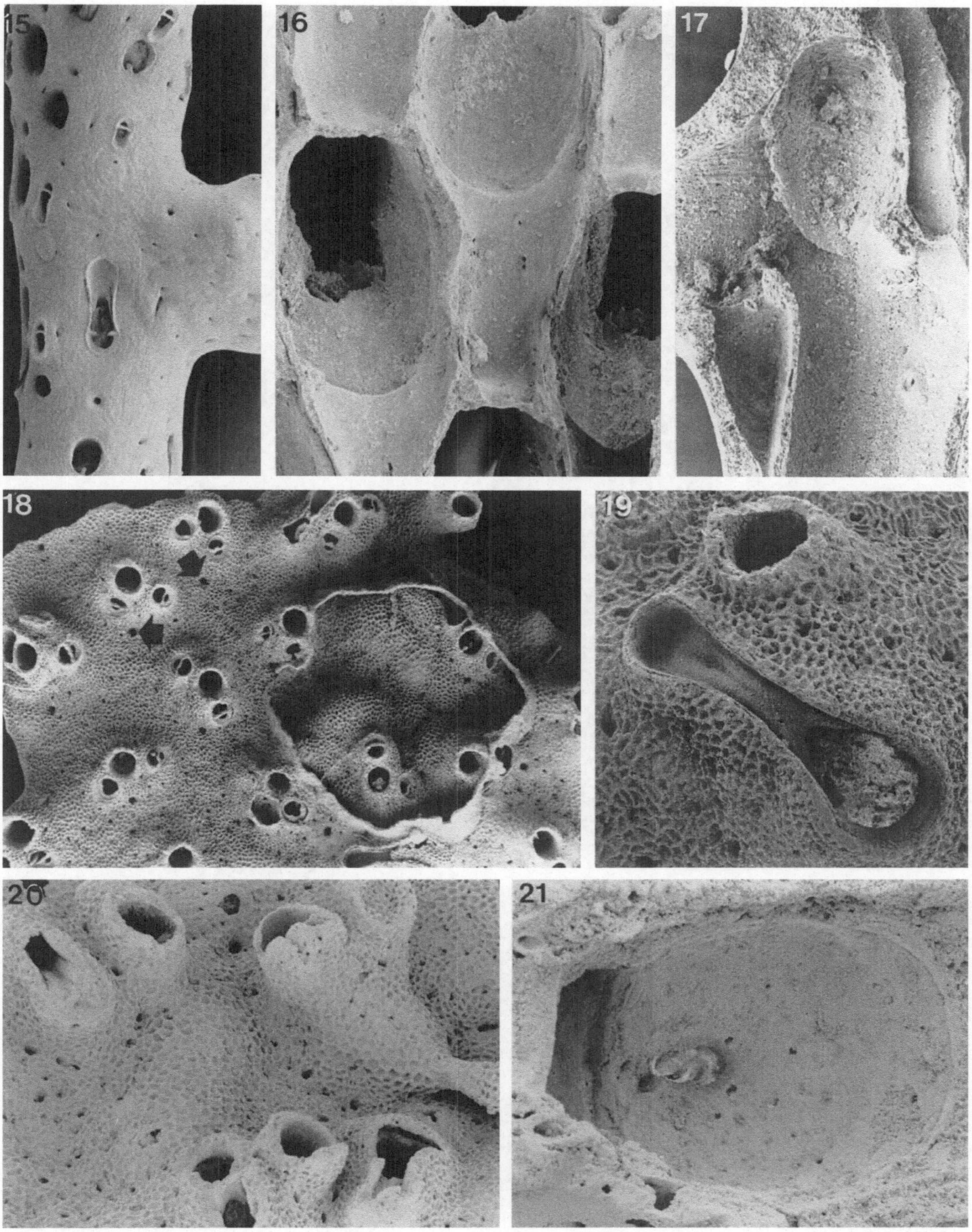

Figures 15-16. *Frurionella cylindrica* (Voigt) n. comb., Danian, from glacial drift boulder, Fehmarn, Staberhuk, Germany (Voigt Collection): 15. Branch fragment (x 26). 16. Interior of zooidal frontal shields (x 91). Figure 17. *Frurionella variegata* (d'Orbigny) n. comb., Upper Campanian, Meudon, France (Voigt Collection): Longitudinal section of zooid showing avicularium chamber and ovicell (x 113). Figures 18-21. *Pachydermopora pachyderma* (Marsson) n. comb., Lower Maastrichtian, Saturn Quarry near Kronsmoor (18, 19) and Rügen (20, 21), Voigt Collection: 18. Colony fragment with relatively less secondary calcification, showing peristomial orifices, small circular spiraminal pores (two arrowed), irregularly disposed small oval avicularia, and two large bispatulate avicularia (x 25). 19. Bispatulate large avicularium (x 91). 20. Heavily calcified colony (x 49). 21. Interior of frontal shield showing opening of spiramen tube in umbonuloid shield (x 96).

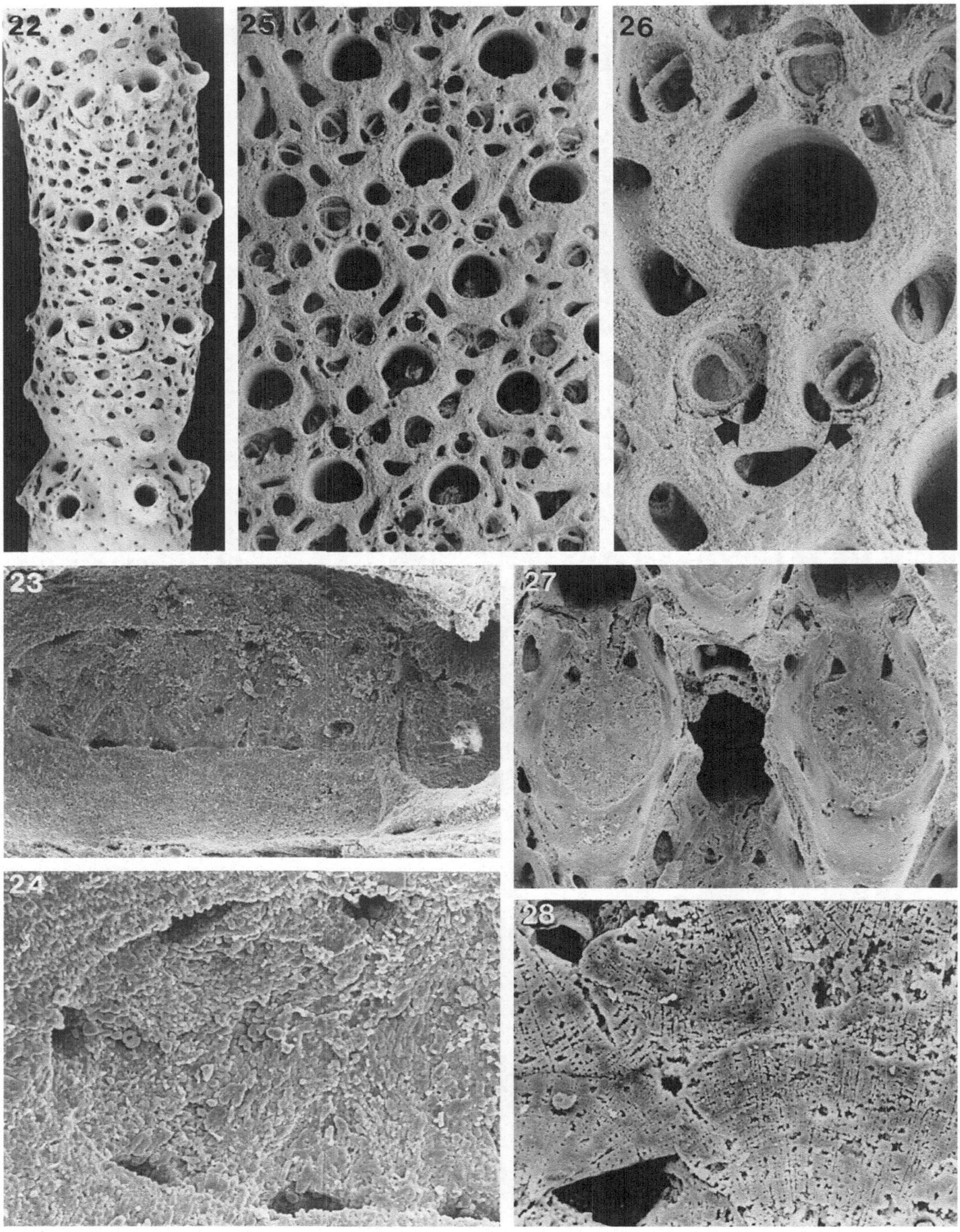

Figures 22-24. *Columnotheca cribrosa* Marsson, Upper Maastrichtian, Stevns Klint, Denmark (Voigt Collection): 22. Cylindrical branch showing whorled peristomial openings and foramina of various sizes depending on degree of secondary calcification (x 23). 23. Interior of frontal shield showing narrow costal field and, at right, the internal openings of the spiramina (x 16). 24. Close-up of same (x 434). Figures 25-28. *Diplobeisselina nobilis* (Levinsen), Dano-Montian, Maastricht (Voigt Collection): 25. Frontal view of branch showing more-or-less regular disposition of peristomial orifices, paired suboral avicularia, adjacent spiraminal openings, and areolar-septular openings (x 53). 26. Close-up of same, spiraminal openings arrowed (x 152). 27. Interior view of frontal shields showing costal fields with paired spiraminal openings (x 108). 28. Close-up of same, showing planar-spherulitic microstructure of costae (x 633).

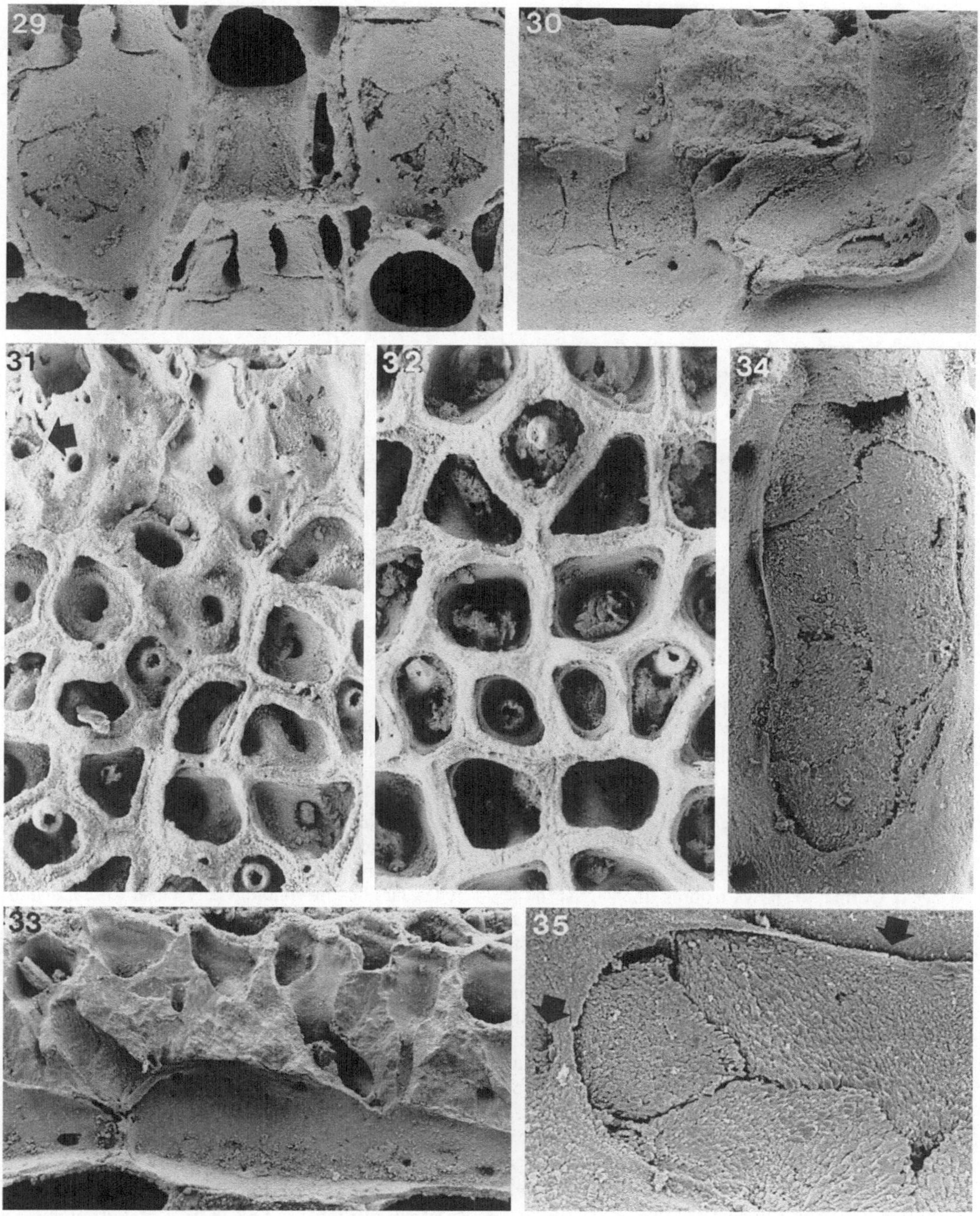

Figures 29, 30. *Diplobeisselina salebrosa* (Marsson) n. comb., Upper Maastrichtian, Stevns Klint, Denmark (Voigt Collection): 29. Zooidal interiors, showing costal fields and spiramen openings (x 89). 30. Longitudinal section through zooid showing costal spines (one apertural), spiramen tube, ovicell, and peristome (x 104). 31-34. *Scoriopora nodensis* (Brydone) n. comb., Upper Maastrichtian, Basbeck, Germany (Voigt Collection): 31. Zooidal surfaces near branch tip; two zooids at top with little secondary calcification, spiramen openings visible as simple holes, not yet tubiform, adventitious avicularium (arrowed) adjacent to one peristome; proximally, other zooids have frontal ridges and spiramen tubes (x 74). 32. Symmetrical arrangement of calyciform depressions in relation to spiramen tubes; actual peristomes not identifiable (x 74). 33. Longitudinal section through zooid and peristome in relation to adjacent ridges and depressions (x 60). 34. Close-up of proximal end of another costal field; entrances to costal lumina arrowed – small pore at left, broad slit at top (x 347). 35. Interior of zooid showing costal field with spiramen (x 161).

the distal end of the umbonuloid field and is longitudinally oval in shape (Fig. 21).

Eschara disticha Goldfuss, 1826 is a very closely related species that can be confused with *P. pachyderma*. External distinguishing features include the proportionately wider spatulate avicularia and single peristomial avicularia (see Voigt 1967 for a full discussion). One way to discriminate between the two species is to fracture colony fragments to view zooidal interiors – the spiramen in *E. disticha* opens in the centre of the umbonuloid field instead of near the distal end. The two species together share characters that are so distinct they may be regarded as congeneric but not coinciding with any existing genus. We therefore propose a new genus for them, *Pachydermopora*, with type species *P. pachyderma* Marsson.

2.8 *Porina nodensis* Brydone, 1930

Porina nodensis (Campanian, Isle of Wight, U.K.) and *P. scoriacea* Brydone, 1930 (Campanian, Meudon, France) are two similar species whose external characters almost defy interpretation. Colonies were erect, bilamellar, and branching, and somewhat lensoidal in cross section rather than flattened. The mature colony surface is made up of a reticulum of ridges circumscribing calyciform depressions that correspond to loci of orifices, spiramina, avicularia, and areolar-septular pore openings. It is not externally obvious which is which except for spiramina (Figs 31, 32); these open at the tips of narrow tubes that occur in some depressions. The ridges comprising the reticulum are narrowest towards the branch apex, wider and of roughly equal thickness proximally.

An exceptionally well-preserved colony from Basbeck, Germany shows zooids that must have been very near a branch tip. They have peristomial orifices, developing reticulum ridges, spiraminal pores, and small oval avicularia with complete crossbars (Fig. 31), thus clarifying the spatial relationships between these elements in a single zooid. Further back from the growing edge, however, it remains difficult to discern peristomial orifices. One longitudinal section (Fig. 33) gives evidence that some of the depressions must correspond to open orifices but, as seems likely, many are probably closed by calcification. In addition to small oval avicularia, large oval avicularia, also with crossbars, are occasionally present, notably along branch margins.

The interior of the frontal shows a costal field (Fig. 34). The costae, about seven in number, are relatively large and flabellate; lumen pores, a small opening in the case of the most proximal costa, broad slits in the case of the large costae, are in evidence adjacent to the outer perimeter of the costal field (Fig. 35). An additional pair of spine elements, medially fused, is separated from the costal field by a large hole representing the internal opening of the spiramen. Ovicells have not yet been encountered in fractured specimens.

Brydone (1930) regarded *P. nodensis* as ancestral to *P. scoriacea*; the latter species has narrower ridges and shallower depressions, with "tiny pits here and there in the walls [ridges], especially at their junctions". Both species are highly distinctive and clearly congeneric, but do not fit in any existing genus. Accordingly, we propose *Scoriopora*, new genus, with *P. nodensis* as type species.

3 DISCUSSION AND CONCLUSIONS

3.1 *Systematic relationships*

It is clear that none of the putative porinids from the Cretaceous and Paleocene considered here correspond to either *Porina* or the Porinidae. Recent *Porina gracilis* has a pseudoporous lepralioid frontal shield and typical ascus sac. On the other hand, among the genera included in the family by Bassler (1935, 1953), *Beisselina*, *Gastropella*, *Pachydermopora*, and *Pachythecella* have umbonuloid shields (inferred in the latter genus) and *Columnotheca* has a primary spinocystal shield overlain by a secondary shield layer that in life would have had an outer hypostegal coelom. Additional "porinid" genera considered here – *Diplobeisselina* Voigt and new genus *Scoriopora* – also have primary spinocystal shields. Thus the family Porinidae as presently conceived is polyphyletic.

How may these genera be classified? Umbonuloid frontal shields are currently distributed among the superfamilies Arachnopusioidea, Adeonoidea, Pseudolepralioidea, Lepralielloidea, and Chlidoniopsoidea (Gordon 2001). Only the Lepralielloidea includes families morphologically close to *Beisselina*, *Gastropella*, *Pachydermopora*, and *Pachythecella* in their suites of characters. Species of *Pachythecella* here moved to *Frurionella* because of lack of a spiramen should be included, with that genus, in the family Lepraliellidae.

Pachythecella s.s., *Gastropella*, *Beisselina*, and *Pachydermopora*, on the other hand, have spiraminal tubes, a character found in the Lepralielloidea only in the Tessaradomidae. In the type genus *Tessaradoma* Norman, 1869 the spiramen opens at the base of the peristome above the primary orifice and operculum, a situation that obtains for *Gastropella*, which may be readily included in this family. In *Beisselina* and *Pachydermopora* the spiramen opens somewhat more proximally, in the umbonuloid shield itself. Ovicells in both *Tessaradoma* and *Beisselina* are concealed when fully formed, opening into the base of the peristome above the primary orifice. While colonial morphology in *Beisselina* and especially *Pachydermopora* appears very different from that in *Tessaradoma*, this difference is chiefly in size and robustness rather than in zooidal characters and the spatial relationships of the various zooidal components. If gracile, four-serial forms of *Beisselina* and *Pachydermopora* were possible they would not then be so different from *Tessaradoma*. We prefer to take a conservative approach and include *Beisselina* and *Pachydermopora* in the family Tessaradomidae rather than create yet another family of Lepralielloidea. Admittedly, the stratigraphic range of Tessaradomidae would traverse the K-T boundary, but this is already the case with the umbonuloid-shielded families Arachnopusiidae, Romancheinidae, and Lepraliellidae (see also Voigt 1985, Taylor 1993).

"Porinids" with a primary spinocyst are more difficult to place. The genera examined here, viz. *Columnotheca*, *Diplobeisselina*, and *Scoriopora* appear to belong to a single clade that would have had a cribrimorph ancestor. Among currently recognised cribrimorph (acanthostegan) superfamilies, only one, the Bifaxarioidea, includes species with mixed frontal shields in which a primary spinocyst is overlain by an additional calcified wall layer. The phylogenetic origin of this extra wall layer is not known in all taxa in which it occurs, but one model derives it from kenozooids (Gordon & Voigt 1996, Gordon 2000). In the Bifaxarioidea, the primary costal field is either coextensive with the autoozoecial opesia that it overarches (Bifaxariidae) or is reduced in area and expressed only in the distal/suboral parts of the frontal shield (Platyglenidae and Mixtopeltidae). The outermost shield layer is interpreted as being exterior-walled on the underside (where it abuts on the costal field) and interior-walled (with a hypostegal coelom) on its frontal side, i.e. structurally equivalent to an umbonuloid layer. This layer is smooth and imperforate in the Platyglenidae, dimpled and imperforate in the Mixtopeltidae, and smooth to granular with small perforations (of uncertain homology) in the Bifaxariidae. Avicularia are absent in mixtopeltids, some bifaxariids, and many platyglenids.

In contrast, the outer shield layer in *Columnotheca*, *Diplobeisselina*, and *Scoriopora* is spiraminate and densely perforated by foramina of a variety of sizes that are frontal expressions of areolar-septular pore canals. Adventitious avicularia are generally also present. I propose a new family, Scorioporidae, and superfamily, Scorioporoidea, to accommodate these three genera.

3.2 *Diagnoses of new taxa*

Pachydermopora gen. nov. (Tessaradomidae): Colony erect, bilamellar, branching. Thickly calcified, zooidal boundaries not frontally evident, the surface densely dimpled. Zooidal peristomes arranged quincuncially, associated with one or more adventitious avicularia with crossbars. Larger spatulate avicularia also present. A single proximal spiraminal pore proximal to orifice, small, non-projecting, opening into umbonuloid shield proximal to base of peristomial shaft. In older parts of colonies peristomes and avicularia concealed by secondary calcification, forming nodularities.

Type species: *Porina pachyderma* Marsson, 1887.

Range: Campanian-Maastrichtian.

Superfamily Scorioporoidea superfam. nov.: Colony erect, bilamellar or cylindrical, branching. Zooids whorled or in quincunx, spiraminate or bispiraminate. Colonial surface densely foraminate, the foramina corresponding to peristomes, frequently obvious, as well as spiraminal pores and the frontal expressions of areolar-septular pore canals, some of the latter being loci for small or large avicularia with crossbars. Primary frontal shield comprising a costal field of relatively few, large costae; spiraminal tube(s) opening at distal end of costal field or at base of peristome. Ovicells, where seen, opening into base of peristome.

Range: Campanian-Danian.

Family Scorioporidae, fam. nov.: Diagnosis as for superfamily.
Range: Campanian-Danian.

Scoriopora gen. nov.: Colony erect, bilamellar, branching. Colony surface a reticulate network of raised ridges that conceal zooids boundaries. Calyciform depressions defined by ridges corresponding to peristomial openings, loci for avicularia and spiraminal pores (unispiraminate), and especially areolar-septular pores. Primary frontal shield comprising relatively large and few fused costae. Spiramen opening into distal end of costal field proximal to apertural bar. Ovicells not seen.

Type species: *Porina nodensis* Brydone, 1930.

Range: Campanian-Maastrichtian.

ACKNOWLEDGMENTS

I am extremely grateful to Professor Dr Ehrhard Voigt for the loan of specimens in his collection (Universität Hamburg), for donated specimens, for use of photographs, and for advice and information. Dr Agnès Rage of the MNHN (Muséum national d'Histoire naturelle, Paris) facilitated the loan of d'Orbigny material and Dr Paul Taylor (The Natural History Museum, London) scanned some uncoated specimens.

REFERENCES

Bassler, R.S. 1934. Notes on fossil and Recent Bryozoa. *Journal of the Washington Academy of Sciences* 24: 404-408.

Bassler, R.S. 1935. Bryozoa. *Fossilium catalogus* 1(67): 1-229.

Bassler, R.S. 1953. Bryozoa. In R.C. Moore (ed.) *Treatise on invertebrate paleontology. Part G.* Boulder & Lawrence: Geological Society of America and University of Kansas Press.

Berthelsen, O. 1962. Cheilostome Bryozoa in the Danian deposits of East Denmark. *Geological Survey of Denmark, ser. 2,* 83: 1-290.

Bock, P.E. 1982. Bryozoans (Phylum Bryozoa). In S.A. Shepherd & I.M. Thomas (eds), *Marine Invertebrates of Southern Australia Part I*: 319-394. Adelaide: Government Printer.

Brydone, R.M. 1930. *Further Notes on New and Imperfectly Known Chalk Polyzoa. Part II. (*Vincularia, Onychocella, Rhagasostoma, Porina, *etc.).* London: Dulau & Co.

Buge, E. 1975. Les Bryozoaires de l'Oligocène du Bassin de Paris et leurs relations avec les faunes des bassins oligocènes d'Europe. *Mitteilungen aus dem Geologisch-Paläontologisches Institut der Universität Hamburg* 44: 45-58.

Canu, F. 1913. Études morphologiques sur trois nouvelles familles de Bryozoaires. *Bulletin de la Société Géologique de France,* sér. 4, 13: 132-147.

Canu, F. & Bassler, R.S. 1917. A synopsis of American Early Tertiary cheilostome Bryozoa. *Bulletin U.S. National Museum* 96: 1-87.

Canu, F. & Bassler, R.S. 1925. Molluscoidea, Bryozoa. In B. Wade (ed.), *The Fauna of the Ripley Formation on Coon Creek, Tennessee. Professional Papers of the U.S. Geological Survey* 137: 741-761.

Goldfuss, G.A. 1826-33. *Petrefacta Germaniae.* Düsseldorf: Arnz & Comps.

Gordon, D.P. 1993. Bryozoan frontal shields: studies on umbonulomorphs and impacts on classification. *Zoologica Scripta* 22: 203-221.

Gordon, D.P. 2000. Towards a phylogeny of cheilostomes—morphological models of frontal wall/shield evolution. In A. Herrer-Cubilla & J.B.C. Jackson (eds), *Proceedings of the 11th International Bryozoology Association Conference*: 17-37. Panama: Smithsonian Tropical Research Institution.

Gordon, D.P. 2001. Genera and subgenera of Cheilostomata. Interim classification (working classification for Treatise on Invertebrate Paleontology). Author's compilation. [Unpublished.]

Gordon, D.P. & d'Hondt, J.-L. 1997. Bryozoa: Lepraliomorpha and other Ascophorina, mainly from New Caledonian waters. *Mémoires du Muséum National d'Histoire Naturelle* 176: 9-124.

Gordon, D.P. & Voigt, E. 1996. The kenozooidal origin of the ascophorine hypostegal coelom and associated frontal shield. In D.P. Gordon, A.M. Smith & J.A. Grant-Mackie (eds), *Bryozoans in Space and Time*: 89-107. Wellington, NIWA.

Jullien, J. 1883. Bryozoaires. Espèces draguées dans l'océan Atlantique en 1881. *Bulletin de la Société Zoologique de France* 7: 497-529.

Koschinsky, C. 1885. Ein Beitrag zur Kenntniss der älteren Tertiärschichten des südlichen Bayerns. I. Abtheilung. Cheilostomata. *Palaeontographica* 32: 1-73.

Lamarck, J.B.P.A. de M. de 1816. *Histoire naturelle des Animaux sans Vertèbres.* Paris: Verdière.

Lang, W.D. 1917. The genotypes of certain polyzoan genera. *Geological Magazine, n.s.,* Decade 6, 4: 169-174.

Levinsen, G.M.R. 1909. *Morphological and Systematic Studies on the Cheilostomatous Bryozoa.* Copenhagen: Nationale Forfatteres Forlag.

Levinsen, G.M.R. 1925. Undersøgelser over Bryozoerne I den danske Kridtformation. *Kongelige Danske Videnskabernes Selskabs Skrifter,* ser. 8, 7: 281-445.

Marsson, T. 1887. Die Bryozoen der weissen Schreibkreide der Insel Rügen. *Paläontologische Abhandlungen* 4: 1-112.

Norman, A.M. [Last report on dredging among the Shetland Islands.] Class Polyzoa. *Report of the British Association for the Advancement of Science* 1868: 303-312.

Orbigny, A.D. d' 1850. *Prodrome de Paléontologie stratigraphique universelle des Animaux mollusques & rayonnés.* Volume 1. Paris: Victor Masson.

Orbigny, A.D. d' 1851-54. *Paléontologie française. Description des Mollusques et Rayonées fossiles. Terrain Crétacé, V. Bryozoaires.* Paris: Victor Masson.

Pergens, E. & Meunier, A. 1886. La faune des Bryozoaires garumniens de Faxe. *Annales de la Société royale Malacologique de Belgique* 21: 187-242.

Reuss, A.E. 1848. Die fossilen Polyparien des Wiener Tertiärbeckens. *Haidingers Naturwissenschaftliche Abhandlungen* 2: 1-109.

Taylor, P.D. 1993. Bryozoa. In M.J. Benton (ed.), *The Fossil Record 2*: 465-489. London: Chapman & Hall.

Voigt, E. 1924. Über neue Bryozoen aus Daniengeschieben Anhalts. *Paläontologische Zeitschrift* 6: 1-13.

Voigt, E. 1930. Morphologische und stratigraphische Untersuchungen über die Bryozoenfauna der oberen Kreide. *Leopoldina* 6: 379-579.

Voigt, E. 1951. Das Maastricht-Vorkommen von Ilten bei Hannover und seine Fauna mit besonderer Berücksichtigung der Gross-Foraminiferen und Bryozoen. *Mitteilungen aus dem Geologischen Staatsinstitut in Hamburg* 32: 15-109.

Voigt, E. 1959. Über *Fissuricella* n.g. (Bryozoa foss.). *Neues Jahrbuch für Geologie und Paläontologie. Abhandlungen* 108: 260-269.

Voigt, E. 1967. Oberkreide-Bryozoen aus den asiatische Gebieten der UdSSR. *Mitteilungen aus dem Geologische Staatsinstitut in Hamburg* 36: 1-34.

Voigt, E. 1968. Zwei homoeomorphe Arten der Bryozoengattung *Columnotheca* Marsson. *Lethaia* 1: 382-401.

Voigt, E. 1985. The Bryozoa of the Cretaceous-Tertiary boundary. In C. Nielsen & G.P. Larwood (eds), *Bryozoa: Ordovician to Recent*: 329-342. Fredensborg, Olsen & Olsen.

Voigt, E. 1987. Die Bryozoen des klassischen Dano-Montiens von Mons. *Mémoires pour servir à l'Explication des Cartes Géologiques et Minières de la Belgique* 17: 1-161.

Waters, A.W. 1905. Notes on some Recent Bryozoa in d'Orbigny's collection. *Annals and Magazine of Natural History,* ser. 7, 15: 1-16.

Wiesemann, G. 1963. Untersuchungen an der Gattung *Beisselina* Canu 1913 und ähnlichen Bryozoen (Maastrichtien, Danien, Montien). *Mitteilungen aus dem Geologische Staatsinstitut in Hamburg* 32: 5-70.

Bryozoan Studies 2001, Wyse Jackson, Buttler & Spencer Jones (eds)
© 2002 Swets & Zeitlinger, Lisse, ISBN 90 5809 388 3

Kubaninella: a new genus of Adeonidae (Bryozoa: Cheilostomata) from the Western Kamchatka shelf of the Sea of Okhotsk

Andrei V. Grischenko & Shunsuke F. Mawatari
Systematics and Evolution, Division of Biological Sciences, Graduate School of Science, Hokkaido University, Sapporo 060–0810, Japan

ABSTRACT: A new encrusting adeonid bryozoan *Kubaninella relicta* gen. et sp. nov. (Bryozoa: Cheilostomata) is described from the Western Kamchatka shelf of the Sea of Okhotsk. The new species clearly differs from species of the encrusting genus *Reptadeonella* Busk, 1884 in having an avicularium asymmetrically associated with a suboral mucro, a tatiform ancestrula, and no gonozooid. It is the second most northerly species of adeonid known, and the first record of the family in the Sea of Okhotsk. Early astogeny suggests that *Kubaninella* may be more closely related to *Adeonellopsis* than to *Reptadeonella*, which it resembles in colonial characters.

1 INTRODUCTION

The cheilostome family Adeonidae is very distinctive. In common with other families of umbonuloid-shielded ascophorans (see Gordon 1993), all species have calcareous frontal shields that overarch the membranous frontal walls of zooids. In contrast with other umbonuloid families, however, adeonids have a suite of characters that readily separates them visually during taxonomic sorting (Harmer 1957, Cook 1973). Most genera and species are rigidly erect and dichotomously branching, although some are encrusting. Zooidal frontal shields are well calcified, with well-developed series of areolar septular channels along the margins, a central simple or compound spiramen, single or paired suboral avicularia, and often vicarious and small adventitious avicularia. Brooding is internal and morphologically distinctive gonozooids may occur. Additionally, adeonids have a distinctive blocky crystal ultrastructure (Lidgard 1996). This is readily seen by SEM in the frontal shield at moderate to high magnifications, but is also evident as an obvious granularity of the skeletal surface under the light microscope. Interestingly, this type of crystal ultrastructure is also present in Adeonella (family Adeonellidae) (Sandberg 1983, Lidgard 1996), a genus that shares many of the features of typical adeonids but has instead a lepralioid frontal shield (Cook 1973). The status of some fossil genera attributed to the Adeonidae needs re-evaluating; nominally, there are 14 genera, of which three are two-dimensionally encrusting (D.P. Gordon 2001, pers. comm.).

In September 1992, the first author collected many samples of bryozoans during a short voyage of the Middle Fishery Refrigerator Trawler *Rodino* to the Western Kamchatka shelf of the Sea of Okhotsk. Two papers have already been published describing new endemic taxa from the collection (Grischenko et al. 1999, Taylor & Grischenko 1999). Here we describe a notable new genus and species of encrusting adeonid that not only augments our understanding of the range of expression of morphological characters within the family, but also increases its known geographic range.

2 MATERIALS AND METHODS

Specimens collected were initially preserved by drying. Colonies were cleaned in sodium hypochlorite solution, rinsed with tap-water and dried in air prior to their measurement under a binocular microscope (NIKON: SNZ–10). The dried colonies were coated with Pd–Pt by an ion sputter (HITACHI: E–1030) and observed under a scanning electron microscope (HITACHI: S–2380N) at 10–20 kV accelerating voltage.

The specimens described here are deposited in the Zoological Institute, Russian Academy of Sciences (ZIRAS), Saint Petersburg; the Zoological Institute, Faculty of Science, Hokkaido University (ZIHU), Sapporo; The Natural History Museum (NHM), London; the National Institute of Water and Atmospheric Research (NIWA), Wellington; and the National Science Museum (NSMT), Tokyo.

3 SYSTEMATICS

Order Cheilostomata Busk, 1852
Superfamily Adeonoidea Busk, 1884
Family Adeonidae Busk, 1884
Genus *Kubaninella* gen. nov.

Diagnosis: Colony encrusting, multiserial, sheet-like. Zooids with umbonuloid frontal shields. Avicularia small, suboral, on distolateral slope of suboral mucro. Numerous marginal tubular areolar pore canals centripetally encroaching on the frontal shield. Frontal spiramen in a central depression, initially single, remaining single or becoming compound in many zooids. Vicarious avicularia, oral spines and kenozooids absent. Basal pore chambers present. Ancestrula tatiform with three periancestrular zooids budding proximolaterally.

Type species: *Kubaninella relicta* sp. nov.

Etymology: The genus is named in memory of Russian bryozoologist, Dr Andrei A. Kubanin (1949–1999), who made a significant contribution towards investigation of fouling bryozoans in the Russian Far Eastern Seas.

Kubaninella relicta sp. nov.
(Figure 1A–M)

Material examined: **Holotype**: ZIRAS 1/49296 (colony on a pebble), sorted from a crab trap by A.V. Grischenko on the Middle Fishery Refrigerator Trawler *Rodino*, 12 September 1992. Type locality 57°36.2'N, 156°09.0'E, Sea of Okhotsk, Western Kamchatka Shelf, about 32 km from Cape Hayryuzova, depth 78–81 m, bottom silt, sand, gravel. **Paratypes**: ZIRAS 2/49297 (colony on pebble), same data as for holotype. ZIRAS 3/49298 (colony on a pebble), same data as for holotype. NHM 1997.7.28.1. (colony on gravel), same data as for holotype. NIWA P–1214 (one intact colony), same data as for holotype; P–1215 (several pieces from a large colony on a shell of *Chlamys*), same data as for holotype. ZIHU–02012 (several fragments of intact colonies), ZIHU–02013 (ancestrulate colony), same data as for holotype. NSMT Te–676 (one colony on a pebble), same data as for holotype.

Additional material examined: several colonies stored in AVG personal collection.

Measured specimens: ZIHU–02012, 02013.

Description: Colony encrusting, multiserial, sheet-like, bright yellow with pale growing margin of developing zooids when alive, and beige, light yellow or white when dry. Colony of irregular form, maximally attaining 33 mm length in any direction, usually not exceeding 15–20 mm in size.

Zooids hexagonal, oval, or pyriform, relatively large, arranged somewhat regularly (Fig. 1A). Newly budded zooids demarcated (Figs 1K, L). Boundaries between zooids occluded in older regions of colony by secondary calcification (Fig. 1B).

Frontal shield umbonuloid, finely granular, convex in young zooids, flattened in ontogenetically older zooids. Frontal pores 39–54 in number, arranged in several rows, representing openings of tubular areolar pore channels which connect main coelom to hypostegal coelom (Figs 1L, M). Frontal spiramen in a central depression, initially single; either remaining single or becoming compound by development of cross partitions, usually subdivided into two, or sometimes three, pores (Figs 1D–G). Secondary orifice usually semicircular with straight proximal margin, sometimes rounded or oval. Suboral mucro forming frontal bulge. Adventitious avicularium usually present but can be absent in old zooids, small, suborally situated on left or right distolateral slope of mucro, directed frontally (Fig. 1C). Avicularian mandible elongate-triangular with blunt tip. Vicarious avicularia absent.

Basal pore-chambers relatively large, 7–11 in number, in the lateral and transverse walls of zooid (Fig. 1M). Two basal pore plates present in distal half of zooid. No oral spines, except on ancestrula and primary zooids, or other heterozooids. No ovicells or obvious gonozooids. Operculum brown. Preserved polypides light brown.

Ancestrula and early astogeny: The ancestrula (Fig. 1H) is tatiform, subcircular, with a narrow gymnocyst and large oval opesia, surrounded by eight jointed hollow vertical spines up to 0.43 mm long. The colour of the frontal membrane is light brown in dry specimens. Three periancestrular zooids (Fig. 1I) differentiate proximally and proximolaterally from the ancestrula. They are small, oval, and clearly demarcated, with thin convex frontal walls and rounded orifices. One to two pairs of oral spines are present at the distal margin of the periancestrular zooids and of zooids of the second-budded generation (Figs 1I, J). Spines are hollow and directed distolaterally. Several other zooids are budded in the opposite direction later, distally and distolaterally from the ancestrula (Fig. 1J). These zooids differ from the first three periancestrular buds in being as large as autozooids. The ancestrula is unrecognizable in older regions of the colony, partly because of secondary calcification (Fig. 1J). Prior to this, however, the ancestrular zooid is modified by regeneration, during which it evidently acquires a frontal shield and loses its spines, thereby resembling periancestrular zooids.

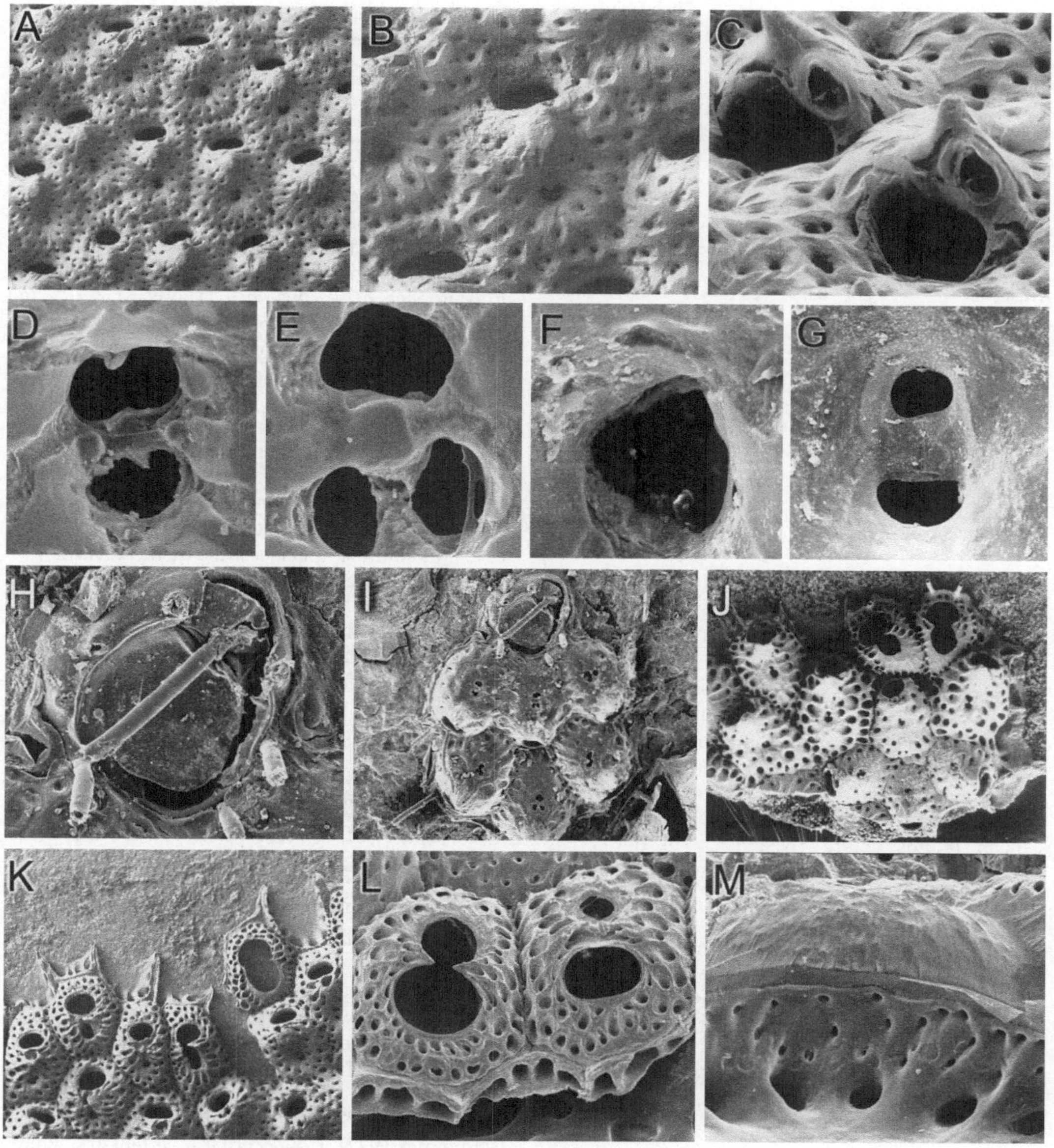

Figure 1A–M: *Kubaninella relicta* gen. et sp. nov. (NHM 1997.7.28.1. for A, B, K; NIWA, P–1215 for C, G; NIWA, P–1214 for J; ZIHU–02012 for D-F, L, M; ZIHU–02013 for H, I). A, zooids in mature region of a colony, x 31; B, enlarged view of mature zooids, showing numerous series of marginal areolar pores, and confluent frontal shields of adjacent zooids, x 67; C, two mature zooids with typical shape of the orifice, mucro, and adventitious suboral avicularium, x 101; D–G, frontal view of spiramina: D, E, compound spiramina x 450; F, simple, x 600; G, spiramen with intact epithecal membrane, x 257; H, ancestrula, partly obscured by a broken spine, x 113; I–J, early astogeny: I, ancestrula with periancestrular zooids, bearing a hollow spines, differentiated proximally and proximolaterally, x 37; J, ancestrula with apparently regenerated frontal shield (zooid in central position) surrounded by three first proximal and proximolateral periancestrular buds and several succeeding zooids developing distally, in the opposite direction, x 26; K, growing edge of a colony, showing developing zooids at different stages of differentiation, x 23; L, two young zooids at the growing margin shoving development of the umbonuloid frontal shield with numerous tubular areolar pore channels encroaching on the frontal shield to open into the hypostegal coelom, x 60; M, inner view of zooidal lateral and frontal walls, showing epistegal cavity, ring scar, basal pore-chambers, and numerous openings of tubular areolar pore channels from the hypostegal coelom, x 195.

Table 1. Measurements on *Kubaninella relicta.* (in mm, except for Nz and N).The following abbreviations are used for measurements: L: Length; W: width; N: number; S.D.: standard deviation; Nz: number of zooids.

		Mean	Range	S.D.	Nz
Zooids	L	0.77	0.62–0.95	0.071	30
	W	0.65	0.40–0.82	0.099	30
Orifice	L	0.21	0.17–0.26	0.015	30
	W	0.15	0.10–0.21	0.021	30
Avicularia	L	0.08	0.06–0.10	0.012	10
Ancestrula	L	0.51	-	-	1
	W	0.57	-	-	1
Ancestrular aperture	L	0.38	-	-	1
	W	0.32	-	-	1
Priancestrular zooids	L	0.57	0.53–0.62	0.037	6
	W	0.45	0.41–0.50	0.033	6
Priancestrular zooids orifice	L	0.17	0.15–0.20	0.020	6
	W	0.13	0.12–0.14	0.005	6
Priancestrular zooids spines	L	0.25	0.09–0.37	0.117	3

Remarks: *Kubaninella* gen. nov. is assigned to the family Adeonidae based on the following combination of characters: umbonuloid frontal shield with single or compound spiramen in a central depression; suboral avicularium situated in the distolateral slope of the mucro; and characteristic granular-blocky skeletal microstructure (best seen in developing zooids). Among the 14 genera provisionally classified in the family Adeonidae (D.P. Gordon 2001, pers. comm.), most representatives have erect colonies with diverse zooidal polymorphism. Only three genera have encrusting species — *Anarthropora* Smitt, 1868, *Triporula* Canu & Bassler, 1927, and *Reptadeonella* Busk, 1884. At least one species of *Adeonellopsis* [*A. baccata* (Hutton, 1878)] is encrusting but can also develop erect lobes (Gordon & Parker 1991). *Anarthropora* and *Triporula* both lack a spiramen, having instead scattered stellate pores. In colonial characters, *Kubaninella* most resembles *Reptadeonella*, a genus with eight Recent species ranging in the tropical Indo–Pacific to the Philippines, western Mexico to Galapagos, Gulf of Mexico–Florida, Brasil, and amphi–North Atlantic–Mediteranean. *Reptadeonella* and *Kubaninella* share such zooidal characters as: one or more spiraminal pores in a central depression, a suboral aviculiferous mucro, numerous areolar septular pores that encroach on the centre of the frontal shield, and the absence of large vicarious avicularia.

On the other hand, *Kubaninella* is clearly distinguished from *Reptadeonella* in many features. In *K. relicta*, the avicularium is asymmetrically set and frontally concealed in the strongly calcified umbo, its palate facing distally while the mandible is directed frontally, whereas in *Reptadeonella* species the mucro is relatively small and the avicularium is situated on its proximal side. The zooids in many species of *Reptadeonella* are clearly demarcated by distinct grooves between neighbouring zooids even in old central region of the colony. In contrast, the boundaries between mature zooids of *K. relicta* tend to be indistinguishable because of secondary calcification. Ephemeral kenozooids, found in the differentiating colony margin of which is characteristic for growing edge of *R. insidiosa* (Hayward & Ryland 1979), are absent in *Kubaninella*. Another difference between the two genera concerns the brooding chambers. In *Reptadeonella*, embryos are brooded in enlarged gonozooids, which are apparently absent in *Kubaninella*.

Early astogeny is significantly different in the two genera. Where known, the larva in *Reptadeonella* species metamorphoses into a complex of six, symmetrically grouped, ancestrular (Hayward & Ryland 1979). In contrast, the ancestrula of *K. relicta* is tatiform with three periancestrular zooids budded proximally and proximolaterally. Further, periancestrular zooids have ephemeral, articulated, distal oral spines — such spines have rarely, if ever, been reported in adeonid zooids.

Ecology and distribution: *Kubaninella relicta* is endemic to the Sea of Okhotsk, and presently known only from a single locality (57°36.2'N, 156°09.0'E) in the Western Kamchatka shelf at a depth of 78–81 m. In the area of occurence of *K. relicta*, the annual temperature range is –2.5°C in winter to 0.4°C in summer. Salinity ranges from 33.0–33.2 ppt at 50 m to 33.25 ppt at 200 m. Current velocities at the seafloor are generally moderate to strong. *Kubaninella relicta* occurs in an area of mixed hard and soft substrata, in a biocenosis of the sponges *Haliclona gracilis* and *Myxilla incrustans incrustans* at 40–85 m depth (Tsalkina 1969). A rich encrusting benthos occupies up to 11% of the sea bottom square in the sampled habitat (Savilov 1961). Bryozoan biomass here comprises 8.2% of total benthic biomass (Izyumova 1977). Co–occuring invertebrates mainly include sponges, hydroids, barnacles, serpulid and spirorbid polychaetes, asteroids, and numerous other bryozoan species. Examination of about 60 colonies collected, indicates that *K. relicta* settles on a limited range of hard substrata including gravel (86%) and broken shells of the bivalve *Chlamys* (14%).

Owing to solid attachment to the substratum by all zooids, and the apparent impossibility of elevating its growing margin, the shape of *K. relicta* colonies accurately follows irregularities of surface microrelief. As a result, in case of boundary interactions, several other bryozoan species not only arrest colony development of *K. relicta* but also overgrow its marginal zooids. These species include

Alcyonidium sp., *Bidenkapia alaskensis* Osburn, 1950, *Cheilopora sincera* (Smitt, 1868), *Escharella dijmphnae* (Kluge, 1929), *Parasmittina trispinosa* (Johnston, 1838), *Porella compressa* (Sowerby, 1805), *Rhamphostomella scabra* (Fabricius, 1780), and *Smittina majuscula* (Smitt, 1868). Mutual arrest of growth was seen in an encounter with *Electra arctica* Borg, 1931.

4 DISCUSSION

The family Adeonidae is most abundantly (but not exclusively) represented in tropical and warm-temperate waters of the world, especially the Indo-Pacific. The most speciose genera are *Adeona* Lamouroux, 1812 and *Adeonellopsis* MacGillivray, 1886. In the Pacific Ocean, species of the latter range from southern Australia and New Zealand through Indonesia and the Philippines to Japan; the most northerly record of any adeonid is that of *Adeonellopsis tuberculata* (Ortmann, 1890) from Kolyuchinskaya Inlet in the Chukchi Sea (67°N) (Kluge 1962). Adeonids are rare in the eastern Pacific, although two species of *Reptadeonella* are known in the area of western Mexico to the Galapagos (Banta 1990). No Antarctic adeonids are known. In the Atlantic Ocean, *Bracebridgia subsulcata* (Smitt, 1873) is known from the Burdwood Bank (54°S); the most northerly occurring Atlantic adeonid is *Reptadeonella violacea*, which ranges to the Isle of Man (ca 54–55°N) (Hayward & Ryland 1999). *Kubaninella relicta* is the second most northerly occurring adeonid species, and the first record of the family for the Sea of Okhotsk.

The ancestrular stage is scarcely known in adeonids, with the exception of some encrusting species of *Reptadeonella* (e.g., Hayward & Ryland 1999, figs 70, 71). Erect adeonids become proximally calcified early in astogeny so that the ancestrula is never normally seen. Bock and Cook (2000) reported erect, proximally attached ancestrular regions in very small colonies of *Adeona* sp.; in contrast, they observed a tiny encrusting phase in an unidentified *Adeonellopsis* colony in which the ancestrula is encrusting and resembles post-ancestrular zooids. *Kubaninella* is the only known adeonid with a tatiform ancestrula and spines. In forming encrusting colonies and in some features of zooidal morphology it resembles *Reptadeonella*. On the other hand, early astogeny of *K. relicta* more closely resembles that in *Adeonellopsis*, i.e., a non-compound ancestrula that buds daughter zooids to one side — proximally in *Kubaninella*, distally in *Adeonellopsis*. The ancestrula in *Adeonellopsis* resembles later zooids (Bock & Cook 2000, fig. 14), but an earlier tatiform morphology followed by regenerative formation of an ancestrular frontal shield (as in *K. relicta*) cannot be ruled out. The earliest-known records of fossil adeonids from the northwestern Pacific are three Pliocene species of *Adeonellopsis* from Japan (Hayami 1971); a fourth species, and two species of *Adeona* are known from the Japanese Pleistocene (Kataoka 1961, Sakagami et al. 1980). Given the evidence of early astogeny, it seems more likely that the ancestry of *Kubaninella* may be found within *Adeonellopsis* than within *Reptadeonella*.

ACKNOWLEDGEMENTS

We are grateful to Dr Dennis P. Gordon, National Institute of Water & Atmospheric Research, Wellington, for his critical reading of the manuscript and discussion. Thanks are also extended to Dr Yoshinobu Nodasaka, School of Dentistry, Hokkaido University, for his kind assistance in scanning electron microscopy.

REFERENCES

Banta, W.C. 1990. A checklist of the Bryozoa of the Galapagos. *Proceedings of the Biological Society of Washington* 103: 789–802.

Bock, P.E. & Cook, P.L. 2000. Early astogeny of *Adeona* colonies. In A. Herrera Cubilla & J.B.C. Jackson (eds) *Proceedings of the 11th International Bryozoology Association Conference*: 161–167. Balboa, Republic of Panama: Smithsonian Tropical Research Institute.

Borg, F. 1931. On some species of *Membranipora*. *Arkiv för Zoologi* 22(4): 1–35.

Busk, G. 1852. An account of the Polyzoa, and sertularian zoophytes, collected in the voyage of the Rattlesnake, on the coasts of Australia and the Louisiade Archipelago. In J. MacGillivray (ed.), *Narrative of the Voyage of H.M.S. Rattlesnake.* Appendix 4: 343–402. London: T & W. Boone.

Busk, G. 1884. Report on the Polyzoa collected by H.M.S. Challenger during the years 1873–76. Part 1. The Cheilostomata. *Report on the Scientific Results of the Voyage of H.M.S. Challenger, Zoology* 10(30): 1–216.

Canu, F. & Bassler, R.S. 1927. Classification of the cheilostomatous Bryozoa. *Proceedings of the U.S. National Museum* 69(14): 1–42.

Cook, P.L. 1973. Preliminary notes on the ontogeny of the frontal body wall in the Adeonidae and Adeonellidae (Bryozoa, Cheilostomata). *Bulletin of the British Museum (Natural History), Zoology*, 25: 243–263.

Fabricius, O. 1780. *Fauna Groenlandica.* Hafniae & Lipsiae: I.G. Rothe.

Gordon, D.P. 1989. The marine fauna of New Zealand: Bryozoa: Gymnolaemata (Cheilostomida Ascophorina) from the western South Island continental shelf and slope. *New Zealand Oceanographic Institute Memoir* 97: 1–158.

Gordon, D.P. 1993. Bryozoan frontal shields: studies on umbonulomorphs and impacts on classification. *Zoologica Scripta* 22: 203–221.

Gordon, D.P. & Parker, S.A. 1991. Discovery and identity of 110-year-old Hutton Collection of South Australian Bryozoa. *Records of the South Australian Museum* 25: 121–128.

Grischenko, A.V., Gordon, D.P. & Taylor, P.D. 1999. A unique new genus of Cheilostomate Bryozoan with re-

versed-polarity zooidal budding. *Asian Marine Biology* 15: 105–117.
Harmer, S.F. 1957. The Polyzoa of the Siboga Expedition. Part 4. Cheilostomata Ascophora. II. *Siboga-Expeditie* 28D: 641–1147.
Hayami, T. 1971. Some Neogene Cheilostomata (Bryozoa) from Okinawa-jima. *Transactions and Proceedings of the Palaeontological Society of Japan, n.s.*, 82: 73–92.
Hayward, P.J. & Ryland, J.S. 1999. Cheilostomatous Bryozoa Part 2. Hippothooidea–Celleporoidea. Notes for the identification of British species. *Synopses of the British Fauna, n.s.* 14: 1–416.
Izyumova, E.A. 1977. Data on the ecology of Western Kamchatka Shelf Cheilostomata. Abstract of report. *1st All-Russian Marine Biology Conference*: 59–60. Vladivostok.
Johnston, G. 1838. *A History of British Zoophites*. Edinburgh, London & Dublin: W.H. Lizars.
Johnston, G. 1847. *A History of British Zoophytes*. 2nd edn. London: John van Voorst.
Kataoka, J. 1961. Bryozoan fauna from the "Ryukyu Limestone" of Kikai-jima, Kagoshima Prefecture, Japan. *Science Reports of the Tohoku University, ser. 2, Geology*, 32: 213–272.
Kluge, H. 1929. Die Bryozoen des sibirischen Eismeers. *Raboty Murmanskoi Biologicheskoi Stantsii* 3: 1–33.
Lamouroux, J.V.F. 1812. Extrait d'un mémoire sur la classification des Polypiers coralligènes non entièrement pierreux. *Nouveau bulletin scientifique de la Société Philosophique* 3: 181–188.
Lidgard, S. 1996. Zooid skeletal morphogenesis of some Australian and New Zealand *Adeonellopsis* (Cheilostomatida). In D.P. Gordon, A.M. Smith & J.A. Grant-Mackie (eds), *Bryozoans in Space and Time*: 167–177. Wellington: NIWA.
MacGillivray, P.H. 1886. Descriptions of new, or little-known, Polyzoa. Part IX. *Transactions and Proceedings of the Royal Society of Victoria* 22: 128–139.
Ortmann, A. 1890. Die japanische Bryozoenfauna. Bericht über die von Herrn Dr. L. Döderlein in Jahre 1880–81 gemachten Sammlungen. *Archiv für Naturgeschichte* 54: 1–74.
Osburn, R.C. 1950. Bryozoa of the Pacific coast of America. Part 1. Cheilostomata Anasca. *Allan Hancock Pacific Expeditions* 14: 1–269.
Sakagami, S., Arakawa, S. & Hayami, T. 1980. Check list and bibliography of Japanese Cenozoic fossil Ectoprocta (Bryozoa), 1935–1978. In *Professor Saburo Kanno Memorial Volume*: 317–338.
Sandberg, P.A. 1983. Ultrastructure, mineralogy, and development of bryozoan skeletons. In Robison, R.A. et al. (eds), *Treatise on Invertebrate Paleontology. Part G, Bryozoa Revised*: 238–286. The Geological Society of America & The University of Kansas: Boulder & Lawrence.
Savilov, A.N. 1961. Ecological description of Sea of Okhotsk bottom communities. *Trudy Instituta Okeanologii* 46: 3–34.
Smitt, F.A. 1868. Kritisk förteckning öfver Skandinaviens Hafs-Bryozoer, IV. *Öfversigt af Kongliga Vetenskaps-Akademiens Förhandlingar* 25: 3–230.
Smitt, F.A. 1873. Floridan Bryozoa, collected by Count L. F. de Pourtales. Part II. *Kongliga svenska Vetenskaps-Akademiens Handlingar* 11(4): 1–83.
Sowerby, J. 1805. *The British Miscellany* 8(1): 1–131. [Privately published.]
Taylor, P.D. & Grischenko, A.V. 1999. *Rodinopora* gen. nov. and the taxonomy of fungiform cyclostome bryozoans. *Species Diversity* 4 (1): 9–33.
Tsalkina, A. V. 1969. Characteristices of the epifauna of the west Kamchatka shelf In *Problems of commercial hydrobiology. Proceedings of All-Union Scientific Research Institute of Marine Fishery Farm and Oceanography (VNIRO)*, 65: 248–257.

Bryozoan Studies 2001, Wyse Jackson, Buttler & Spencer Jones (eds)
© 2002 Swets & Zeitlinger, Lisse, ISBN 90 5809 388 3

Partitioning phenotypic variation: implications for morphometric analyses (Bryozoa)

Steven J. Hageman
Department of Geology, Appalachian State University, Boone, NC 28608, USA

Micha Bayer
Department of Systematics and Evolution, Royal Botanic Garden Edinburgh, 20A Inverleith Row, Edinburgh EH3 5LR, UK

Christopher D. Todd
Gatty Marine Laboratory, School of Environmental and Evolutionary Biology, University of St. Andrews, St. Andrews, Fife KY16 8LB, UK

ABSTRACT: Using clonal replicates of *Electra pilosa* (L.) colonies grown in controlled environmental conditions, it is possible to partition sources of morphological variation of skeletal hard parts into their Genotype, Environment and GxE interaction components. Preliminary studies showed these factors to account for only about 40% of variation among measured zooecial characteristics, resulting in a Residual of 60%. Here we accounted for additional, within colony sources of variation in an ANOVA models with a nested factorial design: Patches of zooecia within replicate colonies and Columns and Rows of zooecia within Patches. We analyzed data for five zooecial characteristics (opesia and zooecia width and length and spine number). The large Residual variance (among zooecia – within colonies) was reduced by two-thirds, to < 20% when sources of within colony variation were accounted for in ANOVA models. Significant environmental variation can be detected (Tank effect ~ 2–3%), even in experiments designed to minimize it as a main factor. Patch effect within colonies was highly significant, and often was greater than Genotype, ~ 25–35%. Failure to account for Patch effects can bias conclusions regarding other factors. If zooecia are sampled randomly and widely across a colony and are pooled, Patch effect variance will be allocated to the Residual. However, if zooecia are all collected from a single, small patch on a colony, or as single small colony fragment (e.g. fossil), the variance attributable to Patch is incorrectly and differentially allocated to Genotype, Environment, and GxE interaction. Row and Column effects within patches were small, ~ 5–10%, but significant. If Row and Column are omitted as factors in ANOVA, their variance does not affect the outcome of other factors and is allocated to the Residual. Morphometric data should not be collected from single patches or small fragments of bryozoans, because the within colony Patch effect will artifactually and significantly inflate the apparent among Colony factor, frequently used by palaeontologists and neontologists as a proxy for genetic differences.

1 INTRODUCTION

1.1 *Morphometric analyses in bryozoans*

Traditional morphometric methods involve quantitative measurements from biological features, which are used as proxies for the shape and size of key parts of an organism. Variation in those characteristics is controlled by both internal (genetic and developmental) and external (environmental) factors. Morphometric data typically are treated with univariate or multivariate statistical analyses, but in recent decades more specialized, sophisticated techniques have been developed, particularly to quantify changes in overall shape (e.g. Bookstein 1996). Nevertheless, traditional morphometric methods often provide results, the biological significance of which can be more directly interpreted than can complete analyses of shape.

To the naked eye, the subtle differences among the multitude of zooecia within a bryozoan colony, combined with variation among conspecific colonies, strongly suggests that minimal, ordered distribution of zooecial characteristics could be present or recognized. However, even early application of quantified ranges for zooecial characters proved valuable in making taxonomic distinctions among bryozoan species (e.g. Busk 1884, Ulrich 1890). Modern computational and systematic analysis of morphometric zooecial data within and among colonies has yielded a surprising level of clarity among the superficial chaos. This has been convincingly demonstrated for taxonomic (Cheetham 1968, Cook 1977, Jackson & Cheetham 1990, Hageman 1991, Holdener 1994), ecological (Abbott 1973, Farmer & Rowell 1973, Schopf 1976, Schopf & Dutton 1976, Taylor & Furness 1978, Pachut & Anstey 1979, Pachut 1982, Key 1987, Okamura 1987, Okamura & Bishop 1988, Lidgard & Buckley 1994, Hageman 1994, 1995, O'Dea & Okamura 1999a, 1999b, Okamura & Partridge 1999) and microevolutionary studies (Cheetham 1986, Pachut & Cuffey 1991, Cheetham et al. 1993, 1994, 1995).

The experimental design of many of these studies only is possible with colonial organisms such as bryozoans due to the inherent hierarchical arrangement of their modular units (Boardman et al. 1970, Schopf 1976). That is, on the assumption of no somatic mutations, zooids within a colony can be considered as genetic replicates of each other. Moreover, various patches or branches within a colony (or even individual zooids) may represent growth under different spatial or temporal microenvironmental conditions. The ability to partition morphological variance into Genotypic, Environment and Genotype x Environment interaction sources using ANOVA has been demonstrated by Jackson & Cheetham (1990) and Cheetham et al. (1993, 1994, 1995) using techniques of common garden plots with maternal inheritance, and Hunter & Hughes (1994) and Hageman et al. (1999) using techniques of clonal replicates grown in controlled laboratory environments.

In an ANOVA, the Residual is the amount of variance that is not accounted for by the other factors in the study, whether it be random variation or attributable to additional, unidentified factors. Previous studies have shown approximately the same degree of Residual variance (40–65% of the total variation) for zooids within colonies (Jackson & Cheetham 1990, Cheetham et al. 1993, 1994, 1995, Hunter & Hughes 1994, Hageman et al. 1999). Thus, although it has been important to show that Genotype and other factors are statistically significant, two major questions arise from these findings:

(1) what is the source of the large Residual variance (i.e. unaccounted factors)?

(2) what are the implications for traditional morphometric studies of ignoring such a large percentage of the total variance, the within colony variance in these examples being approximately twice that attributable to Genotype?

Hageman et al. (1999) proposed the possible systematic sources of within colony variation: (1) temporal microenvironmental differences, i.e. budding of zooids in different parts of the colony at different times under different ambient conditions; (2) spatial microenvironmental differences, i.e. growth of different parts of the colony affected by substratum or other local heterogeneities; and (3) packing constraints based on budding interactions (space filling): nearest neighbor effects caused by growth being restricted to lineal budding columns or by position along a budding series (= zooid rows).

1.2 *Goals of this study*

(1) Grow clonal replicates of *Electra pilosa* (L.) in a laboratory environment under which the Environment effect is minimized. That is, how small can one ever expect the Environment effect to be among and within colonies?

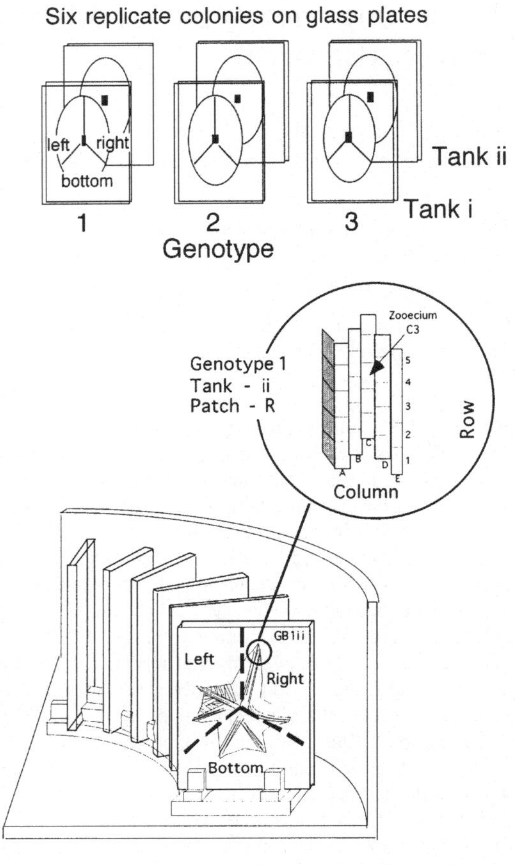

Figure 1. Methods and experimental design. (Top) Six replicate colonies growing on glass plates from three genotypes allocated to two tanks in a common water bath. (Middle) Patch of 5 Rows by 5 Columns sampled from an echelon. (Bottom) Replicate colony illustrating Left, Right, and Bottom sectors. Patches measured from primary growth axes.

(2) Partition sources of phenotypic variation, reproducing the results of Hageman et al. (1999) for Genotype, Environment, GxE Interaction, but to further partition the Residual variance into within colony variance of Patch effects (due to differences among distinct clusters of zooids analytically nested within colonies), and local Row and Column effects (variance among zooids due to their position in adjacent rows and budding columns nested within patches, within colonies).

(3) Evaluate the implications of these analyses on broader morphometric studies in colonial organisms, both living and fossil.

No genetic analyses have been undertaken to quantify the degree of genetic difference among genotypes in this study. The original colonies collected were, however, all complete with their ancestrulae. They therefore had separate sexual origins, and un-

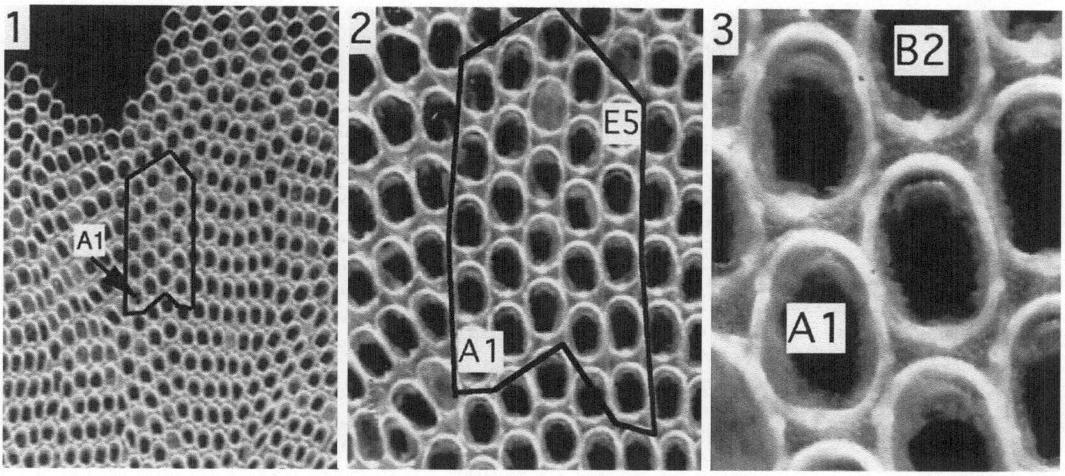

Figure 2. Specimen of *Electra pilosa* illustrating (1) position of 5x5 zooid "right" patch for Genotype 5, Tank i, (2) intermediate view (map) of 5x5 patch with Columns (A-E) and Rows (1-5), (3) close-up of zooids A1 and B2 at scale for measurement.

doubtedly were separate genotypes with allelic differences.

2 MATERIALS AND METHODS

2.1 *Laboratory procedure for generating clonal replicates*

Single colonies of *Electra pilosa* were physically cut into separate, small patches of zooids. Due to the remarkable regenerative abilities of Bryozoa (and *E. pilosa* in particular), damaged zooids within these starter patches heal readily and budding (colonial growth) proceeds as normal. Multiple colonies can be grown separately and all have the same Genotype (i.e. they are clonal replicates). Bayer & Todd, (1996) and Hageman et al. (1999, p. 1720, fig. 2) have provided the experimental details.

2.1.1 *Bryozoan specimens for this study*

Colonies of *Electra pilosa*, complete with ancestrulae, were collected from Gruinard Bay, NW Scotland. Colonies were transferred to the laboratory and maintained in a single circular 6 l glass tank with 0.2 μm filtered seawater at the optimal experimental growth temperature of 18°C ± 0.1°. Three colonies with unique genotypes (different ancestrulae) were allowed to grow onto a pre-scoured coverslip and then split with a scalpel. The resulting fragments were clipped onto larger glass plates and allowed to grow onto these, which served as the starter clusters for the replicate colonies.

Clonal replicates grew on substrata of separate 5 x 6 cm glass plates, each beginning as a starter group of nine zooids. The glass plates were allocated to two 6 l tanks (Tank-i and Tank-ii) for the 10 week growth phase of the experiment. Colonies were maintained upright on a circular Perspex rack, with each glass plate oriented in the same way and spaced at equal distances (Fig. 1). Throughout the experiment, the flagellate *Rhodomonas* sp. (100 cells·μl^{-1}) was given as food. At the end of the 10 week growth period the colonies were removed, cleaned in freshwater and air-dried for subsequent examination and measurement. Specimens representing these genotypes and tanks were randomly selected as a subset from a larger experimental design that included twenty genotypes with three replicate colonies each, assigned to three tanks in a common waterbath.

2.2 *Experimental design*

The experimental design was of one replicate colony of each of three Genotypes (1–3) allocated to each Tank (i and ii) for a total of six replicate colonies (plus other genotypes not employed in this study). In this study, the Tank factor is equivalent to the Environment factor (microenvironmental differences among colonies due to their unique positions in space). The biological, physical and chemical conditions associated with these differences are not identified here and under ideal circumstances of the experimental design should approach zero. The experimental design accounted for six possible sources of morphological variation: Genotype, Tank (= Environment), GxE interaction, Patch within colony and two types of local (Row and Column) variation within patches (see below and Fig. 1).

2.2.1 *Morphometric measurements*

We used morphological characters for zooecia that were reliably measurable at the available resolution: opesia width (OpsW); opesia length (OpsL); zooecium width (ZW); zooecium length (ZL), all maximum measurements, and the number of spines about each opesia (Sp#). We intentionally chose a small number of simple characters from a relatively simple organism, with the assumption that any significant patterns observed would be analytically and interpretatively robust and have the potential to be more pronounced for other, more complex systems. Zooecia were photographed using a video camera on a WILD M8 stereomicroscope, all set to the same magnification, selected such that at least one, but no more than two zooecia filled a screen. Measurements (to 0.1µm, based on six separate calibrations from a 10 µm unit micrometer) were made on the resulting images using NIH Image 1.61 software.

The position from which zooecia were selected for measurement from a colony was chosen carefully. Three patches of 25 zooecia were selected from each replicate colony using the following protocol (Figure 1): (1) etched slide label positioned upper right, (2) three sectors running 120 degrees each, 12 o'clock to 4 o'clock = "Right", 4 to 8 = "Bottom", and 8 to 12 = "Left", (3) patch location within the sector selected as the first suitable cluster clockwise around the periphery of sector to primary axis, then inwards, then accept or reject the patch selection on criteria below. The objective was to both ensure that patches represented different parts of colony (i.e. they were not clustered), and to allow for evaluation of any positional effect within the Perspex racks.

Patches were chevron in shape in order to account for the shape of the leading (budding) edge of a colony. Primary growth axes form as a small number of adjacent, serial columns expanding from the primary colony in a phalanx, with space between axes filled later by lateral budding from the primary column. Thus, zooids A5, B5, C5, D5, E5 (Figure 1) were more likely budded at the same time (of the same age) than were zooids straight across a row.

All patches were measured from well developed primary growth axes, well away from lateral budding effects and crowding effects. We made a conscious decision not to represent the complete amount of variability within each colony, but only the variability among zooids that budded free of any within colony crowding effects.

Column-A in every patch was a margin of a primary growth axis from which lateral budding had occurred. Thus patches were all taken from primary growth axes at least five zooids wide (Fig. 2).

All patches were placed as close as possible to the distal growing edge to obtain a complete patch fulfilling the other criteria. Therefore, zooecia from Row-1 do not represent exactly the same time of origin as those of Row-1 in other patches (or other replicate colonies), but overall, all of the Patches were formed at approximately the same time assuming more or less continual growth up to termination of the experiment.

2.3 *Statistical methods*

In order to appraise the relative contribution of various factors (main effects) on the overall variation for each character, we partitioned the overall (= phenotypic) variation into variance component estimates from ANOVA (Sokal & Rohlf 1981, p. 216). A variance component is the relative proportion that the variance of each factor contributes to the overall = total variance. Estimates were calculated as s^2_A = $(MS_{\text{Groups}} - MS_{\text{Within}})/n_0$ and converted to percentages of all estimates summed (Table 1). Prior to analysis, all data were inspected for normality but no transformation was found to be necessary. In some cases, the MS_{Within} component exceeded the value of the MS_{Groups} component, resulting in a negative contributory term, which, accordingly, was assumed to be a zero contribution (Brown & Mosteller 1991, p. 210).

The relative amount of variance for each factor (given as a percentage) was estimated by calculating the mean for each factor across all five characters. Because some small percentages were involved, we performed an arcsine transformation prior to calculating averages and then "back-transformed" the results (Sokal & Rohlf 1981, p. 427). As a result of this back-transformation, reported percentages are not additive for a given factor (total $\neq$ 100%). All ANOVAs were performed with SuperANOVA v1.11 (Abacus Concepts).

Factors available for ANOVA from this data set can be summarized as follows:

Genotype effect: a = 3 genotypes, n = 150 zooecia per genotype.

Tank effect (= Environment factor): a = 2 tanks, n = 225 zooecia per tank.

Genotype by Tank (GxE) interaction: a = 6 replicate colonies, n = 75 zooecia per replicate. This is an independent response to Environment for each Genotype, also called "reaction norm" in some studies (Falconer 1981).

Patch effect (nested within replicate colony): a = 18 patches, n = 25 zooecia per patch.

Row effect (nested within patch, and within replicate colony): a = 5 rows, n = 5 zooecia per row.

Column effect (nested within patch, and within replicate colony): a = 5 columns, n = 5 zooecia per column.

The Residual is the remaining variation not accounted for by the five factors and GxE interaction in the model, i.e. other unmeasured factors or independent variation among zooecia (n = 450 zooecia in the study).

2.4 *ANOVA methods*

Four separate ANOVA models were assessed. The rationale for multiple analyses was that traditionally, researchers have not employed as complete a factorial model as illustrated here. We wished to determine both the relative importance of factors, and the consequences of ignoring factors (i.e. is the extra variance simply placed in the Residual, or does is it artifactually become allocated to other measured factors?)

2.4.1 *Full within colony factorial model*

Here, all five factors and interaction were evaluated simultaneously in a Two-way Nested Type III model. For nested factors, the order represents their depth of nesting; e.g. Column (P, T, G) = Column within Patch, within Tank, within Genotype.

2.4.2 *Complete pooling within colony*

This model simulated collecting morphometric data for zooecia across a broad area of a colony (to fully document variation within a colony), whether it be (1) measuring every zooid in the colony, (2) randomly sampling zooecia across the colony, or (3) collecting long lineal or lateral transects that extend across gradients within colonies. In any case, this model assumes no control for within colony local effects (Patch, Row, or Column).

2.4.3 *Single Patch per replicate colony, accounting for Row and Column effects*

This simulated a study whereby a worker measured all of the zooecia from a single area on each colony *and* kept track of the relative Row and Column position of nearest neighbor zooecia. For both this and the foregoing analysis that utilize single, independent patches, in this study, the results are reported as averages from three independent ANOVAs for each of the three patches separately.

2.4.4 *Single Patch per replicate colony, pooled Row and Column effects*

This simulated a study in which all of the measured zooecia were from a single area of the colony, with no control for nearest neighbor (Row/Column) positioning. This approach to data collection is quite common when taphonomic effects limit the available material, as for example when only fragments of fossil colonies are available. Furthermore, it is common for workers to sample from a single region of a colony even when more material is available, and few workers routinely document local Row/Column effects.

3 RESULTS

An example of a Sum of Squares table (for opesia width, OpsW) is provided in Table 1. Similar tables were generated for each morphological character (OpsW, OpsL, ZW, ZL, Sp#), for every ANOVA model presented. Much of the results presented in this study have been summarized to the level of the relative variance (percentage) for which each factor accounts, or the average variance percentage for a factor across all five characters (Table 2).

3.1 *Full within colony model (Table 3, column 1)*

All factors demonstrated significant differences ($p < 0.001$), with the exception of Tank, which lacked significance for zooecia length ($p = 0.94$): GxE interaction effect on Spine number was also highly significant ($p = 0.004$). It is noteworthy that the relative magnitude of factors was comparable across all characters (Table 2). Factors explaining the greatest amount of variance were Patch (nested within replicate colonies) and Genotype, which accounted for 25-38% and 18-38% of the total variance, respectively. Row (2-7%) and Column position (3-13%) and Genotype x Tank interaction (2-14%) typically explained more variance than Tank alone (2–3%). The percent averages for each component are presented both in Table 2, and in Table 3, column 1, for comparison with results for other models.

Many of the apparent differences in Table 4 between common factors from different models were due to the inevitable scaling that results from the inclusion of a different number of factors in the respective models. For example, in comparing Table 3, columns 1 and 2, each corresponding value in the 3-factor model was 1.14 times larger than their counterpart in the 5-factor model (scaling value derived prior to arcsine back transformation). This principle became important when we examined where the variance of known factors would be allocated following their hypothetical omission from analysis (i.e. allocation to residual, versus the artificial inflation of other factors, see *Discussion*).

3.2 *Complete pooling within colony (Table 3, column 2)*

These results (G = 31.7%, T = 2.6%, GxT interaction = 5.5%, Residual = 58.4%) are closely comparable to our previous results using the same ANOVA model on a different, but similar *Electra pilosa* data set (G = 26.5%, T = 6.9%, GxT = 0.8%, and Residual = 55.9%; Hageman et al. (1999)). In the present ANOVA model, the variance due to within colony variation (i.e. 31.7, 7.6, 4.4 from full factorial Table 3, column 1:) was allocated to the Residual (column 2: 58.4% versus column 1 = 18.0%).

3.3 *Single patch per colony models (Table 3, columns 3, 4)*

When only a single patch is measured for each colony replicate — that is the 5x5 adjacent zooecia are employed as a proxy for the entire colony replicate — the GxT interaction was exaggerated nearly fourfold. That is, the variation among patches within

Table 1. Example of a Sum of Squares table for a full factorial ANOVA for the character opesia width (OpsW).

Source	df	SS	MS	*F*	*p*	n_o	S^2	Var %
Genotype	2	533.12	266.56	273.74	0.0001	150	1.77	31.5
Tank	1	32.48	32.48	33.36	0.0001	225	0.14	2.5
Genotype x Tank	2	68.53	34.27	35.19	0.0001	75	0.44	7.9
Patch (T, G)	12	433.06	36.09	37.06	0.0001	25	1.40	25.0
Row (P, T, G)	72	162.49	2.26	2.32	0.0001	5	0.26	4.6
Column (P, T, G)	72	294.38	4.09	4.20	0.0001	5	0.62	11.1
Residual	288	280.45	0.97				0.97	17.4
						total	5.61	100.0

Patch is nested within Tank and Genotype, Row and Column are nested within Patch, which is nested within Tank and Genotype.

Table 2. Percent of variance within morphological characters accounted for by each factor and mean values (± 95% c.i.) for each factor across all five morphological characters.

Source	Opesia width	Zooecia width	Opesia length	Zooecia length	Spine number	neg.	Mean	pos.
Genotype	31.5	18.1	26.7	37.5	25.6	-6.1	27.7	6.6
Tank	2.5	2.4	8.4	0.0**	2.5	-2.0	2.3	3.5
Genotype x Tank	7.9	2.5	3.6	14.3	1.7*	-3.3	5.2	4.8
Patch (T, G)	25.0	38.2	36.1	33.5	26.2	-5.1	31.7	5.3
Row (P,T,G)	4.6	3.8	6.9	2.2	5.2	-1.4	4.4	1.7
Column (P,T,G)	11.1	12.7	3.3	3.0	10.9	-3.7	7.6	4.9
Residual	17.4	22.3	15.1	9.4	27.9	-5.8	18.0	6.6
	100.0%	100.0%	100.0%	100.0%	100.0%		96.9%	

All values are significant (F-test) at < 0.001 except (* = 0.004) and (** = 0.94). Patch is nested within Tank and Genotype, Row and Column are nested within Patch, which is nested within Tank and Genotype. Local Effects (Patch, Row, Column) total = 43.7%. Patch is nested within Tank and Genotype, Row and Column are nested within Patch, which is nested within Tank and Genotype.

Table 3. Average variance percent for all characters with ANOVA models accounting for various levels of local effects.

	1	2	3	4
	Full within colony factorial	Complete pooling within colony	Single Patch per replicate colony accounting for Row/Column	Single Patch per replicate colony pooled Row/Column
Genotype	27.7	31.7	33.0	33.7
Tank	2.3	2.6	6.4	6.2
GxT	5.2	5.5	21.2	22.0
Patch (T, G)	31.7	-	-	-
Row (P,T,G)	4.4	-	4.3	-
Column (P,T,G)	7.6	-	7.4	-
Residual	18.0	58.4	17.9	30.2

Single patch data represent averages of three independent trials. Variances within model are not additive (totals ≠ 100%) due to back-transformation from the arcsine.

Table 4. Average percent of variance contributed by each factor using Colony factor only.

	1	2	3	4			
	Full	Patches	Patches	Single Patch	Single patches		
	Factorial	Nested	Pooled	Average	Left	Right	Bottom
Colony	–	30.7	34.6	59.9	46.9	60.5	72.2
Genotype	*27.7*	-	-	-	-	-	-
Tank	*2.3*	-	-	-	-	-	-
GxT	*5.2*	-	-	-	-	-	-
Patch	31.7	36.1	-	-	-	-	-
Residual	18.0	33.1	65.4	40.1	53.1	39.5	27.8
*other factors	12.0	-	-	-	-	-	-

*Row and Column positional effects nested within patches and replicate colonies.

colonies was not allocated to the Residual, but was misleadingly assigned to the GxT interaction. Variance due to positional effects of Rows and Columns was, however, allocated to the Residual when appropriate factors are not considered in the model.

4 DISCUSSION

4.1 *Reproducibility of the results*

The overall results from this study, both in terms of the general conclusions and the specific values of variance percentages for individual factors, were very comparable to Hageman et al. (1999), which employed different genotypes of *Electra pilosa*, collected from a different locality, and grown during different experimental trials. This indicates a satisfactory degree of generality and analytical and interpretative robustness of approach.

The precision of the morphometric image analysis system (± 0.1 μm) and the ability of ANOVA to recognize relatively small differences among the means of populations with large sample sizes, raised the question of whether observed differences were meaningful or an artifact of the methodology. Figure 3 shows that clear differences do exist — the ranges of confidence limits consistently were ~ 1μm, the differences among means were up to 5μm, and features measured were ~ 25-30 μm.

4.2 *Expanded model accounting for within colony factors*

This expanded factorial model, explained a large amount (~ 2/3) of the Residual that we set out to explore (18.0% *versus* 58.4%, Table 3, columns 1-2).Much of this residual variance reported in our previous study, was accounted for in this model by variation among Patches nested within colonies. These patches do not represent simple astogenetic positional effects (i.e. ordinal distance from ancestrulae), nor do they represent regions of communality necessarily discrete from other patches. Rather, these Patches are sub-samples from a colony-wide continuum of morphological variation that reflects strong nearest-neighbor effects, as also observed in some Palaeozoic stenolaemates (Hageman 1995, Holdener & Hageman 1998). The magnitude of variance explained by these within colony patches is as great or greater than that explained by Genotype (Table 2). Clearly variation of Patches nested within colonies is a significant source of morphological variation among zooecia in these bryozoans.

4.2.1 *Consequences of ignoring patch variation nested within colonies*

If the nested Patch factor is omitted from ANOVA models, the results will depend greatly on from where the individual zooecia were measured for a colony. If the zooecia are scattered across the colony such that the sample represents a pooling of among colony variance, most of the Patch variance unaccounted for will be allocated to the Residual (Table 3, column 2). However, if zooecia were all to be measured from a small patch on the colony (or a colony fragment), then the Patch variance will be misallocated to Genotype, Tank and the GxT interaction. In the present case, the increase was 19%, 178% and 308% respectively (Table 3, columns 1-2).

4.2.2 *Row and Column local effects*

The local position of zooecia within adjacent rows and columns of a single Patch nested within a colony accounted for a significant amount of variance for all morphological characters, with variance components ranging from 2.2% – 12.7%. If row and column data were omitted from ANOVA models, however, all of their unaccounted variance was allocated to the Residual. Thus, failing to account for Row and Column factors will not artificially affect the interpretation of the significance of other factors.

4.3 *Patches nested within colonies versus systematic effect of patch position*

In this study, it was assumed originally that there would be no systematic variation of Patches among colonies. That is, the orientation of Patches on their glass plates (Left, Right or Bottom) would not have a common influence on corresponding positions for all of the replicate colonies in both of the tanks. However, when this was later tested, the "systematic patch position" factor displayed a small (2.9%), but significant ($p < 0.0001$) effect. It is clear that significant Patch differences do occur *within* colonies and thus our initial ANOVA model is meaningful. A fuller interpretation of these differences would perhaps best be achieved by the investigation of simulated data.

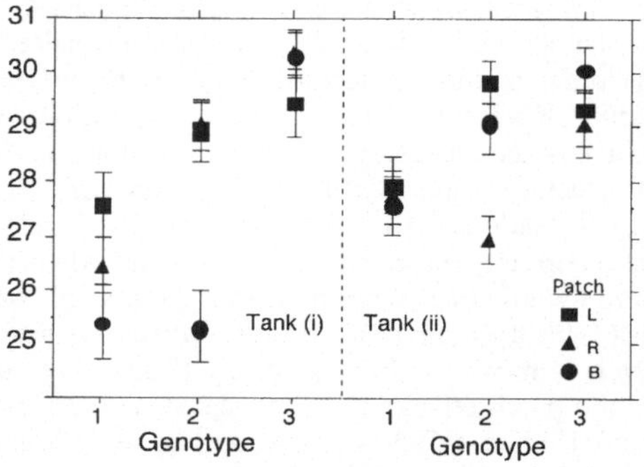

Figure 3. Mean values and 95% c.i. for Patches measured from each of the six replicate colonies representing three genotypes and two tanks.

4.4 *Implications for fragmentary fossil material*

Many morphometric studies, especially on fossil material, have employed single patches within a colony (e.g. fragments, Hageman 1994), despite our developing knowledge from fossil material of significant morphological gradients (patches) within colonies (Anstey et al. 1976, Hageman 1995, Holdener & Hageman 1998). In addition, all palaeontologists and most zoologists will never be able to recognize the degree of genotypic difference among separate colonies. That is, all colonies have to be treated as unique genotypes, without colony replicates. In order to investigate the consequences and limitations for interpretation of data, here we approached this problem in two ways. First we treated our colony replicates as unique genotypes, and second, we modeled the system.

For both fossil and living material, observers work with single colonies. Because clonal replicates of known genotype differences are not available, the amount of observed variance among colonies, which here could be allocated to Genotype, Environment (= Tank), and GxE interaction, all would otherwise have to be allocated to the single factor "Colony". Many workers have made assumptions regarding the relative contribution of Genotype versus Environment to the Colony factor (Schopf 1976, Schopf & Dutton 1976, Pachut & Anstey 1979, Brande & Bretsky 1982, Pachut 1982, Key 1987, Cheetham et al. 1994, Cheetham et al. 2001), but given our present and previous results (Hageman et al. 1999), even with a highly controlled environment, such assumptions seem unwarranted.

Data from this study were analyzed by re-coding each of the six replicate colonies as "Colony" (1–6). All of the ANOVA models presented in Table 3, were reanalyzed with this coding and results are presented in Table 4. Note that column 1 in Table 4 is the result from the full factorial model (Table 3, column 1). The results are not surprising. With "multiple Patches nested within colonies", and the "Patches pooled to simulate measurement from zooecia scattered widely across the colony", the variance attributed to Colony (30.7% and 34.6%) is indeed on the order of the full factorial models (Genotype + Tank + GxT interaction = 35.2%). The differences probably are attributable to scaling for total number of factors and artifacts of arcsine back-transformations. All within colony Patch effects were correctly allocated to the Residual when they were hypothetically omitted from the pooled analysis (Table 4, column 3). However, when only a single patch is measured from a colony (Table 4, column 4), the patch effect was misallocated to the Colony factor (Figure 4, 59.9% is greater than twice the true value of Genotype at 27.7%).

Therefore, if a single fragment or small patch is used to represent a colony, which in turn is assumed to represent a "genotype" in an ANOVA model, results will be very strongly biased by within colony patch variation.

5 SUMMARY

The present study shows that morphometric data should not be collected from single patches or small fragments of bryozoans. If it is omitted from analyses, the within colony Patch effect will artifactually and significantly inflate the apparent among Colony differences, which are frequently used by palaeontologists and neontologists as a proxy for genetic differences.

(1) Sources of phenotypic variation can be partitioned in *Electra pilosa*, into their Genotype and Environment main effects and GxE interaction in repeated treatments. The large Residual (among zooecia – within colonies), reported in previous studies (~55–65%), was reduced by two-thirds when sources of within colony variation were accounted for in ANOVA models, as Patches within colonies and Row and Column positions within Patches.

(2) Even under controlled conditions, designed to minimize environmental differences among replicate colonies, significant environmental variation still can be detected (Tank effect ~ 2–3% of variance). Failure to recognize significance of environmental factors in future studies may be attributable to inadequate sample size.

(3) Patch effect within colonies was highly significant, and contributed as much or more variance as did the main effect of Genotype (~ 25–35%). The magnitude of the Patch effect has major implications for procedural protocols in the collection and analysis of morphometric data.

(4) If zooecia are sampled randomly and widely across a colony and pooled, all of the variance associated with Patch will be allocated to the Residual. That is, values obtained for other factors will not be affected relative to each other (note the proportional scaling of all effects due to the number of factors included in the ANOVA and their finite sum of 100%).

(5) If zooecia all are measured from a single, small patch on a colony, or from a single small colony fragment, the variance from the Patch effect is not simply allocated to the Residual and will artifactually and differentially inflate the values of Genotype, Environment, and their interaction.

(6) Failure to account for Patch effect will make all other results misleading. Therefore, in all morphometric analyses zooecia must adequately represent variance of patches within colonies by including this effect as a separate factor, or by pooling zooecia sampled across a wide range of the colony (Patch effect variance allocated to Residual).

(7) Local, that is Row and Column effects, are small (~ 5–10%), but nonetheless statistically sig-

nificant. If Row and Column effects are ignored in ANOVA, their variance will be allocated to the Residual, and will not affect the outcome for other observed factors.

ACKNOWLEDGEMENTS

We thank A. Cheethman (USNM) and F.K. McKinney (Appalachian State) for comments and suggestions that improved this paper. SJH thanks the University Research Council (Appalachian State) and National Science Foundation (EAR0073648) for grants in support of this work.

REFERENCES

Abbott, M.B. 1973. Intra- and intercolony variation in populations of *Hippoporina neviani* (Bryozoa-Cheilostomata). In R.S. Boardman & A.H. Cheetham (eds.), *Animal colonies, development and function through time*: 223-245. Stroudsburg, Pennsylvania: Dowden, Hutchinson & Ross.

Anstey, R.L, Pachut, J.F. & Prezbindowski, D.R. 1976. Morphogenetic gradients in Paleozoic bryozoan colonies. *Paleobiology* 2: 131-146.

Bayer, M.M. & Todd, C.D. 1996. Effect of polypide regression and other parameters on colony growth in the cheilostomate *Electra pilosa* (L.). In D.P. Gordon, A.M. Smith, & J.A. Grant-Mackie (Des), *Bryozoans in space and time*: 29-38. Wellington, NZ: National Institute of Water and Atmospheric Research.

Boardman, R.S, Cheetham, A.H. & Cook, P.L. 1970. Intracolony variation and the genus concept in Bryozoa. In E.L. Yochelson (ed.), *Proceedings of the North American Paleontological Convention, September 1969, Part C*: 294-320. Chicago.

Bookstein, F.L. 1996. Biometrics, biomathematics and the morphometric synthesis. *Bulletin of Mathematical Biology* 58: 313-365.

Brande, S. & Bretsky, S.S. 1982. Avoid improper statistical analysis in bryozoans: analysis of variance is suitable for study of hierarchical variation. *Journal of Paleontology* 56: 1207-1212.

Brown, C. & Mosteller, F. 1991. Components of variance. In D.C. Hoaglin, F. Mosteller & J.W. Tukey (eds), *Fundamentals of Exploratory Analysis of Variance*: 193-251. New York: John Wiley & Sons Inc.

Busk, G. 1884. Report on the scientific results of the voyage of HMS Challenger during the years 1873-76. Zoology. volume 10, part 30, Report on the Polyzoa – The Cheilostomata. Her Majesty's Stationary Office: London.

Cheetham, A.H. 1968. Morphology and systematics of the bryozoan genus *Metrarabdotos*. *Smithsonian Miscellaneous Collections* 153: 1-121.

Cheetham, A.H. 1986. Tempo of evolution in a Neogene bryozoan: rates of morphologic change within and across species boundaries. *Paleobiology* 12: 190-202.

Cheetham, A.H, Jackson, J.B.C. & Hayek, L.A.C. 1993. Quantitative genetics of bryozoan phenotypic evolution. I. Rate tests for random change versus selection in differentiation of living species. *Evolution* 47: 1526-1538.

Cheetham, A.H, Jackson, J.B.C. & Hayek, L.A.C. 1994. Quantitative genetics of bryozoan phenotypic evolution. II. Analysis of selection and random change in fossil species using reconstructed genetic parameters. *Evolution* 48: 360-375.

Cheetham, A.H, Jackson, J.B.C. & Hayek, L.A.C. 1995. Quantitative genetics of bryozoan phenotypic evolution. III. Phenotypic plasticity and the maintenance of genetic variation. *Evolution* 49: 290-296.

Cheetham, A.H., Jackson, J.B.C. & Sanner, J. 2001. Evolutionary significance of sexual and asexual modes of propagation in Neogene species of the bryozoan *Metrarabdotos* in tropical America. *Journal of Paleontology* 75: 564-577.

Cook, P.L. 1977. The genus *Tremogasterina* Canu (Bryozoa, Cheilostomata). *Bulletin of the British Museum (Natural History) Zoology Series* 32: 103-165.

Falconer, D.S. 1981. *Introduction to quantitative genetics*, . (Second edition). London: Longman Group Ltd.

Farmer, J.D. & Rowell, A.J. 1973. Variation in the bryozoan *Fistulipora decora* (Moore & Dudley) from the Beil Limestone of Kansas. In R.S. Boardman & A.H. Cheetham (eds), *Animal colonies, development and function through time*: 377-394. Stroudsburg, Pennsylvania: Dowden, Hutchinson & Ross.

Hageman, S.J. 1991a. Approaches to systematic and evolutionary studies of perplexing groups: an example using fenestrate Bryozoa. *Journal of Paleontology* 65: 630-647.

Hageman, S.J. 1994. Microevolutionary implications of clinal variation in the Paleozoic bryozoan *Streblotrypa*. *Lethaia* 27: 209-222.

Hageman, S.J. 1995. Observed phenotypic variation in a Paleozoic bryozoan: implications for microevolutionary and systematic studies. *Paleobiology* 21: 314-328.

Hageman, S.J., Bayer, M.M. & Todd, D.T. 1999. Partitioning phenotypic variation: genotypic, environmental and residual components from bryozoan skeletal morphology. Journal of Natural History, 33: 1713-1735.

Holdener, E.J. 1994. Numerical Taxonomy of Fenestrate Bryozoans: Evaluation of Methodology and Applicability to Analyses of Microevolution. *Journal of Paleontology* 68: 1201-1214.

Holdener, E.J. & Hageman, S.J. 1998. Implications of intracolonial variation in a Paleozoic bryozoan. *Journal of Paleontology* 72: 809-818.

Hunter, E. & Hughes, R.N. 1994. The influence of temperature, food ration and Genotype on zooid size in *Celleporella hyalina* (L.). In P.J. Hayward, J.S. Ryland, & P.D. Taylor (eds.), *Biology and palaeobiology of bryozoans*: 83-86. Fredensborg, Denmark: Olsen & Olsen.

Jackson, J.B.C. & Cheetham, A.H. 1990. Evolutionary significance of morphospecies: a test with cheilostome Bryozoa. *Science* 248: 579-583.

Key, M.M., Jr. 1987. Partition of morphologic variation across stability gradients in Upper Ordovician trepostomes. In J.R.P. Ross (ed.), *Bryozoa: past and present*: 145-152. Bellingham: Western Washington University.

Lidgard, S. & Buckley, G.A. 1994. Toward a morphological species concept in cheilostomes: phenotypic variation in *Adeonellopsis yarraensis* (Waters) . In P.J. Hayward, J.S. Ryland, & P.D. Taylor (eds.), *Biology and palaeobiology of bryozoans*: 109-113. Fredensborg, Denmark: Olsen & Olsen.

O'Dea, A. & Okamura, B. 1999a. Intracolony variation in zooid size in cheilostome bryozoans as a new technique for intestigating palaeoseasonality. *Paleaogeography, Palaeoclimatology, Palaeoecology* 162: 319-332.

O'Dea, A. & Okamura, B. 1999b. Influence of seasonal variation in temperature, salinity and food on module size and colony growth of the estuarine bryozoan *Conopeum serati*. *Marine Biology* 135: 581-588.

Okamura, B. 1987. Seasonal changes in zooid size and feeding activity in epifaunal colonies of *Electra pilosa*. In J.R.P.

Ross (ed.), *Bryozoa: past and present*: 197-204. Bellingham: Western Washington University.

Okamura, B. & Bishop, J.D.D. 1988. Zooid size in cheilostome bryozoans as an indicator of relative paleotemperature. *Palaeogeography, Palaeoclimatology, Palaeoecology* 66: 145-152.

Okamura, B. & Partridge, J.C. 1999. Suspension feeding adaptations to extreme flow environments in a marine bryozoan. *Biological Bulletin* 196: 205-215.

Pachut, J.F. 1982. Morphologic variation within and among genotypes in two Devonian bryozoan species: an independent indicator of paleostability. *Journal of Paleontology* 56: 703-716.

Pachut, J.F. & Anstey, R.L. 1979. A developmental explanation of stability-diversity-variation hypotheses: morphogenetic regulation in Ordovician bryozoan colonies. *Paleobiology* 5: 168-187.

Pachut, J.F. & Cuffey, R.J. 1991. Clinal variation, intraspecific heterochrony, and microevolution in the Permian bryozoan *Tabulipora carbonaria*. *Lethaia* 24: 165-185.

Schopf, T.J.M. 1976. Environmental versus genetic causes of morphologic variability in bryozoan colonies from the deep sea. *Paleobiology* 2: 156-165.

Schopf, T.J.M. & Dutton, A.R. 1976. Parallel clines in morphologic and genetic differentiation in a coastal zone marine invertebrate: the bryozoan *Schizoporella errata*. *Paleobiology* 2: 255-264.

Sokal, R.R. & Rohlf, F.J. 1981. *Biometry: the principles and practice of statistics in biological research, 2nd ed.* New York: W.H. Freeman & Company 859 pp.

Taylor, P.D. & R.W. Furness. 1978. Astogenetic and environmental variation of zooid size within colonies of Jurassic *Stomatopora* (Bryozoa, Cyclostomata). *Journal of Paleontology* 52: 1093-1102.

Ulrich, E.O. 1890. Palaeozoic Bryozoa. *Illinois Geological Survey Bulletin* 8: 268-68

Bryozoan Studies 2001, Wyse Jackson, Buttler & Spencer Jones (eds)
© 2002 Swets & Zeitlinger, Lisse, ISBN 90 5809 388 3

Bryozoan distribution in a Messinian coral reef complex of western Algeria

Assia Hamdane
Département des Sciences de la Terre, Université Houari Boumediene, BP 32 El Alia, Bab Ezzouar, Algeria
Pierre Moissette
Université de Lyon I, UFR Sciences de la Terre/UMR CNRS PEPS, 27 Bd du 11 Novembre, 69622 Villeurbanne Cedex, France

ABSTRACT: In north-western Algeria, the Djebel Tzioua is an upper Miocene coral reef complex. The distribution of the associated bryozoans (59 species) reveals a zonation corresponding to an ecological succession. A small number of zoarial growth forms are observed at the base of the main bioherm, corresponding to the *stabilization* of a shoal by oysters and corals. It is followed by a *colonization*, characterised by an increase in the number of zoarial forms, with predominating massive and lamellar species, together with some erect colonies. A *diversification* of species belonging to various zoarial forms then occurs at the top of the bioherm with an increase of the space available for reef dwellers. Other constructions above the bioherm and also laterally are elaborated by vermetids and serpulids bound to the coral reefs. The bryozoans associated with these secondary frame builders show a higher number of zoarial forms with a predominance of massive and encrusting species.

1 INTRODUCTION

Bryozoans are frequent primary reef-builders during the Palaeozoic, but are generally considered as less significant components, but always present, in reefs of the Mesozoic to the Recent (Cuffey 1971, 1974, 1977, 1985). The past decades have seen a marked growth of interest in the diverse communities of sessile organisms found on sheltered hard substrates within coral reefs. These cryptic organisms inhabit crevices, caves and tunnels inside the reef framework. They are dominated by sponges, serpulid worms, encrusting foraminifers, and bryozoans (Choi & Ginsburg 1983, Garrett et al. 1971, Hughes & Jackson 1992, Jackson & Winston 1982, Kobluk 1988, Logan 1981).

The Messinian (uppermost Miocene) calcareous platforms of the Mediterranean are spectacular in their development (Esteban 1979, Saint-Martin 1990). They comprise widespread coral reefs with sometimes abundant bryozoans (Moissette 1993, 1997, Moissette & Saint Martin 1995). This is the case in the Chelif Basin (north-western Algeria) where the small patch reef of the Djebel Tzioua yielded 59 bryozoan species. The main purpose of this paper is to study the bryozoan communities contained in this Miocene reef structure. It is also an attempt to describe the cryptic faunas inside a coral build-up.

The analysis of the distribution of the bryozoan zoarial growth forms and species within the reef reveals a vertical and horizontal zonation in accordance with the model of Alberstadt et al. (1974) based on corals and other organisms. This zonation or ecological succession is in direct relationship with the ecological factors governing the structure of the reef as well as the variations in water depth and energy.

2 GEOLOGICAL SETTING

In the north-western part of Algeria, the Chelif basin is a Neogene sedimentary basin predominantly east-west oriented (Perrodon 1957). The marine sedimentary succession of the Messinian comprises sandy marl or marl deposits followed by an alternation of marl with diatomite beds, succeeded by sands and sandstones and/or coral/algal limestones associated with foraminifers, molluscs, serpulid worms, bryozoans and echinoderms. (Moissette 1988, 2000, Neurdin-Trescartes 1992, Saint Martin 1990).

The Djebel Tzioua is a Quaternary volcano (Fig. 1) with associated Messinian fossiliferous sediments. A quarry was exploited on its southern flank where the Messinian deposits have been studied (Fig. 2). The marly base is not visible. The section begins with a sandy, more or less argillaceous level, containing numerous large oyster shells. Above this

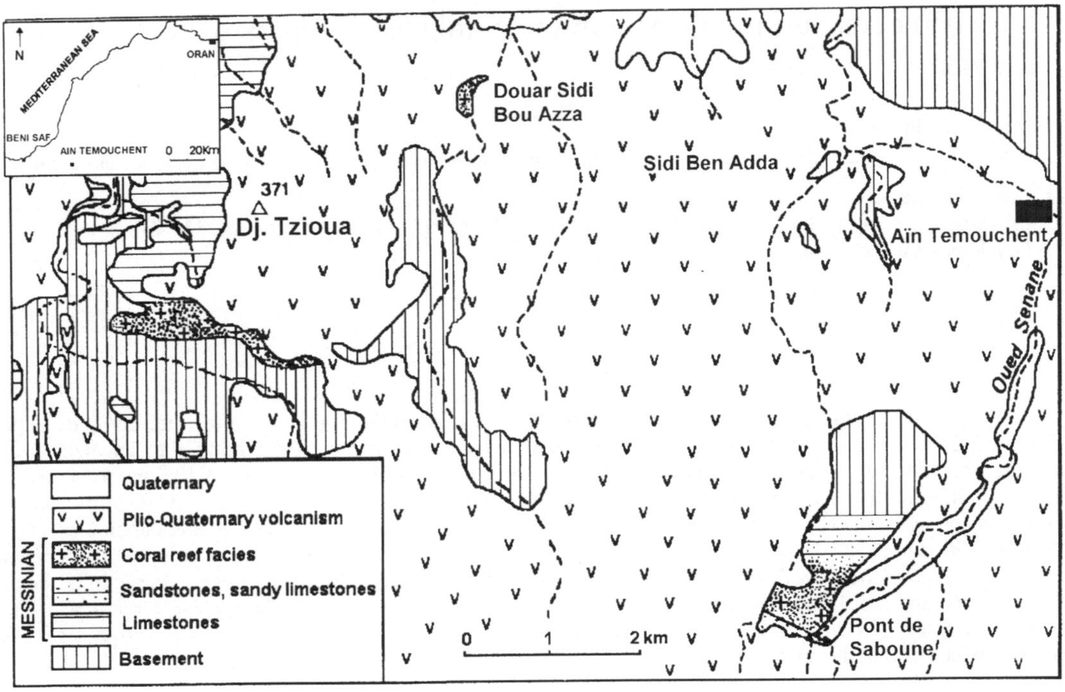

Figure 1. Location map of the Djebel Tzioua in north-western Algeria. (From Guardia 1975).

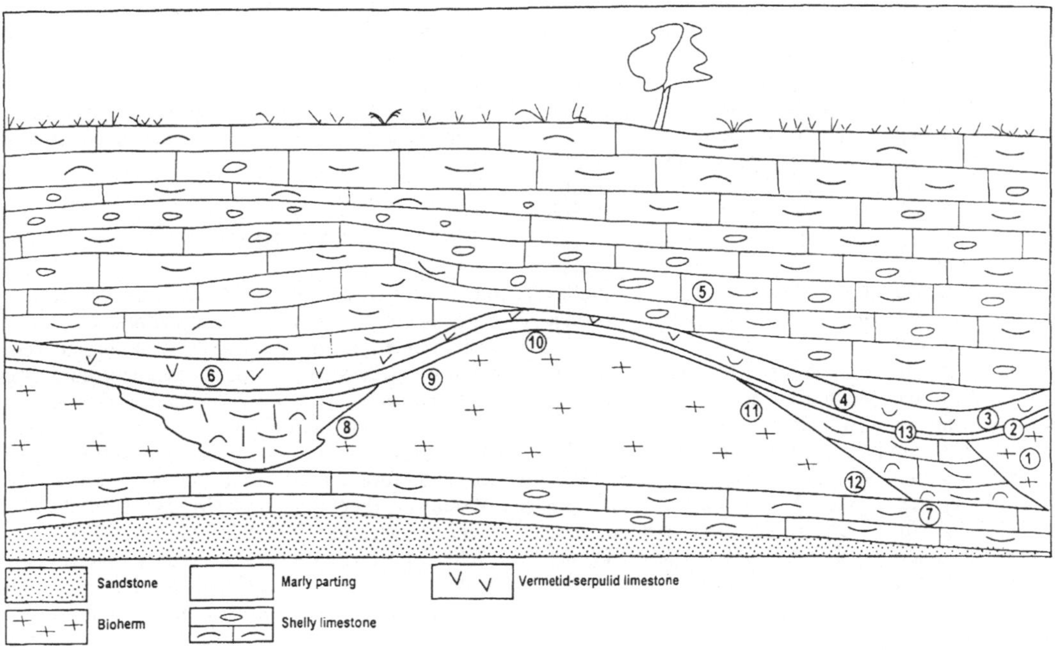

Figure 2. Schematic cross-section of the Djebel Tzioua coral reef complex with indication of the sampling locations.

bed, the reef complex consists of several lenses built by the scleractinian coral genera *Tarbellastraea* and *Porites*. Each lens is separated from the neighbouring ones by bioclastic limestones with abundant bivalves. The main bioherm is about 20 m long and 3 m high. The coral build-ups are covered by a vermetid and serpulid level isolated from the bioherms by a thin marly parting. This unit is overlain by a thick bioclastic limestone containing reefal blocks.

3 MATERIALS AND METHODS

After logging the section, samples were taken from 13 levels likely to contain bryozoans (Fig. 2). Samples were taken from two fractions. (Moissette 2000). The first one from the coarse fraction (> 5 cm): corals and associated macrofauna (mostly bivalve shells), ideal substrates for the bryozoans, and second from the fine material (< 5 cm).

The coarse fraction was prepared by simply washing and brushing the surfaces of the rock slabs. The fine material was sieved in a column of six sieves with diminishing mesh size (2, 1, 0.5, 0.25, 0.125, 0.063 mm). Bryozoans were found mostly in the four coarser sieves. The 0.125 mm sieve contained principally foraminifers and ostracods, associated with very rare erect flexible bryozoans (cellariiforms and catenicelliforms).

The samples thus prepared were observed under a stereo microscope, the species identified and the fragments corresponding to the various zoarial forms were then counted. Whatever its size, each fragment is counted as one unit. In order to compare samples corresponding to fragments of various dimensions or even whole colonies, we used a semi-quantitative grouping: very abundant (more than 100 fragments), abundant (50 to 100 fragments), frequent (11 to 50 fragments) and rare (1 to 10 fragments).

4 ZOARIAL FORMS, EXTANT SPECIES AND PALAEOECOLOGY

4.1 *Zoarial Forms*

Bryozoan colonies from several zoarial types. Their growth has been correlated with various factors of the environment such as water depth, substrate, water energy and sedimentation rate (Hageman et al. 1997, 1998, Harmelin 1988, Lagaaij & Gautier 1965, McKinney & Jackson 1989, Moissette 2000, Nelson et al. 1988, Schopf 1969, Smith 1995, Stach 1936, 1937). Seven distinct bryozoan growth forms have been identified in this study (Table. 1).

4.1.1 *Membraniporiform*

Unilaminar colonies encrusting a hard substrate. This morphotype indicates shallow depths, moderate to high water energy and relatively slow sedimentation rates.

In the section studied, the oyster shells, but also the coral colonies and the vermetid tubes offer numerous hard substrates. The majority of the species (39) thus belong to this zoarial form. The most abundant are: *Annectocyma major, Onychocella angulosa, Cribrilaria innominata, Cribrilaria radiata, Plagioecia sarniensis, Microporella ciliata*, and *Escharoides coccinea.*

4.1.2 *Celleporiform*

The zooids are piled up without order in successive layers on various types of substrates and constitute massive, branching or multilamellar colonies. This zoarial form also indicates shallow depths with a maximum around -30 metres (Moissette 2000).

Three species belonging to this morphotype are abundant to very abundant in the studied levels: *Celleporaria palmata, Celleporina costazii*, and *Turbicellepora coronopus.*

4.1.3 *Cellariiform*

Erect flexible morphotype with long segments articulated by chitinous joints. The zoarium is attached on loose or solid substrates by its chitinous rootlets. This zoarial form is especially abundant in shallow waters (around 10 m) and, due to its flexibility, is adapted to moderate to high sedimentation rate and energy. It is also found deeper.

This morphotype is represented here by seven species. Among them, *Crisia eburnea* and *Scrupocellaria scruposa* are the more abundant.

4.1.4 *Vinculariiform*

Erect rigid colonies with sub-cylindrical dichotomous branches, attached to a solid substrate. This morphotype is found in relatively deep water (with an optimum depth around 40-70 m), but occupies also various sheltered habitats such as cavities (< 10 m).

Five species belonging to this zoarial type have been found in the section under study; they are never very abundant.

4.1.5 *Adeoniform*

Erect rigid bilaminate colonies attached to a solid

	Lower limestone	Bioherm basis	Bioherm middle	Bioherm summit	Lateral reef	Marly parting	Vermetids serpulids	Upper limestone
Membraniporiforms								
Annectocyma major	*	*	**	**	**		**	
Annectocyma indistincta		*					*	
Annectocyma sp.			*	*	*			
Bitubigera gardettensis			*					
Bitubigera sp.			*	*				
Buffonellodes incisa			*					
Calpensia gracilis	*				*			
Calpensia nobilis					*		*	
Chorizopora brongniartii					*		*	
Crepidacantha odontostoma			*	*				
Cribrilaria innominata		*	*	*	*		*	
Cribrilaria radiata		*	*	*	**		**	
Dakaria goniostoma					*			
Desmeplagioecia sp.		*	**	**	**		*	
Disporella hispida		*	*	*		*	*	
Distansescharella seguenzaii			*	*				
Electra monostachys			*	*				
Escharina vulgaris			*	*			*	
Escharoides coccinea					*	*	**	
Herentia montenatii			*					
Hincksina loxopora			*	*	*			
Hippopodinella lata			*	*				
Hippopleurifera surgens								
Hippoporella pauper			*				**	
Lagenipora lepralioides							*	
Lichenopora prolifera		*	*	*	*	*	*	*
Lichenopora fungicula							*	
Lichenopora sp.			*					
Microporella ciliata	*	*	*	*	*			*
Onychocella angulosa	*	*	**	**	**	*		
Plagioecia platydiscus							*	
Plagioecia sarniensis		*	**	*		*	*	
Schizotheca fissa				*	*			
Schizoporella longirostris	*				*		**	*
Schizoporella linearis	*							
Steginoporella altavillae								*
Steginoporella iberica iberica								*
Stomatopora parvula							*	
Tubulipora sp.							*	
Celleporiforms								
Celeporina costazii		*		*	*		**	
Celleporaria palmata		*	*	*	*	*	**	
Turbicellepora coronopus		*	**	**		*	****	
Cellariiforms								
Cellaria salicornioides						*		
Crisia eburnea	*	**	***	**	*	**	**	
Crisia hauerii		*	*		*			
Crisia hoernesii					*		*	
Margaretta cereoides						*		
Scrupocellaria elliptica				*				
Scrupocellaria scruposa	*		*	*		*	*	
Vinculariiforms								
Entalophorecia subverticellata							*	
Idmidronea atlantica						*	*	
Mecynoecia proboscidea			*	*		*	*	
Myriapora truncata			*		*	*	*	
Ybselosoecia typica		*	**	**	**	*	*	
Adeoniforms								
Schizotheca serratimargo			*	**			*	
Reteporiforms								
Sertella cellulosa		*	*	**	*		**	
Catenicelliforms								
Halysisis diaphana	*				*			
Savignyella lafontii		*						
Vasignyella otophora			*		*			

Table 1. Distribution and abundance of bryozoan species in the facies of the Djebel Tzioua reef complex.
* : rare ; ** : frequent ; ** : abundant ; **** : very abundant

substrate. This zoarial morphotype is characteristic of rocky shores, low water energy and relatively slow sedimentation rate. It has an optimum depth of about 40-50 m, but can also live in much shallower water.

In the Djebel Tzioua only one species, *Schizotheca serratimargo,* belongs to this zoarial form. According to Gautier (1962) it has two very different depth ranges: 5-10 m and 60-150 m.

4.1.6 *Reteporiform*

This colonial morphotype is characterized by an erect rigid and fenestrate zoarium, attached to hard substrates at depths of about 20 to 80 m.

In the Djebel Tzioua reef, the only species of this zoarial form, *Sertella cellulosa*, is represented by many fragments.

4.1.7 *Catenicelliform*

This is an erect articulated morphotype attached to a hard substrate by chitinous rootlets. Each very short segment is constituted only of one or two zooids. This zoarial form, adapted to high water energy, is abundant at very shallow depths (0-40 m).

Three species of this form are represented here by very rare fragments found among coral limestones.

4.2 *Extant species*

Most of the species identified in this study, 36 out of 59, are extant. Thus, data on the distribution of the bryozoans living on the Mediterranean continental shelf (Gautier 1962, Harmelin 1976, Zabala 1986) can be used in palaeoecological and palaeobathymetric reconstructions. Many species are almost eurybathic (0 to 100-300 m), but the stenobathic species (Table 2) give more precise information, especially when they show an optimum bathymetric range (indicated in parentheses). Many stenobathic species collected in the Djebel Tzioua present a maximum of frequency around _-30 m. They thus indicate very shallow depths (Moissette 1988, 2000).

4.3 *Palaeoecology*

An assemblage showing the zoarial forms, their relative abundance and the distribution of each species is a good palaeobathymetric indicator (Harmelin 1988, McKinney & Jackson 1989, Smith 1995). In this study, the assemblage is characterized by a great number of species (39) of the membraniporiform morphotype. Numerous other species belonging to various zoarial forms also occur: cellariiform (7), catenicelliform (3), celleporiform (3), vinculariiform (5), adeoniform (1), and reteporiform (1). The composition of this assemblage indicates shallow depths.

The membraniporiform species are represented on average by 10 fragments of colonies per sample and by 20 fragments for the cellariiform, celleporiform and adeoniform morphotypes. This assemblage and the presence of many shallow-water extant species suggests a bathymetric interval of 0 to 20-30 m.

The analysis of the benthic foraminifers (*Borelis melo, Ammonia, Cibicides, Elphidium...*) from the sandy levels at the base of the section (Saint Martin, 1990) also indicates a very littoral environment propitious to the development of the corals.

The abundance and diversity of the encrusting and erect forms reflect the availability of hard substrates. The hard substrates in the Djebel Tzioua are the calcareous skeleton of corals (*Porites* and *Tarbellastraea*), bivalve shells (oysters), as well as vermetid and serpulid tubes. Many encrusting species, but also erect forms find favorable hydrodynamic conditions inside the numerous cavities contained within the reef framework.

Table 2. List of some stenobathic species found in the Djebel Tzioua coral reef. Numbers between parentheses correspond to the bathymetric optimum of the species.

Species	Depth range (in metres)
Annectocyma major	3-50
Calpensia nobilis	0-60
Cellaria salicornioides	30-100 (30-80)
Crisia eburnea	10-30
Escharina vulgaris	0-100 (40-80)
Escharoides coccinea	30-80
Halysisis diaphana	0-60
Hippopodinella lata	0-30
Lagenipora lepralioides	40-80
Margaretta cereoides	10-90 (10-50)
Myriapora truncata	0-130 (30-60)
Onychocella angulosa	0-80 (30-50)
Plagioecia platydiscus	20-40 ?
Plagioecia sarniensis	25-80
Schizoporella longirostris	0-150 (20-60)
Schizotheca serratimargo	5-10
Sertella cellulosa	20-130 (30-80)
Turbicellepora coronopus	40-60
Vasignyella otophora	0-60

5 BRYOZOAN DISTRIBUTION IN THE DJEBEL TZIOUA REEF FACIES

Different units have been identified in the coral reef complex (Fig. 2). Each is characterized by a bryozoan assemblage of species and zoarial forms.

5.1 *Bioclastic limestones*

Mostly membraniporiforms are represented by rare fragments belonging to six species in the lower limestone and five in the upper one. They are cemented on bivalve shells or on coral fragments. A few fragments belonging to two cellariiform species were also found in the lower bioclastic level.

5.2 *Bioherm (base)*

Membraniporiform types predominates (10 species). Other zoarial forms also occur: celleporiform (3 species), cellariiform (2 species), vinculariiform (1 species), reteporiform (1 species), and catenicelliform (1 species). They are however all represented by rare fragments. The establishment of the bryozoans seems to correspond to that of the corals. The bryozoans find here the adequate substrates for their attachment within the reef framework. The very small number of species of the erect rigid type (vinculariiform and reteporiform) indicates adverse conditions.

5.3 *Bioherm (middle part)*

Compared to the previous level, there is an increase in the number of fragments and species of membraniporiforms (22). The number of cellariiform (3) and vinculariiform (2) species increases slightly. The quantity of fragments of celleporiform, cellariiform and vinculariiform colonies rises as well. In addition, a few fragments of one species of adeoniform and one species of catenicelliform occur. The rise in diversity and number of fragments indicates that the area for attachment (abundant branching corals) increased correspondingly.

5.4 *Bioherm (summit)*

The number of species of the membraniporiform type decreases (from 22 to 18), although this morphotype is still represented by a great number of colony fragments. There are more fragments of adeoniforms, reteporiforms, celleporiforms and cellariiforms. The greater number of erect morphotypes indicates a possible increase in the development of cavities where they are attached, sheltered from strong currents.

5.5 *Lateral reef*

The bryozoans found in this facies (with lamellar colonies of *Tarbellastraea* and *Porites* and many bivalve shells) are almost the same as those occurring in the main bioherm: membraniporiform (16), celleporiform (2), cellariiform (3), vinculariiform (2), reteporiform (1), and catenicelliform (2). However, a slight decrease in the number of species is noticed for the membraniporiforms and celleporiforms and the adeoniform is absent.

5.6 *Marly parting*

In this facies, membraniporiform (5), celleporiform (2), cellariiform (4), and vinculariiform (3) species occur. The rare fragments of membraniporiform and vinculariiform species indicate some small hard substrates (shells, corals, etc.), whereas the cellariiforms were at least partly attached to the soft substrate by their chitinous rootlets.

5.7 *Vermetid and serpulid level*

Compared to the main bioherm, there is an increase in the number of species and/or fragments of the following zoarial forms: membraniporiform (19), celleporiform (3), cellariiform (3), vinculariiform (4), adeoniform (1), and reteporiform (1). There is a high number of fragmented colonies belonging to a celleporiform species (*Turbicellepora coronopus*). This hard substrate offered space necessary to the attachment of these various species.

6 ECOLOGICAL SUCCESSION IN THE REEF COMPLEX

6.1 *Corals in the Djebel Tzioua coral reef complex*

The reef framework of the Djebel Tzioua is mostly formed by coral colonies of *Tarbellastraea* locally associated with *Siderastraea* and *Porites* cf. *collegniana.* The principal frame-builder develops in very tight columnar colonies leaving only little space for *Porites* but also for the sediment fillers (Chaix et al. 1986, Saint Martin 1990). However, these branching forms offer sheltered and dark zones favourable to the epilith fauna (bivalves, annelids, and vermetids) and the cryptic fauna (bryozoans). Bryozoans encrust and cover the cavity walls or are attached to the bivalve shells (*Ostrea lamellosa),* themselves cemented on the coral substrate.

6.2 *The Alberstadt et al. model*

The composition, the diversity and the organisation of the biological assemblages in Ordovician reefs changes during the growth of the structure (Alberstadt et al. 1974). This is due to the interactions between organisms, under the control of various physical factors (water depth, sedimentation rate and currents). These modifications lead to an ecological succession characterized by an increase in the diversity and in the complexity of the reef structure and its stability. Four ecological stages are recognised during the growth of a reef (Alberstadt et al. 1974, James 1979a, b, James & Bourque 1992).

Stabilization: corals require hard substrates to settle. This stage is characterized by the sedimentation of bioclastic sands, followed by the installation of sessile epibenthic organisms: corals, coralline algae, encrusting bryozoans, and sponges.

Colonization: at this stage the growth of the coral reef is controlled by the substrate type and the sedimentation rate. The coral assemblage is characterized by few species of massive or lamellar forms. Some delicate branching forms appear and create niches and cavities which will be colonized by other organisms.

Diversification: an increase in the number of coral species is observed and a great variety of growth forms appears: delicate branching, lamellar, encrusting, massive, etc. The result is a greater area of fixing and the structure of the reef rises above the sea floor. During this phase, the diversification of the coral community has an influence on the environment.

Domination: at this stage large encrusting and laminar growth forms predominate, but the diversity decreases. This abrupt change in growth forms seems to correspond to the effects of the wave surge as the reef comes near the surface. The expansion phase of the reef results in a mature and endemic community.

6.3 *Application of the model to the bryozoans of the Djebel Tzioua coral reef complex*

The distribution of the zoarial forms in the main bioherm reveals a vertical zonation. The ecological succession previously established for the corals seems also suitable for the bryozoans within the coral reef.

Stabilization: corresponds to the base of the bioherm. The bryozoans begin their colonization with a few membraniporiform colonies cemented on oyster shells and under some isolated corals. At this stage the erect morphotypes are very rare.

Colonization: this phase is restricted to the middle part of the bioherm. The colonization by membraniporiform, cellariiform and vinculariiform colonies occurs through an increase in the surfaces available for their attachment.

Diversification: this stage corresponds to the highest part of the coral formation. The great number of species and colonial forms (membraniporiform, celleporiform, cellariiform, catenicelliform, vinculariiform, adeoniform, and reteporiform) indicates favourable environmental conditions: numerous cavities, sedimentation rate and water energy.

Domination: this phase is not really distinct from the diversification phase. The membraniporiform morphotype is still abundant. There is nevertheless a slight increase in the number of fragments of adeoniform and reteporiform colonies.

7 DISCUSSION

In his study of the corals of the Djebel Tzioua, Saint Martin (1990) also recognized a vertical zonation. Although the stabilization and colonization stages are well developed (with branching and massive colonies of *Tarbellastraea* and *Porites*), the small size of the reef builders indicates that the construction has not reached the same stage of evolution as the modern reefs. Diversification and domination phases for the main frame-builders are thus missing in the Djebel Tzioua (Saint-Martin, 1990).

For the bryozoans contained in the same reef, all phases of reef construction except the domination phase have been identified. It was made possible because colonization and diversification of the zoarial forms were controlled by the coral framework offering hard substrates, cavities, crevices, fissures and tunnels where shaded cavities are used as permanent habitat (Harmelin 1969, 1980, 1985, Kobluk, 1988).

Membraniporiform colonies cover and cement the roof, the floor and the walls of the cavities on several layers. Some species close the opening of the cavity and seem to protect the fragile erect forms which are attached slightly deeper inside the cavity, sheltered from currents and predators. In this environment, shallow and deep-water species meet and mix. Some encrusting and erect forms also live under coral colonies and bivalve shells.

The reef community in the bioherm reached its maximum of growth and expansion towards the water surface. This is corroborated by the development above the bioherm of the vermetid and serpulid facies, which requires shallow environments. In this facies the membraniporiforms find the substrates favourable to their installation and the celleporiforms are very abundant. Rare erect colonies (cellariiforms, adeoniforms, vinculariiforms and reteporiforms) also occur.

8 CONCLUSION

The Messinian reef environment of the Djebel Tzioua is characterized by a high bryozoan diversity and the relative abundance of their various zoarial forms. Corals are the main frame-builders and bryozoans the accessory frame-builders. Separately they display a vertical zonation within the reefal mosaic. The zonations developed by corals and bryozoans have been identified using a theoretical model based on the construction of a reef by corals and other invertebrates (Alberstadt et al. 1974).

The coral zonation depends on the nature of the substrate and on the biotic changes in the ecological succession throughout the reef growth (Alberstadt et al. 1974, James & Bourque 1992, Perrin 2000).

The bryozoan zonation within the reef also depends on the substrate and mostly on the space available for attachment. This space is created by corals of branching or lamellar forms. Thus, the phases of colonization, diversification and domination can be distinguished. Other environmental factors also contributed to their installation in the reef: sedimentation rate and hydrodynamism.

The variations observed in the Djebel Tzioua are to be placed in a wider Mediterranean context. After the major Tortonian transgression (7 Ma) the installation of the carbonate platforms, with coralline algal limestones followed by coral reefs, indicates a regressive episode. The downward migration of the coral constructions (Rouchy 1982, Rouchy et al. 1986, Saint Martin & Rouchy 1986) between 6.1 and 5.9 Ma is completed by the development of stromatolithic and annelid constructions. The significant lowering of the water depth also led to the deposition of evaporites at the end of the Messinian (Rouchy 1982, Rouchy & Saint Martin 1992).

ACKNOWLEDGMENTS

We would like to thank Dr. J.-P. Saint Martin (University of Marseilles) for his help during the preparation of this paper and A. Armand and P. Desvignes (University of Lyon) for preparing respectively figure 1 and the fossil material.

REFERENCES

Alberstadt, L.-P. Walker, K.-R.& Zurawski, R.-P. 1974. Patch reefs in the Carters limestones (Middle Ordovician) in Tennessee, and vertical zonation in Ordovician Reefs. *Geological Society of America Bulletin* 85(7): 1171-1182.

Chaix, C., Moissette, P. & Saint Martin, J.-P. 1986. Réflexions sur les biocénoses et paléobiocénoses en milieu récifal (Messinien d'Algérie). *Bulletin du Muséum national d'Histoire Naturelle Paris* 2: 219-230.

Choi, D.R. & Ginsburg, R.N. 1983. Distribution of coelobites (cavity-dwellers) across the Florida reef tract. *Coral Reefs* 2: 165-172.

Cuffey, R.J. 1971. Bryozoan distribution in the modern reefs of Eniwetok atoll and the Bermuda platform. *Pacific Geology* 6: 25-50.

Cuffey, R.J. 1974. Delineation of bryozoan constructional roles in reefs from comparison of fossil bioherms and living reefs. *Proceedings of the Second International Coral reef Symposium* 1: 357-364.

Cuffey, R.J. 1977. Bryozoan contributions to reefs and bioherms through geologic time. *American Association of Petroleum Geologists, Studies in Geology* 4: 181-194.

Cuffey, R.J. 1985. Expanded reef-rock textural classification and the geologic history of bryozoan reefs. *Geology* 13: 307-310.

Esteban, M. 1979. Significance of the upper Miocene reefs in the western Mediterranean. *Palaeogeography, Palaeoclimatology, Palaeoecology* 29: 169-188.

Garrett, P., Smith, D.L., Wilson, A.O. & Patriquin, D.1971. Physiography, ecology and sediments of two Bermuda patch reefs. *Journal of Geology* 79: 647-668.

Gautier, Y.V. 1962. Recherches écologiques sur les Bryozoaires Chilostomes en Méditerranée occidentale. *Recueil des Travaux de la Station Marine d'Endoume* 24 (38): 1-434.

Guardia, P. 1975. *Géodynamique de la marge alpine du continent africain d'après l'étude de l'Oranie nord occidentale.* Unpublished thesis, University of Nice.

Hageman, S.J., Bone, Y., McGowran, B. & James, N.P. 1997. Bryozoan colonial growth-forms as paleoenvironmental indicators: evaluation of methodology. *Palaios* 12: 405-419.

Hageman, S.J., Bock, P.E., Bone, Y. & McGowran, B. 1998. Bryozoan growth habits: classification and analysis. *Journal of Paleontology* 72(3): 418-436.

Harmelin, J.-G. 1969. Bryozoaires des grottes sous-marines obscures de la région marseillaise. Faunistique et écologie. *Tethys* 1(3): 793-806.

Harmelin, J.-G., 1976. Le sous-ordre des Tubuliporina (Bryozoaires cyclostomes) en Méditerranée. *Mémoires de l'Institut Océanographique de Monaco* 10: 1-326.

Harmelin, J.-G., 1980. Établissement des communautés des substrats durs en milieu obscur, résultats préliminaires d'une expérience à long terme en Méditerranée. *Memorie de Biologia Marina e Oceanografia* 10: 29-52.

Harmelin, J.-G. 1985. Bryozoan dominated assemblages in Mediterranean cryptic environments. In C. Nielsen & G.P. Larwood (eds), *Bryozoan: Ordovician to Recent*: 135-143. Fredensborg: Olsen and Olsen.

Harmelin, J.-G. 1988. Les bryozoaires, de bons indicateurs bathymétriques en paléoécologie? *Géologie Méditerranéenne* 15(1): 49-63.

Hughes, D.J. & Jackson, J.B.C. 1992. Distribution and abundance of cheilostome bryozoans on the Caribbean reefs of central Panama. *Bulletin of Marine Science* 5: 443-465.

Jackson, J.B.C. & Winston, J.E. 1982. Ecology of cryptic coral reef communities. I: Distribution and abundance of major groups of encrusting organisms. *Journal of Experimental Marine Biology and Ecology* 57: 135-147.

James, N.P. 1979a. Introduction to carbonate facies models. In R.G. Walker (ed), *Facies models*: 105-107. Geoscience Canada, 1.

James, N.P. 1979b. Reefs. In R.G. Walker (ed), *Facies models*: 121-132. Geoscience Canada, 1.

James, N.P. & Bourque, P.-A. 1992. *Sédimentologie, diagenèse et porosité des carbonates.* Groupe Interuniversitaire de Recherches géologiques en Analyse de Bassins. Département de Géologie, University of Laval, Québec (unpublished).

Kobluk, D.R. 1988. Cryptic faunas in reefs. Ecology and geologic importance. *Palaios* 3: 379-390.

Lagaaij, R. & Gautier, Y.V. 1965. Bryozoan assemblages from marine sediments of the Rhône delta, France. *Micropaleontology* 11(1): 39-58.

Logan, A. 1981. Sessile invertebrate coelobite communities from shallow reef tunnels, Grand Cayman. *Proceedings of the 4th International Coral reef Symposium* 2: 735-744.

McKinney, F.K. & Jackson, J.B.C. 1989. *Bryozoan Evolution.* Boston: Hyman.

Moissette, P. 1988. Faunes de bryozoaires du Messinien d'Algérie occidentale. *Documents des Laboratoires de Géologie de Lyon* 102: 1-351.

Moissette, P. 1993. Bryozoan assemblages in Messinian deposits of western Algeria. *Lethaia* 26: 247-259.

Moissette, P. 1997. Bryozoaires récoltés dans les unités messiniennes de sondages offshore dans le golfe de Gabès (Tunisie). *Revue de Micropaléontologie* 40(2): 181-203.

Moissette, P. 2000. Changes in bryozoan assemblages and bathymetric variations. Examples from the Messinian of northwest Algeria. *Palaeogeography, Palaeoclimatology, Palaeoecology* 155: 305-326.

Moissette, P. & Saint Martin, J.-P. 1995. Bryozoaires des milieux récifaux miocènes du sillon sud-Rifain au Maroc. *Lethaia* 28: 271-283.

Nelson, C.S., Hyden, F.M., Keane, S.L., Leask, W.L. & Gordon, D.P. 1988. Application of bryozoan zoarial growth-form studies in facies analysis of non-tropical carbonate deposits in New Zealand. *Sedimentary Geology* 60: 301-322.

Neurdin-Trescartes, J. 1992. *Le remplissage sédimentaire du bassin néogène du Chélif, modèle de référence de bassins intramontagneux*. Unpublished thesis, University of Pau.

Perrin, C. 2000. Changes of palaeozonation patterns within Miocene coral reefs, Gebel Abu Shaar, Gulf of Suez, Egypt. *Lethaia* 33: 253-268.

Perrodon, A., 1957. Etude géologique des bassins Néogènes littoraux de l'Algérie occidentale *Bulletin du Service de la Carte Géologique de l'Algérie* 12: 1-323.

Rouchy, J.-M. 1982. La genèse des évaporites Messiniennes de Méditerranée. *Mémoires du Muséum national d'Histoire Naturelle Paris* 50, 1-267.

Rouchy, J.-M., Saint Martin, J.-P., Maurin, A. & Bernet-Rollande, M.-C. 1986. Évolution et antagonisme des communautés bioconstructrices animales et végétales à la fin du Miocène en Méditerranée occidentale: Biologie et Sédimentologie. *Bulletin des Centres de Recherche et d'Exploration-Production Elf Aquitaine* 10(2): 333-348.

Rouchy, J.-M. & Saint Martin, J.-P. 1992. Late Miocene events in the Mediterranean as recorded by carbonate-evaporite relations. *Geology* 20: 629-632.

Saint Martin, J.-P. 1990. Les formations récifales coralliennes du Miocène supérieur d'Algérie et du Maroc. Aspects paléoécologiques et paléogéographiques. *Mémoires du Muséum national d'Histoire Naturelle Paris* 56: 1-373.

Saint Martin, J.-P & Rouchy, J.-M. 1986. Intérêt du complexe récifal du Cap des trois fourches (Bassin de Nador, Maroc septentrional) pour l'interprétation paléogéographique des événements messiniens en Méditerranée occidentale. *Comptes-Rendus Académie des Sciences Paris* 2, 302(15): 957-962.

Schopf, T.J.M. 1969. Paleoecology of Ectoprocts (bryozoans). *Journal of Paleontology* 43: 234-244.

Smith, A.M. 1995. Palaeoenvironmental interpretation using bryozoans: a review. In D.W.J. Bosence & P.A. Allison (eds), *Marine Palaeoenvironmental Analysis from Fossils*: 231-243. Geological Society Special Publications, 83.

Stach, L.W. 1936. Correlation of zoarial form with habitat. *Journal of Geology* 44: 60-66.

Stach, L.W. 1937. The application of the Bryozoa in Cainozoic stratigraphy. *Report of the 23rd Meeting, Australia-New Zealand Association of Advanced Science* 23: 80-83.

Zabala, M. 1986. Fauna dels Briozous del països Catalans. *Institut d'Estudis Catalans, Arxius de la Seccio de Ciencies* 84: 1-294.

Bryozoan Studies 2001, Wyse Jackson, Buttler & Spencer Jones (eds)
© 2002 Swets & Zeitlinger, Lisse, ISBN 90 5809 388 3

Effects of sewage on bryozoan diversity in Mediterranean rocky bottoms

J.G. Harmelin & S. Capo
Centre d'Océanologie de Marseille, CNRS-UMR 6540, Station Marine d'Endoume, 13007 Marseille, France

ABSTRACT: Impacts of sewage from Marseilles (France, Mediterranean Sea) on sublittoral bryozoans were assessed in the rocky area including the sewer outlet, which is predominantly submitted to westward flow. The assemblage features were assessed on both sides of the sewer outlet at short (0.4 km), intermediate (0.7-1.3 km) and long (2.4-5 km) distances from it. Three assemblage categories were considered: the cryptic fauna of block undersides, the large, erect bryozoans, and the 5 and 12 month-old assemblages colonising artificial cryptic habitats. Each of these fauna categories showed abrupt drop in diversity and abundance at the nearest sites and intermediate west site. A set of species highly frequent in remote sites including a large proportion of cyclostomes were excluded from the most polluted areas, where a few species (*Schizoporella dunkeri*, *Tubulipora aperta*) were particularly successful. Large, erect cheilostome species proved to be highly sensitive to pollution, except *Myriapora truncata*, which was relatively tolerant.

1 INTRODUCTION

The rise of urbanisation of coastal zones during the last decades has considerably increased the pressure of pollution on nearshore benthic communities and subsequently the need for reliable bioindicators that could give early warnings of the alteration of communities and loss of biodiversity (Underwood & Peterson 1988, Noss 1990). Bioassessment studies of sublittoral hard-substrate communities are relatively few in comparison to those devoted to soft-bottom communities (e.g. Pearson & Rosenberg 1978, Reish & Bellan 1996). As stressed by Chapman et al. (1995) impacts of sewage on filter-feeding invertebrates are little documented. This applies to bryozoans, particularly at assemblage level. Most information about impact of pollution on bryozoans comes from harbour communities (e.g. Geraci & Relini 1970, Scholz 1990, 1991, Soule & Soule 1981) or effects of contaminants on particular species (e.g. *Bugula neritina*: review in Gordon & Mawatari 1992, *Schizoporella unicornis*: Powell et al. 1970). Evidence that bryozoans can be most sensitive to metal contamination was given by Ward & Young (1984) in a field transplant experiment on epifauna-bearing *Pinna* shells. Some indications on the impoverishment of the bryozoan assemblages in a region impacted by industrial wastes are found in Harmelin & Hong (1979) and Hong (1980, 1983).

Alteration of marine environment from sewage is a major concern for the Mediterranean countries. This type of contaminant source and the resulting impacts are particularly heterogeneous among local situations, especially because of the diversity of wastewater composition and volume. On the other hand, in most cases the point source nature of sewage discharge generates clear spatial gradients on which assessment studies can be easily managed. The discharge of sewage from Marseilles and its suburbs (France, Mediterranean Sea) on a rocky shore apart from any other proximate source of pollution gives the opportunity to test the impact of this pollution on bryozoan assemblages in a zone which has received many monitoring studies (review by Arfi et al. 2000) and where the distribution of bryozoans is well-known (e.g. Gautier 1962, Harmelin 1976).

In this study, we chose to test the effects of Marseilles' sewage on rocky bottom bryozoans through the response of three types of assemblages: (i) encrusting species of small cryptic habitats, (ii) large erect species, and (iii) colonists of artificial cryptic habitats. Small cryptic habitats in rocky bottoms usually gather large proportions of the local bryozoan species pool (Harmelin 2000) and thus can supply relevant information on the local state of bryozoan diversity. Undersides of small blocks are particularly suitable to the bioassessment of rocky bottoms because this habitat is widespread, relatively homogeneous among sites (flat structure and

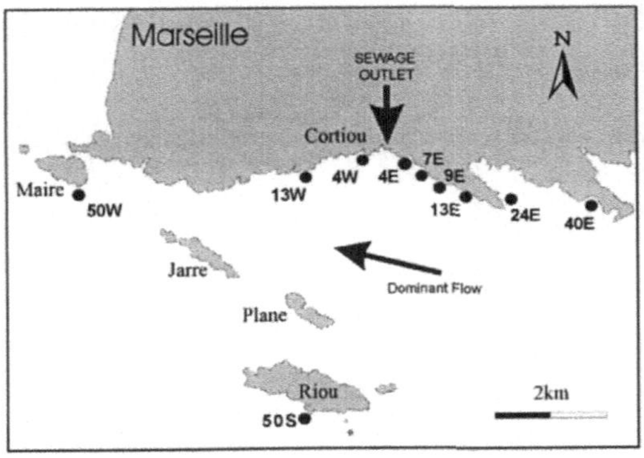

Figure 1. Location of the study sites.

sheltered from light), and close to bottom where sediments may store pollutants. Large, rigidly erect bryozoans are commonly found in the Mediterranean sublittoral shadowed rocky habitats. They are important descriptors of landscape biodiversity, which can be identified and monitored underwater (Cocito et al. 1998). Another reason for selecting large erect bryozoans was their permanent exposure to water-borne contaminants due to their three-dimensional growth in the water column and large size. Successional stages in cryptic environment usually include many bryozoan species (e.g. Harmelin 1976). Artificial habitats can be useful tools for testing the variability of bryozoan colonisation under various disturbance regimes. When appropriately designed they allow easy replication of standardised microhabitats within and among sites.

2 MATERIAL AND METHODS

2.1 *Study sites*

The studied coast is located a few kilometres south of Marseilles (France). It is characterised by steep limestone rocky shores with a general east-west orientation and includes a group of small islands (Riou archipelago) (Fig. 1). The hydrology of this region is typically variable but is predominantly influenced by (i) nearshore branches of the Northern Current (= Liguro-Provençal Current), which flows westward, and (ii) wind-induced mesoscale events resulting in down- and upwellings and counter-currents that modify sporadically the dominant E-W circulation, especially when the Mistral wind blows from NNW (Castelbon 1972, Millot & Wald 1980). The environment of this region is markedly contrasted at small geographical scale with on the one hand relatively preserved bottoms presenting many biological components of high patrimonial value and on the other hand a highly impacted zone around the outlet of the main sewer of Marseille and neighbouring small cities (about 1.5 million inhabitants). This sewer collector opens directly at the surface on the mainland rocky shore (Cortiou cove). Sewage discharge has occurred there since 1886 but the resulting impacts increased considerably during the seventies following the rise of detergent use and urbanisation and, after 1983, with the permanent diversion of the polluted Huveaune river in the main sewer. During the last decade, the environmental quality improved thanks to the treatment of industrial wastes at the source and of sewage waters in an advanced primary treatment plant in operation since 1987 (Bellan et al. 1999). Previous studies on patterns of water circulation (Castelbon 1972), spatial distribution of contaminants and of sandy bottom fauna (Arfi et al. 2000) indicate a zoning of effects and a dominant westward spreading of sewage waters. Therefore, we chose to locate our study sites on both east and west sides of Cortiou cove except for one southern, remote site. These study sites were selected according to three main categories of distance from the outlet: 4-5 km for remote sites, expected to be control sites, 1.3 km for intermediate sites, and 0.4 km for the nearest sites (Table 1). When selecting these sites, particular attention was paid to similarity in their intrinsic features (e.g. topography, rugosity, depth, orientation).

Table 1. List of study sites and location (distance and orientation) relative to Cortiou sewer outlet and types of bryozoan assemblages studied in each site.

Site	Name	Distance (km)	Side	Fauna*
50W	Maire Island	5.0	west	2
13W	Cirque Cove	1.3	west	1-2-3
4W	Melette Islet	0.4	west	1-2-3
4E	Dark Cove	0.4	east	1-2-3
7E	Vaisseau A	0.7	east	2
9E	Vaisseau B	0.9	east	2
13E	Vaisseau C	1.3	east	1-3
24E	Sormiou	2.4	east	1-2
40E	Figuier	4.0	east	3
50S	Riou Island	5.0	south	1-2-3

* 1: fauna of cryptic microhabitats, 2: large, erect species, 3: colonists of artificial cryptic habitats.

2.2 *First approach: bryozoans from small cavities*

Between-site comparisons of bryozoan diversity of cryptic habitats were based on assemblages living underneath blocks lying on rocky substrates at similar depth. In each site, 30 quadrats of 25 cm^2 were sampled at random on undersides of boulders collected at 17-19m depth. Species and the whole percentage cover of bryozoans were recorded in each quadrat. For highlighting between-site differences the stock of the most frequent species was considered by listing species having the four first ranks of occurrence frequency in each site. The percentage cover was estimated under the microscope by point-intercept method (grid of 10 points in eyepiece, 300 points per quadrat) (Sutherland & Karlson 1977).

Six sites were studied at three distances from the sewer outlet (Table 1).

2.3 *Second approach: Large erect bryozoans*

Impacts of pollution on the guild of the large rigidly erect bryozoans were assessed through the local occurrence of six species common in the Provence region at relatively shallow depth: *Adeonella calveti* Canu & Bassler, 1930, *Myriapora truncata* (Pallas, 1766), *Pentapora fascialis* (Pallas, 1766), *Reteporella septentrionalis* (Harmer, 1933), *Smittina cervicornis* (Pallas, 1766), and *Turbicellepora avicularis* (Hincks, 1862). Colonies occurring in 0.25 m^2 quadrats were identified and counted *in situ*. In each site 30 to 35 quadrats randomly distributed onto potentially favourable shadowed rocky walls were sampled at 11-28 m depth. Eight sites were sampled at various distances from the sewer outlet (Table 1).

2.4 *Third approach: Colonists of small artificial cryptic habitats.*

Artificial habitats designed for this study were composed of six 8 x 10.6 cm PVC plates fixed in two piles within a mesh box (32 x 12 x 9 cm) made of plastified wire. Each plate was flanked by two vertical, 3.5 cm high, sides oriented downwards. In each study site two of these mini-artificial reefs (MARs) were attached on heavy breezeblocks and placed at 17-19 m depth among boulders and rocks. MARs were immersed for five months (12/97 – 04/98) and 12 months (07/99 – 07/00) in six sites on the east, west and south sides of the sewer outlet at short, intermediate and long distances from it (Table 1). Species composition, species richness, percentage cover of whole assemblage and of dominant species were considered only on the plate undersides. The within-site occurrence frequency of species was measured in terms of presence-absence data on plates (range: 0-12). This was preferred to abundance measured by colony numbers because of bias introduced by aggregations of juvenile colonies often observed on these substrates. A list of the most frequent species in each site was established by deleting species which occurred on less than 50% of plates. The percentage cover of bryozoans was estimated following the same procedure as for block undersides.

Statistical analyses were made using Statistica © software. Data were transformed when necessary before ANOVA analyses.

3 RESULTS

3.1 *Bryozoans from small cavities*

The mean species richness observed on 25 cm^2 quadrats was significantly different among sites (1-way ANOVA on Log (x) transformed data: $F_{5\text{-}161}$ = 28.69, p = 0.000). SNK post-hoc tests confirm the evidence (Fig. 2) that species richness was very unequally distributed between two groups of three sites : (i) a group of sites with high species diversity, comprising the two furthest sites (50S, 24E) and the eastern intermediate site (13E), and (ii) a group of sites with much lower species richness, comprising the closest sites (4E, 4W) and the western intermediate site (13W). These two groups of sites were particularly well discriminated by species-sample size curves, i.e. species richness on substrates of increasing size (Fig. 3). In the three "poor" sites, richness rapidly reached a plateau and the value achieved when cumulating the whole 30 quadrats (750 cm^2) was similar or lower to that observed on only two cumulated quadrats (50 cm^2) in the three "rich" sites.

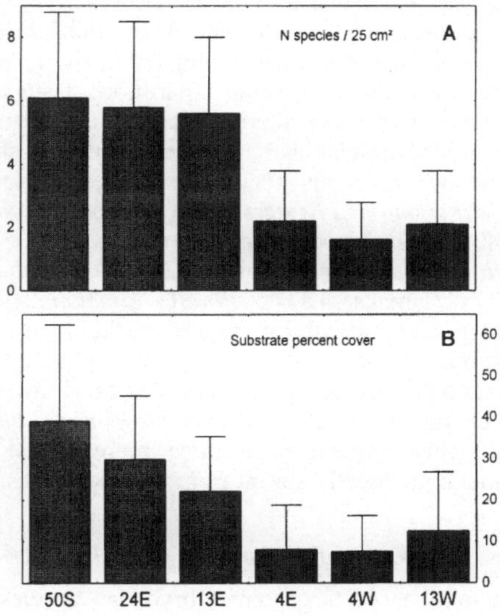

Figure 2. Bryozoans from small cavities in sites variously impacted by sewage. A: mean number of species per 25 cm^2 + SD. B: mean percentage cover (+ SD) in the same samples (30 quadrats per site).

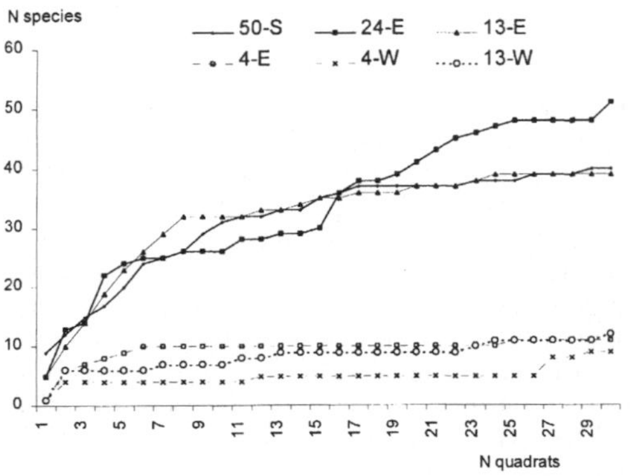

Figure 3. Bryozoans from small cavities: species-sample size curves (1-30 quadrats of 25 cm²) in sites variously impacted by sewage.

The rate of substrate cover by bryozoans was also significantly different between sites (1-way ANOVA on arcsine transformed data: $F_{5.174}$ = 26.21, p = 0.000) and presented the same geographic pattern than species richness (Fig. 2). However, post-hoc tests indicated that among the three "rich" sites the cover rate was significantly higher in the most remote site (50S). The same bipartition of sites also appeared when considering occurrence frequency data in the assemblage restricted to the most frequent species. In this stock of 13 species, six species [*Crisia sigmoidea* Waters, 1916, *Diplosolen obelia*, 1838, *Entalophoroecia deflexa* (Couch, 1844), *Fenestrulina malusii* (Audouin, 1826), *Hincksina flustroides* (Hincks, 1877), *Porella concinna* (Busk, 1854)] were particularly frequent in the three "rich" sites but were absent (5 species) or very poorly represented (1 species) in the "poor" sites. Conversely, the encrusting cheilostomate *Schizoporella dunkeri* was highly frequent in the three impoverished sites although practically absent from the "rich" sites.

3.2 *Distribution of the large, erect bryozoans*

The number of large, erect bryozoan species observed in each site showed a clear rarefaction pattern in relation with decreasing distance from the sewer outlet and orientation (Fig. 4). The whole set (6 species) occurred in the furthest sites, Riou and Maire (50S, 50W), but the southern site was significantly richer than the western site (2.5 ± 1.4 vs 1.5 ± 1.3 species/quadrat, $F_{1\text{-}135}$ = 19.23, p = 0.000). On the east side of Cortiou species disappeared successively when approaching the outlet: five species were present at 2.4 km, four at 0.9 km, one at 0.7 km, and zero at 0.4 km. On the western side, all species were excluded from the closest (4W) and intermediate (13W) sites.

The mean abundance of colonies showed the same geographical pattern as species richness (Fig. 4) with the highest value at Riou (50S) and clear impoverishment in the eastern sites when distance to the outlet decreased. Univariate analysis confirmed that colony densities were significantly different between sites (ANOVA on log (x + 1) transformed data: $F_{4\text{-}159}$ = 12.58, p = 0.000) and SNK post-hoc tests indicated that density in the furthest western site (50W) was not significantly different from that observed in the two closer eastern sites 24E and 9E.

Changes in colony density along these gradients were not uniform among species. The most tolerant species to the situation induced by drainage from Cortiou was *Myriapora truncata*, the only species recorded at site 7E and clearly dominant at sites 9E and 24E. Moreover, in these three sites the density of this species (respectively 0.9, 1.9 and 2.4 colonies per quadrat) was higher than in the most remote sites (50S and 50W: 0.5 and 0.3 colonies per quadrat). Conversely, the other species appeared to be much more severely affected. *Adeonella calveti* appeared to be the most sensitive large erect species. It reached the first rank of abundance at site 50S with a relatively high density (2.4 col./quadrat) and the second rank at site 50W with a lower density (0.7 col./quadrat) but was excluded from the other sites, except site 9E where its occurrence was vestigial. Densities in colonies of *Smittina cervicornis*, *Reteporella septentrionalis* and *Turbicellepora avicularis* also collapsed abruptly when approaching the sewer outlet. The same pattern was observed in *Pentapora fascialis*, but the gradiental decrease was less pronounced.

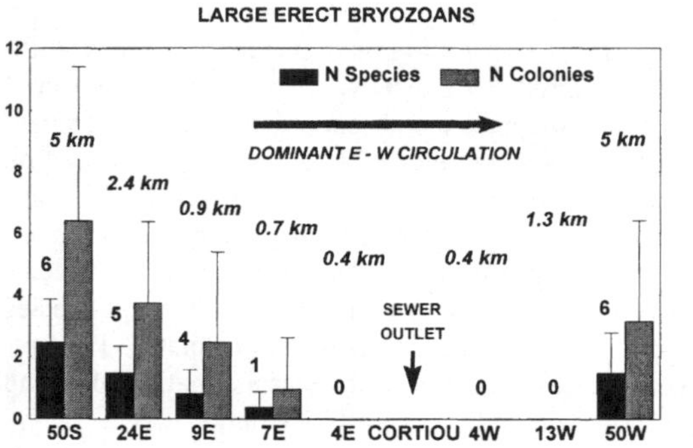

Figure 4. Mean species richness and mean density of large, erect bryozoans in relation with Cortiou sewer outlet. Numerals above histobars of species numbers are the total number of large erect species recorded in site samples.

3.3 *Fauna of artificial cryptic habitats*

At both immersion times, 5 and 12 months, species richness and composition of bryozoan assemblages in the small, artificial, cryptic habitats indicated the same clear-cut discrimination between "rich" and "poor" sites than in the former approaches.

Assemblages observed in MARs from sites 50S, 40E and 13E gathered high proportions of the whole fauna recorded after 5 months (respectively 67.6, 78.4, 56.8% of 37 species) and 12 months (respectively 51.5, 57.4 and 66.2% of 68 species). In sites 13W, 4W and 4E, relative species richness was markedly lower: respectively 35.1, 21.6 and 24.3% of the whole assemblage at 5 months and 23.5, 22.1 and 26.5% at 12 months (Fig. 5). The highest species number recorded on individual plates was 15 species at 5 months (40E) and 26 species at 12 months (13E). The lowest values were found in sites 4E and 4W (2 species) at 5 months and in sites 4E and 13W (6 species) at 12 months.

Differences in mean number of species per plate (Table 2) were highly significant (ANOVA on Log x transformed data) at 5 months ($F_{5,66}$ = 43.11, p = 0.000) and 12 months ($F_{5,66}$ = 48.08, p = 0.000). Scheffé post-hoc tests identified the same bipartite association of sites as above at both immersion times (50S = 40E = 13E > 13W = 4W = 4E) while SNK tests gave more hierarchical results: 50S = 40E = 13E > 13W > 4W = 4E at 5 months and 13E > 50S = 40E > 4W > 4E = 13W at 12 months. Clustering of sites based on species occurrence frequency (Fig. 6) was the same at both colonisation times either considering the whole assemblage or that restricted to frequent species. Two pairs of tightly associated sites are identified, 50S-40E and 13W-4W, the first being linked to site 13E and the second to site 4E. Interestingly, the same site clustering is obtained from simple site presence-absence data.

In 1-yr old assemblages the differences in species richness among the two groups of sites were much more pronounced among cyclostomes than among cheilostomes (Fig. 7). The average number of cyclostome species per plate ranged from 7.6 to 9.7 in sites 50S, 40E and 13E and only from 2.1 to 3.8 in sites 13W, 4W and 4E. The mean cyclostome / cheilostome ratio was particularly high (≥ 1) on plates from the three "rich" sites and nearest east site 4E, and was much lower (around 0.4) on plates from the nearest and intermediate western sites (4W, 13W).

As in small cavities of natural rocky habitats, *Schizoporella dunkeri* was particularly frequent at both colonization times in the "poor" sites (13W-4W-4E). At five months it occurred also in the "rich" sites, relatively frequently in the intermediate east site but only occasionally in the remote east and south sites. At 12 months *S. dunkeri* was not recorded in "rich" sites, except an anecdotal occurrence in the intermediate east site.

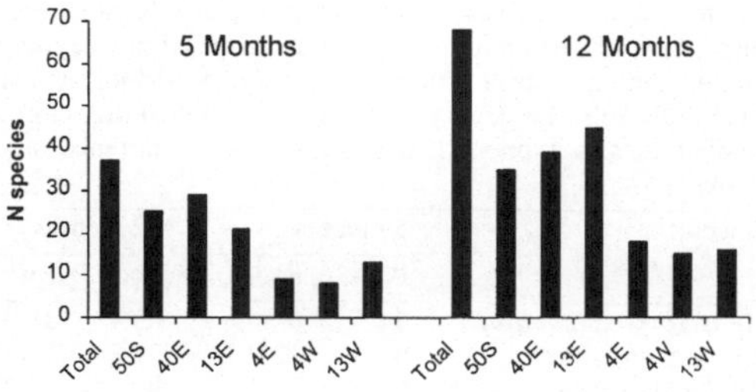

Figure 5. Species richness recorded on artificial substrates (2 MARs = 12 plates / site) after 5 and 12 months of immersion.

Conversely, many species (e.g. *Diplosolen obelia*, *Disporella* sp., *Fenestrulina malusii*, *Schizoporella unicornis*) occurred exclusively or predominantly in the "rich" sites with high frequency at both immersion times (Table 3).

Table 2. Mean species richness and mean percentage cover of bryozoans on plates (85 cm²) of MARs immersed during 5 and 12 months.

Time (months)	Site	Species richness*	Cover**
5	50S	9.0 ± 1.9 (25) 7-12	10.9 ± 6.7
5	40E	10.8 ± 2.8 (29) 7-15	15.6 ± 9.3
5	13E	8.4 ± 1.8 (21) 6-12	6.8 ± 2.5
5	4E	3.3 ± 1.0 (9) 2-5	1.3 ± 0.7
5	4W	3.7 ± 1.2 (8) 2-5	2.0 ± 1.5
5	13W	4.9 ± 1.2 (13) 3-7	7.1 ± 3.3
12	50S	15.1 ± 2.8 (35) 10-20	32.9 ± 9.1
12	40E	16.1 ± 2.9 (39) 9-19	46.7 ± 14.4
12	13E	19.6 ± 8.2 (45) 14-26	28.1 ± 5.1
12	4E	8.4 ± 2.0 (18) 6-12	26.9 ± 10.5
12	4W	10.3 ± 1.3 (15) 7-12	33.3 ± 13.3
12	13W	7.7 ± 1.3 (16) 6-11	34.4 ± 9.9

* Mean species number per plate ± SD (site richness) range.
** Mean percentage cover ± SD.

On 5-month old plates the mean percentage cover rate of plates (Table 2) was particularly low at nearest sites (4W, 4E) moderate at intermediate sites (13W, 13E) and higher at remote sites (40E, 50S).

After one year of colonization, between-site differences in cover were much less contrasted, although overall difference among sites was still highly significant (ANOVA on Arcsine transformed data: $F_{5,66}$ = 4.61, p = 0.0001). SNK post-hoc tests showed that only site 13E differed from the other sites in higher cover of plates by bryozoans.

Table 3. List of species frequent on artificial substrates after 5 and 12 months of immersion. Sites are grouped in two categories according to species richness: poor (13W-4W-4E) and rich (13E-40E-50S). Frequency of species on plates of MARs in the pooled sites is expressed in percentage of occurrence on 36 plates.

Species	5 months		12 months	
	Poor	Rich	Poor	Rich
Annectocyma tubulosa	0	0	69.4	38.9
Crisia sigmoidea	0	13.9	27.8	36.1
Diplosolen obelia	0	75.0	0	91.7
Entalophoroecia deflexa	5.6	91.7	27.8	55.6
Eurystrotos compacta	0	0	0	44.4
Disporella hispida	0	0	13.9	16.7
Disporella sp.	2.8	66.7	0	83.3
Lichenopora radiata	0	0	0	30.6
Mecynoecia delicatula	0	0	0	61.1
Plagioecia sarniensis	0	8.3	0	61.1
Tubulipora aperta	94.4	69.4	100	36.1
T. hemiphragmata	0	0	0	41.7
T. phalangea	0	0	0	63.9
T. plumosa	0	11.1	0	58.3
T. ziczac	0	0	36.1	72.2
Aetea sica	0	41.7	83.3	75.0
Callopora dumerilli	30.6	16.7	41.7	75.0
Chorizopora brongnati	50.0	100	97.2	100
Cellepora pumicosa	2.8	0	36.1	11.1
Celleporina canariensis	0	5.6	61.1	22.2
Escharoides coccinea	0	44.4	0	2.8
Fenestrulina malusii	0	41.7	0	36.1
Hincksina flustroides	0	22.2	0	47.2
Microporella ciliata	25.0	30.6	38.9	30.6
Parasmittina rouvillei	0	0	5.6	25.0
Prenantia cheilostoma	41.7	2.8	0	0
Puellina hincksi	0	8.3	13.9	27.8
Reptadeonella violacea	5.6	11.1	5.6	83.3
Schizomavella cuspidata	50.0	75.0	83.3	50.0
Schizoporella dunkeri	63.9	25.0	91.7	2.8
S. unicornis	5.6	33.3	0	72.2

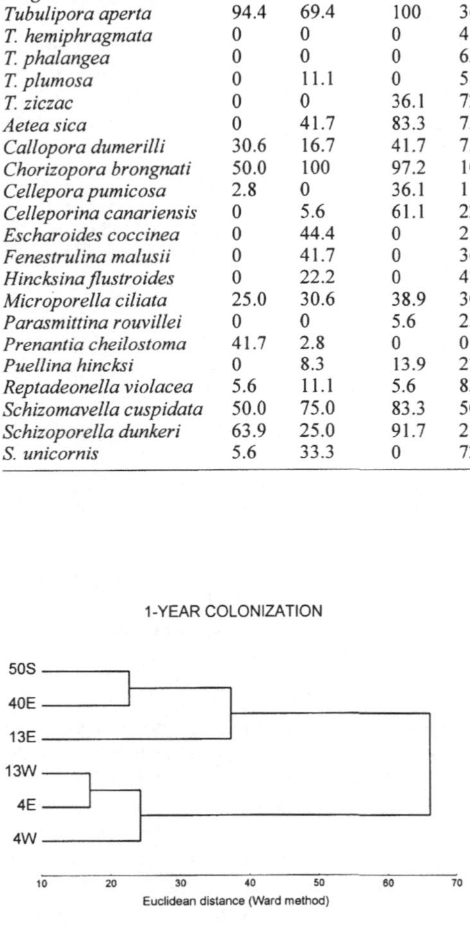

Figure 6. Hierarchical classification of sites based on occurrence frequency of species on MAR plates immersed during one year.

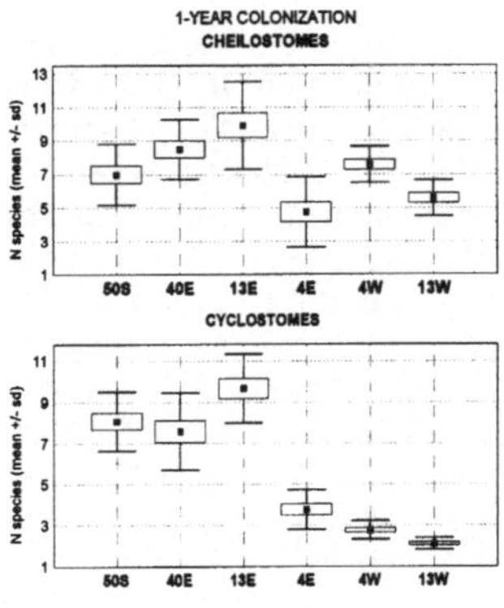

Figure 7. Mean numbers of cheilostome and cyclostome species recorded on artificial substrates of MARs immersed during one year at various distances from the Cortiou sewer outlet.

In the 1-yr old whole assemblage, 12 species occupied the first rank of cover on individual plates. *Chorizopora brongnarti*, the most efficient colonist, was first-ranked on 57% of the whole set of plates (72) without any salient geographical trend. Cyclostomes and cheilostomes were equally represented among these first-ranked species (6 species in both groups), but their ratio differed greatly according sites. In the pool of "rich" sites (50S-40E-13E) five cyclostomes were first-ranked against two cheilostomes while the reverse ratio (2 cyclostomes vs 4 cheilostomes) was observed in pooled "poor" sites (13W-4W-4E).

4 DISCUSSION

The three categories of bryozoan fauna assessed in the zone including the sewage source supplied similar indications on the abrupt decay of bryozoan diversity when pollution increased. The condition of bryozoan assemblages reflected particularly well the spatial near-bottom repartition of stressors spread by Cortiou sewage as it was evidenced by previous studies on water circulation, contaminant deposits and distribution of benthic fauna (e.g. Arfi et al. 2000, Arnoux & Bellan-Santini 1972, Bellan et al. 1999, Bellan-Santini et al. 1992, Leung Tack Kit 1975, Muricy 1991, Verdenal et al. 1990). These studies stressed the dominant westward spreading of sewage waters induced by coastal branches of the general east flow (Millot & Wald 1980) and the layout of islands and mainland coast. Most features of bryozoan assemblages testify to strongest alteration at both sites nearest the outfall (0.4 km: 4E and 4W), i.e. in the hot spot of the drainage spreading in the open sea, but also at the downstream intermediate site 13W, about 1 km west of the sewer outlet. Conversely, at the same distance on the opposite, upstream side (site 13E), the bryozoan assemblages were quite diverse and similar in most features to those recorded in the remote sites (24E, 40E, 50S).

Besides inherent variability in larval supply, reduced recruitment may be an early indication of a pollution event (Fairweather 1991, Menzie 1984). The low rate of substrate colonization observed nearby Cortiou sewer, either in MARs or in natural cryptic microhabitats, contrasts with the particular success of recruitment in the 'rich' sites (e.g. 26 species on a single 85-cm² plate at 13E, Table 2), which attests to the suitability of this substrate type for bryozoans. The decay of bryozoan colonization in the sites highly impacted by sewage may result from the combination of three major sources of alteration: (i) upstream larval pool impacted before reaching places suitable for settlement, (ii) habitats made unsuitable for the settlement of sensitive species, and (iii) impoverished recruitment sources. The habitat resource of rocky bottoms nearby the sewer outlet is obviously altered, chiefly by heavy deposits of fine particles associated to pollutants and by the monopolization of substrate by the encrusting Corallinacae *Lithophyllum incrustans* Philippi (Harmelin et al. 1981). Although undersides of both natural substrates and MAR plates are not directly affected by these factors, they are contiguous to polluted sediments. The relative short larval life of most bryozoans and the limitation of larvae dispersal out of cryptic habitats (Harmelin 2000), as well as similarities in structure and species composition observed within sites between assemblages of MAR and natural cryptic habitats, suggest that the recruitment sources of MAR assemblages are predominantly local. This also indicates that the study of late succession stages in cryptic artificial habitats can thus advantageously replace that of natural habitat assemblages.

Similar indications on the dramatic decline of the diversity of cryptic bryozoans along pollution gradients in the Provence region were given by Hong (1980, 1983) in a thorough study of coralligenous bioherm communities in a zone impacted by industrial discharges (Gulf of Fos) and in a control area (Riou Island). Bryozoans were shown to be among the most sensitive higher taxa to this type of pollution, together with crustaceans, sponges and echinoderms.

The large, erect bryozoans proved to be strongly affected by pollution from sewage. Their particular sensitivity may result from their constant exposure to water-borne pollutants. The spectacular exclusion of these bryozoans successively along the pollution gradient contrasts with the persistence on overhanging walls near the sewer outfall of some other conspicuous sessile invertebrates (e.g. the gorgonian *Eunicella cavolinii*) with which large, erect bryozoans are commonly associated. The disappearance of large, erect bryozoans had also been noticed in the altered parts of the Gulf of Fos and considered as an indication of industrial pollution (Harmelin & Hong 1979). Interestingly, in this situation as in the area impacted by Cortiou drainage, *Myriapora truncata* was the most resistant species. Monitoring of the same area in the Gulf of Fos in 1994 after significant abatement of industrial pollution had shown that the populations of these large erect bryozoans were recovering (Harmelin, unpublished data). In the Marseilles area, large, erect bryozoans were formerly very abundant on coarse detritic bottoms at 30-60 m depth ("graviers à Coralliaires et à Bryozoaires", Marion 1883). They disappeared a long time ago from these bottoms following increase of sedimentation and pollution (Picard 1965).

The distribution of cyclostome species in both natural and artificial cryptic habitats from Cortiou area indicates that many of them are particularly sensitive to stressors associated to sewage. The lack of common species, such as *Diplosolen obelia*, in the impacted sites appears to be a significant indication of benthos alteration. Studies in other types of habitats and pollution are needed to verify if the particular sensitivity of cyclostomes to pollution is a general feature of this group.

5 CONCLUSIONS

The responses of the bryozoan assemblages in cryptic habitats, either natural or artificially exposed to sewage, are clear and can be considered as reliable indications of the general alteration of benthic communities. The particular sensitivity of numerous species otherwise common in less impacted areas suggests that assessment of the condition of communities can be limited to information from simple presence-absence of frequent species.

Considering that cryptic habitats may supply some protection against the direct effects of sewage stressors, one may expect that bryozoans from more open environments are capable of providing earlier warnings of alteration. Three-dimensional growth in the water column may explain the strong response to pollution of the large, erect bryozoans.

These conspicuous, easy to identify, erect bryozoans appear to be most valuable bioindicators of the condition of hard bottom communities in the Mediterranean.

The use of standardized artificial microhabitats, such as MARs, for monitoring rocky bottoms assemblages could be a valuable alternative to natural substrate sampling, which needs much more efforts, particularly to obtain true replicates in among-site comparisons.

Acknowledgements. This study was part of LITEAU programme funded by the French ministry in charge of the environment (MATE).

REFERENCES

Arfi, R., Arnoux, A., Bellan, G., Bellan-Santini, D., Bourcier, M., Dukan, S., Durbec, J.P., Laubier, L., Marinopulos, J., Millot, C., Moutin, T., Patriti, G., Pergent-Martini, C. & Petrenko, A. 2000. *Impact du grand émissaire de Marseille et de l'Huveaune détournée sur l'environnement marin de Cortiou. Etude bibliographique raisonnée 1960-2000*. Unpublished report, Centre d'Océanologie de Marseille & Ville de Marseille, Marseille.

Arnoux, A. & Bellan-Santini, D. 1972. Relations entre la pollution du secteur de Cortiou par les détergents anioniques et les modifications des peuplements de *Cystoseira stricta*. *Téthys* 4(3): 583-586.

Bellan, G., Bourcier, M., Salen-Picard, C., Arnoux, A. & Casserley, S. 1999. Benthic ecosystem changes associated with wastewater treatment at Marseille: implications for the protection and restoration of the Mediterranean coastal shelf ecosystem. *Water Environment Research* 71(4): 483-493.

Bellan-Santini, D., Arnoux, A., Bellan, G., Arnaud, P.M. & Marinopoulos, J. 1992. Spatio-temporal ecological evolution of a hard substratum biocenosis under the impact of the untreated outflow of the main sewer of Marseille. *Mésogée* 52: 52.

Castelbon, C. 1972. Etude de la circulation des masses d'eau dans le golfe de Marseille. *Téthys* 4(2): 269-312.

Chapman, M.G., Underwood, A.J. & Skilleter, G.A. 1995. Variability at different scales between a subtidal assemblage exposed to the discharge of sewage and two control assemblages. *Journal of Experimental Marine Biology and Ecology* 189: 103-122.

Cocito, S., Sgorbini, S. & Bianchi, C.N. 1998. Aspects of the biology of the bryozoan *Pentapora fascialis* in the northwestern Mediterranean. *Marine Biology* 131: 73-82.

Fairweather, P.G. 1991. Implications of "supply-side" ecology for environmental assessment and management. *Trends in Ecology and Evolution* 6(2): 60-63.

Gautier, Y.V. 1962. Recherches écologiques sur les Bryozoaires chilostomes en Méditerranée occidentale. *Recueil des Travaux de la Station Marine d'Endoume* 38(24): 1-434.

Geraci, S. & Relini, G. 1970. Fouling di zone inquinate. Osservazioni nel porto di Genova. I. Briozoi. *Pubblicazioni della Stazione Zoologica di Napoli* 38(suppl.): 21-32.

Gordon, D.P. & Mawatari, S.F. 1992. Atlas of marine-fouling Bryozoa of New Zealand ports and harbours. *Miscellaneous Publications New Zealand Oceanographic Institute* 107: 1-52.

Harmelin, J.G. 1976. Le sous-ordre des Tubuliporina (Bryozoaires Cyclostomes) en Méditerranée. Ecologie et systématique. *Mémoires de l'Institut océanographique, Monaco* 10: 1-326.

Harmelin, J.G. 2000. Ecology of cave and cavity dwelling bryozoans. In A. Herrera Cubilla & J.B.C. Jackson (eds) *Proceedings of the 11th International Bryozoology Association Conference*: 38-53. Balboa, Republic of Panama: Smithsonian Tropical Research Institute.

Harmelin, J.G. & Hong, J.S. 1979. Données préliminaires sur le peuplement d'un fond de concrétionnement soumis à un gradient de pollution. 2 – Faune bryozoologique. *Rapports et Procès-Verbaux des Réunions Commission Internationale pour l'Exploration Scientifique de la Mer Méditerranée* 25/26(4): 175-177.

Harmelin, J.G., Bouchon, C. & Hong, J.S. 1981. Impact de la pollution sur la distribution des échinodermes des substrats durs en Provence (Méditerranée nord-occidentale). *Téthys* 10(1): 13-36.

Hong, J.S. 1980. *Etude faunistique d'un fond de concrétionnement de type coralligène soumis à un gradient de pollution en Méditerranée nord-occidentale (Golfe de Fos)*. Unpublished 3rd cycle thesis, University of Aix-Marseille II, Marseille.

Hong, J.S. 1983. Impact of the pollution on the benthic community. Environmental impact of the pollution on the benthic coralligenous community in the Gulf of Fos, northwestern Mediterranean. *Bulletin of the Korean Fisheries Society* 16(3): 273-290.

Leung Tack Kit, D. 1975. Etude de la faune marine de substrat rocheux dans la zone de rejets des eaux résiduaires de la ville de Marseille (France). *Téthys* 7(2-3): 191-212.

Marion, A.F. 1883. Esquisse d'une topographie zoologique du golfe de Marseille. *Annales du Muséum d'Histoire Naturelle de Marseille* 1: 1-108.

Menzie, C.A., 1984. Diminishment of recruitment: a hypothesis concerning impacts on benthic communities. *Marine Pollution Bulletin* 15(4): 127-128.

Millot, C. & Wald, L. 1980. The effect of Mistral wind on the Ligurian current near Provence. *Oceanologica Acta* 3(4): 399-402.

Muricy, G. 1991. Structure des peuplements de spongiaires autour de l'égout de Cortiou (Marseille, France). *Vie Milieu* 41(4): 205-221.

Noss, R.F. 1990. Indicators for monitoring biodiversity: a hierarchical approach. *Conservation Biology* 4(4): 355-364.

Pearson, T.H. & Rosenberg, R. 1978. Macrobenthic succession in relation to organic enrichment and pollution of the marine environment. *Oceanography and Marine Biology Annual Review* 16: 229-311.

Picard, J. 1965. Recherches qualitatives sur les biocénoses marines des substrats meubles dragables de la région marseillaise. *Recueil des Travaux de la Station Marine d'Endoume* 36(52): 1-160.

Powell, N.A., Sayce, C.S. & Tufts, D.F. 1970. Hyperplasia in an estuarine bryozoan attributable to coal tar derivatives. *Journal of the Fisheries Research Board of Canada* 27(11): 2095-2098.

Reish, D.J. & Bellan, G. 1996. The long-term effects of municipal discharges from outfall sewers from urban areas on the marine environment: a review. In D. Bellan-Santini, G. Bonin & C. Emig (eds), *Functioning and dynamics of natural and perturbed ecosystems*: 701-743. Paris: Lavoisier Printers.

Scholz, J. 1990. The significance of Philippine reef and harbor Bryozoa as indicators of environmental stress. *The Philippine Scientist* 27: 19-26.

Scholz, J. 1991. Bryozoan response to "pollution events". *N.eues Jahrbuch Geologie und Paläontologie. Mh.* H.9: 543-548.

Soule, J.D. & Soule, D.F. 1981. Heavy metals uptake in bryozoans. In G.P. Larwood & C. Nielsen (eds), *Recent and fossil Bryozoa*: 227-233. Fredensborg: Olsen & Olsen.

Sutherland, J.P. & Karlson, R.H. 1977. Development and stability of the fouling community at Beaufort, North Carolina. *Ecological Monograph* 47: 425-446.

Underwood, A.J., Peterson, C.H. 1988. Towards an ecological framework for investigating pollution. *Marine Ecology Progress Series* 46: 227-234.

Verdenal, B., Diana, C., Arnoux, A. & Vacelet, J., 1990. Pollutant levels in Mediterranean commercial sponges. In K. Rützler (ed.), *New Perspectives in Sponge Biology*: 516-524. Washington, D.C.: Smithsonian Institution Press.

Ward, T.W. & Young, P.C. 1984. Effects of metals and sediment particle size on the species composition of the epifauna on *Pinna bicolor* near a lead smelter, Spencer Gulf, South Australia. *Estuarine and Coastal Shelf Science* 18: 79-85.

Bryozoan Studies 2001, Wyse Jackson, Buttler & Spencer Jones (eds)

Predominance of obligate outbreeding in the simultaneous hermaphrodite *Celleporella hyalina* sensu lato

Roger N. Hughes, Peter Wright & Patricio H. Manríquez†
School of Biological Sciences, University of Wales, Bangor, Gwynedd, LL57 2UW, U.K
†current address: *Estacion Costera Investigaciones Marinas, Casilla 114-D, Santiago, Chile*

John D.D. Bishop
Marine Biological Association of the UK, The Laboratory, Citadel Hill, Plymouth PL1 2PB, UK and Department of Biological Sciences, University of Plymouth, Drake Circus, Plymouth PL4 8AA.

ABSTRACT: Simultaneous hermaphroditism affords the potential for self-fertilization, which could be selectively advantageous under restricted opportunity for outbreeding. Such restriction may apply to free-spawning sessile invertebrates, such as bryozoans, when the vagaries of settlement place colonies at a distance from potential partners. Contrary to this scenario, however, most populations studied within the *Celleporella hyalina* species complex showed little or no evidence of self-fertilization. Most colonies kept in reproductive isolation failed to produce embryos or produced few, abortive embryos, whereas controls kept together in culture vessels copiously produced viable embryos. These results, together with previous work demonstrating severe inbreeding depression and the capacity of juvenile colonies to store allosperm, suggest that *Celleporella hyalina* is typically, although not exclusively, an obligate outbreeder.

1 INTRODUCTION

The first evolutionary account of simultaneous hermaphroditism is probably attributable to Darwin (1876, 1877). Reflecting upon the form and function of flowers, Darwin proposed that hermaphroditism is selectively advantageous because, through self-fertilization, it enables seed production to occur even in the absence of cross-pollination. Having become known as the "reproductive assurance" principle (Stebbins 1950), Darwin's proposal relies upon gametic self-compatibility and a lack of severe inbreeding depression. These conditions are frequently met, but are by no means universal, among flowering plants (Barrett 1998). The mating system of benthic invertebrates that rely on water-borne sperm to fertilize retained eggs is analogous to pollination in plants. The reproductive assurance principle therefore could reasonably be expected to apply to such invertebrates, provided that the necessary conditions of gametic self-compatibility and moderate or zero inbreeding depression are met. Fertilization of retained eggs by water-borne sperm is a universal feature of Bryozoa, among which the cheilostome *Celleporella hyalina* (L.) is a convenient model for experimental investigation (Yund & McCartney 1994, Manríquez et al. 2001). We tested for gametic self-compatibility in *C. hyalina* sensu lato by rearing colonies in reproductive isolation, using a common-garden rearing environment as a control for reproductive competence through outbreeding.

2 METHODS

2.1 *Population from Beaumaris, Anglesey, North Wales*

Colonies of *C. hyalina* were established in April 1996 from larvae released by colonies collected from the Menai Strait at Beaumaris (53° 16'N, 04° 06'W, Ordnance Survey co-ordinates SH 609767) and propagated as cuttings (ramets) to generate a corresponding set of clones. For further details of cloning and culturing protocols see Manríquez (1999). Each clone was isolated from sources of allosperm. Seven ramets per clone were housed separately in 300 ml glass jars two-thirds filled with 0.2 μm-filtered, UV-irradiated seawater at 16 ± 2°C and fed daily with *Rhinomonas reticulata* at an ambient concentration of about 100 cells μl^{-1}. To test self-fertilizing capability, one ramet was kept in reproductive isolation and another was exposed to a suspension of autosperm for 2h (for methods of preparing sperm suspensions see Manríquez et al. 2001). To test outcrossing capability, the remaining ramets were each exposed to allosperm suspension obtained from a different member of the set of clones. Ramets, each initially at a size of 900-1300 zooids, were examined daily over a period of 3 months, using a 1.2 x 1.2 mm graticule and 40 x 10 magnification. All gynozooids were examined for the presence of ovulated coelomic oocytes, which were then followed through embryogenesis. Mating success was scored as the proportion of coelomic oocytes resulting in viable embryos. Random larvae released during this period were allowed to settle on acetate and the young colonies grown on for 3 weeks. Viability of progeny was expressed as the proportion of larvae successfully completing settlement, metamorphosis and astogeny to the 3-week stage.

2.2 *Geographically distinct populations*

Colonies were collected from *Laminaria* spp. (Ría de Ferrol, Spain; Amlwch Harbour, North Wales; Lough Hyne, South-west Ireland; Reykjavik, Iceland; Damariscotta Estuary, Maine; North-West Arm, Halifax, Nova Scotia) or *Chondrus crispus* (Nobska Point, Woods Hole, Massachusetts) and transported to the Bangor laboratory where they were cultured, in a separate common-garden environment per population, and induced to release larvae. Newly formed ancestrulae were immediately removed by excision of the acetate substratum, so preventing the precocious uptake and storage of allosperm (Hughes et al. in prep.). To test self-fertilizing capability within each population, 4 ancestrulae were placed in separate 300 ml jars. To test for outcrossing ability 10 ancestrulae were placed in a 2 l plastic drinks bottle, serving as a common-garden environment. Young colonies were grown on for 3 months and inspected weekly for the presence of embryos and larval release. All culture vessels were furnished with aeration and food ports designed to prevent accidental transmission of allosperm (Manríquez et al. 2001). In a subsequent experiment designed to test reproductive compatibility between populations (unpublished), the previously isolated colonies were placed in a common-garden environment presenting the opportunity for outcrossing.

3 RESULTS

3.1 *Population from Beaumaris, Anglesey, North Wales*

Neither the reproductively isolated colonies nor those exposed to suspensions of autosperm produced embryos (Table 1) and indeed no coelomic oocytes could be seen in any of these colonies. In contrast, all colonies exposed to suspensions of allosperm copiously produced embryos and viable progeny (Table 1). Moreover, all potential outcrossing partners were reproductively compatible whether acting in male or female roles and all produced equally viable progeny (Table 1).

3.2 *Geographically distinct populations*

In all but one population most isolated colonies failed to produce embryos Table 2), but abundantly produced embryos when subsequently transferred to a common-garden environment. Output in those isolated colonies that did produce embryos was 1-2 orders of magnitude less than in grouped colonies. In the Nova Scotian population, progeny produced by isolated colonies showed reduced viability. The Woods Hole population was exceptional by sustaining comparable embryo production in isolated and grouped colonies. In all populations, grouped colonies produced embryos of normal viability (Table 2).

Table 1. Mating compatibility between clones A1-Q1 of *C. hyalina* derived from a population at Beaumaris, North Wales. Data are the proportion of coelomic oocytes that developed into embryos and, in parentheses, the proportion of embryos that developed into viable colonies. Clones in the top row are sperm donors and those in the first column are recipients. Zeros in the first row of data represent lack of self-fertilization in reproductively isolated ramets, zeros in the diagonal represent the absence of self-fertilization in ramets exposed to autosperm suspension.

Clones	A_1	D_1	E_1	J_1	M_1	Q_1
	0	0	0	0	0	0
A_1	0	0.63 (0.95)	0.59 (0.92)	0.43 (0.88)	0.53 (0.96)	0.39 (0.94)
D_1	0.52 (0.90)	0	0.43 (1.00)	0.57 (0.88)	0.49 (0.94)	0.36 (0.89)
E_1	0.55 (0.87)	0.64 (0.89)	0	0.49 (0.89)	0.52 (1.00)	0.62 (0.80)
J_1	0.66 (0.96)	0.55 (1.00)	0.48 (0.98)	0	0.58 (0.94)	0.63 (0.97)
M_1	0.55 (0.97)	0.48 (0.98)	0.51 (0.94)	0.60 (0.89)	0	0.63 (0.84)
Q_1	0.66 (0.87)	0.50 (1.00)	0.67 (0.86)	0.64 (0.79)	0.54 (0.94)	0

Table 2. Embryo production by isolated and grouped colonies of *C. hyalina* from each of 7 geographically distinct populations. Data are the fraction of isolated colonies (N = 4 or 10 per population) and grouped colonies (N = 10 per population) producing embryos. Nominal abundance (abundant = several thousand, rare = less than 100, occasional = less than 50) and viability of progeny (normal or impaired) are given in parentheses.

Population	Embryo production	
	Isolated colonies	Grouped colonies
Spain	0/4	10/10 (abundant, normal)
Wales	1/4 (rare, normal)	10/10 (abundant, normal)
Ireland	1/4 (rare, normal)	10/10 (abundant, normal)
Iceland	0/4	10/10 (abundant, normal)
Nova Scotia	2/4 (occasional, impaired)	10/10 (abundant, normal)
Maine	0/10	10/10 (abundant, normal)
Massachusetts	4/4 (abundant, normal)	10/10 (abundant, normal)

The effect of container (300 ml glass jar or 2 l plastic bottle) on embryo production may be discounted for the following reasons. In preparation for inter-population mating trials (in progress), each isolated colony was removed from its jar once having grown to about 3 cm^2 in area and propagated into a number of ramets that were then placed in a bottle. Over subsequent months, ramets derived from maternally non-reproductive colonies remained free of embryos. Moreover, the same ramets began producing embryos when transferred to jars and paired with compatible partners.

4 DISCUSSION AND CONCLUSIONS

From the breeding experiments above, we predict that in 6 of the 7 geographically distinct populations studied (Beaumaris and Amlwch are regarded as sharing the same geographical location), the majority of natural colonies (genotypes) will be incapable of inbreeding. This condition was first reported by Cancino et al. (1991) for a Chilean population of *C. hyalina* sensu lato, in which reproductively isolated colonies produced few ovicells and no viable embryos. Sample sizes in the present study are insufficient to precisely estimate the frequency of inbreeding capability within populations, but adequately indicate that this frequency is relatively low. To the 6 obligately outbreeding genotypes cloned from the Beaumaris population may be added a further 9 that were studied for other purposes but again proved to be incapable of selfing (Manríquez 1999). Drawing from among the same clones Hoare & Hughes (2001) found that severe inbreeding depression results from full-sib mating, as would be expected in a population that is normally outbreeding. Opportunity for outbreeding in *C. hyalina* is promoted by the relatively prolonged fertility of water-borne sperm (Manríquez et al. 2001), the efficient capture of sperm by active lophophores (Temkin 1996, Manríquez 1999), the ability to use stored allosperm to fertilize ova and the precocious capture and storage of allosperm by newly formed, sexually immature colonies (Hughes et al. in prep.).

Despite the predominance of obligate outbreeding in *C. hyalina*, there would seem to be widespread genetic variation for inbreeding capability. Data in Table 2 tentatively suggest that the frequency of inbreeding capability is variable among populations and may be associated with variation in the severity of inbreeding depression. In the Woods Hole population, inbreeding capability appears to be genetically fixed and no signs of inbreeding depression were detected. The native population, epiphytic on *Chondrus crispus*, occurred at a relatively high density that would present ample opportunity for outbreeding. Possibly, inbreeding capability and resistance to inbreeding depression have been driven to genetic fixation by reduced opportunity for outbreeding associated with recurrent reductions in the local population or in a source population occupying some other habitat. Alternatively, fixation of inbreeding capability may represent a phylogenetic trait with deeper history, since the Woods Hole population is cladistically distinct within *C. hyalina* sensu lato (A. Gomez pers. comm.).

Darwin's reproductive-assurance principle, with its implicit assumptions of self-compatibility and of resistance to severe inbreeding depression, clearly is inapplicable to many extant populations of *C. hyalina* sensu lato. Sex ratio among colonies of *C. hyalina* is known to vary within populations, the extremes ranging from all-male (Hunter & Hughes 1995, McCartney 1997) but apparently never to all-female. Consideration of fitness gains resulting from competitive increments in resource allocation (Charnov 1982, Yund et al. 1997) to androzooids, gynozooids and autozooids is likely to play a pivotal role in understanding the expression of hermaphroditism in this bryozoan model system.

ACKNOWLEDGEMENTS

The National Environmental Research Council funded this work through grants GR3/12896 (RNH, PW) and GR3/10226 (RNH, JDDB, PHM). We thank Juan Cancino for his helpful review of the manuscript.

REFERENCES

Barrett, S.C.H. 1998. The evolution of mating strategies in flowering plants. *Trends in Plant Science* 3: 335-341.

Cancino, J.M., Castañeda, B. & Orellana, M.C. 1991. Reproductive strategies in bryozoans: experimental test of the effects on conspecific neighbours. In F.P. Bigey (ed), *Bryozoaires Actuels et Fossiles: Bryozoa Living and Fossil. Bulletin de la Société des Sciences Naturelles de l'Ouest de la France. Memoires Hors Série 1*, Nantes (France) 1991: 81-88.

Charnov, E.L. 1982. *The Theory of Sex Allocation.* Princeton University Press, New Jersey: Princeton.

Darwin, C. 1876. *The Effects of Cross- and Self-Fertilization in the Vegetable Kingdom.* London: John Murray.

Darwin, C. 1877. *The Different Forms of Flowers on Plants of the Same Species.* London: John Murray.

Hoare, K. & Hughes, R.N. 2001. Inbreeding and hermaphroditism in the sessile brooding bryozoan *Celleporella hyalina. Marine Biology* 139: 147-162.

Hughes, R.N., Manríquez, P.H. & Bishop J.D.D. in prep. Female investment is enhanced by reception of allosperm in the sessile hermaphrodite *Celleporella hyalina* (Bryozoa).

Hunter, E. & Hughes, R.N. 1995. Environmental and genetic control of variation in sexual allocation in an epialgal bryozoan. *Marine Ecology Progress Series* 120: 193-201.

Manríquez PH 1999. *Mate choice and reproductive investment in the cheilostome bryozoan* Celleporella hyalina *(L.).* Ph.D. thesis, University of Wales, Bangor.

Manríquez, P.H. Hughes R.N. & Bishop J.D.D. 2001. Age-dependent loss of fertility in water-borne sperm of the bryozoan *Celleporella hyalina* (L.). *Marine Ecology Progress Series* in press.

McCartney, M.A. 1997. Sex allocation and male fitness gain in a colonial, hermaphroditic marine invertebrate. *Evolution* 51: 127-140.

Stebbins, G.L. 1950. *Variation and Evolution in Plants*. New York: Columbia University Press.

Temkin, M.H. 1996. Comparative fertilization biology of gymnolaemate bryozoans. *Marine Biology* 127: 329-339.

Yund, P.O. & McCartney, M.A. 1994. Male reproductive success in sessile invertebrates: competition for fertilizations. *Ecology* 75: 2151-2167.

Yund, P.O., Marcum, Y. & Stewart-Savage, J. 1997. Life-history variation in a colonial ascidian: Broad-sense heritabilities and tradeoffs in allocation to asexual growth and male and female reproduction. *Biological Bulletin* 192: 290-299.

Bryozoan Studies 2001, Wyse Jackson, Buttler & Spencer Jones (eds)

Functional morphology of maculae in a giant ramose bryozoan from the Permian of Greenland

Marcus M. Key, Jr.
Dickinson College, Department of Geology, Carlisle, PA 17013-2896, USA

Lena Thrane (née Madsen)
Baker Atlas Geoscience, Jorcks Passage, Opg. A, 4. DK-1162 Copenhagen K, Denmark

Julie A. Collins
Dickinson College, Department of Geology, Carlisle, PA 17013-2896, USA

ABSTRACT: The need for colony-wide feeding currents should be greater in ramose bryozoan colonies with larger diameter branches due to their greater and more planar surface area. This situation was magnified in a giant branch of *Tabulipora* from the Permian of North Greenland. The 8 mm thick exozone allowed the making of 20 serial tangential peels which intersected two maculae. Macular spacing, size, and shape were measured to see how they varied through the exozone in response to increasing branch size. As the branch grew, the spacing and size of the maculae increased, and the maculae became more stellate. These trends continued until the outer exozone, where they reversed, probably in response to the initiation of a new macula. This indicates that macular excurrent chimneys are dynamic in their position, size, and shape in response to changing hydrodynamics of colony-generated, colony-wide currents on a growing branch of a ramose colony.

1 INTRODUCTION

The goals of this project were to quantify the effects of increasing colony branch diameter on macular spacing, size, and shape. These effects should be accentuated in a giant-sized colony. It was hypothesized that as a colony increases its size (i.e. increasing branch diameter) the macular chimneys should become more widely spaced, larger, and more stellate. This is in response to increasing incurrent area as the colony branch expands laterally. For this paper, a macula was defined as a small cluster of nonfeeding polymorphs and/or extrazooecial skeleton surrounded by autozooecia that may be depressed below, level with, or elevated above the colony surface (Boardman & Cheetham 1983).

2 COLONY-WIDE FEEDING CURRENTS

Bryozoans are active filter feeders (sensu Jørgensen [1966]) in that their autozooids create their own feeding currents to draw food particles toward their mouths. Autozooids have an inverted cone-shaped lophophore. Lateral cilia on each tentacle of the lophophore beat outwards causing water to be evacuated from the center of the tentacle crown and to pass down toward the mouth and away laterally between the tentacles. This evacuated water is replaced by water flowing into the open end of the lophophore from above. Thus an inhalant current (i.e. incurrent) approaches the autozooid from above and an exhalent current (i.e. excurrent) departs laterally between the fixed part of the colony surface and the canopy of protruded lophophores.

In extant species polypides feed simultaneously over large areas of the colony so that when lophophores are protruded, they cover the colony surface as densely as possible with minimal overlapping (McKinney 1986a, 1990). In robust ramose colonies with their resulting tightly packed canopy of lophophores, the water that is being pumped by the lophophores toward the colony surface can not be vented from the colony surface between the lophophores (McKinney & Jackson 1989, Taylor 1999). In ramose colonies with small diameter branches with their inherently more curved branch surfaces, the filtered water can exit between the lophophores as they are more widely spaced, and so chimneys are unnecessary (Banta et al. 1974). In ramose colonies with larger diameter branches, the filtered water spreads laterally towards lower pressure regions marked by the absence of incurrents. These regions are called excurrent chimneys as they are sites where filtered water is expelled back out away from the colony surface (Banta et al. 1974, Cook 1977, Winston 1978, 1979, 1981). Chimneys are required to relieve the hydrodynamic resistance to incurrents in larger colonies where surface areas are large relative to peripheral areas (Winston 1979, McKinney 1986b, Dick 1987, McKinney 1990). Thus with an extensive lophophore canopy, efficient performance in food gathering is secured by creating a common, colonial system of feeding currents which effectively separates incurrent (i.e. unfiltered) and excurrent (i.e. filtered) water masses.

3 MACULAR CHIMNEYS

An excurrent chimney can form in any region that is less densely packed with or completely absent of autozooids with their feeding lophophores (Taylor 1975, 1979), as long as the region is large enough so that the excurrent is unopposed by surrounding incurrents (Boardman 1983). Gymnolaemates can produce a chimney in a group of autozooids simply by titling their lophophores away from one another (Cook 1977, Winston 1978, Cook & Chimonides 1980, Lidgard 1981). Stenolaemates have lophophores that are situated directly above the autozooecial chambers and can not be tilted (McKinney 1988). This may explain why it has never been demonstrated that stenolaemates can form chimneys without involving the skeleton.

Banta et al. (1974) were the first to hypothesize that maculae are two-dimensional skeletal representations of a three-dimensional water exchange phenomenon (Anstey 1987). The hypothesized function of maculae as sites of excurrent chimneys is supported by the observation that skeletal maculae in living bryozoans have chimneys centered on them (Cook 1977, Winston 1978, 1979) and by the observation that many may have been areas of converging flow lines (Anstey 1981, 1987, Patzkowsky 1987).

Thus colony-wide feeding currents centered on chimneys can be inferred from skeletal evidence in fossils. Incurrent areas can be identified as aggregations of autozooecia, whereas macular chimneys can be identified as areas of less densely packed autozooecia or areas devoid of autozooecia (Banta et al. 1974, Taylor 1975, 1979, 1999, Winston 1981). Autozooids are less densely packed in maculae due to more abundant nonfeeding polymorphs and/or buildups of extrazooidal skeleton (McKinney 1986a, Taylor 1999).

In colonies where it has been measured, the total colony surface area dedicated to incurrents is greater than the area of excurrent chimneys (Lidgard 1981, McKinney 1989). Based on this and the principle of continuity, excurrent velocities in chimneys are greater than the incurrent velocities of the individual autozooids supplying the chimneys (Lidgard 1981, McKinney 1986b, 1990, Dick 1987). The principle of continuity states that for every volume of fluid which enters a conduit per unit time, an equal volume must leave (Vogel 1981). These enhanced excurrent flow velocities in the chimneys jet the previously filtered water away from the colony surface, through the boundary layer, and into the overlying flow. Thus, the colony-wide feeding currents keep the incurrents separate from the excurrents which reduces refiltering of water and improves feeding efficiency (Banta et al. 1974, Winston 1978, 1979, Taylor 1979, Lidgard 1981, Dick 1987). Mathematical modeling of water flow in encrusting bryozoan colonies by Grünbaum (1995, 1997) and Eckman and Okamura (1998) has shown that with or without interzooidal feeding interference, colony-wide feeding currents are beneficial to the colony especially in low current velocities where flow remains laminar.

Previous workers have hypothesized several additional benefits of colony-wide feeding currents. Colony-wide currents may: 1) remove waste materials (Cook 1977, Winston 1979), 2) keep sediment from settling on the colony surface (Cook 1977, Taylor 1979, Boardman 1983), 3) aid gamete and larval dispersal especially for those reproductive polymorphs centered in or near the excurrent areas (Taylor 1979), 4) prevent larval recruitment of spatial competitors on the adjacent substrate (Buss 1979), and 5) mediate competition among adjacent benthic filter feeders in the low velocity fluid boundary layer (Buss 1980, 1981, Lidgard 1981, McKinney 1992).

Most of these functions have also been inferred for colony-wide feeding currents associated with maculae in fossils and in particular for maculae in trepostomes that are elevated above the colony surface (i.e. monticules). There are several additional inferred functions for maculae in trepostomes. It has been suggested that maculae were centers of zooecial budding (Anstey & Delmet 1972, Delmet & Anstey 1974, Anstey et al. 1976, Pachut & Anstey 1979, Anstey & Pachut 1980). Maculae may have also been the centers of cormidia (i.e. subcolonies) involved in morphogenetic regulation of the colony (Anstey et al. 1976, Pachut & Anstey 1979, Patzkowsky 1987). As with living colonies, it has been suggested that maculae in fossil colonies served a reproductive role as many maculae contained large polymorphs similar to reproductive gonozooids in living stenolaemates (Ulrich 1890, Astrova 1973, Anstey et al. 1976).

4 MATERIAL

This study was based on a single Geological Survey of Greenland specimen (GGU 196054-1) of the stenoporid trepostome stenolaemate bryozoan *Tabulipora*. The branch fragment had a length of 135 mm and a diameter of 37.5 mm with an endozone diameter of 21.5 mm and an exozone width of 8.0 mm. Due to its immense size, it was possible to cut a large (1800 mm^3) block out of the exozone that was 15 mm wide, 15 mm long, and 8 mm deep. From this block, 20 serial tangential acetate peels were made. Smaller branch fragments were available for study, but these were not utilized as their thinner exozones prohibited the making of enough serial peels to draw statistically significant conclusions. The macular pat-

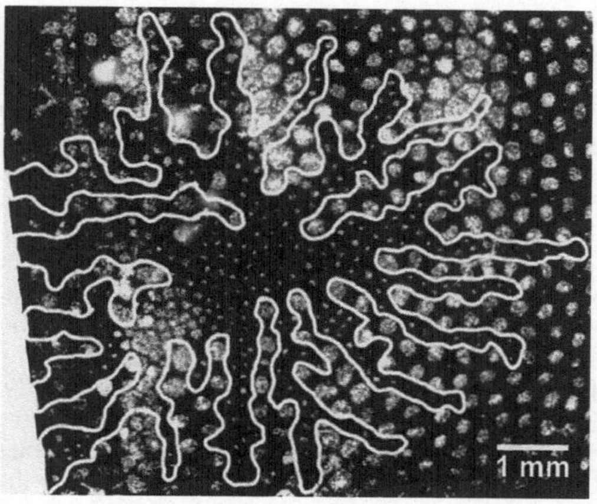

Figure 1. Photomicrograph of a tangential section of the exozone of *Tabulipora* sp. (GGU 196054-1) showing Macula 1 (outlined in white) at 6.51 mm from endozone. Note the stellate macula defined by contiguous exilazooecia. The left side of the macular channels are truncated.

terns described below are known from smaller specimens and other species and are not simply a phenomenon of this giant specimen.

Other giant stenolaemate colonies are known from elsewhere (e.g. Taylor & Voigt 1999), but these Greenland colonies are at least an order of magnitude larger than other stenolaemates specifically and bryozoans in general (Madsen 1991). Their large size has been attributed to symbiotic, photosynthetic, zooxanthellae algae living intracellularly within the zooids (Håkansson & Madsen 1991).

The shallowest peel was made at a depth from the surface of 0.62 mm, and the deepest was at the endozone/exozone boundary at a depth of 8.00 mm. The spacing between the serial peels ranged from 0.05 mm to 1.23 mm with a mean of 0.39 mm. These same peels were used in previous studies to document exozonal budding (Madsen 1991, 1994) and skeletal space-filling (Key et al. 2001).

The specimen was collected during the 1980 expedition of the regional geological mapping project of eastern North Greenland (Håkansson et al. 1981). The specimen, previously figured by Håkansson & Madsen (1991, Pl. 1, Fig. 4), and its peels are housed in the Geological Museum at the University of Copenhagen (MGUH 25.988-26.008). The sample was collected from Midnatfjeld in the Kim Fjelde area in eastern Peary Land in eastern North Greenland (Stemmerik & Håkansson 1989: Fig. 16, locality 11c). The Kim Fjelde Formation is Early Permian (late Artinskian to Kungurian stages) in age (Stemmerik et al. 1996).

With a branch diameter or 37.5 mm, this branch fragment is wider than typical *Tabulipora* colony branch fragments of equal length (Håkansson & Madsen 1991: Figs. 1-4) and is classified as having an erect robust maculate radial branch growth form

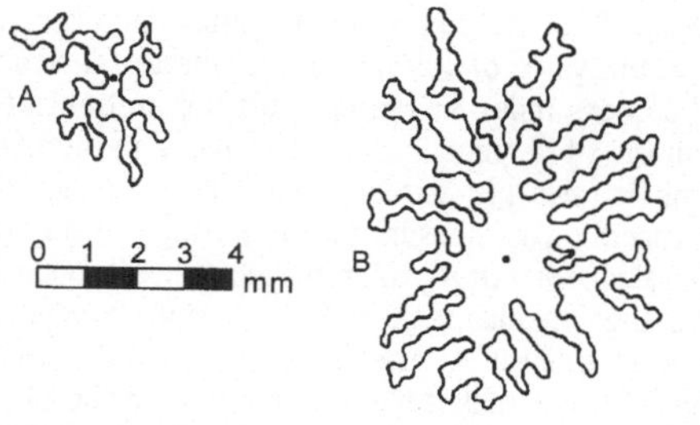

Figure 2. Outlines of Macula 2 of *Tabulipora* sp. (GGU 196054-1) at A) 1.02 mm from endozone and B) 7.38 mm from endozone.

(sensu McKinney 1986a, b, 1990). This growth form usually contains regions of feeding zooids (autozooids) that form incurrent regions that surround regions of excurrent flow (maculae) where autozooids are absent. The maculae in this colony were composed of clusters of contiguous exilazooecia similar to those in cystoporates (e.g. Boardman 1983: Fig. 59.5). The intermacular autozooecial chamber cross-sectional areas in this colony were on average 10 times larger than the macular exilazooecial chamber cross-sectional areas (Key et al. 2001). As the macular exilazooecia were so small, they probably did not house feeding lophophores like the intermacular autozooecia. This inability of kenozooecia (e.g. exilazooecia and mesozooecia) to feed has been suggested by previous workers (Ulrich 1890, Boardman 1983, McKinney & Jackson 1989, Taylor 1999). Without feeding, the exilazooecia would not have been able to create inflow. Without inflow, the macular clusters of contiguous exilazooecia would have acted as excurrent chimneys.

5 CHARACTERS MEASURED

In the specimen, maculae were practically invisible on the colony surface due to their lack of relief, but in magnified tangential section they were very prominent with their stellate shape formed by radiating bifurcating channels (Fig. 1). The block of exozone contained two maculae (herein referred to as Macula 1 and Macula 2). Macular outlines in the peels were determined by the distribution of contiguous exilazooecia (Fig. 1). Macular outlines were drawn for 18 of the 20 peels. The maculae could not be found in peel 13, because it was of poor quality, and in peel 20, because it was too close to the endozone for the maculae to have developed.

Five characters were measured as follows: 1) Intermacular distance was measured as the linear distance between macular centers; 2) Macular area was measured as the area within the macular outline; 3) Macular shape was quantified using a stellateness index defined as the macular outline perimeter length squared divided by the product of the macular area and 4π. This yielded a dimensionless number with a minimum value of 1 for a perfect circle. The macular stellateness index increased with the amount of convolution of a macula's shape; 4) The number of channels (arm-like extensions of the maculae; Fig. 1) per macula was measured by counting the number of separate projections of contiguous exilazooecia out from the macular center. For example Macula 2 in the peel closest to the endozone had 14 channels (Fig. 2A), whereas it had 43 channels in the peel farthest from the endozone (Fig. 2B); 5) Channel length was measured as the mean curvilinear distance from the macular center to the end of the five longest channels.

All of the characters were measured using digitized video images of the tangential serial peels at 50x or 100x magnification. All of the characters were measured on each of the 20 peels except for peels 13 and 20. All of the measurements were repeated ten times per macula per peel except for the number of macular channels which was measured once per peel per macula and the intermacular distance which was measured once per peel.

Tabulipora has distinctive moniliform (i.e. alternating thin-and thick-walled) zooecial walls in the exozone. Tangential sections intersecting both thin- and thick-walled zones reveal multiple bands of thin- and thick-walled zooecia oriented parallel to the branch axis. All macular characters were measured in the thick-walled zones.

6 SOURCES OF ERROR

One problem with tangential sections of cylindrical colony branches is that the depth of the sections below the colony surface increases laterally from the edges of the section (parallel to the growth axis of the branch) to the center along the proximal-distal axis. At the lateral edges, the section intersects the branch at the colony surface. Along the center, the section intersects the branch deeper in the exozone. As a result, there is a systematic lateral variation in the depth of tangential sections of cylindrical surfaces. This is more of a problem in colonies with small diameters, and less in giant colonies like the one in this study. The amount of this variation depends on both the lateral width of the tangential section and the radius of the branch. In this colony branch fragment, the effect of making tangential sections of a cylindrical surface had a 2.3 % maximum error on the macular characters (Key et al. 2001).

There was another potential error in the macular characters due to the truncation of the lateral margin of maculae in some peels. This error was noticeable in Macula 1 in the nine outermost peels and in Macula 2 in the three outermost peels (e.g. left side Fig. 1).

7 RESULTS AND DISCUSSION

Intermacular distance increased significantly (R^2 = 0.614, $p < 0.001$) through the exozone from 9.36 mm at 1.02 mm from the endozone to 10.01 mm at 7.38 mm from the endozone (Fig. 3). If these maculae functioned as excurrent chimneys as originally proposed by Banta et al. (1974), then they may have had a geometric regularity in their spatial distribution. Previous workers have noted the rhombic/hexagonal close-packed array of chimneys over colony surfaces (Banta et al. 1974, Delmet & Anstey 1974, Anstey et al. 1976, Pachut & Anstey 1979, Lidgard 1981, Boardman 1983). Chimney spacing should be relatively constant in encrusting colonies that grow peripherally. In encrusting colonies, the peripheral growth does not affect the flow dynamics of the pre-existing chimneys which are distant from the colony margin. This is supported by the observations of previous workers who have reported that chimneys have a relatively constant spacing of 1-5 mm (Taylor 1975, Cook 1977, Lidgard 1981, McKinney 1986b). This regular spacing may not be the case in ramose colonies that grow by expansion of the entire branch. As the branch expands in size, previous workers have inferred or shown that macular chimneys migrate apart (Taylor 1975, Anstey et al. 1976, Pachut & Anstey 1979, Anstey 1981). Chimney spacing also varies with chimney size as bigger chimneys tend to be more widely spaced (Pachut & Anstey 1979). The only way to maintain constant chimney spacing in ramose colonies is to bud new maculae in the intermacular regions.

The surface areas occupied by the two maculae increased significantly through the exozone (Fig. 4). Macula 1 experienced a three fold increase in area (R^2 = 0.929, $p < 0.001$) while Macula 2 grew six fold (R^2 = 0.945, $p < 0.001$). These increases are minimums as the maculae actually increased from a smaller size at the endozone/exozone boundary to a larger size at the colony surface. The maculae increased their size by budding new macular exilazooecia (Key et al. 2001).

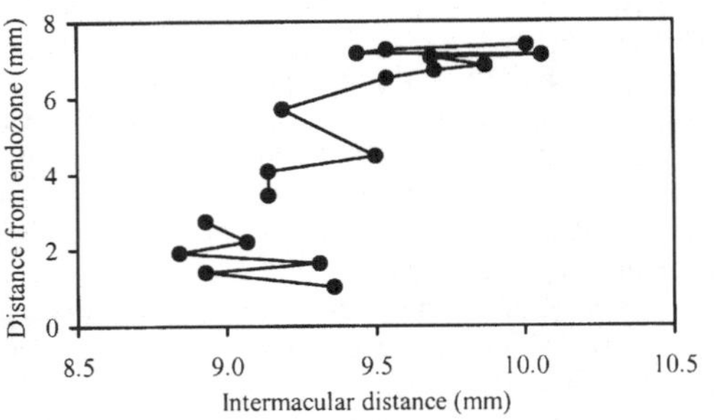

Figure 3. Plot of intermacular distance versus distance from endozone.

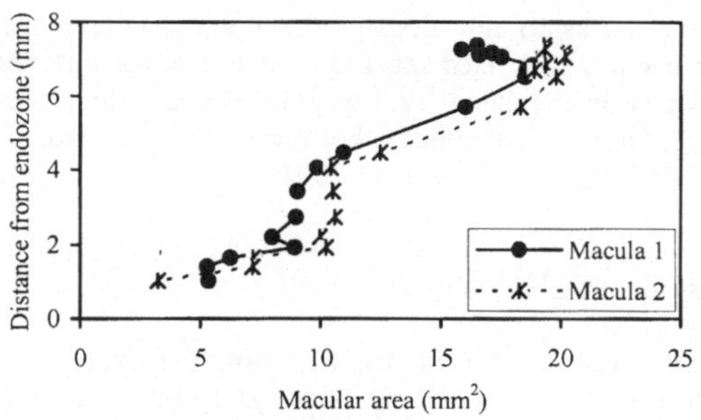

Figure 4. Plot of macular area versus distance from endozone.

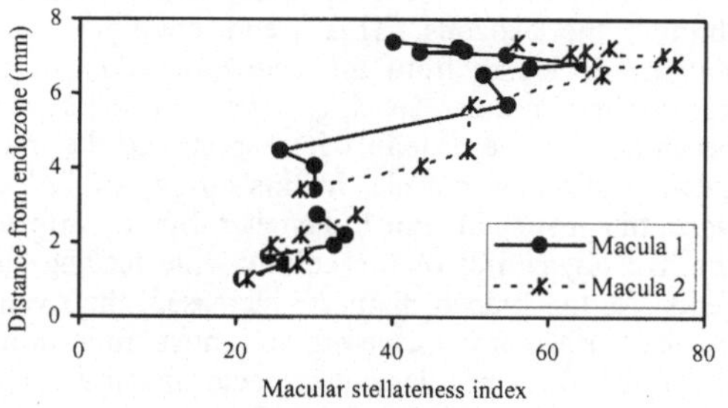

Figure 5. Plot of macular stellateness index versus distance from endozone.

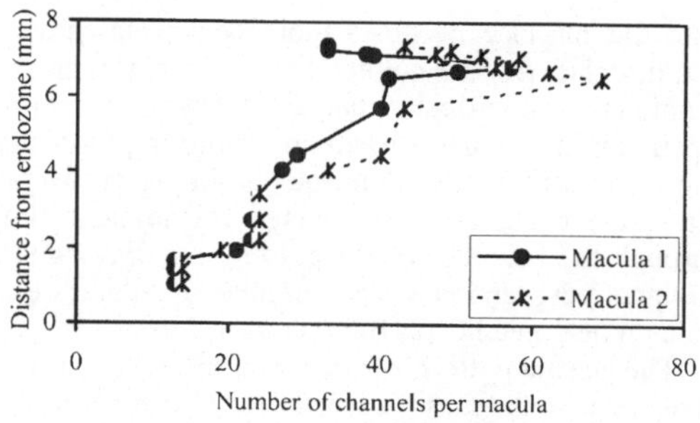

Figure 6. Plot of number of channels per macula versus distance from endozone.

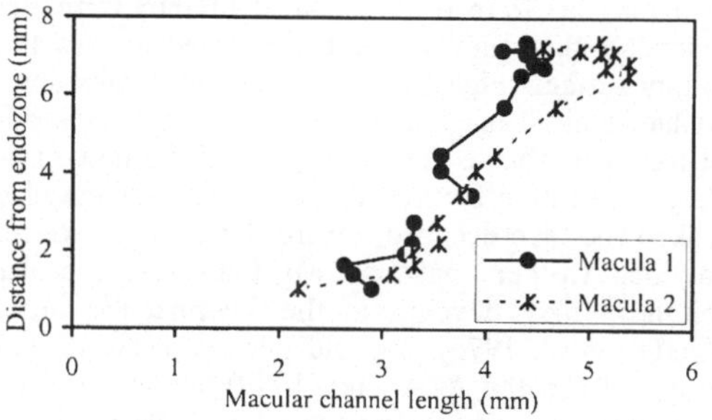

Figure 7. Plot of macular channel length versus distance from endozone.

Other workers have measured macular chimney size in both living and fossil bryozoans. Macular chimney size should be relatively constant in encrusting colonies that grow peripherally. In encrusting colonies, the peripheral growth does not affect the flow dynamics of the established macular chimneys which are distant from the colony margin. This is supported by the observations of previous workers who have reported that macular chimneys have a relatively constant diameter of 0.6 to 2.0 mm (Banta et al. 1974, Cook 1977, Cook & Chimonides 1980, Lidgard 1981). This should not be the case in ramose colonies that grow by expansion of the entire branch. As the colony branch expands in size, previous workers have inferred or shown that macular chimneys increase in size as well (Anstey et al. 1976, Pachut & Anstey 1979, Podell & Anstey 1979, Anstey 1981). This is probably due to increasing incurrent area requiring a larger excurrent flow through the macular chimney (McKinney 1986b). Thus, as a colony branch expands by increasing the width of its exozone, more maculae should develop. In fact, maculae rarely occur on branches with diameters less than 2 mm (McKinney 1986a).

The shape of the maculae in *Tabulipora* became significantly more stellate through the exozone (Macula 1: $R^2 = 0.687$, $p < 0.001$; Macula 2: $R^2 = 0.897$, $p < 0.001$) (Fig. 5). Many excurrent maculae are star-shaped (Anstey 1987), for example those in the cystoporate *Constellaria* (Boardman 1983, Fig. 59.1) and the trepostome *Heterotrypa* (Anstey & Perry 1973, Plate 17). As first noticed by Anstey & Pachut (1976) and Anstey et al. (1976), stellate maculae are similar to centripetal flow structures in other organisms such as astrorhizae in living sclerosponges (e.g. Hartman & Goreau 1970: Fig. 5) and fossil stromatoporoids (e.g. LaBarbera & Boyajian 1991: Fig. 1) as well as inorganic centripetal flow structures such as star dunes (e.g. Nielson & Kocurek 1987: Fig. 2A).

The number of channels per macula and the length of the channels also increased significantly through the exozone (Figs. 6 and 7, respectively). The number of channels in Macula 1 more than doubled ($R^2 = 0.761$, $p < 0.001$), while the number in Macula 2 tripled ($R^2 = 0.848$, $p < 0.001$). Macula 1's mean channel length increased 52 % ($R^2 = 0.916$, $p < 0.001$), while Macula 2's mean channel length increased 132 % ($R^2 = 0.904$, $p < 0.001$).

These last three characters, stellateness index, number of channels, and channel length all increased in response to increasing colony branch size. This probably reflects the need for the macular chimneys to handle more water flow as the incurrent surface area increases. With a more complex stellate shape with more and longer channels, the more efficiently the maculae can collect filtered water into the center of the maculae for ultimate expulsion from the colony surface. This is supported by the work of Patzkowsky (1987) that showed that in some bryozoans as the maculae increase in size, the water flow into the maculae becomes more centripetal around the maculae. As the colony branch increases in circumference, the macular channels act as rivers cutting upstream in a drainage basin and capturing more and more "runoff" as the "drainage" basin grows. Macular shape could also be affected by ambient flow through the branches of the colony in a way similar to the effect seen in elongate maculae in colonies epizoic on nektonic hosts (Baird et al. 1989).

The general pattern of increasing macular spacing, size, and stellateness through the exozone masks some smaller scale changes. All five of the macular characters measured reached a maximum value in both maculae from 6.51 mm to 7.11 mm from the endozone; then they generally decreased toward the colony surface (Figs. 3-7). This indicates something fundamental changed in the exozone in this depth range. It is probable that the initiation of a new macula caused this effect, as the sizes of both maculae peaked and then declined toward the colony surface. Previous workers have shown that new macular chimneys may develop in the intermacular areas (Anstey et al. 1976), and old ones can shrink and disappear (Pachut & Anstey 1979). What happens to the exilazooecia in a macula when it decreases in size? They transform ontogenetically into autozooecia, pass into the intermacular areas, and are not replaced by new exilazooecia.

The decrease in macular size was more pronounced in Macula 1 than Macula 2. Macula 1's size definitely decreased while Macula 2's remained relatively constant (Fig. 4). This suggests a new macula had been initiated closer to Macula 1 than Macula 2. There was no evidence in the peels of a new macula developing in the exozone from 6.51 mm to 7.11 mm from the endozone. This may reflect that the new macula was on an adjacent part of the exozone and not in the block of exozone sampled in this study. The fact that macular size decreased and did not stay constant at some optimum size suggests that some of the incurrent autozooecia supplying the previous maculae had been pirated by the new macula (sensu stream piracy). Taylor (1975) argued that as a colony branch grows and its circumference increases, there will be an increase in the spacing between adjacent macular chimneys and a deviation from the optimum spacing. Hence, unless new macular chimneys are added and/or pre-existing ones expand in size, the feeding efficiency of the colony will decrease as the colony branch grows in circumference.

These significant changes through the exozone in the macular characters were in marked contrast to the lack of statistically significant changes through the exozone in the intermacular regions of this colony (Madsen 1994, Key et al. 2001). Autozooecial packing did not change significantly through the exozone in the incurrent areas. Consistent packing of autozooecia in the intermacular areas was needed to keep the optimal spacing of lophophores in the canopy so as to not disrupt the colony-wide feeding currents. This fixed spacing of intermacular autozooecia was regulated by the exilazooecia which were randomly distributed between the autozooecia (Madsen 1991, Key et al. 2001).

8 CONCLUSIONS

This study represents the first time the dynamic nature of macular spacing, size, and shape has been documented serially through the exozone of a ramose bryozoan colony. It was shown that maculae significantly increased in spacing, size, and stellateness through the exozone. This trend continued until roughly 6.81 mm from the endozone when it reversed, and macular spacing, size, and stellateness began to decrease probably in response to the intercalation of a new macula. In this colony of *Tabulipora*, the growth in branch diameter directly affected the hydrodynamics of the colony-wide feeding currents. As the branch diameter increased, the overall branch surface area increased, and most importantly the area of intermacular autozooecia increased. This increased the area of incurrent flow. As a result of the principle of continuity, the macular chimneys increased in spacing, size, and stellateness presumably to accommodate the increasing incurrent flow with a higher excurrent flow rate. The distance between the maculae increased as the branch grew so the maculae had to increase in size and stellateness to maintain their functionality until a new macula budded, and they began to decrease in size. If maculae did not increase in spacing, size, and stellateness and/or if new maculae did not develop, the feeding efficiency of the colony would have decreased as the branch diameter increased.

In smaller colonies of *Tabulipora* with 3-4 mm diameters, macular chimneys may have functioned with only a cluster of three exilazooecia. In this large colony branch, larger more stellate maculae were needed to create functional excurrent chimneys. This type of excurrent chimney which can change size and shape by adding or removing exilazooecia (Key et al. 2001) may be very flexible to accommodate a range of colony branch sizes.

Do macular chimneys have an optimal spacing, size, and shape? These results suggest not, because intermacular distance, macular area, and stellateness index never reached stasis. Macular position in ramose colonies is not fixed in an absolute sense as maculae migrate away from one another as branch diameter increases and toward one another as the number of adjacent maculae increases. Maculae are fixed in their position relative to one another in a radial growth trajectory sense. The spacing, size, and stellateness of maculae increase with increasing branch diameter and then decrease when an adjacent macula is budded. Macular spacing appears to be a function of macular size, and size appears to be a function of incurrent area. In contrast, maculae may have an optimal spacing, size, and shape in encrusting colonies where colony growth is restricted to the colony margins.

Efficient colony-wide feeding currents centered on macular chimneys may be an intrinsic part of bryozoan colony function and shape, but they have a cost to the colony. These currents are maintained at some metabolic expense to the colony as the chimneys occupy colony surface area which could be filled with feeding zooids (Lidgard 1981, Taylor 1999). In this colony, the combined mean macular areas of 12.7 mm^2 and 14.3 mm^2 occupied 12 % of the area of the peel. This number was representative of the entire colony only if the packing of maculae in this block of exozone was representative of the general surface of the colony. McKinney (1986b) measured the sizes of macular chimneys in Paleozoic bryozoans and found 5-25 % of the colony surface area being occupied with most estimates being 10-15 %. Our value of 12 % supports these estimates.

Despite the loss of feeding zooids to macular area, colony-wide feeding currents have evolved independently in numerous bryozoan clades (Banta et al. 1974). This suggests there must be a metabolic energy gain in colonies that have colony-wide feeding currents. In fact, maculae become a virtual necessity in large colonies to provide a flow rate sufficient to sustain the colony (McKinney 1991). The generation and maintenance of efficient colony-wide feeding currents and their resulting decrease in metabolic cost of expelling filtered water has had an evolutionary impact throughout the history of the phylum (Banta et al. 1974, McKinney 1986a, b, Jackson & McKinney 1990). The development of chimneys for outflow of filtered water was critical in the evolution of the bryozoans as chimneys removed the hydrodynamic constraint on branch width in ramose bryozoans (McKinney 1986a), and branch width in turn affected the structural strength of bryozoan colony branches (Key 1991).

ACKNOWLEDGEMENTS

This research was made possible by grants to M.M. Key from the Knight Foundation (with J.A. Collins) and from the National Science Foundation (BSR-8917108) and to L. Thrane from the Danish Natural Science Research Council. We thank E. Håkansson for placing the specimen at our disposal. This paper benefited greatly from discussions with C.J. Buttler, P.D. Taylor, and P.N. Wyse Jackson as well as from the comments of S.J. Hageman, H.A. Nakrem, and an anonymous reviewer.

REFERENCES

Anstey, R.L. 1981. Zooid orientation structures and water flow patterns in Paleozoic bryozoan colonies. *Lethaia* 14: 287-302.

Anstey, R.L. 1987. Colony patterning and functional morphology of water flow in Paleozoic stenolaemate bryozoans. In J.R.P. Ross (ed.), *Bryozoa: present and past*: 1-8. Bellingham: Western Washington University.

Anstey, R.L. & Delmet, D.A. 1972. Genetic meaning of zooecial chamber shapes in fossil bryozoans. *Science* 177: 1000-1002.

Anstey, R.L. & Pachut, J.F. 1976. Functional morphology of animal colonies by comparison to sand dune paradigms. *Geological Society of America Abstracts with Programs* 8: 124-125.

Anstey, R.L. & Pachut, J.F. 1980. Fourier packing ordinate: A univariate size-independent measurement of the polygonal packing variation in Paleozoic bryozoans. *Mathematical Geology* 12: 139-156.

Anstey, R.L., Pachut, J.F. & Prezbindowski, D.R. 1976. Morphogenetic gradients in Paleozoic bryozoan colonies. *Paleobiology* 2: 131-146.

Anstey, R.L. & Perry, T.G. 1973. Eden Shale bryozoans: a numerical study (Ordovician, Ohio Valley). *Publications of the Museum – Michigan State University Paleontological Series* 1(1): 1-80.

Astrova, G.G. 1973. Polymorphism and its development in the trepostomatous Bryozoa. In G.P. Larwood (ed.), *Living and Fossil Bryozoa*: 1-10. London: Academic Press.

Baird, G.C., Brett, C.E. & Frey, R.C. 1989. "Hitchhiking" epizoans on orthoconic cephalopods: preliminary review of the evidence and its implications. *Senckenbergiana Lethaea* 69: 439-465.

Banta, W.C., McKinney, F.K. & Zimmer, R.L. 1974. Bryozoan monticules: excurrent water outlets? *Science* 185: 783-784.

Boardman, R.S. 1983. General features of the Class Stenolaemata. In R.A.Robison (ed.), *Treatise on invertebrate paleontology. Part G. Revised. Volume 1*: 49-137. Boulder and Lawrence: Geological Society of America and University of Kansas Press.

Boardman, R.S. & Cheetham, A.H. 1983. Glossary of morphological terms. In R.A.Robison (ed.), *Treatise on invertebrate paleontology. Part G. Revised. Volume 1*: 304-320. Boulder and Lawrence: Geological Society of America and University of Kansas Press.

Buss, L.W. 1979. Habitat selection, directional growth and spatial refuges: Why colonial animals have more hiding places. In G. Larwood & B.R. Rosen (eds), *Biology and systematics of colonial organisms*: 459-497. New York: Academic Press.

Buss, L.W. 1980. Bryozoan overgrowth interactions - the interdependence of competition for space and food. *Nature* 281: 475-477.

Buss, L.W. 1981. Mechanisms of competition between *Onychocella alula* (Hastings) and *Antropora tincta* (Hastings) on an eastern Pacific rocky shoreline. In G. Larwood & C. Nielsen (eds), *Recent and fossil Bryozoa*: 39-49. Fredensborg: Olsen and Olsen.

Cook, P.L. 1977. Colony-wide water currents in living bryozoans. *Cahiers de Biologie Marine* 18: 31-47.

Cook, P.L. & Chimonides, P.J. 1980. Further observations on water current patterns in living Bryozoa. *Cahiers de Biologie Marine* 21: 393-402.

Delmet, D.A. & Anstey, R.L. 1974. Fourier analysis of morphological plasticity within an Ordovician bryozoan colony. *Journal of Paleontology* 48: 217-226.

Dick, M.H. 1987. A proposed mechanism for chimney formation in encrusting bryozoan colonies. In J.R.P. Ross (ed.), *Bryozoa: present and past*: 73-80. Bellingham: Western Washington University.

Eckman, J.E. & Okamura, B. 1998. A model of particle capture by bryozoans in turbulent flow: significance of colony form. *American Naturalist* 152: 861-880.

Grünbaum, D. 1995. A model of feeding currents in encrusting bryozoans shows interference between zooids within a colony. *Journal of Theoretical Biology* 174: 409-425.

Grünbaum, D. 1997. Hydromechanical mechanisms of colony organization and cost of defense in an encrusting bryozoan, *Membranipora membranacea. Limnology and Oceanography* 42: 741-752.

Hartman, W.D. & Goreau, T.F. 1970. Jamaican coralline sponges: Their morphology, ecology and fossil relatives. *Symposia of the Zoological Society of London* 25: 205-243.

Håkansson, E., Heinberg, C. & Stemmerik, L. 1981. The Wandel Sea Basin from Holm Land to Lockwood Ø, eastern North Greenland. *Rapport Grønlands Geologiske Undersøgelse* 106: 47-63.

Håkansson, E. & Madsen, L. 1991. Symbiosis - a plausible explanation of gigantism in Permian trepostome bryozoans. In F.P. Bigey (ed.), *Bryozoa: Living and fossil*: 151-159. Nantes: Société des Sciences Naturelles de l'Ouest de la France, Mémoire hors série.

Jackson, J.B.C. & McKinney, F.K. 1990. Ecological processes and progressive macroevolution of marine clonal benthos. In R.M. Ross & W.D. Allmon (eds), *Causes of evolution: A paleontological perspective*: 173-209. Chicago: University of Chicago Press.

Jørgensen, C.B. 1966. *Biology of suspension feeding*. Oxford: Pergamon.

Key, M.M., Jr. 1991. How to build a ramose trepostome. In F.P. Bigey (ed.), *Bryozoa: living and fossil*: 201-207. Nantes: Société des Sciences Naturelles de l'Ouest de la France, Mémoire hors série.

Key, M.M., Jr., Thrane, L. & Collins, J.A. 2001. Space-filling problems in ramose trepostome bryozoans as exemplified in a giant colony from the Permian of Greenland. *Lethaia* 34: 125-135.

LaBarbera, M. & Boyajian, G.E. 1991. The function of astrorhizae in stromatoporoids: Quantitative tests. *Paleobiology* 17: 121-132.

Lidgard, S. 1981. Water flow, feeding, and colony form in an encrusting cheilostome. In G. Larwood & C. Nielsen (eds), *Recent and fossil Bryozoa*: 135-142. Fredensborg: Olsen and Olsen.

Madsen, L. 1991. *The species concept in trepostome bryozoans - a study of phenotypical and genotypical variability within the genus* Tabulipora *in North Greenland*. Unpublished Ph.D. thesis, University of Copenhagen, Copenhagen.

Madsen, L. 1994. Exozonal budding in trepostome Bryozoa. In P.J. Hayward, J.S. Ryland & P.D. Taylor (eds), *Biology and palaeobiology of bryozoans*: 113-115. Fredensborg: Olsen and Olsen.

McKinney, F.K. 1986a. Historical record of erect bryozoan growth forms. *Proceedings of the Royal Society of London* (Series B) 228: 133-148.

McKinney, F.K. 1986b. Evolution of erect marine bryozoan faunas: Repeated success of unilaminate species. *American Naturalist* 128: 795-809.

McKinney, F.K. 1988. Elevation of lophophores by exposed introverts in Bryozoa: A gymnolaemate character recorded in some stenolaemate species. *Bulletin of Marine Science* 43: 317-322.

McKinney, F.K. 1989. Two patterns of colonial water flow in an erect bilaminate bryozoan, the cheilostome *Schizotheca serratimargo* (Hincks, 1886). *Cahiers de Biologie Marine* 30: 35-48.

McKinney, F.K. 1990. Feeding and associated colonial morphology in marine bryozoans. *Critical Reviews in Aquatic Sciences* 2: 255-280.

McKinney, F.K. 1991. *Exercises in invertebrate paleontology*. Boston: Blackwell Scientific.

McKinney, F.K. 1992. Competitive interactions between related clades: Evolutionary implications of overgrowth interactions between encrusting cyclostome and cheilostomes bryozoans. *Marine Biology* 114: 645-652.

McKinney, F.K. & Jackson, J.B.C. 1989. *Bryozoan evolution*. London: Unwin-Hyman.

Nielson, J. & Kocurek, G. 1987. Surface processes, deposits, and development of star dunes: Dumont dune field, California. *Geological Society of America Bulletin* 99: 177-186.

Pachut, J.F. & Anstey, R.L. 1979. A developmental explanation of stability-diversity-variation hypotheses: Morphogenetic regulation in Ordovician bryozoan colonies. *Paleobiology* 5: 168-187.

Patzkowsky, M.E. 1987. Inferred water flow patterns in the fossil *Fistulipora* M'Coy (Cystoporata, Bryozoa). In J.R.P. Ross (ed.), *Bryozoa: present and past*: 213-219. Bellingham: Western Washington University.

Podell, M.E. & Anstey, R.L. 1979. The interrelationship of early colony development, monticules, and branches in Palaeozoic bryozoans. *Palaeontology* 22: 965-982.

Stemmerik, L. & Håkansson, E. 1989. Stratigraphy and depositional history of the Upper Palaeozoic and Triassic sediments in the Wandel Sea Basin, central and eastern North Greenland. *Rapport Grønlands Geologiske Undersøgelse* 143: 21-45.

Stemmerik, L., Håkansson, E., Madsen, L., Nilsson, I., Piasecki, S., Pinard, S. & Rasmussen, J.A. 1996. Stratigraphy and depositional evolution of the Upper Palaeozoic sedimentary succession in eastern Peary Land, North Greenland. *Bulletin Grønlands Geologiske Undersøgelse* 171: 45-71.

Taylor, P.D. 1975. Monticules in a Jurassic cyclostomatous bryozoan. *Geological Magazine* 112: 601-606.

Taylor, P.D. 1979. The inference of extrazooidal feeding currents in fossil bryozoan colonies. *Lethaia* 12: 47-56.

Taylor, P.D. 1999. Bryozoans. In E. Savazzi (ed.), *Functional morphology of the invertebrate skeleton*: 623-645. New York: Wiley.

Taylor, P.D. & Voigt, E. 1999. An unusually large cyclostome bryozoan (*Pennipora anomalopora*) from the Upper Cretaceous of Maastricht. *Bulletin de l'Institut Royal des Sciences Naturalles de Belgique, Sciences de la Terre* 69: 165-171.

Ulrich, E.O. 1890. Paleozoic Bryozoa. In A.H. Worthen, C. Wachsmuth, F. Springer, E.O. Ulrich & O. Everett (eds), *Palaeontology* in *Geology and Palaeontology* 8: 283-688. Springfield: Illinois Geological Survey.

Vogel, S. 1981. *Life in moving fluids*. Boston: Willard Grant.

Winston, J.E. 1978. Polypide morphology and feeding behavior in marine ectoprocts. *Bulletin of Marine Science* 28: 1-31.

Winston, J.E. 1979. Current-related morphology and behaviour in some Pacific Coast bryozoans. In G.P. Larwood & M.B. Abbott (eds), *Advances in bryozoology*: 247-268. London: Academic Press.

Winston, J.E. 1981. Feeding behavior of modern bryozoans. In J.T. Dutro & R.S. Boardman (eds), *Lophophorates: notes for a short course* in *Studies in Geology* 5: 1-21. Knoxville: University of Tennessee.

Bryozoan Studies 2001, Wyse Jackson, Buttler & Spencer Jones (eds)
© 2002 Swets & Zeitlinger, Lisse, ISBN 90 5809 388 3

Freshwater bryozoan remains from the Molteno Formation (Upper Triassic) of South Africa

R. Kohring & A. Hörnig
Institut für Geologische Wissenschaften, Freie Universität Berlin, Germany

ABSTRACT: From the Upper Triassic Molteno Formation of South Africa more than 1000 specimens of small convex and circular-shaped statoblasts have been discovered. They are of two size classes (0.6 mm and 1 mm in diameter), reflecting at least two different taxa. Nearly all specimens are concentrated on bedding planes at the locality Birds River, which is of lacustrine origin. The other Molteno locality to have yielded statoblast is Kapokkraal, which was deposited in an abandoned channel within a braided-river system. Due to their size and shape, some of these statoblasts can be assigned to the Family Pectinatellidae suggesting this group to be a primitive one within the order Phylactolaemata Allman 1856. Up to now, fossil statoblasts of freshwater Bryozoa have been known only from Quaternary, Oligocene, Eocene and Lower Cretaceous deposits. Since the marine orders of Bryozoa are known from the Early Palaeozoic, a much older origin of the Phylactolaemata can be suggested.

1 INTRODUCTION

In contrast to marine Bryozoa, the limnetic Phylactolaemata develop no calcified or other mineralized hard parts. Hence, the soft-bodied freshwater Bryozoa are not to be expected to become fossilized, and are, therefore, generally unknown in the fossil record. Bassler (1953) in the first edition of the bryozoan part of the *Treatise on Invertebrate Paleontology* mentioned doubtful freshwater bryozoans from the Cretaceous of Bohemia. Ehrhard Voigt (1990, pers. comm.) described these fossils as "dubious matter".

Like many invertebrates inhabiting unstable lakes and ponds, freshwater bryozoans form asexually little hard-shelled buds, so-called statoblasts (Bushnell & Rao 1979, Wood 1979). Statoblasts can at least be divided into two different groups, buoyant statoblasts, called floatoblasts, and sessile statoblasts, called sessoblasts.

In general, all statoblasts are made of chitinious material, enclose the yolky germinal mass and consist of a dorsal and a ventral valve, which are separated when the statoblast germinates (Oda & Mukai 1989). Since freshwater bryozoans are not mineralised, it is only their chitinuous statoblasts that can become fossilised.

Ovoid or circular-shaped, biconvex objects, interpreted as statoblasts, have been recorded extremely rarely in the fossil locality. Jell & Duncan (1986) described statoblasts from lake sediments of the Lower Cretaceous Korumburra Group (Victoria/Australia). They are ovally formed, 0.25 to 0.35 mm in diameter and exhibiting a dark central area and a ligther coloured flat annulus. They occur in clusters. Jell & Duncan (1986) interpreted these fossils as statoblasts and assigned them tentatively to the genus *Plumatella*.

The statoblasts from the Middle Eocene of the Eckfeld maar-sediments (Germany) are not yet described. A problematic fossil from the Oligocene of Rott (Germany) is described by Kohring & Reitner (1991). It is more or less circularshaped with a diameter of 0.58 to 0.6 mm. It is one single disk and resembles a floatoblast more than a sessoblast. A small dark bown peripheral annulus displays no special features. The central area, however, shows a reticular network. It is regular and somewhat spiralic, and all polygonal cells have nearly the same dimensions. It is, therefore, quite similar to *Plumatella*. From interglacial and postglacial deposits of the Canadian Arctic, Kuc (1973) described and figured extremely well-preserved statoblasts, assigned to *Cristatella mucedo*.

The purpose of the present study is the description of statoblasts from the Upper Triassic of the Molteno Formation (South Africa), where mass occurrences of fossil statoblasts are common at two localities (Kohring & Hörnig 1998).

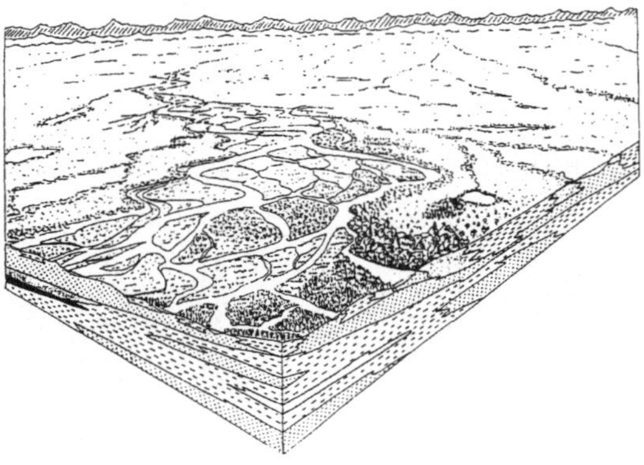

Figure 1. Regional palaeoenvironemtal reconstruction of the Upper Trassic Molteno Formation with particular emphasis of the river systems and the vegetation. Modified after Cairncross et al. (1995). The locality type of Bir III (Birds River) is indicated by a circle.

2 LOCALITY AND MATERIAL

The Molteno Formation is a part of the Karoo supergroup in South Africa. Its depositional environment is characterised as a northerly thinning intracratonic clastic wedge extending over an area of about 25,000 km^2 which attains a maximum thickness of some 460 m in the southeast (Anderson 1981, Cairncross et al. 1995).

The rich palaeoentomofauna (Anderson et al. 1998) derives from extensive flood-plain deposits of the Gondwana interior (Fig. 1). The stratigraphic age of the Molteno Formation has been assigned to the Carnian within the Upper Triassic (Anderson 1981) (220-230 ma).

The great majority of Molteno specimens are concentrated on a few bedding planes at one locality (Bir III, Birds river). These are sediments of lacustrine origin, where permanent changes of water quality, temperature or nutrient supply can be expected, often promoting the formation of statoblasts.

Nearly 99% of all specimens have been sampled from this locality, sometimes they have been found in clusters up to 100 specimens. The only other Molteno locality to have yielded statoblasts is Kapokkraal.

In contrast to Bir III this site was deposited in an abandoned channel within a braided-river system. Only a few isolated individuals up to now have been found here.

3 SYSTEMATIC DESCRIPTION

In the Molteno Formation, the statoblasts are of two distinct size classes (Type A and Type B).

Type A: Diameter 0.5-0.7 mm. Central area about 0.4-0.5 mm in diameter. Shelf not very broad (about 0.1 mm) (Fig. 2).

Type B: Diameter 0.9-1.0mm. Central area about 0.5 mm. Shelf relatively broad (0.4 mm).

All statoblasts are convex and circularshaped or slightly ovoid (Figs 2a-d). In all specimens a peripheral annulus is developed, sometimes darker in colour than the central area. No spines are visible. However, in some specimens outside the annulus small holes are visible, surrounding at least some parts of the statoblast. Possibly these are imprints of denser and harder parts of spines, when valves have been deposited on their convex side. Obviously all specimens are isolated valves.

Type A is dominating with 90 % of all observed specimens and is common in both localities, mainly accumulated in big clusters up to more than 100 specimens (Fig. 2e). In Kapokkraal only isolated and few specimen of both types could be found. Distribution pattern of isolated valves in both localities is simlar to modern lake environments after germination (Fig. 2f) (T. Wood, pers. comm.).

4 DISCUSSION

The two types of statoblasts found in the Molteno sediments probably reflecting at least two different taxa (genera?, species?). In modern environments, the diversity of the Phylactolaemata is highly variable, ranging from 1 to 12 species per site.

Due to the poor visible characters of these specimens and the age of the deposits, a certain assignment of the Molteno statoblasts to any modern genus is impossible. However, size and shape as well as the short spines as indicated by the tiny imprints in the sediments is typical for statoblasts of the family Pectinatellidae (T. Marsh and T. Wood, 2001, pers. comm.). This assignment of the (up to now) earliest record of the Phylcatolaemata agrees with the sequence coding for ribosomal RNA (M. Lore, 2001, pers. comm.) in which the Pectinatellidae has been isolated longer than any other group of modern freshwater Bryozoa.

The statoblasts of the Molteno Formation are at present the oldest evidence of the class Phylactolaemata. However, due the history of discovery of fossil statoblasts, two suggestions can be made: the evolution of the freshwater Bryozoa began in the Triassic and is, therefore, an outcome of the "Triassic explosion", or statoblasts have been overlooked in the pre-Triassic deposits till now. Since many fossil statoblasts have been found in association with fossil insects, older statoblasts than of Triassic age can be expected in Carboniferous and Permian insect localities. Because the other groups of Bryozoa, all

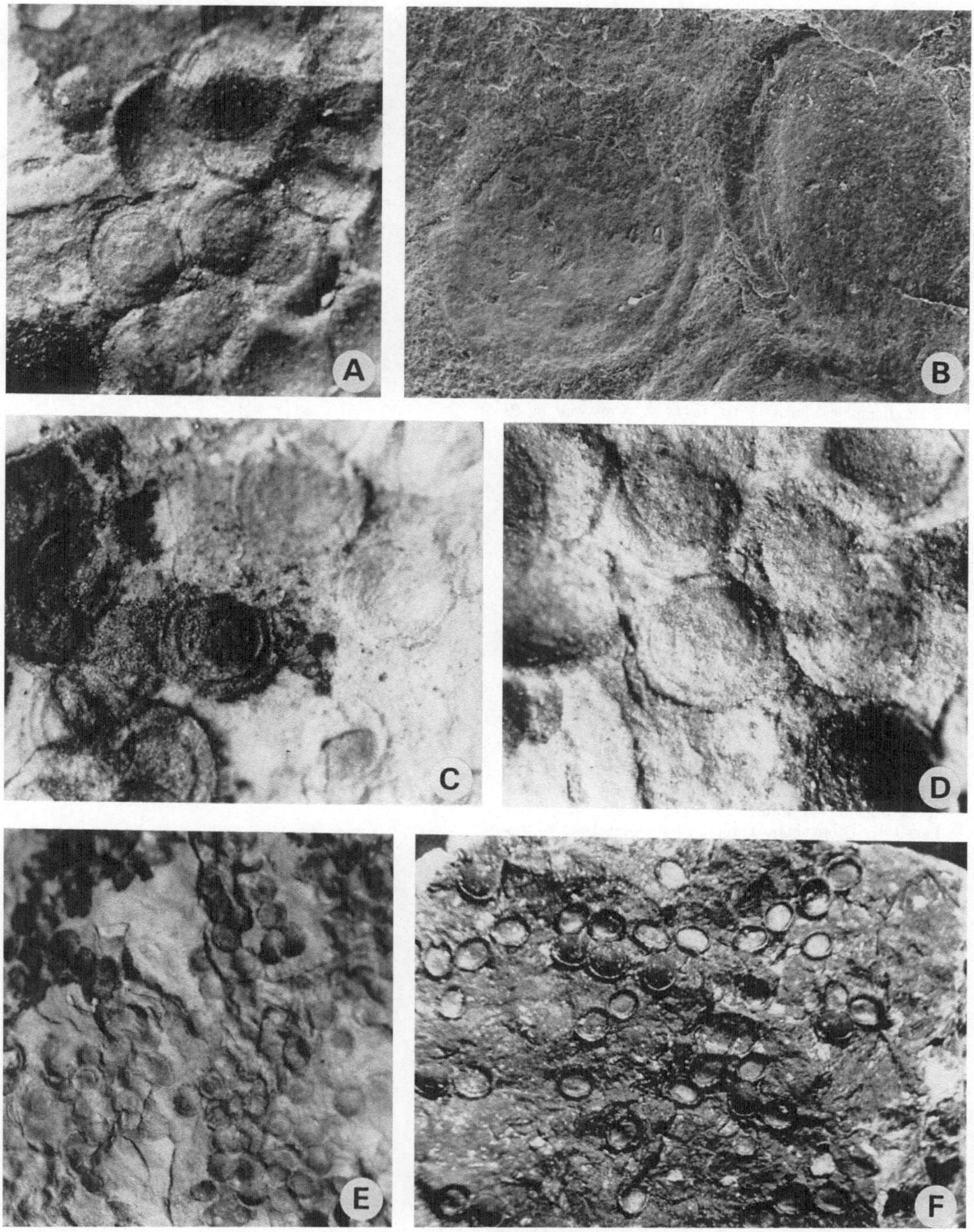

Figure 2. a-e: Freshwater bryozoan statoblasts from the Upper Triassic Molteno Formation of South Africa (locality: Birds River). a-d: Type A, diameter of specimens is 0.5-0.7 mm. e: Clusters of statoblasts in the Molteno Formation (Birds River). f: Modern statoblasts (*Pectinatella* sp.) found on rocks in Lough Anilaun at Cleggan, County Galway, Ireland. Many of the sessoblasts have germinated, leaving behind the concave basal valve attached to the rock. Other sessoblasts are still intact, as seen by the convex frontal valve.

living in marine environments, are known from the Early Palaeozoic, a much older origin of the Phylactolaemata is not inconceivable.

ACKNOWLEDGEMENTS

The authors thank John Anderson (Pretoria, RSA) for loan of the specimens, and Michael Lore (Dayton, Ohio), Terrence Marsh (Naperville, Illinois), Timothy Wood (Dayton, Ohio) and Emmy Wöss (Vienna, Austria) for discussion.

REFERENCES

Anderson, J.M. 1981. World Permo-Triassic correlations: their biostratigraphical basis. 3-10. In M.M. Cresswell & P. Vella (eds.) *Gondwana Five (Proceedings of the Fifth International Gondwana Symposium, Wellington New Zealand, 1980)* Rotterdam: Balkema.

Anderson, J.M., Kohring, R. & Schlüter, T. 1998. Was insect biodiversity in the Triassic akin to today? - a case study from the Molteno Formation. *Entomologia generalis* 23(1/2): 15-26.

Bassler, R.S. 1953. Bryozoa. In R.C. Moore (ed.) *Treatise of Invertebrate Paleontology*, Part G. University of Kansas Press, Boulder. 254 pp.

Bushnell, J.H. & Rao, K.S. 1979. Freshwater Bryozoa: Microarchitecture of statoblasts and some aufwuchs animal associations. In G.P. Larwood & M.B. Abbott (eds) *Advances in Bryozoology*: 75-92. Academic Press.

Cairncross, B., Anderson, J.M. & Anderson, H.M. 1995. Palaeoecology of the Triassic Moilteno Formation, Karoo Basin, South Africa - sedimentological and palaeontological evidence. *South African Journal of Geology* 98: 452-478.

Jell, P.A. & Duncan, P.M. 1986. Invertebrates, mainly insects, from the freshwater, Lower Cretaceous, Koonwarra Fossil Bed (Korumburra Group), South Gippsland, Australia. *Mem. Ass. Australasian Palaeontols* 3: 111-205.

Kohring, R. & Hörnig, A.C.F. 1998. The earliest freshwater Bryozoa: evidence from the Upper Triassic Molteno Formation (South Africa). *Journal of African Earth Scinces* 27(1A): 125.

Kohring, R. & Reitner, J. 1991. A problematic microfossil from the Oligocene of Rott (Siebengebirge, Germany) with affinities to freshwater bryozoan statoblasts (Phylocatolaemata). *Berliner geowissenschaftliche Abhandlungen A* 134: 221-225.

Kuc, M. 1973. Fossil statoblasts of Cristatella mucedo Cuvier in the Beaufort Formation and in interglacial and postglacial deposits of the Canadian Arctic. *Geological Survey of Canada*, Paper 72-28: 1-12.

Oda, S. & Mukai, H. 1989. Systematic position and biology of Pectinatella gelatinosa Oka (Bryoozoa: Phylactolaemata) with the description of a new genus. *Zoological Science* 6: 401-408.

Wood, T. 1979. Significance of morphological features in bryozoan statoblasts. In G.P. Larwood & M.B. Abbott (eds) *Advances in Bryozoology*: 59-74. Academic Press.

Bryozoan Studies 2001, Wyse Jackson, Buttler & Spencer Jones (eds)
© 2002 Swets & Zeitlinger, Lisse, ISBN 90 5809 388 3

Bryozoa of the high Arctic fjord – a preliminary study

P. Kuklinski
Institute of Oceanology, Polish Academy of Science, ul.Powstanców Warszawy 55, Sopot 81-712, Poland (mailing address) & UNIS, P.O.Box 156, N-9171 Longyearbyean, Norway

ABSTRACT: A bryozoan faunal investigation of Kongsfjorden (79°N and 12E° - western coast of Spitsbergen) based on a collection made in the summer of 1998 is presented. In 19 samples, obtained by grab and SCUBA diving, 89 taxa were recognized; 69 of them were determined to species, 17 to genus and 3 to family level. Collection stations ranged from 4 to 258 m depth. 12 of the species recognized were recorded for the first time in Svalbard waters. Depth, salinity and temperature data was analysed. Species composition depends on the depth of the station, as confirmed by MDS (multidimensional scaling) and cluster analysis. *Electra crustulenta* var. *arctica, Cylindroporella tubulosa, Celloporella hyalina, Callopora craticula* and *Tegella arctica* were the most frequent taxa found in the fjord. Biodiversity increases with the depth. The most probable factor responsible for that is lower stability of environment and disturbance caused by ice and icebergs in shallow part of the fjord.

1 INTRODUCTION

Fjords are defined as a deep, high-latitude estuaries which has been (or are presently being) excavated or modified by land-based ice (Syvitski et al. 1987). All estuaries are ephemeral geologic features, but fjords are the youngest of all. Products of the general retreat of ice and sea-level fluctuations that have occurred since last glacial maximum some 17,000 years BP. Fjords are non-steady-state systems, evolving and changing over relatively short time scales (Syvitski et al. 1987). They are transition regions between the land and open oceans, regions of strong physical and chemical gradients where fresh and salt waters mix and react (Syvitski et al. 1987). Therefore they are very dynamic systems.

There are no studies about bryozoan ecology looking at the a fjord as a whole system. This research is part of study started in 1998 examining fjord as one system. The main purpose of this study is to follow the patterns of bryozoan distribution and community structure along the depth gradient in one of the Svalbard Archipelago fjords. These are preliminary results and further research will probably enrich the knowledge of fjordic bryozoan communities.

Bryozoans play a major role in association with other sessile arctic organisms (Rózycki & Gruszczynski 1986, Lippert 1998). A recently produced check list of Svalbard fauna (total: 1645 species) revealed that bryozoans are among one of the most numerous groups of animals (196 species) in that area (Gulliksen et al. 1999).

2 MATERIALS AND METHODS

2.1 *Study area*

Kongsfjorden is situated on the Western coast of Spitsbergen on 79°N and 12°E. The fjord is 26 km long and on average 8 km wide. The area of the fjord is 208 km^2. Maximum depth is 428 m while average is 141 m. Length of coast line of the fjord is 90 km, 16 km is covered by glaciers. The basic features responsible for hydrological and sedimentary conditions in a fjord include bathymetry, vicinity of the ocean, presence and activity of tidal glaciers and climate. Kongsfjorden does not have a typical fjord sill at the entrance, which results in strong influence of open ocean waters on hydrological conditions (Weslawski et al. 1991, Ito & Kudoh 1997).

There are two main water masses which influence Kongsfjorden. The first one is West Spitsbergen Current, which is a branch of the warm (4°C) and highly saline (35 psu) Norwegian Current. This current flows along the western and northern parts of Spitsbergen (Loeng 1991). Transportation of this warm water masses causes a milder climate compared to other areas at the same latitude (Gammelsrod & Rudels 1983). The second water mass influencing this area is the East Spitsbergen Current. The cold, dense and highly saline water of this cur-

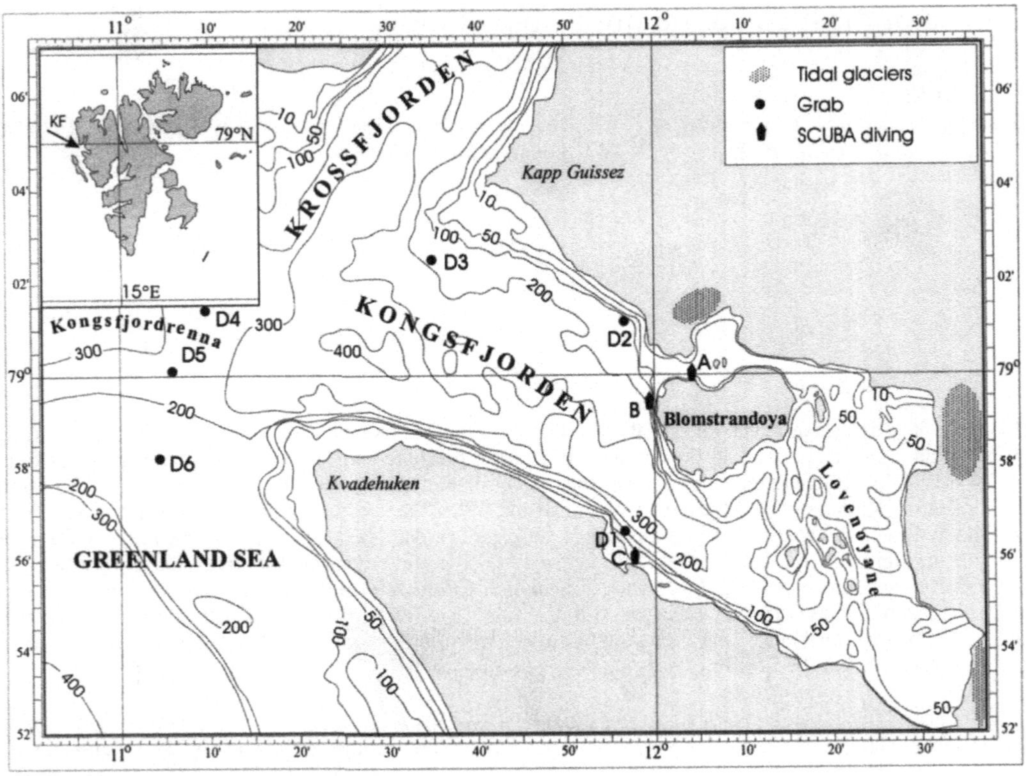

Figure 1. Area of study and location of sampling stations (KF - Kongsfjorden)

rent originates from the Arctic Ocean. It flows along the East coast of Spitsbergen down to the southern tip of the island where it becomes the Sorkapp Current. The Sorkapp Current brings to western coast of Spitsbergen waters which range in temperature form –1.5 to 1°C and salinity from 34 to 35 psu (Loeng 1991, Beszczynska-Moller et al. 1997).

Suspension concentration close to glaciers can reach 300-500 mg/dm^3, decreasing in the central and outer parts of the fjord in summer. In the inner basin concentrations in intermediate water layer may reach 20-25 mg/dm^3 and 2-3 mg/dm^3 in the central parts, while relative clear water in the outer parts of the fjord contains 0.5 mg/dm^3 (Elverhoi et al. 1983, Zajaczkowski 2000).

2.2 *Methods*

Samples were taken during the 1998 cruise of r/v Oceania. There were two techniques used to collect the samples: Van Veen grab and SCUBA diving. Temperature and salinity was measured by a CTD probe. Of the 19 samples collected 13 were by SCUBA diving and 6 by grab. Bryozoa were found on rocks, stones, algae, and barnacles. Bryozoans from shallow sample sites occurred on algae, rocks, stones, barnacles while deep ones only occur on stones and rocks. Location of the sampling stations is shown on Figure 1. SCUBA diving technique was used to collect samples down to 20 m, while all the other samples were collected by grab. Collected material was dried in case of rocks, stones and barnacles. Algae and bushy bryozoans were fixed in 4% formalin. The volume of substratum on which the bryozoan occurred was estimated by the volume of water displaced from a glass cylinder by a given item. For stones this was a rough estimate. Species determination was made using a stereomicroscope. Frequency of occurrence were calculated by using equation:

$F= n_i * n^{-1}$

F – frequency of given taxa [%]

n_i – number of samples where given taxa was present

n – number of all samples

The multivariate analysis was applied to species presence-absence data using the PRIMER package. The Bray-Curtis similarities were calculated. Ordination (non-metric multidimentional scaling – MDS) and classification (using group average linking) of samples were performed. Groups of samples were distinguished based on the resulting dendrogram and

MDS plot. The diversity was measured as: number of taxa per sample and Shannon-Wiener index. Shannon-Wiener (log base e) was calculated for data standardized to 150 cm^3.

3 RESULTS

The salinity varied from 31.2 (B1) to 34.93 (D4) psu. Within shallow stations (4-20m) salinity ranged from 31.2 (B1) to 34.2 (C) psu. Salinity among deep stations (114-258 m) varied from 34.59 (D2) to 34.93 (D4) psu. Temperature ranged between stations from minus 0.69°C (C) to 2.58°C (D5). In the shallow areas temperature ranged from minus 0.69°C (C) to 2.38°C (A2). In the deep areas temperature varied from 0.63°C (D1) to 2.58°C (D5).

89 taxa were found. 69 taxa were identified to species, 17 to genus and 3 to family level (Table 1). Identified taxa belonged to three orders: Cyclostomatida, Ctenostomatida and Cheilostomatida. 80 % of taxa belonged to order Cheilostomatida, 14% to Cyclostomatida and 6% to the order Ctenostomatida. The number of taxa per sample varied from 1 (B4) to 28 (D5, D6), while on average for all samples it was 15 taxa. For the shallow samples the average number of taxa was 13, while for the deep samples it was 20 taxa. Shannon-Wiener index varied from 0.92 (D4) to 2.97 (D5), average for all samples 1.95. For the shallow samples the average value of the index was 1.74, while for the deep samples it was 2.36. Cluster analysis showed differences between deep and shallow samples, what was confirmed by MDS analyses as well (Figure. 2). *Electra crustulenta* var. *arctica* (65 %), *Cylindroporella tubulosa* (60 %), *Callopora craticula* (55%), *Celleporella hyalina* (55%) were the most common species (Table 1). 25 species occurred only in the shallow samples. The most frequent were *Celleporella hyalina, Harmeria scutulata, Tegella arctica* and *Tegella armifera*. In the deep samples 33 species were present, the most frequent were *Hippodiplosia obesa, Schizoporella pachystega, Schizoporella costata* and *Escharelloides spinulifera.* 11 species were present within a wide range of depth. Here the most frequent were *Electra crustulenta* var. *arctica, Cylindroporella tubulosa, Hippothoa divaricata* var. *arctica* and *Callopora craticula*. Compared with Kluge (1975), Lippert (1998) and Gulliksen et al. (1999) 12 species (marked by + sign in the Table 1) were noted for the first time in Svalbard area.

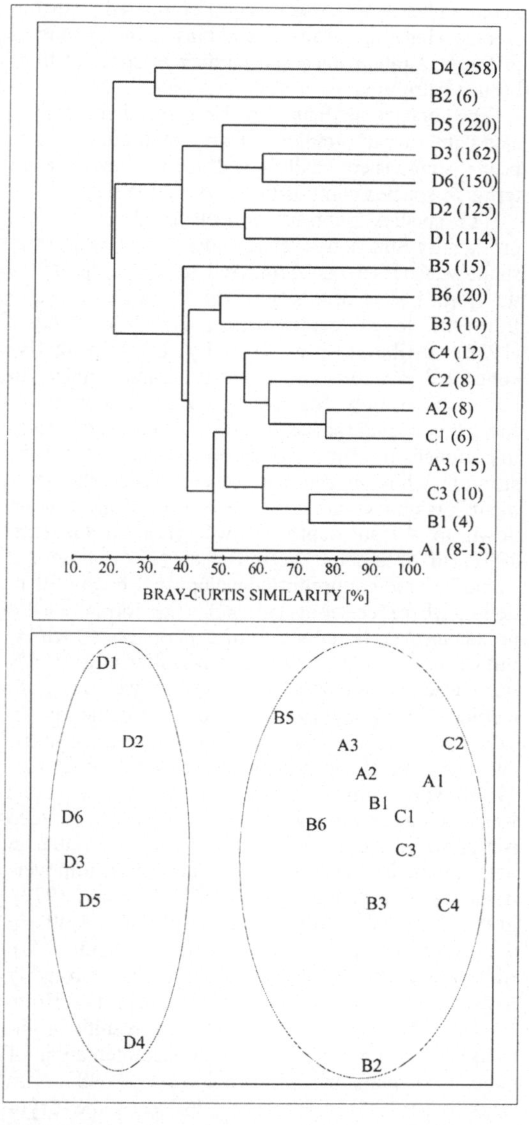

Figure 2. Cluster and MDS based on data from Table 1 (depth in brackets, shallow and deep samples separately circled).

4 DISCUSSION

There are difficulties in obtaining quantitative data when majority of bryozoans live on a hard substratum. Shallow sampling can be done by a SCUBA diver using frame of a known surface, while for deep sampling a combination of grab and video techniques seems to be the best solution to obtain quantitative data (Rumohr 1995). However in this study as a substitute, abundance and number of taxa per volume of substratum was used (to calculate Shannon-Wiener index). The aim was to provide the quantitative comparability of data collected by Van Veen grab in deep sites and by a diver (without a frame) in shallow waters. Certainly there are some drawbacks to the method used. In the samples from shallow waters the substrata examined comprised algae and stones while in the deep sites the bryozoans were found only on stones. Surface/volume ratio will be much higher in the case of algae compared with the stones. That might cause some bias in interpretation of the data (abundance and number of species) from both substratums.

The bryozoan fauna in Kongsfjorden changes along the depth gradient. Two communities have been distinguished: shallow and deep. They differ in terms of species composition and biodiversity.

The species richness as well as diversity (expressed by Shannon-Wienner index) was lower in the shallow waters when compared to deeper parts of the fjord. Lower diversity in shallow waters results from the lower environmental stability. Sanders (1958) attributed low diversity to "physically-controlled environments", where animals have to adapt to strongly fluctuating, stressful conditions. Mechanical disturbance of the bottom is of major importance. In winter ice cover scratches the substratum, while in summer icebergs scour the sediments, which could be observed in Kongsfjorden down to 40 m depth (Dowdeswell & Forsberg 1992). In Antarctica Dayton et al. (1974) observed stable benthic community development below 30 m depth, where ice scour and anchor ice formation did not occur. Additionally a fjord experiences a high input of fresh water (Beszczynska-Moller et al. 1997) from glacier meltwater discharge as well as from melting of snow and ice. This induces fluctuations in salinity as well as changes in concentrations of nutrients and mineral suspended matter (Weslawski et al. 1988). The salinity measured at the sampling sites was much more variable in shallow sites then in deep water, while the temperature was very variable in both zones. Reduced salinity results in an impoverished bryozoan fauna (Ryland 1970, Winston 1977). In the Skagerrak lower salinity (>30 psu) results in the occurrence of about 130 species (Ryland 1970), while along the Norwegian coast (salinity roughly 35psu) 268 species were noted (Brattegard & Holte 1997). The vicinity of the glaciers results in increased concentration and rate of sedimentation of mineral suspensions (Elverhoi et al. 1983, Zajaczkowski 2000). This might influence the bryozoan diversity in shallow sites situated close to active glacier fronts. Inorganic suspensions are especially stressful to suspension feeding and sedentary bryozoans as they may clog their filtering organs and hinder their settlement (Moore 1977). The biological factors- including predation might also be important (McKinney & Jackson 1991). The perennial macroalge – the main substratum of bryozoans in shallow waters are consumed by abundant sea urchins - *Strongylocentrotus droebachiensis* (Hop et al., in press).

The much higher heterogeneity of substratum in shallow sites has not resulted in higher biodiversity, which is usually the case (Begon et al. 1996). This suggests that disturbance caused by ice and the instability of the environment are the decisive factors here. A similar pattern of bryozoan diversity (increasing with depth), was observed by Rosso (1991, 1992) in Terra Nova Bay (Ross Sea, Antarctica) and Winston & Hayward (1994) in the Ross Sea.

Published data on Arctic bryozoan faunas are scarce. Powell (1968) showed that in Arctic Canada bryozoans mostly predominate at a depth of 30 m or below. *Discoporella hispida, Eucratea loricata, Cribrilina annulata, Celleporella hyalina* were also found in the shallow waters of Kongsfjorden. *D. hispida* and *E. loricata* also occurred in samples from deep parts of the fjord. *Cribrilina spitsbergensis, Escharoides jacksoni, Posterula sarsi, Celleporina ventricosa* indicate a deeper environment (Powell 1968). In case of Kongsfjorden that was observed only for the *C. spitsbergensis,* while three others do not occur at all in presented study.

In the vicinity of Svalbard in shallow waters of Franz Josef Land Androsova (1977) observed 27 species of bryozoa. Among all the shallow samples investigated in Kongsfjorden sample B1 contains 25 taxa which is close to the number noted by Androsova (1977). Only 7 taxa found in Kongsfjorden are present in Androsova's list of species as well. Among those 7 taxa only *Dendrobeania fruticosa* and *Microporella arctica* occured exclusively in the shallow parts of the fjord, the other 5 taxa had a wider range of occurrence.

Further investigations of bryozoan ecology from arctic fjord should complete our knowledge concerning the species distribution in the fjord, structure of community, productivity and level of influence of various environmental factors.

5 CONCLUSIONS

In Kongsfjorden the bryozoan community species composition as well as biodiversity change along the depth gradient. The multivariate analysis shows two communities – a shallow (down to 20 m) and a deep one (114-256 m). The fauna is more diverse in the deeper parts of the fjord than the shallow one. The lower stability of shallow environment (ice cover scratch, icebergs scour, glacier meltwater discharge, sedimentation) is most likely responsible for that kind of pattern.

ACKNOWLEDGEMENT

I would like to thank very much Dr Maria Wlodarska-Kowalczuk for her great support and valuable comments on the manuscript.

Table 1. List of taxa found in Kongsfjorden with their distribution among stations (+ species noted for first time to Svalbard waters, * lack of data) salinity, temperature, Shannon-Wiener index for each station

Taxa	A1 (8-15m)	A2 (8m)	A3 (15m)	B1 (4m)	B2 (6m)	B3 (10m)	B4 (12m)	B5 (15m)	B6 (20m)	C1 (6m)	C2 (8m)	C3 (10m)	C4 (12m)	D1 (114m)	D2 (125m)	D3 (162m)	D4 (258m)	D5 (220m)	D6 (150m)
Size of the sample (cm^3)	1124	548	469	1974	*	430	*	57.5	150	833	742	509	361	1701	816	2101	1097	2152	2789
Oncousoeciidae																			
Oncousoecia canadensis Osburn, 1933																x		x	x
Oncousoecia diastoporides (Norman, 1869)																x		x	x
Oncousoecia sp. Canu, 1918																			x
Tubuliporidae																			
Tubulipora flabellaris (Fabicius, 1780)									x		x		x						
Tubulipora sp. Lamarck, 1816									x										
Tubuliporidae indet.					x			x	x				x		x	x	x	x	x
Crisiidae																			
+*Crisiella producta* (Smitt, 1865)												x							
Crisia eburnea (Linnaeus, 1758)				x															
Crisia sp. Lamouroux, 1812								x											
Crisidae indet.				x															
Lichenoporidae																			
Disporella hispida (Fleming, 1828)									x										x
Lichenopora sp. Defrance, 1823	x		x	x	x				x	x	x	x	x			x	x	x	x
Alcyonidiidae																			
Alcyonidium mamillatum Alder, 1857									x										
+*Alcyonidium mamillatum* var. *erectum* Andersson, 1902															x				
Alcyonidium mytili Dalyell, 1847				x		x						x							
Alcyonidium sp. Lamouroux, 1813													x						
Flustrellidae																			
Flustrellidra corniculata (Smitt, 1872)									x										
Eucrateidae																			
Eucratea loricata (Linnaeus, 1758)				x				x		x	x	x	x		x				
Membraniporidae																			
Electra crustulenta var. *arctica* Borg, 1931	x	x	x	x				x	x	x	x	x			x	x		x	x
Electra crustulenta var. *catenularia-similis* Kluge, 1962															x				x
Calloporidae																			
Callopora aurita (Hincks, 1877)				x	x							x							
Callopora craticula (Alder, 1857)	x		x	x		x		x			x	x		x	x			x	
+*Callopora sedovi* Kluge, 1962	x			x															
Callopora lata (Kluge, 1907)									x										
Callopora sp. Gray, 1848			x	x						x	x	x		x	x				x
Cauloramphus intermedius Kluge, 1955	x	x	x	x															
Amphiblestrum sp. Gray, 1848														x					
Tegella arctica (d'Orbigny, 1850)		x	x	x		x		x	x	x	x	x	x						
Tegella armifera (Hincks, 1880)	x	x		x		x			x	x		x	x						
Tegella nigrans (Hincks, 1882)												x							
Tegella sp. Levinsen, 1909				x				x		x			x						

Table 1. (continued)

Taxa	A1 (8-15m)	A2 (8m)	A3 (15m)	B1 (4m)	B2 (6m)	B3 (10m)	B4 (12m)	B5 (15m)	B6 (20m)	C1 (6m)	C2 (8m)	C3 (10m)	C4 (12m)	D1 (114m)	D2 (125m)	D3 (162m)	D4 (258m)	D5 (220m)	D6 (150m)
Size of the sample (cm^3)	1124	548	469	1974	*	430	*	57.5	150	833	742	509	361	1701	816	2101	1097	2152	2789
Scrupocellariidae																			
Scrupocellaria arctica (Busk, 1855)									x										
Tricelaria ternata (Ellis and Solander, 1786)				x	x	x			x	x									
Scrupocellariidae indet.			x	x		x		x			x	x			x			x	
Microporidae																			
Dendrobeania fruticosa (Packard, 1863)				x		x	x	x	x										
Dendrobeania murrayana (Johnston, 1847)			x	x	x			x											
Cribrilinidae																			
Cribrilina annulata (Fabricius, 1780)			x	x				x	x			x							
Cribrilina sp. Gray, 1848	x																		
Cribrilina spitsbergensis Norman, 1903															x				
Hippothoidae																			
Hippothoa divaricata var. *arctica* Kluge, 1906						x		x	x					x	x	x	x	x	x
Hippothoa expansa Dawson, 1859			x	x								x				x			x
Celleporella hyalina (Linnaeus, 1767)	x	x	x	x	x	x			x	x	x	x	x						
Harmeria scutulata (Busk, 1855)	x	x	x	x						x	x	x	x						
Escharellidae																			
Escharella abyssicola (Norman, 1869)																x			
+*Escharella latodonta* Kluge, 1962																		x	
Escharella sp. Gray, 1848														x	x	x			
Escharelloides spinulifera (Hincks, 1889)															x	x		x	x
Lepralioides nordlandica (Nordgaard, 1905)																		x	
Smittinidae																			
Smittina majuscula (Smitt, 1868)									x					x					
Smittina minuscula (Smitt, 1868)			x					x	x					x	x				x
Smittina rigida Lorenz, 1886	x														x	x	x	x	x
Smittina sp. Norman, 1903														x					
Parasmittina trispinosa (Johnston, 1838)														x	x				
Porella concinna (Busk, 1854)																	x		
Porella concinna var. *belli* (Dawson, 1859)																		x	
Porella minuta (Norman, 1869)		x		x				x		x	x			x				x	
+*Porella princeps* Norman, 1903																			x
Porella proboscidea Hincks, 1888																x		x	x
Porella sp. Gray, 1848																		x	
+*Porella tumida* Kluge, 1955																x			
Palmicellaria skenei var. *bicornis* (Busk, 1859)																		x	

Table 1. (continued)

Taxa	A1 (8-15m)	A2 (8m)	A3 (15m)	B1 (4m)	B2 (6m)	B3 (10m)	B4 (12m)	B5 (15m)	B6 (20m)	C1 (6m)	C2 (8m)	C3 (10m)	C4 (12m)	D1 (114m)	D2 (125m)	D3 (162m)	D4 (258m)	D5 (220m)	D6 (150m)
Size of the sample (cm^3)	1124	548	469	1974	*	430	*	57.5	150	833	742	509	361	1701	816	2101	1097	2152	2789
Umbonulidae																			
Umbonula arctica (Sars, 1851)															x				
Umbonula patens (Smitt, 1868)																			x
Schizoporellidae																			
Schizoporella pachystega Kluge, 1929														x	x	x		x	x
+*Schizoporella auriculata* var. *lineata* Nordgaard, 1896									x										
Schizoporella biaperta (Michelin, 1841)																			x
Schizoporella costata Kluge, 1962														x	x	x			x
+*Schizoporella elmwoodiae* Waters, 1900																		x	
+*Schizoporella hexagona* Nordgaard, 1905															x				x
Schizoporella limbata Lorenz, 1886																		x	x
Schizoporella magniporata Nordgaard, 1906																		x	
+*Schizoporella ortmanni* Kluge, 1955									x										
Schizoporella sp. Hincks, 1877															x				x
+*Schizoporella stylifera* (Levinsen, 1887)																		x	
Hippodiplosia obesa (Waters, 1900)														x	x	x	x	x	x
Stomachetosellidae																			
Stomachetosella cruenta (Busk, 1854)			x					x						x				x	
+*Stomachetosella producta* (Packard, 1863)														x				x	x
Ragionula rosacea (Busk, 1856)									x										
Microporellidae																			
Microporella ciliata (Pallas, 1766)								x											
Microporella arctica Norman, 1903																	x	x	
Microporella sp. Hincks, 1877				x															
Tessaradomidae																			
Cylindroporella tubulosa (Norman, 1868)	x	x	x	x				x	x	x	x			x	x	x			x
Cleidochasmatidae																			
Hippoporella hippopus (Smitt, 1868)																x			
Hipponella sp. (Canu and Bassler, 1920)														x	x				
Rhamphostomellidae																			
Rhamphostomella scabra (Fabricius, 1780)																			x
Rhamphostomella bilaminata (Hincks, 1877)						x													
Rhamphostomella sp. Lorenz, 1886																			x
Celleporidae																			
Cellepora sp. Linnaeus, 1767																x		x	
Hippopodinidae																			
Cheilopora sincera (Smitt,1868)	x														x			x	
salinity (psu)	33,09-33,7	33.09	33.7	31.2	31.2	32.6	32.6	32.9	33.8	34.2	34.2	34.2	34.2	34.61	34.59	34.74	34.93	34.91	34.82
teperature (C)	1,05-2,38	2.38	1.05	0.02	0.02	0	0.5	0.7	1.02	-0.69	-0.69	-0.69	-0.69	0.63	0.82	1.49	2.41	2.58	2.12
number of taxa	12	8	15	25	6	10	1	17	23	12	12	16	10	17	24	19	7	28	28
number of colonies/150cm^3	8	11	56	20	*	33	*	73	331	32	34	49	76	5	12	7	5	6	13
Shannon-Wiener index	1.49	1.69	1.69	2.22	*	1.29	*	2.65	2.07	1.23	1.37	1.93	1.52	2.53	2.53	2.45	0.92	2.97	2.77

REFERENCES

Androsova, E.I. 1977. Bryozoa in the biocoenoses of the Arctic Ocean (Franz Josef Land). In: *Explorations of the fauna of the seas XIV (XXII).* (in Russian) Leningrad: Izdatielstvo "Nauka", 471: 194-205.

Begon, M., Harper, J.L., Townsend, C.R. 1996. *Ecology: individuals, populations and communities.* Oxford: Blackwell Science Ltd.

Beszczynska-Moller, A., Weslawski, J.M., Walczowski, W. & Zajàczkowski, M. 1997. Estimation of glacial meltwater discharge into Svalbard coastal waters. - *Oceanologia,* 39(3): 289-298.

Brattegard, T. & Holte, T. 1997. Distribution of marine, benthic macro-organisms in Norway. *Research Report for DN 1997-1.* Directorate for Nature Management.

Dayton, P.K, Robilliard, G.A., Paine, R.T. & Dayton, L.B. 1974. Biological accommodation in the benthic community at McMurdo Sound, Antarctica. *Ecological Monographs,* 44: 105-125.

Dowdeswell, J.A. & Forsberg, C.F. 1992. The size and frequency of icebergs and bergy bits derived from tidewater glaciers in Kongsjorden, northwest Spitsbergen. *Polar Research* 11(2): 81-91.

Elverhoi, A., Lonne, O. & Seland, R. 1983. Glaciomarine sedimentation in a modern fjord environment, Spitsbergen. *Polar Research,* 1 n.s.: 127-149.

Gammelsrod, T. & Rudels, B. 1983. Hydrographic and current measurements in the Fram Strait, August 1981. *Polar Research* 1 n.s.: 115-126.

Gulliksen, B., Palerud, R., Brattegard, T. & Sneli, J. (eds), 1999. *Distribution of marine benthic macro-organisms at Svalbard (including Bear Island) and Jan Mayen.* Research Report for DN 1999-4. Directorate for Nature Management.

Hop, H., Pearson, T., Hegseth, E.N., Kovacs, K.M, Wiencke, C., Kwasniewski, S., Eiane, K., Weslawski, J.M., Cochrane, S., Leakey, R., Gulliksen, B., Mehlum, F., Wlodarska-Kowalczuk, M., Lydersen, C., Lønne, O.J., Gabrielsen, G.W., Zajaczkowski, M., Falk-Petersen, S., Kendall, M., Wängberg, S.A., Kai Bischof, Voronkov, A.Y., Kovaltchouk, N.A., Wiktor, J., Poltermann, M., Prisco, G., Estoppey, A., Papucci, C., Gerland, S. & Lippert, H. 2001. The ecosystem of Kongsfjorden, Svalbard. *Polar Research.* (in press)

Ito, H. & Kudoh, S. 1997. Characteristics of water in Kongsfjorden, Svalbard. Proc. of NIPR Symp - *Polar Meteorology and Glaciology,* 11: 211-232.

Kluge, G.A. 1975. *Bryozoa of the northern seas of the USSR.* New Delhi: Amerind Publishing Co.Pvt.

Lippert, H. 1998. *Die Algenassoziierte Makrofauna des Kongsfjordes (Spitsbergen) –Artenzusammensetzung und Verteilung auf verschiedenen Arten von Makroalgen.* - Studiengang Diplombiologie Diplomarbeit. Carl von Ossietzky Universitat Oldenburg.

Loeng, H. 1991. Features of the physical oceanographic conditions of the Barents Sea. In Sakshaug, E., Hopkins, C. C. E. & Oritsland, N. A. (eds.): *Proceedings of the Pro Mare Symposium on Polar Marine Ecology, Trondheim, 12-16 May 1990.* Polar Research, 10(1): 5-18.

McKinney, F.K. & Jackson, J.B.C. 1991. *Bryozoan evolution.* Chicago: The University of Chicago Press.

Moore, P.G. 1977. Inorganic particulate suspensions in the sea and their effects on marine animals. *Oceanography and Marine Biololgy Annual Review,* 15: 225-363.

Powell, N.A. 1968. Bryozoa (Polyzoa) of Arctic Canada. Journal of the Fisheries Research Board of Canada, 25: 2269-2320.

Rosso, A. 1991. Infralittoral bryozoa assiociated to macroalgae from the first Italian Antarctic Oceanographic Expedition (Terra Nova Bay, Ross Sea). *Bol. Soc. Biol. Concepcion, Chile,* 62: 179-186.

Rosso, A. 1992. Bryozoa from Terra Nova Bay (Ross Sea, Antarctica). In V.A. Gallardo, O.Ferretti & H.I. Moyano (eds), *Oceanografia en Antartica*: 359-369. Chile: U de Concepepcion.

Rózycki, O. & Gruszczynski, M. 1986. Macrofauna associated with laminarians in the coastal waters of West Spitsbergen. *Polish Polar Research,* 7(4): 337-351.

Rumohr, H. 1995. Monitoring the marine environment with imaging methods. *Scientia Marina,* 59: 129-138.

Ryland, J.S. 1970. *Bryozoans.* London: Hutchinson&CO Ltd.

Sanders, H.L. 1958. Marine benthic diversity: a comparative study. *Am. Nat.,* 102: 243-282.

Syvitski, J.M., Burrell, D.C. & Skei, J.M. 1987. *Fjords: processes and products.* New York: Springer-Verlag.

Weslawski, J.M., Jankowski, A., Kwasniewski, S., Swerpel, S. & Ryg, M. 1991. Summer hydrology and zooplankton in two Svalbard fiords. *Polish Polar Research,* 12(3): 445-460.

Weslawski, J.M., Zajaczkowski, M., Kwasniewski, S., Jezierski, J. & Moskal, W. 1988. Seasonality in an Arctic fjord ecosystem: Horsund, Spitsbergen. *Polar Research,* 6: 185-189.

Winston, J.E. 1977. Distribution and ecology of estuarine ectoprocts: a critical review. *Chesapeake Science,* 18 (1): 34-57.

Winston, J.E. & Hayward, P.J. 1994. Bryozoa of the U.S. Antarctic Research Program: Preliminary Report. In P.J. Hayward, J.S. Ryland & P.D. Taylor (eds.), *Biology and Palaeobiology of Bryozoans*: 205-210. Fredensborg: Olsen and Olsen.

Zajaczkowski, M. 2000. *Inflow and rate of sedimentation of chosen fiords of Western Spitsbergen.* (in Polish). Doctoral thesis. Gdansk University.

Bryozoan Studies 2001, Wyse Jackson, Buttler & Spencer Jones (eds)
© 2002 Swets & Zeitlinger, Lisse, ISBN 90 5809 388 3

Chesterian (Carboniferous) *Septopora* (Order Fenestrida), eastern North America

Frank K. McKinney
Department of Geology, Appalachian State University, Boone, NC 28608, USA

ABSTRACT: The acanthocladiid fenestrate bryozoan *Septopora* is widespread but only locally common in diverse Chesterian (Carboniferous; Upper Mississippian) quiet water carbonate platform and basin deposits in the eastern United States. The genus is characterized by colonies with fenestrate meshworks of main branches and short, fused pinnate, usually with two rows of autozooecia per branch; and by hemispherical polymorphs (cyclozooecia) of unknown function opening on reverse sides of junctions of main and lateral branches and between autozooecial apertures on frontal branch surfaces. Species characteristics include distance between branch centers, diameter of the distal tube (equivalent to apertural diameter on the branch surface), apertural spacing along rows, fenestrule size, and size of cyclozooecia. *S. subquadrans* Ulrich is the most abundant, widespread and apparently long-ranged species. Other species, including the type species, *S. cestriensis* Prout, are distinct but less abundant or temporally long-ranging, including the new species *S. gracilis*.

1 INTRODUCTION

For at least the last half century, E.O. Ulrich has been most widely known among geologists for his discarded concepts of stratigraphy in the mid-continent of North America. However, he is more highly regarded among bryozoologists. If one addresses any aspect of the morphology of Paleozoic bryozoans that occur in North America, it is worth searching through Ulrich's writings to see what he had to say about it. Regarding *Septopora*, Ulrich (1890, p. 626) wrote, 'This genus includes some very interesting but, at the same time, exceedingly difficult forms. They required much study, and had I not had an abundance of good material to work upon, the following classification into species and varieties would have been nearly impossible.' Ulrich's statements still hold.

H.A. Prout established the genus *Septopora* in 1859 for a single species, *S. cestriensis*, from the 'Upper Archimedes Limestone, Carboniferous series, Chester", Illinois, U.S.A., i.e. from the uppermost Mississippian Chesterian Series (equivalent to Namurian A and B of the Upper Carboniferous. *S. cestriensis* and the genus were characterized by two rows of zooecia on main branches and also zooecia on 'dissepiments' (i.e. pinnae). The species *Synocladia biserialis* Meek, 1872, from younger Carboniferous beds, was transferred to *Septopora* by Foerste in 1887, but the first systematic comparison of many specimens and evaluation of species characteristics was in Ulrich's 1890 work. The genus has since been recognized as a wide-spread member of the Permo-Carboniferous bryozoan fauna, being reported from many parts of the world and with species confidently placed in the genus ranging from Viséan (highest Lower Carboniferous) through the Upper Permian.

Septopora is less abundant than *Archimedes* and some other fenestellid and polyporid species in the eastern North American Chesterian, and as noted by Ulrich (1890) many specimens are required to determine consistent suites of characteristics. He considered that main branches in Chesterian *Septopora* increased in number mainly by branch bifurcation and secondarily by transformation of pinnae into main branches. Ulrich divided them into five species: *S. cestriensis* Prout, two new species, *S. subquadrans* Ulrich., 1890 and *S. decipiens* Ulrich, 1890, and two varieties of other species, *S. robusta* var. *intermedia* Ulrich, 1890 and *S. biserialis* var. *nervata* Ulrich, 1890.

2 PURPOSE

The purpose of this paper is to re-examine the fenestrated *Septopora* from Chesterian deposits of eastern North America, including redescription of species, summary of their known stratigraphic and geographic ranges, and interpretation of their paleoenvironmental distribution. *S. decipiens* Ulrich was based on specimens from which pinnae diverge from single main branches, with only local, occasional lateral linkage of pinnae. The species is not included in this study pending further study to determine if it belongs within the genus.

3 METHODS

Specimens were embedded in epoxy, polished, etched, and acetate peels were made from the etched surfaces, following the methods of Boardman and Utgaard (1964) and Merida and Boardman (1967). Measurements of zoarial and zooecial characters were made manually from projected images of the peels, and they include distance between center of adjacent main branches (BRS), distance between bases of successive pinnae (PS), branch width (BRW), pinna width (PW), angle between pinnae and main branches (PA), distance between centers of exozonal portions of zooecial chambers within a longitudinal row (AS), (diagonal) distance between centers of exozonal portions of zooecial chambers on opposite sides of a branch (ASD), and diameter of exozonal portion of zooecial chamber (AD). Measurements involving exozonal portions of zooecial chambers are considered surrogates for measurements of zooecial apertures, inasmuch as exozonal portions of zooecial chambers are essentially perpendicular to the frontal surface and chamber cross sections represent former positions and sizes of apertures. Fenestrule length and width were also measured but are not included in analysis because they are, respectively, equal to pinna and branch spacing minus width. Initially, specimens were grouped morphologically by unweighted pair group clustering using SPSS and Statistica software.

Almost all colonies yielded 10 measurements per character. Results of measurements are reported in μm (except for PA) and are given as mean ± standard deviation, followed by parentheses enclosing number of colonies measured. All characters indicated as useful in discriminating between species are significant at $p < 0.05$ (Mann Whitney U Test).

4 DISTRIBUTION

Septopora is widespread in Chesterian deposits of eastern North America (Fig. 1), especially in the mid-Chesterian Glen Dean Limestone and its approximate calcareous equivalents.

Well over 90% of specimens of the three previously named species of *Septopora* recognized here (*S. cestriensis*, *S. intermedia*, *S. subquadrans*) occurred in interbedded calcareous shales and thin-bedded skeletal packstones, secondarily in calcareous shales and lime wackestones/mudstones, and much less commonly in grainstones. The interbedded calcareous shales and thin-bedded skeletal packstones are typically adjacent to or within a few meters stratigraphically from coarser, cleaner limestones that commonly are cross-bedded. Thick-bedded lime wackestones/mudstones of the Goreville member of the Kinkaid Formation of southern Illinois contain scattered *S. subquadrans*.

The favored environment of *Septopora*, in which the interbedded shales and packstones were deposited, is interpreted to have been intermittently agitated, typically quiet, near leeward sides of barriers. *Septopora* specimens collected from grainstones have all been transported some unknown distance, perhaps from local protected pockets within the barrier system. The wackestones/mudstones of the Goreville contain robust *Archimedes* colonies up to 1 m high (McKinney and Gault, 1980) and apparently were deposited below normal wave base, an open-water but kinetically similar environment to the back-barrier environments where *Septopora* was most common.

The new species *S. gracilis* was found at only one locality, in terrigenous clay-rich boundaries between thick, non-cross bedded crinoidal packstone beds totaling about 3 m thick. The crinoidal packstones are underlain by fenestrate bryozoan packstones and ooid grainstones and are overlain by cross-bedded fenestrate packstones. This general suite of lithologies, and the cross-bedding, suggest deposition in a shallow-water barrier system. However, the presence of clay-sized sediments in all but the ooid grainstones indicates that all the other lithologies were deposited in areas of less vigorous water movement.

Branch spacing and robustness vary with ambient water motion for at least some delicate, cemented, rigidly erect bryozoans (e.g., Haas, 1948; Harmelin, 1973). In order to test for possible correlation of morphological attributes of *Septopora* with inferred kinetic energy of the environment, I grouped together all specimens occurring in mud-dominated lithologies (calcareous shales, interbedded shales and thin packstones, lime mudstones/wackestones) that are interpreted to have been deposited in quite water and compared them with specimens occurring in apparently higher-energy, coarser-grained deposits (packstones and grainstones). The two groups are grossly unequal in size. Among measured specimens that were collected *in situ*, 99 are from mud- dominated lithologies and 10 from coarser-grained deposits.

Only two characters, branch width and pinna width, were significantly different at $p \leq 0.05$ with the Student t-test ($p = 0.04$ and 0.01, respectively). They are marginally not significant with the non-parametric Mann-Whitney U test (0.06 for both). Interestingly, branch widths average 500 μm in mud-dominated lithologies versus 450 μm in coarser-

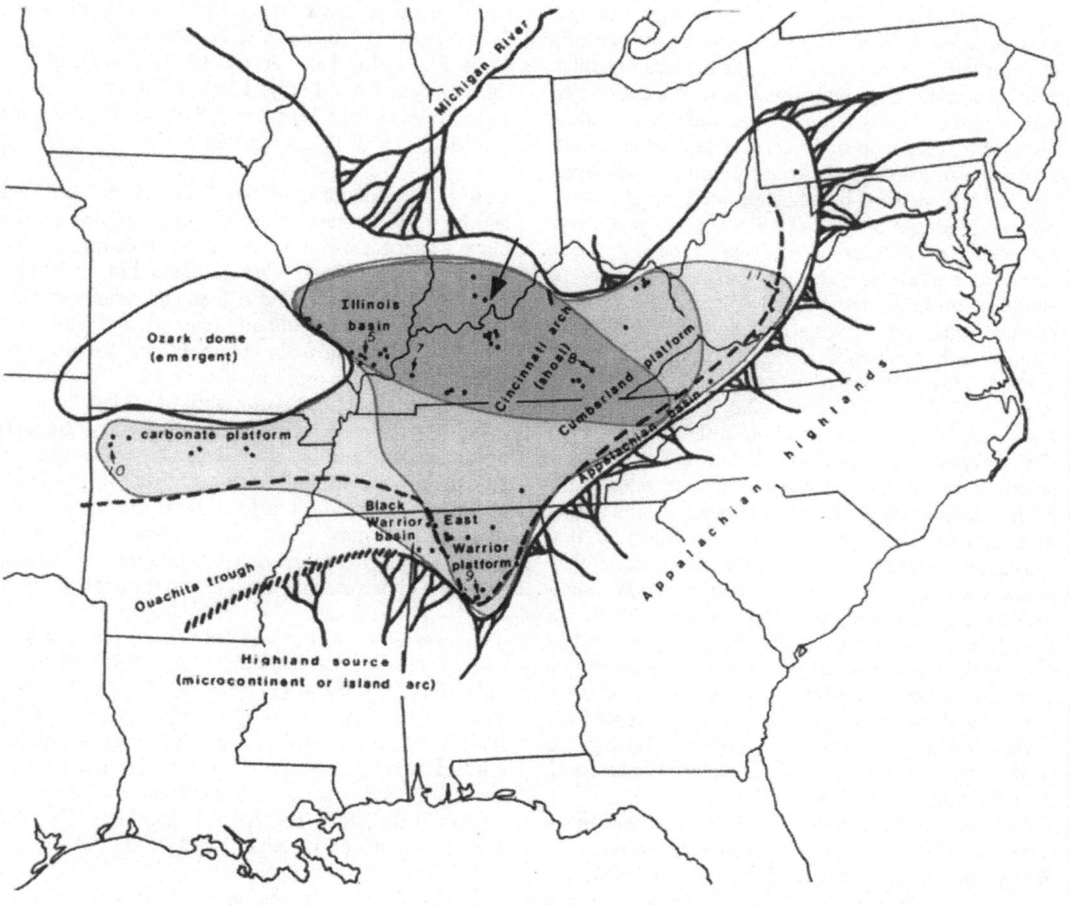

Figure 1. Map of Chesterian collecting localities (represented by dots) in eastern North America. Lightest shading shows geographic distribution of *Septopora subquadrans*, intermediate shading shows distribution of *S. intermedia*, darkest shading represents distribution of *S. cestriensis*, and arrow points to type locality of *S. gracilis*. Not every locality encircled by the distribution boundary yielded *Septopora*. Base map after McKinney and Gault (1980).

grained deposits, and corresponding pinna widths average 360 μm and 321 μm. This poorly expressed correlation is opposite that of coarser branches in more agitated waters for the living cheilostome *Reteporella* (Haas, 1948) and the living cyclostome *Exidmonea* (Harmelin, 1973), and it defies interpretation.

5 TAXONOMY

Genus *Septopora* Prout, 1859

Type species: *Septopora cestriensis* Prout, 1859

Diagnosis: Zoarium typically planar or curved fan or cone-shaped, obverse surface innermost in conical zoaria. Main branches straight, bearing lateral pinnae, with most pinnae short but some developing as main branches that also develop lateral pinnae; main branches also originating dichotomously; main branches and pinnae with two (occasionally one or three) rows autozooecia. Dissepiments absent; pinnae from adjacent branches typically in contact and fused skeletally. Zooecia budded from laterally curved plate; endozonal portions of autozooecial chambers elongate pentagonal to rhombic in deep tangential sections; length of endozonal portion of chamber greater than height or width, topped by elongate distal tubes with diameter equal to width of endozonal chamber. Hemisepta and diaphragms absent. Axial wall extending as median keel on obverse side of branch, commonly height increased by laminar skeleton deposited above core of granular skeleton, bearing robust spines that may ramify distally into several short, lateral branches. Granular skeleton present in thick basal plate that curves up laterally to form lateral endozonal walls, in moderately thick axial and transverse endozonal autozooecial walls, and in thin exozonal autozooecial walls surrounding distal tube. Extrazooecial skeleton laminated. Autozooecial chambers locally lined by thin laminar skeleton. Reverse sides of branches striated longitudinally because of ridges on reverse side of basal plate. Cyclozooecia most common at main branch – pinnae junctions on reverse sides of branches but may be abundant on both obverse and reverse sides of main branches and pinnae. A narrow, low peristome surrounds each cyclozooecium. Stratigraphical range: Viséan – Upper Permian.

Remarks: The most conspicuous zoarial characteristics of *Septopora*, as noted by Prout, are a) the characteristic two rows of zooecia per main branch and b) the presence of short, zooecia-bearing pinnae that meet and fuse between the main branches, so that zoaria are fenestrate. In one of the first generic diagnoses, Ulrich (1886, p. 6) described 'a variable number of scattered dimorphic pores' on reverse sides of branches. These hemispherical pits also occur less commonly between autozooecial apertures on frontal surfaces and are were named cyclozooecia by Morozova (1974, p. 65). Cyclozooecia apparently are limited to a few fenestrate genera, although as phylogeny of the fenestrates is currently conceived, the genera are not all closely related.

Silvaseptopora Chronic, 1953 and *Lanarkopora* Graham, 1975 are junior objective synonyms of *Septopora*. Chronic (1953, p. 125) noted that the type species of *Silvaseptopora*, *S. incaica* Chronic 1953, differed from species of *Septopora* only in that the tips of spines extending from frontal keels are divided into multiple lateral spines. However, such spines are only of species significance in *Archimedes* (McKinney. 1992) and are considered here to be only of species significance in *Septopora*. *Synocladia carbonaria* Etheridge jun., 1873, the type species of *Lanarkopora* Graham (1975, p. 9), is a typical *Septopora* in having main branches linked by short, fused pinnae, all branches bearing two rows of zooecia. Graham (1975, p. 10) noted that Prout's (1859) illustration of *Septopora cestriensis* did not show median carinae on pinnae, whereas they are present in the type specimens of *Synocladia carbonaria*. However, carinae on the pinnae are variably well developed and preserved within all Chesterian species of *Septopora* in eastern North America, and their presence or absence is insufficient in itself to discriminate a different genus.

Septopora cestriensis Prout, 1859
Figure 2, A-C

Septopora cestriensis Prout, 1859, p. 448, Pl. 8, fig. 2-2b. Ulrich, 1890, p. 628, Pl. 64, 1-1b. ?Morgan, 1924, Pl. 27, fig. 16. Coryell, 1942, p. 454, Pl. 67, fig. 4. Utgaard and Perry, 1960, p. 28, Pl. 6, figs. 6, 7. (non Meek and Worthen, 1870, p. 15; Worthen and Meek, 1873, Pl. 24, fig. 14a-c)

Description: Zoaria fan-shaped, less than 2 cm high and broad in present material. Branches and pinnae closely spaced, width of branches almost that of fenestrule width, pinnae wide at bases but with pronounced distal taper. Main branches originating by bifurcation or from distally curved and elongated pinnae. Zooecia in main branches and pinnae relatively small, pentagonal to less commonly quadrangular in basal section, width about half length in endozone in main branches but relatively less elongate in pinnae, distal tubes about equal to width and orientated toward frontal surface, about 60° with respect to branch axis. Up to at least five small apertural stylets may occur locally, not seen in all zooecia. Cyclozooecia small, common on reverse surface and between distal tubes on frontal surface.

Measurements: AD = 118±11 µm (13); AS = 237±16 (13); ASD = 237±21 (13); BRS = 943±121 (13); BRW = 408±46 (12); CD = 62±15 (12); NS = 794±141 (7); PA = 79±6 ° (13); PS = 889±140 (13); PW = 352±35 (12). This species may be distinguished from *S. subquadrans* by its smaller BRS, BRW, PS, and larger PA; from *S. intermedia* by its smaller BRS, BRW, PS, AS, CD and larger PA; and from *S. gracilis* by its larger PW and AD.

Remarks: The specimens on which Prout based the species have been lost, and the topotype specimen illustrated by Ulrich, 1890, Plate 64, fig. 1 (USNM 518285) is here designated as neotype of *Septopora cestriensis*. *S. cestriensis* has the smallest, most dense meshwork of the Chesterian species of the genus in eastern North America.

Distribution: Although McFarlan (1942) indicated that *S. cestriensis* ranges throughout the Chesterian Stage, I have encountered it only from the lower and middle Chesterian (Paint Creek through Menard limestones and equivalents) from the Illinois Basin to northeastern Tennessee and have not found it at all in strata further south and southwest.

Septopora gracilis, sp. nov.
Figure 2, D-F

Etymology: From the Latin, *gracilis* (slender, thin), in reference to the slender pinnae.

Type material: Holotype, USNM 518287; paratypes, USMN 518288-518290.

Diagnosis: Branches and pinnae thin and closely spaced; pinnae commonly bearing single row of zooecia; cyclozooecia very small.

Description: Zoaria fan-shaped, less than 2 cm high and broad in present material. Branches and pinnae closely spaced, width of branches almost that of fenestrule width, pinnae very narrow, commonly bearing single row of zooecia. Style of origination of main branches not seen. Zooecia in main branches and pinnae relatively small, pentagonal or rhombic in basal section, width less than half length in endozone in main branches but conspicuously less elongate in pinnae, with width about two-thirds length; distal tubes about equal to endozonal width. Apertural stylets not seen. Cyclozooecia very small, present on reverse surface and between distal tubes on frontal surface.

Measurements: AD = 107±4 μm (4); AS = 248±14 (4); ASD = 222±30 (4); BRS = 889±16 (4); BRW = 402±79 (4); CD = 56 (1); NS = not available; PA = 74±8° (4); PS = 862±56 (4); PW = 276±37 (4). This species differs from *S. cestriensis* in having smaller PW and AD, and from both *S. intermedia* and *S. quadrans* in its smaller BRS, BRW, PS, PW and AD.

Distribution: Mid-Chesterian (Haney Limestone), junction of Indiana Highway 37 and Interstate Highway 64, Sulphur, Indiana.

Septopora intermedia Ulrich, 1890
Figure 2, G-I

Septopora robusta var. *intermedia* Ulrich, 1890, p. 634, Pl. 56, fig. 10, Pl. 64, fig. 4-4a. McFarlan, 1942, p. 454.
Septopora subquadrans Ulrich. Utgaard and Perry, 1960, p. 29 (part), Pl. 6, fig. 1. McKinney, 1998, Fig. 2-5.
Septopora sp. McKinney and Gault, 1980, fig. 4c.

Description: Zoaria fan-shaped, up to at least 4 cm high and 3 cm broad in present material. Branches and pinnae widely spaced, width of branches about equal to fenestrule width, pinnae conspicuously narrower and less thickened on reverse side than are main branches. Main branches originating most commonly from distally curved and elongated pinnae. Zooecia in main branches and pinnae large, commonly elongate rectangular in basal section, width about half length in endozone in main branches but relatively less elongate in pinnae, distal tube diameters about equal to endozonal chamber width and orientated at high angle, commonly 60-70° with respect to branch axis and commonly tilted out about 30° laterally. Pinnae usually with two rows of zooecia but may have three rows. Apertural stylets not seen. Cyclozooecia on reverse surface very large, diameters slightly greater than distal tubes of autozooecia; those between distal tubes on frontal surface typically small, less than 50% distal tube diameter.

Measurements: AD = 120±10 μm (30); AS = 252±14 (30); ASD = 261±21 (30); BRS = 1089±150 (31); BRW = 536±63 (28); CD = 133±18 (31); NS = 838±212 (14); PA = 68±6 ° (31); PS = 1514±301 (28); PW = 372±37 (26). This species is distinguished from *S. cestriensis* by its greater BRS, BRW, PS, AS, CD and smaller PA; from *S. subquadrans* by its larger BRS, BRW, PS, PW, and CD; and from *S. gracilis* by its greater BRS, BRW, PS, PW and AD.

Remarks: The specimen illustrated by Ulrich, 1890, Plate 64, fig. 4a (USNM 43300) is here designated as lectotype of *Septopora intermedia*. This species has not been recorded or redescribed since Ulrich named it as a variety of *Septopora robusta*, except for a brief note in McFarlan (1942) that it is reported from the Glen Dean and Clore limestones. However, it is common in the Chesterian of eastern North America and is easily distinguished by its overall robust meshwork and by the very large cyclozooecia on the reverse side, which are about twice the diameter of cyclozooecia in all co-occurring species of the genus.

Distribution: Mid-Chesterian, from the Golconda through Glen Dean limestones in the Illinois Basin and in approximate Golconda and Glen Dean equivalents to eastern Kentucky and northern Alabama.

Septopora subquadrans Ulrich, 1890
Figure 2, J-L

Septopora cestriensis Prout. Meek and Worthen, 1870, p. 15; Worthen and Meek, 1873, Pl. 24, fig. 14a-c.
Septopora subquadrans Ulrich, 1890, p. 629, Pl. 56, figs. 7, 8, Pl. 64, fig. 2-2c. McFarlan, 1942, p. 454, Pl. 67, figs. 1-3. Utgaard and Perry, 1960, p. 29 (part), pl. 6, ?figs. 2-5 (not 1).
Septopora biserialis var. *nervata* Ulrich, 1890, p. 632, Pl. 64, fig. 6. MarFarlan, 1942, p. 454.

Description: Zoaria fan-shaped, up to at least 3.5 cm high and 4.5 cm broad in present material. Branches and pinnae relatively widely spaced, width of branches one-third to two-thirds that of fenestrule width, pinnae somewhat narrower and less thickened on reverse side than are main branches. Main branches originating by bifurcation or from distally curved and elongated pinnae. Zooecia in main branches and pinnae relatively large, pentagonal in basal section, width usually less than half length in

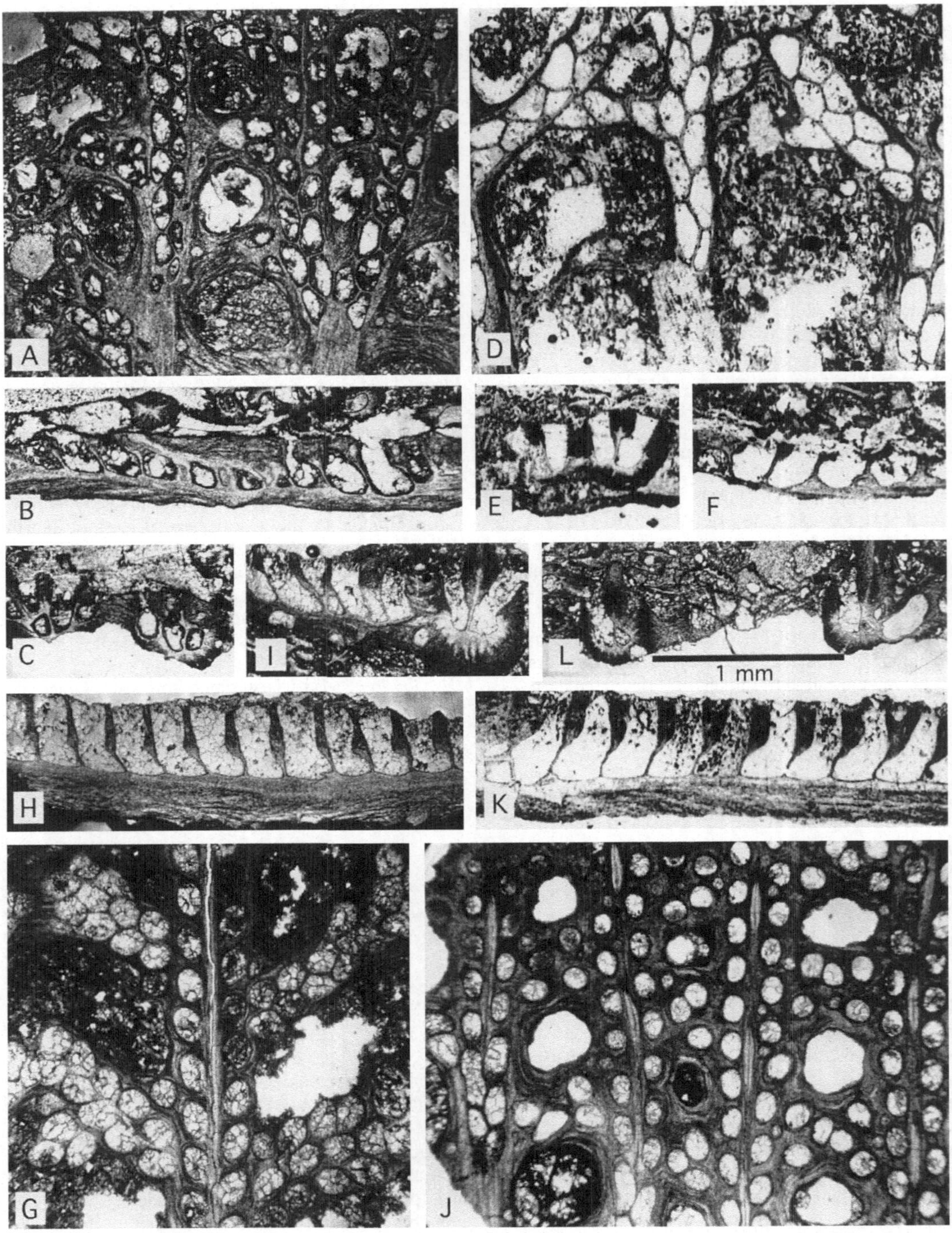

Figure 2. Chesterian *Septopora* species of eastern North America. A-C, *S. cestriensis* Ulrich, topotype USNM 518286, Chester, Illinois; A, tangential; B, longitudinal; C, branch cross sections. D-F, *S. gracilis, sp. nov.*, Sulphur, Indiana; D, tangential, holotype, USNM 518287; E, branch cross sections and F, longitudinal, paratype, USNM 518288. G-I, *S. intermedia* Ulrich, paralectotype, USNM 496175, Litchfield, Kentucky; G, tangential; H, longitudinal and I, branch cross section with longitudinal of pinna. J-L, *S. subquadrans* Ulrich; J, tangential, paralectotype, USNM 518292, Caldwell County, Kentucky; K, longitudinal, paralectotype, USNM 518293, Sloans Valley, Kentucky; branch cross sections, paralectotype, USNM 518294, Sloans Valley, Kentucky. All illustrations at same magnification.

endozone in main branches but relatively less elongate in pinnae, distal tube diameters about equal to endozonal chamber width and orientated at a very high angle, commonly 90° with respect to branch axis but commonly tilted up to 30° laterally. Up to at least five small apertural stylets may occur locally, not seen in all zooecia. Cyclozooecia on reverse surface and between distal tubes on frontal surface typically small, less than 50% distal tube diameter.

Measurements: AD = 119±11 μm (54); AS = 247±21 (54); ASD = 249±17 (54); BRS = 1030±112 (30); BRW = 490±60 (53); CD = 70±16 (46); NS = 864±226 (25); PA = 72±7° (54); PS = 1363±473 (53); PW = 353±39 (52). This species is distinguished from *S. cestriensis* by its larger BRS, BRW, PS, and smaller PA; from *S. intermedia* by its smaller BRS, BRW, PS, PW, and CD; and from *S. gracilis* by its greater BRS, BRW, PS, PW, and AD.

Remarks: The specimen illustrated by Ulrich, 1890, Plate 64, figs. 2, 2a (USNM 518291) is here designated as lectotype of *Septopora subquadrans*. Tangential sections demonstrate that new main branches arise at least as frequently by from extension of pinnae, which are laterally budded branches, as from dichotomy of pre-existing main branches. Ulrich (1890) considered that main branches of *S. subquadrans* proliferated primarily by dichotomy and therefore split off as *S. biserialis* var. *nervata* a few Chesterian specimens in which main branches increased mainly by transformation of pinnae but that otherwise had characteristics of *S. subquadrans*. I consider them to belong together in a single species.

Utgaard and Perry (1960) indicated that their specimens have larger 'accessory pores' (cyclozooecia) than the specimen illustrated by Ulrich (1890) as Plate 64, figure 26. The specimen illustrated from the reverse side by Utgaard and Perry as *S. subquadrans* in Plate 6, figure 1 belongs in *Septopora intermedia* Ulrich because of the large cyclozooecia and robust branch size. Other specimens illustrated and measured by Utgaard and Perry could belong to either species, both of which occur in the localities that they studied.

Distribution: This is the most abundant and widespread *Septopora* species in the eastern North American Chesterian. It ranges from the lowest (Paint Creek) through the highest (Goreville) limestones in the Illinois Basin, east at least to the Virginia/West Virginia border, south into northern Alabama, and southwest into northern Arkansas.

ACKNOWLEDGEMENTS

This research was supported by grants from the National Science Foundation and the Petroleum Research Foundation, administered by the American Chemical Society. In addition, I thank Marg McKinney, who recorded approximately 14000 measurements called out monotonously in a crepuscular room.

REFERENCES

Boardman, R.S. & Utgaard, J. 1964. Modifications of study methods for Paleozoic Bryozoa. *Journal of Paleontology* 38: 768-770.

Chronic, J. 1953. Phylum Bryozoa. In N. D. Newell, J. Chronic & T. G. Roberts (eds), Upper Paleozoic of Peru. *Geological Society of America Memoir* 58: 108-129.

Coryell, H.N. 1924. Bryozoa. In Morgan, G.D. *Geology of the Stonewall Quadrangle, Oklahoma. Oklahoma Bureau of Geology Bulletin* 2: 176-184.

Foerste, A.M. 1887. Flint Ridge Bryozoa. *Bulletin of the Scientific Laboratories of Denison University* 2: 71-88.

Graham, D.K. 1975. A review of Scottish Carboniferous acanthocladiid Bryozoa. *Bulletin of the Geological Survey of Great Britain* 49: 1-21.

Haas, H. 1948. Beitrag zur Kenntnis der Reteporiden. *Zoologica* 101: 1-138.

Harmelin, J.-G. 1973. Morphological variations and ecology of the Recent cyclostome bryozoan "Idmonea" atlantica from the Mediterranean. In G.P. Larwood (ed.), *Living and fossil Bryozoa*: 95-106. London: Academic Press.

McFarlan, A.C. 1942. Chester Bryozoa of Illinois and western Kentucky. *Journal of Paleontology* 16: 437-458.

McKinney, F.K. 1992. Intraspecific variability in axial morphology of *Archimedes*, Pitkin Limestone (Chesterian), Arkansas. *Oklahoma Geological Survey Circular* 94: 23-37.

McKinney, F. K. 1998. Avicularia-like structures in a Paleozoic fenestrate bryozoan. *Journal of Paleontology* 72:819-826.

McKinney, F.K. & Gault, H.W. 1980. Paleoenvironment of Late Mississippian fenestrate bryozoans, eastern United States. *Lethaia* 13: 127-146.

Meek, F.B. & Worthen, A.H. 1870a. Note on the relations of *Synocladia* King, 1849, to the proposed genus *Septopora* Prout, 1858. *Proceedings of the Academy of Natural Sciences of Philadelphia* 1870: 15-18.

Merida, J.E. & Boardman, R.S. 1967. The use of Paleozoic Bryozoa from well cuttings. *Journal of Paleontology* 41: 763-765.

Morozova, I.P. 1974. Reviziya roda *Fenestella. Paleontologicheskiy Zhurnal* 1974 (2): 54-67.

Prout, H.A. 1859. Third series of descriptions of Bryozoa from the Palaeozoic rocks of the western States and Territories. *Transactions of the St Louis Academy of Sciences (3rd series)* 1: 443-451.

Ulrich, E.O. 1886. Descriptions of new Silurian and Devonian fossils. *Contributions to American Palaeontology* 1: 3-35.

Ulrich, E.O. 1890. Paleozoic Bryozoa. *Geological Survey of Illinois* 8: 285-688.

Utgaard, J. & Perry, T.G. 1960. Fenestrate bryozoans from the Glen Dean Limestone (middle Chester) of southern Indiana. *Indiana Department of Conservation Geological Survey Bulletin* 19: 1-32.

Worthen, A.H. & Meek, F.B. 1873. Descriptions of invertebrates from the Carboniferous system. *Geological Survey of Illinois* 5: 321-619.

Bryozoan Studies 2001, Wyse Jackson, Buttler & Spencer Jones (eds)
© 2002 Swets & Zeitlinger, Lisse, ISBN 90 5809 388 3

The ctenostome collar – an enigmatic structure

Marjorie J. McKinney
Department of Geology, Appalachian State University, Boone, NC 28608, USA

Ruth Ann Dewel
Department of Biology, Appalachian State University, Boone, NC 28608, USA

ABSTRACT: Setigerous, or pleated, collars are characteristic of ctenostome bryozoans, yet little has been recorded describing their form, microstructure, behavior, function, or possible taxonomic significance. Observations of living ctenostomes in the northern Adriatic in Croatia and Bogue Sound, North Carolina and a study of the ctenostome literature suggest that there are four collar forms. Forms 1 and 2 are short, unflared, simple cuffs. Form 3 is short and flared. Form 4 is long and flared. Rods support the collar membrane in all but Form 1. Collar edges may be scalloped in Forms 2, 3, and 4. Shorter collars fold inward when they retract. Longer collars twist. Collars may serve to reduce the amount of debris encountered by the lophophores. When the collar protrudes, it pushes this surficial debris aside, preventing the lophophore from becoming fouled.

1 INTRODUCTION

Setigerous, or pleated, collars are characteristic of the gymnolaemate bryozoan order Ctenostomata. The term ctenostome comes from the resemblance that the jagged distal margin of many collars has to the teeth of a comb (Prenant & Bobin 1956). Collars have also been reported in the cheilostome genera *Membranipora*, *Aetea*, and *Scruparia* (Banta et al. 1995) and are here reported in a colony of *Aetea truncata* (Landsborough).

Collars are easily overlooked. Often the first indication one has of their presence is a flash of iridescence when light strikes the structure. The membrane and rods are very thin and clear. The rod diameter is distinctly smaller than the tentacle diameter, making them especially difficult to discern in unflared collars. Even when one expects to see them, collars can be missed unless the light and microscope optics are very good. Little has been described about the structure of bryozoan collars or of the rods and their relationship to the membrane. The purpose of this paper is to systematically summarize existing literature on collars, supplemented by new observations. Particular attention is paid to shape and relative size of the fully extended collars, to movement of the collar, and to its position and shape when retracted.

The collar is a thin membrane attached to the atrial sphincter (diaphragm). When protruded, the collars may be simple cuff-like structures or they may be flared. Regularly spaced, delicate rods within membrane may extend beyond the free edge, giving it a jagged margin. The collar emerges prior to the lophophore and completely encircles its base. Generally, the collar is folded or pleated when not fully open. When the lophophore retracts into the vestibule, the collar may follow, effectively closing the atrium (vestibule).

The nature of the ctenostome collar is described and illustrated in the earliest literature of that group, although various names have been applied to it. In the original description of *Bowerbankia*, the genus commonly used to illustrate ctenostome characteristics, Farre (1837) described the collar as a "horny cell" consisting of delicate bristle-shaped processes, or setae, that are restricted in their expansion by a delicate membrane that encircles the lophophore. He also thought the notched "opercular margin" in *Farella* might be formed by short setae. Johnson (1847) characterized the collar as a short tube with the appearance of a truncated cone that preceded the lophophore's protrusion and encircled its base. The tube was reported to have very fine bristles connected by a translucent membrane at their base. Hincks (1880) described the "setose operculum" of *Bowerbankia pustulosa* (Ellis & Solander) as having tall, delicate bristles united by a delicate membranous sheath. He observed that the aperture of *Mimosella* is closed by a membranous covering, and he illustrated several species with "setose opercula". Jullien (1888) characterized the ctenostomes as those bryozoans whose tentacle sheath ends in a collar like a creased bundle of bristles at the aperture of the zooecium. In living zooids, the collar formed a funnel from which the tentacles emerge.

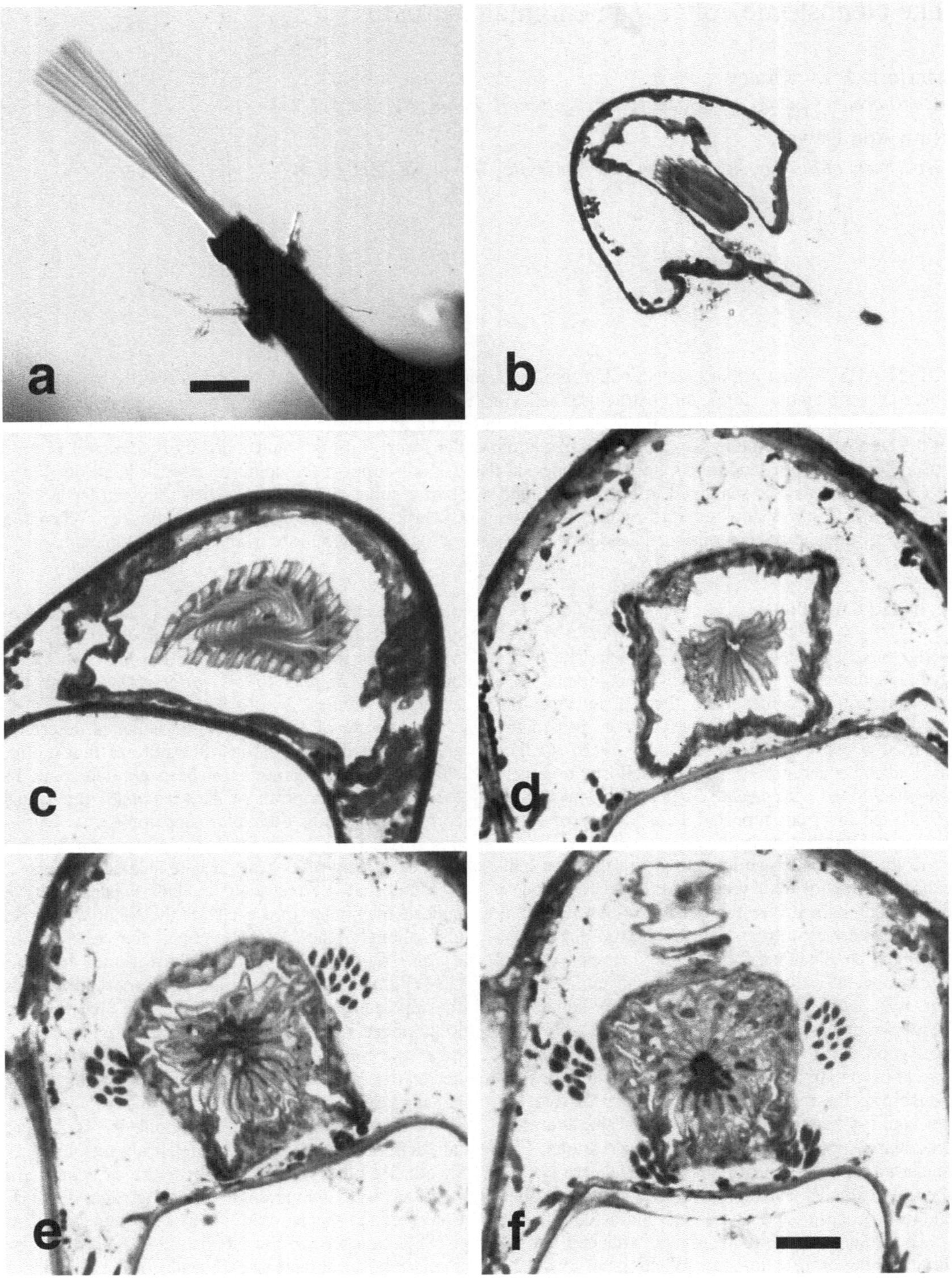
a
b
c
d
e
f

Figure 1. (a) Twisted partially retracted collar of a living *Aeverrilla setigera* zooid. (b) Slightly oblique cross-section through an *A. setigera* zooid. (c) Cross-section through an *A. setigera* zooid to show the retracted twisted collar, the rhombic to rectangular cross-section of the rods, and the membrane enclosing them. (d) Distal portion of the retracted collar of *Amathia convoluta* showing the membrane enclosing rods of circular cross-section. (e) Retracted collar of *A. convoluta* in a more proximal section than (d) showing well-defined membrane creases. (f) Basal portion of an *A. convoluta* collar showing cuticle-secreting cells in most of the collar folds. Bars = 20 μm.

In more recent literature, Prenant & Bobin (1956) described the collar as a circular, pliant membrane, commonly held up by bristles that extend beyond the free margin. They reported that some bristles are rudimentary and that their transparency makes them difficult to see. Hyman (1959) used the term orificial collar or pleated collar in describing the pleated structure that closes the orifice and protects the zooid. She considered the rods as being thickened chitin. Franzén (1960) described a rib-like thickening of cuticle that added support to the collar. Kluge (1962) described a pecten in the form of a chitinous membrane like a collar. The longitudinal folds of the structure of thickened chitin provided tight closure of the orifice. He reported that the collar is developed differentially in the ctenostome families but did not describe that difference. Ryland (1970) described the collar in ctenostomes as a membrane that encircles the lower part of the tentacle sheath and is withdrawn after the lophophore to block the orifice. Hayward (1985) considered the pleated collar at the base of the lophophore to be a special characteristic of the ctenostomes. He described it as sometimes being thickened by rigid elements that make it appear spiny. Mukai et al. (1997) described the collar as folding into a cone that blocks the vestibule.

Collar descriptions have not been included systematically in taxonomic studies. A notable exception is the work of Prenant & Bobin (1956), which contains at least simple descriptions of the collar for most of the taxa. Occasionally, collars have been included in ctenostome species descriptions, and they are commonly illustrated in ctenostome literature, both in the retracted and protruded positions, with no explanation of them. The attachment of the base of the collar at the diaphragm has been frequently reported in the literature.

2 METHODS

Ctenostome colonies were collected in the northern Adriatic off Rovinj, Croatia (45°5' N, 13°40' E) and in Bogue Sound, near Morehead City, North Carolina (34°43' N, 76°43' W). Colonies of *Walkeria tuberosa* (Heller) were collected by a diver June 1997 off the island of Banjole near Rovinj and November 1997 in the Rovinj harbor. *Amathia prutovi* Calvet and *Nolella stipata* Gosse (November 1997) and *Bowerbankia gracilis* Leidy and *Amathia vidovici* (Heller) (July 1998) were collected off the island of Figarole near Rovinj.

In Bogue Sound, *Nolella stipata* and *Aeverrilla armata* (Verrill) were collected from a floating dock at the University of North Carolina Institute of Marine Sciences during September 1997. Dredging at the Turning Basin and at the edge of the salt marsh north and east of Beaufort Inlet during September 1997 provided colonies of *Amathia convoluta* Lamouroux, *Nolella stipata*, *Aeverrilla armata* and *A.* cf. *armata*. Subsequently *A. convoluta*, ?*Bowerbankia* sp., and *Aeverrilla setigera* (Hincks) were collected in April 2000; while *Nolella dilatata* (Hincks), *A. convoluta*, *Aeverrilla* sp., and *Anguinella palmata* Van Beneden were collected in September 2000. Colonies were kept alive in seawater and observed after a few hours to 2 - 3 days after collection.

Representative colonies of *Aeverrilla setigera* and *Amathia convoluta* with actively feeding zooids were brought back to the laboratory at Appalachian State University, where small clumps of active zooids were separated and prepared for light microscopy. Specimens were fixed after Stricker & Reed (1981) in 2.5% glutaraldehyde in 0.2 M phosphate buffer and 0.14M NaCl for 2 hours at room temperature and rinsed in buffer before being postfixed at 4°C in 2% OsO_4 in 1.25% sodium bicarbonate adjusted to pH 7.4 with 1N HCl. After dehydration in acetone, specimens were embedded in Spurr's resin, serially sectioned at 1 μm with glass knives, and stained with 1% toluidine blue in 1% borax.

3 RESULTS

Once protruded fully, collars were either simple parallel-sided sleeves or flared. Generally, collars remained stationary once they protruded even when lophophores within them moved. Collars in some *Aeverrilla setigera* and *A. armata* zooids twisted slightly and moved from side to side.

Just prior to eversion of the collar and lophophore in the species with tubular zooecia (all but *Aeverrilla* in this study), the distal portion of the atrium usually extended to look like an elongated bump with the diameter of the extension smaller than that of the zooecium. That extension could appear long before the collar everted and remain long after retraction of the collar. Some collars of *Walkeria tuberosa*, a *Bowerbankia* sp., *Amathia vidovici*, and *A. convoluta* remained closed at their free edge as they everted, retaining their conical retracted outline and tentacles could be seen within them. As the lophophore retracted, the distal portion of the collar was drawn together once more, closing to form a cone. The entire structure retracted until what looked like a single rod or spine was left exposed. The collar of some zooids of *Amathia pruvoti*, *A. convoluta*, and *Walkeria tuberosa* behaved similarly except that as the collar initially protruded and as it finally retracted, the distal third became pinched together and caused the outline of the collar to appear biconcave. The tentacles could be seen within the collar and seemed to expand the distal portion of the collar as the lophophore everted. When the lophophore retracted, the distal part of the collar appeared to collapse laterally. In some zooids of *W. tuberosa* and in a few zooids of *Bowerbankia* sp., the collar initially looked like a single, short rod or spine. This single "spine" could be seen for some time at the orifice before the collar fully protruded and after the lophophore and most of the collar retracted. Collars in the above-mentioned species were slightly flared as they protruded and as they retracted. Collars of *Aeverrilla setigera* twisted at their midpoint before retracting and untwisted upon eversion (Fig. 1a). The distal margin did not pinch together but remained slightly flared. Some zooids of *Aeverrilla armata* were observed to have both the long collar and the lophophore twisted together, but the collar untwisted and expanded before the lophophore fully protruded.

It was common to see everted collars in zooids with retracted lophophores in *Amathia convoluta*, *Nolella stipata*, *Walkeria tuberosa*, *Aeverrilla armata*, *A. setigera*, *Amathia pruvoti*, and *Anguinella palmata*. Some of the collars were deflated or partially flared and in varied stages of retraction. Dead zooids of *W. tuberosa*, *A. setigera*, and *A. armata* were observed with fully protruded collars, some flared and others deflated. Some were at an angle to the zooid instead of parallel to the long axis of the zooecium as is the case in living zooids.

All collars observed in this study were composed of membranes that appeared to have their basal attachment at the atrial sphincter (diaphragm). Although collars appeared to be delicate, none were observed torn or damaged. Some were wrinkled as they unfolded during protrusion, but quickly spread tautly between the rods. The delicate, evenly spaced rods noted in all zooids but those of the *Nolella* species collected provided visible support to the membrane. Both the membrane and rods were clear to translucent and most appeared iridescent. Rods in collars of *Aeverrilla setigera* zooids have a rhombic cross-section visible in 1 μm thick sections (Fig. 1b, c). They appear to be embedded in the membrane with the membrane thinned on the outer contact with the rod. Rods in *Aeverrilla armata* collars had 1/3 to 1/5 the diameter of the tentacles, generally comparable to rod diameter in collars of most other living species observed. Rods *of Amathia convoluta* collars were circular in cross-section (Fig. 1d). The membrane made contact with the outer edge of the rod. It was thinner along the outer edge and appeared thickened where it met the inner part of the rod.

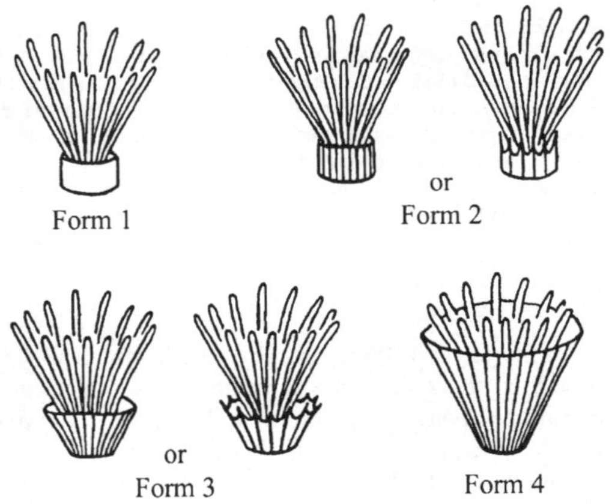

Figure 2. Setigerous collar forms.

In 1 μm thick sections of *Amathia convoluta* and *Aeverrilla setigera* the midline of the membrane between the rods was drawn inward as if along a crease (Fig. 1d-f). Observations of the living zooids gave the same impression of membranes pleated like a well-organized umbrella. In *A. convoluta*, there was slight localized twisting of the folded membrane but, for the most part, the midlines were drawn to the center of the retracted collar, touching to produce a radial pattern (Fig. 1d-f). The rods were encircled by the more pliant part of the membrane. Sections of *A. setigera* showed that retracted collars are gathered in tight swirls within the ring of rods at the zone of twisting (Fig. 1c). The rods do not always form a complete circle or oval around the folded membrane but tend to have a slight gap and offset at the narrower end (Fig. 1c), a condition that may be an artifact.

Four collar plans are recognized here based on collar shape when fully extended, presence or absence of rods, and length of the collar relative to that of the lophophore (Fig. 2). Two types (here illustrated as Forms 1 and 2 in Fig. 2) have the appearance of a short, unflared, simple cuff surrounding the proximal portion of the extended lophophore. In both of these forms, the distal collar margin is smooth or nearly so. Form 1 collars appear to have no rods and may be smooth or pleated. *Nolella stipata*, *N. dilatata*, *Walkeria tuberosa*, and *Bowerbankia gracilis* have Form 1 collars. Form 2 collars are conspicuously setigerous with the rods sometimes extending slightly beyond the distal margin. All species of *Amathia*, *Anguinella palmata*, and some zooids of *B. gracilis*, and the cheilostome *Aetea truncata* had Form 2 collars. Form 3 collars are relatively short, extending no more than 1/3 the length of the extended lophophore. They flare from the base and the rods may extend well beyond the distal collar edge to produce a serrated margin in which the membrane may be scalloped. Many zooids of *W. tuberosa*, and *Amathia convoluta* had Form 3 collars. Form 4 collars are broadly flared and the collar margin may extend to just below the tentacle tips. *Aeverrilla setigera* and *A. armata* were the only species collected to have this type of collar.

Collars do not appear to be as sensitive to mechanical stimulation as do the lophophores. Grazing by a pynchnogonid on *Aeverrilla armata* produced no response from the collar although the lophophore retracted. Collars of a *Walkeria tuberosa* colony were not affected by the stimulation that caused the immediate retraction of the lophophores of the colony. Collars in an *Amathia pruvoti* colony retracted slowly when a brush was applied to the distal edges.

Ctenostome colonies are often at least partially covered with fine detritus and, presumably, bacterial films. However, the lophophores, tentacles, and collars are clean. Commonly, a concentration of particulate material forms a ring at the base of the collar. Sometimes, the distal collar margins have detritus adhering. A collar of *Aeverrilla armata* was seen to push away detritus as it protruded

Some fecal pellets were observed leaving the anus. Of 35 disposal events recorded for 7 species, 21% (7 events) were along Pathway 3 described by McKinney (1997). Species that followed that track all had short collars. All other observed pellets (89%) were carried above the level of the anus and either moved into the lophophore and out distally (Pathway 1: 51%, 18 events), out laterally between the tentacles (Pathway 2: 18%, 6 events), or moved up along the extended lophophore (Pathway 4: 10%, 3 events).

Of the 24 living zooids in the suborder Carnosa collected, the majority (91%) had either collar Form 1 or 2. Of 38 living zooids in the suborder Stolonifera, the collars were more diverse with over half (63%) being Form 1 or 2 and most of those remaining being Form 3.

4 DISCUSSION

Although setigerous collars are a characteristic of ctenostome bryozoans, little detailed information has been recorded about their form, microstructure, behavior, function, or possible taxonomic significance. Collars, often partially everted ones that look like spines at the orifice, are commonly illustrated in the literature, but there is generally no description or even mention of their presence. Lack of detailed information is easily understood. The delicate appearance and transparency of the collar allow it to be easily overlooked. Even when one expects to see this structure, it can be missed if lighting and magnification are insufficient. A flash of iridescence may be the only indication of a collar. Another difficulty is that the ctenostome genera observed in this study, and presumably other ctenostomes, tend to be lethargic in a laboratory setting. Many hours of waiting for the animals to acclimate and begin to feed were required. However, all of the living colonies collected had active zooids 2 - 3 days after collection and life in non-circulating sea water.

The delicate appearance of the collar membrane suggests fragility, but no damaged collars were observed. The even distribution of the rods in the collar may help to protect it as well as provide support. The membrane appears to be a uniform sheet. However, geometrically precise folding of the retracting collar in *Amathia convoluta* and *Aeverrilla setigera* indicate that there are linear elements in the membrane that are not readily visible. The collars fold sharply along the midline between rods in sharp folds as if creased.

The linear supports of the membrane have been called rods, bristles or setae by some authors and have been described as thickened chitinous areas by others. The term "rod" is being used here for consistency in this paper although "bristle" is very descriptive of the hair-like structures seen in some collars. The term "seta" has specific biological implications and more detailed study is needed to determine if its use is appropriate. "Setigerous" is retained here as a descriptor of collars with obvious rods because of its common use in the literature. The supports in *Amathia convoluta* and *Aeverrilla setigera* are seen to be definite rods in serial 1 μm sections and are not merely thickened cuticle. Flexibility of the rods is suggested by the pinched distal portions noted in several species, the conical shape of closed collars, and twisting in *Aeverrilla* collars. The mechanism for retracting the collar as a whole has been described by many authors including Hyman (1959), Ryland (1970), and Hayward (1985), but the mechanism for pinching rods together distally in some collars and twisting them in others is not presently evident.

The four collar types suggested by this study (Fig. 2) have been observed in living bryozoans. It is felt that drawings in the literature that were used as corroborative information are generally accurate representations of what was observed. Even with limited microscopic capability, earlier workers tended to be good observers and illustrators. In determining collar type, only collars associated with fully extended lophophores were used. Even though collars protrude before the lophophore, they may be deflated before the lophophore protrudes and might cause confusion in determining the type. Collar Form 1, which has no rods, may not be as common as it seems. Rods often become apparent after a long period of observation and are easily missed. Pleating in Form 1 collars is difficult to understand without the presence of supportive linear elements.

Collars might be expected to restrict feeding currents and affect waste removal pathways. No feeding currents were evaluated although minute particles were observed being carried in currents passing between the tentacles with no apparent interruption. The very long collars (Form 4) seen in *Aeverrilla* species might be expected to restrict feeding currents since they extend to just below the distal tips of the tentacles. Winston (1978) illustrated *Aeverilla armata* and showed that some of the tentacles bend at the level of the collar margin. That species also exhibits scanning behavior and the collar has been observed in this study to move from side to side, all of which may overcome the difficulty expected by a collar that would seem to restrict the lophophore.

In 89% of the total waste removal events noted, fecal pellets traveled away from the zooid at a level above that of the anus. Of those, 78% of the pellets exited at the distal portion of the lophophore, either having traveled through the lophophore or up along the side of it. In the remaining 22% events, the pellet moved into the bell approximately 1/3 the length of the lophophore before exiting. Of the 21% of the total waste removal events remaining, the zooids had short collars and the pellets were not impeded as they moved away from the collar. Thus it would seem that the zooids are able to remove pellets without restriction caused by the collar.

Collars of live zooids remained everted in several species. Maturo (1957) reported that fully protruded but unfurled collars of *Amathia armata* were common and that lophophores could be seen in the collars. Collars of *Aeverrilla setigera* and *A. armata* in this study remained extended, either partially or fully, after retraction of the lophophore. In *Walkeria tuberosa*, the lophophore periodically everted and retracted through the extended collar. Jullien (1888) noted that one of the characteristics of ctenostomes was that the collar of dead zooids remains extended. In obviously dead zooids of all species observed in this study, the collars often remained protruded and often tilted limply. Since so many collars remain everted long after the lophophore retracts, that they are extended after death of the zooid is not unexpected.

Lack of rigid skeletal elements in non-calcified ctenostomes increases the vulnerability of the zooid. The orifice is constricted by the longitudinal parietal muscles (Hayward 1985) when the lophophore retracts. The quadrate arrangement of the muscles produces the commonly observed squared orifice. Closing of the collar at the distal end of the zooecium and its retraction after that of the lophophore provides an effective defensive closure as has been recognized by many authors.

The collar may serve another function – that of keeping the lophophore clean. Ctenostomes, especially the bushy or clumped colonies tend to collect fine detritus, yet the lophophore and most of the collars observed were free of detritus and able to function. A ring of concentrated debris was observed at the bases of collars in *Walkeria tuberosa*, and detritus was sometimes seen along the distal edges of some collars in several species. The collar always emerges before the lophophore and as such would move any detritus out of the way of the emerging lophophore.

Suggestions that collars may have taxonomic significance made by Kluge (1962) and in this paper need more data in order to be adequately evaluated. At present, descriptions and illustrations are too incomplete to determine with confidence if correlation exists between any taxonomic group and collar type. The results of a tally of living zooids and extended collars illustrated in the literature indicate correlation. Ninety two percent of the 24 carnosans scored had simple cuff collars (Forms 1 or 2 in Fig. 2). Of the individual carnosan genera scored which had multiple species represented, *Nolella* and *Victorella* exhibited only Form 1. *Alcyonidium* exhibited more diversity with the majority of species having Forms 1 or 2 and one appeared to be 3. Sixty three percent of the 38 stoloniferans scored had cuff collars of Forms 1 or 2. Of the stoloniferan genera that had multiple species, collar form was variable. Variation might be real or, more likely, due to inability to see rods or to discern flare.

The report of Banta et al. (1995) of a setigerous collar in the cheilostome species *Membranipora chesapeakensis* Banta, Perez & Santagata was illustrated in a retracted individual. It has the appearance of the conical "pleated" collar typical of many ctenostomes. They reported that collars had been reported in the cheilostome genera *Aetea* and *Scruparia* as well. A simple collar (Form 2) in zooids of *Aetea truncata* was observed in this study. Due to the erratic movements of the lophophores in *A. truncata* as they evert and retract, it was very difficult to identify collars, and they were not seen in all active zooids. Banta et al. suggested that presence of a setigerous collar in *Membranipora* supports the hypothesis that most, if not all, cheilostomes evolved from ctenostomes but Prenant & Bobin (1956) and Jebram (1992) have suggested that convergent evolution may be responsible. Collars are especially difficult to see in encrusting cheilostomes, which may require sectioning of many species in order to determine their distribution in this group.

ACKNOWLEDGEMENTS

Data and observations were collected during visits to the Ruder Boskovic Institute Center for Marine Research in Rovinj, Croatia. Andrej Jaklin was instrumental in obtaining specimens for study in Rovinj. Facilities at the University of North Carolina's Institute of Marine Sciences in Morehead City, North Carolina were made available for collection and observation of other specimens. William Dobson loaned an aerated seawater box that enabled us to transport the living colonies back to the laboratory. We thank F.K. McKinney for his guidance and for his critical reading of and suggestions for improving this manuscript.

REFERENCES

Banta, W.C., Perez, F.M. & Santagata, S. 1995. A setigerous collar in *Membranipora chesapeakensis* n. sp. (Bryozoa): implications for the evolution of cheilostomes from ctenostomes. *Invertebrate Biology* 114(1): 83-88.

Farre, A. 1837. Observations on the minute structure of some of the higher forms of polypi with views of a more natural arrangement of the class. *Philosophical Transactions of the Royal Society of London* 127: 387-426.

Franzén, A. 1960. *Monobryozoon limicola* n. sp., a ctenostomatous bryozoan from the detritus layer on soft sediment. *Zoologiska Bidrag från Uppsala* 33: 135-151.

Hayward, P.J. 1985. *Ctenostome Bryozoans*. Synopses of the British Fauna, vol. 33. London: E.J. Brill/Dr. W. Backhuys.

Hincks, T. 1880. *A history of the British marine Polyzoa.* London: John Van Voorst.

Hyman, L.H. 1959. *The Invertebrates: Smaller Coelomate Groups.* Vol. V. New York, London, Toronto: McGraw-Hill.

Jebram, D.H.A. 1992. The polyphyletic origin of the "Cheilostomata" (Bryozoa). *Zeitschrift für Zoologische Systematik und Evolutions-forschungs* 30: 46-52.

Johnston, G. 1847. *A history of the British zoophytes.* London: John Van Voorst.

Jullien, J. 1888. *Mission Scientifique du Cap Horn 1882-83, Tom VI Zoologia Bryozoaires.* Paris: Gauthier-Villars et Fills.

Kluge, G.A. 1962. *Bryozoa of the Northern Seas of the USSR.* Moscow, Leningrad: Akademiya Nauk SSSR. (English translation 1975. New Delhi: Amerind Publishing Co.)

Maturo, F. 1957. A study of the bryozoa of Beaufort, North Carolina, and vicinity. *Journal of the Elisha Mitchell Scientific Society* 73(1): 11-68.

McKinney, M. 1997. Fecal pellet disposal in marine bryozoans. *Invertebrate Biology* 116(2): 151-160.

Mukai, H., Terakado, K. & Reed, C.G. 1997. *Microscopic Anatomy of Invertebrates, vol.13: Lophophorates, Entoprocts & Cycliophora.* New York: Wiley-Liss, Inc.

Prenant, M. & Bobin, G. 1956. *Bryozoaires, pt. 1, Entoproctes, Phylactolemes, Ctenostomes. Faune de France 60.* Paris: Paul Lechevalier.

Ryland, J.S. 1970. *Bryozoans.* London: Hutchinson University Library.

Stricker, S.A. & Reed, C.G. 1981. Larval morphology of the nemertean *Carcinonemertes epialti* (Nemertea: Hoplonemertea). *Journal of Morphology* 169: 61-70.

Winston, J. 1978. Polypide morphology and feeding behavior in marine ectoprocts. *Bulletin of Marine Science* 28: 1-31

Bryozoan Studies 2001, Wyse Jackson, Buttler & Spencer Jones (eds)
© 2002 Swets & Zeitlinger, Lisse, ISBN 90 5809 388 3

Mitochondrial evidence of geographical isolation within *Bugula dentata* Lamouroux

J.A. Mackie,[1,2] & M.J. Keough
[1]*Zoology Department, University of Melbourne, Australia*

J.A. Norman & L. Christidis
[2]*Museum Victoria, Australia*

ABSTRACT: Levels of divergence in the mitochondrial *cytochrome c oxidase* subunit I gene were assessed from geographically widespread populations of *Bugula dentata* Lamouroux. An average pairwise distance of 21% (Kimura-2P corrected) was found between colonies from the eastern Australian seaboard, Albany in south west Australia, and Oahu in Hawaii. Assuming COI mutation rates are comparable with other organisms, cladogenisis leading to these taxa would have occurred in the Miocene. Morphological differences between specimens from eastern Australia and Hawaii are consistent with regional taxonomic differences identified previously in the literature, while phylogenetic division between the southern Australian specimens has not been diagnosed taxonomically. Examination of population-level differences within south eastern Australia revealed a shallow phylogenetic structure consistent with a recent colonisation event by a small number of founders. Genetic differences between populations are structured (F_{ST} = 1.0), suggesting small scale barriers to gene flow.

1 INTRODUCTION

The occurrence of cryptic species complexes, and the likelihood that human activities have caused an unnoticed expansion of the range of some species are alternative explanations for the observation of widely cosmopolitan distributions (Knowlton 1993, Carlton 1996). As bryozoans have potentially many cryptic species, and in shallow water species there is as a high potential for movement on artificial vectors, the coupling of these factors poses a serious problem for the definition of evolutionarily and ecologically relevant units.

Watts et al. (1998) found a relationship in cheilostomes between the ecological ability of species to foul artificial substrata and global range suggesting dispersal mechanisms involving fouling, such as the movement of colonies on ship hulls (Allen 1953; Ryland 1965, Carlton and Hodder 1995) and drift plastics (Winston 1982, Stevens et al. 1996), have increased the range of many species (Watts et al. 1998).

Genetic studies have shown the presence of cryptic species, e.g. in *Bugula neritina* (Davidson & Haygood 1999) *Alcyonidium* sp. (Thorpe & Ryland 1979), and *Parasmittina* and *Stylopoma* (Jackson & Cheetham 1990), The breeding and allozyme experiments of Jackson and Cheetham (1990) supported that the usefulness of 'morphospecies' concept to detect species divisions. The extension of this concept to the finding of a significant geographic correlation of morphometric variation in *Adeonellopsis yarraensis* (Waters), in populations from New Zealand and Australia suggests species –level endemism may be much more finely grained than the nominative distribution suggests (Lidgard and Waters 1994).

Hoare et al. (2001) found evidence of mitochondrial divergences between populations of the cosmopolitan Hippothoid *Celleporella hyalina* on different continents which are higher than would be expected for a single species in other groups. Subtle morphological differences have been reported between populations. An exception to this pattern was indicated by finding of relatively lower differentiation was between haplotypes of colonies from Sweden and USA, suggesting a recent translocation of genetic variation possibly due to human activities. Similarly, in *Membranipora membranacea* identical haplotypes have been reported from both sides of the Atlantic.

In the genus *Bugula*, many species have widely cosmopolitan distributions. Some distributions may reflect long term processes of endemism, while others may be the result of anthropogenic dispersal. *B. neritina*, *B. stolonifera*, *B. simplex* and *B. flabellata* are potentially invasive species as they occur in port environments and are globally widespread occurring in southern Australia, the United Kingdom, and other centres (Keough & Ross 1999). Extreme cosmopolitanism, as is exemplified by *B. neritina* L.

which is found in all areas except cold polar and subarctic/antarctic waters, may itself be considered good *a priori* evidence that ocean-going vessels have contributed in some part to the present range (Ryland & Hayward 1977). Alternatively, cosmopolitanism in anascans which are generally characterised by lower levels of calcification and skeletal morphology compared with ascophorans, may reflect the presence of highly generalised species complexes (in Schopf 1979). Existing quantitative evidence, however, does not support average higher cosmopolitanism of nominative anascan species (Watts et al. 1998).

Davidson and Haygood (1999) detected the presence of two mitochondrially diverged cryptic variants of *B. neritina*.on the U.S. west coast in California. These variants, designated type *D* and *S*, were first detected because they exhibit produce different of chemotypes for bryostatin molecules. The sampling from the west coast of a single colony with a haplotype corresponding to one of the type *S* haplotypes suggests shipping may have resulted in at least a physical reconnection of ecologically and genetically diverged populations.

Is there actually such a thing as a non-introduced cosmopolitan bryozoan species? We present results of a preliminary study of the mitochondrial population structure of widespread populations of the temperate-tropical shallow water anascan, *Bugula dentata* Lamouroux (1816) to determine whether phylogenetic connections reflect a model of primary marine endemism or recent spread corresponding to anthropogenic dispersal.

1.1 *Distribution of* Bugula dentata *Lamouroux*

Bugula dentata Lamouroux (1816) is readily distinguished from other bugulids in the field due to its unique green colouration. An account of zoarial characters is provided in Ryland (1974): The nominal species has a pan temperate-tropical distribution, having been collected from various localities on the Australia-New Guinea continent including Victoria, Heron Island in Queensland and the Torres Strait; as well as the Celebes Sea, South Africa, New Zealand, Japan and Madeira (Ryland, 1974).

Populations in south eastern Australian have been described both as 'native' (Keough & Ross 1999) and 'cryptogenic' (Parry et al. 1997). This uncertainty is due in part to the ecological distribution which crosses the interface between natural and artificial environments. In southern Australia populations occur on shallow rocky reefs, a habitat in which other *Bugula* species have not been reported, as well as artificial structures, with other *Bugula* species. For example, *B. neritina, B. flabellata, B. simplex* and *B. denata* co-occur on the Williamstown Gasloading Pier near Melbourne (pers. obs.).

Ryland (1974) reported the occurrence of morphological variation throughout the range of *B. dentata.* The avicularium of specimens from Victoria and New Zealand is positioned proximally on the zooid, while in specimens from Heron Island in Queensland this structure is closer to the third spine, thus medially positioned. As concordance in all other characters was observed, Ryland considered the Heron Island *B. dentata* conspecific with *B. dentata* from Victoria and New Zealand, adding that this taxon 'is perhaps distinguishable at some infraspecific level'.

Ryland found greater morphological basis to doubt the conspecificity of specimens of *B. dentata* collected from the central IndoPacific on the Siboga expedition. These exhibit avicularial dimorphism: 'Giant' avicularia are found in the place of the standard avicularium on some of the zooids (Harmer 1926, Plate XXXII, Fig. 21,22) In addition, the ovicells of this morph are 'oblique, inwardly orientated'. In specimens from Heron Island, Victoria and New Zealand these have a more horizontal orientation. (Ryland 1974, Fig. 2A).

1.2 *Phylogenetic expectations of marine endemism*

Recently, studies of mitochondrial variation encompassing the entire range to globally distributed temperate species complexes have been published. Lessios et al. (2001) dated the basal divergence leading to pan tropical species of *Diadema* to between 6.5 - 14.0 MyBP on the basis of mitochondrial divergences. Basal divergences of other temperate-tropical 'ring species' complexes are also of Miocene age, e.g. within the Trumpetfishes, *Aulostomus* (Bowen et al. 2001); and between Atlantic and Pacific populations of the algivorourous reef fish *Ophioblennius* (Muss et al. 2001).

Given comparable rates of mutation, deep (Miocene-age) divergences would be expected within *Bugula dentata* if the isolation of geographically widespread extant populations occurred simultaneously with the cladogenetic events leading to the present distributions of other taxa. Lower divergence is expected if cosmopolitanism is due to more recent spreading.

2 METHODS

2.1 *Samples*

Whole colonies of *B. dentata* were collected from four sites in south eastern Australia: from the piers of Williamstown and Flinders in Victoria and from shallow reef habitat at Spring Beach in Tasmania, and Twin Reef, Wonthaggi in Victoria (approximately 50 km from Flinders). In southern Western Australia we collected colonies from the town jetty at Albany. At piers we sampled single colonies from separate pylons. At Spring Beach colonies were

Table 1. Collection localities of *B. dentata* specimens.

Locality	Grid Reference	Museum Victoria voucher
Flinders Pier, VIC	38.48S 145.06E	F89210-14
Twin Reef, Wonthaggi, VIC	38.78S 145.62E	F89215
Spring Beach Reef, TAS	42.35S 147.54E	F89216-20
Gasloading Pier, Williamstown,VIC	37.52S 144.53E	F89221-25
Wistari Reef, Heron Island, QLD	23.29S 151.53E	F89226
Jetty,Albany WA	35.05S 117.55E	F89227-29
Waiahole Reef, Oahu, Hawaii	20.96N 156.43W	F89230

sampled from discrete patches of reef, along a 400 m stretch of coast line. We also obtained single colonies from, Wistari Reef, Heron Island in Queensland, and Waiahole Reef in Kaneohe Bay on the island of Oahu, Hawaii. Collection localities are shown in Table 1.

We noted the presence or absence of giant avicularia, the zooidal position of avicularia, and the orientation of ovicells.

Twenty-one colonies were used for sequencing: sixteen from sites in south eastern Australia, three from Albany, and single colonies from Heron Island and Oahu.

Sequences of *B. neritina* haplotypes *D* (AF061428); *S1,* (AF06146); and *S2,* (AF061425); and *B. pacifica* (AF061433) published by Davidson and Haygood (1999) from GenBank were used. We generated sequences for colonies of *B. cucullata* Busk (F89231) collected near Portsea, Victoria, and *B. flabellata* Thompson (F89232) collected at St Kilda, Victoria. These sequences establish a establis[1]h a reference level of interspecific variation. As the outgroup we generated a sequence for *Tricellaria inopinata* d'Hondt and Occhipinti Ambrogi (from Poole Harbour, UK, F89233).

2.2 *DNA extraction*

Colonies were preserved in 75% ethanol at collection. One or two zooids (0.4 – 1.5 mg wet wt) were isolated from one tip or preceding bifurcation per colony. DNA was extracted in 200 mL of 12% Chelex (BioRad Laboratories) prepared with Milli Q H_2O. Samples were incubated at 50 °C with periodic vortexing for 10-12 hours followed by 15 minutes at 95 °C.

2.3 *Amplification and sequencing*

The universal LCO1490 and HCO2198 primers of Folmer et al. (1994) which were designed for the amplification of COI in diverse metazoans were used for initial amplification and sequencing of a 700 bp product in a subset of samples. Generally, this product was used as a template for re-amplification with LCOI490 and a bryozoan-specific internal non-coding primer as this increased the DNA yield. The bryozoan-specific primers were designed based on consensus of sequences of *Bugula* and *Watersipora* spp generated using the universal primers. Used in combination with the LCO1490 primer, bryozoan-specific primers: BryCOIintNC:
5'GATAAGGGAGGGTATACGTTTG; and Bry-COIH:
5' TGTTGGTATAGAATAGGATCCC produce fragments of 330 bp and 630 bp respectively. Sequences for Albany and Hawaiian colonies were generated by re-amplifying the template DNA using BryCOIL: 5' CATAAACAGGAA-GAGGTTTAAG; (designed as above), with a non-coding primer
BryCOIHIV: 5'CCCCCTCCAGCAGGATCAAA, designed based on consensus data including sequences of *Mucropetraliella ellerii,* as this combination was found to increase the reliability of PCRs. This combination produced 540 nucleotide data. Sequences were compared across 271 nucleotides of the shorter (330 bp) sequence amplified previously in three colonies from Albany and found to be identical.

PCR reactions consisted of 6.25 μL of DNA, 1x Reaction Buffer, 0.875 mM MgCl2, 250 μM dNTPs, 0.5 μM of each primer, and 1 U Taq DNA poly erase (Geneworks) in a 12.5-μL volume. Thermocycler parameters were 93° for 30 seconds, 50° for 30 seconds and 72° for 40 seconds for 30 cycles, followed by a finishing step at 72 °C for 3 minutes. Products were checked on 1.2 % agarose gels. Two 12.5 μL reactions were pooled and cleaned using the Ultra Clean purification kit (MoBio Laboratories, Inc.). Sequencing reactions were performed using the Promega *f*mol thermal cycle-sequencing kit with gamma ^{33}P-ATP end-labeled primers. Sequences were run manually on 6% denaturing polyacrylamide gels and visualised using autoradiograph. Representative sequences have been submitted to GenBank.

2.4 *Sequence analysis*

Sequences were translated using the *Drosophila* mitochondrial code in DNA Strider 1.2 (Marck 1994), revealing no stop codons. Maximum parsimony trees were constructed for the full inter-intraspecific sequence set and for *B. dentata* sequences only using the heuristic search and tree-bisection-reconnection options in PAUP* 4.0b1a (Swofford 2000).

We constructed a distance-based tree using the maximum likelihood procedure of PAUP. The choice of DNA substitution model was determined by the hierarchical maximum likelihood method of Heulsenbeck & Crandall (1997) implemented in Modeltest 3.06 (Posada and Crandall 1998). Bootstrap support for nodes was based on 1000 replicates using the fast-heuristic search option in PAUP and maximum parsimony due to computational restrictions of maximum likelihood.

Percentage differences between haplotypes discussed in the text are corrected using the Kimura (1980) two parameter correction, for ease of comparison with other studies.

The geographical distribution of mutations separating haplotypes of *B. dentata* was shown using a parsimony network, constructed in PAUP. A measure of the genetic structure of the three populations in which we sampled five colonies was provided by the standardised variance statistic, F_{ST} (Wright 1951). The AMOVA procedure implemented in Arlequin 2.0 with 10,000 randomisations, was used to test the significance of genetic structure (Schneider et al. 2000).

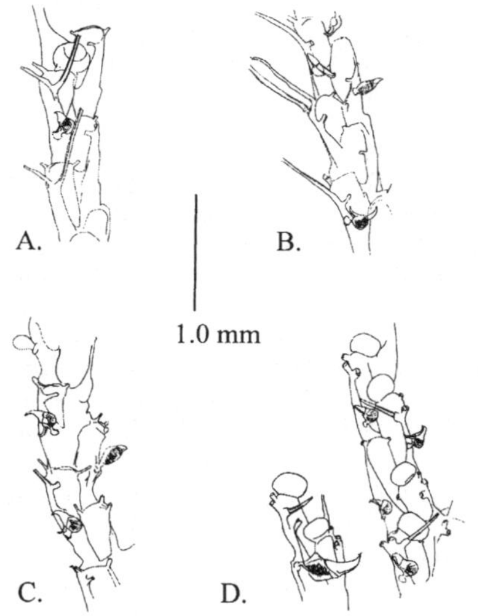

Figure 1. *Bugula dentata* zooid structure: A. Williamstown; B. Heron Island; C. Albany; D. Oahu (a zooid with a giant avicularium to the left).

3 RESULTS

3.1 *Morphological characters*

Examination of three Oahu colonies revealed standard-sized avicularia, and secondary giant avicularia on some zooids at colony bifurcations. The giant avicularium mandible is tripartite, matching the structure described by Harmer (1926). Ovicells of Oahu colonies are strongly inwardly orientated (Fig. 1).

Colonies from other localities had standard avicularia only. These were positioned proximally, in colonies from Albany and south eastern Australia, and medially in the colony from Heron Island. (Fig. 1). Ovicells found in Albany colonies are horizontally orientated, not sharply oblique, resembling the south eastern Australian morphology.

3.2 *Sequence data*

We obtained between 360 and 640 nucleotide data from 21 colonies of *B. dentata*. Amplification of the short product only was successful in the specimen from Heron Island. This is presumably because of difficulties in the recovery of DNA. Most zooids appeared to have lost their tissue in vivo.

Sequences from *B. dentata* formed three lineages with divergences ranging from 16.9 % between Albany and Oahu, and 22.7 % between Albany and Williamstown. Comparatively lower divergence (9.7%) separates *B. neritina* type *D* and *S* sequences. The minimum divergence found so far between non-disputed *Bugula* species is 25.2 % between *B. flabellata* and *B. pacifica*.

Unambiguous sequence data totaling 472 nucleotides were used for phylogenetic analysis. In parsimony analysis six trees were found, all of which included *B. dentata* sequences as a single clade. Accordingly, bootstrap support for the monophyly *B. dentata* sequences is high (90%). Monophyly within sequences of congeners, and the absence of mutations prohibiting coding functionality, suggests divergences are likely to be in homologous genes rather than due to the amplification of contaminant COI or non-coding paralagous copies.

The Tajima and Nei (1984) model of nucleotide evolution provided the best fit with sequence data. This model assumes a single evolutionary rate for transition-type mutations, and an independent rate for each of the four remaining classes of transversion mutations (α =1.3327). The maximum likelihood tree is shown in Figure 3. The short sequence from Heron Island was found to group unambiguously with the south eastern Australian clade.

Sequences sampled from south eastern Australian populations exhibit low diversity, (the average pairwise difference, π , is 0.003). Comparison of sixteen colonies revealed two mutations only. Sequences sampled from the same population were identical. Colonies from Flinders and the colony from Twin Reef also shared an identical haplotype. Single differences separate the three population samples, with the medial haplotype corresponding to the Flinders colonies The Heron Island haplotype is separated from Williamstown by two differences (see Appendix).

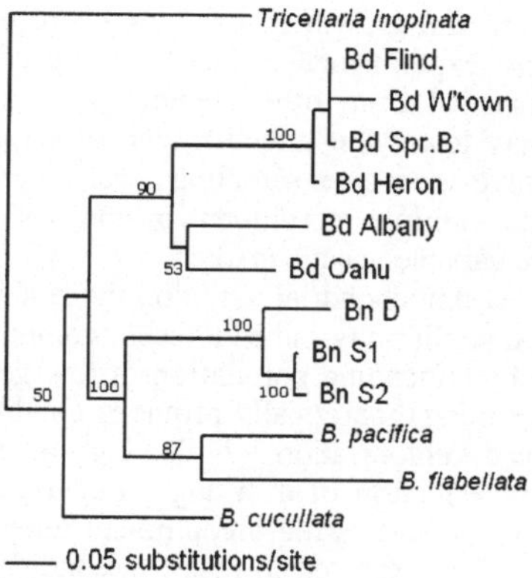

Figure 2. Maximum likelihood tree of *Bugula* COI sequences with *Tricellaria inopinata* as the outgroup. Bootstrap values shown for nodes which received >50% majority support were derived using parisimony. *B. dentata* (*Bd*) locality abbreviations refer to Table 1., *B. neritina* (*Bn*) haplotypes are discussed in the text.

The grouping of unique mitochondrial alleles at the three south eastern Australian localities at which multiple colonies were sampled corresponds to the maximum coefficient of population differentiation, or F_{ST} value equal to 1.0 (with a probability <0.000 of being generated by chance).

4 DISCUSSION

The sampled *B. dentata* colonies correspond to three divergent lineages. Colonies from Oahu, Albany and eastern Australia are separated on the basis of major divergences, therefore it is unlikely that the distribution of *B. dentata* at these centres is due to recent common dispersal.

The range of major divergences found in *B. dentata* (16.9 - 22.7 %) corresponds to the upper range of COI divergences reported in cryptic complexes of other invertebrates (listed in Table 5 of Rocha-Olivares et al. 2001).

As estimates of the rate of COI evolution based on independent geologic data are presently unavailable in bryozoans, dating of cladogenisis in *B. dentata* relies on the assumption that a standard estimate derived from other organisms can be applied. Rate calibrations are commonly based on the level of divergence observed between geminate lineages formed by the closure of the Panama Isthmus (3.2 MyBP). Estimates of COI divergence per million years for invertebrates include 1.66 % for grapsid crabs, 2.4 % for *Alpheus* shrimps (in Rocha-Olivares et al. 2001), and 1.6 – 2.6 % for four echinoid genera (Lessios et al. 2001). Based on the echinoid range of rates, which brackets other estimates, divergence of the three *B. dentata* lineages occurred 6.5 – 14.MyBP.

Our estimate suggests separation of the three lineages occurred in the middle to late Miocene. Ongoing fluctuations in sea level commencing with a drop in sea level of approximately 80 m 10 MyBP (Haq et al. 1987), have been proposed as a possible trigger of the basal cladogenisis of genera including the urchin *Diadema* (Lessios et al. 2001). The punctuated radiation of the Atlantic neotropical ascophoran bryozan *Metrarabdotos* 7 MyBP (Cheetham 1986), a well studied cladogenetic event, also fits within the assumed window of cladogenisis of these lineages. Separation of these lineages may have occurred after the end of the Miocene period (5 MyBP) if the mutation rate of COI in *B. dentata* is substantially faster than in other organisms, (i.e. >3% / My).

4.1 *Concordance between mitochondrial lineages and taxonomic characters*

The three *B. dentata* lineages are approximately equidistant genetically (Fig. 3). As colonies from Albany and south eastern Australia are not distinguished taxonomically based on variable characters in the literature, they demonstrate the potential for cryptic species.

The grouping of closely-related sequences within south eastern Australia, and an established close relatedness of the Heron Island morph, suggests a single phylogenetic species occupies a contiguous range on the eastern Australian coast.

Oahu colonies, which possess giant avicularia and oblique ovicells, correspond morphologically with the variant illustrated by Harmer (1926) from the Celebes Sea. There is not a strict correspondence between the occurrence of avicularial dimorphism and

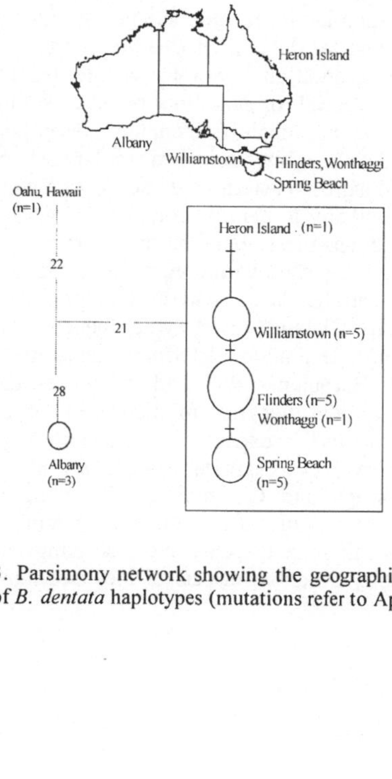

Figure 3. Parsimony network showing the geographical distribution of *B. dentata* haplotypes (mutations refer to Appendix).

phylogenetic strains, as Ryland (1974) found that although a colony collected from north west of New Guinea on the Siboga expedition possesses giant avicularia, other aspects of morphology are identical to specimens from the Torres Strait and Heron Island.

The repositioning of the avicularium, which defines the Heron Island morph, occurred relatively recently, as indicated by the minimal divergence between the Heron Island and south eastern Australian haplotypes. Nothing of the genetic basis of this trait is known, i.e., whether repositioning involves mutation at a single or multiple loci. Shifts in avicularial characters may be dependent upon epistatic interactions with loci which maintain the plasticity of response to changing environmental cues (e.g., Cheetham et al. 1995). The occurence of equivalent character configurations in Albany and southeastern Australian samples, points to a potential influence of stabilizing selection. Phenotypic states may be maintained despite the accumulation of genetic differentiation between populations under common selective regimes (Lande 1985). The absence of reports of avicularial dimorphism in *B. dentata* from temperate localities, and its widespread occurrence in the tropics suggests selective pressures on avicularial traits may be divided between these latitudinal zones.

4.2 *Population structure*

Schopf (1979) showed that empirical patterns of endemism in extant shelf Bryozoa agree well with expectations of isolation based on the present day position of continents and major patterns of ocean circulation. Genetic divergence apparent between colonies from Albany and eastern Australia provides tentative preliminary support for a suggested zoogeographic break on the southern Australian shelf however samples from intermediate localities are required to test for the existence of this pattern.

Similarly, additional samples are needed to estimate the potential for gene flow between south eastern Australia and localities including Heron Island to the north. Although we searched similar reef and pier habitats along the eastern coast south of Sydney we failed to find any *B. dentata* colonies, suggesting it is absent or at least less abundant in this area.

Structured genetic variation, as detected in south eastern Australia, is consistent with predictions of Isolation by Distance under which genetic drift and short range migration determine gene frequency variation (Barbujani 1987 and references therein). Probable limited dispersal of the lecithotrophic larvae of *B. dentata* is expected to maintain population structure over even smaller spatial scales (Schopf 1974, Keough and Chernoff 1987). Although the sampling of colonies from discrete patches is intended to minimise the chance of sampling variation produced by potential effects of localised inbreeding due to potential growth of colonies by regeneration of stolons, experimental isolation of the effects of larval dispersal from other demographic processes which may lower the effective size of populations will involve extensive sampling over finer spatial scales and multiple recruitment events, and the use of a more variable genetic marker.

The low mitochondrial variation evident in south eastern Australia may indicate recent colonisation of this area by a founding population of low genetic diversity. Rafting theoretically provides conditions for widespread colonisation. Following a founding 'pulse' introduction of a rafting group of colonies, reversion to larvae as the predominant intermediates of dispersal, is expected to result in conditions leading to a rapid increase in F_{ST} (Avise et al. 1988). Alternatively, the sampled area may have been colonised recently by the steady diffusion of populations from beyond this range according to a model of linear genetic expansion (e.g. Hengeveld 1989), which we have been unable to detect due to our small sample size and the perceivably slow rate of COI evolution. These two genetic models of colonisation can be distinguished through fine scale analysis of population structure (Le Corre & Kremer 1998).

Initial evidence of deep divergences between three geographically widespread localities suggests the cosmopolitan distribution of *B. dentata* reflects the presence of cryptic species. Sampling of the entire range of the nominative species is required.. Determination of the time of arrival of *B. dentata* populations in the Hawaiian Islands will be useful for assessing the roles of rafting and human introduction in bryozoan dispersal.

ACKNOWLEDGEMENTS

We are grateful to Nerida Wilson, Chela Zabin and Peter Dyrynda for providing specimens. We also thank Abigail Smith and an anonymous reviewer who read an early draft of this paper and provided helpful suggestions. This work was supported by a Melbourne University Collaborative Research Grant and Museum Victoria.

REFERENCES

Allen, F.E. 1953. Distribution of marine invertebrates by ships. *Aust. J. Mar. Fw. Res.* 4: 306-316.

Avise, J.C., Ball, M.R. & Arnold, J. 1988. Current versus historical population sizes in vertebrate species with high gene flow: a comparison based on mitochondrial DNA lineages and inbreeding theory for neutral mutations, *Mol. Biol. Evol.* 5: 331-344.

Barbujani, G. 1987. Autocorrelation of gene frequencies under Isolation by Distance. Genetics 117:777-782.

Bowen, B.W., Bass, A.L., Rocha, L.A., Grant, W.S. & Robertson, D.R. 2001. Phlogeography of the Trumpetfishes

(Aulostomus): Ring species complex on a global scale. *Evol:* 55: 1029-1039.

Carlton, J.T. 1996. Biological invasions and cryptogenic species. *Ecology* 77: 1654-55.

Carlton, J.T. & Hodder, J. 1995. Biogeography and dispersal of coastal marine organisms: experimental studies on a replica of a 16th-century sailing vessel. *Mar. Biol.* 121: 721-730.

Clague, D.A. & Dalyrmple, G.G. 1987. Tectonics, geochronology and origin of the Hawaiian-Emperor volcanic chain. In: Decker, R.W., Wright, T.L. & Stauffer, P.H. (eds), *Volcanism in Hawaii*: 1-54. US Geological Survey Professional Paper 1350. Washington, DC: US Governement Printing Office.

Clary, D.O. & Wolstenholme, D.R. 1985. The mitochondrial DNA molecule of *Drosophila yakuba*: nucleotide sequence, gene organization and genetic code. *J. Mol. Evol.* 22: 252-271.

Cheetham, A.H. 1986. Tempo of evolution in a Neogene bryozoan: rates of morphologic change within and across species boundaries, *Palaeobiology* 12: 190-202.

Cheetham, A.H., Jackson, J.B.C. & Hayek, L.C. 1995. Quantitative genetics of bryozoan phenotypic plasticity and the maintenance of genetic variation. *Evol.* 49: 290-296.

Davidson, S.K. & Haygood, M. G.1999. Identification of sibling species of the bryozoan *Bugula neritina* that produce different Anticancer Bryostatins and Harbour distinct strains of the bacterial symbiont: *Candidatus Endobugula sertula. Biol. Bull.* 196: 273-380.

Folmer, O., Black, M., Hoeh, W., Lutz, R., and Vrijenhoek, R. 1994. DNA primers for the amplification of mitochondrial cytochrome c oxidase subunit I from diverse metazoan invertebrates *Mol. Mar. Biol. Biotechnol.* 3: 294-299.

Harmer, S. F. 1926. The Polyzoa of the Siboga Expedition. Part 2. Cheilostomata Anasca. *Siboga Exped.* 28A: 1-180.

Hoare, K., Goldson, A.J., Giannasi, N., & Hughes, R.N. 2001. Molecular Phylogeography of the Cosmopolitan Bryozoan *Celloporella hyalina*: Cryptic Speciation?. *Mol. Phyog. Evol* 18: 488-492.

Huelsenbeck, J.P. & Crandall, K.A. 1997. Phylogeny estimation and hypothesis testing using maximum likelihood. *A. Rev. Ecol. Syst.* 28: 437-466.

Haq, B.U., Hardenbol, J. & Vail, P.R. 1987. Chronology of fluctuating sea level since the Triassic. *Science* 235: 1156-166.

Hengeveld, R. 1989. *Dynamics of Biological Invasions,* London: Chapman and Hall.

Jackson, J.B.C. & Cheetham, A.H. 1990. Evolutionary significance of morphospecies: a test with Cheilostome Bryozoa. *Science* 248: 578-583.

Keough, M.J. & Chernoff, H. 1987. Dispersal and population variation in the bryozoan *Bugula neritina. Ecology* 68: 199-210.

Keough, M.J. & Ross, J. 1999. Introduced fouling species in Port Phillip Bay. In C. L. Hewitt., M. L. Campbell, R. E. Thresher & R. B. Martin (eds) *Marine Biological Invasions of Port Phillip Bay Victoria*: 193-226. Hobart: CSIRO Marine Research.

Kimura, M. 1980. A simple method for estimating evolutionary rates of base substitutions through comparative studies of nucleotide sequences. *J. Mol. Evol.* 16: 111-120.

Knowlton, N. 1993. Sibling species in the sea. *A. Rev. Ecol. Syst.* 24: 189-216.

Lamouroux, J.V.F. 1816. *Histoire des Polypiers Coralligénes flexibles*. Caen.

Lande, R. 1985. Expected time random genetic drift of a population between stable phenotypic states. *Proc. Natl. Acad. Sci. USA* 82: 7641-7645.

Le Corre, V. & Kremer, A. 1998. Cumulative effects of founding events during colonisation on genetic diversity and differentiation in an island and stepping-stone model. *J. Evol. Biol.* 11: 495-512.

Lessios, H.A., Kessing, B.D. & Pearse, J.D. 2001. Population structure and speciation in tropical seas: global phylogeography of the sea urchin *Diadema*. *Evol.* 55: 955-975.

Lidgard, S. & Buckley, G.A. 1994. Toward a morphological species concept in cheilostomates: phenotypic variation in Adeonellopsis yarraensis (Waters). In P.J Haywood, J.S. Ryland & P.D. Taylor (eds), *Biology and palaeobiology of bryozoans:* 113-115. Fredensborg: Olsen and Olsen.

Marck, C. 1994. DNA Strider. Version 1.2. *A Program for DNA and Protein Sequences Analysis.* Sevice de Biochemie et de Genetique Moleculaire Department de Biologie Cellulaire et Moleculaire. France.

Muss, A., Robertson, D.R., Stepien, C.A., Wirtz, P. & Bowen, B.W. 2001. Phylogeography of *Ophioblennius*: the role of ocean currents and geography in reef fish evolution. *Evol.* 55: 561-572.

Posada, D. & Crandall, K.A. 1998. MODELTEST: testing the model of DNA substitution. *Bioinformatics.* 14: 817-818.

Palumbi, S.R. 1994. Genetic divergence, reproductive isolation, and marine speciation. *A.. Rev. Ecol. Syst.* 25: 547-572.

Parry, G.D., Currie, D.R. & Crookes, D.P. 1997. *Exotic marine pests in Portland Harbour and Environs.* Marine and Freshwater Resources Institute Report No. 1. pp. 42. Queenscliff: MAFRI.

Rocha-Olivares, A., Fleeger, J.W. & Foltz, D.W. 2001. Decoupling of molecular and morphological evolution in deep lineages of a meiobenthic Harpacticoid copepod. *Mol. Biol. Evol.* 18: 1088-1102.

Ryland, J.S. 1965. *Catalogue of Main Marine Fouling Organisms (Found on ships coming into European waters).* 2. *Polyzoa.* Paris: Organisation for Economic Cooperation and Development.

Ryland, J.S. 1974. Bryozoa in the Great Barrier Reef Province. *Proceedings of the Second International Coral Reef Symposium 1.* Great Barrier Reef Committee, Brisbane, pp.341-348.

Ryland, J.S. & Hayward, P.J. 1977. *British Anascan Bryozoans*. London: Academic Press.

Schneider, S., Keuffer, J.-M., Roessli, D., & Excoffier, L. 2000. *Arlequin. Ver. 2.0: a software for population genetic analysis.* Genetics and Biometry Laboratory, University of Geneva, Switzerland.

Schopf, T.J.M. 1974. Long term (3 to 5 year) records of gene frequencies in natural populations of an abundant subtidal species (the bryozoan *Schizoporella errata*). *Biol. Bull.* 147: 498.

Schopf, T.J.M. 1979. The role of biogeographical provinces in regulating marine faunal diversity through geologic time. In J. Gray & A.J. Boucot (eds), *Historical Biogeography, Plate Tectonics, and the Changing Environment*: 449-557. Corvallis: Oregan State Univ. Press.

Schwaninger, H.R. 1999. Population structure of the widely-dispersing marine bryozoan *Membranipora membranacea* (Cheilostomata): implications for population history, biogeography, and taxonomy *J. Mar. Biol.* 135: 411-423.

Stevens L.M., Gregory, M.R. & Foster, B.A. 1996. Fouling Bryozoa on pelagic and moored plastics from northern New Zealand. In Gordon, D.P., Smith, A.M., & Grant-Mackie, J.A. (eds), *Bryozoans in Space and Time*: 321-340. Wellington: NIWA.

Swofford, D.L. 2000. *PAUP* Phylogenetic analysis using parsimony (*and other methods).* Vers. 4. Sinauer, Sunderland: MA.

Tajima, F. & Nei, M. 1984. Estimation of evolutionary distance between nucleotide sequences. *Mol. Biol. Evol.* 3: 269-285.

Thorpe, J.P. & Ryland, J.S. 1979. Cryptic speciation detected by biochemical genetics in three ecologically important intertidal bryozoans. *Estuar. Coastl Mar. Sci.* 8: 395-398.

Watts, P.C., Thorpe, J.P. & Taylor, P.D. 1998. Natural and anthropogenic dispersal mechanisms in the marine environment: a study using cheilostome Bryozoa. *Phil. Trans. R. Soc. Lond. B.* 353: 453-464.

Winston, J.E. 1982. Drift plastic - an expanding niche for a marine invertebrate? *Mar Poll. Biol.* 13: 348-351.

Wright, S.J. 1951. The genetical structure of populations. *Ann. Eugen.* 15: 323-354.

APPENDIX

Matrix of variable sites in COI used to construct the parsimony network in Figure 3. Position 1 corresponds to position 1517 of the published *Drosophila yakuba* mitochondrial genome (Clary & Wolstenholme 1985).

```
                          Variable Position
Haplotype                 111111111111111111111122
identifier   667777788999001112233555667788899900
             391235818014251470928036251703625617
Williamstown TAAACTTGGGCTTCTTTGTTTCTTGATCGGCCGTAA
Flinders     ....................................
Spring Beach ....................................
Heron Island ..G.................................
Albany       CG.GTAAAAATCCTACCA.AATC.TG..AATTACT.
Oahu, Hawaii C...TAAAGA....A..AACAT.CC.CTAC..A.TG

             22222222222222222222222223333333
             11133444455677778888888990001122
             13614016928703690123589120135817
             CACACGCCTATAAATCATACATCAAATAGAAC
             .......................C........
             ..............C........C........
             ...........................T....
             T.AT.CTTC...T.CTTCT.GAT..T..AC..
             .GGTTATTCGCT.T.TTC.T....CTC.T.CA
```

Bryozoan Studies 2001, Wyse Jackson, Buttler & Spencer Jones (eds)
© 2002 Swets & Zeitlinger, Lisse, ISBN 90 5809 388 3

Results of a freshwater bryozoan survey in the Pacific Northwestern United States

Terrence Marsh
Department of Biology, North Central College, Naperville, IL 60540-4690, USA

Timothy S. Wood
Department of Biological Sciences, Wright State University, Dayton, OH 45435, USA

ABSTRACT: This study represents the first known attempt to gather region-wide information on bryozoans in the Pacific Northwestern United States. For four weeks in July and August 1998, we traveled through Oregon and Washington as well as part of adjacent Idaho collecting colonies and sieving shoreline debris for statoblasts. We narcotized live colonies and then preserved them in 70% ethanol. We stored statoblasts, dried, in sealed plastic bags. Many of the species we encountered are common in eastern North America. Here we report finding *Fredericella sultana* and *Plumatella mukaii* in North America for the first time. We did not expect the widespread occurrence of *Stephanella hina*, a species well known in western Asia and previously known from only a few populations in New England and Virginia. We also found a species closely resembling *Plumatella fungosa*, but with unusually large tubercles over all of the floatoblast.

1 INTRODUCTION

The first freshwater bryozoologist to survey a large area of North America was Bushnell (1965a, b, c) with his study in Michigan. Subsequent surveys have been conducted in Massachusetts (Smith 1989), Ohio (Wood 1989), eastern Canada (Ricciardi & Reiswig 1994), and Illinois (Marsh & Wood, in prep.). Documentation of species known to occur outside the northeastern quadrant of North America has only been made when researchers reported findings of one or more species in relatively few localities (Wood 2001a).

To begin to address this lack of information on the freshwater Bryozoa of other parts of North America, we undertook a survey of the Pacific Northwest, especially the states of Oregon and Washington. The Pacific Northwest is generally understood to include part or all of adjacent Idaho, although we did not include a significant portion of this state in our survey due to time constraints. We reasoned that an initial survey would reveal whether a more extensive survey would be warranted, e.g., if we discovered new species or species not known to occur in North America.

2 METHODS

We did most of the sampling during a one-month trip to the Pacific Northwest in the summer 1998. In the early spring prior to the trip one of us (Marsh) had spent a week sampling primarily lakes and reservoirs in southeastern Oregon. At that time of year few colonies were growing so almost all of the material collected was sieve samples containing statoblasts. On the summer trip we tried to collect from different areas of the states of Oregon and Washington. We sampled lentic and lotic sites in each state along a meandering east-west route as well as sites along a meandering north-south route from southern Oregon to northern Washington. We chose sites based on their accessibility from highways, the type of habitat, and physiographic location (Fig. 1).

We collected bryozoan material in two ways: by direct sampling of colonies and by examining sediment for statoblasts. We normally collected colonies by wading in water less than 1 m deep and looking at the undersides of submerged logs, rocks, or aquatic plants. Occasionally one of us would dive to 2-3 m to collect rocks and examine the undersides. When we found colonies we would cut or chip them from the substrate to which they were attached to include the attaching structures of the colony with the collected material. Sampling in this manner typically required 30-90 minutes per site to examine all of the substrates that were available and accessible. We stored specimens in 1-litre polyethylene bottles until the end of the day, when we narcotized them with menthol flakes then preserved them in 70% ethyl alcohol (Wood 2001a).

Our second method of collecting involved collecting shoreline debris or a thin layer of upper bottom sediment from lentic sites and sieving these ma-

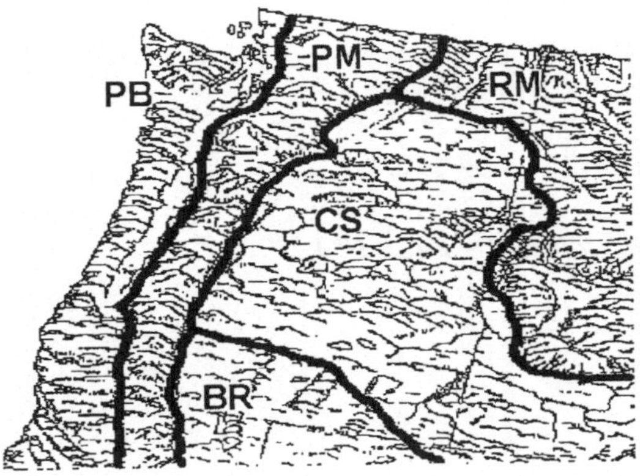

Figure 1. Physiographic regions of the Pacific Northwestern US (from Hunt 1974). BR = Basin & Range, CS = Columbia-Sanke River Plateau, PB = Pacific Border, PM = Pacific Mountain System, and RM = Rocky Mountains

terials for statoblasts. We used a three-screen method (Jones, et al. 2000) to separate statoblasts from as much of the mineral and other organic material as possible. The uppermost screen (mesh opening of 1 mm) held the sample and prevented material larger than any statoblast from being included with the statoblasts. The middle screen (mesh opening of 500 μm) caught larger statoblasts (e.g. those of *Pectinatella magnifica*), and the bottom screen (mesh opening of 150 μm) caught the smaller statoblasts (e.g. those of *Plumatella* spp.) Subsequent separation of the mineral fraction on the latter two screens resulted in a prepared sample consisting of statoblasts and a much smaller amount of mineral and other organic material. We dried the prepared sample on a paper coffee filter and stored each sample, labeled, in a self-sealing plastic bag. Sometimes we handpicked large statoblasts out of the screens and preserved and labeled them.

For species identifiable only by scanning electron microscopy we isolated statoblasts in deionized water and removed the adhering membrane, if necessary. Statoblasts were then freeze dried, mounted on aluminum stubs, and sputtered with gold palladium alloy for examination with a Philips 500 scanning electron microscope.

3 RESULTS

Figures 2 and 3 show the locations of the collecting sites in Oregon and Washington/Idaho respectively. Table 1 shows all of the bryozoan species that we found at the collecting localities. Some of our identifications are only to the genus level pending further study by us or because the material did not include statoblasts that are necessary for species determination.

For analysis we divided the Pacific Northwest into "desert" and "mountain" regions where "desert" includes all lentic sites in Hunt's (1974) Columbia-Snake River Plateau and Basin and Range physiographic provinces and "mountain" includes his Pacific Border, Pacific Mountain System and Rocky Mountains physiographic provinces (Fig. 1). Table 2 shows the pH and conductivity of lentic collection sites in the desert region of the Pacific Northwest and the bryozoan species we found at those sites.

We included Omak Lake, a site in Hunt's mountain region, in this group. While Omak Lake is several kilometers north of Hunt's Columbia River boundary for the desert region, it is an internally draining lake of high conductivity that is more typical of the desert region. The vegetation surrounding Omak Lake includes sagebrush (*Artemisia* sp.) and related desert plants. Three of the seven distinguished bryozoan species on this list are unique to the desert region in our survey. Using Microsoft Excel's t-test on the means with unequal variances we compared the pH and conductivity values between desert and mountain sites (Table 3) and found highly significant differences in both pH ($p < 0.008$) and conductivity ($p < 0.002$).

Cristatella mucedo. Colonies and/or floatoblasts occurred in 14 mountain localities, all lentic. Colonies

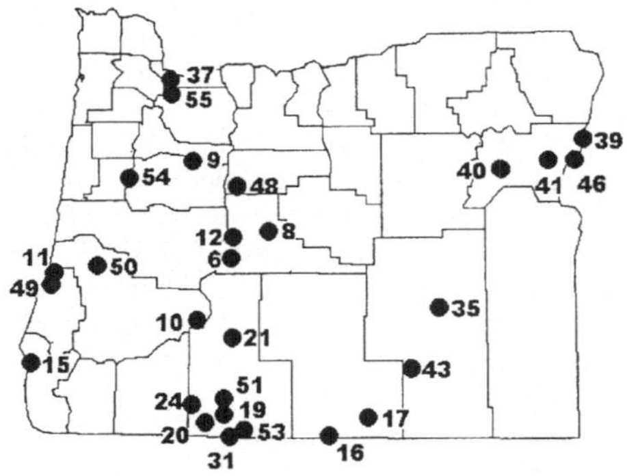

Figure 2. Collecting sites in Oregon.

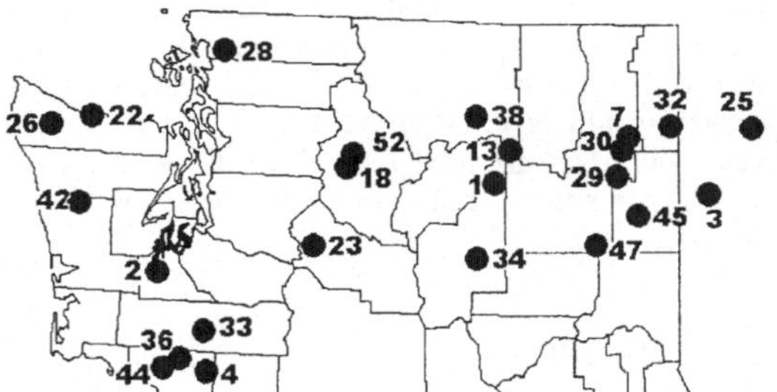

Figure 3. Collecting sites in Washington & adjacent Idaho.

Table 1. Bryozoan species of the Pacific Northwestern US.

NO	ST	COUNTY	BODYWATER	TAX1	TAX2	TAX3	TAX4	TAX5	TAX6
1	WA	Grant	Banks Lake	*H.pun.*	*P.art.*	*P.mag.*	*P.nit.*		
2	WA	Thurston	Black Lake	*C.muc.*	*P.mag.*	*Plum.*	*S.hin.*		
3	ID	Kootenai	Coeur d'Alene Lake	*H.pun.*	*P.mag.*	*Plum.*	*S.hin.*		
4	WA	Cowlitz	Coldwater Lake	*Plum.*	*S.hin.*				
5	OR	Lane	Cottage Grove Lake	*C.muc.*	*H.pun.*	*L.car.*	*P.mag.*	*Plum.*	
6	OR	Deschutes	Crane Prairie Reservoir at dam	*C.muc.*	*P.muk.*	*S.hin.*			
7	WA	Stevens	Deer Lake	*C.muc.*	*P.mag.*	*P.muk.*	*P.nit.*	*P.vai.*	*S.hin.*
8	OR	Deschutes	Deschutes River	*F.ind.*					
9	OR	Linn	Detroit Lake	*C.muc.*	*P.ema.*	*Plum.*			
10	OR	Douglas	Diamond Lake	*P.nit.*					
11	OR	Coos	Eel Lake	*P.art.*	*P.mag.*	*P.rug.*	*S.hin.*		
12	OR	Deschutes	Elk Lake	*P.fru.*	*P.rug.*	*S.hin.*			
13	WA	Okanogan	Franklin D. Roosevelt Lake	*C.muc.*	*H.pun.*				
14	WA	Skamania	Franz Lake	*P.mag.*	*P.muk.*	*P.sim.*	*S.hin.*		
15	OR	Curry	Garrison Lake	*F.sul.*	*Plum.*				
16	OR	Lake	Goose Lake	*P.fun.*					
17	OR	Lake	Greaser Reservoir	*P.nod.*					
18	WA	Chelan	Hidden Lake Outlet	*C.muc.*					
19	OR	Klamath	irrigation canal off Klamath River	*L.car.*	*P.ema.*	*Plum.*			
20	OR	Klamath	John Boyle Reservoir	*L.car.*	*P.mag.*	*P.sim.*	*S.hin.*		
21	OR	Klamath	Klamath Marsh	*Plum.*					
22	WA	Clallum	Lake Crescent	*C.muc.*	*P.mag.*	*Plum.*			
23	WA	Kittitas	Lake Easton	*C.muc.*	*P.rug.*	*S.hin.*			
24	OR	Klamath	Lake of the Woods	*F.ind.*	*P.rug.*				
25	ID	Bonner	Lake Pend Oreille	*C.muc.*	*H.pun.*	*P.mag.*	*P.ema.*		
26	WA	Clallum	Lake Pleasant	*C.muc.*	*F.ind.*	*P.art.*	*P.mag.*	*Plum.*	*S.hin.*
27	WA	King	Lake Washington	*C.muc.*	*P.mag.*	*Prug.*	*S.hin.*		
28	WA	Whatcom	Lake Whatcom	*C.muc.*	*Fred.*	*P.mag.*	*Plum.*	*S.hin.*	
29	WA	Stevens	Long Lake	H.pun.	*P.mag.*	*P.ret.*	*S.hin.*		
30	WA	Stevens	Loon Lake	*P. nit.*	*Plum.*				
31	CA	Siskiyou	Lower Klamath Lake	*Plum.*					
32	WA	Pend Oreille	Marshall Lake	*Fred.*	*S.hin.*				
33	WA	Lewis	Mayfield Lake	*C.muc.*	*Fred.*	*P.mag.*	*P.cas.*	*Plum.*	
34	WA	Grant	Moses Lake	*P.ret.*					
35	OR	Harney	Mud Lake	*Plum.*					
36	WA	Cowlitz	N. Fork Toutle River	*Fred.*	*P.ema.*				
37	OR	Multnomah	Oaks Bottom slough	*P.mag.*	*Pvai.*	*Plum.*			
38	WA	Okanogan	Omak Lake	*P.sim.*					
39	OR	Baker	Snake River-Oxbow Reservoir	*Plum.*					
40	OR	Baker	Phillips Lake	*P.ema.*	*P.rug.*				
41	OR	Baker	Powder River	*Plum.*					
42	WA	Grays Harbor	Quinault Lake	*F.ind.*	*P.mag.*	*Plum.*			
43	OR	Harney	Rock Creek Reservoir	*Plum.*					
44	WA	Cowlitz	Silver Lake	*P.mag.*	*Plum.*	*S.hin.*			
45	WA	Spokane	Silver Lake	*P.nit.*	*P.rug.*				
46	OR	Baker	Snake River below Brownlee Dam	*P.art.*	*Plum.*				
47	WA	Lincoln	Sprague Lake	*P.art.*	*P.nit.*				
48	OR	Jefferson	Suttle Lake	*C.muc.*	*P.nit.*	*S.hin.*			
49	OR	Coos	Tenmile Lake	*P.art.*	*P.mag.*	*Plum.*	*S.hin.*		
50	OR	Douglas	Umpqua River	*Fred.*	*P.rug.*				
51	OR	Klamath	Upper Klamath Lake	*Plum.*					
52	WA	Chelan	Wenatchee Lake	*S.hin.*					
53	CA	Siskiyou	White Lake	*Plum.*					
54	OR	Benton	Willamette River	*P.muk.*					
55	OR	Clackamas	Willamette River below locks	*F.bro.*	*P.mag.*	*P.ema.*			

C.muc. = Cristatella mucedo; F.bro = Fredericella browni; F.ind. = Fredericella indica; Fred. = Fredericella unidentified; *F.sul = Fredericella sultana; H.pun. = Hyalinella punctata; L.car. = Lophopodella carteri; P.art. = Paludicella articulata; P.mag. = Pectinatella magnifica; P.cas. = Plumatella casmiana; P.ema. = Plumatella emarginata; P.fun. = Plumatella fungosa; P.fru. = Plumatella fruticosa; Plum. = Plumatella* unidentified; *P.muk. = Plumatella mukaii; P.nit. = Plumatella nitens; P.nod. = Plumatella nodulosa; P.ret. = Plumatella reticulata; P.rug. = Plumatella rugosa; P.sim. = Plumatella semilirepens; P.vai. = Plumatella vaihiriae; S.hin. = Stephanella hina.*

Table 2. Bryozoan species found in lentic sites in "desert" regions of the Pacific Northwestern US.

Site	State	County	Body of water	Taxon	pH	Conduc-Tivity
16	OR	Lake	Goose Lake	*P.fun.*	9.1	1230
17	OR	Lake	Greaser Res.	*P.nod.*	8.4	280
31	CA	Siskiyou	Lower Klamath Lake	*Plum.*	8.0	810
34	WA	Grant	Moses Lake	*P.ret.*	9.8	340
35	OR	Harney	Mud Lake	*Plum.*	9.0	1740
38	WA	Okanogan	Omak Lake	*P.sim.*	9.8	1610
40	OR	Baker	Phillips Lake	*P.ema.* *P.rug.*	9.3	80
43	OR	Harney	Rock Creek Res.	*Plum.*	8.6	150
45	WA	Spokane	Silver Lake	*P.ema.* *P.rug.*	9.2	680
47	WA	Lincoln	Sprague Lake	*P.art.* *P.nit.*	9.7	380
53	CA	Siskiyou	White Lake	*Plum.*	8.9	1450

P.art. = *Paludicella articulata; P.ema.* = *Plumatella emarginata; P.fun.* = *Plumatella fungosa; Plum.* = *Plumatella* unidentified; *P.nit.* = *Plumatella nitens; P.nod.* = *Plumatella nodulosa; P.ret.* = *Plumatella reticulata; P.rug.* = *Plumatella rugosa; P.sim.* = *Plumatella semilirepens.*

Table 3. The pH and conductivity in "desert" vs. "mountain" lentic sites in the Pacific Northwestern US.

Site	Statistic	pH	Conductivity (μS)
desert	average	9.1	795
	standard deviation	0.6	613
mountain	average	8.5	63
	standard deviation	0.5	38
desert vs mountain	t-test	$p < 0.008$	$p < 0.002$

of this species were abundant on the undersides of logs several hundred meters above the dam in Crane Prairie Reservoir in Oregon. They were also abundant on the undersides of rocks in the small outlet stream from Hidden Lake, Washington. There was no current in Crane Prairie Reservoir, whereas the current was moderate in the small outlet stream from Hidden Lake.

Fredericella browni colonies occurred on the undersides of logs below the locks (and falls) on the Willamette River at West Linn, Oregon. Here it occurred with *P. magnifica.* There was no current at this site.

Fredericella indica. We found colonies of this species in four localities, all in mountain sites. Water flow at these sites varied from none to sluggish.

Fredericella sultana. We collected colonies of this species in Garrison Lake in southwestern Oregon. This freshwater lake is just inland of the Pacific Ocean at Port Orford. This species was abundant on sticks in a slow moving channel under the Arizona Street bridge that divides the lake into two parts. We also found colonies of an unidentified *Plumatella* at this site.

We made several other collections of *Fredericella* sp., all in lentic or lotic mountain sites. Because these colonies lacked statoblasts, we could not identify them to species.

Hyalinella punctata. We found colonies in two localities, one mountain and the other bordering the desert region. In Franklin D. Roosevelt Lake, an impoundment of the Columbia River formed by the Grand Coulee Dam, we found the colonies on sticks we collected from a couple of meters below the surface. In Lake Pend Oreille we collected colonies from concrete riprap and a branch at the lake edge. We also sieved statoblasts at four mountain lakes and a lake that is fed by Franklin D. Roosevelt Lake.

Lophopodella carteri. We found evidence of this species in only three mountain localities, all in western Oregon. None of our collections included colonies, only floatoblasts.

Paludicella articulata. We collected colonies of this ctenostome species in both still and flowing water throughout the Pacific Northwest. Colony growth was luxuriant in Tenmile Lake.

Pectinatella magnifica. We collected colonies or statoblasts of this species in 18 localities in this survey, none of which were lentic desert localities. Banks Lake is a reservoir that receives and returns water to the Columbia River for electricity generation. We collected only statoblasts in Quinault Lake, Lake Crescent, and Lake Washington, all in western Washington; Deer Lake in northeastern Washington; and Lake Pend Oreille in northern Idaho. Sallie Jones, a biologist at the Bonneville Lock and Dam, (pers. comm.) documented the occurrence of colonies of this species passing through the Bonneville Dam facility on the Columbia River (Multnomah County, Oregon, and Skamania Co., Washington) in the fall seasons for several years prior to our survey. She also documented this species in Drano Lake, an impoundment created by fill for a railroad bed along the Washington side of the Columbia River several km upstream of Bonneville Dam.

Plumatella casmiana. We encountered colonies only in Mayfield Lake in Washington. They were abundant on logs in 1-2 m deep water underneath a bridge at the upper end of the lake.

Plumatella emarginata colonies occurred in four localities in our survey; we found statoblasts in two additional localities. Colonies were common in all four sites growing on logs and/or rocks. At two of

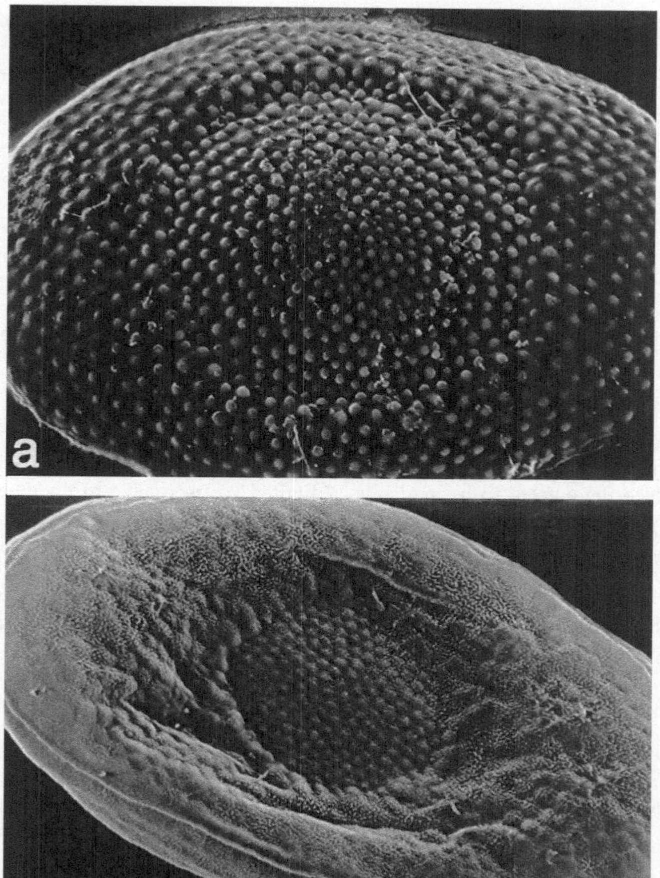

Figure 4. Selected statoblasts encountered in this study. a) Floatoblast of *Plumatella fungosa* from Greaser Lake showing unusually strong tuberculation. b) Floatoblast of *Plumatella mukaii* from Crane Prairie Reservoir showing the wrinkled surface membrane characteristic of this species. c) Sessoblast of *Stephanella hina* from Eel Lake showing sharp, conical projections characteristic of this species. Scale bar = 0.1 mm.

the sites the current was judged to be "moderate" or "swift," vs. the other two sites where the water was still.

Plumatella fungosa. We sieved floatoblasts of this species from Goose Lake, a desert locality in southern Oregon (Fig. 4a). Access to the lake was gained in Goose Lake Recreation Area at Pine Creek, Oregon.

Plumatella fruticosa colonies occurred in Elk Lake in the central Oregon Cascade Mountains. At this lake we found the colonies growing on logs with *P. rugosa* in a protected bay at the southern end of the lake.

Plumatella mukaii. We collected colonies in two lotic, mountain localities, both in western Oregon. At these localities we judged the current to be moderate and swift, respectively. We sieved statoblasts in two lentic, mountain localities, both in Washington (Fig. 4b).

Plumatella nitens. We collected colonies in four localities, all lentic. We collected statoblasts at three more lentic localities. The first three localities where we collected colonies are mountain sites while the fourth, Sprague Lake, is a desert lake which had been badly fouled by fishermen leaving remains of their catches in the water. Silver Lake, where we collected statoblasts, is also a desert locality.

Plumatella nodulosa. We isolated statoblasts of this species from a sieve sample of lakeshore debris that we collected at Greaser Lake, a desert lake in south central Oregon.

Plumatella reticulata. We collected colonies at two lentic localities. At Moses Lake, a desert lake, the colonies were growing on rocks, while at Long Lake, a mountain lake, the colonies were growing on logs.

Plumatella rugosa. Colonies occurred in four mountain localities in our survey; we found statoblasts at an additional four localities – two mountain and two desert localities. This species was abundant only at Eel Lake, where the colonies were small (1-2 cm in diameter). There was adequate substrate to support colony growth at all four of the colony sites except Elk Lake.

Plumatella similirepens. We found colonies or statoblasts in two mountain and one desert lentic localities.

Plumatella vaihiriae. We recovered statoblasts of this species in two mountain lentic localities.

Stephanella hina. Casual sampling by one of us (Marsh) in 1998 had revealed the presence of *S. hina* in Silver Lake in western Washington. Heretofore the species had been known only from New England (Smith 1988) and Virginia (Wood unpubl.) since its original description from Japan (Oka 1908). On every subsequent visit to Silver Lake we have found statoblasts of this species but have yet to find the translucent colonies. The sites where we collected this species were all mountain lentic sites with little or no current. Despite intensive searching, we encountered colonies of *Stephanella* only at Marshall Lake in northeastern Washington (Fig. 4c).

4 DISCUSSION

While our collections in the Pacific Northwest were limited, we were able to document a number of species in this under-studied region. Even the large and unique *Pectinatella magnifica* had not previously been noted in the literature, although colleagues familiar with the region had reported its occurrence to us. Previously published records of this species in North America note only that it has been found as far west as Texas.

The number of bryozoan species appearing in this survey (20) is comparable with the diversity reported in other regional or state surveys. Bushnell (1965a, b, c) reported 13 species in Michigan, compared to 14 in Massachusetts (Smith 1989), 14 in eastern Canada (Ricciardi & Reiswig 1994), and 13 in Ohio (Wood 1989). The high diversity we found in Washington and Oregon reflects both the relatively large area of the region and recent advances in the taxonomy of this bryozoan group. More than one third of our species have been described only since 1988. Other surveys have not included sieving for statoblasts. In this survey four species were found only as statoblasts.

Freshwater bryozoans are likely disseminated over long distances on the feathers and in the guts of migratory waterfowl (Bushnell 1973). Thus it is reasonable that all but two of the species we found have been reported elsewhere in North America. *Fredericella sultana*, however, has not previously been recorded in North America but is common in Europe. *Plumatella mukaii* is know in this hemisphere from Chile (Orellana-Liebbe 1999) but otherwise has been reported only from a variety of localities in eastern Asia and India (Wood 2001c). We predict that future surveys will reveal additional North American localities where these species occur. Waterfowl that migrate to Arctic nesting grounds could bring statoblasts to local water bodies where colonies could germinate and grow. Statoblasts produced by the colonies in these arctic bodies of water could be subsequently dispersed into other parts of the continent by birds using various migratory routes.

This same dispersal mechanism could account for the distribution of *P. nodulosa*. Greaser Lake in Oregon, the only site where this species was found, is frequented by migratory waterfowl. In mid-March, for example, we noted Canada geese and other unidentified waterfowl on the lake and killdeers along the shoreline. Thus we are not surprised that this species also occurs at several other sites in northeastern North America (Wood 2001b).

Nearly 60% of the bryozoan species in this study occurred only in mountain localities in western Oregon, western Washington, and northeastern Washington, in what Hunt (1974) refers to as the Pacific Border, Pacific Mountain System, and Rocky Mountains physiographic regions (Fig. 1). The remaining species occurred in the desert regions of southeastern Washington and the central and eastern portions of Oregon: Hunt's Columbia-Snake River Plateau and the Basin and Range physiographic regions.

The species encountered the most frequently, *Stephanella hina*, occurred exclusively in the mountain regions. We found only statoblasts in all localities except Marshall Lake, Washington, where we unknowingly collected colony material. Colonies of this translucent species are very difficult to see on the typical dark substrate we examine in the field. This species has not been found as frequently in any other survey in North America. From Smith's (1988) first record of *Stephanella* in the Swift River in Massachusetts and our findings, it is tempting to describe this species as restricted to holarctic regions, as is the case for *C. mucedo* (Wood 2001a). However Smith (pers. comm.) has subsequently found this species in a second locality nearer the coast of Massachusetts and Wood has found it in a reservoir in southwestern Virginia. Both findings suggest that this species is not necessarily restricted to a more northern, cooler climate.

The three sites from which we recovered statoblasts of *L. carteri* are all in the mountain region. This species is known from scattered sites in a number of states in northeastern North America (Wood 2001a, Ricciardi & Reiswig 1994) and Panama (Marsh & Wood, unpubl.). Wood (2001a) has noted that this species has been found in a growing number of localities in North America following its introduction in the 1930s.

Lentic sites in the desert region are home for only several species of *Plumatella* and *P. articulata* based on our collections (Table 2). We found three of the nine taxa, *P. fungosa*, *P. nodulosa*, and an undescribed species of *Plumatella*, only in desert localities. Of these three taxa, *P. nodulosa* has been found elsewhere in North America (Wood, 2001b). There is a highly significant difference between the conductivity and pH in lentic sites in the desert region vs. the mountain region. We attribute this difference to the limited drainage and high rates of evaporation in lentic sites in the desert region. Our findings suggest that *Plumatella* spp. and *P. articulata* are the only species able to cope with the differences in water conductivity and pH that exist among mountain and desert localities.

We collected statoblasts of *P. similirepens* at Omak Lake in Washington, a desert region site, and Franz Lake in Washington, a mountain region site, as well as colonies in John Boyle Reservoir in Oregon, a mountain region site. Previously this species was only known from two collections in Illinois (Wood 2001b).

The undescribed plumatellids are known so far only from their floatoblasts. In one, the floatoblast annulus is paved, as in *P. emarginata*, but the ventral fenestra is entirely smooth. The other species resembles that of *P. repens*, but the annulus bears no nodules. We anticipate new collections very soon, to be followed by published descriptions of both species.

Statoblast records alone are useful for indicating the presence of colonies in the drainage area. We recognize that statoblasts do not necessarily prove the occurrence of colonies in the body of water where they are found.

No single plumatellid species occurred frequently in this survey. This may be due to the relatively few sites surveyed, compared to our three-year survey of Illinois bryozoans (Marsh & Wood, in prep.). *P. reticulata*, occurring at 22% of all Illinois sites, was found at only one mountain and one desert locality in the Pacific Northwest. We were surprised that we did not find the ctenostome, *P. articulata*, in more lotic (1) than lentic (6) sites. In our survey of Illinois (Marsh & Wood, unpubl.) we have found this species in 34 localities, 28 of which are lotic. Another ctenostome bryozoan, *Pottsiella erecta*, has been reported infrequently but consistently in surveys of northeastern North America (Smith 1989, Wood 1989, Ricciardi & Reiswig 1994, Marsh & Wood, unpubl.) but we did not find this species in this survey.

ACKNOWLEDGEMENTS

We thank Jennifer Winkelmann and Sabina Zuk for their assistance in the field in the summer 1998. Kimberly Jones and Katharine Marsh also helped collect field samples. Sallie Jones, Biological Technician, Bonneville Fisheries Office, Bonneville Lock and Dam, Cascade Locks, Oregon, provided information on *P. magnifica* at localities along the Columbia River and access to Franz Lake. Dr Richard Wilders provided help with the statistics. Laura Marsh helped us edit the manuscript. Dr Ann Taylor assisted with the scanning electron microscopy. The study was supported in part by a grant from North Central College. Emmy Wöss reviewed the manuscript and provided a number of improvements it.

REFERENCES

Bushnell, J.H. Jr. 1965a. On the Taxonomy and Distribution of Freshwater Ectoprocta in Michigan. Part I. *Transactions of the American Microscopical Society* 84(2): 231-244.

Bushnell, J.H. Jr. 1965b. On the Taxonomy and Distribution of Freshwater Ectoprocta in Michigan. Part II. *Transactions of the American Microscopical Society* 84(3): 339-358.

Bushnell, J.H. Jr. 1965c. On the Taxonomy and Distribution of Freshwater Ectoprocta in Michigan. III. *Transactions of the American Microscopical Society* 84(4): 529-548.

Bushnell, J.H. Jr. 1973. The Freshwater Ectoprocta: A Zoogeographical Discussion. In G.P. Larwood, (ed.), *Living and Fossil Bryozoa: Recent Advances in Research*: 503-521. London: Academic Press.

Hunt, C.B. 1974. *Natural Regions of the United States and Canada.* San Francisco: Freeman.

Jones, K.E., Marsh, T.G., & Wood, T.S. 2000. Surveying for phylactolaemate bryozoans by sieving lentic sites for their statoblasts. In A. Herrera Cubilla & J.B.C. Jackson (eds.), *Proceedings of the 11th International Bryozoology Association Conference:* 259-264. Panama: Smithsonian Tropical Research Institute.

Oka, A. Über eine neue Gattung von Susswasserbryozoen. *Annotationes Zooloogicae Japonenses.* 6: 277-285.

Orellana-Liebbe, M. C. 1999. Briozoos de agua dulce en Chile central (Bryozoa, Phylactolaemata). *Masters Thesis,* Universidad de Conceptión. 67 pp.

Ricciardi, A. & Reiswig, H.M. 1994. Taxonomy, distribution and ecology of the freshwater bryozoans (Ectoprocta) of eastern Canada. *Canadian Journal of Zoology*, 72: 339-359.

Smith, D.G. 1988. *Stephanella hina* (Ectoprocta: Phylactolaemata) in North America, with Notes on its Morphology and Systematics. *Journal of the North American Benthological Society*, 7(3): 253-259.

Smith, D.G. 1989. Keys to the freshwater macroinvertebrates of Massachusetts. No. 4: Benthic colonial phyla, including the Cnidaria, Entoprocta, and Ectoprocta (colonial hydrozoans, moss animals). Westborough, MA: *Massachusetts Division of Water Pollution Control.*

Wood, T.S. 1989. Ectoproct Bryozoans of Ohio. *Bulletin of the Ohio Academy of Science,* New Series, 8: 1-66.

Wood, T.S. 2001a. Bryozoans. In J. Thorp & A. Covich, (eds), *Ecology and Classification of North American Freshwater Invertebrates*: 505-525. 2nd Ed. New York: Academic Press.

Wood, T.S. 2001b. Three new species of plumatellid bryozoans (Ectoprocta: Phylactolaemata) defined by statoblast nodules. Journal of the North American Benthological Society. 20(1): 133-143.

Wood, T.S. 2001c. Plumatella mukaii, a new phylactolaemate bryozoan from Asia and South America. Hydrobiologia, 445: 51-55.

Bryozoan Studies 2001, Wyse Jackson, Buttler & Spencer Jones (eds)
© 2002 Swets & Zeitlinger, Lisse, ISBN 90 5809 388 3

Highly diverse bryozoan faunas from the Plio-Pleistocene of the Greek island of Rhodes

Pierre Moissette
Université de Lyon I, UFR Sciences de la Terre/UMR CNRS PEPS, 27 Bd du 11 novembre, 69622 Villeurbanne Cedex, France

Nils Spjeldnaes
Universitet i Oslo, Institutt for Geologi, BOKS 1047, Blindern, 0316 Oslo, Norway

Efrosyni Georgiades-Dikeoulia
University of Athens, Department of Historical Geology-Paleontology, Panepistimiopolis, 15784 Athens, Greece

ABSTRACT: With an estimated 300 species, the Plio-Pleistocene bryozoan faunas of Rhodes are very diverse, especially when compared with the about 300 potentially fossilizable species described in the whole present-day Mediterranean. The studied bryozoans are classified into several palaeobiogeographic groups. The largest one consists of species with an eastern Atlantic/Mediterranean distribution. Smaller groups are the Mediterranean endemics, the cosmopolitan species, and the rare Indo-Pacific species. Two other groups are lacking from the present-day Mediterranean: the boreal-Atlantic and the tropical species. Their presence and the high biodiversity could be due to several factors. During colder periods some species entered the Mediterranean from the North Atlantic, but the warmer and stable climate of the relatively isolated Aegean also permitted tropical species to survive. Due to tectonic movements and glacio-eustatic variations, several cycles of regression and transgression occurred. Together with temperature oscillations this could have made more habitats available and thus explain most of the observed uncommon biodiversity.

1 INTRODUCTION

The Hellenic Arc corresponds to the subduction zone of the African plate sliding under the Aegean plate. At the front of the subduction zone is the Hellenic Trench, followed northwards by a sedimentary arc, and then a volcanic arc. The island of Rhodes is situated at the eastern end of the sedimentary arc (Fig. 1).

Upper Pliocene to lower Pleistocene marine sediments largely outcrop in the north and east of Rhodes (Meulenkamp et al. 1972, Løvlie et al. 1989, Hanken et al. 1996). These beds contain rich invertebrate faunas, generally well preserved: foraminifers, corals (both shallow- and deep-water), serpulid worms, molluscs, brachiopods, decapods, ostracods, echinoderms. Coralline algae and fish remains are very often also associated. Among these organisms, the bryozoans are practically always present (and abundant to very abundant) in most samples and sediment types.

With an estimated number of species of about 300, the bryozoan faunas we collected and started identifying in these Plio-Pleistocene deposits show an unusually high diversity.

To understand this high species richness we consider first the biogeographic and climatic affinities of the identified taxa. On this basis, the aim of this paper is then to put forward a number of environmental factors explaining a phenomenon that can apparently, among the benthic invertebrates of Rhodes, be observed only in the bryozoans.

2 BRYOZOAN DIVERSITY

During several years of field work in Rhodes we collected a total of about 340 samples from 23 sections. The corresponding marine deposits comprise mostly conglomerates, biogenic sands and sandstones, clays, and various types of limestones.

The analysis of about 90 of these samples allowed the more or less precise identification of 250 bryozoan species (some still in open nomenclature, amid them probably a number of new species). These taxa are distributed among the ctenostomes (3 species), the cyclostomes (52 species) and the cheilostomes (195 species). Approximately 80% of the identified species are extant. The zoarial forms represented in the analysed sedimentary deposits are also diversified: membraniporiform (148 species), celleporiform (14 species), cellariiform (21 species), catenicelliform (2 species), vinculariiform (39 species), adeoniform (11 species), reteporiform (7 species), lunulitiform (5 species), as well as 3 boring ctenostomate species. Only the conescharelliniform

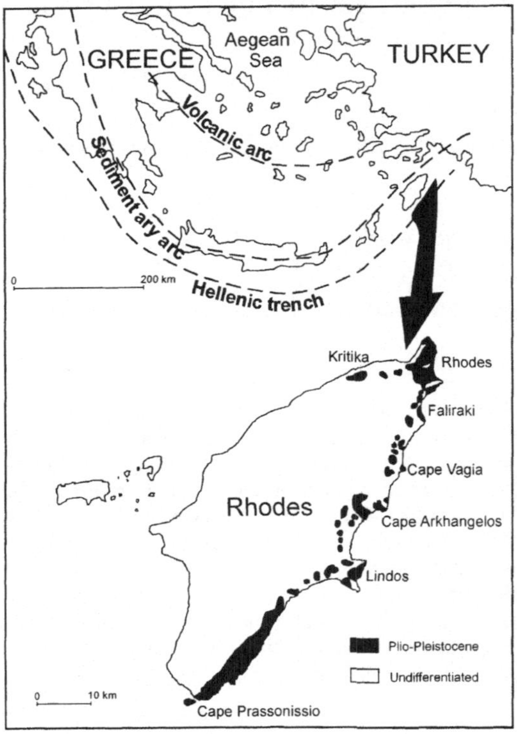

Figure 1. Map of the eastern Aegean with its tectonic features and the main Plio-Pleistocene outcrops on the island of Rhodes. (From Hanken et al. 1996).

morphotype is missing (Moissette & Spjeldnaes 1995).

Although the study of the bryozoan faunas from the Plio-Pleistocene of Rhodes is far from finished (Moissette & Spjeldnaes 1995, Spjeldnaes & Moissette 1997) we estimate that the total number of identified species at the completion of our work will probably reach 300. This diversity is particularly high when compared with the about 300 potentially fossilizable (entirely calcified) species described until now in the whole present-day Mediterranean (Zabala & Maluquer 1988, Harmelin 1992).

Even when taking into account the longer time span involved by our sampling (slightly more than 1 Ma) the diversity of the bryozoan faunas is still remarkable, especially when considering that in the marine Plio-Pleistocene deposits of Rhodes only some habitats and depth ranges (from sea level to a maximum of not more than 600 m) are represented (Moissette & Spjeldnaes 1995).

3 BIOGEOGRAPHIC AND CLIMATIC AFFINITIES OF THE BRYOZOAN SPECIES

On the basis of their palaeobiogeographic and palaeoclimatic distributions, six main groups are distinguished among the bryozoan species already identified in Rhodes.

3.1 *Eastern Atlantic/Mediterranean species*

The species with an eastern Atlantic range are by far preponderant (about 59% of the total). In the Recent, these species are mostly found between Madeira or Morocco and the British Isles and they also occur in the Mediterranean. They live in the cool-temperate to warm-temperate waters of this region. A much smaller number of fossil species (mostly Miocene to Pliocene) also belong to this group.

Among many others, some of the more frequent and characteristic extant species of this abundant group are: *Adeonella polystomella*, *Annectocyma major*, *Cribrilaria innominata*, *Cryptosula pallasiana*, *Electra monostachys*, *Escharina vulgaris*, *Figularia figularis*, *Hornera frondiculata*, *Lichenopora hispida*, *Reptadeonella violacea*, *Schizoporella longirostris*, *Sertella septentrionalis*, *Smittina cervicornis*, and *Turbicellepora coronopus*.

3.2 *Mediterranean species*

The Mediterranean endemics (around 22% of the fauna) are represented by extant or fossil species never or rarely found outside the Straits of Gibraltar. They may correspond to relics of Tethyan faunas, or they may have evolved in situ from immigrants of Atlantic or Indo-Pacific origin.

The following Recent species are typical examples of this group: *Annectocyma arcuata*, *Cribrilaria pedunculata*, *Distansescharella seguenzai*, *Myriapora truncata*, *Onychocella vibraculifera*, *Rhynchozoon neapolitanum*, and *Schizomavella rudis*. The fossil Mediterranean (and often also Paratethys) endemics are only slightly less numerous than the Recent ones; they are represented by species such as: *Anoteropora persimplex*, *Aplousina bobiesi*, *Hippaliosina clavula*, *Hornera subannulata*, *Smittina canavarii*, etc.

3.3 *Cosmopolitan species*

The species with an almost cosmopolitan distribution are relatively abundant (approximately 9%). Outside the eastern Atlantic/Mediterranean, they are also found in the western Atlantic, the Indian ocean and sometimes also in the Pacific. Although largely eurythermal they are however generally absent from polar waters.

The cyclostome species belonging to this group are notably *Idmidronea atlantica*, *Plagioecia patina*, *Plagioecia sarniensis*, and *Tervia irregularis*. Among many other cheilostomes, some of the most typical examples are probably the following: *Aetea sica*, *Arthropoma cecilii*, *Caberea boryi*, *Chorizopora brongniartii*, *Copidozoum tenuirostre*, *Escharina dutertrei*, *Micropora coriacea*, *Microporella ciliata*, and *Parellisina curvirostris*.

3.4 *Indo-Pacific species*

Some rare species (about 2% of the total) are considered as having originated in the Indo-Pacific. Among them *Annectocyma indistincta*, *Margaretta cereoides*, and *Savignyella lafontii* are often cited as relics from a time when the Mediterranean was connected to the Indian ocean (till the end of the Early Miocene).

We do not know much however about the Recent bryozoans of the Indian ocean and still less about the Tertiary faunas of this region. A better knowledge of these faunas is needed to assess how such taxa could have remained in the Mediterranean, particularly during the Messinian salinity crisis (Moissette & Pouyet 1987). As noted by Hayward (1974) the similarity between the Mediterranean and the Indo-Pacific is more likely to be found when taking into account the distribution of genera instead of species. This is the case with mostly Indo-Pacific genera such as *Cosciniopsis*, *Hippaliosina*, *Margaretta* and *Monoporella*, also represented in Rhodes.

3.5 *Tropical species*

True warm-water, tropical taxa are generally lacking from the present-day Mediterranean, excepted perhaps for a small number of species from the eastern Mediterranean, notably the Aegean sea. This group of mostly shallow-water taxa is represented in the Plio-Pleistocene of Rhodes by stenothermal extant species or genera only found in the Recent in the tropical Atlantic and sometimes also in the Indo-Pacific.

Among the circumtropical species belonging to this group (around 3% of the total), the most remarkable example (Cheetham 1967) is *Metrarabdotos moniliferum* occurring in great number of colonies or fragments in many deposits in Rhodes. With this last representative, the tropical genus *Metrarabdotos* completely disappears from the Mediterranean at the beginning of the Pleistocene.

Other warm-water species are less conspicuous but often present: *Alderina peduncolata*, *Biflustra savartii*, *Chaperia annulus*, *Tremopora radicifera*, and *Exechonella antillea*.

3.6 *Boreal-Atlantic species*

The last group, also absent from the modern Mediterranean (or only rarely found in some sampling stations), comprises a small number (approximately 4 %) of cold-water species of boreal-Atlantic origin (Harmelin & d'Hondt 1993). The present-day distribution of these stenothermal taxa generally ranges from the Arctic Ocean south to Nova Scotia (western Atlantic) or the British Isles (eastern Atlantic).

Among the shallow-water bryozoans belonging to this group are notably *Amphiblestrum solidum*, *Callopora lineata* and *Cribrilina cryptoecium*.

Some boreal deep-water species are also represented in Rhodes: *Escharella immersa*, *Hornera lichenoides*, *Euginoma vermiformis*, *Gemellipora eburnea*, and *Tessaradoma boreale*. Together with other Boreal Guests, this group has also been repeatedly reported from various Pleistocene deposits of Southern Italy (Di Geronimo et al. 1997, Rosso & Di Geronimo 1998).

4 CLIMATE VS. EUSTATISM VS. TECTONICS

Among the factors that may explain the observed high bryozoan diversity, the most obvious are the temperature oscillations, the eustatic variations and the tectonic movements recorded in the Plio-Pleistocene marine deposits of Rhodes.

Two oceanographic features of Rhodes may – directly or indirectly – influence the bryozoan diversity. The lack of a continental shelf and the proximity of an oceanic trench create special conditions, which are reflected in the unusually high percentage of planktonic foraminifers even in rather shallow-water sediments in the Plio-Pleistocene of Rhodes.

4.1 *Climate*

During colder periods, temperature stratification and water circulation may have been different at the end of the Pliocene and the beginning of the Pleistocene. This allowed some species to enter the Mediterranean, and even the eastern Mediterranean, from the colder North Atlantic through the Straits of Gibraltar. These so-called Boreal Guests have now disappeared from the Mediterranean but their presence in the studied levels testifies to cooling episodes and easier eastward oceanic circulation. During these lower temperature episodes, massive cold-water influxes from the Atlantic were permitted through a deeper and wider Straits of Gibraltar. Bathyal species were thus introduced into the western and eastern Mediterranean. They later disappeared from the deep-sea benthos when warmer conditions were established. The uplift of the Gibraltar sill (now at about 280 m) cut off the Mediterranean from the deep oceanic circulation and played a role in the disappearance of many cold stenothermal species, thus leading to the impoverishment of the bathyal bryozoan faunas (Harmelin & d'Hondt 1993, Di Geronimo et al. 2001).

However, some Boreal Guests remained in the Mediterranean after the glacial phases, their distribution being limited to the coldest waters. Among the bryozoans found in Rhodes *Anarthropora monodon*, *Buskea dichotoma*, *Hornera lichenoides*, *Palmicellaria elegans*, and *Palmiskenea skenei* are probable representatives of this group of Residual Boreal Guests (Rosso & Di Geronimo 1998, Di Geronimo et al. 2000).

At the same time, the warmer climate of the relatively isolated Aegean region also allowed warm-water species to survive and even thrive in some areas in spite of cold-water influx. The surface temperature in the Aegean is indeed among the highest in the Mediterranean, and as shown by Thiede (1979) the difference between the highest and lowest temperatures is low. These conditions are documented (Thiede 1979) also for the last glaciation and are assumed to have prevailed in the area during the whole Plio-Pleistocene (since the end of the Messinian crisis). The result is that some species live only in the Aegean Sea and in the southern Atlantic or in the warmest waters of the western Mediterranean (Harmelin 1968, 1969, Hayward 1974). The cheilostomes occurring also in the Plio-Pleistocene of Rhodes include *Chleidochasma porcellanum*, *Exechonella antillea*, *Hippomenella mucronelliformis*, *Rectonychocella disjuncta*, *Smittina remotorostrata*, and *Smittina tubulifera*.

Even more important than high temperature in itself is the stability. The difference between highest (July) and lowest (January) surface water temperatures is smaller than in most of the Mediterranean (Thiede 1979). The difference between the present (interglacial) and maximum glacial temperatures is also smaller than in other parts of the Mediterranean. Such stable conditions, resembling on a smaller scale those found in tropical regions such as the Indo-Pacific, are likely to induce high diversity. Some bryozoans are know to survive outside their normal range, if the temperature is stable, with small annual or longer fluctuations. One example of that is *Kinetoskias arborescens*, an arctic species normally living in subzero temperatures, which is found in deep water with warmer, but very stable temperature (up to 6-7 degrees) as an arctic relict (Spjeldnaes 1964).

This combines to indicate that the climatic conditions in Rhodes during the Plio-Pleistocene were optimal for the development of high diversity bryozoan faunas.

4.2 *Eustatism*

The organization of the Plio-Pleistocene marine deposits of the island of Rhodes reflects numerous cycles of regression and transgression due to glacio-eustatic oscillations (Hanken et al. 1996, Kovacs & Spjeldnaes 1999).

The glacio-eustatic cycles are however overshadowed by the tectonically induced isostatic movements linked to the subduction zone (Kovacs & Spjeldnaes 1999). Whereas the isostatic variations have an amplitude of about 600 m or more, the glacio-eustatic curves have an amplitude of less than 100 m (Fig. 2).

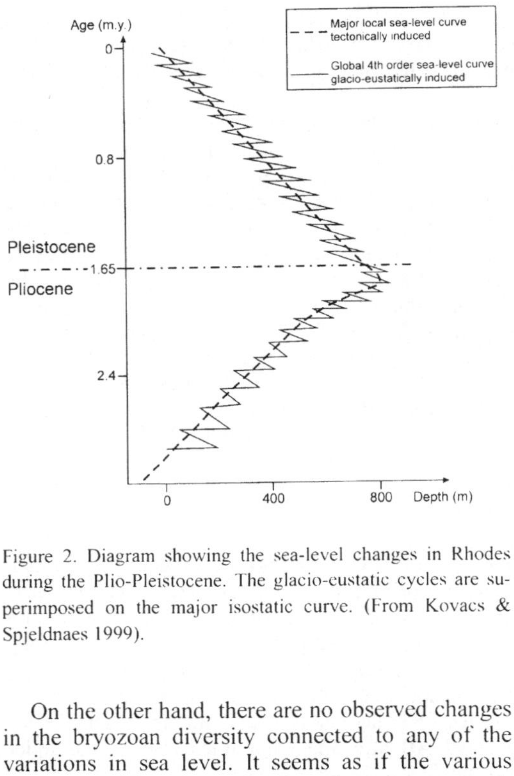

Figure 2. Diagram showing the sea-level changes in Rhodes during the Plio-Pleistocene. The glacio-eustatic cycles are superimposed on the major isostatic curve. (From Kovacs & Spjeldnaes 1999).

On the other hand, there are no observed changes in the bryozoan diversity connected to any of the variations in sea level. It seems as if the various depth assemblages moved only a short distance horizontally (some hundred metres or a few kilometres at most). Even if there were several small, isolated basins, the distance between them was small, and this, and the prevailing long-shore currents secured a rapid transfer of larvae in the region. Even if the eustatic and isostatic movements were rapid, and gave dramatic stratigraphic and sedimentological effects, they probably did not influence the diversity of the bryozoans (and other benthic faunal and floral elements) except very locally.

4.3 *Tectonics*

The Mesozoic basement of Rhodes was tilted eastward and faulted before the late Pliocene, thus resulting in a number of horsts and grabens bounded by normal faults (Meulenkamp et al. 1972, Pirazzoli et al. 1989, Hanken et al. 1996). This tectonically generated palaeogeography later controlled the formation and evolution of small, relatively independent, sedimentary basins on the east coast. The main local event is a major transgression reaching its maximum (with depths of 400 to 600 m) in the late Pliocene, followed by a regression which still continues (Moissette & Spjeldnaes 1995, Kovacs & Spjeldnaes 1999).

Intense Holocene tectonic movements (uplift and subsidence) are recorded on the coasts of Rhodes where erosional notches corresponding to various sea-level stands are carved into the Jurassic rocks (Pirazzoli et al. 1989, Dermitzakis & Georgiades-Dikeoulia 1992). Other evidence of former sea-level positions are given by organic constructions consisting of coralline algae, vermetid gastropods, serpulid worms, and encrusting foraminifers, but also by submerged coastal archaeological remains (Pirazzoli et al. 1989) and raised beachrocks (Dermitzakis & Theodoropoulos 1975).

Late Pliocene higher sea levels are also indicated by numerous bivalve and sponge borings observable up to several metres on sub-vertical Mesozoic limestone cliffs (Pirazzoli et al. 1989, Bromley & Aasgard 1993). In addition, the different terraces and abrasion platforms seen around the eastern coast of the island point to successive temporary Plio-Pleistocene sea-level standstills (Hanken et al. 1996).

The tectonic position of Rhodes, at the southern edge of the Aegean Arc, with a very steep slope towards a 4000 m deep oceanic trench (Woodside et al. 2000) may also be relevant to the bryozoan diversity. Studies of the recent benthic marine fauna (Vamvakas 1970 and references in that paper) indicate very sparse faunas, contrasting dramatically with those found in the Plio-Pleistocene of Rhodes. The earliest study of the marine benthic faunas in the Aegean (Forbes 1843), in contrast, reported rich faunas and assemblages, most of which are similar to those found in the Plio-Pleistocene of Southern Rhodes. This difference, and the impoverished modern fauna may either be due to special causes (possibly pollution) or to recurrent changes, such as the sapropelitic events seen in many marine sections in the Mediterranean. Because of the position of Rhodes, its southern coast may have been protected against such events due to the increased vertical movements of water-masses and sediments, securing a more stable ecological environment than in both the rest of the Aegean Sea and other parts of the Mediterranean.

5 CONCLUSION

The net results of these climatic, paleoceanographic and tectonic variations are probably to have made more habitats available for bryozoans and may explain at least some of the high diversity. Apart from the environments also occurring in the present-day Mediterranean, several biotopes were well represented around Rhodes during the Plio-Pleistocene (e.g. deep-water corals, microbial calcareous crusts and mounds, rapidly cemented beachrocks, etc.).

The reason for the unusually high bryozoan diversity in the Plio-Pleistocene of Rhodes is certainly complex and remains a partial enigma, requiring much more detailed analysis. Not only is the fauna diverse, it is also dense, as the number of specimens in the majority of samples is very high. In most cases there are also many abundant species. The samples where one or a few species dominate quantitatively are comparatively rare. Biogeographically this is similar to regions with warm and stable ecological conditions. The present analysis indicates that the two most important factors contributing to the high diversity and density is the warm climate, being stable, with small fluctuations, both annually and over longer periods without climatic changes. In addition, the special position of southern Rhodes at the edge of a deep ocean trench may have secured greater ecological stability, in spite of high tectonic activity.

No comparable phenomenon is known to occur in the other invertebrates of Rhodes. Groups such as foraminifers, molluscs, brachiopods, ostracods, and echinoderms are also diverse in the Plio-Pleistocene of Rhodes, but quantitative data both from Rhodes and the Mediterranean in general are not available to conclusively show if this diversity is as exceptional as that seen in bryozoans.

REFERENCES

Bromley, R.G. & Asgaard, U. 1993. Endolithic community replacement on a Pliocene rocky coast. *Ichnos* 2: 93-116.

Cheetham, A.H. 1967. Paleoclimatic significance of the bryozoan *Metrarabdotos*. *Transactions of the Gulf Coast Association of Geological Societies* 17: 400-407.

Dermitzakis, M. & Theodoropoulos, D. 1975. Study of beachrocks in the Aegean sea. Observations on occurrences in SE Crete, Rhodes, and Metopi. *Annales géologiques des Pays Helléniques* 26: 275-305 (in Greek).

Dermitzakis, M. & Georgiades-Dikeoulia, E. 1992. Contribution à l'étude de la géologie côtière de l'île de Rhodes. *Annales géologiques des Pays Helléniques* 35: 103-123.

Di Geronimo, I., D'Atri, A., La Perna, R., Rosso, A., Sanfilippo, R. & Violanti, D. 1997. The Pleistocene bathyal section of Archi (Southern Italy). *Bollettino della Società Paleontogica Italiana* 36 (1-2): 189-212.

Di Geronimo, I., Di Geronimo, R., La Perna, R., Rosso, A. & Sanfilippo, R. 2000. Cooling evidence from Pleistocene shelf assemblages in SE Sicily. In M.B. Hart (ed.), *Cli-*

mates: Past and Present: 113-120. London: Geological Society, Special Publications, 181.
Di Geronimo, I., Rosso, A, La Perna, R. & Sanfilippo. 2001. Deep-sea (250-1,550 m) benthic thanatocoenoses from the Southern Tyrrhenian Sea. In E.M. Faranda, I. Gugliemo & G. Spezie (eds), *Mediterranean Ecosystems: Structures and Processes*: 277-287. Milano: Springer-Verlag Italia.
Forbes, E. 1843. Report on the Mollusca and Radiata of the Aegean Sea, and their distribution, considered as bearing on Geology. *Annual Report of the British Association for the Advancement of Science* 1843: 130-193.
Hanken, N.-M., Bromley, R.G. & Miller, J. 1996. Plio-Pleistocene sedimentation in coastal grabens, north-east Rhodes, Greece. *Geological Journal* 31: 271-296.
Hansen, K.S. 1999. Development of a prograding carbonate wedge during sea level fall: Lower Pleistocene of Rhodes, Greece. *Sedimentology* 46: 559-576.
Harmelin, J.G. 1968. Bryozoaires récoltés au cours de la campagne du Jean Charcot en Méditerranée orientale (août-septembre 1967). I. Dragages. *Bulletin du Muséum d'Histoire Naturelle de Paris* 40 (6): 1179-1208.
Harmelin, J.G. 1969. Bryozoaires récoltés au cours de la campagne du Jean Charcot en Méditerranée orientale (août-septembre 1967). I. Dragages. *Bulletin du Muséum d'Histoire Naturelle de Paris* 41 (1): 295-311.
Harmelin, J.G. 1992. Facteurs historiques et environnementaux de la biodiversité de la Méditerranée: l'exemple des bryozoaires. *Revue de Paléobiologie* 11 (2): 503-511.
Harmelin, J.G. & d'Hondt, J.L. 1993. Transfers of bryozoan species between the Atlantic Ocean and the Mediterranean Sea via the Strait of Gibraltar. *Oceanologica Acta* 16 (1): 63-72.
Hayward, P.J. 1974. Studies on the cheilostome bryozoan fauna of the Aegean island of Chios. *Journal of Natural History* 8: 369-402.
Kovacs, E. & Spjeldnaes, N. 1999. Pliocene-Pleistocene stratigraphy of Rhodes, Greece. *Newsletter on Stratigraphy* 37 (3): 191-208.
Løvlie, R., Støle, G. & Spjeldnaes, N. 1989. Magnetic polarity stratigraphy of Plio-Pleistocene marine sediments from Rhodos, eastern Mediterranean. *Physics of the Earth and Planetary Interiors* 54: 340-352.
Meulenkamp, E., De Mulder, E.F.J. & Van De Weerd, A. 1972. Sedimentary history and Paleogeography of the Late Cenozoic of the Island of Rhodos. *Zeitschrift der Deutsche Geologische Gesellschaft* 123: 541-553.
Moissette, P. & Pouyet, S. 1987. Bryozoan faunas and the Messinian salinity crisis. *Proceedings of the VIIIth RCMNS Congress, Annales Instituti Geologici Publici Hungarici* 70: 447-453.
Moissette, P. & Spjeldnaes, N. 1995. Plio-Pleistocene deep-water bryozoans from Rhodes, Greece. *Palaeontology* 38: 771-799.
Pirazzoli, P.A., Montaggioni, L.F., Saliege, J.F., Segonzac, G., Thommeret, Y. & Vergnaud-Grazzini, C. 1989. Crustal block movements from Holocene shorelines: Rhodes Island (Greece). *Tectonophysics* 170: 89-114.
Rosso, A. & Di Geronimo, I. 1998. Deep-sea Pleistocene Bryozoa of Southern Italy. *Geobios* 30 (3): 303-317.
Spjeldnaes, N. 1964. Climatically induced faunal migrations: examples from the littoral fauna of the Late Pleistocene of Norway. In A.E.M. Nairn (ed.), *Problems in Palaeoclimatology*: 353-356. Proceedings of the NATO Palaeoclimates Conference, Newcastle upon Tyne, 1963. Interscience Publishers.
Spjeldnaes, N. & Moissette, P. 1997. Celleporid (bryozoan) thickets from the upper Pliocene of the island of Rhodes, Greece. *SEPM Special Publication, Cool-water Carbonates* 56: 263-270.
Thiede, J. 1979. The glacial Mediterranean and Bay of Biscay. In *Évolution des atmosphères planétaires et climatologie de la Terre*: 121-132. Colloque International, Nice, 1978. CNES (Centre National d'Études Spatiales).
Vamvakas, C.N.E. 1970. Peuplements benthiques des substrats meubles du sud de la mer Égée. *Tethys* 2: 89-130.
Woodside, J., Mascle, J., Huguen, C. & Volkonskaia, A. 2000. The Rhodes basin, a post-Miocene tectonic trough. *Marine Geology* 165: 1-12.
Zabala, M. & Maluquer, P. 1988. Illustrated keys for the classification of Mediterranean Bryozoa. *Treballs del Museu de Zoologia de Barcelona* 4: 1-294.

Bryozoan Studies 2001, Wyse Jackson, Buttler & Spencer Jones (eds)
© 2002 Swets & Zeitlinger, Lisse, ISBN 90 5809 388 3

Interactions of bryozoans and microbes in a chemosynthetic hydrothermal vent system: Big Cove Formation (Lower Codroy Group, Lower Carboniferous, Middle Viséan/Arundian), Port au Port Peninsula, Western Newfoundland, Canada

Penny A. Morris
Department of Natural Science, University of Houston Downtown, 1 Main St., Houston, Tx. 77002, USA

Peter H. von Bitter
Department of Palaeobiology, Royal Ontario Museum, Toronto, Canada, M5S, 2C6; Department of Geology University of Toronto, Toronto, Canada, M51 1A1

Paul E. Schenk
Department of Earth Sciences, Dalhousie University, Halifax, Canada, B3H3J1 (Deceased)

Susan J. Wentworth
Lockheed Martin, NASA/Johnson Space Center, C23, Houston, Tx. 77058, USA

ABSTRACT: In the lower Carboniferous (330 Ma), a hydrothermal vent system developed within an extensive fault setting over a fractured Ordovician limestone basement in western Newfoundland. The vent system supported a complex biological community, which ultimately accumulated to form carbonate mounds. The most likely source of energy for the mounds was chemosynthesis. These mounds are unique because they contain the first recorded fossil bryozoan remains associated with a vent community, and possibly the only fossilized record of the interactions of bryozoans with abundant microbial activity, probably promoted by sulfides emitted by the vent system. The microbial deposits and, to a lesser extent, the sulfides, coated the invertebrate population of monospecific trepostomatous bryozoans, brachiopods, and probable vestimentiferan tubeworms. Various extant sessile invertebrates, including bryozoans, secrete antimicrobial substances that prevent excessive microbial growth. Based on evidence from the fossil record, i.e., the general absence of excessive microbes in association with bryozoans, it appears that the same was probably true in the past. High abundances of materials of probable microbial origin associated with the Carboniferous bryozoans in western Newfoundland are enigmatic.

1 INTRODUCTION

During the Middle Viséan (lower Carboniferous, lower Codroy Group) the subaerial climate associated with the Port au Port Peninsula marine sedimentary environment was predominantly arid. Extensive fracturing of the limestone basement from Newfoundland to Nova Scotia formed a series of steep-sided basins.

Biologically, the Port au Port Peninsula basins were dominated by microbial activity. This resulted in the incrustation and restricted diversity of other organisms such as brachiopods, probable vestimentiferan tube worms, and bryozoans. Organisms associated with mounds, particularly the bryozoans, were made primarily of crystalline calcite (Schenk et al. 1994a, b). The primary energy source for the mounds was probably chemosynthesis, driven by a sulfide-dominated hydrothermal vent system; oxygen levels may have fluctuated (von Bitter et al. 1992, Schenk et al. 1994a, b).

The present study focuses on only a small portion of this complex biological community. It reveals a number of significant details about the geological and paleontological history of the region, such as: (a) an unusual occurrence of a bryozoan associated with a vent system (von Bitter et al.1992); (b) the first reported example of an extensive fossilized bryozoan-microbial association; (c) the effects of microbial activity on bryozoan development and colony form; (d) the characteristics of the fossilized microbial deposits; and (e) the effects of microbes on mineral deposition.

2 MATERIALS AND METHODS

Collections were made on the Port au Port Peninsula in Lead Cove, Mistaken Cove, Bellman's Cove, Aguathuna Quarry, Boswarlos, and Big Cove (Fig. 1).

This study is based on material collected and sectioned by the authors, thin sections deposited by previous investigators at the Royal Ontario Museum (ROM) (Schenk et al. 1990, von Bitter et al. 1992),

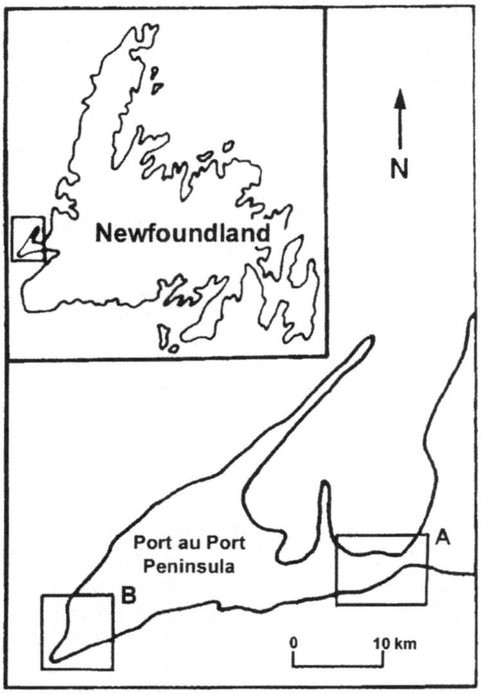

Figure 1. Map of Newfoundland; A and B designate collecting areas; (A) area containing Boswarlos, Lead Cove, Mistaken Cove, Bellman's Cove, and Aguathuna Quarry; (B) area containing Big Cove; figure modified from von Bitter et al. (1992).

and on material kindly loaned by George Dix of Carleton University, Ottawa.

Petrographic thin sections were examined by optical microscopy. Freshly fractured fragments from interiors of specimens were examined with a JEOL 6340F field emission scanning electron microscope (FE-SEM). An IXRF energy dispersive x-ray spectroscopy (EDS) system was used to determine qualitative elemental compositions.

3 GEOLOGICAL AND FOSSIL DATA

3.1 *Bryozoa-Microbe development and colony form*

The bryozoan colonies within the mounds are discrete but difficult to detect with the naked eye. Colonies form the nuclei of spherical microbial deposits. The subsequent layers surrounding the bryozoan nucleus are composed of alternating bryozoan and microbial layers. The spheroids range up to 10 mm or more in diameter. Some bryozoan colonies initially incrust small gastropod shells and subsequently form alternating bryozoan-microbial layers. The balls or spheres contain up to 10 alternating layers (Fig. 2A). Colonies incrusting *Beecheria* brachiopods lack substantial microbial deposits.

Dix and James (1989a, b) described the bryozoan-microbial association as consisting of vertically-oriented branching thickets with each layer representing a new colony that is attached to a previous colony or to another sessile invertebrate mound inhabitant. Our samples indicate that each bryozoan layer represents the regeneration or rejuvenation of existing colonies (Fig. 2B). Growth and regeneration of the colony apparently stopped if the microbial layer exceeded ~1 micrometer in thickness. Further growth of the spheres usually consisted of alternating microbial and inorganic micritic layers (Fig. 2C).

3.2 *Bryozoan identification*

Dix and James (1989a, b) listed two bryozoan genera, *Stenoporella*? and *Diploporaria*, from Port au Port localities. Of the two, *Stenoporella*? is by far the most common within the vent mounds, and it is the only genus which we report herein. The trepostome bryozoan has been assigned different names, *Stenoporella*? (Dix & James, 1987, 1989a, b; Dix & Edwards, 1996; von Bitter et al.1992) and *Anisotrypa* (Morris et al. 1997, 1998). The differences in names may be the result of trepostome systematics, sampling, or the poor preservation of many specimens.

Trepostome identification is generally based on repetitive skeletal characteristics (Madsen, 1991). Within the Stenoporidae, which includes *Stenoporella* and *Anisotrypa*, there is extreme variability within characteristics normally used for species identification (Madsen, 1991).

Our Port au Port bryozoan samples are, in general, more typical of *Anisotrypa* than *Stenoporella*? of Dix and James (1989a, b). For instance, the presence of centrally perforate diaphragms, albeit rare, and the absence of exilazooecia are indicative of *Anisotrypa*. The small spots in zooecial angles vary from those illustrated in Figure 2D, to dark spots that are similar to those described by McKinney (1972) for *Anisotrypa symmetrica*. The presence of scattered hemiphragms is an anomaly as they are absent from *Anisotrypa* but present in *Stenoporella*. The exozone in our Port au Port fossils is absent to poorly developed, and when it is present, the zooecia are shorter and the walls are only marginally thicker than the endozone walls. McKinney (1972, p. 34) described similar morphological developments in his *Anisotrypa* material from the Bangor Limestone, Alabama.

R. Cuffey identified the Port au Port bryozoan material for Dix and James (1987, p. 792). Cuffey based his identification on a limited sample composed of poorly preserved material (pers. comm. Sept. 2001); he also believes that *Anisotrypa* is the best generic assignment at this time. P.N. Wyse

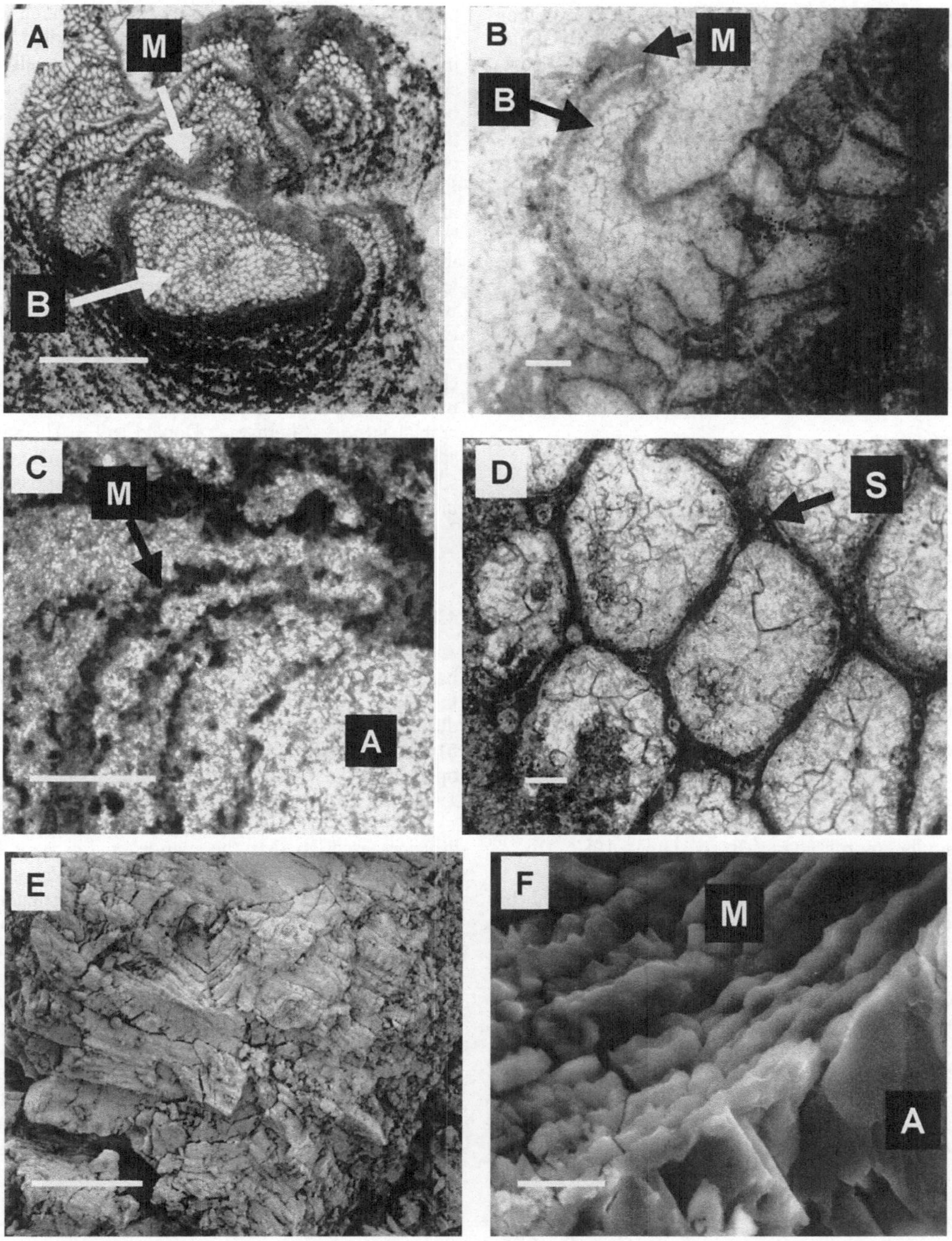

Figure 2A. Cross section of Bryozoa (*Anisotrypa?*) forming small balls or spheres (B); bryozoans form altenating layers with microbial carbonate deposits (M); west wall, Bellman's Cove. Scale=1 mm. ROM 93-1362. Figure 2B. Regeneration or rejuvenation of existing colony as indicated by budding, hook-like zooecia (B); thin microbial layer (M) is present; central Aguathuna Quarry. Scale=.1mm. ROM BeC-18. Figure 2C. Microbial layer (M) exceeds 1 μm; outer layers of bryozoan "sphere" consist of alternating micrite (A) and microbial (M) layers; west wall, Bellman's Cove. Scale=.1mm. ROM BeC-18. Figure 2D. *Anisotrypa*(?) showing zooecial corner dark spots similar to McKinney, 1972, p. 34, pl. 8; central Aguathuna Quarry. Scale=.1mm. ROM 93-1708. Figure 2E. Fine-grained calcium carbonate aggregates like layered logs; west wall, Bellman's Cove. Scale=50 μm. Figure 2F. Microbial layers (M) on micritic deposits (A); microbial surfaces possess rippled texture probably caused by weathering; west wall, Bellman's Cove. Scale=5μm.

Jackson (pers. comm. Aug. 2001) is of the opinion that this species more closely resembles *Stenoporella*. E. Gilmour (written comm. July 2001) believes that a generic identification is impossible to determine because of the poor development and absence of an exozone.

All of the above factors could possibly confirm Madsen's (1991) conclusions regarding the variability within the family, or indicate the existence of another variation within *Anisotrypa*. With the present evidence, the second possibility may be the best, particularly because there are several instances in which members of the Stenoporidae, possessing a mixture of diagnostic generic features, have been assigned to *Anisotrypa* (Perry & Horowitz 1963, McKinney 1972, Karklins 1986).

3.3 *Bryozoan preservation*

Preservation of the mound bryozoans is generally poor and, as a result, some petrographic thin sections of the colonies show only "ghost-like" remnants. Other possible signs of poor preservation include exozone walls with variable thickness, and perforated diaphragms that are collapsed. The poor preservation may, in part, account for the difficulty in the generic assignment.

3.4 *Microbes-metallic deposits*

Petrographic thin-sections indicate that the microbial layers are dense, clotted to laminar or linear, and that they appear to accumulate irregularly on the bryozoan colony (Figs 2A-C). Previous authors described a similar morphology (Dix & James 1987, Dix & Edwards, 1996, von Bitter et al.1992).

FE-SEM and EDS indicate that the microbial (biotic) materials are dominated by fine-grained calcium carbonate aggregates (Figs 2E-F). The microbial layers either radiate from, or lie parallel to, the bryozoan surface. The biotic layer has a rippled texture in association with perpendicular fractures that follow cleavage planes. The rippled texture may account, in part, for the clotted optical images of the microbially mediated carbonates (Fig. 2C).

The micrite is composed of larger rhombohedral shaped crystals that do not exhibit rippled textures (Fig. 2F). The differences between the abiotic and biotic calcium carbonate (rippled vs. non-rippled) suggest that microbial surfaces weather more easily than the coarser micritic or abiotic surfaces (Fig. 2F).

Pyrite is abundant in our mound samples. It is found in large pockets, thin layers, and as small, scattered grains. The pockets (1-2 m) consist of aggregates of cm-sized pyrite crystals. It is probable that the pyrite pockets were the direct result of hydrothermal vent activity because pyrite, along with other sulfides, is commonly found at both modern and fossil hydrothermal vents (Lavoie 1997). The thin pyrite layers (or possibly veins), which can vary from a few millimeters to several hundred millimeters wide, contain crystals that vary equally in size, and appear to follow bedding planes in the carbonate. It seems likely that hydrothermal fluids from the vent system deposited the layers and the pyrite pockets. Small, scattered grains, ranging in size from 1 μm to 2 μm are intermixed with microbially-mediated calcite and the mound-inhabiting invertebrates (including the bryozoans). The origin of the third type of pyrite (the small, scattered grains) is less clear. Further study is needed to determine whether this pyrite is closely related to the other two types, or whether it might be biogenic.

Other sulfide minerals found in the mounds include galena (PbS) and sphalerite (ZnS). Sphalerite, but not galena, is known to be associated with microbial activity (Southam & Donald 1999). The small sphalerite grains (1 μm or less) associated with fossilized microbial layers in our samples are possibly biotic in origin.

4 DISCUSSION

4.1 *Basin environment*

A complex fault system formed steep walled, marine basins that influenced the deposition of sediments in the Big Cove Formation (lower Codroy Group) during Viséan/Arundian times (Schenk et al. 1994a, b). Marine springs and hydrothermal vents developed in association with the faulting (Schenk et al. 1994a, b).

Salinity in the basins was variable, but it gradually increased, so that the marine environment became increasingly restricted with time (Schenk et al. 1994a, b). The steep walled basins, and particularly the sulfide-producing vent system, formed an environment that was dominated by a restricted number of taxa that occurred in great abundance. The organisms thrived in a relatively deep-sea environment with a food-web base ultimately controlled by vent-produced sulfides and sulfur reducing microbes (von Bitter et al. 1992). The microbes and their organic products were the primary food source for filter-feeding invertebrates such as probable vestimentiferan tube worms, brachiopods, and bryozoans. The organisms formed a complex of intimately intertwined prokaryotes and eukaryotes.

4.2 *Prokaryote-Eukaryote complex and the vent community*

Presumably, 330 Ma ago, vent-inhabiting invertebrates lived in much the same way as modern ones, i.e., in a narrow zone where vent fluids mix with oxygenated seawater. This type of environment is unstable, and the survival of organisms is controlled by migrating vent systems, which are ultimately controlled by events within the crust (Tunnicliffe 1992). According to Tunnicliffe (1992), prokaryotic microbes within this unusual environment play an important mediating role in converting the sulfides, which are common in hydrothermal vents, to organic components within their cells, using a process called chemosynthesis. The microbes can serve as an important base of the food web for filter feeders, can develop symbiotic relations with other vent inhabitants, and can form various microbial deposits, such as mats (Belkin & Jannasch 1989).

The vent system that formed the lower Carboniferous mounds was probably similar to the vent environment described by Tunnicliffe (1992). Fossilized microbial deposits and the remains of filter-feeding invertebrates dominated the mounds. Bryozoans, although not included in Tunnicliffe's (1992) generalized macrofaunal vent inhabitants, were an important component of the Port au Port ecosystem. Microbes formed thick coatings and/or mats, flakes, arborescent growths, and peloids (von Bitter et al. 1992). The microbial material was deposited on the bryozoans, brachiopods, and probable vestimentiferan tube worms.

The microbes, along with their organic products, most likely were an important energy source for bryozoans within this extreme environment. Microbial mats and their associated biofilms may also have been important for bryozoan survival strategies (Scholtz, 1993). However, the microbes that formed the excessive microbial mats on the Port au Port bryozoan colonies affected bryozoan development. Examples are the absence or poor development of the exozone layer, the absence of any colony branching within a group that normally develops this mode of growth, and the rarity of internal support structures, such as hemisepta and perforated diaphragms.

Living bryozoans secrete antimicrobial agents to restrict extensive microbial growth (Pettit et al. 1983, Walls et al. 1993, 1995). We presume that this also occurred in the geological past, as there are no literature reports on extensive microbial deposits associated with bryozoans. There are at least two possible explanations to account for the excessive microbial deposits on the bryozoans. The first is that fluctuating vent conditions alternately favored bryozoan and microbial growth, and the second is that the bryozoan colonies secretion of antimicrobial products was diminished.

4.3 *Bryozoan colony morphology*

Dix and James (1989a, b) described the mounds of Big Cove Formation as being composed primarily of bryozoan-microbial communities, with the bryozoans forming branching thickets. In this study we have not found branching thickets. All the bryozoans were incrusting; none of them had features indicative of a branching or arborescent morphology. The bryozoans formed layered spheres reaching 10 mm in diameter (Fig. 2A). Bryozoan colonies in the mounds were not clumped and did not form frameworks or baffles to trap sediments.

Dix and James (1987, 1989a, b) and Dix & Edwards (1996) interpreted Big Cove Formation deposition as representing a high energy, subtidal environment with a water depth less than 10 m. This model contradicts the deeper ($\geq$100 m) water, low energy environment proposed by von Bitter et al. (1990, 1992). Dix and James' hypothesized depositional environment would support the ball-like bryozoan-microbe association. A near shore, high energy environment would result in tumbling action, turning over the colonies repeatedly and allowing spherical growth. In a deeper water, low-energy environment, tumbling would have resulted from vent gases. This alternative explanation is supported by evidence presented by von Bitter et al. (1990, 1992).

4.4 *Microbial fossilization and the role of metallic ions*

Microbial cells and filaments attract, precipitate, and nucleate metallic ions by either passive or enzyme-mediated methods (Ferris et al. 1994; Southam & Donald 1999). Many microbes also produce biofilms that are subject to mineralization (Southam & Donald 1999).

Within the Big Cove Formation, hydrothermal mounds, calcium carbonate, sphalerite (ZnS), galena (PbS) and pyrite (FeS_2) are present. All four minerals can be produced by hydrothermal fluids (von Bitter et al. 1990), and with the exception of galena, can be formed by microbial activity (Southam & Donald 1999).

Fossilized microbial layers composed of calcium carbonate from the Big Cove Formation appear optically clotted (von Bitter et al. 1992, Dix & James 1987, Dix & Edwards 1996) (Figure 2C). Higher resolution FE-SEM imaging indicates a fine-grained texture (Figure 2E-F). The clotted structures and fine-grained texture could suggest that the clotted carbonates of the Big Cove Formation are microbially precipitated. For instance, many ancient stromatolites, when viewed optically, possess a clotted texture, and although there is no evidence of fossil bacteria, they are regarded as microbial precipitates (Chafetz & Buczynski 1992). Additionally, studies and laboratory experiments confirm the fine-grained texture of microbially precipitated carbonate minerals (Krumbein et al. 1977, Chafetz 1994, Ferris et al. 1994). We believe that both of these lines of evidence indicate a strong possibility for the microbially precipitated carbonates.

The fine-grained carbonates appear to weather more easily than the coarser, micritic layers that are present in the bryozoan microbial spheres. A reason may be that there is a larger surface area in relation to volume for fine-grained carbonates, and therefore more area is exposed to weathering.

Pyrite mineralization is extensive in the Port au Port mounds, and large pockets of pyrite are apparently the result of hydrothermal venting. Smaller pyrite occurrences are found in or on sedimentary layers and invertebrates associated with microbial deposits. "Black smoker" hydrothermal vents commonly produce sulfides that are not directly attributed to bacterial activity. It is possible, however, that the smaller-sized grains are related to microbial activity or fossilization as they are intimately intermixed with known biogenic materials. The latter may be a possible source as pyrite can be formed during early diagenesis and microbial decomposition of organic matter (Fortin et al. 1997, Liebig 1996, 1998).

5 CONCLUSIONS

We conclude that:

- The bryozoan spheres common in the lowest Carboniferous carbonate mounds in western Newfoundland belong to the family Stenoporidae, and have traits similar to *Anisotrypa.* Some of the taxonomic problems may be attributable to poor preservation or incomplete understanding of diagnostic traits.
- Environmental factors, including microbial deposits and a sulfide rich environment may have affected morphological development of the bryozoans i.e., the general absence of exozones, and rarity of exozonal structures.
- The vents and their sulfide rich fluids were closely associated with the carbonate mounds and were probably responsible for most of the sulfide minerals that are present today.
- Fine-grained pyrite in the Big Cove Formation samples should be investigated further for possible biogenicity because the pyrite occurs within biologically-mediated carbonates, and calcium and iron are commonly associated with microbial fossilization.
- The presence of extensive microbial deposits on the Big Cove bryozoan colonies is puzzling because modern bryozoans secrete antimicrobial substances, and it is presumed that ancient bryozoans also did so.

ACKNOWLEDGEMENTS

We wish to acknowledge support from the following: Royal Ontario Museum Foundation (Peter H. von Bitter); NSERC (Paul E. Schenk), NASA Grant NAG9-1113, NAG9-1114 (Penny A. Morris). The senior author, Penny A. Morris, wishes to take this opportunity to thank John D. Soule for his support and encouragement to study bryozoans. The relationship began while Penny was an undergraduate student. John died on June 30, 2001, from Parkinson's disease.

REFERENCES

Belkin, S. & Jannasch, W. 1989. Microbial mats at deep-sea hydrothermal vents: new observations. In Y. Cohen & E. Rosenberg (eds.), *Microbial mats: physiological ecology of benthic microbial communities,* Washington, D. C: 16-21.

Belt, E.S. 1968. Carboniferous continental sedimentation, Atlantic Provinces, Canada. *Geological Society of America, Special Paper* 106: 128-176.

Chafetz. H.S. 1994. Bacterially induced precipitates of calcium carbonate and lithification of microbial mats. In: W. E. Krumbein & D. M. Patterson (eds.), *Biostabilization of Sediments, Bibliotheks und Informationssystem der Universität Oldenburg,* Oldenburg (Germany) p. 149-163.

Chafetz, H.S. & Buczynski, C. 1992. Bacterially induced lithification of microbial mats. *Palaios* 7: 277-293.

Chafetz, H.S. & Folk, R.L. 1984. Travertines: depositional morphology and the bacterially constructed constituents. *Journal Sedimentary Petrology* 54: 289-316.

Dix, G.R. & Edwards, C. 1996. Carbonate hosted shallow-submarine and burial hydrothermal mineralization in the Upper Mississippian Big Cove Formation, Port au Port Peninsula, western Newfoundland. *Economic Geology* 91: 180-203.

Dix, G.R. & James, N.P. 1987. Late Mississippian bryozoan/microbial build-ups on a drowned karst terrain: Port au Port Peninsula, western Newfoundland. *Sedimentology* 34: 779-793.

Dix, G.R. & James, N.P. 1989a. Stratigraphy and depositional environments of the Upper Mississippian Codroy Group: Port au Port Peninsula, western Newfoundland. *Canadian Journal Earth Sciences* 26: 1089-1100.

Dix, G.R. & James, N.P. 1989b. Upper Mississippian bryozoan/microbial bioherms, western Newfoundland. *Canadian Society of Petroleum Geologists Memoir* 13: 667-671.

Ferris, F.G., Wiese, R.G. & Fyfe, W.S. 1994. Precipitation of carbonate minerals by microorganisms: implications for silicate weathering and the global carbon dioxide budget. *Geomicrobiology Journal* 12: 1-13.

Fortin, D., Ferris, F.G. & Beveridge, T.J. 1997. Surface-mediated mineral development by bacteria. In: J.F. Banfield & K. H. Nealson (eds.), *Geomicrobiology: Interactions between microbes and minerals. Reviews in Mineralogy* 35: 161-180.

Karklins, O.L. 1986. Chesterian (Late Mississippian) bryozoans from the upper Chainman Shale and the lowermost Ely Limestone of western Utah. *Journal of Paleontology Memoir 17,* 60: 1-48.

Krumbein, W.E., Cohen, Y. & Shilo, M. 1977. Solar Lake (Sinai). 4. Stromatolitic cyanobacterial mats. *Limnology and Oceanography* 22: 635-656.

Lavoie, D., 1997. Hydrothermal vent bacterial community in Ordovician ophicalcite, southern Quebec Appalachians. *Journal of Sedimentary Research* 67: 47-53.

Liebig, K. 1998. Fossil microorganisms from the Eocene Messel Oil Shale of southern Hesse, Germany. *Kraupia: Darmstadter Beiträge Naturgeschichte* 7: 1-98.

Liebig, K., Frey, E., Theobald, G. & Willis, P. 1996. Taphonomy of crocodilians from the Eocene Lake Messel (Darmstadt, Germany) with implications for the sedimen-

tary environment. *Neues Jahrbuch für Geologie und Palaeontologie Abhandlungen* 4: 218-231.
Madsen, L. 1991. Species differentiation in trepostome bryozoans-a discussion based on morphological variability in *Tabulipora*. In: F. P. Bigey (ed.), *Bryozoaires Actuels et Fossiles: Bryozoans Living and Fossil. Bulletin de la Societé des Sciences Naturelles de l'ouest de la France Hors Serie 1 Nantes (France)*, p. 229-263.
McKinney, F.K. 1972. Nonfenestrate Ectoprocta (Bryozoa) of the Bangor Limestone (Chester) of Alabama. *Geological Survey of Alabama Bulletin* 98: 1-144.
Morris, P.A., von Bitter, P.H. & Schenk, P.E. 1997. Effects of microbes on bryozoan colony development. *Abstracts, Geological Society of America, 1997, Annual Meeting, Salt Lake City, Utah*: 194.
Morris, P.A., von Bitter, P.H. & Schenk, P.E. 1998. Microbial Activity and Possible microbial biofilms in marine tufa mounds of the Lower Codroy Group (Lower Carboniferous, Middle Visean/Arundian), Port au Port Peninsula, southwest Newfoundland. *Abstracts, Canadian Paleontological Conference, Corner Brook, Newfoundland, October, 1998*: 17
Perry, T.G. & Horowitz, A.S. 1963. Bryozoans from the Glen Dean Limestone (middle Chester) of southern Indiana and Kentucky. *Indiana Geological Survey Bulletin* 26: 1-51.
Pettit, G.R., Herald, C.L., Kamano, Y., Gust, D., & Aoyagi, R. 1983. The structure of bryostatin 2 from the marine bryozoan *Bugula neritina*. *Journal of Natural Products* 46: 528-531.
Schenk, P.E., Matsumoto, R. & von Bitter, P.H. 1990. Deep-water bacterial precipitated peloidal laminite containing chemosynthetic, microbial mounds, Carboniferous, Atlantic, Canada. *Abstracts, Geological Association of Canada* 15: 177.
Schenk, P.E., von Bitter, P.H. & Matsumoto, R. 1994a. Deep-basin/deep-water carbonate-evaporite deposition of a saline giant: Loch Macumber (Viséan), Atlantic Canada. *Carbonates and Evaporites* 9: 187-210.
Schenk, P.E., Matsumoto, R. & von Bitter, P.H. 1994b. Loch Macumber (early Carboniferous) of Atlantic Canada. *Journal of Paleolimnology* 11: 151-172.
Scholz, J. 1993. Indications for microbial clues for bryozoans when settling. *Facies* 29: 107-118.
Southam, G. & Donald, R. 1999. A structural comparison of bacterial microfossils vs. 'nanobacteria' and nanofossils. *Earth Science Reviews* 48: 251-264.
Tunnicliffe, V. 1992. The nature and origin of the modern hydrothermal vent fauna. *Palaios* 7: 338-350.
von Bitter, P.H., Scott, S.D. & Schenk, P.E. 1990. Early Carboniferous low-temperature hydrothermal vent communities from Newfoundland. *Nature* 344: 145-147.
von Bitter, P.H., Scott, S.D. & Schenk, P.E. 1992. Chemosynthesis: an alternate hypothesis for Carboniferous biotas in bryozoan/microbial mounds. Newfoundland, Canada. *Palaios* 7: 466-484.
Walls, J.T., Blackman, A. J. & Ritz, D. A. 1993. Fouling, surface bacteria and antibacterial agents of four bryozoan species found in Tasmania, Australia. *Journal of Experimental Marine Biology, Ecology* 169: 1-13.
Walls, J T., Blackman, A J. & Ritz, D.A. 1995. Localization of the amathimide alkaloids in surface bacteria of *Amathia wilsoni* Kirkpatrick, 1888 (Bryozoan: Ctenostomata). *Hydrobiologia* 297: 163-172

Bryozoan Studies 2001, Wyse Jackson, Buttler & Spencer Jones (eds)
© 2002 Swets & Zeitlinger, Lisse, ISBN 90 5809 388 3

Bryozoa from oceanic south eastern Pacific islands: diversity and zoogeography

Hugo I. Moyano G.
Departamento de Zoología, Universidad de Concepción, Casilla 160-C, Concepción, Chile

ABSTRACT: This is a twofold comparative report on the already known bryozoan faunas from Galápagos, Juan Fernández and Easter Island and on the hitherto unpublished new bryozoan data from Sala y Gómez and San Félix islands. Data on the bryozoan diversity of the bryozoan fauna of islands and archipelagos over the Nazca plate in terms of genera and species number are analyzed comparatively showing the proportionally large number of cyclostomate bryozoans. Twenty and 23 species from Sala y Gómez and San Félix respectively are reported for the first time. The presence of the genera *Gigantopora* and *Rhynchozoon* in Sala y Gómez and San Félix extends the Indo-pacific Bryozoan fauna component within *c.* 500 km in front of the south American coast. Four species are described as new in the genera *Micropora, Opaeophora, Hippoporina* and *Gigantopora.*

1 INTRODUCTION

As part of extensive and linear submarine ridges within the Nazca plate there are several relatively small, very isolated islands like Easter (Pascua) and Sala y Gómez, and archipelagos like Galápagos, Juan Fernández (Róbinson Crusoe, Alejandro Selkirk and Santa Clara) and Las Desventuradas (San Félix and San Ambrosio). All of them are situated out direct influence of the cold Humboldt current, being principally affected by tropical and subtropical southeastern Pacific waters.

This oceanographic and geographic situation suggests that they should have a different bryozoological fauna in comparison with the Chilean-Peruvian one between 42°S and *c.* 6°S along the western Pacific coast of South America. Published reports on the bryozoans of these remote islands clearly indicate that Easter Island (20 families, 27 genera, 37 species: Moyano 1995) and Juan Fernández archipelago (22 families, 24 genera, 43 species: Marcus 1921; Moyano 1983) constitute two tropical and subtropical bryozoogeographical provinces, respectively, with little or no zoogeographical connections with the Chilean-Peruvian province (Briggs 1974).

The extreme isolation and lack of fresh water in Sala y Gómez and in Las Desventuradas has precluded human colonization and therefore exploration and study. The islands are, however, important stepping stones in the dispersal of benthic species across the vastness of the Pacific Ocean (Parin et al. 1997). They represent the visible and emergent western part of a large submarine guyot chain encompassing the submersed Sala y Gómez and Nazca Ridges. Very close to the eastern end of the Sala y Gómez Ridge emerge the Desventuradas islands, and, six degrees latitude further south, facing central Chile, lies Juan Fernández archipelago.

The current study of some bryozoan samples collected in October 1999 in intertidal places around Easter island yielded a new record for *Jellyella eburnea* (Hincks) growing on plastic debris. A similar study carried out on samples gathered at 50-100 m depth around the smallest Salas y Gómez island (only 2.5 km^2) has already yielded 20 species. Some of them belong to families and genera not previously recorded in Chilean waters, namely *Gigantopora, Mollia* and *Rhynchozoon.* At least one species, *Microporella orientalis* Harmer, extends its known range from the western to the eastern tropical Pacific.

New samples collected in October 2000 around San Félix island - one of the two Desventuradas - confirm the presence there of the genera *Gigantopora* and *Rhynchozoon* and also the occurrence of *Celleporella muricata,* previously found in Juan Fernández. The aim of this paper is to describe the new taxa and new records, and provide a preliminary overview of the bryozoan fauna of the oceanic islands situated on the Nazca plate.

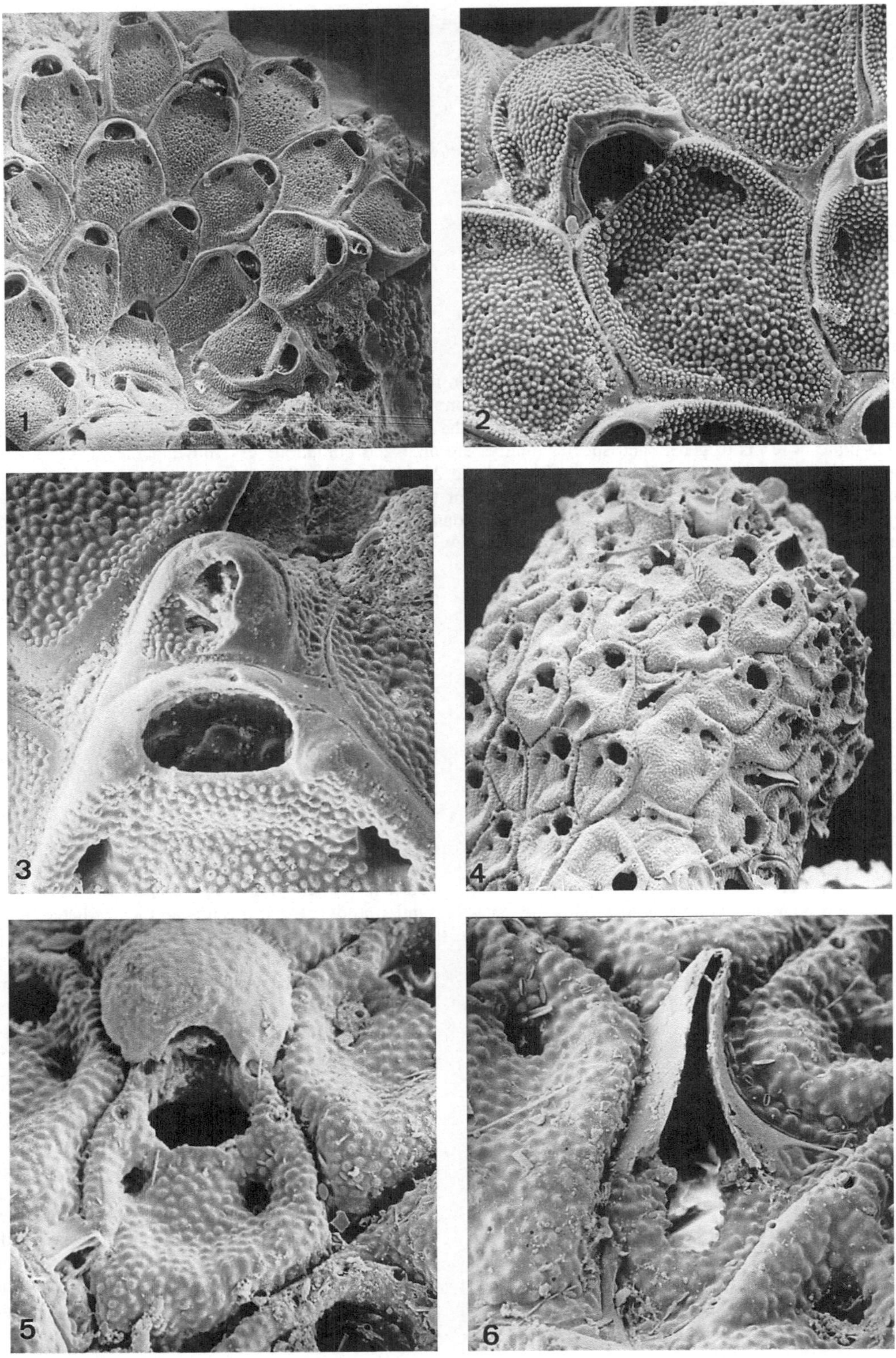
1
2
3
4
5
6

Figure 1, 1-3. *Micropora inexpectata* sp. nov. (San Félix Island). 1. A small colony encrusting a little stone, x30; 2. Fertile zooid showing a kenozooidal ovicell with the same frontal structure as ordinary zooids, x80; 3. Distal view of a zooid exhibiting latero-proximal opesiules, an oral aperture without spines, and a distal interzooidal avicularium provided with a straight articular bar, x160. 4-6. *Opaeophora occulta* sp. nov. (Sala y Gómez Island). 4. View of a colony showing autozooids and avicularia, x30; 5.Fertile zooid with a globular hyperstomial ovicell and the bases of four oral spines, x90; 6. Elongated avicularium with incomplete articular bar and slightly asymmetrical rostrum, x160.

2 MATERIALS AND METHODS

Data on Easter Island bryozoans were obtained from previously published material (Moyano 1973, 1983), including from Juan Fernández (Marcus 1921, Moyano 1983, 1987) and the of Galapagos (Banta & Redden 1990).

Bryozoan samples analysed for this paper were gathered from the following collecting stations of the Chilean Navy R/V *Vidal Gormaz* around San Félix (Desventuradas) and Sala y Gómez Islands.

The bryozoan samples were studied following conventional methods of treatment including washing, rinsing with tap water and fixing in 70% ethanol. Plates were prepared with SEM microphotographs obtained at the Concepción University Electron Microscopy Laboratory.

Table 1. Collecting stations around San Félix island during the CIMAR-6 cruise to the oceanic archipelagos of Juan Fernández and Las Desventuradas.

Site	Latitude	Longitude	Depth (m)	Place	Date	Gear
57	26°25.35	80°00.43	1900	S.Félix	22.10.00	Trawl
51	26°16.34	80°06.00	209	S.Félix	23.10.00	Trawl-Dredge
61	26°17.72	80°08.00	100	S.Félix	24.10.00	Trawl-Dredge
59	26°19.08	80°02.89	151	S.Félix	24.10.00	Trawl
61a	26°17.37	80°08.06	83	S.Félix	24.10.00	Trawl

3 RESULTS

3.1 *Galápagos bryozoans*

The Galapagos islands were investigated by the R.V. *Albatross* of the U.S. Fisheries Commission in 1888 and 1891, collecting 50 species studied by Canu & Bassler (1930). Hastings (1930) added other 37 taxa and Osburn (1950, 1952, 1953) and Osburn & Soule (1953) aggregated several new records issued from the collecting activities of the *Velero III* expeditions of the Allan Hancock Foundation. Finally, Banta & Redden (1990) listed 184 bryozoan species. Table 2 summarices data on the Galápagos bryozoan fauna. According to Banta and Redden (1990) this is an isolated tropical fauna with Panamanian affinities, with a disproportionately large number of cooler water Northern Hemisphere species and also with an "abnormally" high number of cyclostomate species (18%). The overall endemism is only 18%, but is 42% for the Cyclostomatida.

3.2 *Easter Island bryozoans*

Easter Island bryozoans were first surveyed by Moyano 1973. All the 10 recorded species were collected manually by him in the intertidal zone around the island: some from rocky beaches and some from sand grains of Anakena beach. Subsequent collections have elevated the initial number to 37 species (Moyano 1983) (Table 2). This figure is very modest in comparison with that of the Galápagos and very likely does not represent the actual bryozoan diversity of the island which is likely to be higher because of the irregular relief of the seafloor around the island, which is very difficult to survey. The non-littoral samples previously studied came from the undersides and from stomachs of fishes that eat corals or coralline algae but not from soft or hard bottoms. Successful collecting from similar environmental conditions around Sala y Gómez Island however, yielded many unknown families, genera and species not previously recorded at Easter Island. (see Tables 3 and 4), which probably inhabit shallow and deep waters there.

During the 1999 CIMAR-5 expedition to the western Chilean oceanic islands, only a few new samples were successfully taken from Easter Island. These mainly comprised plastic debris hand-collected intertidally, but containing a new generic and specific record for the island and for the eastern Pacific: *Jellyella eburnea* (Hincks).

Table 2. Total number of genera and species of bryozoan orders Cyclostomatida, Ctenostomatida and Cheliostomatida from islands and archipelagos of the Nazca plate.

Orders/ Suborders	GAL N(%)	EAS N(%)	J.F. N(%)	S. y G. N(%)	DES N(%)
Cyclostomatida					
Genera	**13**(14.8)	**8**(29.6)	**9**(26.5)	**4**(28.6)	**5**(23.8)
Species	**33**(21.4)	**12**(32.4)	**13**(30.2)	**4**(20.0)	**5**(21.7)
Ctenostomatida					
Genera	**2**(2.3)	**0**(0)	**1**(2.9)	**0**(0)	**0**(0)
Species	**2**(1.3)	**0**(0)	**1**(2.3)	**0**(0)	**0**(0)
Cheliostomatida (Anasca)					
Genera	**22**(25.0)	**9**(33.3)	**9**(26.5)	**4**(28.6)	**4**(19.0)
Species	**21**(13.6)	**10**(27.0)	**10**(23.3)	**5**(25.0)	**4**(17.4)
Cheliostomatida (Ascophora)					
Genera	**51**(58.0)	**10**(37.0)	**15**(44.1)	**10**(71.4)	**12**(57.1)
Species	**98**(63.6)	**15**(40.5)	**19**(44.2)	**11**(55.0)	**14**(60.9)
Total					
Genera	**88**	**27**	**34**	**14**	**21**
Species	**154**	**37**	**43**	**20**	**23**

GAL = Galápagos; EAS = Easter Island; J.F. = Juan Fernández; S. y G = Sala y Gómez Island; DES = Desventuradas (San Félix and San Ambrosio Islands)

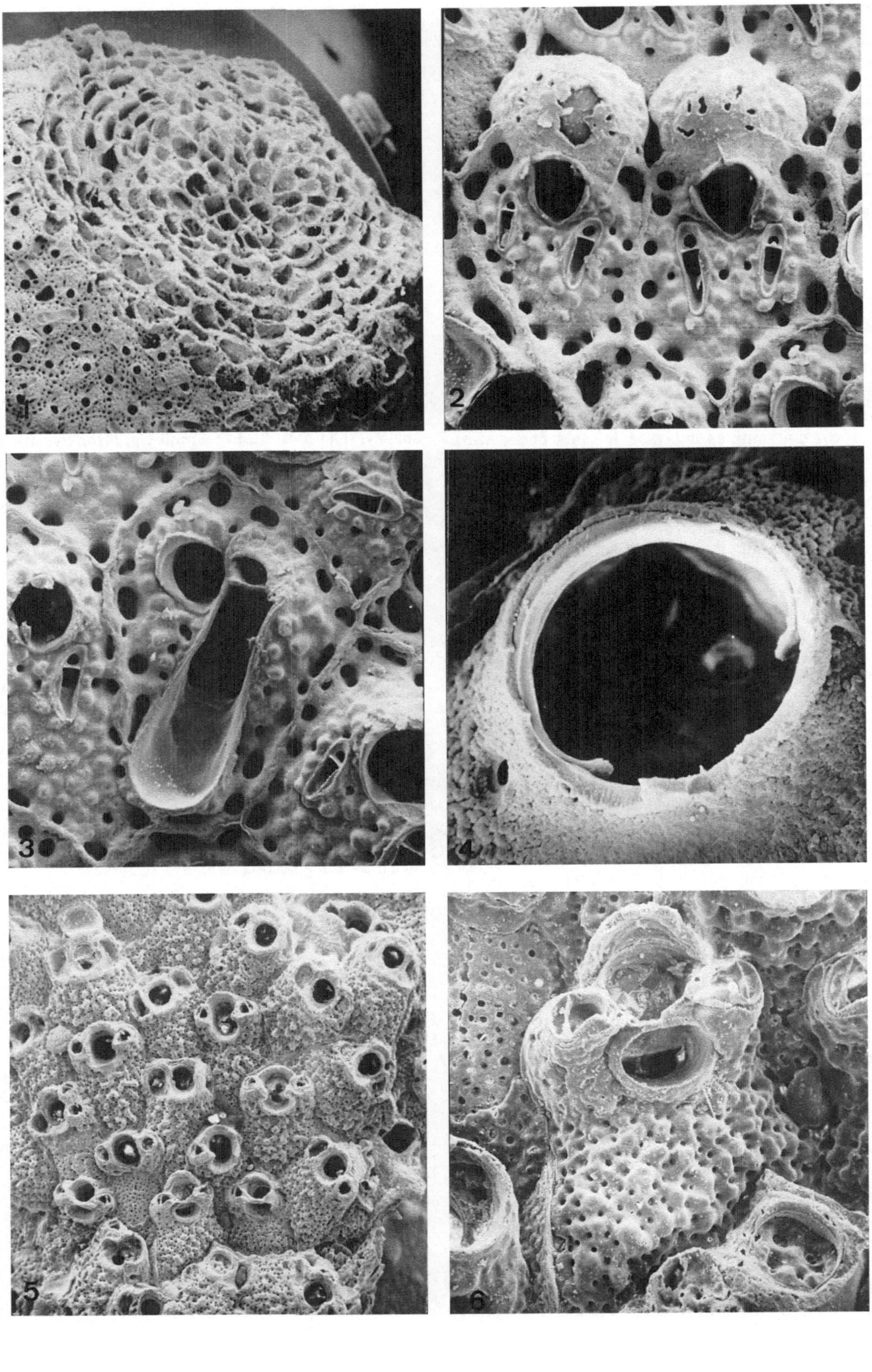
1
2
3
4
5
6

Figure 2, 1-4. *Pleurocodonellina ingens* sp. nov. (Sala y Gómez Island). 1. A broken cylindrical branch showing its multilaminar structure, x10; 2. Ovicellate zooecia having two latero-oral acute avicularia, x70; 3. Infertile zooid with a huge linguiform avicularium, x72; 4. Oral aperture showing the two latero-proximal articulating condyles; x300. 5-6. *Gigantopora verrucosissima* sp. nov. (San Félix Island). 5. General view of colony, x23; 6. A zooid with an incipient ovicell and an extremely verrucose frontal shield, x83.

3.3 *Juan Fernández bryozoans*

The first survey of bryozoans from the Juan Fernández area, obtained by several German collectors was published by Marcus (1921). The 19 species reported were collected during the Swedish Juan Fernández Expedition 1916-1917, commanded by Carl Skottsberg. Although only two new species were described, later studies (Moyano 1983) demonstrated that so-called *Haswellia auriculata* Busk and *Fasciculipora ramosa* d'Orbigny represented new species of the genera *Galeopsis* and *Frondipora* respectively, and that many new records and new taxa remained to be published.

The total number of Juan Fernández Bryozoa (Table 2) is slightly higher than that of Easter Island, and the number of cyclostome genera and species is rather similar to and almost twofold that of Galápagos islands (Banta & Redden 1990). Among the Cyclostomatida, the commonest species appearing in bottom samples are *Frondipora masatierrensis* Moyano and *Annectocyma* sp., which are usually accompanied by two or three species of *Galeopsis*.

3.4 *Sala y Gómez bryozoans*

The genera and species in Table 3 represent the first bryozoan collection ever made at this very small and isolated volcanic island.

According to this list, the genera, *Mollia*, *Rhynchozoon* and *Gigantopora* are new to the southeastern Pacific subtropical islands. *Opaeophora* was previously known from the Juan Fernández archipelago where is represented by *O. browni,* a species very akin to *O. monopia* from New Zealand. *Pleurocodonellina ingens* sp. n., described elsewhere in this paper, is a species with large celleporiform zoaria producing irregularly cylindrical branches 1 cm diameter.

3.5 *Desventuradas bryozoans*

The samples collected around San Félix Island, although partially sorted and studied, have already yielded 23 species shown in Table 4. The most interesting feature of this collection is the presence of the genera *Rhynchozoon* and *Gigantopora* and the species *Celleporella muricata.* *Rhynchozoon* zoaria from this island look different from those of Sala y Gómez and might represent a new taxon. The zoaria of *Gigantopora* correspond to the same species collected in Sala y Gómez. *Celleporella muricata* is shared with Juan Fernández. *Annectocyma sp.* seems to be present in the Juan Fernández archipelago also. It is also possible that *Aplousina* sp., *Crassimarginatella* sp., *Chorizopora* sp., *Puellina* sp. 1 and *Galeopsis* sp. might be the same species found in the Juan Fernández archipelago. *Micropora inexpectata* n. sp. is different from *Micropora mortenseni* previously known from the western Pacific and Juan Fernández.

There are at least two species of the genus *Puellina* differing in the number of oral spines and pericystal costae from species of this genus present in Easter Islands and Sala y Gómez. Island. These and other species encrust large pieces of eroded *Porites*-like scleractinians indicating the tropical nature of the Desventuradas Islands and confirming their zoological links with the more tropical islands of the central and western Pacific.

Table 3. Bryozoans from Sala y Gómez islands collected by the CIMAR-5 expedition in October 1999.

Species	S. y G.	Easter	J.F.
Cheliostomatida (Anasca)			
?Bugula prismatica Gray	-	-	-
Bugula sp.	x	-	-
Crassimarginatella sp.	x	x	x
Mollia sp. nov.	x	-	-
Opaeophora occulta sp. nov.	x	-	-
Cheliostomatida (Ascophora)			
Chorizopora sp.	x	-	-
Crepidacantha anakenensis Moyano	x	x	-
Crepidacantha sp. nov.	x	-	-
Puellina paschalis (Moyano)	x	x	-
Exochella sp.	x	-	-
Fenestrulina sp.	x	-	-
Gigantopora verrucosissima sp. nov.	x	-	-
Pleurocodonellina ingens sp. nov.	x	-	-
Microporella orientalis Harmer	x	-	-
Parasmittina sp.	x	-	-
Rhynchozoon sp. nov.	x	-	-
Cyclostomatida			
Annectocyma sp.	x	-	x
Crisia sp.	x	-	-
?Eryostratos ridleyi Borg	x	x	-
Mesonea radians Lamarck	x	x	-

S. y G = Sala y Gómez; J.F. = Juan Fernández.

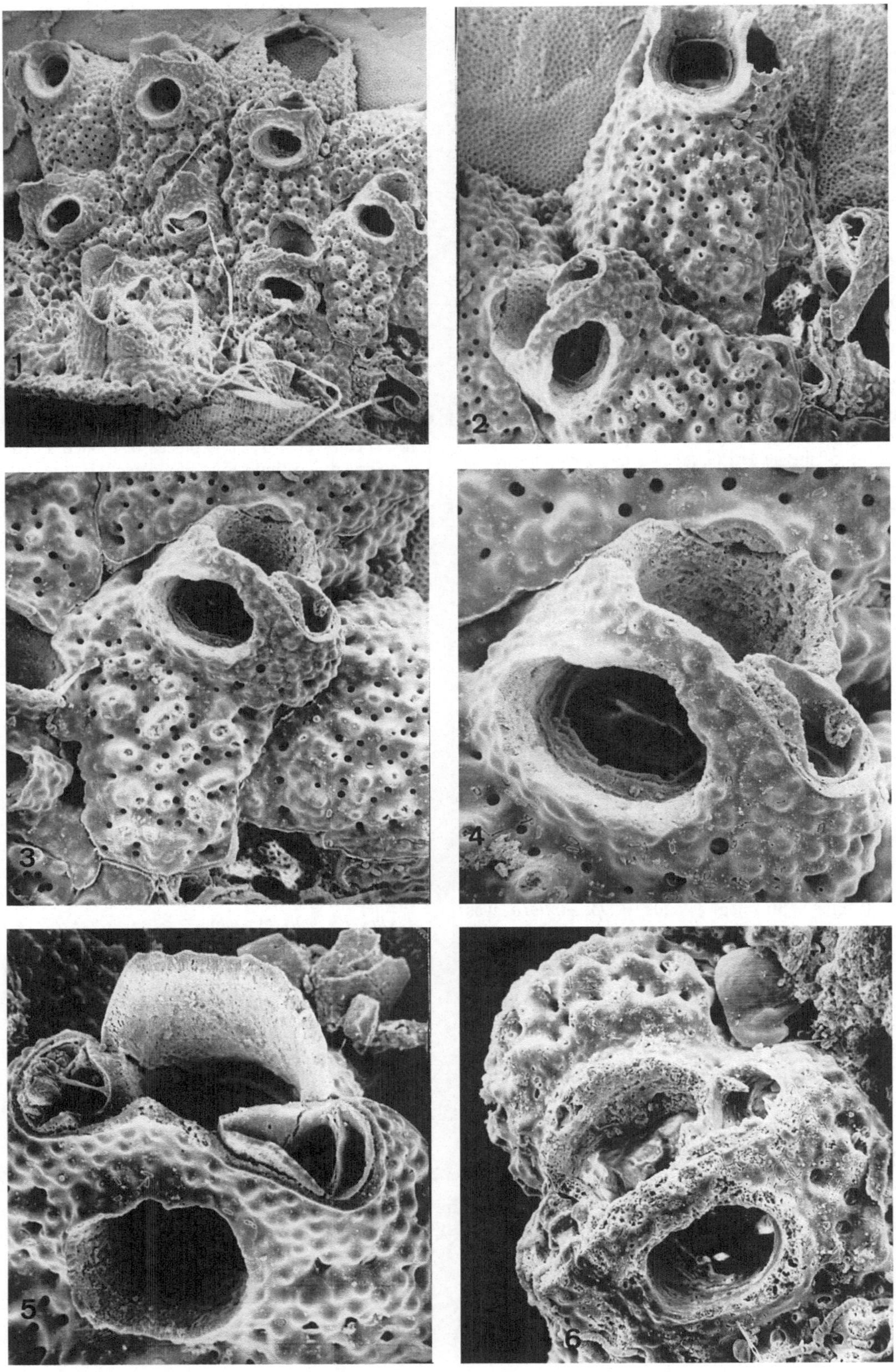
1
2
3
4
5
6

Figure 3, 1-6. *Gigantopora verrucosissima* sp. nov. (Sala y Gómez Island). 1. Frontal view of a juvenile colony, x38; 2. Two zooids at the distal margin of a colony, x68; 3. Extremely verrucose zooid with one lateral avicularium and the complete frontal arcuate bridge, x80; 4. Close-up of the same zooid in 3, x160, 5. Frontal superior view of a zooid with two avicularia, x143; 6. Frontal superior view of an old fertile zooid, x120.

Table 4. A preliminary list of bryozoan species collected at San Félix island (26°17'S, 80°05'W).

Species	J.F.	S.F.	S.yG.	Easter
Cheliostomatida (Anasca) (17.4%)				
Aplousina sp.	x	x	-	-
Caberea sp.	x	x	-	-
Crassimarginatella sp.	x	x	-	-
Micropora inexpectata sp. nov.	x	x	-	-
Cheliostomatida (Ascophora) (60.9%)				
Celleporella muricata Busk	x	x	-	-
Chorizopora sp.	x	x	-	-
Crepidacantha sp. 1	-	x	x	-
Crepidacantha sp. 2	-	x	-	-
Puellina sp. 1	?	x	-	-
Puellina sp. 2	-	x	-	-
Galeopsis sp.	x	x	-	-
Gigantopora verrucosissima sp. nov.	-	x	x	-
Hippothoa sp.	-	x	-	-
?Mamilloporidae	-	x	-	-
Escharina pesanseris	-	x	-	x
Microporella sp.	x	x	-	-
Rhynchozoon sp.	-	x	x	-
Schizoporella sp.	-	x	-	-
Cyclostomatida (21.7%)				
Annectocyma sp.	x	x	-	-
Crisia parvinternodata Moyano	x	x	-	-
Eryostratos sp.	-	x	-	x
Idmidronea sp.	x	x	-	-
Tubulipora sp.	x	x	-	-

J.F. = Juan Fernández; S.F. = San Félix; S.yG = Sala y Gómez.

3.6 *Some new taxa*

Micropora inexpectata sp. nov.
(Figure 1, 1-3)

Diagnosis: Colony encrusting, umultiserial, unilaminar, forming white sheets on hard substrata. Zooids well calcified, nearly flat except at the oral area which is slightly elevated; orifice-opesia roughly semicircular without spines; slightly complex, complete granulose cryptocyst evenly and sparsely perforated by small pores; 2 distolateral infundibular reniform opesiules situated proximally to the poorly developed latero-oral knobs, without additional proximal opesiules. Ovicells roughly semicircular, convex, surface granular like the cryptocyst, sharply delimited from adjacent zooids, with submarginal elongated opesiule-like openings. Interzooidal avicularia scarce, small, roughly semicircular, distal to an autozooid, mandible triangular, diagonally directed, with a short, triangular mandible articulated on a complete straight bar.

Etymology: From Latin *inexpectatus*, meaning unexpected.

Material: Holotype: UCCC 26069. San Félix island: Paratypes: UCCC 26070, San Félix island.

Remarks: The specific name alludes to the unexpected finding of a *Micropora* species similar to Magellanic *Micropora karukinkaensis* Moyano, 1994. Both are the only ones known to have a kenozooidal ovicell and to lack oral spines. Although similar, they differ in that the avicularium has a cross bar and poorly developed suboral knobs in the new species.

Distribution: San Félix, one of the two principal islands of Las Desventuradas archipelago.

Opaeophora occulta sp. nov.
(Figure 1, 4-6)

Diagnosis: Colony encrusting, unilaminar. Zooids white, heavily calcified, separated by deep furrows. Primary frontal wall membranous, overlying a complete granular cryptocyst with a pair of round lateral opesiules. Zooecial aperture subcircular with five spines in non-ovicellate zooids and 4 in ovicellate ones. Ovicell hyperstomial, large, granular, globular. Large interzooecial avicularia with long, pointed, slightly asymmetrical, laterally curved mandible.

Etymology: From the Latin *occultus*, meaning hidden, alluding to the hidden place on a large bryozoan colony of genus *Pleurocodonellina* where it was found.

Material: Holotype UCCC 26071, Sala y Gómez Island; Paratypes UCCC 26072, Sala y Gómez Island.

Remarks: This new species is akin to both *O. monopia* from New Zealand and *O. browni* from Juan Fernández Archipelago (Moyano 1983) but differs from both in the pointed avicularian mandibles instead of the linguiform ones of those species (Gordon 1984). It differs also from *O. lepida* from Australia, New Zealand and Magellanic Chile (Moyano 1983) which has 6-8 opesiules instead of 2 of the new species.

Distribution: Sala y Gomez Island.

Pleurocodonellina ingens sp. nov.
(Figure 2, 1-4)

Diagnosis: Colony large, celleporiform, producing large, irregularly cylindrical branches, pluriserial, plurilaminar, growing parricidally. Zooids longer than wide, irregular in form and position on the zoarial surface; frontal shield heavily granular with large, irregular marginal pores; zooidal aperture with a poorly developed peristome, widely pear-shaped, primary aperture without lyrule or central denticle, and two lateroproximal hinging condyles delimiting a wide concave poster. Avicularia variable in form, size and situation; normally two latero-oral per zooid directed proximally; the non-giant ones elongate, round tipped and somewhat distally curved; sporadically giant epizooecial avicularia occupying a great part of the zooidal frontal shield, with long spatulate mandible. Ovicell hyperstomial, globular, perforated irregularly by irregular pores.

Etymology: From the Latin *ingens* meaning enormous, giant, alluding the large size of both colony and avicularia.

Material: Holotype: UCCC 26073, Sala y Gómez Island; Paratypes: UCCC 26074, Sala y Gómez Island.

Remarks: Species forming a large parricidal colony suggesting similar ones of the genera *Parasmittina* or *Osthimosia*. It also agrees with some smittinids and celleporids in having several different avicularia including giant spatulate ones. It is here provisionally included in *Pleurocodonellina* owing to the lack of a lyrule or a well-developed peristome.

Distribution: Sala y Gómez Island.

Gigantopora verrucosissima sp. nov.
(Figure 2, 5-6; Figure 3, 1-6)

Diagnosis: Colony encrusting, unilaminar. Zooids medium sized, 0.75 mm long x 0.4-0.5 mm width, frontal shield coarse, becoming extremely verrucose with increasing calcification, with many connate or isolated calcareous elevations with a summit pore. Primary orifice longer than wide with a well-marked poster. Latero-oral avicularia triangular, directed almost straight each other, forming a strongly arcuate verrucose bridge delimiting an ovate wide spiramen. Ovicell hyperstomial, granulose and evenly perforated.

Etymology: From the Latin superlative of *verrucosa* meaning highly warty, alluding to the frontal surface structure.

Material: Holotype: UCCC 26075, San Félix island: Paratypes: UCCC 26076, Sala y Gómez Island.

Remarks: The species resembles *Gigantopora proximalis* Gordon, 1984 from the Kermadec islands, but is larger and more warty. It differs from *G. pupa* (Jullien) and *G. polymorpha* (Busk) by its smaller zooids and spiramen.

Distribution: Sala y Gómez and San Félix Islands.

4 DISCUSSION AND CONCLUSION

4.1 *Nazca plate insular bryozoan faunas*

The bryozoan lists from Sala y Gómez and San Félix (Desventuradas) are very preliminary because of the incomplete taxonomic determinations of many species, although four species are described as new and several others will be described elsewhere. The ascription of species to known genera allows, however, a comparison of the bryozoan faunas of the different islands, their species richness, and the relative importance of the different orders. In this sense these islands are characterized by a high number of cyclostomate species and by lacking or having poor representation of the ctenostomes (see Table 2). Banta and Redden (1990) noted similar proportions in an analysis the Galápagos bryozoan fauna, where cyclostomates attain 21%. This figure is even higher in Juan Fernández where it reaches up to 30%. The reason for this is currently unknown.

Although better known, Easter Island bryozoans are probably incompletely surveyed and reported owing to the very irregular volcanic bottom topography which makes it extremely difficult to sample. Nevertheless, successful sampling around Sala y Gómez yielded families, genera and species not yet known from Easter Island but presumably living there.

In the future the bryozoan fauna from Juan Fernández might increase because of new samples obtained from Alexander Selkirk, the most isolated of its three islands. The analysis of these sample may aid in resolving the true zoogeographical links between Juan Fernández and Las Desventuradas.

4.2 *New taxa and endemism*

The extreme isolation of these islands suggests *a priori* the existence of high endemism at the species level. This feature has been demonstrated in Easter Island and Juan Fernandez archipelago where endemism accounts for more than 30% (Moyano 1991), whereas endemism of the Galápagos Bryozoa does not exceed 20% (Banta & Redden 1990). Nevertheless, stenolaemate endemism at the Galápagos is very high (42%) (Banta and Redden 1990). Levels of endemism at the other islands and archipelagos have not yet been completely evaluated owing to the incompleteness of species determinations.

4.3 *Zoogeographical status of Las Desventuradas archipelago*

The Galápagos, Easter Island and Juan Fernandez archipelago have been recognized as tropical warm-temperate zoogeogeographical provinces (Briggs 1974; Moyano 1991). The proximity of Sala y Gómez (*c.* 600 km) to Easter Island has suggested the possibility of integrating it in the same Rapanuian province (Parin *et al.* 1997). Nevertheless, the current bryozoan knowledge of these islands is not yet sufficient to say that they are one or two separated insular provinces. On the other hand, the zoogeographical situation of Las Desventuradas (San Félix and San Ambrosio) is open to discussion. Its situation on a submerged rise also containing the Juan Fernández Islands suggests one zoogeographical province for them. Nonetheless, the presence in Las Desventuradas of the bryozoan genera *Gigantopora* and *Rhynchozoon*, of the irregular sea urchin *Clypeaster isolatus* (Larraín 1995), of scleractinians corals such as *Porites* (in which many of the Bryozoa from San Félix were found; personal observations) and of the fishes *Gymnothorax bathyphylus, Synodus capricornis, Exocoetus volitans, Physiculus luminosa, Trachypoma macracanthus, Decapterus macarelus, Pseudocaranx dentex* and *Entomacrodus chapmani* (Pequeño & Lamilla 1996) make this small archipelago, and the guyot chain linking it with Easter Islands and Sala y Gómez Island, the easternmost outpost of the Indo-West Pacific zoogeographic realm, as previously suggested by Parin (1991), Parin et al. (1997) and Pequeño y Lamilla (1995). It remains for future research to solve the actual zoogeogeographical relationships between Las Desventuradas and Juan Fernández archipelagos and the actual extention of the bathyal Nazcaplatensis province (Parin et al. 1997) that would join Easter Island, Sala y Gómez Island and the guyot chains of the the Sala y Gomez and Nazca submarine ridges such as sugested by Parin et al. (1997).

ACKNOWLEDGEMENTS

The author sincerely thanks the Chilean Navy National Oceanographic Committee (CONA) for sponsoring the sampling activities during the CIMAR-5-1999 and CIMAR-6-2000 cruises respectively to the western and eastern Chilean oceanic islands. It is also thanked the Dirección de Investigación de la Universidad de Concepción for sponsoring the microphotographies used here (Grant 99113048-1). The following persons that actually made the sampling: Margarita Marchant and Victor Hugo Ruiz (Universidad de Concepción), Susana Giglio (Universidad Católica de Valparaíso) and Bladimiro López (Universidad de Magallanes) are also especially acknowledged.

REFERENCES

Banta, W.C. & Redden, J.C. 1990. A checklist of the Bryozoa of the Galapagos. *Proceedings of the Biological Society of Washington* 103(4): 789-802.

Briggs, J.C. 1974. *Marine Zoogeography*. New York: MacGraw-Hill.

Canu, F. & Bassler, R.S. 1930. The bryozoan fauna of the Galápagos Islands. *Proceedings of the U.S. National Museum* 76(13): 1-78.

Gordon, D.P. 1984. The Marine Fauna of New Zealand: Bryozoa: Gymnolaemata from the Kermadec Ridge. *New Zealand Oceanographic Institute Memoir* 91: 1-198

Hastings, A.B. 1930. Cheilostomatous Bryozoa from the vicinity of the Panamá Canal collected by Dr. C. Crossland on the cruise of the S.Y. "St. George". *Proceedings of the Zoological Society of London* 1929(4): 1-17.

Larrain, P.A. 1995. Biodiversidad de equinodermos chilenos: Estado actual del conocimiento y sinopsis biosistemática. *Gayana Zoología* 59(1): 73-96.

Marcus, E. 1921. Bryozoa von den Juan Fernández-Inseln, págs. 93-124. In K. Skottsberg (ed.) *The Natural History of Juan Fernández and Easter Island* 3: 93-124.

Moyano G., H.I. 1973. Briozoos Marinos Chilenos I. Briozoos de la Isla de Pascua I. *Gayana Zoología* 26: 1-22.

Moyano G., H.I. 1983. Southern Pacific Bryozoa: A General View with emphasis on Chilean species. *Gayana Zoología* 46: 1-45.

Moyano G., H.I. 1987. Bryozoa Marinos Chilenos VI. Cheilostomata Hippothoidae: South Eastern Pacific Species. *Boletín de la Sociedad de Biología de Concepción*, 57: 89-135.

Moyano G., H.I. 1991. Bryozoa Marinos Chilenos VIII: Una síntesis zoogeográfica con consideraciones sistemáticas y la descripción de diez especies y dos géneros nuevos. *Gayana Zoología* 55(4): 305-389.

Moyano G., H.I. 1995. Bryozoa: 163-173. In J.A. Simonetti et al. (eds) *Diversidad Biológica de Chile.* Santiago de Chile: Conicyt.

Osburn, R.C. 1950. Bryozoa of the Pacific coast of North America. Part I, Cheilostomata-Anasca. *Allan Hancock Pacific Expeditions* 14(1): 1-269.

Osburn, R.C. 1952. Bryozoa of the Pacific coast of North America. Part II, Cheilostomata-Ascophora. *Allan Hancock Pacific Expeditions* 14(2): 271-611.

Osburn, R.C. 1953. Bryozoa of the Pacific coast of North America. Part III, Cyclostomata, Ctenostomata, Entoprocta and Addenda. *Allan Hancock Pacific Expeditions* 14(3): 613-841

Osburn, R.C. & Soule, J.D. 1953. Order Ectoprocta, Suborder Ctenostomata. In R.C. Osburn, Bryozoa of the Pacific coast of North America. Part III, Cyclostomata, Ctenostomata, Entoprocta and Addenda. *Allan Hancock Pacific Expeditions* 14(3): 726-758

Parin, N.V. 1991. Fish fauna of the Nazca and Sala y Gómez submarine ridges, the easternmost outpost of the Indo-West zoogeographic region. *Bulletin of Marine Science* 49: 671-683.

Parin, N.V., Mironov, A.N. & Nesis K.N. 1997. Biology of the Nazca and Sala y Gómez submarine ridges, an outpost of the Indo-West Pacific Fauna in the Eastern Pacific Ocean: composition and distribution of the Fauna, its

communities and history. *Advances in Marine Biology* 32: 145-242.
Pequeño, G. & Lamilla, J. 1996. Desventuradas Islands, Chile: the easternmost outpost of the Indo-West Pacific zoogeographic region. *Revista de Biología Tropical* 44(2): 929-931.

Bryozoan Studies 2001, Wyse Jackson, Buttler & Spencer Jones (eds)
© 2002 Swets & Zeitlinger, Lisse, ISBN 90 5809 388 3

A Moscovian (Carboniferous) bryozoan buildup from Svalbard

Hans Arne Nakrem
Paleontological Museum, University of Oslo, P.O. Box 1172 Blindern, N-0318 Oslo, Norway

ABSTRACT: A mid-Carboniferous (pre-latest Moscovian) buildup in central Spitsbergen is composed almost entirely of fenestrate bryozoans (13 of 18 species, which contribute almost 100% of bryozoan biomass). Identified species are generally of early to late Carboniferous age, but some have been reported from Permian strata. Fenestrate sheets acted as sediment traps, and fenestrules are in varying degree closed off by initial marine cementation. Buildup development (sedimentation, diagenesis) is like that described from adjacent basins in Canada and eastern North Greenland, but in Spitsbergen the buildups are smaller (20 m thick, 200 m wide).

1 MATERIAL

The material studied was collected during fieldwork in 1992 at the Nordströmfjellet locality, central Spitsbergen, Svalbard (Fig. 1). Around 400 samples were collected and more than 200 thin sections have been investigated. Very few acetate peels could be made due to extensive dolomitisation of the rocks. Conodonts were processed from a crinoid limestone above the bryozoan buildup for dating purposes. Illustrated material is catalogued and housed at the Palaeontological Museum, University of Oslo (PMO).

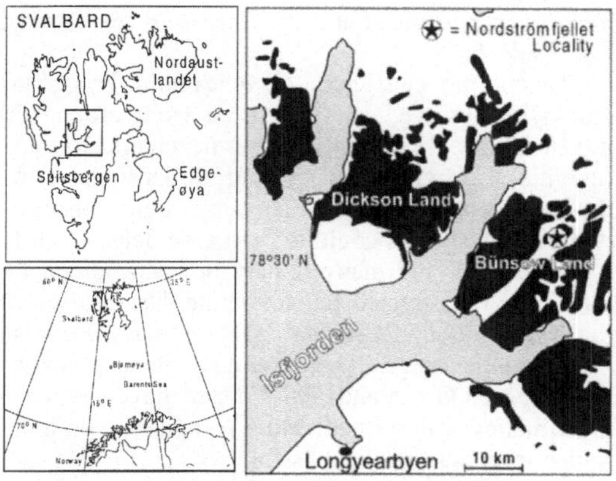

Figure 1. Locality map showing Svalbard's position in the Barents Sea and Carboniferous-Permian outcrops (black) in the investigated area.

2 GEOLOGICAL SETTING

In Late Carboniferous through Early Permian time Svalbard was situated in a sub-tropical position and drifted northwards from 25°N to 35°N. Deposition of dominantly carbonates took place under warm and humid, and later arid, conditions. The Minkinfjellet Formation (Fig. 2), of Moscovian age, consists mainly of thinly bedded dolomites, shales and gypsum beds.

The bryozoan buildups are developed in the uppermost part of the Minkinfjellet Formation in central Bünsow Land (Fig. 1), in what is tectonically known as the Billefjorden Trough.

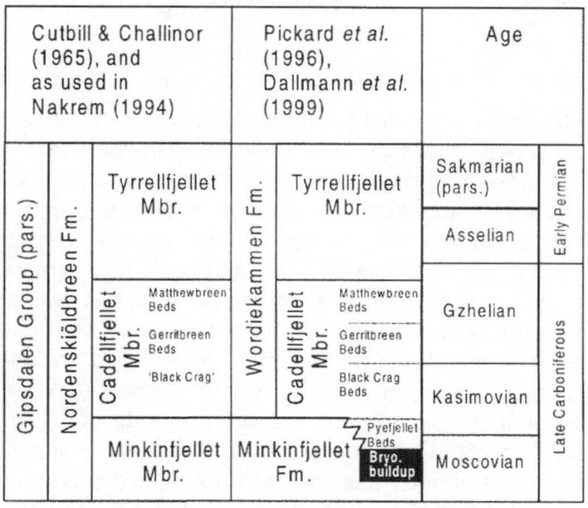

Figure 2. Lithostratigraphic units and corresponding ages in the investigated area.

3 BUILDUP DESCRIPTION

The buildup description is based on the material collected at Nordströmfjellet, as well as data in Pickard *et al.* (1994) and separate papers in Samuelsberg (2000) which cover buildups from Nordströmfjellet and Stenhousebreen.

The buildups are more than 20 m thick with a 200 m wide lateral development, and rest on dolostones with rare gypsum vugs which are generally devoid of bioclastic material. Core facies are composed mainly of a light grey, massive, fenestrate bryozoan wackestone that is commonly replaced by fabric-preserving dolomite. Locally fenestrate cementstone fabrics dominate, where initial porosity is replaced by pervasive calcite cement. Complex stromatactoid cavity systems are developed within the wackestones, and are often "roofed" by large fenestrate bryozoan sheets. This sediment trapping has resulted in visible macro-porosity in the rocks. Some cavities (now filled with sparry calcite) may have resulted from dissolved sponges, and spicules are commonly observed in thin sections. Sediment stabilisation was most probably a combination of marine cementation and binding algae (*Tubiphytes*).

Buildup flanks can occasionally be traced laterally. These flanks dip away from buildup cores (around 20°) and consist of coarse-grained crinoid-bryozoan grainstone and packstone. Intraclasts of fenestrate wackestone and crinoid particles are common in these detrital flank beds.

There is at least one palaeokarst surface near the top of the buildup indicating subaerial exposure (Fig. 3). The uppermost part of the buildup has also yielded phylloid algae in some thin sections, thus indicating that growth took place within the photic zone. The eroded top of the buildup is overlain by *c.* 15 m of barren dolostones marking the uppermost part of the Minkinfjellet Formation.

3.1 *Bryozoan components*

The bryozoan constituents are dominated by different groups of fenestrates (Table 1). Fenestellids form the most important group, both as skeletal components in wackestones as well as sediment trappers and substrate for calcite cement growth. Large sheets of "*Ptylopora*" are also quite common, whereas fragments of *Penniretepora* and *Polypora* are only occasionally seen in thin sections. Encrusting cystoporates (*Fistulipora* cf. *incrustans*) are observed in more than 30 thin sections where they are encrusting both reverse and obverse side of fenestrate sheets. They occur also on brachiopod shells. Encrustation took place on both living and dead fenestrate colonies. Cryptostomes and hexagoneliid cystoporates (e.g. *Goniocladia*) are only subordinate faunal elements. In addition to fistuliporid encrustations fenestrate colonies are in varying numbers encrusted by foraminifera, e.g. *Archaegonus* (or the cyclostomatous bryozoan *Hederella*), worms and the enigmatic algae(?) *Tubiphytes*.

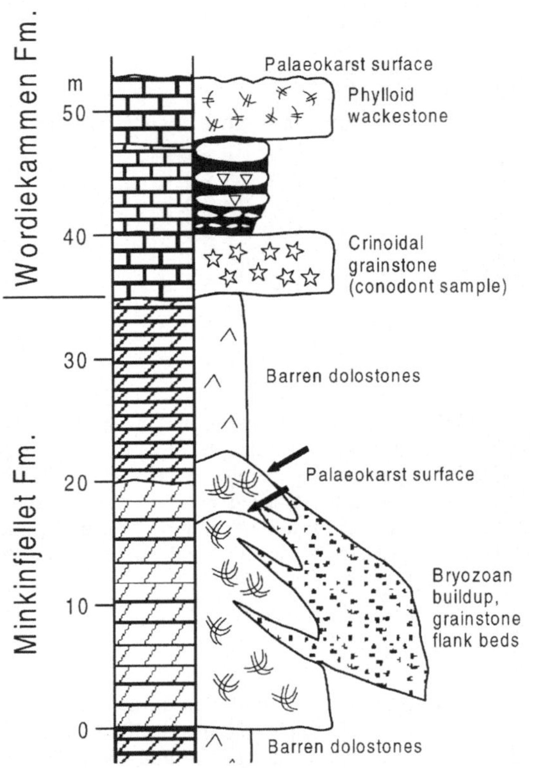

Figure 3. Lithological section through the Nordströmfjellet buildup. Arrows indicate palaeokarst surfaces.

The crinoidal pack/grainstone above contains bryozoan fragments of a complete different nature. Fenestrates are present as abraded fragments, in addition to rare specimens of *Archimedes*, *Goniocladia*, cryptostomes and an encrusting trepostome, see Table 2.

Taxonomic characters of fenestellid bryozoans could be observed and measured from around 70 thin sections. Important taxonomic characters, e.g. internal zooecial geometry and dimensions were, however, often lost due to pervasive dolomitisation. Only a few species could therefore be determined to species level, and many remain in open nomenclature. The investigated fauna contains species which have a wide mid-Carboniferous to early Permian stratigraphic range. The time-equivalent bryozoan buildups of Greenland and Canada have not been systematically described, and a comparison on a detailed level is so far not possible.

The overlying crinoid pack/grainstone, defining the base of the overlying lithostratigraphic unit (the Cadellfjellet Member of the Wordiekammen Formation) contains a different bryozoan fauna with a distinct Moscovian affinity. This interval is composed

Table 1. Bryozoans in the buildup core and flanks

Taxon	Distribution
Fistulipora cf. *incrustans* (Phillips 1836), Figure 5D	Early Carboniferous of the British Isles, Russia and North America
Goniocladia sp.	
Rhombotrypella cf. *dvinensis* Shul'ga-Nesterenko 1955	Moscovian-Kasimovian of Russia
Indet. encrusting trepostome	
Clausotrypa cf. *ramosa* (Owen 1973)	Early Carboniferous of Ireland
Alternifenestella bifida (Eichwald 1860), Figure 4E, G	Late Carboniferous-Early Permian of Russia
A. pseudovirgosa (Nikiforova 1938)	Late Carboniferous of Russia, Early Carboniferous of Ireland
A. cf. *pulcherrima* (Shul'ga-Nesterenko 1941)	Late Carboniferous-Permian of Russia
A. cf. *basloensis* (Bassler 1929)	Permian
Exfenestella(?) sp. A, B, C, Figure 4C,D	
Flexifenestella cf. *foraminosa* (Eichwald 1860), Figure 4F	Late Carboniferous - Early Permian of Russia, Late Carboniferous of China
F. sp. A	
Laxifenestella(?) sp. A, B, C, Figure 4H	
Rectifenestella rudis (*Cubifenestella* by Snyder 1991a)	Early Carboniferous, North America and Ireland
Penniretepora spp.	
"*Ptylopora*" sp. A (?new genus, Figure 4A, 5A-C)	
Polypora cf. *sulaensis* Nikiforova 1938	Late Carboniferous and Early Permian, Russia
Polyporella cf. *gracilis* (Nikiforova 1938)	Early Carboniferous, North America
P. cf. *rhombocellata* (Nikiforova 1938)	Late Carboniferous and Early Permian, Russia
(?)*Hederella*	Mainly Carboniferous

of carbonates that are not affected much by dolomitisation, and the taxonomic characters of the bryozoa are more readily obtained from acetate peels.

Fenestrate bryozoans dominate in the buildup, with a composition largely of previously described taxa. There may, however, be a new taxon, - genus or species, denoted "*Ptylopora*" herein. Its colony form (Fig. 5C) is identical to *Ptylopora* as revised from type material from the Carboniferous of Ireland (Bancroft 1985). The zooecial features differ in that the Svalbard specimens have a clearly defined hemiseptum (Fig. B), which is also weakly developed in *Ptylopora* in Snyder (1991b). A systematic description of this fauna is in preparation by the author..

Additional faunal components in the buildup facies include algae, - *Bersella* sp. (cf. Mamet & Stemmerik 2000: fig. 5), *Stacheoides* sp. and "calcispheres", rare trilobites like *Ditomopyge* sp., rare ammonoids, tubular and other non-fusulinid foraminiferans like *Archaeogonus* sp. and *Tetrataxis*, crinoid grains (often strongly micritised and/or encrusted by marine cement), rare solitary corals, locally abundant ostracods and sponge spicules. The non-skeletal matrix is a mixture of mud, often with a clotted structure, porefilling cement (fibrous or sparry calcite) and peloidal grains (in the upper part of the buildup).

Table 2. Bryozoans in the basal Cadellfjellet Member (crinoid pack/grainstone)

Taxon	Distribution
Fistulipora sp.	
Goniocladia sp.	
Rhombotrypella cf. *rectangulata* Shul'ga-Nesterenko 1955	Late Carboniferous of Russia
Archimedes sp. A	
Rectifenestella cf. *veneris* (Fischer 1866)	Middle Carboniferous (Moscovian) to Early Permian of Russia, North America and China
Penniretepora cf. *longicellata* (Morozova 1955)	Early Carboniferous of Russia
Rhombopora sp.	
Ascopora cf. *oblonga* (Nikiforova 1933)	Moscovian of Russia

Samples were collected from the overlying crinoidal pack/grainstone above the bryozoan buildup for dating purposes. Fusulinid foraminiferans include *Wedekindellina dutkevichi* (Rauser-Chernousova 1951), *Fusuliniella* aff. *eopulchra* (Rauser-Chernousova 1951), *Ozawainella* sp. and *Quasifusulinoides*? spp. (I. Nilsson in Pickard *et al.* 1996), and a fairly rich conodont fauna was extracted. The latter includes:

Streptognathodus cancellosus (Gunnell 1933)
Neognathodus cf. *medexultimus* Merrill 1972
Neognathodus cf. *medadultimus* Merrill 1972
Idiognathodus delicatus Gunnell 1931
Idiognathodus obliquus Kossenko & Kozitskaya1978
Idiognathodus aff. *podolskensis* Goreva 1984
Idiognathodus aff. *sinuosus* Ellison & Graves 1941
Hindeodus sp.

This fauna places the investigated interval in the Myakhovian Horizon (uppermost Moscovian) of the Russian Platform and is correlative to Desmoinesian faunas of North America and Canada, see e.g. Beauchamp *et al.* (1989).

3.2 *Diagenesis*

Most of the diagenetic modifications seen in the buildup and flank depostits are believed to have taken place shortly after deposition. The most important phases include pervasive, porosity-destructive early cementation of the buildup cores and subsequent dolomitisation of the entire succession.

Epitaxial and radial-fibrous calcite growth into fenestrule openings was followed by entire filling of such porosity by microspar cement. This early cementation took place in varying degree, and some

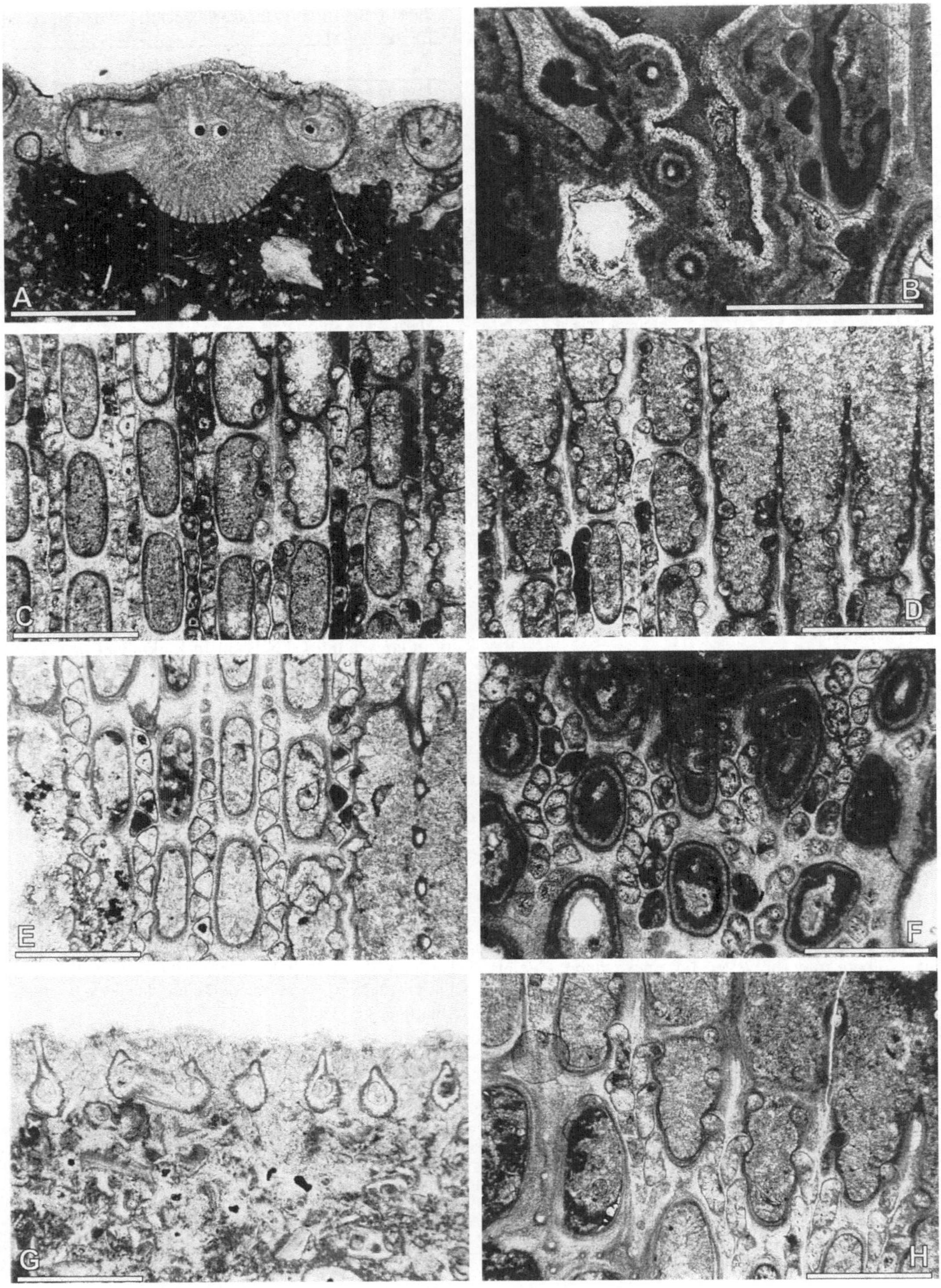
A
B
C
D
E
F
G
H

Figure 4 (facing page). A. "*Ptylopora*" sp., transverse section through main stem and lateral branches, PMO 136.507B. B. *Alternifenestella* sp., tangential section, with several generations of cement, PMO 136.596B. C. *Exfenestella*? sp., deep (left) to shallow tangential section, PMO 136.522B. D. *Exfenestella*? sp., shallow tangential section, PMO 136.522B. E. *Alternifenestella bifida*, deep (left) to shallow tangential section, PMO 136.623A. F. *Flexifenestella* cf. *foraminosa*, deep tangential section, PMO 136.646B. G. *Alternifenestella bifida*, transverse section PMO 136.623C. H. *Laxifenestella*? sp., deep (bottom) to shallow tangential section, PMO 136.636E. Scale bar 1 mm

fenestrules seems to be unaffected (Fig. 5E). Multiple generations of cement, and local dissolution horizons are visible in some colonies (Fig. 4B), and may be after initial aragonite cements.

Dolomitisation, including preservation of dolomite rhombs (Fig. 5I) apparently took place early in the post-depositional history. Karst surfaces are observed near the top of the buildup, and subaerial exposure and meteoric alteration most likely took place. Dolomitising fluids may have been a mixture of freshwater or evaporitic brines, as evidenced from presence of gypsum nodules in adjacent dolostones. Aragonite dissolution and dolomitisation may have regained some initial destroyed porosity. Silicification played a minor role, although some bryozoan skeletal microstructures are destroyed by silica growth. Chert nodules are present in layers in the overlying Cadellfjellet Member. Abundant sponge spicules as observed in thin sections may be the source for silicification.

4 DEPOSITIONAL ENVIRONMENT

The bryozoan buildup grew below fairweather wave base on the outer shelf or near the shelf edge. The water depth was sufficient for vertical aggregation, but presence of phylloid algae in the top of the buildup indicates growth into the photic zone. The presence of karst surfaces in the upper part indicates periods of very shallow water or even subaerial exposure. Growth into very shallow water most probably started laterally and can be followed in the flank deposits. Peloidal and clotted muds comprise parts of the buildup matrix similar to those present in other Carboniferous non-skeletal buildups or mudmounds (Pickard 1992). There is no evidence that bryozoans or other skeletal elements, such as corals, formed a true reef framework, and the growth of the buildup was more controlled by sea level changes than an ecological succession in the sense of Walker & Alberstadt (1975)

5 REGIONAL COMPARISON

The investigated bryozoan buildups are time-equivalent to buildups reported from adjacent areas in the palaeo-Arctic. Reef-mounds more than 150 m in height and hundreds of metres in width are reported from the Canadian Arctic Sverdrup Basin (Beauchamp 1992). A succession of structures, from deeper distal bryozoan buildups, to proximal algal reefs is considered to reflect a depth-controlled biological gradient. Temperature is believed to control the large-scale presence of organisms, and warm-water biota disappeared successively in the Early to Late Permian as a reaction to the significant cooling that took place. Moscovian bryozoan-dominated buildups in eastern North Greenland are up to 40 m thick and several hundred metres wide (Stemmerik 1989, 1992, 2000, Stemmerik & Elvebakk 1994). Frequent occurrences of algae in both buildup cores and flanks indicate that the buildups formed in shallow water (more shallow than the Sverdup buildups) well within the photic zone. Buildups (mainly algal bioherms) are reported from the Moscovian Lavrovskaya Formation in Novaya Zemlya, but bryozoans occur abundantly in these strata as well (Sobolev & Nakrem 1997: p. 18). Time-equivalent buildups may be present in the Barents Shelf, between deeper basins (Stemmerik 2000, p. 109, fig. 9).

ACKNOWLEDGEMENTS

The material was collected during field work in 1992 together with Nils-Martin Hanken, Neil A. H. Pickard, and Tommy J. Samuelsberg. The project is part of a study of Upper Carboniferous - Lower Permian successions in Svalbard led by the "Carbonate Research Group" based at the University of Tromsø, and supported by several oil companies. I express my thanks to this group of people/companies for making data and information available for the current study. Thin sections were prepared by Johny Skrårudseter at the Geological Museum, University of Oslo. Patrick Wyse Jackson is thanked for his constructive review.

REFERENCES

Bancroft, A.J. 1985. The Carboniferous fenestrate bryozoan *Ptylopora pluma* M'Coy. *Irish Journal of Earth Sciences* 7: 35-45.

Beauchamp, B., Harrison, J.C. & Henderson, C.M. 1989. Upper Paleozoic stratigraphy and basin analysis of the Sverdrup Basin, Canadian Arctic Archipelago: Part 1, time frame and tectonic evolution. *Current Research, Part G, Geological Survey of Canada, Paper* 89-16: 105-113.

Beauchamp, B. 1992. Carboniferous and Permian reefs of Sverdurp Basin, Canadian Arctic: an aid to Barents Sea exploration. In T.O. Vorren, E. Bergsager, Ø.A. Dahl-Stamnes, E. Holter, B. Johansen, E. Lie & T.B. Lund (eds) *Arctic Geology and Petroleum Potential*. Norwegian Petroleum Society (NPF) Special Publication 2: 217-241. Amsterdam: Elsevier.

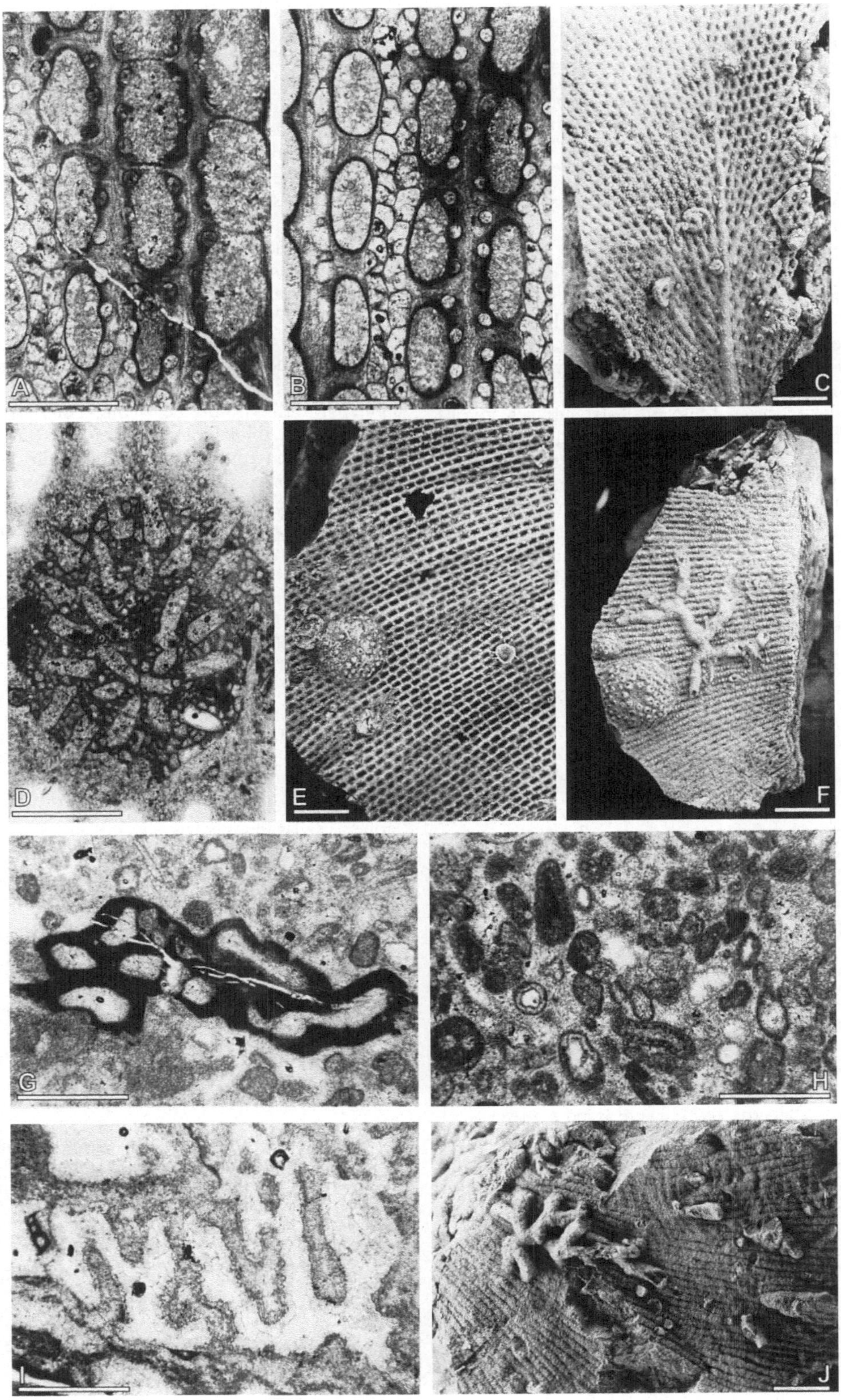
A
B
C
D
E
F
G
H
I
J

Figure 5 (facing page). A. "*Ptylopora*" sp., deep (left) to shallow tangential section, PMO 136.292A. B. "*Ptylopora*" sp., deep (left) to shallow tangential section, note well developed hemisepta, PMO 136.292A. C. "*Ptylopora*" sp., external features, PMO 136.612. D. *Fistulipora* cf. *incrustans*, deep tangential section, PMO 136.345B. E. Colony of *Fistulipora* cf. *incrustans* encrusting a fenestellid sheet, PMO 136.333. F. Colony of *Fistulipora* cf. *incrustans* and *Archaeogonus* sp. (a tubular foraminifera) (or the cyclostomatous bryozoan *Hederella*) encrusting the reverse side of a fenestellid sheet, PMO 136.630. G. *Tubiphytes*, PMO 136.483A. H. Peloid grains and *Bersella* (an algae, central, lower) PMO 136.483A. I. Indet fossil, possibly phylloid algae or *Palaeoaplysina*, PMO 136.483C. J. A tubular foraminifera (or ?*Hederella*) encrusting a fenestellid sheet. Note intense calcification filling fenestellid fenestrules, PMO 136.661. Scale bar 1 mm; except C, E and F: 10 mm

REFERENCES

Cutbill, J.L. & Challinor, A. 1965. Revision of the stratigraphical scheme for the Carboniferous and Permian rocks of Spitsbergen and Bjørnøya. *Geological Magazine* 102: 418-439.

Dallmann, W.K. (ed.) 1999. *Lithostratigraphic Lexicon of Svalbard. Review and recommendations for nomenclature use. Upper Palaeozoic to Quaternary Bedrock.* 318 pp. Tromsø: Norsk Polarinstitutt.

Mamet, B.L. & Stemmerik, L. 2000. Carboniferous algal microflora, Kap Jungersen and Foldedal Formations, Holm Land and Amdrup Land, eastern North Greenland. *Geology of Greenland Survey Bulletin* 187: 79-101.

Nakrem, H.A. 1994. Middle Carboniferous-Lower Permian bryozoans from Spitsbergen. *Acta Palaeontologica Polonica* 39(1): 45-116.

Pickard, N.A.H. 1992. Depositional controls on Lower Carboniferous microbial buildups, eastern Midland Valley of Scotland. *Sedimentology* 39: 1081-1100.

Pickard, N.A.H., Hanken, M.-M. & Dickson, T. 1994. *Porosity evolution project. Year end report 1994*. Unpublished report, University of Tromsø, 68 pp.

Pickard, N.A.H., Eilertsen, F., Hanken, N.-M., Johansen, T.A., Lønøy, A., Nakrem, H.A., Nilsson, I., Samuelsberg, T.J. & Somerville, I.D. 1996. Stratigraphic framework of Upper Carboniferous (Moscovian-Kasimovian) strata in Bünsow Land, central Spitsbergen: palaeogeographic implications. *Norsk Geologisk Tidsskrift* 76: 169-185.

Samuelsberg, T.J. 2000. *Late Palaeozoic Carbonates on the Norwegian Barents Shelf.* Unpublished Dr. Scient. thesis, University of Tromsø.

Snyder, E.M. 1991a. Revised taxonomic procedures and paleoecological applications for some North American Mississippian Fenestellidae and Polyporida (Bryozoa). *Palæontographica Americana* 57: 1-275.

Snyder, E.M. 1991b. Revised taxonomic approach to Acanthocladiid Bryozoa. In F. P. Bigey (ed.) Bryozoaires actuels et fossiles: Bryozoa living and fossil. *Bulletin de la Société des Sciences Naturelles de l'Oust de la France, Mémoire HS* 1: 431-445.

Sobolev, N.N. & Nakrem, H.A. 1997. Middle Carboniferous - Lower Permian conodonts of Novaya Zemlya. *Norsk Polarinstitutt Skrifter* 199: 128 pp. [Dated 1996]

Stemmerik, L. 1989. Crinoid-bryozoan reef mounds, Upper Carboniferous, Amdrup Land, eastern North Greenland. In H. H. J. Geldsetzer, N. P. James & G. E. Tebbutt (eds) *Reefs – Canada and adjacent area.* Canadian Society of Petroleum Geologists Memoir 13: 690-694.

Stemmerik, L. 1992. Moscovian bryozoan-dominated buildups, northern Amdrup Land, eastern North Greenland. In T.O. Vorren, E. Bergsager, Ø. A. Dahl-Stamnes, E. Holter, B. Johansen, E. Lie & T. B. Lund (eds) *Arctic Geology and Petroleum Potential.* Norwegian Petroleum Society (NPF) Special Publication 2: 99-106. Amsterdam: Elsevier.

Stemmerik, L. 2000. Late Palaeozoic evolution of the North Atlantic margin of Pangea. *Palaeogeography, Palaeoclimatology, Palaeoecology* 161: 95-126.

Stemmerik, L. & Elvebakk, G. 1994. A newly discovered mid-Carboniferous – ?early Permian reef complex in the Wandel Sea Basin, eastern North Greenland. *Rapport Grønlands Geologiske Undersøgelse* 161: 39-44.

Walker, K.R. & Alberstadt, L.P. 1975. Ecological succession as an aspect of structure in fossil communities. *Paleobiology* 1: 238-257.

Bryozoan Studies 2001, Wyse Jackson, Buttler & Spencer Jones (eds)
© 2002 Swets & Zeitlinger, Lisse, ISBN 90 5809 388 3

Brood chambers in cribrimorphs evolved by fusion of costae: further arguments

Andrew N. Ostrovsky
Department of Invertebrate Zoology, Faculty of Biology & Soil Science, St. Petersburg State University, Universitetskaja nab. 7/9, St. Petersburg, 199034, Russia. Current address: *Institut für Geowissenschaften, Christian-Albrechts-Universität zu Kiel, Olshausenstr. 40, 24118 Kiel, Germany*

ABSTRACT: Comparative study of the ovicell anatomy and morphology in three cribrimorph bryozoans belonging to the genus *Puellina* revealed new facts in favour of Lang's (1921) hypothesis of the costate origin of the brood chambers in Cribrimorpha. Among other findings, a closed horizontal slit in the distal ooecial wall was found that can be considered as a rudiment of the former passage between brooding space and epistege present in ancestral forms. Possible scenaria of common as well as independent evolution of the ovicells among Calloporoidea and Cribrimorpha are considered.

1 INTRODUCTION

Similar to the majority of cheilostomes, cribrimorph bryozoans brood their young in the special receptacles or chambers called ovicells. In this clade they are hyperstomial and cleithral (closed by an operculum), and ooecial (inner) vesicle is not known (Boardman et al. 1983, Ostrovsky 1998).

There are three ovicellar categories in Cribrilinidae. The ooecial fold can be a proximal component of either the distal (daughter) autozooid (category A) or kenozooid (category B). In category C the ooecium is formed by the maternal autozooid (Bishop & Househam 1987, see also Zabala & Maluquer 1988). Ostrovsky (1998) assigned categories A and B to the 1st, and category C to the 2nd cheilostome ovicellar types. The latter may perhaps have evolved from the former by a gradual reduction of the distal zooid. Thus, ovicells of category A are considered as plesiomorphic (see Bishop & Househam 1987, and Ostrovsky 1998 for further discussion).

Another suggested trend that could characterize an evolution of brood chambers in cribrimorphs, is the formation of an ovicellar roof (ooecium) by fusion of marginal costae in the proximal part of the daughter autozooid. This hypothesis was proposed first by Lang (1921, p. xxxv), who stated that the "costal origin [of ovicells] is evident" in some Cretaceous cribrimoprphs (for example, in *Leptocheilopora tenuilabrosa* Lang). Larwood (1962) made similar suggestion but was less categorical, while Ryland (1979, p. 214) wrote that in the species mentioned "the ooecium was either formed by or covered with a series of spines".

Lang's hypothesis was an alternative to the opinion of Harmer (1902), who studied the Recent cribrimorph *Euthyroides episcopalis* (Busk). Harmer speculated that the ooecium is formed by "the fusion of a pair of greatly expanded oral spines, the bases of which should communicate with the fertile [maternal] zooecium on each side of the operculum" (1902, p. 283), but could not prove this. Silén (1977) supported Harmer's point of view, but gave no real evidence in its favour (criticized by Ostrovsky 1998). In the same work Harmer (1902) stressed the strong resemblance between the development of the ovicell and that of frontal costae in this species.

The cheilostomes *Tendra zostericola* Nordmann and *Heteroecium amplectens* (Hincks) (Tendridae, Calloporoidea) possess so-called acanthostegous "ovicells" (Nordmann 1839, Repiachoff 1875, Hincks 1881, Ostroumoff 1886, Levinsen 1902, 1909, Paltchikova-Ostroumova 1925, Braiko 1967, Hyman 1959, Occhipinti Ambrogi & d'Hondt 1981). In these species the space between a frontal shield of tightly fitted marginal spines and frontal membrane is used for brooding, and Braiko (1967) and Santagata and Banta (1996, p. 178) supposed that the acanthostegous brood chambers of tendrids "might represent a primitive stage in the development of the ovicell complexes like those in [cribrilinid] *Figularia*". In other words, the space between a "spinocyst" and frontal membrane could be divided into two parts - brooding cavity and epistege – during an earlier stage of evolution (see also fig. 25 in Ostrovsky 1998).

A study of the external as well as internal ovicellar morphology of three Recent cribrilinids belonging

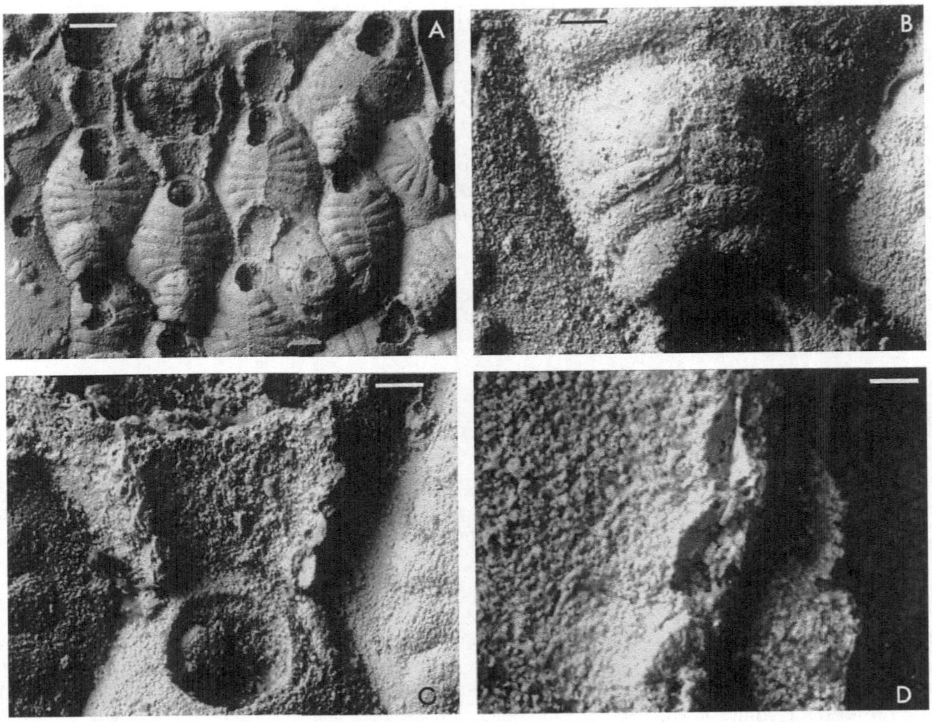

Figure 1. A-D, *Leptocheilopora tenuilabrosa* Lang (D28892 holotype, Santonian, England). A, fragment of the colony with complete and damaged ovicells; B, complete ovicell; C, damaged ovicell; D, detail of the ooecium, showing broken costae (ovicell floor at the left). Scale bars: A = 167 µm; B = 38.5 µm; C = 40 µm; D = 16.7 µm.

to the genus *Puellina* was undertaken. Further evidences of ovicellar origin from costae were found. The presented paper deals with the new data obtained, and discusses relationships of cribrimorphs with their possible ancestors in this connection. SEM-illustrations of the Cretaceous cribrilinid *Leptocheilopora tenuilabrosa* are used for comparative purposes.

2 MATERIAL AND METHODS

Colonies of *Puellina denticulata* Harmelin & Aristegui were collected on 27.03.1930 by dredging from 27-36 m depth near La Luz (Las Palmas, Canary Islands) during the expedition of T. Mortensen, and fixed in 70% alcohol (collection of the Zoological Museum, University of Copenhagen). Colonies of *P. hincksi* (Friedl) and *P. radiata* (Moll) were collected on 12.06.1997 by SCUBA from 22 m depth near Riou Island (Western Mediterranean). Specimens of the two latter species were fixed in Bouin's fluid without acetic acid and 70% alcohol. For light microscopy, colonies decalcified in Bouin's fluid were embedded in plastic (epon), sectioned (1-2.5 µm thick) and stained with toluidine blue using standard methods. For SEM studies, alcohol fixed colonies were cleaned in a 7.5 % solution of sodium hypoclorite, dehydrated in absolute alcohol, air-dried and coated with gold. Colonies of *Leptocheilopora tenuilabrosa* Lang (D28892 holotype; Santonian, *Marsupites* Zone; Brighton, Sussex, England) from the collection of The Natural History Museum (London) were photographed using an Environmental SEM.

3 RESULTS

Ovicells in *Leptocheilopora tenuilabrosa* are hyperstomial. The ooecia are formed by daughter autozooids (1st type, category A). Their external surface is distinctly costate, with a longitudinal medial suture that is aligned with the medial suture of the distal zooid (Fig. 1A, B). The appearance of the ovicellar costae is identical to that of the zooidal ones. Broken ooecia clearly demonstrate that their wall is completely formed by costae, rather than being covered by them (Fig. 1C, D).

In all three *Puellina* species studied ovicells are hyperstomial, cleithral, and formed by daughter autozooids (1st type, category A). The roof of each ovicell is continuous with the spinocyst of the distal zooid (Fig. 2A-D, G, I). The ooecium is divided into

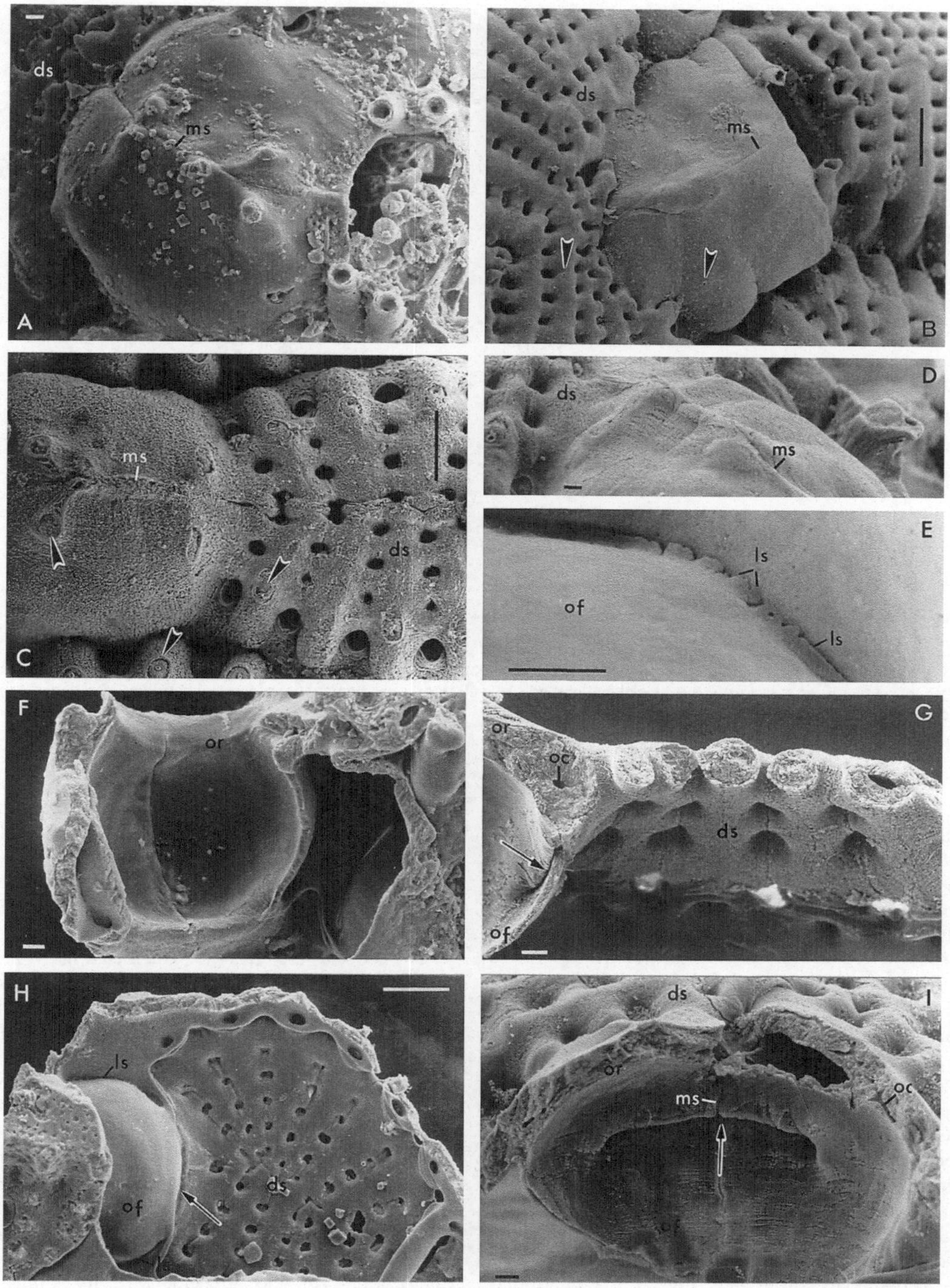

Figure 2. A, E-H, *Puellina denticulata* Harmelin & Aristegui (Recent, Canary Islands), B, *P. radiata* (Moll) (Recent, Mediterranean), C-D, I, *P. hincksi* (Friedl) (Recent, Mediterranean). A-D: ovicellar external view, E-I, details of inner morphology. Abbreviations: ds = spinocyst of the distal zooid, ls = lateral communicative slit, ms = medial ooecial suture, oc = lacuna of the ooecial coelom, of = ovicell floor, or = ovicellar roof (ooecium). Rudimentary ooecial slit arrowed. Pelmatidia shown by arrowheads. Scale bars: A, D-G, I = 10 μm; B-C, H = 50 μm.

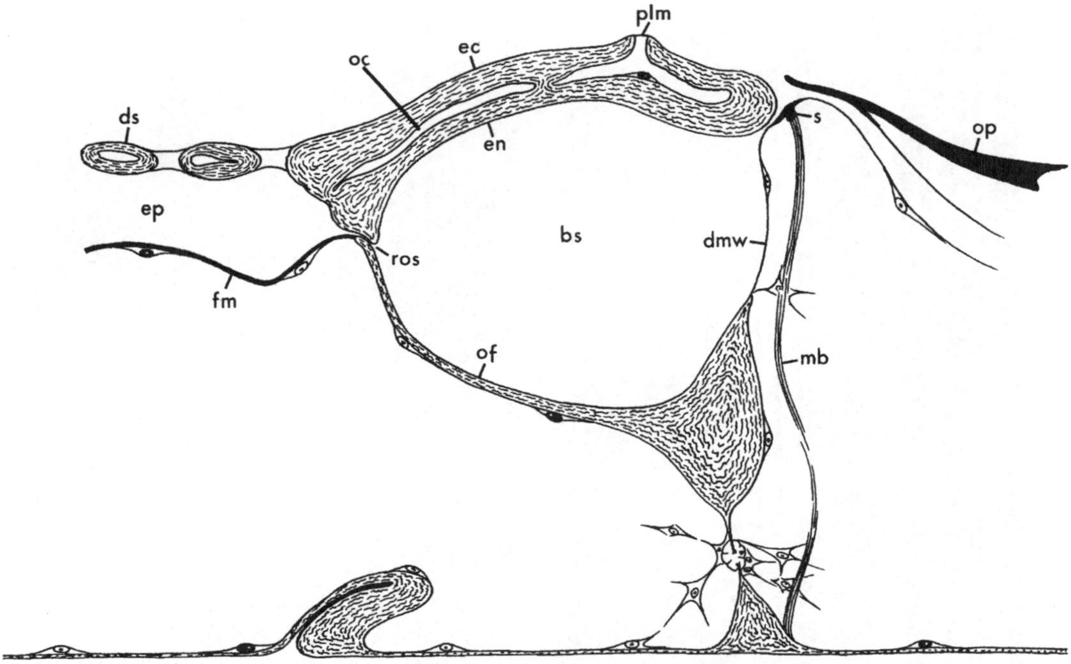

Figure 3. Schematic longitudinal section of *Puellina hincksi* (Friedl). Abbreviations: bs = brooding space, dmw = distal wall of the maternal zooid, ds = spinocyst of the distal zooid, ec = ectooecium, en = entooecium, ep = epystege, fm = frontal membrane, mb = muscle bands, oc = lacuna of the ooecial coelom, of = ovicell floor, op = operculum, plm = pelmatidium, ros = rudimentary ooecial slit, s = sclerite. Scale bars: A, D-G, I = 10 μm; B-C, H = 50 μm.

two parts by a medial suture, which is well seen on the external as well as the internal surface of the ooecium (Fig. 2A-D, F, I). The ooecial suture and the medial suture between the left and the right sides of the frontal shield of the distal zooid are aligned. Most ovicells also possess a variably developed longitudinal medial keel coinciding with the external suture.

In *Puellina denticulata* the ectooecium normally has more or less symmetrically positioned ridges near the medial suture (Fig. 2A), and similar ridges are seen on the proximal parts of the frontal costae. The ooecia of *P. radiata* are often laterally folded (Fig. 2B). In *P. hincksi* several ovicells with a relief resembling costae, were found (Fig. 2D). In all three species radiating wrinkles are common on the inner surface of the entooecium.

All species studied have pelmatidia on the ooecium as well as on the spinocyst. In *Puellina hincksi* each pelmatidium is placed in the centre of the drop-shaped border area (Fig. 2C). Each frontal costa bears one proximal pelmatidium. On the ovicells the pelmatidia are always elevated a little above the ooecial surface, and usually positioned bilaterally near the medial suture, 2-3 on each half of the ooecium (Fig. 2C). In *P. radiata* and *P. denticulata* the pelmatidia are small short tubes. There are 1-3 on each frontal costa (proximally, centrally and/or distally), and they have the same appearance and patterning on the ooecial surface (Fig. 2B). Similarly, rows of pores were described and pictured by Lang (1921) on both the frontal zooidal costae and the costate ooecia in the Cretaceous cribrimorph *Pliophloea striata* Lang.

The ooecia are heavily calcified. In younger zooids an ooecial coelomic cavity still exists, but the ectooecium and entooecium are fused in older zooids, and the cavity is reduced to a few canals and lacunae confluent with cavities of pelmatidia (Figs 2G, I, 3). A border between fused ooecial walls is seen in some broken ooecia, but normally it is not present (Fig. 2G, I). Ooecial lacunae communicate with the coelomic cavity of the daughter zooid through two curved lateral slits (1.5-2 μm wide) (Fig. 2H). Careful examination showed that in older zooids each slit is a row of narrow pores of different sizes (Fig. 2E), and it is the result of the slit closure by ooecial wall calcification. The ooecial coelomic cavity is lined by epithelial cells. I was not able to find any signs of cell obstruction closing the communication slits; however, their narrowness implies that they may be plugged at least in some places.

The ovicell aperture is closed by the membranous part of the distal wall and the operculum of the mother autozooid (Fig. 3). In the species sectioned (*Puellina radiata* and *P. hincksi*) parietal muscles are found that probably provide the opening of the ovicell aperture. Proximally, they are fixed onto the basal zooidal wall, distally either in the middle part of the distal wall (*P. radiata*) or in its upper part (*P. hincksi*) where cuticular sclerite was found in the latter species (Fig. 3). Additional study is necessary to specify the details of their structure and positioning.

The inner surface of the ooecial chamber displays a distinct horizontal slit or groove (Fig. 2G, I), that is perpendicular to the medial suture and separates ovicellar floor and roof. The slit is completely closed, but its edges are not fused in the majority of cases observed (Fig. 2G), so it is also seen from the side of the daughter autozooid (Fig. 2H). The frontal membrane of this zooid is attached to the lower edge of the slit (Fig. 3), thus ovicellar floor and the frontal membranous wall are parts of a continuous structural unit.

4 DISCUSSION

Structure of the ooecia in the *Leptocheilopora tenuilabrosa* permits the conclusion that they were formed by the costae (modified mural spines) of proximal gymnocyst of the daughter autozooid. One may suggest that this variant was initial in cribrimorphs that acquired solid ooecia later on in their evolution.

Costate ovicells are known in several cribrimorphs from the Upper Cretaceous (Lang 1921, Larwood 1962, Voigt 1994). However, the ovicells in some Recent cribrimorphs also bear evidence of their costate origin in one form or another. There are only a few species whose ooecia have either an identical or a very similar morphology to their frontal shields, for instance, *Figularia tahitiensis* (Waters) (as *Figulina*), *Figularia pulcherrima* Tilbrook, Hayward & Gordon, *Bellulopora bellula* (Osburn) (as *Colletosia*), *Puellina modica* Bishop & Househam (see Waters 1923, Osburn 1950, Moyano 1984, Winston & Håkansson 1986, Bishop & Househam 1987, Reverter & Fernandez 1996, Tilbrook et al. 2001). Several species have ridges in a radiating pattern or wrinkles resembling fused costae on their ovicells: *Puellina bifida* (d'Hondt), *P. minima* (Harmelin) (as *Cribrilaria*), *P. arrecta* Bishop & Househam, *P. corbula* Bishop & Househam, *P. denticulata* Harmelin & Aristegui, *P. hincksi* (see Harmelin 1984, Bishop & Househam 1987, Harmelin & Aristegui 1988, Reverter & Fernandez 1996, herein). Also, the "radial" pattern of ectooecial calcification during ovicellogenesis has been interpreted as an indication of ooecial origination from fused costae (Ostrovsky 1998). Such a pattern was found in *Puellina harmeri* Ristedt (as *Cribrilaria*) and *Cribrilina annulata* (Fabricius) (see Ristedt 1985, and Ostrovsky 1998).

Ovicells in the majority of Recent cribrimorph bryozoans (from the genera *Membraniporella, Cribrilina, Puellina, Collarina, Reginella*) and at least some fossil ones (from the *Pliophloea, Anaptopora, Monoceratopora, Lagynopora, Castanopora*) are characterized by either a longitudinal medial suture or a carina (sometimes both) that may be a trace of the coalescence of the left and right sides of the proximal spinocyst of the daughter zooid (for Recent species see descriptions in Prenant & Bobin 1966, Harmelin 1970, 1976(1978), 1984, 2001, Harmelin & Aristegui 1988, Hayward 1988, Harmelin & d'Hondt 1993, Hayward & Ryland 1979, 1998, Moyano 1984, Gordon 1984, Winston 1984, Ristedt 1985, Zabala 1986, Zabala & Maluquer 1988, Bishop 1986, 1994, Bishop & Househam 1987, Àlvarez 1988, Mawatari 1988, Ryland & Hayward 1992, Soule et al. 1995, Reverter & Fernandez 1996, Tilbrook et al. 2001; for fossil species see Lang 1921, Canu & Bassler 1933, Larwood 1962, 1981, 1985). The ooecial medial suture is aligned with the medial suture when it exists of the frontal shield of the distal zooid. Sometimes, the suture is seen mainly on the inner ovicellar surface, as in the case of *Cribrilina annulata*. It divides the distal edge of the ooecium into two parts, and there is an inner double septum corresponding to the suture (see Ostrovsky 1998). In the Recent genera *Euthyroides* (see Harmer 1902, Gordon 1989), *Figularia* (Harmer 1926, Boardman et al. 1983, Wass & Yoo 1983, Moyano 1984, Hayward 1988, Àlvarez 1988, Zabala & Maluquer 1988, Gordon 1989, Soule et al. 1995, Hayward 1995, Tilbrook et al. 2001) and *Filaguria* (Moyano 1991, Hayward 1995), the precence of the medial suture (and consequently the bipartite appearance of the ovicells) is a generic feature (see also Bock & Cook 1994, 1998, and Gordon 1986 for *Corbulipora*). If the partition is complete, each half of such ooecia should be communicated with the coelomic cavity of the daughter zooid through a separate slit, as in *Puellina* species studied (see above). The Cretaceous cribrimorph *Lagynopora amphora* Lang possibly displays a similar structure (Larwood 1962, fig. 29).

Like the costal lumina, the coelomic cavity of the ooecium is connected with a coelom of the distal zooid (ovicells of categories A and B) (Boardman et al. 1983, herein). Similar to the costae, the ooecial cavity is gradually reduced as a result of wall calcification in ontogenesis.

Thus, in the light of the hypothesis under discussion (Lang 1921), the cribrimorph ooecium can be considered as derived from fused costae (and their lumina) of the left and right sides of the frontal shield of the distal zooid. Recent findings of the ovicellar closed slit and the continuity of the ovicell floor and the frontal membranous wall in *Puellina* species strongly support this hypothesis. Communication that existed between the brooding space and the epistege shows that the former could be a part of the latter in some ancestral taxa. The presence of some additional structures (as pelma, fenestrae, pelmatidia, ridges, etc.) on the frontal costae and on the ectooecia may indicate that two latter structures are homologous.

Evolutionary steps of such a transformation, resulting in a separation of brooding and hydrostatic functions of the epistege, can be imagined as follows: (1) incidental brooding in the epistegal space; (2) obligatory brooding in the proximal part of the epistege (much as within the acanthostegous brood chambers of Recent tendrids); (3) gradual fusion of several of the most proximal marginal costae on each side of the daughter autozooid to give the ooecial roof, where the ooecium consists of two hollow plates whose cavities are connected with zooidal coelom through slits; (4) further extension of the proximal gymnocyst to form the ovicell floor; (5) further fusion of the costae and formation of the distal part of the semispherical ooecial capsule; (6) gradual separation of the brooding space from the epistege: reduction of the passage between the cavity of the brood chamber and the epistege, sometimes with complete coalescence of the inner ooecial wall and the ovicell floor (fusion of the lateral communication slits gives the appearance of a common arch-shaped slit, as in *Cribrilina punctata* (Hassall), Ostrovsky, unpubl.); (7) closure of the ooecial aperture by the distal wall (or its outgrowth) of the proximal zooid; (8) modification of the muscle system of the distal wall of the maternal autozooid, and formation of a sclerite; (9) further strengthening of the ovicellar walls by calcification. These evolutionary events must have occuried relatively quickly since the majority of Upper Cretaceous cribrimorphs possess solid ovicells (Lang 1921, Larwood 1962).

In Calloporoidea (supposed ancestral group for Cribrimorpha) the earliest ovicells are known in *Wilbertopora mutabilis* Cheetham from Lower Cenomanian (Cheetham 1954) and Upper Albian (Taylor & McKinney, this volume). These are non-costate, and possess a longitudinal "suture-like" line on the ectooecial surface (Cheetham 1975, p. 553). Bilaterally simmetrical ovicells with a medial suture or keel are known among fossil as well as Recent Calloporidae (*Alderina, Callopora, Crassimarginatella, Copidozoum, Retevirgula, Leptinatella, Bryocalyx*) (see Canu & Bassler 1933, Prenaunt & Bobin 1966, Harmelin 1973a, Zabala & Maluquer 1988, Tilbrook 1998, Cook & Bock 2000). Ovicells in the genus *Valdemunitella* are very similar to those in cribrimorphs *Figularia* and *Euthyroides* (see illustrations in Gordon 1986, 1989), and Hayward (1995, p. 80) considered the "bipartite, bifenestrate ooecia" of this calloporid as a character of generic level. Similar to cribrimorphs, calloporids possess all three categories of ovicells (Harmelin 1973b, Bishop & Househam 1987, Ostrovsky 1998), and a similar trend (reduction of the distal zooid) during ovicell evolution can be supposed in this clade. Bipartite calcification of the future ovicell floor during early ovicellogenesis is known for both taxa as well (Levinsen 1909, Ostrovsky 1998, Ostrovsky & Schäfer, in prep.). Finally, findings in both calloporids (Ostrovsky & Schäfer, in prep.) and cribrilinids (see above) of the sclerite to which the parietal muscles of the distal wall of the maternal zooid are attached provide additional evidence of the suggested common origin of their brood chambers.

However, the above arguments are still speculative since costate ovicells are unknown in Calloporidae, and *Wilbertopora* stratigraphically precedes the earliest Cribrimorpha (Santonian) (see also Ostrovsky 1998 for discussion). Two evolutionary scenaria appear possible. (1) Common origin of ovicells in calloporids and cribrimorphs: the brood chamber of several most proximal non-fused mural spines was the earliest skeletal embryonal receptacle that originated in primitive Calloporoidea; it evolved into a costate and, later on, a solid acleithral ovicells (see above) within this clade; simultaneously, the ooecial vesicle with its special muscle system and the sclerite were formed; cribrimorphs inherited the costate acleithral ovicells with the inner vesicle and sclerite from a calloporid ancestor; calloporids with costate ovicells became extinct; the inner vesicle and sclerite were lost in some cribrimorphs; their ovicells became cleithral. (2) Independent origin of ovicells in calloporids and cribrimorphs: within Calloporoidea brood chambers evolved from spines at least twice; in the family Calloporidae ovicells originated from several or two flattened and fused spines in the proximal part of the distal zooid (Santagata & Banta 1996, see also Cheetham 1975); the inner vesicle was formed to close an ovicellar aperture, and acleithral ovicells evolved; cribrimorphs inherited costate ovicells from one of the clades of Calloporoidea (another than Calloporidae); cleithral ovicells evolved within Cribrimorpha; in both groups discussed sclerites evolved independently.

Acanthostegous brood chambers in which the entire frontal shield is used as an ooecium, are not true ovicells. Brooding in epistege may represent a primitive stage in evolution of the ovicells (see above), but it is unlikely that they derived from the brooding structures as occur in the Recent Tendridae.

Costate ovicells that were recently found in some Lower Cenomanian Microporidae (see Taylor & McKinney, this volume) can be also interpreted as either of common or independent origin in relation to calloporids.

All variants proposed imply an origin of the first brood chambers as protective structures (Hughes 1987). There is no evidence that "vestibular brooding preceded evolution of ovicell among cheilostomes" (Santagata & Banta 1996, p. 178). The calloporid *Antropora tincta* (Hastings) broods its embryos in endozooidal ovicells (Hastings 1930, Tilbrook 1998), not in a diverticulum of the vestibule as Santagata and Banta believed (1996). Further fossil evidence as well as investigations of Recent Cheilostomatida are necessary to test the hypotheses presented.

The great similarity between the ovicell and the frontal shield in some cribrimorphs implies the existence of true pores between the rudiments of the costae in the ooecia of bryozoans such as *Bellulopora bellula* and *Figularia tahitiensis* (see above). However, Silén (1945) noted that embryos of the calloporid *Callopora dumerili* (Audouin) died when in contact with see water. Moyano (1984) described the presence of a membrane (homologous to a frontal membrane?) between the perforated ectooecium and solid entooecium in *B. bellula*. It is unclear what he meant, and additional research is required to understand the ovicell structure in this species.

Ristedt (1985, fig. 10C) presented a schematic longitudinal section of the cribrimorph *Puellina harmeri* (Ristedt) (as *Cribrilaria*). According to this figure, in the case of ovicells belonging to the 1st type (category A), ooecial cavity is a confluent with a coelom of the daughter as well as mother zooids and the ovicell floor consists of two layers. The data herein however clearly show that the ooecial cavity communicates with the daughter zooid only and the lower wall of the ovicell is single.

ACKNOWLEDGEMENTS

I wish to thank sincerely Professor Dr Priska Schäfer (Institut für Geowissenschaften, Christian-Albrechts-Universität zu Kiel) for providing working facilities and for valuable aid, advice and much encouragement. Specimens of *Puellina denticulata* were kindly given by Dr Karen Bille Hansen (Zoological Museum, University of Copenhagen). I thank Dr Jean-George Harmelin (Station Marien d'Endoume, Centre d'Oceanologie Marseille) for collecting and identifying material, and Dr Paul D. Taylor (The Natural History Museum, London) for unpublished SEM-photos of the fossil cribrimorph and discussion. Thanks are given to Ms Mary Spencer Jones and Mr Kevin Tilbrook (The Natural History Museum, London) for kind help with the literature. I also thank Mr Geert Brovad (Zoological Museum, University of Copenhagen) and Ms Ute Schuldt (Institut für Geowissenschaften, Christian-Albrechts-Universität zu Kiel) for assistance with SEM- and photography. The study was completed during a postdoctoral fellowship sponsored by the Humboldt Foundation (Alexander von Humboldt-Stifting). Deutsche Forschungsgemeinschaft is acknowledged for financial support (grant No. SCHA 355/20-1).

REFERENCES

Àlvarez, J.A. 1988. La familia Cribrilinidae Hincks, 1880 (Bryozoa, Cheilostomida) en la Costa Vasca. *Miscelánea Zoologia* 12: 347-352.

Bishop, J.D.D. 1986. The identity of *Cribrilaria innominata* (Couch, 1844) (Bryozoa, Cheilostomata). *Bulletin of the British Museum (Natural History)*, Zoology 50(2): 93-102.

Bishop, J.D.D. 1994. The genera *Cribrilina* and *Collarina* (Bryozoa, Cheilostomatida) in the British Isles and North Sea Basin, Pliocene to present day. *Zoologica Scripta* 23(3): 225-249.

Bishop, J.D.D. & Househam, B.C. 1987. *Puellina* (Bryozoa: Cheilostomata: Cribrilinidae) from British and adjacent waters. *Bulletin of the British Museum (Natural History)*, Zoology 53(1): 1-63.

Boardman, R.S., Cheetham, A.H., Blake, D.B., Utgaard, J., Karklins, O.L., Cook, P.L., Sandberg, P.A., Lutaud, G. & Wood, T.S. 1983. Bryozoa. In R.A. Robison (ed), *Treatise on Invertebrate Paleontology, Part G. Revised. Volume 1*: 1-625. Boulder and Lawrence: Geological Society of America and University of Kansas Press.

Bock, P.E. & Cook, P.L. 1994. Occurence of three phases of growth with taxonomically distinct zooid morphologies. In P.J. Hayward, J.S. Ryland & P.D. Taylor (eds), *Biology and Palaeobiology of Bryozoans*: 33-36. Fredensborg: Olsen and Olsen.

Bock, P.E. & Cook, P.L. 1998. A new species of multiphased *Corbulipora* MacGillivray, 1895 (Bryozoa: Cribriomorpha) from southwestern Australia. *Records of the South Australian Museum* 30(2): 63-68.

Braiko, V.B. 1967. Biology of reproduction of *Membranipora zostericola* Nordm. (Bryozoa). *Zoologicheskiy Zhurnal* 46: 1119-1121 (in Russian).

Canu, F. & Bassler, R.S. 1933. Bryozoan fauna of the Vincetown limesand. *Bulletin of the United States National Museum* 165: i-v, 1-102.

Cheetham, A. 1975. Taxonomic significance of autozooid size and shape in some early multiserial cheilostomes from the Gulf coast of the U.S.A. In S. Pouyet (ed), *Bryozoa 1974, Documents du Laboratoire de Géologie de la Faculté des Sciences de Lyon, Hors Série* 3(2): 547-564.

Cook, P.L. & Bock, P.E. 2000. Two new genera of Bryozoa (Calloporidae) from New Zealand. *Journal of Natural History* 34: 1125-1133.

Gordon, D.P. 1984. The marine fauna of New Zealand: Bryozoa: Gymnolaemata from the Kermadec Ridge. *New Zealand Oceanographic Institute Memoir* 91: 1-198.

Gordon, D.P. 1986. The marine fauna of New Zealand: Bryozoa: Gymnolaemata (Ctenostomata and Cheilostomata Anasca) from the Western South Island continental shelf and slope. *New Zealand Oceanographic Institute Memoir* 95: 1-121.

Gordon, D.P. 1989. The marine fauna of New Zealand: Bryozoa: Gymnolaemata (Cheilostomida Ascophorina) from the Western South Island continental shelf and slope. *New Zealand Oceanographic Institute Memoir* 97: 1-158.

Harmelin, J.-G. 1970. Les *Cribrilaria* (Bryozoaires Chilostomes) de Méditerranée: systématique et écologie. *Cahiers de Biologie Marine* 11: 77-98.

Harmelin, J.-G. 1973a. *Callopora minuta* n. sp., nouvelle espèce de bryozoaire Chilostome (Alderinidae) des côtes françaises de Méditerranée. *Cahiers de Biologie Marine* 14: 29-37.

Harmelin, J.-G. 1973b. Les bryozoaires des peuplement sciaphiles de Méditerranée: le genre *Crassimarginatella* Canu (Chilostomes Anasca). *Cahiers de Biologie Marine* 14: 471-492.

Harmelin, J.-G. 1976(1978). Sur quelques cribrimorphes (Bryozoa, Cheilostomata) de l'Atlantique Oriental. *Tethys* 8(2): 173-192.
Harmelin, J.-G. 1984. Bryozoaires des peuplement sciaphiles de Méditerranée: révision de *Cribrilina setosa* Waters et description de deux nouvelles espèces de *Cribrilaria* (Bryozoa, Cheilostomata). *Zoologica Scripta* 13(2): 81-88.
Harmelin, J.-G. 2001. *Puellina saldanhai* n. sp., a new cribrimorph cheilostome (Bryozoa: Gymnolaemata) from dark cave environment of southern Portugal. *Boletim do Museu Municipal do Funchal* (in press).
Harmelin, J.-G. & Aristegui, J. 1988. New Cribrilinidae (Bryozoa, Cheilostomata) from the upper bathyal of the Atlanto-Mediterranean region. *Journal of Natural History* 22: 507-535.
Harmelin, J.-G. & d'Hondt, J.-L. 1993. Transfers of bryozoan species between the Atlantic Ocean and the Mediterranean Sea *via* the Strait of Gibraltar. *Oceanologica Acta* 16(1): 63-72.
Harmer, S.F. 1902. On the morphology of the Cheilostomata. *Quarterly Journal of Microscopical Science* 46(182): 263-350.
Harmer, S.F. 1926. The Polyzoa of the Siboga Expedition. II. Cheilostomata Anasca. *Reports of the Siboga Expedition* 28b: 181-501. Leiden: E.J. Brill.
Hastings, A.B. 1930. Cheilostomatous Polyzoa from the vicinity of the Panama Canal collected by Dr. C. Crossland on the cruise of the S.Y. "St. George". *Proceedings of the Zoological Society of London* 4(47): 697-740.
Hayward, P.J. 1988. Mauritian cheilostome Bryozoa. *Journal of Zoology (London)* 215: 269-356.
Hayward, P.J. 1995. *Antarctic cheilostomatous Bryozoa.* Oxford University Press.
Hayward, P.J. & Ryland, J.S. 1979. British ascophoran bryozoans. *Synopses of the British Fauna* 14: 1-312.
Hayward, P.J. & Ryland, J.S. 1998. Cheilostomatous Bryozoa, Part 1. Aetoidea-Cribrilinoidea. *Synopses of the British Fauna* 10 (2nd edition): 1-366.
Hincks, T. 1881. Contributions towards a general history of the marine Polyzoa. *The Annals and Magazine of Natural History*, 5 Series, 8: 122-136.
Hughes, D.J. 1987. Gametogenesis and embryonic brooding in the cheilostome bryozoan *Celleporella hyalina. Journal of Zoology, London* 212: 691-711.
Hyman, L. H. 1959. *The Invertebrates: smaller coelomate groups*, vol. 5: VIII. New York: McGraw-Hill.
Lang, W.D. 1921. The cribrimorphs.- Part 1. *Catalogue of the fossil Bryozoa (Polyzoa) in the Department of Geology, British Museum (Natural History). The Cretaceous Bryozoa (Polyzoa)* 3: i-cx, 1-269.
Larwood, G.P. 1962. The morphology and systematics of some Cretaceous cribrimorph Polyzoa (Pelmatoporinae). *Bulletin of the British Museum (Natural History),* Geology 6(1): 1-285.
Larwood, G.P. 1981. Revision of the Cretaceous cribrimorph Bryozoa described by Hennig from the Swedish Chalk. In G.P. Larwood & C. Nielsen (eds), *Recent and Fossil Bryozoa*: 127-134. Fredensborg: Olsen and Olsen.
Larwood, G.P. 1985. Form and evolution of Cretaceous myagromorph Bryozoa. In C. Nielsen & G.P. Larwood (eds) *Bryozoa: Ordovician to Recent*: 169-174. Fredensborg: Olsen and Olsen.
Levinsen, G.M.R. 1902. Studies on Bryozoa. *Videnskabelige Meddelelser fra Dansk Naturhistorisk Forening i Kjøbenhavn* 54: 1-31.
Levinsen, G.M.R. 1909. *Morphological and systematic studies on the Cheilostomatous Bryozoa.* Copenhagen: F. Bagge.
Mawatari, S.F. 1988. Two cheilostomatous bryozoans new to Hokkaido. *Memoires of the National Science Museum, Tokio* 21: 145-151.
Moyano G., H.I. 1984. Chilean Cribrimorpha (Bryozoa Cheilostomata). *Boletín de la Sociedad de Biología de Concepción* 55: 47-72.
Moyano G., H.I. 1991. Bryozoa marinos chilenos VIII: Una síntesis zoogeográfica con consideraciones systemáticas y la decsripción de diez especies y dos géneros nuevos. *Gayana Zoología* 55: 305-389.
Nordmann, M.A. 1839. Recherches microscopiques sur l'anatomie et le développement du *Tendra zostericola*, espèce de polype de la section des Bryozoaires. *Annales des Sciences Naturelles*, Paris: Masson, Serie 2, 11: 185-191.
Occhipinti Ambrogi, A. & d'Hondt, J.-L. 1981. Distribution of bryozoans in brackish waters of Italy. In G.P. Larwood & C. Nielsen (eds), *Recent and Fossil Bryozoa*: 191-198. Fredensborg: Olsen and Olsen.
Osburn, R.C. 1950. Bryozoa of the Pacific coast of America. Part 1, Cheilostomata-Anasca. *Allan Hancock Pacific Expeditions* 14(1): 1-269.
Ostrovsky, A.N. 1998. Comparative studies of ovicell anatomy and reproductive patterns in *Cribrilina annulata* and *Celleporella hyalina* (Bryozoa: Cheilostomatida). *Acta Zoologica* 79 (4): 287-318.
Ostrovsky, A.N. & Schäfer, P. Ovicell structure in *Callopora dumerili* (Audouin) and two related species (Bryozoa: Cheilostomatida) (in preparation).
Ostroumoff, A. 1886. Die Bryozoen der Bucht von Sebastopol. Vollständigere Ausgabe mit einem ganz neuen Theile über die Morphologie der Bryozoen. *Proceedings of the Society of Naturalists of Kazan* 16(2): 1-124 (in Russian).
Paltschikowa-Ostroumowa, M.W. 1926. Kurze Bemerkung über den Ovidukt bei den Bryozoen. *Zoologischer Anzeiger* 65: 100-102.
Prenant, M. & Bobin, G. 1966. Bryozoaires.Deuxième partie. Chilostomes Anasca. *Faune de France* 68: 1-398.
Repiachoff, W. 1875. Zur Entwickelungsgeschichte der *Tendra zostericola. Zeitschrift für wissenschaftliche Zoologie* 25: 129-142.
Reverter, O. & Fernandez, E. 1996. Cribrilinidae (Bryozoa: Cheilostomatida) from the Ría de Ferrol (NW Spain). *Journal of Natural History* 30: 1247-1260.
Ristedt, H. 1985. *Cribrilaria*-Arten (Bryozoa) des Indopazifiks (Rotes Meer, Seychellen, Philippinen). *Mitteilungen aus dem Geologisch-Paläontologischen Institut der Universität Hamburg* 59: 15-38.
Ryland, J.S. 1979. Structural and physiological aspects of coloniality in Bryozoa. In G.P. Larwood & B.R. Rosen (eds), *Biology and systematics of colonial organisms*: 211-242. New York: Academic Press.
Ryland, J.S. & Hayward, P.J. 1992. Bryozoa from Heron Island, Great Barrier Reef. *Memoirs of the Queensland Museum* 32(1): 223-301.
Santagata, S. & Banta, W.C. 1996. Origin of brooding and ovicells in cheilostome bryozoans: interpretive morphology of *Scrupocellaria ferox. Invertebrate Biology* 115(2): 170-180.
Silén, L. 1945. The main features of the development of the ovum, embryo and ooecium in the ooecioferous Bryozoa Gymnolaemata. *Arkiv för Zoologi* 35A(17): 1-34.
Silén, L. 1977. Polymorphism. In R.M. Woollacott & R.L. Zimmer (eds), *Biology of Bryozoans*: 184-232. New York: Academic Press.
Soule, J.D., Soule, D.F. & Chaney, H.W. 1995. The Bryozoa. In J.A. Blake, H.W. Chaney, P.H. Scott & A.L. Lissner (eds), *Taxonomic Atlas of the Santa Maria Basin and Western Santa Barbara Channel*, vol. 13: 1-344. Santa Barbara: Museum of Natural History.

Taylor, P.D. & McKinney, F.K. Brooding in the Cretaceous bryozoan *Stichomicropora* and the origin of ovicells in cheilostomes. This volume

Tilbrook, K.J. 1998. The species of *Antropora* Norman, 1903 (Bryozoa: Cheilostomatida), with the description of a new genus in the Calloporoidea. *Records of the South Australian Museum* 31(1): 25-49.

Tilbrook, K.J., Hayward, P.J. & Gordon, D.P. 2001. Cheilostomatous Bryozoa from Vanuatu. *Zoological Journal of the Linnean Society* 131: 35-109.

Voigt, E. 1994. Zwei neue Bryozoengenera (*Filicisparsa sommerae* n. g. n. sp. und *Craticulacella schneemilchae* n. g. n. sp.) aus der Oberkreide von Misburg und Höver bei Hannover und von Lägerdorf (Holstein). *Bericht der Naturhistorischen Gesellschaft zu Hannover* 136: 7-25.

Wass, R.E. & Yoo J.J. 1983. Cheilostome Bryozoa from the Southern Australian continental shelf. *Australian Journal of Marine and Freshwater Research* 34: 303-354.

Waters, A. 1923. Mediterranean and other Cribrilinidae, together with their relationship to Cretaceous forms. *The Annals and Magazine of Natural History*, Series 9, 12(71): 545-573.

Winston, J.E. 1984. Shallow-water bryozoans of Carrie Bow Cay, Belize. *American Museum of Natural History Novitates* 2799: 1-38.

Winston, J.E. & Håkansson, E. 1986. The interstitial bryozoan fauna from Capron Shoal, Florida. *American Museum of Natural History Novitates* 2865: 1-50.

Zabala, M. 1986. Fauna dels briozous dels pa_sos Catalans. *Arxius de la Secció de Ciències, Institut d'Estudis Catalans* 84: 1-836.

Zabala, M. & Maluquer, P. 1988. Illustrated keys for the classification of Mediterranean Bryozoa. *Treballs del Museu de Zoologia* 4: 1-294.

Bryozoan Studies 2001, Wyse Jackson, Buttler & Spencer Jones (eds)
© 2002 Swets & Zeitlinger, Lisse, ISBN 90 5809 388 3

Neurotoxin found in the freshwater bryozoan *Lophopodella carteri*

N.L. Peterson
North Central College, Naperville, IL, USA

ABSTRACT: Previous studies have shown that fish die in the presence of crushed *Lophopodella carteri*. However, the mechanism of action was not elucidated. This paper reports that *L. carteri* homogenate contains a neurotoxic component. The presence of neurotoxins was tested by bath applying the *L. carteri* homogenate to rat hippocampal slices while monitoring neuroactivity in the Schaffer Collateral-CA1 pathway using extracellular recording techniques. The homogenate was neurotoxic with an IC_{50} = 6 μL for orthodromic responses and an IC_{50} = 10 μL for antidromic responses. These data suggest a nonsynaptic mode of action. The neurotoxic component apparently acts as an antagonist as noted by the lack of population spikes. Further, the inhibition was reversible. The neuronal response returned after removing the toxin. Partial purification of the toxin suggests that the toxin is not a protein, but a small molecule similar to bioactive molecules that have been found in some marine Bryozoa.

1 INTRODUCTION

In 1957, Rogick first reported that the freshwater bryozoan *Lophopodella carteri* possessed a poisonous substance capable of killing fish. While there is no evidence that fish are killed under natural conditions, fish that are confined in small containers die rapidly when crushed *L. carteri* are present (Rogick, 1957). Subsequently, Tenney & Woolcott (1964) showed that the toxin seemed to be specific to fish and salamander larvae as mice, tadpoles and several invertebrates showed no obvious ill effects when subjected to *L. carteri* extract. Further studies have indicated that the toxin may work through asphyxiation of the animals, but the precise mechanism of the toxin has not been elucidated (Collins et al., 1966, Meacham & Woolcott, 1968). While the toxin only seems to affect animals with a functional gill-breathing apparatus, the reason for this is not clear. It is possible that the toxin is more apparent in these animals because of the easy access to vital cells. The gills expose a large surface area of an organ that, when damaged, will cause rapid death.

This paper describes that the toxin in *L. carteri* is a more general neurotoxin not specific to gilled animals. Extract, when applied directly to rat brain slices, inhibits the neuroactivity. To test for neuroactivity, a rat hippocampus preparation was used. Extracellular electrophysiological techniques allowed easy recording of neuroactivity across a glutamatergic synapse when recording orthodromically and to detect a nonsynaptic response when recording antidromically. These techniques also allowed probing of the mechanism of action. It is possible to differentiate between an agonist and an antagonist by the presence or absence of population spikes (Koerner & Cotman, 1982). Agonists increase the post-synaptic end plate potential resulting in action potential formation in the cell body which are recorded as population spikes. In the presence of an antagonist, the response decreases without any population spikes.

2 METHODS

2.1 *Collection of Secretions*

Colonies of *L. carteri* were collected in Lake Shabbona in DeKalb County, Illinois, located at 41°44'45"N latitude and 88°51'30"W longitude. Colonies were homogenized in 10 mM phosphate buffer (pH 7.2) and centrifuged at 4000xg to precipitate cellular debris. Aliquots of the supernatant (protein concentration = 1.4 μg/ml by Bio-Rad protein assay) were stored at –80° C.

2.2 *Rat Hippocampus Slice Assay*

Experiments were performed using transverse slices of hippocampus obtained from male Sprague-Dawley (Harlan) rats (30-100 days old). The animals were anesthetized with urethane (1.5 g/kg, ip) and then decapitated. The forebrain was quickly removed and placed in cold (0°C) isolation medium comprised of 124 mM NaCl, 3.3 mM KCl, 5.0 mM $MgSO_4$, 0.5 mM $CaCl_2$, 1.2 mM KH_2PO_4, 10 mM glucose and 26 mM $NaHCO_3$ equilibrated with 95 % O_2/5% CO_2 to maintain the pH at 7.4. The hippocampus was isolated by blunt dissection and prepared transverse slices (450 μm thick) using a vibrotome (Campden Instruments, Model VSL). The slices were allowed to incubate for 15 minutes in the isolation medium and then transferred into incubation medium comprised of 124 mM NaCl, 3.3 mM KCl, 2.4 mM $MgSO_4$, 2.5 mM $CaCl_2$, 1.2 mM KH_2PO_4, 10 mM glucose and 26 mM $NaHCO_3$ equilibrated with 95 % O_2/5% CO_2 to maintain the pH at 7.4. After at least 1 hour, individual slices were transferred to a 380 μL microperfusion recording chamber (Koerner & Cotman, 1983). Stimulation (0.1 ms, 10-40 V) was delivered by a pair of twisted, Teflon-insulated stainless steel wires (0.003 inches diameter) at a frequency of 0.1 Hz. To record responses, a blunt stainless steel electrode (3-4 MOhms) was used. Orthodromic potentials were recorded by placing the stimulating electrode in the Schaffer Collateral axon bundle and the recording electrode in the CA1 dendritic field. Antidromic potentials were recorded by placing the stimulating electrode in the alveus and the recording electrode in the CA1 pyramidal cell body layer.

After electrode placement, the slice was submerged in oxygenated and rapidly stirred medium. When the electrophysiological response stabilized, toxin was applied to the bath and the extracellular synaptic field potentials were monitored by an oscilloscope. The amplitudes of the field potentials were noted before and after incubation of toxin. The secretion concentration was doubled every 4 minutes which was sufficient time for the response to restabilize.

2.3 *Gel filtration*

Separation of the extract was done using a Sephadex G-50 column. The void volume was determined by Dextran Blue. The inclusion volume was determined using NaCl and measuring conductivity of fractions.

3 RESULTS

Both orthodromic and antidromic responses were inhibited when *L. carteri* extract was bath applied to hippocampus slices (Figs 1-2). No population spikes were evident during the application. The toxin had an IC_{50} of 6 μL for orthodromic responses and 10 μL for antidromic responses. The responses were eliminated after addition of 10-20 μL of secretion corresponding to 14-28 μg of protein in the extract. These tests were performed using hippocampi from at least two different rats. The toxic effects were reversible upon removal of the extract from the slice chamber.

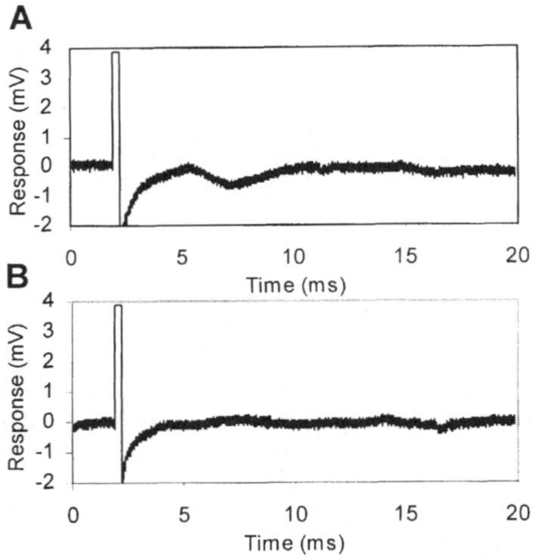

Figure 1. Inhibition of Orthodromic Neuronal Response. A. Response before addition of *L. carteri* extract. B. Response after bath application of 10 μL *L. carteri* extract.

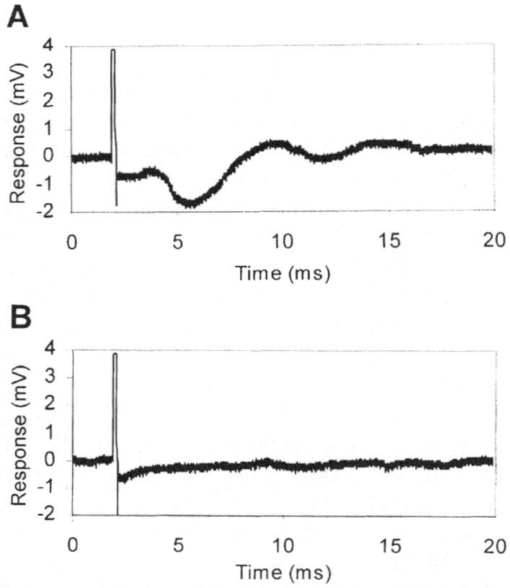

Figure 2. Inhibition of Antidromic Neuronal Response. A. Response before addition of *L. carteri* extract. B. Response after bath application of 20 μL *L. carteri* extract.

The toxic component has been partially purified. The extract was incubated in a boiling water bath for 15 minutes and centrifuged in a microcentrifuge for 5 minutes. The supernatant was then tested for its ability to inhibit neurotransmission. Similar to the original extract, 10-20 μL of heat-treated extract inhibited orthodromic neurotransmission completely.

The heat-treated extract was then run on a Sephadex G-50 gel filtration column. Each fraction was then tested for neuroactivity. The toxic portion of the extract was eluted with the inclusion volume, indicating that the toxin has an apparent molecular weight less than 1500.

4 DISCUSSION

Previous studies have shown that *L. carteri* possess a poisonous substance that is toxic to fish when the Bryozoa are crushed. Here it is shown that the extract may exert a portion of this toxicity through the inhibition of neurotransmission. The lack of population spikes indicate the toxin is working as an antagonist rather than an agonist at the site of action. The extract was shown to inhibit both the orthodromic response and the antidromic response. This suggests that the mode of action is not at the level of synaptic transmission because antidromic responses record neuroactivity along an axon, not across a synapse. This would be an effective antipredator adaptation because the toxin would affect both vertebrates who use acetylcholine as their neuromuscular junction neurotransmitter and invertebrates who use glutamate as their neuromuscular junction neurotransmitter. These data also argue that the specificity seen in fish and salamander larvae is due to the ability of the toxin to gain access to cells that are vital to the animal's survival, not necessarily because the toxin is specific for gill apparati. Invertebrates were previously shown to have no apparent ill effects when exposed to *L. carteri* extract (Tenny & Woolcott, 1964). This would not be unexpected if the toxin acts at motor neurons. The animal would have to get the toxin to the neuron actively by a mechanism such as ingestion whereas a gilled animal would be exposed to the toxin passively by having the toxin in solution. Further, since the gills are vital to life, the effect of the toxin would be noticed much more quickly.

Several marine species of Bryozoa have been shown to have non-protein bioactive components such as bryostatins (Pettit, 1991), alkaloids (Sjoblom & Bohlin, 1983, Prinsep et al., 1991, Holst et al., 1994), and quinones (Choi et al., 1993). The bioactive component of *L. carteri* is most likely a similarly small organic molecule rather than a larger protein. The toxin is heat stable and elutes with the salt peak in a Sephadex G-50 gel filtration column. If it were a larger protein, the heat treatment would have likely denatured the protein rendering it nonfunctional. Further, the gel filtration data points to the molecule being less that 1500 molecule weight. It therefore is suspected that the neurotoxic component will be a non-protein, organic molecule such as an alkaloid. Additional experiments are underway to test this hypothesis and to isolate the neurotoxic component of the extract.

Once the toxin is isolated and identified, it will be possible to determine if the active component exists as a toxin within the animal normally or whether the toxin becomes active upon homogenation. It is likely that the toxin already exists in an active form before homogenation due to the observations that fish have been swimming near, even brushing up against *L. carteri* but have not been observed eating them (Collins et al., 1966). In fact, it has been reported that bluegill sunfish, when offered individual colonies, would mouth the organism than quickly reject it (Meacham & Woolcott, 1968). However, larval caddisflies and turbellarians have been shown to be predators of L. carteri (Ricciardi & Reiswis, 1994). It is possible that these invertebrates have developed a mechanism that allows them to be unaffected by the toxin. Interestingly, several groups have reported insect larvae, snails, turbellarians, several species of fish and even ducks serving as predators of other species of freshwater bryozoa (Applegate, 1966, Bushnell, 1974, Ricciardi & Reiswis, 1994). These observations also indicate that the toxin in *L. carteri* may indeed show antipredator ability under physiological, biologically relevent conditions.

ACKNOWLEDGEMENTS

I would like to thank Dr Terrence Marsh for his assistance in identification and collection of the *L. carteri* colonies. He further served as a source of information about bryozoan biology. Financial support for this project was provided by North Central College Summer Grants.

REFERENCES

Applegate, R.L. 1966. The use of a broyzoan, *Fredericella sultana*, as food by sunfish in Bull Shoals reservoir. *Limnology & Oceanography* 11:129-130.

Bushnell, J.H. 1974. Bryozoans (Ectoprocta). In C.W. Hart, Jr. & S.L.H. Fuller (eds), *Pollution ecology of Freshwater invertebrates*: 157-194. New York: Academic Press.

Choi, Y.H, Park, E. & Schmitz, F.J. 1993. Perfragilins A and B, cytotoxic isoquinolinequinones from the bryozoan *Membranipora perfragilis*. *Journal of Natural Products* 56(8): 1431-1433.

Collins, E.J., Tenney, W.R. & Woolcott, W.S. 1966 Histological effects of the poison of *Lophopodella carteri* (Hyatt) on the gills of *Carassius auratus* (Linnaeus) and larval *Ambystoma opacum* (Gravenhorst). *Virginia Journal of Science* 17: 155-163

Holst, P.B., Anthoni, U., Christophersen, C. & Nielsen, P.H. 1994. Marine alkaloids, 15. Two alkaloids, Flustramine E and Debromoflustramine B, from the marine bryozoan *Flustra foliacea*. *Journal of Natural Products* 57(7): 997-1000.

Koerner, J.F. & Cotman, C.W. 1982. Response of Schaffer Collateral-CA1 pyramidal cell synapses of the hippocampus to analogues of acidic amino acids. *Brain Research* 251: 105-115.

Koerner, J.F. & Cotman, C.W. 1983. A microperfusion chamber for brain slice pharmacology. *Journal of Neuroscience Methods* 7: 243-251.

Meacham, R.H. & Woolcott, W.S. 1968. Studies of the coelomic fluid and isotonic homogenates of the freshwater bryozoan *Lophopodella carteri* (Hyatt) on fish tissues. *Virginia Journal of Science* 19(3): 143-146.

Pettit, G.R. 1991. The Bryostatins. In W. Herz, G.W. Kirby, W. Steglich & C. Tamm (eds), *Progress in the Chemistry of Organic Natural Products*: No. 57, 153-195. New York: Springer-Verlag.

Prinsep, M.R., Blunt, J.W. & Munro, M.H.G. 1991. New cytotoxic β-carboline alkaloids from the marine bryozoan, *Cribricellina cribraria. Journal of Natural Products* 54(4): 1068-1076.

Ricciardi, A. & Reiswis, H.M. 1994. Taxonomy, distribution, and ecology of the freshwater bryozoa (Ectoprocta) of eastern Canada, *Canadian Journal of Zoology* 72:339-359.

Rogick, M.D. 1957. Studies on freshwater Bryozoa, XVIII *Lophopodella carteri* in Kentucky. *Tranactions of the Kentucky Academy Science* 18(4): 85-87.

Sjoblom, T. & Bohlin, L. 1983. Studies of Swedish marine organisms: Muscle-relaxant alkaloids from the marine brozoan *Flustra foliacea. Acta Pharmaceutica Suecica* 20: 415-419.

Tenney, W.R. & Woolcott, W.S. 1964. A comparison of the responses of some species of fishes to the toxic effect of the Bryozoan, *Lophopodella carteri* (Hyatt). *Virginia Journal of Science* 15(1): 16-20.

Bryozoan Studies 2001, Wyse Jackson, Buttler & Spencer Jones (eds)
© 2002 Swets & Zeitlinger, Lisse, ISBN 90 5809 388 3

Bryozoans in Ordovician depositional sequences, Cincinnati Arch region, USA

June R.P. Ross
Department of Biology, Western Washington University, Bellingham, WA 98225, USA
Charles A. Ross
Department of Geology, Western Washington University, Bellingham, WA 98225, USA

ABSTRACT: Various biostratigraphic and lithostratigraphic models are used to interpret the complex facies and age relationships of the Ordovician Lexington Limestone (Trentonian) and Cincinnatian strata of the Cincinnati Arch region. This famous stratigraphic succession is rich in fossils, including bryozoans. Regional unconformities separate the succession into at least four, and perhaps as many as twelve, depositional sequences and provide the framework in which bryozoan assemblages are analyzed. At least five bryozoan assemblages characterize these depositional sequences. The Lexington Limestone rests unconformably on a regional erosional surface which is a widely recognized sequence boundary event throughout the eastern North American craton. The upper Lexington Limestone forms a shallow carbonate bank on the Jessamine Dome and indicates a regional lowering of sea-level. Off the bank's flanks, different facies expand, contract, and change abruptly into a siltstone-calcisiltite-wackestone of the Clays Ferry Formation. Tongues of the upper Lexington Limestone extend northward into the lowest part of the Cincinnatian Kope Formation (i.e. Point Pleasant Member). Most of the Kope Formation is shale and siltstone deposited during high sea-level and has a low bryozoan diversity. General shallowing near the top of the Kope brings a return to more carbonate-rich shales and siltstones (McMicken Member) and increased bryozoan diversity. Gradual shallowing continues through the Maysvillian, accompanied by the introduction of many additional bryozoan species. Most of these range upward into the Arnheim Formation where they become extinct. Only a few bryozoans survived into the overlying Waynesville, Liberty, and Whitewater formations.

1 INTRODUCTION

Rich bryozoan assemblages are common in Upper Ordovician (Blackriveran through Gamachian) strata in the midwestern states (southeastern Minnesota, northeastern Iowa, Illinois, southeastern Indiana, Ohio, Kentucky, and Tennessee). Early studies by numerous geologists, including Ulrich (1890, 1893) reported on the prolific bryozoan faunas. Subsequent studies (Anstey & Perry 1973, Anstey 1986, Brown & Daly 1985, Karklins 1983, 1984, 1985, Utgaard & Perry 1964) have further documented these faunas where they occur in third, fourth-, and smaller-order depositional sequences. This paper presents a preliminary report on how bryozoan distributions in the Upper Ordovician Mohawkian and Cincinnatian series relate to the depositional sequence framework of the region. The Late Ordovician bryozoans exemplify in considerable detail great abundance, diversity, and rapid rates of evolutionary change. Relatively few Ordovician bryozoan species lineages actually survived beyond the end of the Ordovician. The placement of these diverse bryozoan faunas into a framework of depositional sequences provides a means of analyzing their environmental abaptations and their distributions within specific depositional sequences. The latest Ordovician Gamachian Stage was a time of major extinctions and few originations within all bryozoan lineages and the earliest Silurian Ruddanian Stage records only a slow evolutionary recovery in the few surviving bryozoan lineages.

2 REGIONAL RELATIONS

The Upper Ordovician of southwestern Ohio, southeastern Indiana, and northcentral Kentucky consists of the Mohawkian and Cincinnatian Series which form the standard reference sections for the latest Ordovician series in North America. Keith (1985) illustrated a late Trentonian Sebree Trough which passed northeastward across southeastern Indiana through central Ohio. This trough separated the Lexington Platform on the southeast from the Galena Platform on the northwest. During the succeeding Cincinnatian Stage, the Sebree Trough remained a prominent feature, and, as Elias (1995) and

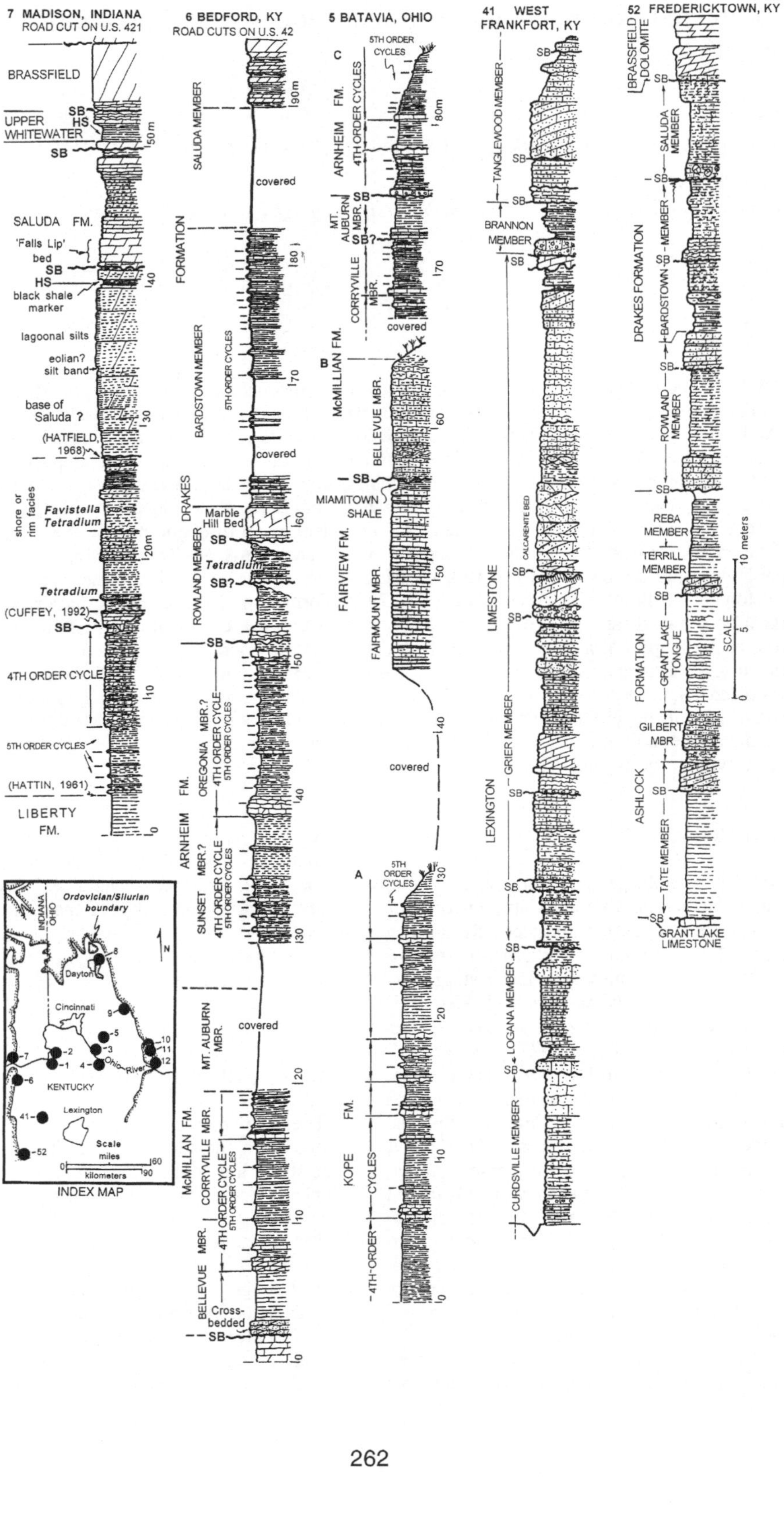
7 MADISON, INDIANA
ROAD CUT ON U.S. 421
BRASSFIELD
UPPER WHITEWATER
SALUDA FM.
'Falls Lip' bed
black shale marker
lagoonal silts
eolian? silt band
base of Saluda ?
(HATFIELD, 1968)
shore or rim facies
Favistella
Tetradium
(CUFFEY, 1992)
4TH ORDER CYCLE
5TH ORDER CYCLES
(HATTIN, 1961)
LIBERTY FM.
6 BEDFORD, KY
ROAD CUTS ON U.S. 42
SALUDA MEMBER
covered
DRAKES FORMATION
BARDSTOWN MEMBER
5TH ORDER CYCLES
Marble Hill Bed
ROWLAND MEMBER
ARNHEIM FM.
OREGONIA MBR.?
SUNSET MBR.?
4TH ORDER CYCLE
MT. AUBURN MBR.
McMILLAN FM.
CORRYVILLE MBR.
BELLEVUE MBR.
Cross-bedded
5 BATAVIA, OHIO
5TH ORDER CYCLES
ARNHEIM FM.
4TH ORDER CYCLES
MT. AUBURN MBR.
CORRYVILLE MBR.
McMILLAN FM.
BELLEVUE MBR.
MIAMITOWN SHALE
FAIRVIEW FM.
FAIRMOUNT MBR.
KOPE FM.
4TH ORDER CYCLES
41 WEST FRANKFORT, KY
TANGLEWOOD MEMBER
BRANNON MEMBER
LEXINGTON LIMESTONE
CALCARENITE BED
GRIER MEMBER
LOGANA MEMBER
CURDSVILLE MEMBER
52 FREDERICKTOWN, KY
BRASSFIELD DOLOMITE
SALUDA MEMBER
DRAKES FORMATION
BARDSTOWN MEMBER
ROWLAND MEMBER
REBA MEMBER
TERRILL MEMBER
ASHLOCK FORMATION
GRANT LAKE TONGUE
GILBERT MBR.
TATE MEMBER
GRANT LAKE LIMESTONE
SCALE
10 meters
Ordovician/Silurian boundary
INDIANA
OHIO
Dayton
Cincinnati
Ohio River
KENTUCKY
Lexington
Scale
miles
kilometers
INDEX MAP

Figure 1 (facing page). Locations of sections. Section 5 just north of Batavia, Ohio, remeasured and modified after Hay et al. 1981: A) along eastbound entry road from Baur Road onto Ohio highway 32; B) along Ohio highway 32 immediately east of A; and C) along Ohio highway 132 at Filagar Road. Section 6 in road cuts on U.S. highway 42, 4 km (2.5 miles) east of Bedford, KY, modified from Hattin et al. 1961. Section 7 in road cuts along U.S. highway 421, north-northeast of Madison, Indiana; remeasured and reinterpreted from Cuffey, 1998. Section 41 road cuts along east side of highway U.S. 127 just north of junction with Kentucky 1005 western side of Frankfort, Kentucky. Section 52 road cuts along highway U.S. 150 2 km west of Fredericktown, Kentucky.

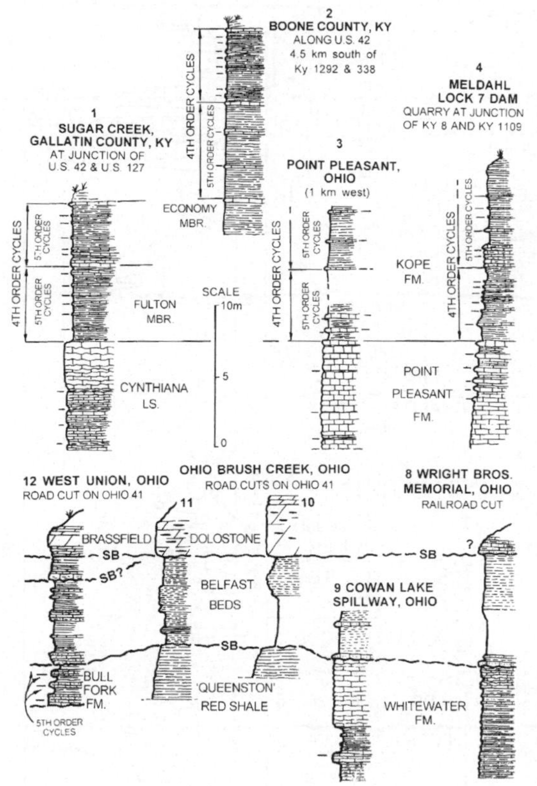

Figure 2. Sections 1 to 4 in the Point Pleasant, Cynthiana, and lower part of Kope Formations. Sections 1 and 2 modified from Hattin et al., 1961. Section 3 from stream outcrops and road cuts in a new housing development at Point Pleasant, OH. Section 4, at Bradford, KY, locality of Weiss et al., 1965, and Stop 1 of Meyer et al., 1981; remeasured in a reopened quarry that shows Point Pleasant–Kope relations. Sections 8 to 12 in the Upper Ordovician along the eastern side of the Cincinnati Arch. Section 8, railroad cut at north side of Wright Brothers Memorial Hill, Dayton, OH. Section 9, Cowan Lake spillway. Section 10, a road cut along Ohio highway 41, south side of Ohio Brush Creek. Section 11, another road cut 0.2 km farther south. Section 12, about 5 km north of West Union, on Ohio highway 41.

Young & Elias (1995) showed from their study of corals, it separates two faunal provinces, the Richmond Province to the east and the Red River-Stony Mountain Province to the west. These differences in faunas make biostratigraphic correlations somewhat difficult and emphasize the need to fully understand the history of depositional sequences in order to include sea-level history in the criteria for establishing detailed correlations.

As Tobin (1986) and Cuffey (1998) have emphasized, many of the Cincinnatian rock units were originally defined on their fossil content and were not strictly based on a lithologic or rock-type definition. However, most of these units do approximate lithostratigraphic units. Cuffey (1998) presented a recent and well-thoughtout correlation scheme for many of the lithologic (or rock) names used in the area. With some modification, this rock unit scheme can be adapted to a depositional sequence classification.

Our bryozoan collections and localities from the Cincinnati Arch are well located in 163 stratigraphic sections from our 1995 and 1998 field work.. Sixteen of these stratigraphic sections are shown in Figures 1 & 2. In 1995 and 1998, we also undertook to place the 134 measured sections from our 1962, 1963 and 1964 field work, along with the many bryozoan collections, in a depositional sequence framework. More than 30,000 thin sections have been prepared from these collections.

The extensive Kentucky mapping project during the 1960s and 1970s by members of the US Geological Survey (Cressman 1973, Weir et al. 1984) has provided a detailed stratigraphic framework in central and northern Kentucky and also resulted in extensive bryozoan studies by Karklins (1983, 1984).

Others also have shown considerable interest in defining depositional sequences in these strata. Holland (1993, 1995) identified five (and in 1995, six) third-order depositional sequences in the Cincinnatian Series and Holland and Patzkowsky (1996 & 1997) extended their studies into the Trentonian and Blackriveran beds of Kentucky and Tennessee. Pope & Read (1995, 1997) also examined these strata as depositional sequences and pointed out their regional significance. They identified only four large depositional sequences and their interpretations differ in several other aspects from those presented by Holland and Patzkowsky. In our studies from Kentucky, Ohio, Indiana, Tennessee, Illinois, Iowa, and Minnesota, we have identified several additional depositional sequences, whether they are fourth- or third-order cycles is not clear (Figs 1 to 4).

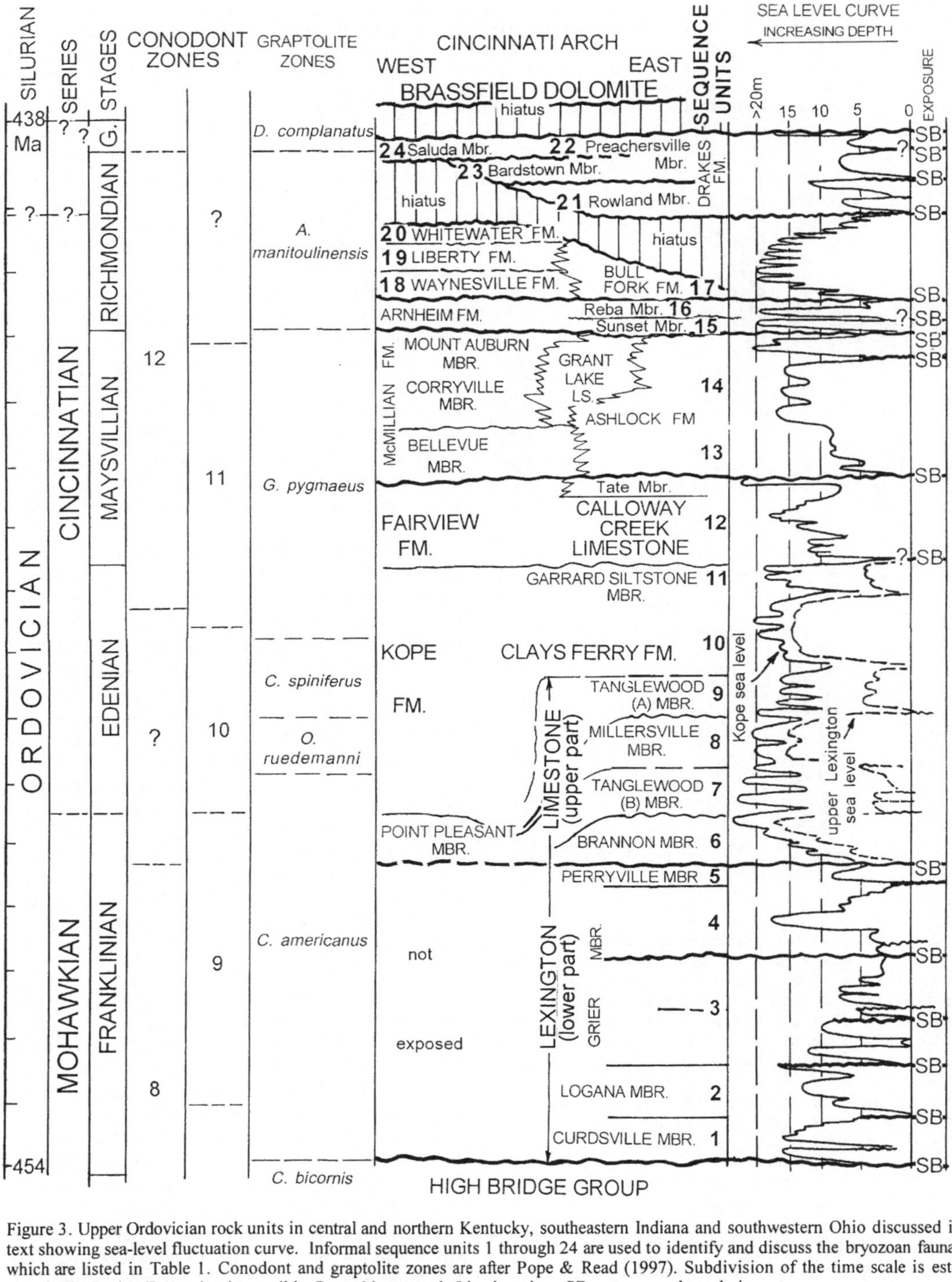

Figure 3. Upper Ordovician rock units in central and northern Kentucky, southeastern Indiana and southwestern Ohio discussed in text showing sea-level fluctuation curve. Informal sequence units 1 through 24 are used to identify and discuss the bryozoan faunas which are listed in Table 1. Conodont and graptolite zones are after Pope & Read (1997). Subdivision of the time scale is estimated. The Drakes Formation is possibly Gamachian or early Llandoverian. SB = sequence boundaries.

3 DEPOSITIONAL SEQUENCES AND BRYOZOAN DISTRIBUTION

More than 190 species of bryozoans have been reported from the Lexington Limestone and the Cincinnatian Series in the Cincinnati Arch region of Kentucky, Ohio, and Indiana (Table 1). These are listed by informal 'sequence units' shown in Figure 3. In the middle Richmondian depositional sequence, we have kept separate the four formations in order to illustrate progressive faunal changes. Species are arranged by first stratigraphic appearance of genera, then by first appearance of species within a genus. Table 1 is only a general approximation of species distributions because of the number of taxonomists involved in identification and stratigraphic locations of the samples are not always precise. No attempt has been made to determine a phylogenetic succession of species within genera.

A regional unconformity and several volcanic ash beds mark the top of the lower part of the Mohawkian Series and form the base of the succession discussed here (Cressman 1973, Weir et al. 1984). The upper Mohawkian (Franklinian Stage) and the overlying Cincinnatian Series have well studied faunas, biostratigraphy and lithostratigraphy. In general, the depositional environments of these rocks suggest a broad, shallow ramp of low relief, except for penecontemporaeous faulting activity (Borella & Osborne 1978, Ettensohn & Kulp 1995) in the Jessamine Dome area of central Kentucky (and the Nashville Dome area, Tennessee).

Pope and Read (1997) recognized four third-order depositional sequences which they traced from the Cincinnati Arch region eastward into southwestern Virginia and made comparison with several earlier studies depicting sea level curves in this stratigraphic interval. Within these four packages of sediments, they also recognized additional fourth-order depositional cyclicity, including well-developed sequence boundaries, however, they did not trace these smaller cycles across the region or extensively correlate them from one stratigraphic section to the next. Figure 2 shows the relations of the four sequences to some of the major stratigraphic units. Holland & Patzkowsky (1996) recognized seven different depositional sequences through this upper Mohawkian and Cincinnatian interval, basing their sequence boundaries in several places at different unconformities in the succession than did Pope & Read (1997).

We generally follow Pope & Read's interpretation for their major sequences: 1) the Franklinian (Lexington Limestone); 2) Edenian (Kope/Clays Ferry); and 3) the Maysvillian into the overlying lower beds of the Richmondian (top of Ft. Ancient Member). However, the cyclicity in the middle and upper parts of the Richmondian are more complicated than shown by either Holland & Patzkowski (1996) or Pope & Read (1997). For these strata, we recognize at least one or two additional third- and fourth-order depositional cycles.

Sequence 1 is about 60 m thick and starts with a relatively rapid sea-level rise and transgression (Curdsville Member) over the post-Blackriveran unconformity at the sequence boundary (SB). Maximum flooding occurred within the Logana Member 5 to 10 m above the SB. The carbonates of this sequence shoal upwards through the Grier Member and overlying tidal flat facies of the Perryville Member.

The top of the Perryville is karsted with infilling down to about 1.5 m. Bryozoans are rare in the beds below the maximum flooding event and include scattered *Orectodictya pansa, Trigonodictya cirrita, Stictopora neglecta, Escharopora eparata, Graptodictya* sp., *Tarphophragma multitabulata* and *Mesotrypa angularis.* After the maximum flooding event, sea level gradually fell and, with shallowing, a succession of bryozoans was able to inhabit the shoaling upwards facies, particularly in the Grier Member. First, *Prasopora falesi, Heterotrypa trentonensis, Eridotrypa trentonensis,* and *Homotrypa granulifera* appear. Five meters higher, *Cyphotrypa acervulosa* appears, which is followed 4 m higher by *Balticoporella tenuimurale* and *Homotrypa cressmani. Eridotrypa mutabilis* and *Dekayia epitrima* appear 4 m higher. The upper 5 m of the Grier reveal a sudden influx of *Cyphotrypa switzeriensis, Stigmatella multispinosa, Crepipora venusta, Ceramophylla alternatum, Amplexopora aff. A. winchelli,* and *Balticopora arcuatilis.* The Perryville tidal flat facies has only rare bryozoans. Bryozoans in sequence 1 belong to the *Tarphophragma multitabulata* assemblage zone of Karklins (1984).

Depositional sequence 2 of Pope & Read (1997) is about 70 m thick and includes the upper part of the Lexington Limestone (Tanglewood carbonate bank) and Clays Ferry Formation in central Kentucky and equivalent beds in the Kope Formation in the Cincinnati area. Sea level rose significantly during deposition of the lower part of the Kope where the maximum flooding event is recorded (Fig. 4). Holland & Patzkowski (1996) use a brief shallowing event in the Point Pleasant Member, which did not shoal to an exposure surface, as an additional boundary to subdivide this sequence into two nearly equal subsequences. Within the Tanglewood Bank we recognize at least two well-developed shallow to deep cycles, which may be sea level events, but alternatively are more likely caused by local tectonic events. Within this diverse set of lithologies, the shallow-water Tanglewood carbonate bank passes laterally and vertically into 2 to 5 m cyclical couplets of calcareous siltstone and shale and thinner shell beds which eventually overstep and cover the bank.

The top of Pope & Read's (1997) sequence 2 is placed in the Garrard Siltstone, a ramp margin wedge (8 to 30 m thick) that is a widely distributed shal-

Table 1. Distribution of bryozoan genera and species in sequence depositional units in upper Mohawkian and Cincinnatian rocks as shown in Figure 2B. Data principally from Cressman (1973), Karklins (1983, 1985), Weir & others (1984) and Brown & Daley (1985).

Sequence units	1	2	3	4	5	6	7	8	9	10	11	12	13	14	15	16	17	18	19	20	21	22	23	24
Bryozoan genera and species ranges	C	L	LG	UG	P	Br	LT	M	UT	CF	Gar	FC	BG	C	Ss	R	BF	Wa	L	Ww	Ro	P	Ba	S
Orectodictya pansa	X																							
Trigonodictya cirrita	X		X	X	X	X		X																
Trigonodictya? spp.													X	X			X							
Stictopora neglecta	X	X	X	X	X	X	X	X	X	X														
Stictopora sp.										X		X												
Stictopora cf. S. lata										X	X						X				X		X	
Stictopora? sp.																					X		X	
Escharopora eparmata	X			X	X	X	X	X	X	X														
Escharopora spp.										X		X					X							
Escharopora falciformis												X		X				X						
Escharopora cf. E. hilli												X												
Escharopora hilli												X												
Tarphophragma multitabula	X	X	X	X	?																			
Homotrypa similis	X																							
Homotrypa granulifera		X	X																					
Homotrypa cressmani				X		X	X	X	X															
Homotrypa spp.										X		X	X	X	X	X								X
Homotrypa obliqua											?	X	X	X										
Homotrypa cincinnatiensis											?	X												X
Homotrypa alta												X												
Homotrypa flabellaris												X	X	X	X	X		X	X	X				
Homotrypa flabellaris spinifera												X	X											
Homotrypa curvata												X	X	X										
Homotrypa cf. H. spinea												X	X	X										
Homotrypa? spp.												X	X	X			X	X	X		X		X	
Homotrypa cf. H. flabellaris spinifera												X	X											
Homotrypa cf. H. cincinnatiensis												X												
Homotrypa pulchra													cf.	X	X	?		cf.						
Homotrypa communis																		X	X					X
Homotrypa nodulosa																		X	X	X				
Homotrypa ramulosa																						X		
Homotrypa wortheni																								X
Graptodictya spp.	X									X		X	X		X		X				X		X	
Prasopora falesi		X	X	X				X																
Mesotrypa angularis (=spinosa)		X	X	X																				
Mesotrypa spp.								X		X			?	?										
Mesotrypa patella																			X	X				

Table 1 continued

Sequence units	1	2	3	4	5	6	7	8	9	10	11	12	13	14	15	16	17	18	19	20	21	22	23	24
Bryozoan genera and species ranges	C	L	LG	UG	P	Br	LT	M	UT	CF	Gar	FC	BG	C	Ss	R	BF	Wa	L	Ww	Ro	P	Ba	S
Heterotrypa trentonensis		X	X	X	X	?																		
Heterotrypa foliacea						X	X	X	X	X														
Heterotrypa spp.							X		X	X	X	X	X	X		X	X							
Heterotrypa ulrichi										X	X													
Heterotrypa frondosa										X	X	X	X	X	X	X								X
Heterotrypa cf. H. frondosa											X	X	X			X	X							
Heterotrypa subfrondosa												X	X											X
Heterotrypa solitaria													X	X										
Heterotrypa inflecta													X	X										
Heterotrypa subramosa																						X		
Eridotrypa trentonensis		X	X	X				X																
Eridotrypa mutabilis			X	X	X	X	X	X	X															
Eridotrypa sadievillensis						X	X	X	X															
Eridotrypa spp.						X	X	X		X							X		X	X			X	
Dekayia appressa		X	X	X																				
Dekayia epitrima				X			X	X																
Dekayia spp.								X				X	X				X				X			
Dekayia? spp.										X			X											
Dekayia multispinosa											?	X	X	X	X									
Dekayia maculata											?	X		X	X									
Dekayia aspera											?	X	X	X	X		X							
Dekayia pelliculata											?	X												
Dekayia cf. D. appressa												X	X	X			X							
Dekayia cf. D. nicklesi												X	X	X			X				X		X	
Dekayia semipilaris												X												
Dekayia magna												X												
Dekayia catenula															X									
Dekayia cf. D. aspera																	X							
Dekayia stidhami																				X				
Stigmatella conica		X	X																					
Stigmatella multispinosa			X	X			X																	
Stigmatella clavis							X			X	X													
Stigmatella spp.							X			X	X			X			X							
Stigmatella? spp.									X	X			X				X							
Stigmatella crenulata																		X	X					
Stigmatella interporosa																		X	X					
Stigmatella cf. S. personata																				X				
Cyphotrypa acervulosa			X	X		X	X	X																
Cyphotyrpa switzeriensis			?	X	?	?																		
Cyphotrypa spp.									X	X			?	?										
Balticopora tenuimurale			X	X	X	X	X		X	X														
Balticopora arcuatilis.						X									X									
Crepipora venusta				X	X	X	X	X	X	X														
Crepipora simulans												X												
Crepipora spp.													?	X			X				X		X	

Table 1 continued

Sequence units	1	2	3	4	5	6	7	8	9	10	11	12	13	14	15	16	17	18	19	20	21	22	23	24
Bryozoan genera and species ranges	C	L	LG	UG	P	Br	LT	M	UT	CF	Gar	FC	BG	C	Ss	R	BF	Wa	L	Ww	Ro	P	Ba	S
Ceramophylla alternatum				X	X	X	X	X	X	X														
Constellaria teres						X	X	X	X	cf.	cf.													
Constellaria cf. C. fisheri							X	X	X	cf.														
Constellaria spp.										X		X	X				X					X		
Constellaria florida												X	X											
Constellaria polystomella																		X	X					X
Peronopora milleri						X	X	X	X	X														
Peronopora vera								X		X														
Peronopora spp.										X		X	X	X	X			X						
Peronopora cf. P. decipiens												X	X	X	X		X				X			
Peronopora? spp.												X		X										
Peronopora hospitalis																		X	X	X				
Ceramoporella distincta						X	X	X	X	X														
Ceramoporella? spp.										X		X	X	X			X							
Ceramoporella spp.												X	X	X			X							
Ceramoporella cf. C. ohioensis															X		X							
Parvohallopora nodulosa						X	X	X	X	X	X													
Parvohallopora onealli							X	X	X	X														
Parvohallopora ramosa								X	X	X	X	X	X	X	X	X	X				X		X	
Parvohallopora subnodosa								?	X		X							X	X	X				
Parvohallopora subplana											X													X
Parvohallopora cf. P. ramosa												X	X	X	X			X	X	X				
Parvohallopora spp.												X	X	X	X		X					X		X
Papillalunaria spatiosa						X																		
Atactoporella newportensis							X	X	X	X														
Atactoporella mundula												X	X			X								
Atactoporella sp.													X											
Atactoporella? sp.													X											
Acanthoceramoporella valliensis								X																
Amplexopora septosa										X	X	X												
Amplexopora persimilis										X		cf.												
Amplexopora filiasa												X	X	X	X									
Amplexopora ampla												X	X					X						
Amplexopora spp.												X	X			X	X				X		X	
Amplexopora cf. A. filiasa													X											
Amplexopora cf. A. robusta													X		X									
Amplexopora robusta																	X							
Amplexopora cf. A. pustulosa																	X							
Amplexopora variabile																								X

Table 1 continued

Sequence units	1	2	3	4	5	6	7	8	9	10	11	12	13	14	15	16	17	18	19	20	21	22	23	24
Bryozoan genera and species ranges	C	L	LG	UG	P	Br	LT	M	UT	CF	Gar	FC	BG	C	Ss	R	BF	Wa	L	Ww	Ro	P	Ba	S
Batostoma jamesi										X														
Batostoma implicatum										X		X												
Batostoma gracilis												X	X	X	X	X								
Batostoma spp.												X	X				X					X	X	
Batostoma? sp.													X											
Monticulipora mammulata										X	X													
Monticulipora spp.												X	X				X							
Monticulipora epidermata																			X	X				
Leptotrypa? sp.										X														
Leptotrypa minima													X	X										
Atactopora sp.										X														
Hemiphragma sp.										X														
Phylloporina sp.										X		X												
Phylloporina? sp.												X												
Bythopora sp.										?														
Bythopora dendrina											X		X				X						X	X
Bythopora aff. B. dendrina													X				X				X		X	
Calloporella? sp.											X													
Calloporella sp.													X											
Homotrypella spp.													X				X	X	X		X		X	
Homotrypella gracilis													X				X						X	X
Homotrypella cf. H. meeki																	X							
Homotrypella varians																		X	X	X				X
Homotrypella delicatula																				X				
Homotrypella cumingsi																				X				
Homotrypella cf. H. delicatula																						X		
Nicholsonella? sp.													X											
Nicholsonella vaupeli													X	X	X	X		X	X					
Gortanipora cf. G. bassleri																	X							
Gortanipora cf. richmondensis																	X							
Gortanipora spp.																	X					?		
Gortanipora bassleri																	X	X	X				X	
Rhombotrypa quadrata																	X	X	X	X				X
Rhombotrypa cf. R. quadrata																	X							
Rhombotrypa multitabulata																				X				
Hallopora congrua																				X				
fenestellid indet.																						X		

low-water marker in central Kentucky. To the northwest, the Garrard Siltstone is overlain by the Fairview Formation. We chose the top of the Garrard Siltstone as our sequence boundary (top of our sequence unit 11) because it is overlain by an easily recognizable sea level rise.

Bryozoans in the lower half of sequence 2, particularly in the shallow-water carbonate bank facies, are common and diverse and form Karklins' (1984) *Constellaria teres* assemblage zone. Some bryozoans range upwards from sequence 1, however, the following are first appearences in this region. These are in order of first appearances in the transgressive deposits (gradual sea-level rise): *Stigmatella* sp., *Constellaria teres,* and *Homotrypa similis;* followed by *Peronopora milleri* and *Heterotrypa foliacea;* and then in short succession, *Ceramoporella distincta, Parvohallopora nodulosa, Acanthoceramoporella valliensis, Papillalunaria spatiosa, Eridotrypa sadievillensis, Parvohallopora onealli, Stigmatella clavis,* and *Peronopora vera.* This succession is followed by an interval of 10 to 12 m that lacks bryozoans. In the upper part of sequence 2 in Kentucky, beds of the Point Pleasant tongue of the Clays Ferry have *Dekayia* sp., and a succession of appearances of *Atactoporella newportensis, Mesotrypa sp.,* and *Amplexopora persimilis,* through about 5 m. About 40 m higher, in the upper part of sequence 2, *Mesotrypa* sp. and *Batostoma* cf. *B. jamesi* appear and indicate a correlation with the upper part of the Kope Formation. Immediately below the Point Pleasant beds are the last regional appearances of *Prasopora simulatrix* and *Constellaria* sp. and within the Point Pleasant is the last appearance of *Eridotrypa mutabilis* in the region. *Peronopora vera* and *Atactoporella* cf. *A. typicalis* are also present in the Point Pleasant and range higher. Immediately above the Point Pleasant beds *Atactoporella typicalis, Heterotrypa ulrichi, Amplexopora septosa* and *Balticoporella whitfieldi* first appear. Higher, 15 m of siltstones lack bryozoans and are overlain by a series of progressively shallowing shale–siltstone–limestone cycles containing the earliest *Batostoma jamesi* , and the last *Balticoporella whitfieldi* and *Amplexopora septosa.* The McMicken Member (or equivalents) at the top of the Kope has *Dekayia aspera* and several variants of *Heterotrypa ulrichi.*

In Kentucky, sequence 3 of Pope & Read (1997) includes the stratigraphic interval from the Calloway Creek Limestone to the top of the Tate Siltstone Member in the lower part of the Ashlock Formation. It extends upward through thin transgressive systems tracts of poorly bedded packstones, calcisiltites, and shales into a maximum flooding event in the lower part of the Calloway Creek Limestone. The highstand systems tract consists of skeletal wackestones and packstones of Calloway Creek and the dolomitic Tate Siltstone Member which locally has a well-marked exposure surface at its top with

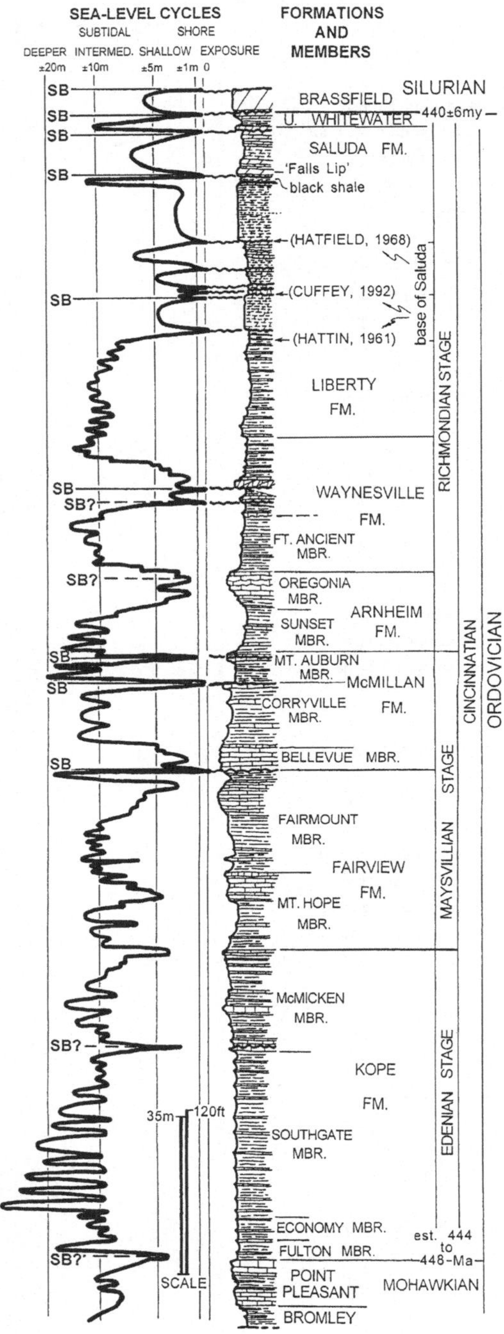

Figure 4. Generalized stratigraphic column of the Cincinnatian Series (with more traditional ages) showing sea-level fluctuation curve, sequence boundaries, maximum flooding surfaces and interpretation of depths of depositional environments. 'SB'= Sequence Boundary, 'HS' = Highstand.

paleosols and subareal diagenetic features. The top of the Tate Member appears to correlate with the unconformity between the Fairmont and Bellevue members in the Cincinnati area which makes sequence 3 equivalent to the Fairview Formation. Elsewhere in their study, Pope & Read (1997) seem to refer this boundary to an unconformity 13 m higher where a well-developed unconformity appears 1 m below the top of the Grant Lake Limestone (Figs 3 & 4).

Again, a succession of bryozoans appear progressively upward through the sequence and relatively few species range up from lower beds. In Kentucky, the new appearances include in ascending order *Homotrypella* spp., *Constellaria* spp., *Peronopora* spp., *Ceramoporella* sp., *Escharopora* sp. *Parvohallopora* sp., *Batostoma implicatum, Homotrypa* spp., *Stictopora* sp., *Heterotrypa* sp., *Heterotrypa frondosa, Monticulipora* spp., *Atactoporella mundula, Heterotrypa inflecta, Dekayia aspera, Graptodictya* sp., and *Calloporella* sp. (a new arrival) at the top of this sequence 6 m below the top of the Grant Lake Limestone.

In their sequence 4, Pope & Read (1997) included the rest of the upper Ordovician strata, however, this interval in southeastern Indiana, southwestern Ohio, and northeastern Kentucky is much more complicated and more complete than in central Kentucky and requires more detailed consideration. We subdivide this interval into sequence units 15 to 24 as shown in Figure 3. Two deepening to shoaling events mark the Arnheim Formation (Fig 4) at the stratigraphic level of the Sunset (sequence unit 15) and Reba (sequence unit 16) members in Kentucky.

Lithologically and faunally the Bull Fork Formation (sequence unit 17) is a lateral facies of the Waynesville, Liberty, and the Whitewater formations (sequence units 18, 19, 20) and make up a major depostional sequence (a part of Pope & Read's, 1997, sequence 4). We listed these formations separately to show the progressive changes in the fauna upwards through this depositional sequence.

A deepening to shoaling event marks the base of the overlying Fort Ancient Member of the lower Waynesville Formation (sequence unit 18). In the Cincinnati area, (Fig. 4) this event is followed by a series of short-event, meter scale, shale-siltstone to silty limestone cycles in the middle and upper parts of the Waynesville Formation. These cycles gradually shallow upward into the Liberty (sequence unit 19) and Whitewater formations (sequence unit 20). In northeastern Kentucky, the equivalent cyclical succession is in the Bull Fork Formation (sequence unit 17). This interval of Richmondian well-developed meter scale cycles is a distinctive sedimentary package. However, the thickness of this depositional sequence is extremely irregular and we propose that its upper surface was extensively eroded, in places down to the Reba and Sunset members to produce a topography of more than 25 m of relief prior to the deposition of the overlying Drakes Formation. Wier & others (1984, p. 49, fig.30) show these general geometric relations but consider the Bull Fork to be a lateral facies of the overlying Drakes Formation rather than being unconformable.

Bryozoans in the lower 6 m of the Bull Fork Formation include *Graptodictya* sp., *Heterotrypa* spp., *Parvohallopora* spp., *Dekayia aspera, Phylloporina* sp., *Batostoma* sp., *Ceramoporella* sp., *Eridotrypa* sp., *Homotrypa* sp. and monticuliporoids. A few meters higher, *Peronopora* cf. *P. decipiens, Homotrypella* spp., *Heterotrypa* cf. *H. frondosa, Gortanipora* cf. *G. bassleri, Homotrypa cincinnatiensis,* and *Amplexopora* cf. *A. pustulosa* are present but most of the earlier bryozoan species are no longer present. About 10 m higher, a thin bed with *Nicholsonella?* sp. appears, another 10 to 14 m higher *Bythopora* sp.and *Stigmatella crenulata* appear, 18 m higher *Gortanipora* cf. *richmondensis* and *Rhombotrypella* cf. *R. quadrata* are present, and the highest Bull Fork record is 15 m higher with *Gortanipora* sp. just below a sharp boundary with the overlying Drakes Formation. In the Cincinnati area, *Homotrypa flabellaris, Batostomella varians, Parvohallopora subnodosa,* and *Nicholsonella vaupeli* range into this interval. *Homotrypa communis, H. nodulosa, Peronopora hospitalis, Constellaria polystomella, Stigmatella interporosa,* and *Monticulipora epidermata* first appear in the Waynesville and range into the Liberty and Whitewater. *Rhombotrypella quadrata, Homotrypa ramulosa, H. wortheni,* and *Mesotrypa patella* first appear in the Liberty and range into the Whitewater. Four species restricted to the Whitewater are *Rhombotrypa multitabulata, Hallopora congrua, Amplexopora variable,* and *Homotrypella cumingsi.*

The Drakes Formation comprises a fifth major depositional sequence and is made up of four members (Fig. 3). The Rowland Member (sequence unit 21) was deposited on an irregularly eroded upper Bull Fork surface and filled valleys with about 15 m of silty dolomitic shales. It is unconformably overlain in the eastern part of the area by the Preachersville Member (sequence unit 22) which eventually covered most of the remaining topography. These beds are unconformably overlain by the Silurian Brassfield Dolomite.

In the western part of the area, the Bardstown Member (sequence unit 23) and Saluda Member (sequence unit 24) (Figure 1, section 52) are considered a facies of the Preachersville Member. Locally, the Saluda rests unconformably on the lower Whitewater Formation where this formation stood in topographic relief above the Rowland and Bardstown valley fills. The post-Saluda unconformity is usually overlain by the Brassfield Dolomite (upper part of the Lower Silurian; Telychian Stage, Llandoverian

Series) although locally the Brassfield may rest on lower or higher beds.

The Drakes Formation's stratigraphic relations of valley incision and subsequent filling are similar to those of the earliest Silurian (Rhuddanian) Mosalem Formation in northwestern Illinois and southeastern Iowa, the Wilhelmi Formation in northeastern Illinois, Edgewood Formation in western and southern Illinois (Willman et al. 1975), and the Medina Group and the lower part of the Lower Clinton Group of western New York State. This comparison leads to speculation that the Drakes Formation is actually post-Ordovician and younger than generally believed. Thus, its relatively rare fauna could be early Llandoverian.

At Madison, Indiana, (Fig. 1, section 7), a 3 to 4 m interval between the top of the Saluda and the Brassfield Dolomite (assigned to the 'Upper Whitewater' by Cuffey & Davis 1998) is a more normal marine, deeper water carbonate and shale. Based on its stratigraphic position as traditionally interpreted, this thin, unconformity-bounded pre-Telychian interval could be Gamachian (latest Ordovician), however, we believe that it is more likely Aeronian (middle Llandoverian, Early Silurian) and, therefore, an age equivalent to the Elwood Formation of northeastern Illinois, the Edgewood Formation of southern Illinois, and the upper part of the Lower Clinton of western New York.

Only a few bryozoans are recorded from this interval and include *Stigmatella* sp., trepostomes indet., and fenestellid indet. Major sequence 5 has many physical and lithologic similarities, including a poor fossil record, and the same general relations to the upper Ordovician Cincinnatian and Silurian Brassfield Dolomite as the Schweizer Member (Rhuddanian) and Birds Member (Aeronian) of the Wilhelmi Formation (lower and middle Llandoverian) which filled valleys cut into the Maquoketa Shale and lie below the Kankakee Limestone of northeastern Illinois.

4 CONCLUSIONS

Bryozoan species' distributions in the late Mohawkian (Lexington Limestone) and Cincinnatian series of the Cincinnati Arch region illustrate the effect of sea-level rise and fall relative to species appearances and disappearances, suggesting that many bryozoans were adapted to fairly narrow water-depth ranges. Relatively few bryozoans (mainly cryptostomes) were able to become established during the rapid sea-level rise at the beginning of the late Mohawkian (Lexington). During the subsequent fall in sea level, apparently at a slower rate of change, progressively increasing abundance and diversity of bryozoans became established. The importance of the unconformity and exposure event (sequence boundary) that separates sequence 1 from sequence 2 is shown by a major loss of bryozoan species followed by a second introduction of gradually increasing numbers and diversity of bryozoans in the lower part of sequence 2. A rapid sea-level rise to a major highstand in the middle of sequence 2 is accompanied by a great reduction in bryozoan numbers and diversity in lower Edenian beds. Only after a series of progressively shallowing fourth- and fifth-order depositional cycles in the late Edenian do bryozoan numbers and diversity recover.

The beginning of Fairview (lower Maysville) deposition has a well-established and varied bryozoan fauna whose over all generic and species composition changes little through the Maysvillian and the Arnheim Formation in the lowest Richmondian. These strata, in contrast to those just below, are relatively shallow-water carbonate ramp facies (with several ramp-wide unconformities) with a rich associated invertebrate fauna. They also stand in contrast to strata above which are much more silty and clayey and have well-developed, 1 to 2 m, silt-shale-limestone cycles. Bryozoans in the Waynesville, Liberty, Whitewater, and Drakes formations, are much reduced in diversity and in number of bryozoan-bearing beds. And nearly two-thirds of the bryozoan species are confined to these middle and upper Richmondian units and of the half dozen species that do range up from lower beds, half of them are not known above the middle part of the Waynesville. Our interpretation is that middle and higher Richmondian environments had much more suspended silts and clays and were a little higher on the ramp or shelf so that they experienced greater environmental changes during sedimentary cycles than the bryozoan-rich beds in the Maysvillian.

Sequence 5, the Drakes Formation, is dolomitic silty shale and siltstone, and has very rare and poorly preserved shelly faunas. Traditionally, these beds are placed in the upper part of the Richmondian (Ordovician), however, their sharp contact on the underlying Bull Fork Formation and the apparent significant erosional topography at the top of the Bull Fork, suggest the Drakes Formation is associated with lowered glacioeustastic sea levels during the Hirnantian/Gamachian (at the end of Ordovician) or with the slow and cyclical recovery of sea level seen in the Rhuddanian and Aeronian stages of the Early Silurian Llandoverian Series.

ACKNOWLEDGEMENTS

June Ross thanks Western Washington University for a Faculty Study Leave in Winter and Spring, 1995, and a Bureau for Faculty Research Grant, Western Washington University, for partial support of field work in 1995, and The American Chemical Society, Petroleum Research Fund Grant 31513-B8 for support of research and field studies in 1997-1999. We also thank Roger Cuffey and Robert Sloan for helpful discussions about the stratigraphy and classification of Cincinnatian strata.

REFERENCES

Anstey, R.L. 1986. Bryozoan provinces and patterns of generic evolution and extinction in the Late Ordovician of North America. *Lethaia* 19: 33-51.

Anstey, R.L. & Perry, T.G. (1973). Eden Shale bryozoans: a numerical study (Ordovician, Ohio Valley): *Publications of the Museum, Michigan State University, Paleontological Series* 1 (1): 1-80.

Borella, P.E. & Osborne, R.H. 1978. Late Middle and Late Ordovician history of the Cincinnati Arch province, central Kentucky to central Tennessee. *Geological Society of America Bulletin* 89: 1559-1573.

Brown, G.D. & Daly, E.J. 1985. Trepostomatous Bryozoa from the Dillsboro Formation (Cincinnatian Series) of southeasten Indiana. *Indiana Geological Survey Special Report* 33: 1-95.

Cressman, E.R. 1973. Lithostratigraphy and depositional environments of the Lexington Limestone (Ordovician) of central Kentucky. *U.S. Geological Survey Professional Paper* 768:1-68.

Cuffey, R.J. 1998. An introduction to the type-Cincinnatian. In R.A. Davis & R.J. Cuffey (eds), Sampling the layer cake that isn't: the stratigraphy and paleontology of the type-Cincinnatian. *Ohio Geological Survey Guidebook* 13: 2-9.

Elias, R.J. 1995, Origin and relationship of the Late Ordovician Red River–Stony Mountain and Richmond solitary rugose coral provinces in North America. In J.D. Cooper, M.L Droser & S.C. Finney (eds), *Ordovician odyssey: short papers for the Seventh International Symposium on the Ordovician System*: 439-442. Fullerton, California: Pacific Section Society for Sedimentary Geology (SEPM).

Ettensohn, F.R. & Kulp, M.A . 1995. Structural-tectonic control on Middle-Late Ordovician deposition of the Lexington Limestone, central Kentucky. In J.D. Cooper, M.L. Droser & S.C. Finney (eds), *Ordovician odyssey: short papers for the Seventh International Symposium on the Ordovician System*: 261-264. Fullerton, California: Pacific Section Society for Sedimentary Geology (SEPM).

Hattin, D.E., Nosow, E., Perkins, R.F., Stumm, E.C., Mound, M.C. & Utgaard, J. 1961. Field excursion to the Falls of the Ohio. *Guidebook for field trips, Cincinnati meeting, 1961, Field Trip* 9: 297-350. New York: Geological Society of America Guidebook Series.

Hay, H.B., Pope, J.K. & Frey, R.C. 1981. Lithostratigraphy, cyclic sedimentation, and paleoecology of the Cincinnatian Series in southwestern Ohio and southeastern Indiana, p. 73-86: In T.G. Roberts (ed.), *Geological Society of America Cincinnati '81 Field Trip Guidebooks, volume 1, Stratigraphy, Sedimentology*. Falls Church, Virginia: American Geological Institute.

Holland, S.M. 1993. Sequence stratigraphy of a carbonate-clastic ramp, the Cincinnatian Series (Upper Ordovician) in its type area. *Geological Society of America Bulletin* 105: 306-322.

Holland, S.M. 1995. The stratigraphic distribution of fossils. *Paleobiology* 21: 92-109.

Holland, S.M. & Patzkowsky, M.E. 1996. Sequence stratigraphy and long-term paleo-oceanographic change in the Middle and Upper Ordovician of the eastern United States. In B.J. Witzke, G.A. Ludvigson & J. Day (eds), Paleozoic sequence stratigraphy: views from the North American craton. *Geological Society of America Special Paper* 306: 117-130.

Holland, S.M. & Patzkowsky, M.E. 1997. Distal orogenic effects on peripheral bulge sedimentation: Middle and Upper Ordovician of the Nashville Dome. *Journal of Sedimentary Research* 67: 250-263.

Holland, S.M., Miller, A.I., Dattilo, B.F. & Meyer, D.L. 1996. Variations in cyclicity within the Cincinnatian Kope Formation: stratigraphic resolution and perceptions of cycle architecture. *Geological Society of America Abstracts with programs, 1996 Annual Meeting*: A-236.

Karklins, O.L. 1983. Ptilodictyoid cryptostome Bryozoa from the Middle and Upper Ordovician rocks of Central Kentucky. *Paleontological Society Memoir* 14: 31p.

Karklins, O.L. 1985. Bryozoan biostratigraphy of the Upper Middle Ordovician and Upper Ordovician of Kentucky. In C. Nielsen & G.P. Larwood (eds), *Bryozoa: Ordovician to Recent*: 346-347. Fredensborg: Olsen & Olsen.

Karklins, O.L. 1984. Trepostome and cystoporate bryozoans from the Lexington Limestone and the Clays Ferry Formation (Middle and Upper Ordovician) of Kentucky. *U.S. Geological Survey Professional Paper* 1066I: 1-105.

Keith, B. D. 1985. Facies, diagenesis, and the upper contact of the Trenton Limestone of northern Indiana. *Michigan Basin. Geological Society, Special Paper* 4: 15-32.

Meyer, D.L., Tobin, R.C., Pryor, W.A., Harrison, W.B. & Osgood, R.G., with contributions by Hinterlong, G.D., Krumpolz, B.J. & Mahan, T.K. 1981. Stratigraphy, sedimentology, and paleoecology of the Cincinnatian .Series (Upper Ordovician) in the vicinity of Cincinnati, Ohio. In T.G. Roberts (ed.), *Geological Society of America Cincinnati '81 Field Trip Guidebooks, volume 1, Stratigraphy, Sedimentology*: 31-71. Falls Church, Virginia: American Geological Institute.

Pope, M. & Read, J.F. 1995. Late Ordovician Tanglewood bank: a highly compartmentalized buildup reflecting tectonics and moderate-amplitude eustasy on the Cincinnati Arch. *Geological Society of America Abstracts with Programs* 27: A450.

Pope, M. & Read, J.F. 1997. High-Resolution surface and subsurface sequence stratigraphy of Late Middle to Late Ordovician (Late Mohawkian-Cincinnatian) Foreland Basin rocks, Kentucky and Virginia. *American Association of Petroleum Geologists Bulletin* 81(11): 1866-1893.

Tobin, R.C. 1986. An assessment of the lithostratigraphic and interpretive value of the traditional "biostratigraphy" of the type Upper Ordovician of North America. *American Journal of Science* 286: 673-701.

Ulrich, E.O. 1890. Paleozoic Bryozoa. Paleontology of Illinois, geology and paleontology. *Illinois Geological Survey*. 8(2), Section 6: 287-688.

Ulrich, E.O. 1893. On lower Silurian Bryozoa of Minnesota. *Geology Minnesota* 3, Part1, Chaper 4: 96-332.

Utgaard, J.E. & Perry, T.G. 1964. Trepostomatous bryozoan fauna of the upper part of the Whitewater Formation (Cincinnatian) of eastern Indiana and western Ohio. *Indiana Geological Survey Bulletin* 33: 1-111.

Weir, G.W. Peterson, W.L. & Swadley, W.C. 1984. Lithostratigraphy of Upper Ordovician strata exposed in Kentucky. *U.S. Geological Survey Professional Paper* 1151E: 1-121.

Weiss, M.P., Edwards, W.R., Norman, C.E. & Sharp, E.R. 1965. The American Upper Ordovician standard. VII. Stratigraphy and petrology of the Cynthiana and Eden Formations of the Ohio Valley. *Geological Society of America Special Paper* 81: 1-76.

Willman, H.B., Atherton, E., Buschbach, T.C., Collinson, C., Frye, J.C., Hopkins, M.E., Lineback, J.A. & Simon,

J.A. 1975. Handbook of Illinois Stratigraphy. *Illinois State Geological Survey Bulletin* 95: 1-261.
Young, G.A. & Elias, R.J. 1995. Latest Ordovician to earliest Silurian colonial corals of the east-central United States: *Bulletins of American Paleontology* 108(347): 1-153.

Bryozoan Studies 2001, Wyse Jackson, Buttler & Spencer Jones (eds)
© 2002 Swets & Zeitlinger, Lisse, ISBN 90 5809 388 3

Bryobaculum carinatum gen. et sp. nov.: a new Mediterranean Pliocene deep-water bryozoan

A. Rosso
Dipartimento di Scienze Geologiche, Università di Catania, Catania, Italy

ABSTRACT: A new species of an articulate branching bryozoan arising from an encrusting base is described from deep-sea Upper Pliocene sediments that crop out in NE Sicily. The colony is formed by quadriserial rod-like branches consisting of twins of back-to-back zooids, each rotated at 90° from the previous one. The zooids have a finely granular, sparsely porous frontal wall and a sinuated orifice, without spines. The species is characterised by paired orificial adventitious avicularia; hyperstomial, finely granular ovicells not closed by the operculum and bearing two small uncalcified symmetrical slits; and frontal kenozooids to articulate lateral branches. The distinctive features of this species do not allow a suitable attribution to the already known generic and familial taxa. Thus a new genus is described and a new family proposed to accommodate the new species. The growth pattern is described. The meaning of the species in the recolonisation of the Mediterranean after the Messinian salinity crisis is discussed.

1 INTRODUCTION

Deep-sea bryozoans from the Mediterranean area are presently poorly known, although efforts in the last decades have increased knowledge about Recent (see Harmelin & d'Hondt 1982, 1992a, b, 1993) and Holocene (Di Geronimo et al. 2001) assemblages. Fragmentary data on Pleistocene species from Italian deep-sea sediments can be found in papers by Seguenza (1880) and Neviani (1900) and only a few assemblages have been recently studied (Rosso 1990, 1998, 1999, Barrier et al. 1996, Di Geronimo et al. 1998, Rosso & Di Geronimo 1998). Nothing is known about the Pliocene, except data on deep-water bryozoans from Rhodes (Greece) by Moissette & Spjeldnaes (1995).

Nevertheless, knowledge of both Upper Miocene and Pliocene deep-sea bryozoan faunas is required to reconstruct the history of deep-sea biotas and to understand the modalities of the recolonisation of the Mediterranean depths after the Messinian salinity crisis. With this aim, Pliocene fossiliferous outcrops containing deep-sea faunas have been searched for and located in north-eastern Sicily. Preliminary studies show that assemblages are rich and diverse, as they comprise, besides the Recent Atlanto-Mediterranean and most of the typical Pleistocene species, other taxa presently living in the Atlantic, or showing Atlantic affinities and/or having counterparts in the Atlantic but yet not recorded from Pleistocene localities. Some other species seem to be very similar to Eocene and Oligocene taxa from northern Italy and central Europe and accurate comparisons are needed to definitely state their attribution. Finally, some species are new and some of them show very distinctive features. The goal of this paper is to describe one particularly interesting taxon.

2 MATERIALS AND METHODS

The studied material was derived from some samples collected at Capo Milazzo peninsula (north-eastern Sicily) within the locally called "yellow calcareous marls". These sediments were deposited in epibathyal environments as testified by brachiopod (Gaetani & Saccà 1984, Gaetani 1986) and foraminiferan (Violanti 1988) faunas, during and after the rapid post-Messinian marine transgression (Fois 1990) and lie directly on the crystalline basement and locally on Miocene sediments. Outcrops are narrow, discontinuous and show variable thickness and erosional basal surfaces usually marked by the presence of breccias as bathyal sediments fill previous depressions in coastal proximal areas (Fois 1989).

The studied succession, cropping out at Cala S. Antonino, along the south-eastern coast of the peninsula, belongs to the Upper Pliocene and according to Violanti (1988) their sedimentation occurred during MP15 and MP16 zones, largely overlapping with the Gelasian Stage of Rio et al. (1994). Sediments consist of silts, sandy silts, silty sands and muddy sands characterised by an extremely high content in $CaCO_3$ due to the almost exclusive presence of calcareous foraminifers, mainly represented by planctonic (globigerinids and *Orbulina*) species. Some coarser bioclastic layers are interbedded, enriched in skeletons of benthic faunas among which isidid fragments are the most obvious element, together with bryozoans, serpuloideans and echinoids.

Bryozoan palaeocommunities comprise up to 60 species as a whole, with a sharp dominance of erect, both rigid and flexible, growth morphotypes. Morphological evidences suggest that most species were able to colonise the soft bottom directly. The most abundant species is nearly always *Tessaradoma boreale* (Busk), often representing about one half of the assemblages. Other common and frequent taxa are *Tervia irregularis* Meneghini, *T. barrieri* Rosso, *Caberea ligata* Jullien, *Cellaria* cf. *salicornioides* (Lamouroux), *Scrupocellaria incurvata* Waters and *Reteporella sparteli* (Calvet). Other species are well represented, such as *Gemellipora eburnea* (Smitt), or dominant, such as *Batopora rosula* (Reuss), although only in one or a few samples.

Samples were routinely treated i.e. washed, using a 63 μm mesh, and subsequently sieved. All the residues coarser than 500 μm were systematically sorted and very small specimens, such as those belonging to the species here described, were picked up from the 250-500 μm sized fraction. Specimens were coated with gold for SEM analyses.

3 SYSTEMATIC ACCOUNT

Order Cheilostomatida Busk, 1852
Suborder Ascophorina Levinsen, 1909
Infraorder Lepraliomorpha Gordon, 1989
Family Bryobaculidae fam. nov.

Diagnosis: As for genus. The new family is monogeneric and it is here proposed to accommodate the newly described genus and species which show distinctive features not allowing a suitable attribution to the already known familial taxa. The explanation is given below.

Genus *Bryobaculum* gen. nov.

Type species: *Bryobaculum carinatum* sp. nov.

Diagnosis: Colony composed of articulated branches from an encrusting base. Branches rod-like consisting of twins of back-to-back zooids, each twin rotated at 90° from the previous one. Frontal wall finely granular, with sparse pores. Zooidal orifice sinuated. Orificial spines absent. Orificial adventitious avicularia paired. Ovicells hyperstomial, conspicuous with a finely granular ectooecium leaving two small uncalcified symmetrical slits; not closed by the operculum. Kenozooids to articulated lateral branches randomly present on the frontal wall of some zooids.

Etymology: the generic name is derived from the name of the phylum Bryozoa and the Latin *baculum*, a neuter noun meaning "rod", "stick", in reference to the general shape of single segments.

Bryobaculum carinatum sp. nov.
(Figs 1-19)

Dignosis: As for genus

Etymology: From the Latin adjective *carinatus* i.e. "having a carina" alluding to the mid-zooidal narrow and delicate carina and to the general appearance of the frontal surfaces of the young zooids resembling the keel of a boat.

Materials: **Holotype**: A long fertile segment composed of 7 pairs of zooids but not containing kenozooids. It belongs to the sample 2 (1998), collected at Cala S. Antonino W, and is deposited in the Palaeontological Museum of the Catania University (PMC B.11a. 27.04.1998) (Fig. 1). **Paratypes**: All the remaining 285 specimens present in the same sample consisting of fertile and sterile branches. One long fragment comprising 8 plus 2 twins of zooids and six additional ones (2 fertile and 4 sterile segments showing arcticular kenozooids and 2 proximal ends) are deposited in the Natural History Museum London (code number NHM BZ4884) whereas the remaining ones are in PMC (PMC B.11b. 27.04.1998). **Additional material**: Cala S. Antonino W, sample 1 (1997): 41 specimens; sample 1 (1998): 7 specimens; Cala S. Antonino Centre: sample 2 (1999): 19 specimens; sample 17 (2000): 4 specimens; Cala S. Antonino E: sample 13 (2000): 5 specimens; sample 14 (2000): 115 specimens; Punta Mazza, a nearby locality: a cobble from the basal part of the "yellow calcareous marls" encrusted by serpuloideans, primnoid bases and bryozoans among which running cyclostomes and ascophorans such as *Sertulipora guttata* Harmelin & d'Hondt and an inferred basal portion of the described species.

Description: Colony known from fragments consisting of very small unbranched rods not exceeding 3 mm in length (Figs 1-4, 8-9, 13-14). No fragment is entire: most of them are broken at the two ends but some show an unbroken tapering and sometimes slightly curved, proximal end (Figs 2, 13-14) and two fragments have a complete distal growing tip (Fig. 7). Stems quadriserial with zooids arranged in superimposed series of back-to-back couples which alternate with each other at 90°. Each zooid buds distally a new zooid in the same series but this remains tubular in its proximal half part (i.e. during the full development of the couple of zooids of the alternating se-

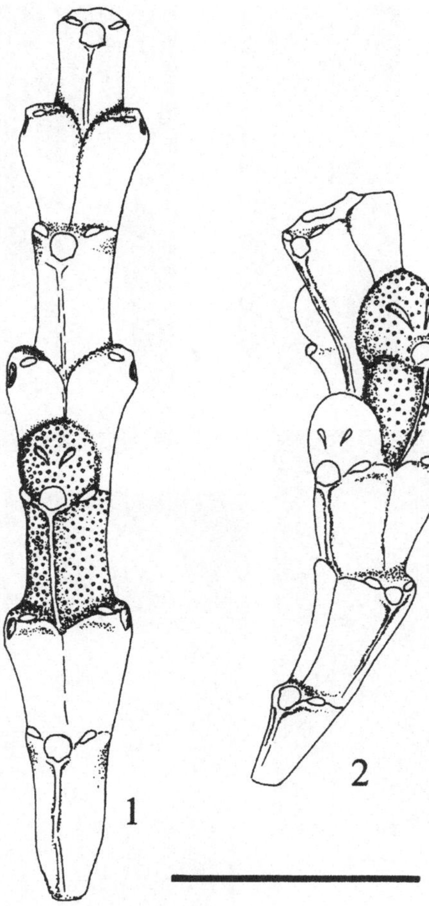

Figure 1-2. *Bryobaculum carinatum* gen. et sp. nov.. Sample 1 (1997). Cala S. Antonino W. Upper Pliocene. Scale bar: 1 mm. 1. Holotype. 2. Paratype: a club-shaped fertile branch.

ries) and thus can be seen only in the juvenile stage (Figs 7, 10) whereas it becomes obscured in later otogeny when laterally overgrown by adjacent zooids which leave uncovered only the median *carina*. Stems with preserved proximal parts begin with a single zooid, proximally elongated into a short tubule, budding two new zooids from the distal dorsal part. Usually 3 to 5 series are present but longest fragments comprise up to 8 series plus at least two other ones at the proximal and distal broken ends. Branch width ranges from 220-250 to 380-450 μm.

Zooids (Figs 1-4, 7) consisting of a hidden tubular proximal part growing adjacent to the enlarged distal parts of the preceding twin and an enlarged distal part, sub-quadrangular, wide one half of the length (about 200 x 400 μm). Total length actually reaching about 800 μm considering the proximal tubular part. Frontal shield finely granular, perforated by isolated, small pores mostly near the raised mid line where a marked but smooth and thin keel is present (Fig. 11). At high magnification the structure of the frontal surface appears honeycomb-like, formed by nearly iso-oriented small prismatic crystals, almost perpendicular to the surface but growing up to different levels so creating depressed and raised zones. Microgranular crystals locally cover the preceding frontal structure forming a smooth surface along the carina and the orificial edge. Frontal surface gradually protruding distally and more depressed proximally where lateral parts extend into wings steeply raising to form the distal edge of the orifice of the preceding pair of zooids. This peculiar morphology of the frontal shields is reflected in the serrated outline of the branches (Figs 1, 4, 13).

Primary orifice (80-90 μm wide and 100-120 μm long) with a shallow subtriangular sinus marked by two minute embayments along the edge and inwardly directed lateral condyles (Figs 15-16). The edge smooth and slightly raised. Orifice distally located, its edge formed only proximally by the same zooid and distally by the proximal projections of the subsequent zooidal shields. The orifice is slightly dimorphic with female zooids characterised by a slightly wider orifice with a wider and shallower sinus (Fig. 4). Oral spines lacking.

Ovicell hyperstomial, recumbent, conspicuous (270-500 but usually about 400 μm long and 250-340 μm wide) but not prominent, overgrowing more than half to all the proximal tubular part of the subsequent zooid in the series and wide almost enough to reach the median carina of the two next alternating zooids (Figs 1-2, 4-6). Ectooecial surface finely granular and sparsely perforated like the zooidal frontal shield, leaving two small narrow symmetrical drop-like fenestrae sometimes joined in the midline and extending proximally, thus resembling a "V" or a "Y". Orifice of the ovicelled zooids with a slightly wider and shallower sinus (Figs 1-2, 4). Ovicells isolated but frequently present on both zooids of a twin and often on all the zooids of two subsequent twins, thus forming swollen parts giving the branch a club appearance.

Adventitious avicularia constant, paired, located laterally but not adjacent to the orifice along the distal zooidal margins, at the level of the half distal part of the orifice. They are small (50 μm long and 30 μm wide) elliptical or slightly spathulate, laterally or disto-laterally directed and persist in the same position also on ovicellate zooids.

Articular kenozooids (Figs 8-9, 14) adventitious, proximo-laterally located in respect to the zooidal orifice, swollen and protruding distally, not well differentiated from the frontal zooidal surface, with a wide, rounded (about 100 μm in diameter), distally inclined orifice. They appear randomly distributed along the segments and usually isolated (only one per stem). Sometimes two articular kenozooids co-

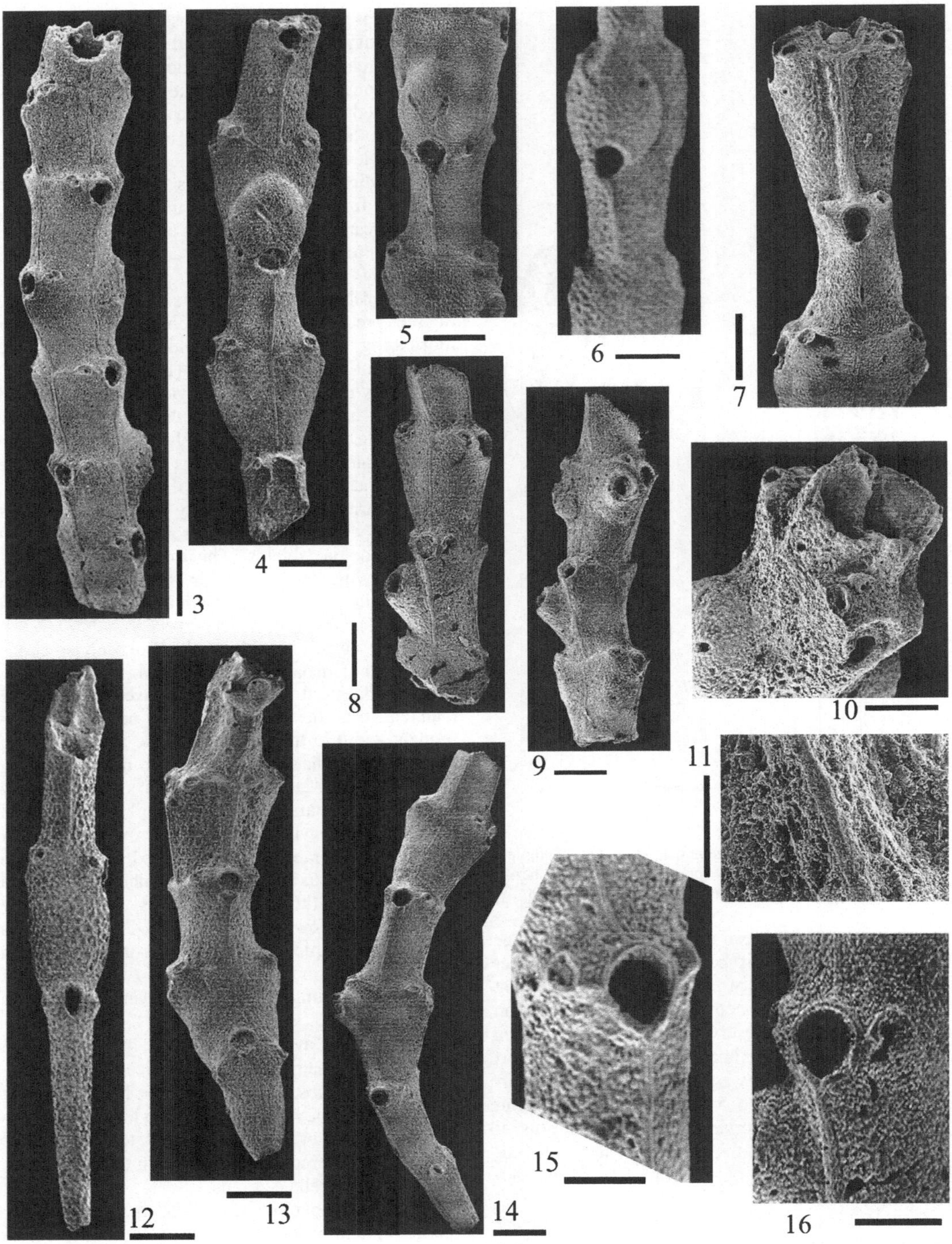
3
4
5
6
7
8
9
10
11
12
13
14
15
16

Figures 3-16 (opposite). *Bryobaculum carinatum* gen. et sp. nov., Upper Pliocene, Milazzo. Figures 3-5. Paratypes. Sample 1 (1997). Cala S. Antonino W. Scale bar: 200 μm. Figure 3. A long segment broken at both ends. Figure 4. A fertile segment showing an ovicell with large frontal fenestrae. Figure 5. Detail of an ovicell with large frontal fenestrae. Figure 6. A very large ovicell completely overgrowing the next alternating couple of zooids. Sample 1 (1998). Cala S. Antonino W. Scale bar: 200 μm. Figure 7. A growing tip showing the proximal tubular parts of the last, still forming twin of zooids. Sample 14 (2000). Cala S. Antonino E. Scale bar: 100 μm. Figure 8. Sample 1 (1997). Cala S. Antonino W. Scale bar: 200 μm. A broken old and eroded branch with an articular kenozooid. Figures 9-10. Sample 14 (2000) Cala S. Antonino Est. Figure 9. An old broken branch, the only one showing three articular kenozooids, two of which in the same twin. Scale bar: 200 μm. Figure 10. A broken segment showing the four serial arrangement of zooids with the two swollen distal parts of a twin and the slender tubular part of the subsequent one. Scale bar: 100 μm. Figures 11-13. Paratypes. Sample 1 (1997). Cala S. Antonino W. Figure 11. Close-up of the median carina, just below an orifice. Scale bar: 50 μm. Figure 12. Proximal end of a young straight segment with frontal walls of zooids coarsely granular. Scale bar: 200 μm. Figure 13. Proximal end of an old segment. Scale bar: 200 μm. Figure 14. Curved proximal end of a segment with an articular kenozooid in the second twin. Sample 1 (1998). Cala S. Antonino W. Scale bar: 200 μm. Figures 15-16. Paratypes. Sample 1 (1997). Cala S. Antonino W. Scale bar: 100 μm. Figure 15. Orifice of a young zooid with oral avicularia at the anter level. Figure 16. Orifice of an old zooid.

exist on the same fragment and in two cases they are located on two contiguous alternating couples, athough in different longitudinal series and zooidal sides; sometimes kenozooids are located along the same longitudinal series; only once they are paired on the same zooid. Articular kenozooids may also be located at the beginning of a stem and usually in the third series (second twin) and only once an articular kenozooid was observed on the first twin of an article.

Inferred basal parts (5 as a whole, but only one satisfactorily preserved) have been found only in 3 samples, two from Cala S. Antonino and a third from a nearby locality (Punta Mazza). Encrusting parts are small (maximum size 1.5 x 3 mm) and irregularly shaped, composed of no more than 25 heterozooids. They are small (from 200 to 400 μm long and 150-250 μm wide); some of them are similar to the autozooids of the erect part possessing two distally located avicularia on projecting tubules and a median carina but have occluded orifices; other smaller ones are subtriangular and have only one avicularium, centrally disposed on a short and slender tubule (Figs. 17-18). Other heterozooids are interspersed mostly in the peripheral parts, raised and swollen, with a large opening, about the same shape and size of the articular kenozooids present on the stems.

Variability: The general appearance of segments changes seemingly with ontogeny as juvenile segments are slender and with a decidedly serrated outline whereas the others show a greater diameter and a less marked serrated outline. At the same time young zooids show a marked boat-keel shape and an evenly and coarsely granular surface (Figs 7, 12) which become lost in senile more globular zooids, characterised by a finely granular to smooth, sometimes vitreous, surface (Fig. 3).

Remarks: The described species has no counterparts within the already known bryozoan fauna of the deep-sea environments of Pliocene and Pleistocene age from the Mediterranean area and the Recent NE Atlantic.

The described segments are, at first sight, reminescent of *Cellaria schreibersi* Reuss, a deep-sea Eocene species from Northern Italy and the Carpatian area. Nevertheless this species differs strongly for the orifice shape (elliptical and without sinus), the nature of the frontal surface (evenly perforated), the absence of kenozooids and the colony morphology involving a vincularian organisation (Reuss 1848, 1969, Zagorsek, in press). This species was for a long time been included in *Tetraplaria* (Braga 1963, Vávra 1977) although Cheetham (1972) suggested separating this and other bifurcating unarticulated branching species from that genus. Only

17

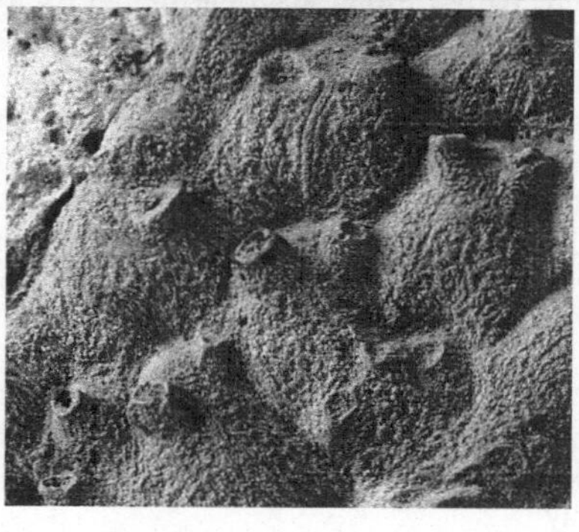

18

Figures 17-18. 17. A complete basal portion. Punta Mazza. Scale bar: 1 mm. 18. A close-up of some basal zooids. Punta Mazza. Scale bar: 200 μm.

recently Zagorsek (in press) created the new genus *Tychinella* for *C. schreibersi* Reuss but left it within the family Tetraplariidae Harmer, considering as a criterion the quadriserial disposition of the zooids along the branches (although some *Tetraplaria* species are multiserial, such as *Tetraplaria orospinea* Gordon & d'Hondt). Moreover *Tychinella* lacks articulation and differs from tetraplariids in further characters such as orifice morphology. A re-evaluation of this attribution seems therefore needed.

Owing to the presence of sinuate orifices and articulate segments, *Tetraplaria* could appear as a suitable genus for the present species. Nevertheless, *Tetraplaria* Tennison-Woods shows evenly perforated frontal and ovicellar shields lacking slit-like fenestrae (see Tennison Woods 1879, Harmer 1957) and possesses quadriserial-to-multiserial branches bifurcating at their tips, each new segment having a biserial proximal part and arising from an interzooidal kenozooid, according to Cheetham (1972). Finally, this genus lacks frontal and oral avicularia. All these features preclude placing of the present species in *Tetraplaria.*

Some characters of *Bryobaculum carinatum* such as the articulation, the presence of a sinuate dimorphic orifice and large ovicells, and the lack of oral spines, are shared with species belonging to the family Phorioppnidae recently erected by Gordon & d'Hondt (1997) for some south-western Pacific genera. Nevertheless the species contained in this family have projecting zooidal margins, a perforated depressed frontal wall showing proximally located paired areolae and interzooidal articular kenozooids in spite of the frontal ones as in *B. carinatum*. On the contrary Phorioppnidae species lack oral avicularia and the frontal slit-like fenestrae of the ovicells which are closed by the operculum of the maternal zooid, unlike *B. carinatum.*

Finally, *Bryobaculum carinatum* shares some characters also with representatives of the family Pasytheidae Davis, such as the flexible joined colony arising from an encrusting base and the elongate zooids with a frontal wall perforated by a few shattered pores and bearing an orifice with a shallow poster and without spines. Nevertheless, both avicularia and ovicells are lacking in this family and zooids are arranged in a different manner (see Gordon 1984).

Thus, to suitably place the new species it has been necessary to describe a new genus and create a new family.

Distribution: *Bryobaculum carinatum* is presently known only from Upper Pliocene (MP15-MP16) deep water bioclastic sediments that crop out in NE Sicily.

From the previous discussion it appears that, notwithstanding its morphological peculiarities, the species shows a certain affinity both with Tertiary species from the Tethys and with the Recent Western Pacific species. Thus the present species seems to represent a Tethysian element but further data are needed to interpret its real meaning. Taking into account the present knowledge and considering that *Bryobaculum carinatum* has been found only in Upper Pliocene sediments whereas it seems to be absent from the Pleistocene of the same region, the species could be considered as a Mediterranean Pliocene palaeoendemic, descended from a Tethysian stock. Based on present data it is puzzling to work out if *Bryobaculum carinatum* evolved or persisted within the Mediterranean area, "surviving" the Messinian salinity crisis possibly within small local basins as suggested for some marine benthic, particularly euriecious species, by David & Pouyet (1984), or if it had an original wide geographical distribution, comprising not only the Mediterranean but also the Atlantic, from where it re-colonised the Mediterranean during the Pliocene. It appears worth of note to outline as the species has not been found in any sample from deep-sea sediments aged from the upper part of the Lower Pleistocene to the Middle Pleistocene. This probably means that *Bryobaculum carinatum* was not able to withstand the changes that happened in the Mediterranean deep-sea hydrology, from about 1 my BP, involving the establishment of an homothermic regime (see Rosso & Di Geronimo 1998).

4 FUNCTIONAL MORPHOLOGY

The observed fragments actually represent parts of articulated segments as it appears obvious from the tapering proximal ends of some of them, constituted by a single, very long and narrowing zooid showing a proximal wide pore which is the evidence of an uncalcified joint. Further evidence of a nodal articulation of the branches is the presence along the stems of kenozooids whose orifices have about the same diameter as the segment proximal ends. These structures have been interpreted as sites of articulation. The presence of articular kenozooids budding from zooidal frontal walls and their irregular location and space-out throughout the internode, points out to an irregularly branching colony with new branches arising at different high along the parental internodes and not bifurcating at their tips as usually found in almost all the erect, articulate species.

Such a peculiar branching pattern appears unusual. Within the Mediterranean area a somewhat similar branching pattern is known in *Cellaria normani* Hastings which shows, besides the normal bifurcating branching at the distal internode tips, also the possibility of forming new branches from frontal articular kenozooids from which orthogonal branches spring (Prenant & Bobin 1966, Rosso, unpublished).

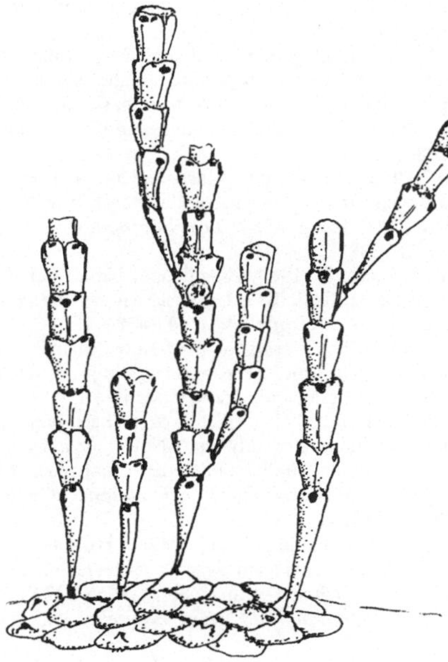

Figure 19. Inferred reconstruction of a living colony of *B. carinatum* with a few, slightly branched internodes articulated on an encrusting basal part.

Within the Ascophorans *Margaretta cereoides* (Ellis & Solander) develops new branches only from some zooids all around the parental branch, often in verticils (Rosso, unpubl.). Chitinous joints born from subcircular orifices of modified autozooids lacking the peristomes typical of the species, and join with septula in the proximal frontal walls of the first zooid in the new branch, thus ensuring articulation. New branches are straight but sometimes slightly curved at the beginning.

In both the discussed cases colonies are fixed to the bottom by means of chitinous rhizoids.

A general colonial morphology, involving the lateral development of new branches has been recently observed in *Quadriscutella papillata* Bock & Cook (Bock & Cook 1993), a living species from off the south Australian coast. Nevertheless, unlike *Bryobaculum carinatum*, this large sized and robust species possesses interzooidal articular kenozooids and the first joint is reinforced by further cuticular tubules arising from above the lateral frontal septulae of the proximal zooids of a branch. Traces of such kind of chitinous tubules, also present in *M. cereoides*, seem to be absent in *B. carinatum*.

Unlike the latter two above described species *M. cereoides* and *Q. papillata*, *Bryobaculum carinatum* would presumably form more delicate, candelabriform erect parts characterised by a loose branching pattern with new slender segments arising sparsely from the parental stem (none-to-one, rarely two branches per segment). New branches were seemingly slightly diverging (straight segments) or subparallel (segments with slightly curved proximal ends) with respect to the main branch.

It is therefore worth of note that the percentage of segments comprising joint sites (kenozooids) is low ranging from about 20 to 30%. This could mean that colonies were rarely ramified and probably often represented by individual unbranched rods. Moreover the number of the basal tapering parts appears low in comparison to the number of the inferred articular kenozooids (about one half), although a certain number of articular kenozooids could have been not active, as observed by Bock & Cook (1993) in *Q. papillata*. The percentage of basal parts is still lower in reference to the total number of fragments (less than 10%). Finally, the basal observed portions are often more slender than most fragments, and formed by zooids showing juvenile features, thus probably representing young growing branches and not broken proximal parts of longer segments. The different total number of broken branches and proximal ends could be related to a breaking off of segments with production of fragments belonging to different size fractions with the proximal-most parts particularly small and easily removed by a selective transport agent. In fact it is not supportable that the skeletal fragments of *B. carinatum* present in the studied samples represent the result of *in situ* disarticulation after death i.e. that they are in the same proportion as in living colonies. Probably they are part of a residual palaeocommunity and the observed proportion of fragments derives from a selective preservation phenomenon.

Finally, it is presumed that colonies of *B. carinatum* did not adhere to the substratum by means of rhizoids. The discovery of five small crusts formed by several "zooids" (showing a high degree of morphological affinities with zooids composing the rod-like segments) and among which are interspersed heteromorphs looking exactly as those present on autozooids, allowed to infer that they represent the encrusting bases of *B. carinatum*. It is thus hypothesised that erect segments were directly joined to basal heteromorphs and *B. carinatum* encrusted bottom clasts large enough to becolonised. A single inferred base has been found still connected with its substratum (represented by a large consolidated block of "yellow marls") but the other ones are detached from their substrata and probably colonised bioclasts originally present on the bottom. This inference is confirmed by the presence of this species in samples with sediments comprising coarse (mainly bioclastic) fractions and its absence from samples exclusively made up by foraminiferan oozes, as expected.

From all the above inferences, the general colonial morphology of *B. carinatum* may be summarised as follows: a small encrusting base comprising some articular kenozooids, each supporting an erect part composed of single rod-like branches and a few branching articulated ones (Fig. 19). Erect articulated branching colonies usually adhere to the substratum by means of basal rootlets and give rise to a bush-like or three-shaped colonies or raise from uni-to-biserial encrusting running portion forming slender erect segments, such as *Gemellipora eburnea* Smith. Small erect unbranched stems can rise from encrusting bases in a very few cheilostome species among which *Phorioppnia cookae* Gordon & d'Hondt (Gordon & d'Hondt 1997) having a basal encrusting part springing 1-3 unbranched erect stems. On the contrary, a colonial morphology similar to that inferred for *B. carinatum* involving two different growth phases (encrusting and erect articulate) has been suggested only for species of *Characodoma* by Cook and Bock (1996).

ACKNOWLEDGEMENTS

I am indebted to Mary Spencer Jones for bibliographic help, to Giampietro Braga and Kamil Zagorsek for useful discussions and to Donata Violanti for dating samples. Paul Taylor, Dennis Gordon and Italo Di Geronimo reviewed the manuscript. A particular appreciation to Eva Skarmoutsou for SEM assistance.

REFERENCES

Barrier, P., Di Geronimo, I., La Perna, R., Rosso, A., Sanfilippo, R. & Zibrowius, H., 1996. Taphonomy of deep-sea hard and soft bottom communities: the Pleistocene of Lazzàro (Southern Italy). *Il Reuniòn de Tafonomia y fossilizacion:* 39-45.

Bock, P.E. & Cook, P.L. 1993. *Quadriscutella* gen. nov. (Bryozoa, Cheilostomata) from the Tertiary and Recent of Victoria and South Australia. *Alcheringa* 17: 159-171.

Braga, G.1963. I Briozoi del Terziario veneto. I Contributo. *Bollettino della Società Paleontologica Italiana* 2: 16-55.

Cheetham, A.H. 1972. Cheilostome Bryozoa of Late Eocene Age from Eua, Tonga. *Geological Survey Professional Paper 640-E.*

Cook, P.L. & Bock, P.E. 1996. Characodoma Maplestone, a senior synonim of Cleidochasma Harmer (Cleidochasmatidae). In D.P. Gordon, A.M. Smith & J.H. Grant-Mackie (eds.). Bryozoans in space and time: 81-88.

David, L. & Pouyet, S. 1984. Les bryozoaires méditerranéens du Miocène à l'Actuel: succession stratigraphique, signification paléoécologique, essai biogéographique. *Annales Géologiques des Pays Helléniques* 32: 129-139.

Di Geronimo, I., D'Atri, A., La Perna, R., Rosso, A., Sanfilippo, R. & Violanti, D. 1998. The Pleistocene bathyal section of Archi (Southern Italy). *Bollettino della Società Paleontologica Italiana* 36(1-2): 189-212.

Di Geronimo, I., Rosso, A., La Perna, R. & Sanfilippo, R. 2001. Deep-sea (250-1,550m) benthic thanatocoenoses from Southern Thyrrenian Sea. In: Faranda F.M., Guglielmo L. & Spezie G. (eds.) *Mediterranean Ecosystems: Structures and Processes*, ch. 36: 277-287. Springer Verlag.

Fois, E. 1989. La successione neogenica di Capo Milazzo (Sicilia NE). *Rivista Italiana di Paleontologia e Stratigrafia* 95(4): 397-440.

Fois, E. 1990. Stratigraphy and palaeogeography of the Capo Milazzo area (NE Sicily, Italy): clues to the evolution of the southern margin of the Tyrrhenian basin during the Neogene. *Palaeogeography, Palaeoclimatology, Palaeoecology* 78: 87-108.

Gaetani, M. 1986. Brachiopod palaecommunities from the Plio/Pleistocene of Calabria and Sicilia (Italy). In P.R. Racheboef & C.C. Emig (eds) *Biostratigraphie du Paléozoïque* 4: 281-288.

Gaetani, M. & Saccà, R. 1984. Brachiopodi batiali nel Pliocene e Pleistocene di Sicilia e Calabria. *Rivista Italiana di Paleontologia e Stratigrafia* 90(3): 407-458.

Gordon, D.P. 1984. The Marine fauna of New Zealand: Bryozoa: Gymnolaemata from the Kermadec Ridge. *New Zealand Oceanographic Institute Memoir* 91: 1-198.

Gordon, D.P. & d'Hondt, J.-L.1997. Bryozoa: Lepraliomorpha and other Ascophorina, mainly from New Caledonian waters. In A. Crosnier (ed) Résultats des Campagnes MUSORSTOM, 18. *Mémoires du Muséum national d'Histoire naturelle* 176: 9-124.

Harmelin, J.-G. & d'Hondt, J.-L. 1982. Bryozoaires Cyclostomes bathyaux des campagnes océanographiques de l'"Atlantis II", du "Chain" et du "Knorr" (1967-1972). *Bulletin du Muséum national d'Histoire naturelle,* Paris, séc. A (1-2), 4, 4: 3-23.

Harmelin, J.-G. & d'Hondt, J.-L. 1992a. Bryozoaires des parages de Gibraltar (campagne océanographique BALGIM, 1984) 1 - Cheilostomes. *Bulletin du Muséum national d'Histoire naturelle*, Paris, 4, 14A, 1: 23-67.

Harmelin, J.-G. & d'Hondt, J.-L. 1992b. Bryozoaires des parages de Gibraltar (campagne océanographique BALGIM, 1984) 1 - Cténostomes et cyclostomes. *Bulletin du Muséum national d'Histoire naturelle*, Paris, 4, 14A, 3-4: 605-621.

Harmelin, J.-G. & d'Hondt, J.-L. 1993. Transfers of bryozoan species between the Atlantic Ocean and the Mediterranean Sea *via* the Strait of Gibraltar. *Oceanologica Acta* 16(1): 63-72.

Harmer, S. F. 1957. The Polyzoa of the Siboga expedition. Part 4. Cheilostomata Ascophora II. *Siboga Expedition* 28: 641-1147.

Moissette, P. & Spjeldnaes, N. 1995. Plio-Pleistocene deep-water bryozoans from Rhodes, Greece. *Palaeontology* 38 (4): 771-799.

Neviani, A. 1900. Briozoi neogenici delle Calabrie. *Paleontographia Italica* 6: 115-266.

Prenant, M. & Bobin, G. 1966. Bryozoaires, 2° partie. Cheilostomes Anasca. *Faune de France* 68: 1-647.

Reuss, A.E. 1848. Die fossilen Polyparien des wiener Tertiarbeckens. Haidinger. naturwissenschaftliche Abhandlungen, 2: 1-109.

Reuss, A.E. 1869. Palaontologische Studien ueber die alteren Tertiarschichten der Alpen. II part, Die Fossilen Anthozoen und Bryozoen der Schichtengruppe von Crosara. Sitzungberichte der K. Akademie der Wissenschaften. 17-36: 215-298.

Rio, D. Sprovieri, R. & Di Stefano, E. 1994. The Gelasian Stage: a proposal of a new chronostratigraphic unit of the Pliocene series. *Rivista Italiana di Paleontologia e Stratigrafia* 100(1): 103-124.

Rosso, A. 1990. Thanatocenose wurmien à Bryozoaires bathyaux en Mer Tyrrhenienne. *Rapports Commission internationale pour l'Exploitation de la Mer Méditerranée* 32(1): 23.

Rosso, A. 1998. New bryozoan species from the deep-sea Pleistocene sediments of Southern Italy. *Rivista Italiana di Paleontologia e Stratigrafia* 104(3): 423-430.

Rosso, A. 1999. Recent and fossil species of *Characodoma* Mapleston, 1900 (Bryozoa) from the Mediterranean with description of two new species. *Journal of Natural History* 33: 415-437.

Rosso, A. & Di Geronimo, I. 1998. Deep-sea Pleistocene Bryozoa of Southern Italy. *Géobios* 30(3): 303-317,

Seguenza, G. 1880. Le formazioni terziarie della Provincia di Reggio (Calabria). *Regia Accademia dei Lincei, Memorie della Classe di Scienze Fisiche, Matematiche e Naturali,* s.3 (6): 416 pp.

Tennison Woods, J.E. 1879. On some Australian tertiary fossil corals and polyzoa. *Journal and Proceedings of the Royal Society of New South Wales* 12: 57-61.

Vávra, N. 1977. *Bryozoa teriaria. Catalogus fossilium Austriae.* Springer Verlag, Wien, New York.

Violanti, D. 1988. I Foraminiferi plio-pleistocenici di Capo Milazzo. *Bollettino Museo Regionale di Scienze Naturali,* Torino 6 (2): 359-392.

Zagorsek, K. (in press). Upper Eocene Bryozoa from Waschberg (Austria). *Beitraege zur Palaeontologie.*

Bryozoan Studies 2001, Wyse Jackson, Buttler & Spencer Jones (eds)
© 2002 Swets & Zeitlinger, Lisse, ISBN 90 5809 388 3

Life history characters and ecology of some encrusting ctenostomates

J.S. Ryland
School of Biological Sciences, University of Wales Swansea, Swansea SA2 8PP, UK

ABSTRACT: The reproduction of four species has been followed in Milford Haven, south-west Wales. Data were numbers of newly settled ancestrulae present on samples of *Fucus serratus* fronds and, for *Alcyonidium hirsutum* and *A. reticulum* only, numbers of embryo-containing brood chambers. Based on 2_ years' results, *A. hirsutum* and *A. reticulum* are shown to breed in winter and *A. polyoum* and *Flustrellidra hispida* in summer, with some variation between sites. Different settling seasons help to limit competition for space on growing *F. serratus*.

1 INTRODUCTION

Four encrusting ctenostomates occur on intertidal *Fucus serratus* in South Wales: *Alcyonidium hirsutum* (Fleming), *A. polyoum* (Hassall) (recently synonymized with *A. gelatinosum* (L.), see Thorpe & Winston (1984)), *A. reticulum* Ryland & Porter, and *Flustrellidra hispida* (Fabricius). At the time of the recent study of settlement seasons (de Putron & Ryland, 1998), following the *Sea Empress* oil spillage at the entrance to Milford Haven, South Wales, in February 1996, *A. reticulum* had neither been recognized as present on *F. serratus* nor had it been properly distinguished from the very similar *A. polyoum* (see Ryland & Porter, 2000, as *A. gelatinosum*). Great difficulty, therefore, was experienced in assigning newly settled ancestrulae to species. To clarify the situation and try to ascertain more clearly the effects of the spillage on the reproduction of *A. polyoum*, a further year's study was undertaken.

The previous study had been based primarily on two rocky shores in Milford Haven, a drowned river valley (ria) in south-west Wales: Dale, which was little affected by either the spillage or subsequent cleaning (SEEEC, 1998), and Angle, which was severely impacted first by crude oil carried into the Haven by the wind, then by fuel oil released after the *Sea Empress* had been towed into the Haven, and finally by the clear-up operations (SEEEC, 1996; SEEEC, 1998). In addition, two unaffected *Alcyonidium* populations were studied in a nearby, but much smaller, ria at Solva. By the end of the study the following had been established: (1) *A. hirsutum* occurred at Dale and at Solva, having a short, winter breeding season (see also Owrid & Ryland, 1991); (2) *A. polyoum* occurred at Solva, Dale and Angle and had a summer breeding season; (3) *A. reticulum* occurred at Dale and Angle and appeared to have an extended winter breeding season, but it was not known how to separate its recent settlement from that of *A. hirsutum*. Subsequently, it was discovered that the early astogeny of the two species differed (Ryland *et al.*, 2002), providing a method for quantifying the settlement of each.

At Angle the reproductive output of *A. polyoum* for 1996 had been annihilated by the spillages and their clear up (de Putron & Ryland, 1998, as *A. gelatinosum*). Further, in 1997 its breeding season was later than at Dale, only 7 km distant. Was this normal or a consequence of the disaster? Could it result at Angle in a confusing overlap between late settlement of *A. polyoum* in summer and early settlement of *A. reticulum* in autumn? Moreover, ecological considerations, such as resource partitioning between cohabiting species, needed to be addressed; and the fourth species, *Flustrellidra hispida*, had not been included in the original investigation. For these reasons a further study, from December 1997 to February 1999, was undertaken to facilitate comparison between years of the reproductive periods of the three *Alcyonidium* species at Dale and at Angle, and to determine that of *F. hispida*.

It has become apparent that there are nomenclatural problems affecting two of the four species discussed in this paper. The details are complex and unlikely to be resolved without recourse to the International Commission of Zoological Nomenclature. The nomenclature adopted here represents a provisional solution pending further clarification.

2 MATERIALS AND METHODS

Shore visits were made to Dale (OS grid SM813061) and Angle Point (SM875032) in Milford Haven, South Wales, fortnightly: i.e. on as many spring tides as was practicable between 14th December 1997 and 1st February 1999 (see Appendix).

Two types of collection were made. In the first, samples of the bryozoans were collected for determination of reproductive state; they were fixed in Bouins Fluid for 24 hours and stored in 70% ethanol. In the second, distal portions of *Fucus serratus* thalli were removed from plants in the close vicinity of clumps bearing the bryozoans; these were later trimmed to a length of ~10 cm, fixed in 4% formaldehyde in sea water, and stored in a 1% freshwater solution of propylene phenoxetol. Fruiting tips were removed as they are avoided by settling larvae (Ryland 1959a) and their inclusion would distort the frond area calculation. These frond lengths stayed flexible, were safe to handle, and any settled ancestrulae remained white and undistorted.

Subsequently, newly settled ancestrulae (defined as having no more than 1-2 incipient buds) on both sides of 20 frond lengths were counted under a stereo-microscope; the fronds were then lightly compressed below a rigid clear acrylic sheet on a lightbox and their outlines drawn onto A3 tracing paper with black ink. The 20 outlines were photo-reduced onto white A4 paper and scanned as .TIF files; the outlines were filled black in Adobe Photoshop, the areas of all the silhouettes determined in Optimas 5 and downloaded into an Excel spreadsheet. Areas were then doubled (for the two sides), basic statistical analysis conducted, and settlement densities expressed per 100 cm^2 of frond surface.

The embryo-containing brood chambers of *Alcyonidium hirsutum* and *A. reticulum* were clearly visible in preserved specimens but, unfortunately, those of *A. polyoum*, so clear in fresh material, were hidden by the fixation-induced opacity of the colony. An area of colony surface 1.41 cm _ 0.94 cm, captured by a monochrome video camera using a zoom lens and extension tubes, was viewed on a computer fitted with a grabber board. The .TIF image was transferred to Adobe Photoshop and its contrast strongly enhanced. It was then easy to count the fertile gonozooids on the monitor screen, 'spotting' each as it was counted. When material was sufficient, counts were made on 20 colonies. The data were expressed as densities per 1 cm^2 and analysed in Excel spreadsheets.

3 RESULTS

3.1 *Settlement (unbudded ancestrulae)*

Sixteen sets of 20 fronds were analysed for settlement The sample mean lengths ranged from 7.87 cm to 12.54 cm (mean 10.16 cm) and the two-sided area means ranged from 36.09 cm^2 to 67.15 cm^2 (mean 53.30 cm^2). The recorded numbers of ancestrulae of the four species *Alcyonidium hirsutum, A. polyoum, A. reticulum* and *Flustrellidra hispida* per sample, with the 95% confidence limits of the mean, are shown in Figures 1-2. To simplify the presentation, data are separated by season as well as by locality. Settlement of all four species occurred at Dale but of only *A. polyoum* and *A. reticulum* at Angle (adult colonies of the other two being present but uncommon). *A. hirsutum* settled at Dale over a short period from late January to mid-March (Fig. 1A). For some reason unknown, but presumably an artifact of sampling (see Discussion), no settlement of *A. reticulum* was found at Dale amongst the heavy settlement of *A. hirsutum*, although the ancestrulae of the two species could reliably be distinguished. At Angle *A. reticulum* settled from December until the end of March, with some settlement still taking place at Dale through April (Fig. 1B). Settlement of *A. polyoum* was first recorded at Dale in May and continued until August (Fig. 2A); at Angle the season was later (Fig. 2B), with heavy settlement during July (but the record is imperfect owing to some lost samples). *Flustrellidra hispida* settled from the end of March until early August, most densely between April and June (Fig. 2A).

3.2 *Presence of embryos in brood chambers*

When the zygotes of *A. hirsutum* are transferred to the brood chamber, they become arranged in a circle (Owrid & Ryland, 1991) and are very visible (and which led Hassall (1841) to create the genus *Cycloum*). These rings first became visible in November (Fig. 1D) and persisted until all larvae had been shed. The zygotes of *A. reticulum* after transfer form a clump in a brood zooid that enlarges to form a prominent sphere (Ryland & Porter, 2000), white while embryos are present, and persisting some time after their discharge. These filled brood chambers were first observed at both sites during November (Fig. 1C,D) and were present throughout winter. The embryos of *A. polyoum* are buff-coloured when live and do not become arranged in rings, nor do the brood chambers differentiate in the manner observed in *A. reticulum*. When the colony becomes opaque following fixation, the embryos are difficult to see and the brood chambers cannot be counted. *Flustrellidra hispida* has thick colonies and the embryos are scarcely visible.

3.3 *Consolidated results*

In view of the gaps present in the 1998 samples, settlement data from 1997 have also been included in the figures. *Alcyonidium hirsutum*, which is abundant at Dale but rare at Angle, displays rings of

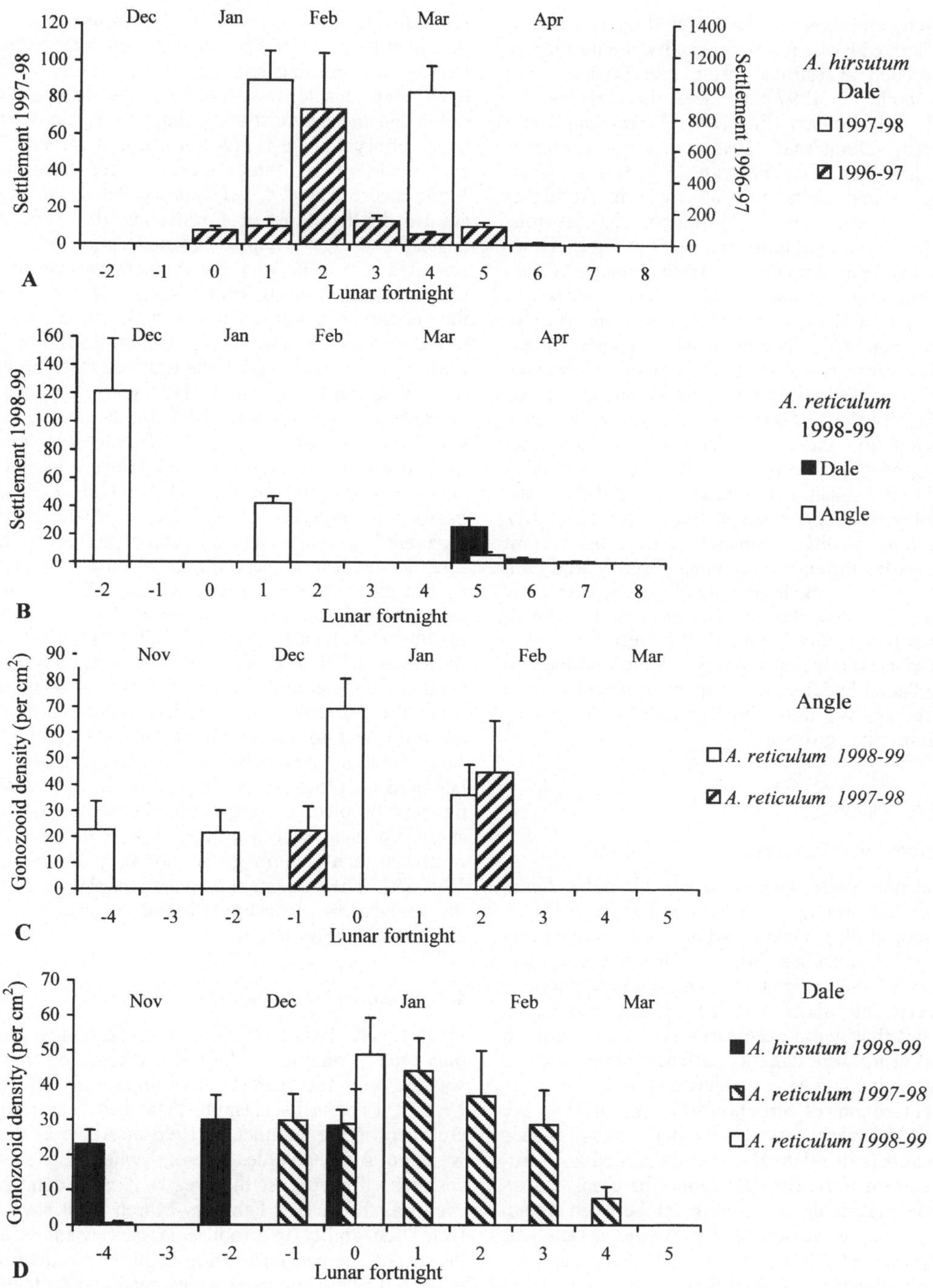

Figure 1. Winter breeding of *Alcyonidium* species at Angle and Dale, 1996-99. A, B. Settlement (ancestrulae 100 cm^{-2} *Fucus serratus*). A. *A. hirsutum*, Dale, 1997-98; 1996-97 on right *y*-axis. B. *A. reticulum*. C, D. Density of brood zooids with embryos (number cm^{-2}); C. Angle; D. Dale. Gaps between bars within series indicate absence of samples, not zero values. Fortnight 0 represents the first new/full moon in January; actual dates vary slightly between years (see Appendix). Error bars are 95% confidence limits.

developing embryos by November (Fig. 1D) and in 1998 liberated larvae (as evidenced by the occurrence of unbudded ancestrulae) from late January until early March. In 1997 there was also some settlement in early January (Fig. 1A). *A. reticulum* has a settlement season that starts earlier and continues longer than that of *A. hirsutum* (Fig. 1B), with ancestrulae found abundantly at Angle in December (and some settlement in November the previous year) but continuing until late April at Dale. However, many Dale samples comprised abundant *A. hirsutum* but no *A. reticulum*. The settlement season of *A. polyoum* at Dale, as in 1997, was from May to August (Fig. 2A), though actual densities of ancestrulae were much lower than those of *A. hirsutum*. Densities at Angle were much higher but the record is marred by the lost samples. However, the data, supported by the 1997 settlement, show that breeding of *A. polyoum* at Angle lags considerably behind that at Dale, not commencing until June and continuing (in 1997) through September (Fig. 2B). *Flustrellidra hispida*, abundant at Dale but rare at Angle, settled throughout summer, from March until August, with the mode in May (Fig. 2A). Its early astogeny was described by Barrois (1877), who illustrates a bilaterally symmetrical pattern. This was observed at Dale in many young colonies but the rest had unilateral budding, rather as in *A. polyoum* (also described and illustrated by Barrois (1877)), leading to asymmetrical growth.

4 DISCUSSION

4.1 *Winter breeding species*

Two ctenostomate species in the intertidal *Fucus serratus* community in Milford Haven (Seed & O'Connor, 1981) breed in winter, *Alcyonidium hirsutum* and *A. reticulum* (Fig. 1). The settlement of *A. hirsutum* in 1998 (Fig. 1A) superficially appears to have been only about one-fifth of that recorded in 1997, but that is not necessarily so. The counts in the two years were done by different observers, and the criteria limiting what budding stages were or were not included almost certainly differed. In 1998 the author used very restrictive criteria and excluded huge numbers of relatively recently settled ancestrulae. Consequently, the data cannot be used to compare the magnitude of recruitment between years. The reproductive season of *A. hirsutum* at Dale was already well established at the time of the *Sea Empress* disaster (Owrid & Ryland, 1991, 1994) but that of *A. polyoum*, then believed to be the only other *Alcyonidium* present, was not. The *Sea Empress* study (de Putron & Ryland, 1998) established that it bred in the summer but that there was another, then apparently un-named, species in the Haven which bred in winter. This was duly described as *A. reticulum* by Ryland & Porter (2000) and distinguished from *A. polyoum* (as *A. gelatinosum*). De Putron & Ryland (1998), even had they realized that two winter breeding species were present, would have been unable to distinguish the ancestrulae. Following the demonstration that *A. reticulum* had larger embryos than *A. polyoum*, several samples of ancestrulae of all three species were measured. While ancestrulae of *A. polyoum* are indeed, on average, smaller than those of *A. reticulum*, the difference is small and the overlap considerable; moreover, ancestrulae of *A. reticulum* and *A. hirsutum* are effectively the same size (Ryland *et al.*, 2002). However, the ancestrular budding pattern in *A. reticulum* was found to resemble that of *A. polyoum*, illustrated by Barrois (1877, as *A. mytili*): the first bud is lateral but the second and third, which arise together, are distolateral and proximo-lateral, above and below the first bud. The early astogeny is therefore asymmetrical. In *A. hirsutum*, on the other hand, the second bud is lateral and opposite the first. The incipient colony passes through a distinctive phase in which, at first, there are large and small bud anlagen beside the ancestrula and, a little later, the initial three zooids are aligned in a transverse row (Ryland, 1962). Interestingly the present results support the earlier ones (de Putron & Ryland, 1998) in failing to register the settlement of *A. reticulum* at Dale. Although the studied shore, known as Black Rocks, is relatively small, the abundances of the species seem to differ to the north and to the south of the rocky outcrop. The result is presumably the result of collections failing to sample *A. reticulum* adequately. There was no such problem at Angle and the season is presumed to have been correctly represented by the combined data (sampling was not very frequent in 1998-99). The settlement data are complemented by the abundances of occupied brood chambers, especially at Dale (Fig. 1C,D).

4.2 *Summer breeding species*

De Putron & Ryland (1998) established that *A. polyoum* bred in summer. They also discovered that it was the only one of the species apparently affected by the oil and/or its cleanup: there had been no recruitment in the summer of 1996 at Angle as there were no small colonies present when the study commenced in August that year. It is evident that events as far back as February-March must have affected the ability to produce larvae several weeks later. The mechanism for this is quite unknown but is likely to have involved widespread and prolonged polypide degeneration which impacted adversely on the onset of gametogenesis. The following summer, *A. polyoum* bred prolifically at Angle but its settlement season was considerably later than at Dale. Was this a delayed effect of the impact or a normal occurrence related to habitat differences between the two shores? Resolution of these alternatives was a

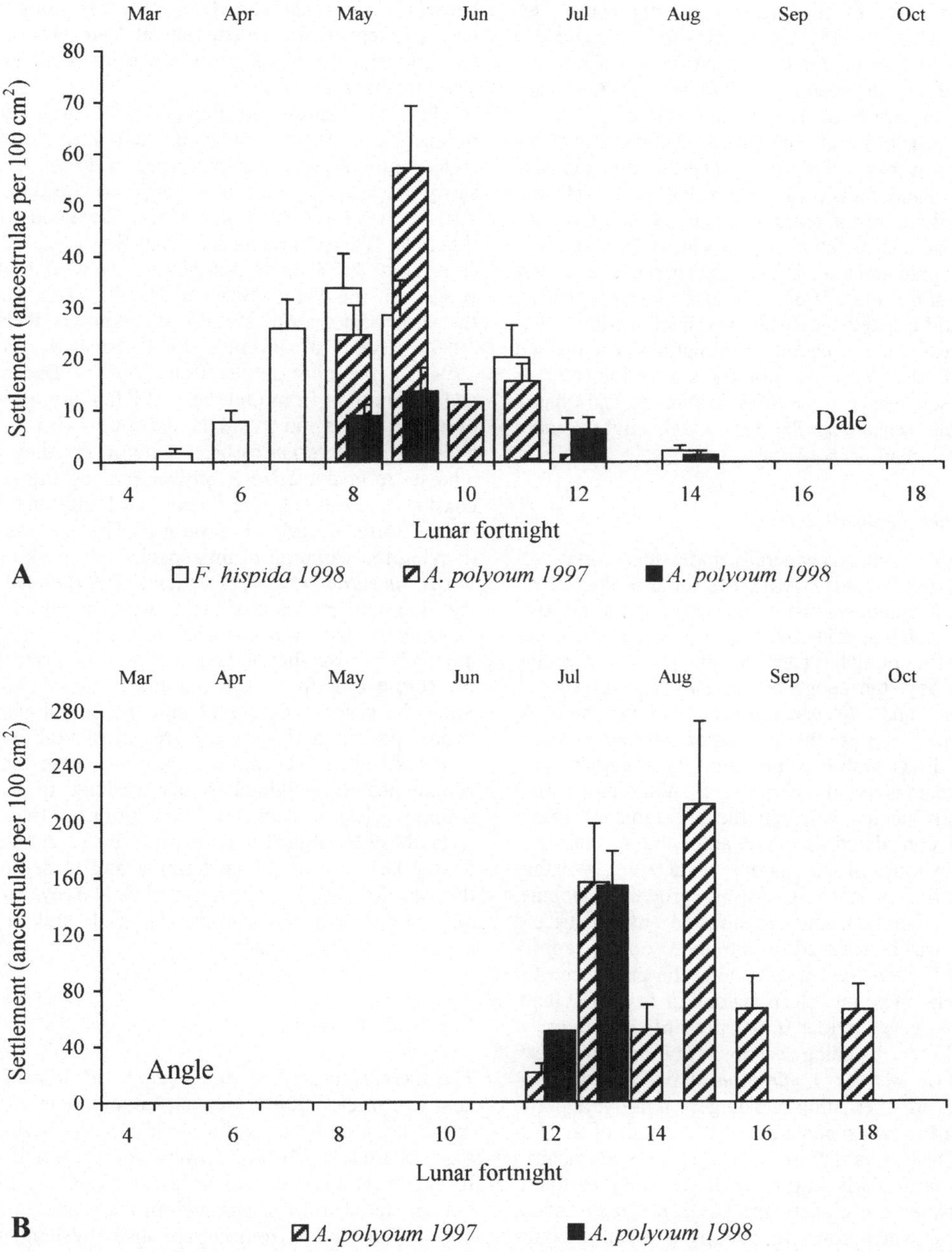

Figure 2. Summer breeding of *Alcyonidium* species and *Flustrellidra hispida* at Dale and Angle, 1997 and 1998, measured as numbers of unbudded ancestrulae 100 cm^{-2} *Fucus serratus* frond. A. Dale; B. Angle. Gaps between bars within series indicate the absence of samples, not zero values. Actual dates vary slightly between years (see Appendix). Error bars are 95% confidence limits. Note that the settlement of *A. polyoum* is later at Angle than at Dale.

major objective of the present investigation. The settlement results (Fig. 2) clearly show, despite the subsequent loss of some relevant samples, conformity between the results of 1997 and 1998: though the seasons are both of the same duration, that at Dale precedes that at Angle (only 7 km distant) by about 10 weeks. The reason(s) for this are unknown: Angle, tucked into a bay within the Haven, is certainly more sheltered than Dale, which faces up the Haven with a fetch of ~15 km, but most other hydrographic and biological characteristics are similar (see Nelson-Smith, 1965; Owrid & Ryland, 1994). It should be noted, however (see next section), that the *Alcyonidium* breeding seasonalities established for south-west Wales are not necessarily the same as those elsewhere. *Flustrellidra hispida* at Dale has a settlement season similar to, though slightly more protracted than, *A. polyoum*.

4.3 *General considerations*

Gause's Principle, as generally understood today, affirms that two species with the same ecological requirement cannot coexist in the same habitat; i.e. two closely related species living together must be occupying different niches (see Stebbing, 1971). Stebbing showed how the congeners *Crisia eburnea* and *C. aculeata*, and *Scrupocellaria reptans* and *S. scruposa*, living on the fronds of *Flustra foliacea*, avoided direct spatial competition by occupying different regions of the frond. Spatial competition manifestly occurs between the epibionts of *Fucus serratus* considered here, and all of the species generally respond to contiguous neighbours, whether conspecifics or not, by ceasing growth (Stebbing, 1973). Competition, manifested by restricted growth, will be reduced or eliminated on the rapidly growing *F. serratus* by combining the preference for settlement on young, clean regions of frond (Ryland, 1959a) — very obvious in the present collections — with different breeding seasons. The spatial consequences of this for *A. hirsutum* and *F. hispida*, for example, are alternating and largely non-overlapping zones of the two species down the length of the *Fucus* thallus (Oswald & Seed, 1986). This mechanism may be particularly important for avoiding competition between the closely similar *A. polyoum* and *A. reticulum*, which come to occupy alternating zones in the manner described for *A. hirsutum* and *F. hispida*. The direct spatial competitor for the latter is *A. polyoum*, but *F. hispida* differs from it in the large size of its lophophore (28-40 tentacles, ~1.4 mm top diameter, compared to 19-20 tentacles, ~0.8 mm top diameter in *A. polyoum* (Ryland, 1976)). This may affect food selection but does not seem to make *F. hispida* a better spatial competitor, simply through having the larger lophophore, as reported for two other bryozoan species with disparate lophophore sizes (Buss, 1980). The nature of any interactions between *A. hirsutum* and *A. reticulum* has not been noted, except to the extent that at Dale, where the two co-occur, larvae of the two species seem not to settle together.

The investigation has highlighted the importance of breeding season in ecology and taxonomy, but discrete populations of the same species do not necessarily display the same phenology, as the differing settlement seasons of *A. polyoum* at Dale and Angle illustrate. The settlement season at Solva (de Putron & Ryland, 1998) is contemporaneous with that at Angle. In the Menai Strait in North Wales, where the settlement seasons are also well known (Ryland, 1959b; Owrid & Ryland, 1994; Seed & Wood, 1994), *A. polyoum* settles from June to December and *A. hirsutum* from October until January: thus *A. polyoum* settles later than in its south-west Wales sites and *A. hirsutum* earlier. Autumn breeding also appears to characterize *A. polyoum* along the south coast of England (e.g. the Solent and Langstone Harbour) (Porter & Ryland, unpubl.). The reasons underlying the variation in this species are not known. For *A. hirsutum*, Owrid & Ryland (1994) concluded that late winter release of larvae was normal and advantageous: the embryos are brooded through the most adverse weather and released just in advance of the spring phytoplankton outburst. In the Menai Strait, however, they found that the conditions favoured prolific and luxuriant growth of both *Fucus serratus* and its *Alcyonidium* epibionts, with the resultant heavily weighted fronds breaking up in the autumn gales. Under these conditions, Owrid & Ryland (1994) argued that it would be advantageous for the larvae to be released earlier and to settle, as they do (Ryland, 1959a; Hayward & Harvey, 1974), on young clean areas of frond that would survive the autumn and winter intact.

5 CONCLUSIONS

The recently described *Alcyonidium reticulum* (Ryland & Porter, 2000) is a major component of the sessile community of epibionts found on *Fucus serratus* at sites in Milford Haven, south-west Wales. It was previously (e.g. in Seed & O'Connor, 1981) not separated from *A. polyoum* in the same community. Quite apart from certain distinguishing morphological characters (Ryland & Porter, 2000; Ryland *et al*., 2002), this study shows that a major ecological difference between these apparently very similar species is the timing of reproduction. In Milford Haven, at least, *A. reticulum* breeds in winter and *A. polyoum* in summer. Since *F. serratus* itself grows, the two congeners come to occupy alternating bands up the *Fucus* thallus, as earlier described for *A. hirsutum* and *Flustrellidra hispida* (Oswald & Seed, 1986). This is likely to reduce competition between them greatly. Evidence is accruing that the *Alcyon-*

idium species discussed here may have quite different breeding seasons elsewhere in their geographic range. Only one specific case has so far been investigated, comparing *A. hirsutum* in Milford Haven and the Menai Strait (Owrid & Ryland, 1994). In general, the factors that control the timing of bryozoan reproduction are unknown.

ACKNOWLEDGEMENTS

The field work conducted in 1998 was made possible by a grant from the British Ecological Society.

REFERENCES

Barrois, J. 1877. *Recherches sur l'embryogénie des Bryozoaires.* Lille: Université de Lille.

Buss, L.W. 1980. Bryozoan overgrowth interactions — the interdependence of competition for space and food. *Nature (London)* 281: 475–477.

De Putron, S. & Ryland, J.S., 1998. Effects of the 'Sea Empress' oil spillage on reproduction and recruitment of *Alcyonidium* (Bryozoa) populations on *Fucus serratus.* In R. Edwards & H. Sime (eds) *The Sea Empress oil spill*: 457–466. Lavenham: Terence Dalton & CIWEM.

Hassall, A.H. 1841. Descriptions of two new genera of Irish zoophytes. *Annals and Magazine of Natural History* 7: 483–486.

Hayward, P.J. & Harvey, P.H. 1974. The distribution of the settled larvae of the bryozoans *Alcyonidium hirsutum* (Fleming) and *Alcyonidium polyoum* (Hassall) on *Fucus serratus* L. *Journal of the Marine Biological Association of the United Kingdom* 54: 665–676.

Nelson-Smith, A. 1965. Marine biology of Milford Haven: the physical environment. *Field Studies* 2: 155–188.

Oswald, R.C. & Seed, R. 1986. Organisation and seasonal progression within the epifaunal communities of coastal macroalgae. *Cahiers de Biologie Marine* 27: 29–40.

Owrid, G.M.A. & Ryland, J.S. 1991. Sexual reproduction in *Alcyonidium hirsutum* (Bryozoa: Ctenostomata). In F.P. Bigey (ed.) *Bryozoaires actuels et fossiles: Bryozoa living and fossil. Bulletin de la Societé des Sciences Naturelles de l'Ouest de la France* HS 1: 317–326.

Owrid, G.M.A. & Ryland, J.S., 1994. Variation in life histories in two *Fucus serratus – Alcyonidium hirsutum* associations. In J. C. Aldrich (ed.) *Quantified phenotypic responses in morphology and physiology*: 93–105. Ashford, Ireland: Japaga.

Ryland, J.S. 1959a. Experiments on the selection of algal substrates by polyzoan larvae. *Journal of Experimental Biology* 36: 613–631.

Ryland, J.S. 1959b. *Settlement behaviour in Polyzoa*, Unpublished PhD thesis, University of Wales.

Ryland, J.S. 1962. Biology and identification of intertidal Polyzoa. *Field Studies* 1(4): 33–51.

Ryland, J.S. 1976. Parameters of the lophophore in relation to population structure in a bryozoan community. *Proceedings of the 9th European Marine Biology Symposium*: 363–393.

Ryland, J.S. & Porter, J.S. 2000. *Alcyonidium reticulum* sp. nov., a common intertidal bryozoan from south-west Britain. *Journal of the Marine Biological Association of the United Kingdom* 80: 563–564.

Ryland, J.S., Porter, J.S. & de Putron, S. 2002. The genus *Alcyonidium* (Bryozoa) in the Bristol Channel and south-west Wales: recognition and ecology of intertidal species. In M.S.C. Havard (ed.) *The coast and estuaries of South Wales.* Coastal Zone Topics: Process, Ecology and Management 9 (in press).

Seed, R. & O'Connor, R.J. 1981. Epifaunal associates of *Fucus serratus* at Dale, south-west Wales. *Holarctic Ecology* 4: 1–11.

SEEEC. 1996. *Initial Report.* Cardiff, *Sea Empress* Environmental Evaluation Committee.

SEEEC. 1998. The environmental impact of the *Sea Empress* oil spill. *Final report of the* Sea Empress *Environmental Evaluation Committee.* London: The Stationery Office.

Stebbing, A.R.D. 1971. The epizoic fauna of *Flustra foliacea* [Bryozoa]. *Journal of the Marine Biological Association of the United Kingdom* 51: 283–300.

Stebbing, A.R.D. 1973. Competition for space between the epiphytes of *Fucus serratus* L. *Journal of the Marine Biological Association of the United Kingdom* 53: 247–261.

Thorpe, J.P. & Winston, J.E. 1984. On the identity of *Alcyonidium gelatinosum* (Linnaeus, 1761) (Bryozoa: Ctenostomata). *Journal of Natural History* 18: 853–860.

APPENDIX

Dates of new and full moons corresponding to lunar fortnights as used in this paper. The last few fortnights in the year preceding the nominal year are indicated by minus values.

Fortnight	Nominal year 1997	Nominal year 1998
-5	26/10/96	31/10/97
-4	11/11/96	14/11/97
-3	25/11/96	30/11/97
-2	10/12/96	14/12/97
-1	24/12/96	29/12/97
0	09/01/97	12/01/98
1	23/01/97	28/01/98
2	07/02/97	11/02/98
3	22/02/97	26/02/98
4	09/03/97	13/03/98
5	24/03/97	28/03/98
6	07/04/97	11/04/98
7	22/04/97	26/04/98
8	06/05/97	11/05/98
9	22/05/97	25/05/98
10	05/06/97	10/06/98
11	20/06/97	24/06/98
12	04/07/97	09/07/98
13	20/07/97	23/07/98
14	03/08/97	08/08/98
15	18/08/97	22/08/98
16	01/09/97	06/09/98
17	16/09/97	20/09/98
18	01/10/97	05/10/98
19	16/10/97	20/10/98
20	31/10/97	04/11/98

Bryozoan Studies 2001, Wyse Jackson, Buttler & Spencer Jones (eds)
© 2002 Swets & Zeitlinger, Lisse, ISBN 90 5809 388 3

Eocene Bryozoan Assemblages of the St Vincent Basin, South Australia

R. Schmidt & Y. Bone
Department of Geology & Geophysics, Adelaide University, Adelaide, South Australia 5005, Australia

ABSTRACT: The first stratigraphically detailed taxonomic study of Eocene Bryozoa in the St Vincent Basin has revealed a faunas ranging from very high abundance and diversity to almost monotypic assemblages. Some formations contain over 200 species of cheilostomes and over 50 species of cyclostomes. This is more diverse than all other macrofauna present combined. The Phidoloporidae are the most diverse and ubiquitous group with over 20 species, at a geological time which is close to their estimated origination. The Cellariidae constitute over 15 species and form distinct biofacies. Lunulitiform growth form is displayed by the disparate families Otionellidae, Calloporidae, Lepraliellidae and Didymosellidae. Biogeographical comparisons exhibit distinct differences with the published (mainly Oligo-Miocene) fossil bryozoan faunas of Victoria, but similarities with the Eocene of Antarctica, New Zealand and even central Europe.

1 INTRODUCTION

This paper reports some of the results from the first extensive and stratigraphically detailed taxonomic study carried out on the fossil Bryozoa occurring in the Eocene sediments of the St. Vincent Basin, South Australia. Understanding these bryozoan faunas is important for geological, evolutionary and taxonomic reasons. Despite being almost ubiquitous in the Tertiary of Australia and frequently rock forming, most of these fossil Bryozoa have not been extensively studied.

During the later Eocene the Earth was still in a Greenhouse state, before the onset of extensive glaciations in the earliest Oligocene. Although the southern margin was positioned at 55°S latitude (20° further south than today), the climate was mesothermal (sub-tropical) with extensive rainforests (Scriven 1994). Sea levels were relatively high and flooded many basins along the western and southern Australian margin. Many of these were initiated during the rifting of Australia and Antarctica in the Cretaceous. Both continents were still connected via Tasmania until the Late Eocene, forming a large gulf (Fig. 1). This resulted in very different ocean currents to today. It is thought, however, that a Proto-Leeuwin Current was already bringing Indo-Pacific topical waters into the southern margin via western Australia (McGowran *et al.* 1997), even though the Indonesian Gateway did not exist.

The St. Vincent Basin is a comparatively small basin (Fig. 1a). It had restricted access to the open ocean for most of its history due to the Kangaroo Island basement high (James & Bone 2000). The internal topography is controlled be ancient fault systems, which create several embayments and subbasins. This geomorphology resulted in diverse and often enigmatic environments. The earliest fossil Bryozoa are of Middle to Late Eocene age, and are the focus of this paper. (Fig. 2)

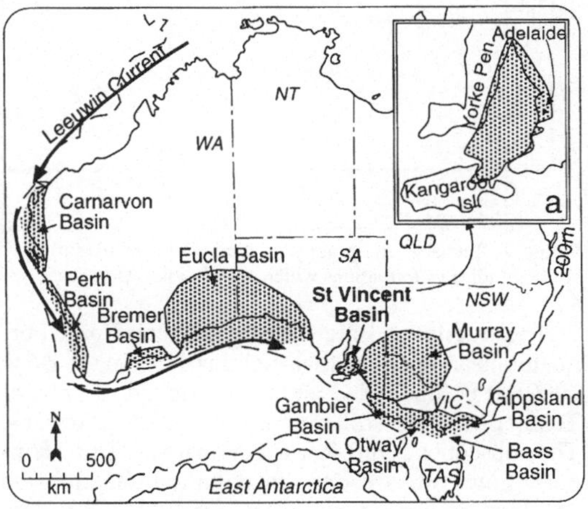

Figure 1. Australia in the Eocene, with major Tertiary Basins schematically outlined in grey (adapted from Palfreyman 1984), modern path of the Leeuwin Current also indicated; inset A: St Vincent Basin.

On the eastern margin of the basin the oldest marine stratum is the late Middle Eocene Tortachilla Limestone, which grades up from the deltaic and goethite rich South Maslin Sand. The basal unit contains abundant goethite pellets and is relatively unconsolidated. This probably represents a sea grass meadow environment. The upper unit is glauconitic and very cemented towards the top, This was probably deposited under more restricted conditions. The top may represent an exposure surface. Bryozoans generally constitute the bulk of the sediment but are more abundant and diverse in the lower unit. Other common macrofossils include turritellid gastropods, pectinid bivalves, infaunal and epifaunal echinoids, rare solitary and even rarer colonial corals.

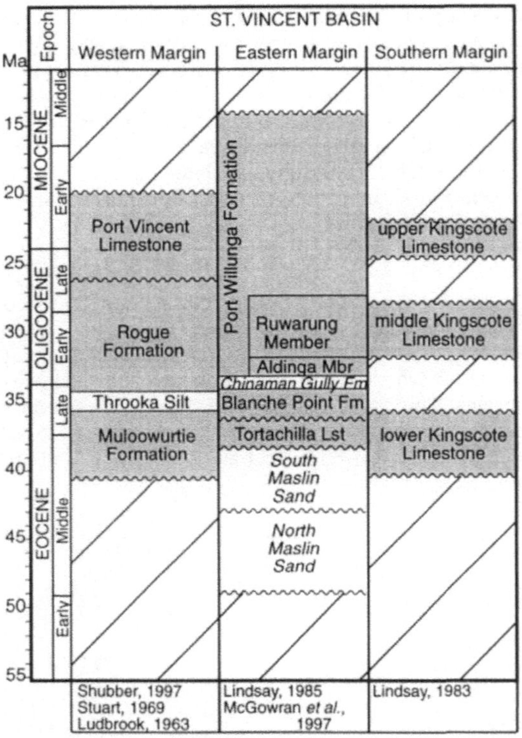

Figure 2. Tertiary stratigraphy of the St Vincent Basin; grey shading indicates formations which contain fossil Bryozoa.

The overlying Late Eocene Blanche Point Formation is an opaline silica-rich marl (James & Bone 2000). It is subdivided into four members: The basal Tuketja Member still contains abundant bryozoans. The subsequent Gull Rock, Perkana and Tuit Members, however, are dark in colour, indicating the presence of unoxidised organic matter due to low oxygen levels (McGowran & Beecroft 1986). The fauna is dominated by infaunal turritellid gastropods and sponge spicules. Bryozoa are present in variable but mostly low abundances and diversity. Only in a few beds do they become common and moderately diverse. The most conspicuous Bryozoa are large rafts of multilaminar *Celleporaria*. This formation was deposited under eutrophic conditions, with no marginal marine influence evident in the coastal outcrops (Cooper 1979).

The thin overlying clay-rich Chinaman Gully Formation is considered to be a regressive, terrestrial deposit, marking the regression that signifies the Eocene/Oligocene boundary.

The Muloowurtie and Rogue Formations on the western margin (east coast of Yorke Peninsula) are mostly siliciclastic and only contain occasional fossiliferous beds (Stuart 1970). These are generally dominated by bivalves. The greatest abundance and diversity of Bryozoa is directly overlying the Cambrian basement rocks, which probably formed a palaeohigh. There was probably a greater terrestrial sediment and freshwater influx, which generally restricts bryozoan faunas.

On the southern margin (near Kingscote on Kangaroo Island) the Eocene Lower Kingscote Limestone is heavily cemented and the predominant macrofossils are infaunal echinoids such as *Monostychia* and *Fibularia*. Bryozoa are common but rarely become dominant.

2 RESULTS AND DISCUSSION

2.1 *Taxonomy*

Despite the rich bryozoan faunas, no taxonomic work has been published on the Bryozoa of the St. Vincent Basin, apart from three dozen species described by Waters (1885). It is likely that most of the Tertiary basins in Australia contain bryozoan assemblages at least as diverse as the well studied Oligo-Miocene Victorian deposits (e.g. MacGillivray 1895, Brown 1958, Bock & Cook 1993a, b, 1995a, b, 1996, 1999, Cook & Bock 1996, Cook & Chimonides 1987, and references therein).

An accurate and up to date taxonomic species list of this fauna is therefore essential for any other aspect of these Bryozoa to be investigated. The primary goal was therefore the identification of as much of this bryozoan fauna as possible, under tight stratigraphic control. Cheilostomes were investigated in greatest detail, as the identification of the fragmented cyclostomes to species was difficult, and corresponding diversity estimates are probably a slight underestimate.

Figure 3. Cheilostomata of the Tortachilla Limestone, Maslin Beach. A. Calloporidae gen. et sp. nov., SAM 39320; B. *Tubulipora magna* (Tenison Woods), SAM 39326; C. *Otionellina* cf. *exigua* (Tenison Woods), SAM 39328; D. *Celleporaria nummularia* (Busk), SAM 39331, three colonies, one from basal view; E. *Cellaria* sp. nov. a, SAM 39335; F. *Cellaria* sp. nov. b, SAM 39336; G. *Cellaria* sp. nov. c, SAM 39337; H. *Cellaria* sp. nov. d, SAM 39338.

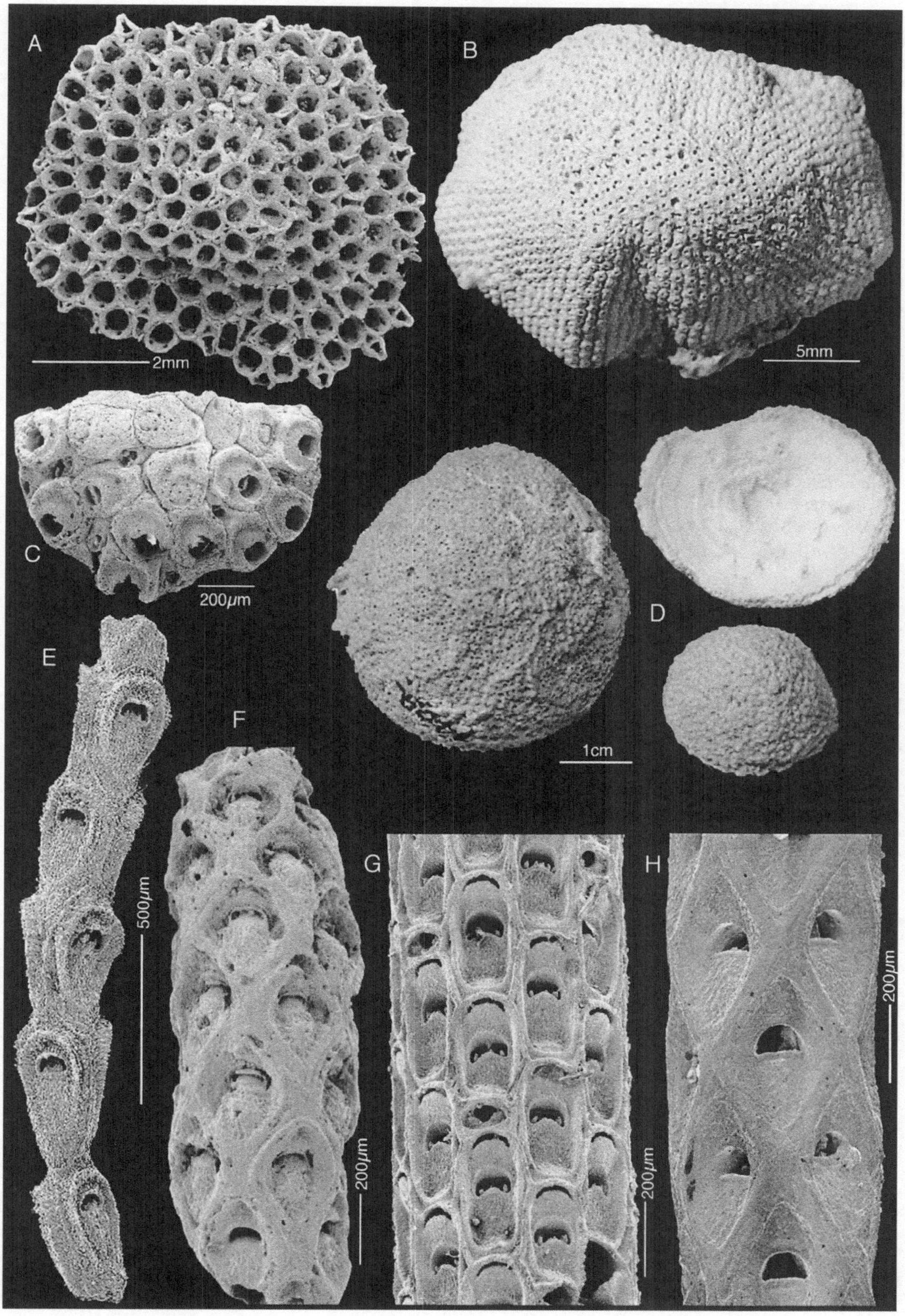
A
B
2mm
5mm
C
200μm
D
1cm
E
F
G
H
500μm
200μm
200μm
200μm

The taxonomic species list clearly shows that there is a high diversity. Current data indicates that there are more than 200 species of Cheilostomata within over 80 genera. It is likely that well over 50 of the species (~20%) are described for the first time. At least five new genera will also need to be introduced. The Cyclostomata are represented with more than 50 species.

The most diverse families are the Phidoloporidae (5 genera, 20 species), Cellariidae (3 genera, 17 species), Smittinidae (4 genera, 15 species), Schizoporellidae (4 genera, 10 species), Catenicellidae (5 genera, 10 species), Lepraliellidae (3 genera, 8 species), Romancheinidae (4 genera, 6 species) and Adeonidae (4 genera, 6 species). The most diverse genera are *Cellaria* (14 species) and *Smittoidea* (8 species)

Most of this diversity occurs within the thin (~2m thick) Tortachilla Limestone. The remaining diversity peaks occur in locally fossiliferous horizons of the basal Muloowurtie Formation, Blanche Point Formation and Kingscote Limestone.

The type material, which has resulted from this investigation, is stored in the South Australian Museum (SAM), Adelaide.

2.2 *Biogeography*

The Australian fauna is of great interest because of the long isolation from other continents and the resultant unique biogeographic relationships. This is particularly interesting for the Bryozoa, which are generally considered to have maintained an endemic character throughout the whole Cenozoic.

Biogeographical comparisons within Australasia are difficult, as the only detailed and up to date taxonomic work is restricted to the Oligo-Miocene of Victoria, and the Tertiary of New Zealand (Gordon & Taylor 1999). This research must therefore be considered as a further piece in the puzzle that will be useful for future research.

Comparisons with existing studies indicate some endemism of this fauna because of the significant proportion of new species and genera. It is significantly distinct from the faunas recorded in the Victorian Tertiary. Many species, genera or even families (most notably the Selenariidae), which are common there, are absent in the St Vincent Basin. Although the Cellariidae are a common and diverse component in both regions, most of the species do not appear to be the same. Figure 3E - H display the range of variation within this group found within a single horizon of the Tortachilla Limestone. All these species are new.

The restricted nature of the St Vincent Basin during much of the Tertiary caused by Kangaroo Island may have resulted in a number of sufficiently unique environments to produce very different faunas. It could also have formed a type of geographic faunal barrier, restricting dispersal.

Such absences or differences of taxa must, however, be seen in the context of two circumstances. Firstly, it must be noted that many of these formations have been heavily altered by diagenesis. Many species with primary aragonitic or high magnesium calcite skeletons may therefore have been removed through dissolution or neomorphism. Only the relatively impermeable Blanche Point Formation contains horizons with unaltered fossils, frequently still aragonitic. Secondly, most of the hundreds of fossil species recorded in Australia stem from Oligo-Miocene deposits, and are usually 10 – 20 million years younger. This is a sufficient 'hiatus' for many taxa to originate or disappear.

The occurrence of several genera, which are also found in Antarctica, indicates its close geographic links with Australia. This is true for both the recent, e.g. the genus *Chondriovelum* (Hayward & Thorpe 1988), as well as the Tertiary, eg. *Reticrisina* cf. *transennata* Waters (Hara, pers. comm. 2001). This is not surprising, as the two continents were still joined at the shelf in the Eocene (Fig. 1).

There are also some links to New Zealand (such as the species *Chondriovelum fossilis* Gordon & Taylor 1999).

Interestingly, some links with Tethyan Europe may be considered. The cheilostome family Phidoloporidae is considered to have originated in the later part of the Eocene in the Australasian region. Their ubiquitous abundance and diversity in the late Middle Eocene of the St Vincent Basin, makes it feasible to consider a local evolution for this group. The Late Eocene of Austria, however, also contains a range of phidoloporid species (Zágorsek, pers. comm.). This opens up the possibility of a Tethyan origin, with a rapid dispersal to these two far reaches of this vanishing ocean. Such a link questions the degree of endemism Australia's faunas maintained during its 'isolation'.

2.3 *Palaeoenvironments*

The presence of many embayments and restricted access to the open ocean, resulted in environments developing during the history of this basin that are diverse and complex. The exact environmental conditions of several Eocene formations are still unclear, and various fossil and sedimentological analyses often result in contradictory interpretations (James & Bone 2000). A combined analysis of the bryozoan species and growth-form assemblages was carried out, in order to interpret these facies.

Significant time-averaging is evident within many of the horizons (especially the Tortachilla

Limestone). Many sedimentologically similar layers, however, contain distinct and sometimes very disparate bryozoan assemblages of both species and growth forms. Most layers within the Tortachilla Limestone, for example, contain abundant fragments of species of Cellariidae. One layer, however, contains almost none, but has higher than usual rates of encrustation. This layer does not appear different to the surrounding layers at the outcrop level to the surrounding ones. It is not clear what palaeoenvironmental implications this has as yet.

The lunulitiform, or free-living cap-shaped growth form was common in the Eocene St Vincent Basin. At least five species from four unrelated families are commonly found in the fossiliferous limestones. *Tubiporella magna* (Tenison-Woods 1880), a Didymosellid, has a large dome-shaped colony with an irregular marginal outline (Fig. 3B). *Celleporaria nummularia* (Busk 1859), a Lepraliellid, forms large multilaminar colonies with basal surfaces concave to various degrees (Fig. 3D). A similar species, *C. emancipata* Gordon (1989), occurs around New Zealand today. *Otionellina* cf. *exigua* (Tenison-Woods 1880), an Otionellid (Fig. 3C), is similar to some of the species which still live on the Australian continental shelf. The final two species are as yet undescribed and of unknown (possibly new) generic affinity, probably within the Calloporidae (Fig. 3A). These unnamed species have only been found in the Tortachilla Limestone and exclusively settle on pellets of goethite (a hydrated iron oxide). This may have been a preference for a hydrodynamically stable substrate, as this mineral is considerably heavier than quartz or carbonate grains.

All these species are common and generally occur together in the Tortachilla Limestone in beds consisting almost exclusively of shell-fragments. These beds were probably within wave base for some time as the fragmentation of fossils is very high.

The consistently high Cheilostome to Cyclostome ratios may initially have been even higher. Cyclostomes generally construct their skeletons from relatively stable low Magnesium calcite, and can be expected to remain intact once they are buried below the taphonomically active zone. Cheilostomes often use meta-stable high Mg calcite or aragonite for their skeletons (Bone & James 1993, 1997, Smith *et al.* 1998), and these groups are more likely to have been removed from the fossil assemblage by diagenetic fluids. Such high ratios are typical of shallow tropical environments. The paucity of azoozanthellate corals, however, makes this interpretation tenuous.

The palaeoenvironments of the Blanche Point Formation have created the most controversy. Previous studies, such as ostracods, have indicated it to be deposited at well over 100m depth (Majoran 1995). The Bryozoa that occur in the narrow diversity peaks within the otherwise bryozoan-barren Blanche Point Formation, are generally the same or similar species to those occurring in the more fossiliferous formations (e.g. *Celleporaria* spp., various phidoloporids and a fenestrate onychocellid). This may indicate that these are opportunistic species and groups, which can exploit a variety of environments. It may also indicate brief periods during which environments in the basin returned to less eutrophic and sub-oxic conditions.

3 CONCLUSIONS

The fossil Bryozoa of the St Vincent Basin were diverse in the Late Eocene. Over 200 species of Cheilostomata and 50 species of Cyclostomata are recognized.

Comparisons with the later Oligo-Miocene of Victoria indicate that the Australian bryozoan fauna has changed significantly through the Tertiary. It may have, however, been less endemic relative to other continents prior to the change in the global ocean current regime at the Eocene-Oligocene boundary. Similarities at both generic and specific levels with the Paleogene faunas of Antarctica, New Zealand and even Central Europe exist.

Cheilostome/cyclostome ratios are typical of shallow tropical environments. The Cellariidae delineate distinct biofacies. The impoverished fauna of the Blanche Point Formation contains species that are similar to those in the fossiliferous formations, suggesting either wide environmental tolerances for these species, or short returns to normal marine conditions.

ACKNOWLEDGMENTS

We acknowledge the support of Australian Research Council for this research. RS would like to thank the Geological Society of Australia (SA Div.) and the Walter & Dorothy Duncan Trust for financial travel support to attend the 12th IBA conference in Dublin, Ireland.

REFERENCES

Bock, P.E. & Cook, P.L. 1993a. *Catenariopsis morningtoniensis* Maplestone (Bryozoa, Anasca) - a Tertiary chaperiid from Victoria. *Alcheringa* 17: 153-157.

Bock, P.E. & Cook, P.L. 1993b. *Quadriscutella* gen. nov. (Bryozoa, Cheilostomata) from the Tertiary and recent of Victoria and South Australia. *Alcheringa* 17: 159-171.

Bock, P.E. & Cook, P.L. 1995a. *Chasmozoon* gen. nov., a new genus of ascophoran bryozoan for *Cellepora abdita*

MacGillivray, from the Tertiary of Victoria. *Alcheringa* 19: 113-117.

Bock, P.E. & Cook, P.L. 1995b. *Pachystomaria parvipuncta* MacGillivray, a 'sand fauna' bryozoan species from the Tertiary of Victoria. *Alcheringa* 19: 171-174.

Bock, P.E. & Cook, P.L. 1996. Concatenellidae, a new family of Catenelloidea (Cheilostomatida) for *Concatenella airensis* (Maplestone) from the Tertiary of Victoria. In D.P. Gordon, A.M. Smith & J.A. Grant-Mackie (eds.) *Bryozoans in Space and Time; Proceedings of the 10th International Bryozoology Conference*: 47-53. Wellington: NIWA.

Bock, P.E. & Cook, P.L. 1999. Notes on Tertiary and Recent 'Lunulite' Bryozoa from Australia. *Memorie di Scienze Geologiche* 51: 415-430.

Bone, Y. & James, N.P. 1993. Bryozoans as carbonate sediment producers on the cool-water Lacepede Shelf, Southern Australia. *Sedimentary Geology* 86(3-4): 247-271.

Bone, Y. & James, N.P. 1996. Bryozoan stable isotope survey from the cool-water Lacepede shelf, southern Australia. In N.P. James and J.D.A. Clarke (eds.) *Cool Water Carbonates*: 93-105. SEPM Special Publication.

Brown, D.A. 1958. Fossil Cheilostomatous Polyzoa from south-west Victoria. *Memoirs of the Geological Survey of Victoria* 20: 1-90.

Cook, P.L. & Chimonides, P.J. 1987. Recent and fossil Lunulitidae (Bryozoa: Cheilostomata), 7. *Selenaria maculata* (Busk) and allied species from Australia. *Journal of Natural History* 21: 933-966.

Cook, P.L. & Bock, P.E. 1996. *Characodoma* Maplestone, a senior synonym of *Chleidochasma* Harmer (Chleidochasmatidae). In D.P. Gordon, A.M. Smith & J.A. Grant-Mackie (eds.) *Bryozoans in Space and Time; Proceedings of the 10th International Bryozoology Conference*: 81-88. Wellington: NIWA.

Cooper, B.J. 1979. Eocene to Miocene stratigraphy of the Willunga Embayment *Report of Investigations of the Geological Survey of South Australia* 50.

Gordon, D.P. and Taylor, P.D. 1999. Latest Paleocene to earliest Eocene bryozoans from Chatham Island, New Zealand. *Bulletin of the Natural History Museum, London (Geology)* 55(1): 1-45.

James, N.P. and Bone, Y. 2000. Eocene cool-water carbonate and biosiliceous sedimentation dynamics, St. Vincent Basin, South Australia. *Sedimentology* 47: 761-786.

Lindsay, J.M. 1983. Late Eocene to Late Oligocene age of the Kingscote Limestone, Kangaroo Island, S.A. *Transactions of the Royal Society of South Australia* 107: 127-128.

Lindsay, J.M. 1985. Aspects of South Australian foraminiferal biostratigraphy, with emphasis on studies of *Massilina* and *Subbotina*. In J.M. Lindsay (ed.) *Stratigraphy, Palaeontology, Malacology, Papers in Honour of Dr Nell Ludbrook.* Department of Mines and Energy, Special Publication 5: 187-214.

Ludbrook, N.H. 1963. Correlation of the Tertiary rocks of South Australia. *Transactions of the Royal Society of South Australia* 87: 5-15.

Majoran, S. 1996. Late Eocene ostracoda of the Blanche Point Formation, South Australia. *Revista Española de Paleontologia* 11(1): 18-34.

McGowran, B. and Beecroft, A.S. 1986. Foraminiferal biofacies in a silica-rich neritic sediment, Late Eocene, South Australia. *Palaeogeography, Palaeoclimatology, Palaeoecology* 52: 321-345.

McGowran, B., Li, Q. & Moss, G.D. 1997. The Cenozoic neritic record in southern Australia: The biogeohistorical framework. In N.P. James & J.D.A. Clarke (eds.) *Cool-Water carbonates in space and time*. SEPM Special Volume 56: 185-203.

Palfreyman, W.D. 1984. Guide to the Geology of Australia *BMR Bulletin* 181.

Scriven, L.J. 1994. *Diversity of the Mid-Eocene Maslin Bay Flora, South Australia.* Ph.D. Thesis, University of Adelaide. 410pp.

Shubber, B. 1996. *Mid Cenozoic cool-water carbonate facies and the diagenetic history*. Ph.D. Thesis, University of Adelaide. 362 pp.

Smith, A.M., Nelson, C.S. and Spencer, H.G. 1998. Skeletal carbonate mineralogy of New Zealand bryozoans. *Marine Geology* 151: 27-46.

Stuart, W.J.J. 1970. The Cainozoic stratigraphy of the eastern coastal area of Yorke Peninsula, South Australia. *Transactions of the Royal Society of South Australia* 94: 151-178.

Waters, A.W. 1885. Chilostomatous Bryozoa from Aldinga and the River-Murray Cliffs, South Australia. *Quarterly Journal of the Geological Society, London* 41: 279-310.

Bryozoan Studies 2001, Wyse Jackson, Buttler & Spencer Jones (eds)
© 2002 Swets & Zeitlinger, Lisse, ISBN 90 5809 388 3

Changing concepts in species diversity in the northeastern Pacific

Dorothy F. Soule & John D. Soule
Allan Hancock Institute of Marine Studies, University of Southern California, Los Angeles, CA 90089-0371, USA, and Santa Barbara Museum of Natural History, Santa Barbara, CA 93105, USA

Penny A. Morris
Natural Sciences Department, University of Houston, Downtown, One Main St. Houston, TX 77002, USA

ABSTRACT: Changing concepts in species diversity are reflected in reevaluation of cosmopolitan species and concomitant increase in new species previously included within wide ranges of variation. This information is important for biodiversity totals but also for determining environmental impacts of coastal developments. If most species are perceived to have near-worldwide distribution with little speciation occurring in local environments, identification of species may be negated and unnoticed rare species lost to development. Examples of diversity hidden by prior cosmopolitan identifications were found in studies off Santa Barbara, California. Current re-examination of '*Parasmittina trispinosa*' collected from Alaska to the Galapagos indicates that this species may not occur in the eastern Pacific; instead a complex of described and undescribed species occurs. Similar reexamination of presumed species could increase species diversity extensively.

1 ENVIRONMENTAL IMPACT EVALUATION AND SPECIES DIVERSITY

1.1 *Federal and State Legislation*

Over three decades ago the United States passed the National Environmental Policy Act, and states followed with similar statutes such as the California Environmental Quality Act. Regulations to implement the legislation required, among other things, that any new coastal development be evaluated by producing environmental impact assessments and/or reports that included listings of species found in the area of development, either from recent literature or from field surveys. The quality of such listings and surveys varied considerably.

1.2 *Species diversity*

Determination of Species Diversity Indices became a common method for comparing areas. However, so-called minor groups of organisms such as bryozoans were often ignored, in part because identifications were difficult and no specialists were available, but also because familiar species in many such groups were considered cosmopolitan in distribution, giving credence to the notion that destroying some populations would not greatly affect continued existence of species. Producing species diversity calculations while ignoring the organisms less convenient to identify defeats the entire concept. To improve the quality of surveys the US Bureau of Land Management Minerals Management Service in about 1990 funded taxonomic studies of invertebrate groups for which there was no current literature, in areas where offshore drilling for oil might occur. This resulted in a series of atlases on the Santa Maria Basin and Western Santa Barbara Channel of which our monograph was part (Soule, Soule & Chaney, 1995). There had been no major monographs produced on northeastern Pacific bryozoans since Osburn (1950, 1952, 1953) and Soule, (in Osburn, 1953). It is unfortunate that funding was not available to study more of the northeastern Pacific in similar detail.

2 FACTORS AFFECTING DISTRIBUTION

2.1 *Topography, faunal zones and regions*

The northeastern Pacific continental shelf differs greatly from the Atlantic in being narrow, with steep drop-offs subtidally and few large embayments. Shorelines are mostly characterized by rugged cliffs, with relatively few sandy beaches and a dearth of wide intertidal rocky flats. Pt. Conception, California, northwest of Santa Barbara, was long considered the southern limits of a boreal (cool temperate) zone extending from the Gulf of Alaska, and the beginning of a warm temperate zone extending down the coast of Baja California and into the northern Gulf of California (Sea of Cortez, Vermillion Sea). The zones were later subdivided into the Aleutian region, extending south from the Gulf of Alaska to somewhat north of Vancouver Island, Canada, the Oregonian region extended south to Point Conception, and south of that the San Diegan Region extending to meet the tropical region at the southern end of Magdalena Bay, Baja California. The area to the north and south of Pt. Conception was considered a transition zone, where the rugged coast turns eastward and heavy upwelling normally occurs. However, bryozoans seldom have been confined to these narrowly defined zones or regions over a long period.

2.2 *The Channel Islands*

Key to the transition zone near Pt. Conception, and to bryozoan speciation and distribution, are the Channel Islands, which lie to the south. Separated from the mainland by deep canyons, they are characterized by steep rocky cliffs and few sandy beaches. Most are uninhabited. Including more southerly subsurface banks, these islands, lying roughly parallel to the northwest-to-southeast-trending southern California coastline, extend to the Mexican border. There are numerous microhabitats, which offer a great variety of exposures, depths, water temperatures, substrates and currents and hence open to very diverse speciation.

2.3 *Currents and coastal transport, La Niña and El Niño events*

The cool temperate California Current travels south from the Gulf of Alaska, but lies offshore sufficiently that locally generated countercurrents often flow northward inshore. The concept that identical fauna found in shallow water in the north descend in depth along a temperature gradient moving south encouraged the concept that many bryozoan species would be cosmopolitan from Alaska to Baja California, and in some cases to Panama, Colombia and the Galapagos Islands. This would require transit across the Gulf of Panama and the tropical current flowing north to reach the southward flowing tropical current. The California Current is not static, but sometimes flows entirely oceanward of the Channel Islands, and at other times has been documented to split up south of Pt. Conception, breaking into a series of counterclockwise gyres between the islands and shore. Crucial to distribution has been the role of the tropical countercurrent (undercurrent) which flows from the tropics northward from equatorial waters, joining local longshore currents. In non-El Niño years (popularly dubbed La Niña), the countercurrent flows most strongly during the winter months, usually surfacing and dissipating in the Southern California Bight south of Pt. Conception. In strong El Niño years, upwelling ceases around Pt. Conception, and large volumes of warm tropical waters extend north, sometimes as far as British Columbia, with warm waters lingering in the Bight for as much as a year after the main event. During El Niño events, it is certainly possible that species may be transported northward, and if they find a suitable microhabitat, such as a solar-warmed bay or a power plant effluent plume, they may survive, become established and possibly form new species when they are isolated by resumption of cool water flow offshore. Of course, transport on ships' hulls and other artificial means is always possible for some species, and natural coastal transport by floating wood or algae or on larval stages of molluscans or crustaceans can account for coastal movement along shore.

3 'COSMOPOLITAN' SPECIES REVISITED

3.1 *Revised distribution of species in the Santa Barbara (CA) area studies*

When the relatively small area, the Santa Barbara Basin and Western Santa Barbara Channel, was sampled by dredge, a total of 118 species were identified, with six new genera and 42 new species described (Soule, Soule & Chaney, 1995). New species thus constituted 37% of the total. However, virtually all of these new species were included in the so-called wide range of variation of previously described species, as reported by Osburn (1950, 1952, 1953). The type localities of the previously described species ranged from Patagonia to the Arctic Circle, from Britain to Africa, and to Australia. Half were species originally described from Britain or the Atlantic Arctic/Boreal, followed by the Mediterranean/ Red Sea, more indicative of the then-available literature than reality. Descriptions of these species were enlarged to make them cosmopolitan or close to it. The 1995 descriptions in some cases restricted previously described species to ranges outside the southern California study area, to the north or south, and species originally described from Britain and the Mediterranean/Red Sea were almost all eliminated from the northeastern Pacific (Table 1).

4 THE NORTHEASTERN PACIFIC '*PARASMITTINA TRISPINOSA*'

4.1 *A complex of identified and unidentified* Parasmittina *species with acute avicularia*

Currently under study are forms originally placed in *Parasmittina trispinosa* (Johnston, 1825) from the northeastern Pacific and the eastern Pacific tropics in the Allan Hancock Foundation collections ranging from Alaska to Venezuela and the Galapagos Islands. Originally described from Britain, and illustrated in 1838 without ovicells or avicularia, *P. trispinosa* apparently ranges from Norway to the Mediterranean (Hayward & Ryland 1979) but was previously considered to be cosmopolitan in the temperate zone by Hastings (1930) and Osburn (1952). However, Osburn remarked that none of the specimens in the

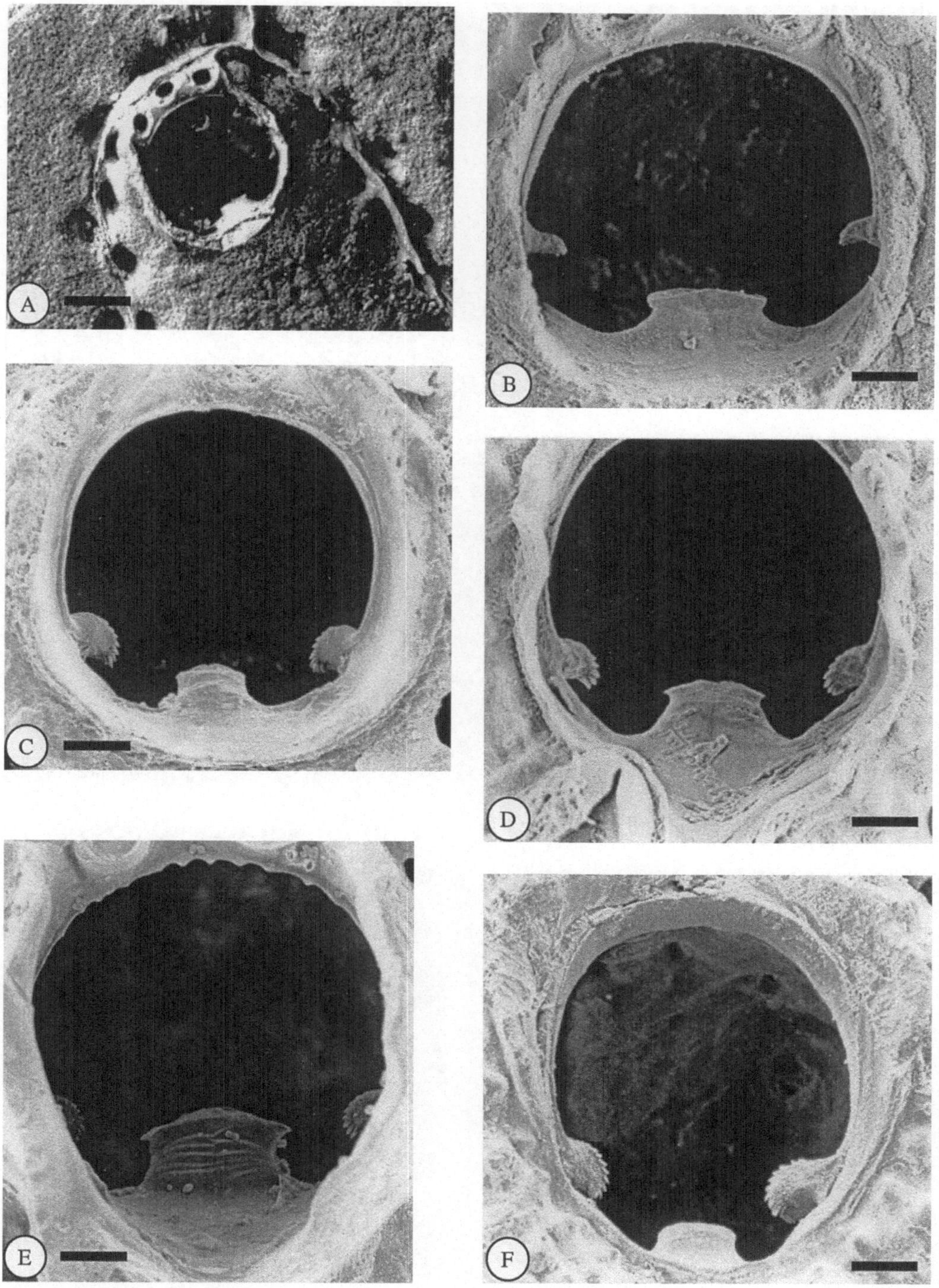

Plate 1. Apertures of eastern Pacific *Parasmittina* spp., compared with *P. trispinosa* lectotype. A. *P. trispinosa* lectotype BMNH 47.9.16.61, courtesy of Dr Paul Taylor, Natural History Museum, London; young zooid with thin condyles, moderate lyrula, 3 spine scars. B. *Parasmittina* sp., Canoe Bay, Alaska, small condyles, wide lyrula, 2 spines. C. *Parasmittina* sp., Santa Rosa Island, CA, large burred condyles, moderate lyrula, 1 spine. D. *Parasmittina* sp., Gulf of California, Mexico, large burred condyles, wider, arched lyrula, suboral sinus, 2-3 spines. E. *Parasmittina* sp., Dillon Beach northern California, large burred condyles, wide, arched lyrula, suboral sinus, 2 spines. F. *Parasmittina* sp., Galapagos Islands, large burred condyles, moderate lyrula, 1-3 spines. Size bars: Figure A, 62.5 μm; Figures B-F, 25.4 μm.

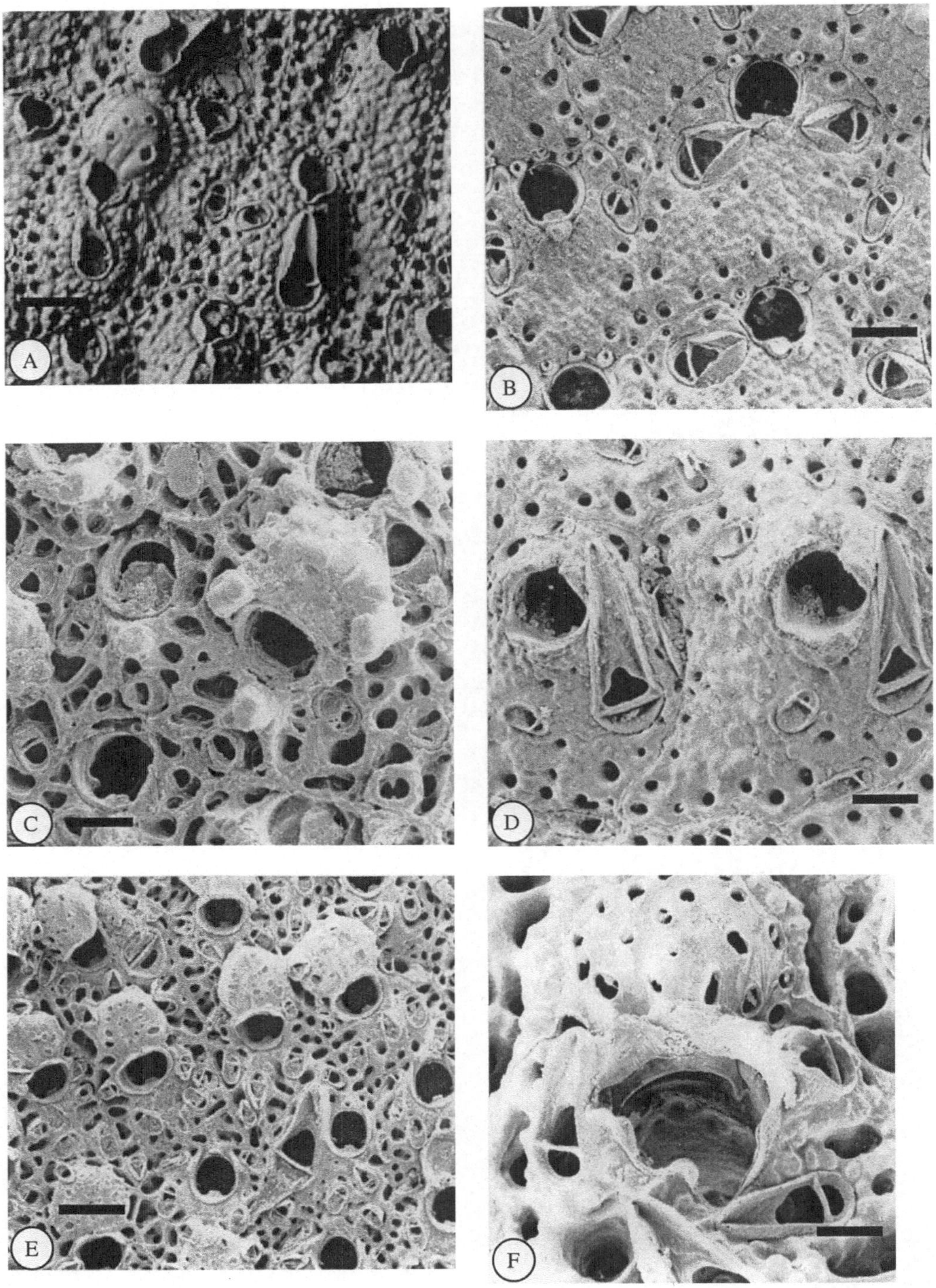

Plate 2. Ovicells, avicularia of eastern Pacific *Parasmittina* spp. compared with *P trispinosa* lectotype. A. *P. trispinosa*, Johnston Collection BMNH 49.9.18.14, courtesy of Dr Paul Taylor, Natural History Museum, London; ovicell with 4 pores, large acute suboral avicularium; B. *Parasmittina* sp., Canoe Bay, Alaska, smaller acute frontal avicularia directed centrally, 3 spines; no ovicells known. C. *Parasmittina* sp., Dillon Beach northern CA, ovicells with 10-12 large and small pores obscured by calcification. D. *Parasmittina* sp., Pt. Dume, southern CA, acute avicularia flanking aperture; ovicells (not shown) with 23-24 pores. E. *Parasmittina* sp. Monterey County, central CA, ovicell with 14-19 small and large pores, with avicularia encroaching; 1-2 spines, moderate lyrula, large condyles. F. *Parasmittina* sp., Gulf of California, ovicell with 19-21 large and small pores, acute avicularia encroaching on peristome. Size bars: Figure A, 250 μm; Figures B-D, 125 μm; Figure E, 250 μm; Figure F, 63 μm.

Table 1. Soule et al. (1995) new species, original and selected former names, prior ( ) and present distribution.

1995 new species	Original and selected former names	Prior ( ) and present distribution
Aetea pseudoanguina	*Sertularia anguina* Linné 1758 *Aetea anguina*, Robertson 1905	(Britain; worldwide except polar) central, S California
Aetea paraligulata	*Aetea ligulata* Busk 1852	(Patagonia, Strait of Magellan, British Columbia to Galapagos Islands); S California Channel Islands to Baja California, Galapagos Islands
Bicrisia robertsonae	*Sertularia cornuta* Linné 1758 *Crisidia cornuta*, Osburn 1953	(Arctic-Boreal, Britain, Atlantic to Mediterranean; British Columbia to Peru);?British Columbia; Monterey, San Pedro Bay, California
Cauloramphus californiensis	*Flustra spinifera* Johnston 1832 *Cauloramphus spiniferum*, Osburn 1950	(Britain; cool temp. Atlantic, NE Pacific); S California Channel Islands
Codonellina hastingsae	?*Codonella granulata* Canu & Bassler	(Galapagos Islands, Gulf of California); Galapagos Islands, Gulf of California, Cortez Banks, Mexico
	Codonellina anatina, Osburn 1952	? S California southern Channel Islands
Codonellina santabarbara	*Schizoporella longirostrata* Hincks 1883 *Hippomonavella longirostrata*, Osburn 1952 (part)	(Pleistocene, Santa Barbara, S California Channel Islands; Pleistocene, Santa Barbara
Conopeum osburni	*Electra crustulenta* var. Osburn 1950	Oregon to Baja California, Mexico
Copidozoum adamantum	*Membranipora tenuirostre* Hincks 1880	(Madera, Naples, temp, trop waters worldwide); British Columbia to Galapagos Islands
	Copidozoum tenuirostre, Osburn 1950	
	Membranipora planum Hincks 1880 *Copidozoum planum*, Osburn 1950	(Australia, Japan, California, Panama, Galapagos Islands, Channel Islands);?British Columbia-Peru
Dengordonia uniporosa	*Lepralia bella* Busk 1860 *Smittina bella*, Osburn 1952 (part)	(Atlantic Boreal to Arctic); S California
Diplosolen harmelini	*Tubulipora obelia* Johnston 1838 *Diplosolen obelium*, Osburn 1953	(Arctic, Britain, Mediterranean, Alaska); Alaska to N. Channel Islands, S California
Fenestrulina farnsworthi	*Cellepora malusii* Savigny & Audouin 1817, 1826 *Fenestrulina malusi*, Osburn 1952 (part)	(Cosmopolitan temp, trop waters; British Co lumbia S California Pleistocene); S California
Fenestruloides blaggae	(See *Cellepora malusi,* above)	(See above); central California
Fenestruloides eopacifica	(See *Cellepora malusi,* above)	(See above); S California
Fenestruloides miramara	(See *Cellepora malusi,* above	(See above); S California
Fenestruloides morrisae	(See *Cellepora malusi,* above) *Fenestrulina malusi*, Osburn 1952 (part)	(See above); S California, Islands off W Baja California, Gulf of California
Haywardipora orbicula	*Mucronella spinosissima* form *major* Hincks 1884 *Mucronella major*, Osburn 1952 (part)	(British Columbia to Galapagos Islands) central California, Gulf of California, ?British Columbia Galapagos Islands
Haywardipora rugosa	*Lepralia ventricosa* Hassall 1842 *Mucronella ventricosa* Osburn 1952 (part)	(Arctic-Boreal Atlantic to Mediterranean; Cape Cod; Alaska to Oregon); Channel Islands, S California, British Columbia to Galapagos Islands);
Haywardipora rylandi	*Mucronella spinosissima* form *major* Hincks 1884 *Mucronella major*, Osburn 1952 (part)	Alaska, ? W Baja California
Hippoporina mexicana	*Schizoporella insculpta* Hincks 1882	(Alaska to Coronados Islands off S California); southern Channel Islands, S California
	Hippodiplosia insculpta Osburn1952 (pt)	southern Channel Islands, California to Cocos Island off Costa Rica
Micropora santacruzana	*Flustra coriacea* Johnston 1847 *Micropora coriacea* (Esper 1791), Osburn 1952 (part)	(Britain, cool temp., warm temp waters worldwide); Br Columbia to S California Channel Islands
Microporella catalinensis	*Eschara ciliata* var.B Pallas 1766 *Microporella ciliata*, Osburn 1952(part)	(Britain, cosmopolitan except polar); S Channel Islands, California
Microporella infundibulipora	*Lepralia californica* Busk 1856 *Microporella californica*, Osburn 1952(pt)	(Oregon to Galapagos Islands, ? British Columbia); S California Channel Islands
Microporella planata	*Eschara ciliata* var.B Pallas 1766 *Microporella ciliata*, Osburn 1952 (part)	(Britain, cosmopolitan except polar); S California Channel Islands
Neodakaria islandica	*Schizoporella ordinata* CH & E O'Donoghue 1923 *Dakaria ordinata*, Osburn 1950 (part)	(British Columbia to Mexican border); S California Channel Islands, ? Baja California
Neodakaria umbonata	*Schizoporella ordinata* CH & E O'Donoghue 1923 *Dakaria ordinata*, Osburn 1950 (part)	(British Columbia to Mexican border); central, S California
Pleurocodonellina californica	*Schizoporella longirostrata* Hincks 1882 *Hippomonavella longirostrata*, Osburn 1952 (part)	(British Columbia, S California, Baja California, Gulf of California); S California, ? Gulf of California

Table 1. continued

1995 new species	Original and selected former names	Prior ( ) and present distribution
Porella taylori	*Porella marsupium* form *porifera* Hincks 1884 *Porella porifera*, Osburn 1952 (part)	(British Columbia; Pleistocene, Recent S California, off Baja California, Costa Rica); Santa Barbara Channel, S. California, ? Baja California
Puellina (?*Cribrilaria*) *perplexa*	*Lepralia innominata* Couch 1884 *Colletosia radiata* form *innominata*, Osburn 1950	(Britain, N Atlantic, Mediterranean); S California
Puellina (*Cribrilaria*) *californiensis*	*Cribrilina setosa* Waters 1899 *Colletosia setosa & C. radiata* form *innominata*, Osburn 1950 (part)	(Madeira, Naples, British Columbia to Colombia); S California Channel Islands, ? Gulf of California
Raymondcia osburni	*Lepralia landsborovi* Johnston 1847 *Smittina landsborovi*, Osburn 1952 (part)	(Cosmopolitan, Atlantic; Oregon to Galapagos); Santa Barbara Channel, S California
Raymondcia macginitei	*Lepralia belli* Busk 1860 *Smittina belli*, Osburn, 1952 (part)	(Boreal-Arctic, Alaska, California); Alaska
Schizomavella robertsonae	*Lepralia auriculata* Hassall 1842 *Schizomavella auriculata*, Osburn 1952 (part)	(Britain, N Atlantic, Mediterranean; British Columbia to Baja California, Gulf of California); S California Channel Islands to? Galapagos Islands
Schizomavella triavicularia	*Schizomavella auriculata*, Osburn 1952 1952 (part)	(See above) S California Channel Islands to Galapagos Islands
Schizosmittina pedicellata	*Lepralia auriculata Hassall* 1842 *Schizomavella auriculata*, Osburn 1952 (part)	(See above) S California Channel Islands
Schizoporella pseudoerrata	*Lepralia errata* Waters 1878 *Schizoporella unicornis*, Osburn 1952(part)	(Warm temp to subtrop Atlantic, Gulf of Mexico); central, S California
Smittina ovirotula	*Smittia spathulifera* Hincks 1884 *Smittina spathulifera*,Osburn 1952	(British Columbia to Baja California, Gulf of California); S California Channel Islands to Baja California
Smittina veleroa	*Smittia landsborovi* Johnston 1847 *Smittina landsborovi*, Osburn 1952 (part)	(Cosmopolitan, Atlantic; Oregon to Galapagos Islands); S California, Gulf of California
Stomachetosella condylata	*Lepralia sinuosa* Busk 1860 ?*Stomachetosella sinuosa*, Osburn 1952 (part)	(N Atlantic to Cape Cod; Alaska, British Columbia, Puget Sound); Puget Sound, S California Santa Barbara Channel
Tegella laruensis	*Membranipora armifera* Hincks 1880,1892 *Tegella armifera*: Osburn, 1950 (part)	(Gulf of St. Lawrence, Cape Cod, Arctic seas of Europe, America); central, S California Channel Islands
Tricellaria circumternata	*Cellaria ternata* Solander, 1786 *Tricellaria* ternata, Osburn, 1950 (part)	(Scotland, northern Atlantic to Arctic, both coasts); British Columbia to central California
Tubulipora aliciae	*Tubulipora pulchra* MacGillivray,1885 *Tubulipora pulchra*, Osburn 1953	(Australia, S California, Islands off Baja California, Galapagos Islands); S California, islands off W Baja California

eastern Pacific were exactly like those from Britain. *P. trispinosa* was recorded in the eastern Pacific by Hincks (1880, 1884), Robertson (1908), O'Donoghue & O'Donoghue (1923, 1926), Hastings (1930), Canu & Bassler (1930), Osburn (1952), J.D. Soule (1961), Soule & Soule (1964) and Dick & Ross (1988) from Alaska and British Columbia south to the Galapagos Islands. Osburn consulted extensively with Hastings while preparing his monograph, and Hastings very much favored including the specimens within the "wide range of variation" considered acceptable at that time, a concept we followed (Soule 1961, Soule & Soule, 1964) but subsequently rejected. Our current research, however, indicates that *P. trispinosa sensu stricto* does not occur in the eastern Pacific at all.

4.2 *Characters that separate northeastern Pacific forms from true* Parasmittina trispinosa*: Condyle size and ovicell pores*

In the Allan Hancock Foundation Collections are numerous specimens tentatively identified by Osburn as *Parasmittina trispinosa* (Johnston, 1838 (actually 1825). Although some final determinations separated out such previously described species as *P. californica* and *P. collifera* (Robertson 1908), and *P dolobrata* (Soule & Soule1973), a number of unidentified species remain. Two characters separate all northwestern and tropical eastern Pacific material of '*P. trispinosa*' examined from south of Alaska: condyle size and the number of ovicell pores. The aperture of *P. trispinosa*, lectotype BMNH 47.9.16.61, has a relatively narrow median denticle and small, simple condyles hooked proximally (Pl 1, Fig A). Specimens from Alaska show similar small condyles, but some have a wide denticle. Dick & Ross (1988) illustrated *P. trispinosa* with a moderately wide lyrula and small condyles, and with few ovicell pores, from Kodiak, Alaska, and Kluge (1965, 1975) illustrated the species with a narrow denticle and thin condyles, and with 2 or 3 pores in the ovicell. Both had relatively small distally directed avicularia but whether these specimens are true *P. trispinosa* has not been determined. All specimens examined from Puget Sound south to the Galapagos Islands have large, burred (layered) condyles, regardless of the width of the denticle (Pl 1, Figs B-F). The lectotype of *P. trispinosa* is a young colony without ovicells, but a specimen from the Johnston Collection on shell, BMNH 49.9.18.14, has only four pores of various sizes (Pl 2, Fig A), although Hayward & Ryland (1979) illustrated seven large or small pores. All specimens examined from Puget Sound to the Galapagos Islands have numerous large and small pores (Pl 2, Figs. C, E, F), although small pores might not have been observed by earlier authors using light microscope examination of material not treated with sodium hypochlorite.

5 CONCLUSIONS

5.1 *Northeastern Pacific coast specific diversity and environmental impact*

The 37% increase in number of species, based on the relatively restricted Santa Barbara Basin and Channel Islands studies, indicates that many more species purported to be of British and European origin, and indeed of Australian or Patagonian origin, need to be examined by SEM comparison with type material. Distribution records of species purported to extend from Alaska to the Galapagos Islands are certainly suspect, and such examination would certainly reveal a much greater diversity. Impacts of habitat alterations in areas of the coast that are under immense development pressure, particularly in southern and central California, must be made more accurate or some species will be irretrievably lost. The increased diversity in the study area was based primarily on revision of existing records and species descriptions.

5.2 *Comparing revised diversity to a little studied area such as the Hawaiian Islands*

Results of the northeastern Pacific coastal diversity studies differ considerably from the situation in the Hawaiian Islands. Soule & Soule (1973) noted that only two smittinid species had been previously described by Busk (1884) from the *Challenger* Expedition, and two by Canu & Bassler (1927) from the *Albatross* Expedition. The Soules described two new genera and 19 new Hawaiian smittinid species out of 23 recorded there, almost 83% of the total. Since that time, a few of these species have been found elsewhere, but the large majority can still be considered endemic.

5.3 *Historic research resources and techniques*

In the nineteenth and early twentieth century, little research existed on species in the Pacific Ocean, except for the *Challenger* and *Albatross* Expeditions, most collections having been of fouling organisms, from settling plates or by hand held dredge. Later came a few offshore expeditions used large biological dredges, and still later, some sampling was by snorkel and SCUBA diving. When Osburn undertook identification of the Allan Hancock Collections during the mid-1940s, he was already retired, and the available literature on eastern Pacific species was very limited: e.g. fossil species from California studied by Conrad (1855), Trask (1857), Gabb & Horn, (1862) and Canu & Bassler (1930), and Recent species from British Columbia (Hincks (1882, 1883, 1884 a, b), O'Donoghue & O'Donoghue (1923, 1925, 1926), all either poorly illustrated or not at all. The works of Robertson (1900, 1905, 1908, 1910) from Alaska to the Mexican border were much better illustrated. Type specimens have disappeared for the Conrad, Trask, Gabb and Horn & Robertson works. The condition of material collected by dredge or substrate scrapings was often poor and the optics used for identification inadequate. The literature available for comparison consisted mostly of descriptions of Recent British and European species and fossils from the Atlantic and Gulf coasts. Osburn's monographs (1950, 1952, 1953) were remarkable considering the relatively few years he had for those studies, retiring again in 1952.

5.4 *Importance of revisiting species diversity*

The advent of Scanning Electron Microscopy (SEM) has changed our ability to distinguish consistent phenotypic differences (morphospecies), as confirmed by Jackson & Cheetham (1990), and eventually techniques such as DNA, (e.g. Dick et al. 2000) will separate species ever more accurately. The changes in techniques and identification emphasize the need for changes in the concept of cosmopolitanism, decreasing its incidence considerably. More important, this would alter our perspective on how many living species might exist, greatly increasing the number of species. The concepts that most species remained stable over long periods of time and that they exist over wide ranges of habitat or climatic variation must be reevaluated. This also applies to the current efforts to synonymize sometimes poorly characterized fossil species with Recent species over large areas in space and time. Furthermore, when habitats are examined for environmental impacts, the impression that species are stable over such wide ranges of habitat and climate exacerbates the assumption that these species are of negligible importance in the analysis of localized project impacts. Revealing the hidden diversity must be made a greater priority than is now accorded to bryozoan species in order to help protect areas from environmental degradation.

ACKNOWLEDGEMENTS

A travel grant was made available from the Kenneth T. and Eileen L. Norris Foundation Fund through the University of Southern California Emeriti College.

REFERENCES

Busk, G. Report on the Polyzoa collected by H.M.S. *Challenger* during the years 1873-76. Pt. 1. Cheilostomata. *Report on the scientific results of H.M.S. Challenger during the years 1873-76, London, Zoology,* vol 10, pt. 30: 1-216.

Canu, F. & Bassler, R.S. 1927. Bryozoaires des Iles Hawaï. *Bulletin de la. Sociétè des Sciences de Seine-&- Oise,* fascicule 7, supplémentaire 1928: 1-66.

Canu, F. &Bassler, R.S. 1930. The bryozoan fauna of the Galapagos Islands. *Proceedings of the United .States National Museum* 76(13): 1-78.

Conrad, T.A. 1855. Note on the Miocene and post-Pliocene deposits of California, with descriptions of two new fossil corals. *Proceedings of the Academy of Natural Sciences of Philadelphia,* 7: 1-441.

Dick, M.H., Freeland, J.R., Williams, LP. & Coggeshall-Burr, M. 2000. Use of 16S ribosomal DNA sequences to investigate sister-group relationships among gymnolaemate bryozoans. In A. Herrera Cubilla. & J.B.C. Jackson (eds), Proceedings of the *11th International Bryozoology Association Conference:* 197-210. Balboa, Republic of Panama: Smithsonian Tropical Research Institute.

Dick, M.H. & Ross, J.R.P. 1988. Intertidal Bryozoa (Cheilostomata) of the Kodiak vicinity, Alaska. Occasional Paper 23. *Center for Pacific Northwest Studies,* Bellingham: Western Washington University, 1-133.

Gabb, W.M. & Horn, G.H. 1862. Monograph of the Fossil Polyzoa of the Secondary and Tertiary Formations of North America, *Journal of the Academy of Natural Sciences of Philadelphia,* New Series 5, Part 2, Article 3: 111-179.

Hastings, A.B. 1930. Cheilostomatous Polyzoa from the vicinity of the Panama Canal, collected by Dr. C. Crossland on the cruise of the S. Y. *St. George. Proceedings of the Zoological Society of London, 1929,* Part 4, Number 47: 697-730.

Hayward, P.J. & Ryland, J.S. 1979. British Ascophoran Bryozoans. In D.M. Kermack and R.S.K. Barnes (eds), *Synopses of British Fauna* (new series). London: Academic Press.

Hincks, T. 1880. *A History of British Marine Polyzoa.* London: Van Voorst.

Hincks, T. 1882. Polyzoa of the Queen Charlotte Islands: Preliminary notice of new species. *Annals and Magazine of Natural History,* Series 5, Volume 10: 248-256; 459-471.

Hincks, T. 1883. Report on the Polyzoa of the Queen Charlotte Islands. *Annals and Magazine of Natural History,* Series 5, Volume 11:442-451.

Hincks, T. 1884a. Report on the Polyzoa of the Queen Charlotte Islands. *Annals and Magazine of Natural History,* Series 5, Volume 13: 49-58; 203-215.

Hincks, T. 1884b. Report on the Polyzoa of the Queen Charlotte Islands. *Geological Survey of Canada*: 1-44.

Jackson, J.B.C. & A.H. Cheetham. 1990. Evolutionary significance of morphospecies: A test with cheilostome Bryozoa. *Science*: 248: 579-583.

Johnston, G. 1838. *A history of British Zoophytes* London, 3-330.

Kluge, G.A. 1962 (English Translation, 1975). *Bryozoa of the Northern Seas.* Academy of Sciences of the USSR, Moscow-Leningrad, Number 76. New Delhi: Amerind Publishing Co.

O'Donoghue, C.H. & O'Donoghue, E. 1923. A preliminary list of Bryozoa (Polyzoa) from the Vancouver Island region. *Contributions to Canadian Biology,* New Series 1 (10):145-201.

O'Donoghue, C.H. & O'Donoghue, E. 1925. List of Bryozoa from the vicinity of Puget Sound. *Transactions of the Puget Sound Biological Station,* 5: 91-108.

O'Donoghue, C.H. & O'Donoghue, E. 1926. A second list of Bryozoa (Polyzoa) from the Vancouver Island region. *Contributions to Canadian Biology and Fisheries,* New Series. 3: 49-131.

Osburn, R.C. 1950. *Bryozoa of the Pacific Coast of America. Part 1, Cheilostomata - Anasca,* Allan Hancock Pacific Expeditions 14 (1): 1-270. Los Angeles: University of Southern California.

Osburn, R. C. 1952. *Bryozoa of the Pacific Coast of America. Part 2, Cheilostomata - Ascophora.* Allan Hancock Pacific Expeditions, 14 (2): 271-611. Los Angeles: University of Southern California.

Osburn, R.C. 1953. *Bryozoa of the Pacific Coast of America. Part 3, Cyclostomata, Ctenostomata, Entoprocta and Addenda.* Allan Hancock Pacific Expeditions, 14 (3): 613-841. Los Angeles: University of Southern California.

Robertson, A. 1905. Non-incrusting chilostomatous Bryozoa of the west coast of North America. *University of California Publications. Zoology* 2(5): 235-322.

Robertson, A. 1908. The incrusting Chilostomatous Bryozoa of the west coast of North America. *University of California Publications in Zoology* 4(5): 253-344.

Robertson, A. 1910. The cyclostomatous Bryozoa of the west coast of North America. *University of California Publications in Zoology* 6(12): 225-284.

Soule, D.F. & Soule, J.D. 1964. The Ectoprocta (Bryozoa) of Scammon's Lagoon, Baja California, Mexico. *Ameican Museum of Natural History Novitates,* Number 2199:1-56.

Soule, D.F. & Soule, J.D. 1973. Morphology and speciation of Hawaiian and eastern Pacific Smittinidae (Bryozoa, Ectoprocta). *Bulletin of the American Museum of Natural History,* 152(6): 367-440.

Soule, D.F., Soule, J.D. & Chaney, H.W. 1995. The Bryozoa. In J.A. Blake, H.W. Chaney, P.H. Scott & A.L. Lissner (eds), *Taxonomic Atlas of the Santa Maria Basin and Western Santa Barbara Channel,* Volume 13: 1-344. Santa Barbara: Santa Barbara Museum of Natural History.

Soule, J.D. 1953. *Ctenostomata.* In R.C. Osburn, *Bryozoa of the Pacific Coast of America, Part. 3, Cyclostomata, Ctenostomata, Entoprocta and Addenda.* Allan Hancock Pacific Expeditions 14(3): 726-784. Los Angeles: University of Southern California.

Soule, J.D. 1961. Results of the *Puritan*-American Museum of Natural History Expeditions to western Mexico, 6, Ascophora-Cheilostomata (Bryozoa). *American Museum of Natural History,* Number 2053: 1-66.

Trask, J.B. 1857. Nine new species of zoophytes from the Bay of San Francisco and adjacent localities. *Proceedings of the California Academy of Sciences* 1: 100-102

Bryozoan Studies 2001, Wyse Jackson, Buttler & Spencer Jones (eds)
© 2002 Swets & Zeitlinger, Lisse, ISBN 90 5809 388 3

Brooding in the Cretaceous bryozoan *Stichomicropora* and the origin of ovicells in cheilostomes

Paul D. Taylor
Department of Palaeontology, The Natural History Museum, Cromwell Road, London, SW7 5BD, UK
Frank K. McKinney
Department of Geology, Appalachian State University, Boone, North Carolina 28608, USA

ABSTRACT: The coilostegan *Stichomicropora* has unusual ovicells normally preserved only as a transverse row of 3-10 small pores located along the proximal mural rim of the distal zooid. Intact ovicells in *S. baccata* show the pores to be bases of flattened, hollow spines overarching proximally and tightly juxtaposed to form a costate roof. Spine bases in some other species of *Stichomicropora* may have been articulated, which would explain their non-fossilization. In all species, however, the ovicell is produced entirely by the distal zooid, the roof from modified mural spines and the floor from proximal gymnocyst. *Stichomicropora* may form a clade (Monoporellidae) with some other coilostegans, including *Monoporella* spp. and '*Macropora' multilamellosa*. Comparison of ovicells in *Stichomicropora* and *Wilbertopora*, the oldest known ovicellate cheilostome, permits the primitive cheilostome ovicell to be reconstructed as consisting of several articulated mural spines belonging to the distal zooid and having no direct structural association with any pore chambers.

1 INTRODUCTION

The great majority of cheilostome bryozoans brood their progeny in small skeletal chambers called ovicells (Ström 1977). Up to a 500-fold increase in volume can occur during brooding as the young embryo matures into a larva ready for release. Primitively, the cheilostome ovicell is a folded, double-layered structure comprising a pair of cuticle-bound exterior walls with a narrow coelomic lumen inbetween. The underside of the lower wall (entooecium) forms the ceiling of the brood chamber, whereas the top surface of the upper wall (ectooecium) is the externally visible roof of the ovicell. Often the ectooecium is not fully calcified but contains cuticle-covered windows, revealing the inner surface of the entooecium, and the ectooecium is an interior wall in some species.

Considerable interest has been focused on the related issues of the development (ontogeny) and evolution (phylogeny) of cheilostome ovicells. An ontogenetic question frequently posed is: are ovicells formed by the maternal zooid, by the zooid distal to it, or by the two zooids in combination? A key evolutionary question is whether ovicells originated through the modification of spines, as posited by several bryozoologists dating back to Harmer (1902, p. 283), or by some other means. The high degree of morphological disparity found among ovicells implies complexities in their ontogenies and phylogenies. Consequently, such questions do not necessarily have a single answer. It is possible, although yet to be demonstrated by rigorous phylogenetic analysis, that ovicells have arisen independently on more than one occasion from ancestors lacking ovicells, in which case not all cheilostome ovicells are homologous and different ovicells could have different patterns of development and evolutionary histories.

Evidence from both modern and fossil cheilostomes can contribute to a better understanding of the ontogeny and phylogeny of ovicells. The study of Recent species (see Ostrovsky 1998 and references therein) has the advantage that soft parts are present, allowing the anatomy of soft tissue connections between ovicells and maternal and other associated zooids to be established, and hence the ontogenetic development of ovicells to be more fully appreciated. On the other hand, because of their age, fossils can provide important evidence on the evolutionary origin of ovicells by revealing primitive morphologies (e.g. Cheetham 1975), including some which may no longer be extant.

Cheilostome ovicells represent a key evolutionary innovation (Taylor 1988; Taylor & Larwood 1990). Ovicellate cheilostomes greatly exceed non-ovicellate cheilostomes in taxonomic diversity at the present day, in spite of the fact that non-ovicellate cheilostomes originated first, appearing in the fossil record some 50 million years before the first species with ovicells. The explosive diversification of cheilostomes beginning in the mid-Cretaceous followed soon after the first appearance of ovicells, and involved only groups with ovicells. If ovicells signal the switch from non-brooded, planktotrophic to brooded, non-planktotrophic larvae, then this radiation may have been powered by increased allopatric speciation rates resulting from the higher probability of genetic isolation in ovicellate groups with short-lived, poorly and spasmodically dispersed larvae.

Here we present the first full description of intact ovicells in the Cretaceous cheilostome *Stichomicropora* Voigt, 1949. This genus dates back to the Early Cenomanian and is consequently one of the earliest ovicellate cheilostome genera in the fossil record. The ovicell in *Stichomicropora* is shown to be formed by a group of modified mural spines belonging to the zooid distal to the maternal zooid. We explore the significance of this finding to general questions of ovicell ontogeny and evolution in cheilostomes.

2 THE GENUS *STICHOMICROPORA*

Stichomicropora was introduced by Voigt (1949, p. 33) who named a new species, *S. sicksi* from the Upper Campanian of northern Germany, as the type species. He assigned a further three species to the genus: *S.* cf. *clathrata* (Reuss, 1872) (= *Homalostega clathrata* (Reuss) sensu Brydone, 1936), *S. punctilla* (Brydone, 1936) and *S. baccata* (Canu & Bassler, 1926). Additional species, all from the Upper Cretaceous, subsequently placed in *Stichomicropora* by Voigt (1967, 1989) and Favorskaya (1987) are: *S. erecta* (v. Hagenow, 1839), *S. marginula* (Brydone, 1914) (see Taylor 1987, fig. 8), *S. sulcata* (Reuss, 1846), *S. oceani* (d'Orbigny, 1852) (see Voigt 1981, fig. 4g), *S. sussexiensis* (Brydone, 1936), *S. glabra* (Voigt, 1924), *S. membranacea* (v. Hagenow, 1839), and *S. subquadrata* Favorskaya, 1987. Taken together these species span the entire Upper Cretaceous, and the range of *Stichomicropora* is therefore Lower Cenomanian to Upper Maastrichtian. The geographical distribution of the genus extends from southeastern USA, through northern and central Europe to central Asia (Turkmenistan and Tadzhikistan).

Voigt (1949) and Bassler (1953) both assigned *Stichomicropora* to the Microporidae but Voigt later (1967) placed the genus in the Aspidostomatidae on account of its hyperstomial ovicell, an assignment with which Favorskaya (1987) concurred. For the reasons noted below, we tentatively place *Stichomicropora* in the Monoporellidae.

All known species of *Stichomicropora* have encrusting multiserial colonies. Autozooids are rhombic in outline shape, with extensive frontal walls (Fig 1A). A narrow marginal gymnocyst surrounds the slightly depressed and flat to convex cryptocyst which is pierced by one or more pairs of opesiules and may have a plain or granular surface texture. In at least one unidentified species (Fig. 1A), small pores occur scattered across the frontal wall. The frontal wall is commonly partly to completely destroyed in colonies of some species of *Stichomicropora*, including *S. clathrata* (see Voigt 1989, pl. 18, fig. 2) and *S. baccata* (Fig. 1G). This may be true even when the fragile ovicells are intact, suggesting that a high organic or magnesium content is responsible for taphonomic frontal wall loss rather than physical or biological abrasion. The small, semicircular orifice/opesia is typically bordered by one to three pairs of oral spine bases arranged around its distal rim. Occasionally in *S. oceani* a median proximal mural spine base is developed (Fig. 1D). Budding is intrazooidal, from distal and paired distolateral pore chambers with oval pore windows. Polymorphs are uncommon. In *S. baccata*, vicarious avicularia with tall, rounded rostra occur rarely in the syntypes (see Canu & Bassler 1926, pl. 6, fig. 4), while clusters of closed, dome-shaped kenozooids are found in *S. sussexiensis* (see Brydone 1936, pl. 41, fig. 3). Small and irregular kenozooids occasionally develop in spatially restricted areas. In some species of *Stichomicropora* the ancestrula resembles later zooids but has proximal mural spine bases. However, the ancestrula in the oldest species (*S. oceani* (d'Orbigny)) appears to be membraniporimorph and possibly spineless.

3 OVICELLS IN *STICHOMICROPORA*

The unusual ovicell morphology which prompted Voigt (1949) to establish *Stichomicropora* is the most characteristic feature of the genus. Normally, all that remains of the ovicell is a transverse row of pores arranged across the frontal wall of the zooid distal of the brooding zooid (Fig. 1A-D). These pores range from 3 to 10 in number according to species, with low numbers (3-5) occurring, for example, in *S. baccata* and high numbers (10) in *S. sulcata*. The row of pores may be straight, gently convex distally or gently concave distally. Sometimes the outermost pores are slightly offset from the others and positioned more proximally (Fig. 1C). The pores are located on a low ridge which is continuous distolaterally with the raised mural rim at the boundary of the inner cryptocystal and outer gymnocystal parts of the frontal wall. Individual pores vary from relatively large (> 30 μm) openings in *S. baccata* to smaller (< 20 μm) openings containing a concentric ring of calcification (Fig. 1B). There is no direct relationship between the ovicell pores and pore chambers in transverse walls: fractured specimens clearly show that ovicell pores are located well distally of the pore chambers, although both open into the same zooidal chamber. Between the pores and the maternal zooid is the proximal gymnocyst of the distal zooid which has a trapezoidal shape and is enlarged relative to that of zooids not bearing ovicells.

Intact ovicells are known in only two species of *Stichomicropora* - *S. baccata* and *S. subquadrata* - both of Maastrichtian age. Canu & Bassler (1926, pl. 6, fig. 5) depicted a colony of *S. baccata*, from the Ripley Fm of Tennessee, with one complete and several broken ovicells, describing the preserved structures as 'spicules protecting the ovicell'. SEM study of Canu & Bassler's specimen (USNM 69954) shows that the complete ovicell (Fig. 1E) is constructed of 5 unequal spines originating from the line of pores, which can therefore be interpreted as spine bases, and directed proximally towards the orifice of the maternal zooid. The spines are initially separate but soon broaden in width so that their edges are closely juxtaposed with those of the adjacent spines. As they converge on the maternal orifice and space becomes restricted, the spines narrow. Their outer ends form a smooth arch, parallel with the distal edge of the maternal zooid orifice and located a little above the level of this orifice. Abrasion of the two spines on the right of the ovicell suggests that adjacent spines retain a dividing wall (or walls) and that there is no continuity between the lumens of the flattened spines. The surface smooth texture of the spines and indications of planar spherulitic fabric imply that the spines are exterior walled, in common with most other cheilostome spines. The two zooids on either side of the ovicell have depressed crescentic facets in their frontal walls laterally of the outermost spines of the ovicell. Gaps – lateral foramina - at the sides of the ovicell open onto these facets.

There are several broken ovicells (Fig. 1F) in the same colony. These show similar facets in the lateral zooids which are deeper in ontogenetically older zooids and often incorporate an opesiule. Most of the broken ovicells in this colony are constructed of 3 or 4 spines in contrast to the 5-spined complete ovicell. The granular cryptocyst of the frontal wall may drape proximally into the gaps between the spine bases. The floor of the ovicell comprises a smooth, slightly convex wall sometimes having a shallow central furrow. In one ovicell minor corrosion reveals planar spherulitic fabric in two units radiating from the proximolateral corners of the ovicell floor. This undoubtedly indicates the direction of wall calcification. Zooids forming ovicells for a proximal zooid may themselves be maternal zooids having their own ovicell formed from the distal zooid. In some instances all of the zooids from patches of the colony are fertile, with ovicells covering the entire frontal surface between the orifices.

Additional examples of complete and broken spinose ovicells have been found in specimens of *S. baccata* from the Prairie Bluff Chalk of Alabama and the Peedee Fm of North Carolina (Fig. 1G). Despite the poorer surface preservation these ovicells show the same characteristics as the Ripley Fm material. In addition, Prairie Bluff ovicells sometimes have spines with open-ended, slightly upturned tips, further supporting the inference that the lumens of the juxtaposed spines are not confluent. One colony (NHM BZ4796) from the Prairie Bluff Chalk is about 20 mm in maximum diameter and contains more than 60 ovicellate zooids. These are usually arranged in longitudinal strips, often with two rows of alternating ovicellate zooids per strip, aligned along the central axes of lobes in the digitate colony.

The small illustrations of ovicells in *S. subquadrata* published by Favorskaya (1987, pl. 3, figs 1a and 1b) show no obvious differences with the ovicells in *S. baccata*. It would, however, be unwise to infer that the ovicells in the other species of *Stichomicropora* were completely identical to those of *S. baccata* and *S. subquadrata*. This is because the relatively large, irregular and ragged-edged spine bases in *S. baccata* differ from the smaller, more regular and smoother-edged spine bases present in the other species. Furthermore, the inner ring of calcification observed in one species (Fig. 1B) is reminiscent of the structures present at the bases of articulated spines in such genera as *Callopora* (e.g. Taylor 1986, fig. 19A). If the ovicell spines of many *Stichomicropora* species were indeed articulated then this would explain their non-preservation.

The apparent lack of facets on laterally adjacent zooids in these species is another difference from *S. baccata*.

In summary, the ovicell in *Stichomicropora* is large and costate, constructed from 3-10 spines belonging to the zooid distal of the maternal zooid. The spines have no direct connection to pore chambers. In the two species in which complete ovicells are known, the spines are non-articulated, although this is unlikely to be the case for the majority of species of *Stichomicropora* which probably had articulated spines In *S. baccata* the spines arch proximally and are closely juxtaposed at their edges, but do not apparently have confluent lumens, and ovicells have lateral foramina opening onto facets on the frontal walls of the two adjacent zooids.

The position of ovicell spines points to their derivation from proximal mural spines of the distal zooid. In support of this inference the existence of proximal mural spines in the ancestrulae of some species, and occasionally in budded zooids in the stratigraphically oldest species *S. oceani* (Fig. 1D) makes it reasonable to suppose that proximal mural spines were present in all zooids in an ancestral taxon. Articulated mural spines encircle the opesia in some genera of primitive, Lower Cretaceous malacostegans (e.g. *Spinicharixa* Taylor, 1986) from which ovicellate cheilostomes may have evolved. The floor of the ovicell in *Stichomicropora* is clearly formed by the proximal gymnocyst of the distal

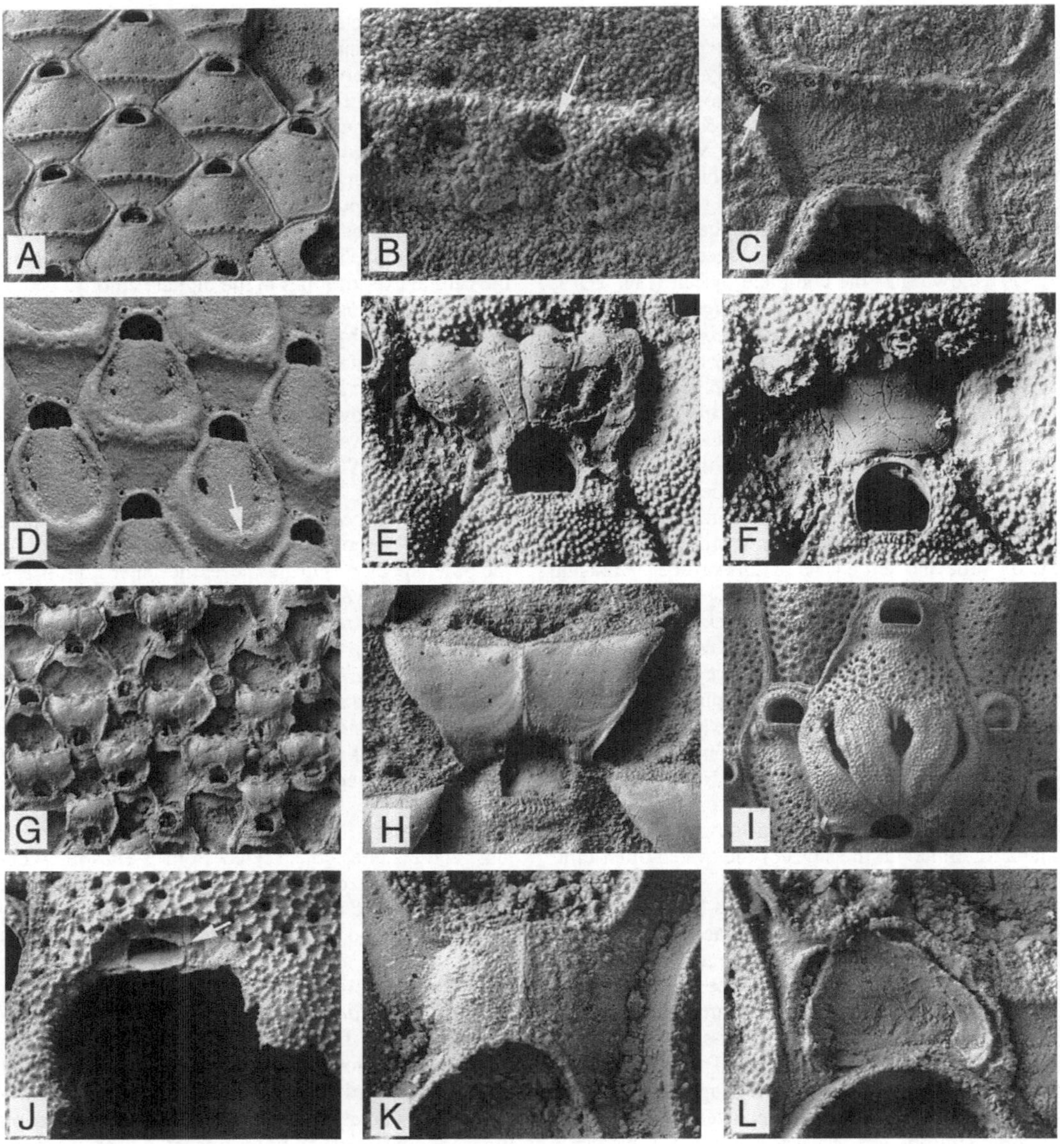

Figure 1. A-B. *Stichomicropora* sp., Campanian, Norwich, England, NHM D42263; A, slightly oblique view of area with most zooids fertile; x 18; B, ovicell spine bases with inner rings, particularly apparent in the arrowed example; x 225. C, *Stichomicropora* sp., Cenomanian, Wilmington, Devon, England, NHM D55618; ovicell with 7 spine bases, the outermost more proximal (arrowed); x 65. D, *S. oceani* (d'Orbigny), Cenomanian, Lamnay, France, NHM D55549; ovicellate and non-ovicellate zooids, the former with 6-7 ovicell spine bases, the latter with one proximal mural spine base (arrowed); x 50. E-F, *S. baccata* (Canu & Bassler), Maastrichtian, Ripley Fm, Tennessee, USA, USNM 69954 (syntype); E, complete ovicell comprising 5 juxtaposed spines; x 80; F, roofless ovicell showing floor formed by proximal gymnocyst of distal zooid; x 90. G, *S. baccata* (Canu & Bassler), Maastrichtian, Peedee Fm, Wilmington, NC, USA, NHM BZ4859; group of zooids with complete ovicells but frontal walls largely destroyed; x 22. H, '*Macropora*' *multilamellosa* Canu & Bassler, Eocene, Castle Hayne Limestone Fm, Wilmington, NC, USA, NHM BZ4860; ovicell comprising two flattened spines; x 55. I-J, *Monoporella* sp., Recent, Alaska, MNHN Canu Collection 2856-7 (b); I, ovicell comprising 4 proximally-growing units of cryptocyst; x 19; J, broken roof of ovicell showing strip lumen (arrowed); x 65. K-L, *Wilbertopora mutabilis* Cheetham, Cenomanian, Grayson Fm, Waco Dam, Texas, USA; K, NHM D58502, complete ovicell with median ridge and suture; x 135; L, NHM D57378, incomplete ovicell with paired crescent-shaped slits, from which the bipartite roof grows, and concave proximal gymnocyst of the distal zooid between, x 150.

zooid which is enlarged in comparison with zooids not bearing ovicells.

These findings contrast with previous views of the ovicell in *Stichomicropora*. Canu & Bassler (1926, p. 36) described the ovicell of *S. baccata* thus: "The ovicell is small, endozooecial, convex, smooth. Above each ovicell and on the cryptocyst of the distal zooecium there are 4 to 6 flat bifurcated spicules forming a roof-like protection." They apparently believed the spicules (spines) to be structures supplementary to the ovicell itself. Voigt (1985, p. 331) stated: "In *Stichomicropora* Voigt there are broad uncalcified hyperstomial ovicells supported by a horizontal row of spines on the proximal part of distally adjacent zooecia." Again, the implication is that the spines are not part of the ovicell but are additional structures.

4 COMPARABLE OVICELLS IN OTHER CHEILOSTOMES

Certain other coilostegans possess ovicells with similarities to those of *Stichomicropora*. *Macropora multilamellosa* Canu & Bassler, 1920 from the Eocene Castle Hayne Fm of North Carolina, USA, has zooids with evenly perforate cryptocystal frontal shields (erroneously interpreted as tuberosities by Canu & Bassler), as in other species of *Macropora*, but the existence of calcified opercula, the defining synapomorphy of *Macropora*, is uncertain. Canu & Bassler's plate 20, figure 2 depicts pairs of horizontal ridges on proximal frontal shields of some of the autozooids, located distolaterally of the orifice of the proximal zooid. These structures were not remarked upon by Canu & Bassler. New material from the Castle Hayne Fm shows that the ridges are the bases of bilateral ovicells comprising two juxtaposed spines overarching from the mural rim of the distal zooid proximally towards the maternal orifice (Fig. 1H). The spines are frontally compressed such that the lumen is very reduced in size. Like the ovicells of *S. baccata*, lateral foramina open onto the facets of the adjacent zooids.

A Maastrichtian species described by Favorskaya (1987) as *Monoporella prisca* also has an evenly perforate cryptocyst and ovicells which, in consisting of 9-10 juxtaposed spines, are even more like those of *Stichomicropora*. Another Cretaceous species assigned to *Monoporella*, *M. exsculpta* (Marsson), was refigured by Voigt (1989). The ovicell in this species is bilobate (pl. 19, fig. 1), but an abraded example (pl. 19, fig. 3) reveals a floor closely similar to that found in *Stichomicropora* as well as facets on the frontal walls of the two adjacent zooids. Similar vicarious avicularia (pl. 19, fig. 4) occur in *M. exsculpta* as those occasionally present in *S. baccata*.

The extant type species of *Monoporella* Hincks, *Monoporella nodulifera* (Hincks, 1881), also has zooids with evenly porous cryptocysts. The large ovicell of *M. nodulifera* has lateral openings (lateral foramina) and a floor formed by the proximal gymnocyst of the distal zooid (Cheetham and Cook 1983, fig. 72). It is very similar in overall size and shape to the ovicell of *Stichomicropora baccata*. A major difference, however, is that the *M. nodulifera* ovicell has a cryptocystal roof. Nevertheless, a line of fissures close to the distal edge of the ovicell could conceivably represent vestiges of intercostal spaces. An undescribed Alaskan species of ?*Monoporella* (represented by Canu Collection specimen 2856-7 in the MNHN, Paris) has even more pronounced fissures in the cryptocystal ovicell roof giving a more obvious costate morphology (Fig. 1I). A fractured ovicell reveals the existence of a lumen in one of the putative costae close to its base (Fig. 1J).

Pending a full phylogenetic study, the tentative conclusion is that *Stichomicropora*, '*Macropora*' *multilamellosa* and *Monoporella* belong to a clade of coilostegans diagnosed by having large ovicells, often with lateral foramina, constructed, at least primitively, from two or more modified proximal mural spines of the distal zooid. The neglected Family Monoporellidae Hincks, 1882 seems to be an appropriate name for this clade.

5 SIGNIFICANCE FOR OVICELL EVOLUTIONARY ORIGIN

It is impossible in this short paper to consider all theories of ovicell evolution in every group of cheilostomes. Instead, we focus on the importance of the *Stichomicropora* ovicell to an understanding of the origin and primitive condition of ovicells in cheilostomes. For this purpose, the most germane comparisons are with ovicells which, on stratigraphical and general morphological grounds, can be regarded as the most primitive known cheilostome ovicells.

The stratigraphically oldest ovicells are found in the calloporid *Wilbertopora mutabilis* Cheetham, 1954 from the late Albian and Cenomanian of Texas. According to Cheetham (1975), ovicells in this species (or species-group) are bilateral, hyperstomial, have their floor formed by the proximal gymnocyst of the distal zooid, and a lumen which originates from two lateral slits in this gymnocyst. Cheetham considered that the ovicell lumen possibly opened directly into the longitudinal pore chamber at the proximal end of this distal zooid. New SEM studies of *W. mutabilis* confirm most of Cheetham's observations. In well-preserved colonies from the Grayson Fm, ovicells have a medial ridge and suture, often marked by a narrow fissure (Fig. 1K). Ovicells appear very close to the growing edge of the colony. The first indication of a new ovicell forming is a pair of embayments in the calcifying proximal gymnocyst

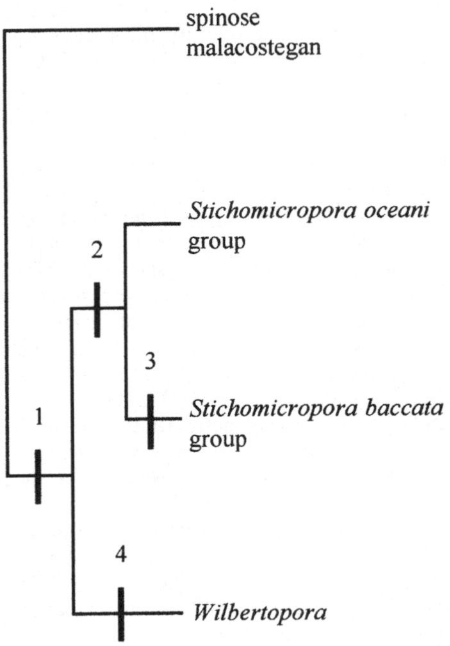

Figure 2. Phylogenetic model of early ovicell evolution showing main character changes and hypothesized relationships between the oldest known ovicellate cheilostome (*Wilbertopora*) and two groups of species of *Stichomicropora*: the *S. oceani* group with inferred articulated ovicell spine bases and the *S. baccata* group with basally fused ovicell spine bases. Principal apomorphies are as follows: 1, clustering of articulated mural spines to form ovicell; 2, extensive cryptocyst with opesiules; 3, fusion of spine bases and juxtaposition of adjacent spines; 4, fusion of spine bases and reduction of spines to two.

of the distal zooid. By the time formation of the gymnocyst is completed these embayments have become long, somewhat crescent-shaped slits (Fig. 1L). The gymnocyst between the slits, which forms the floor of the ovicell, is more concave than it is in zooids not bearing ovicells, and the slits extend almost as far distally as the mural rim where gymnocyst gives way to granular cryptocyst. Upward growth of gymnocystal wall surrounding each slit completes the roof of the ovicell when the two components converge and fuse along the mid-line of the zooid. It is presumed, although not proven, that the lumens of the two components become confluent after meeting. There is no direct link between the ovicell lumen/s and the large pore chamber in the lateral wall between maternal and distal zooids; lumens and pore chambers open separately into the proximal part of the distal zooid.

There are obvious similarities as well as differences between ovicells in *Wilbertopora* and *Stichomicropora*. In addition to general resemblances in size, shape, and location, they share the following features: (1) formation entirely from the zooid distal of the maternal zooid; (2) floor comprising the proximal gymnocyst of the distal zooid; (3) roof a compound structure consisting of two or more components; and (4) each roof component consisting of an exterior (gymnocystal) wall enclosing a hollow space. Differences are (1) the greater number of roof components in the *Stichomicropora* ovicell; (2) subcircular component basal sections in *Stichomicropora* vs crescentic slit-like sections in *Wilbertopora*; (3) apparent non-confluence of the lumens in these components in *Stichomicropora* vs their likely confluence in *Wilbertopora*; and (4) a dominantly proximal growth vector of the components in *Stichomicropora* compared with lateral, inward growth in *Wilbertopora*. While the similarities are more striking than the differences, their status as synapomorphies, symplesiomorphic or homoplasies - and hence their phylogenetic importance - cannot be evaluated without a cladistic analysis incorporating more taxa and a fuller range of characters, which is well beyond the scope of this paper. Provisionally, however, ovicells in *Wilbertopora* and *Stichomicropora* are regarded as homologous structures inherited from a common ancestor rather than structures independently derived from different ancestors. The two clades containing *Wilbertopora* + most other ovicellate cheilostomes, and *Stichomicropora* + other monoporellids are possible sister-groups. Alternatively, one of the two clades may be a monophyletic group nested within the paraphyletic second group. This can be tested by cladistic analysis.

If the proposition is accepted that ovicells in *Wilbertopora* and *Stichomicropora* are homologous then it is possible to make some inferences about ovicell origin in cheilostomes and the primitive morphology of ovicells. Firstly, the primitive cheilostome ovicell is produced by the distal zooid alone, corroborating Nielsen's (1985) contention that such ovicells represent the 'basic' type for cheilostomes. Ovicells incorporating skeletal components of the maternal zooid should be regarded as derived, and terminal ovicells not associated with a distinct distal zooid interpreted as representing extreme reduction of the distal zooid to the ovicell only. There is no evidence for the involvement in primitive ovicell formation of a kenozooid positioned between the maternal and distal zooids (cf. Ostrovsky 1998). Secondly, the primitive cheilostome ovicell is constructed of two or more hollow, gymnocystal components. The likelihood is that these components represent modified mural spines – they are clearly mural spines in *Stichomicropora* and it is easy to interpret the two components of the *Wilbertopora* ovicell as greatly flattened spines. Oral mural spines are found in the zone of astogenetic repetition in many *Wilbertopora mutabilis* colonies (cf. Santagata & Banta 1996, p. 178), so it is not unreasonable to suppose that spines in ancestral species were present on the proximal mural rim. Derivation of ovicells from mural spines lends support to Ostrovsky's (1998) hypothesis of a costal origin for ovicells in cribrimorphs, although not via reduction of an acanthostegous distal zooid. It remains unclear whether the individual spines forming the ovicell are polymorphs or simple outgrowths of autozooidal skeleton. The answer to this question can now be seen to hinge on whether mural spines are polymorphs or outgrowths. A tentative phylogenetic model outlining morphological changes during early ovicell evolution is shown in Figure 2.

Santagata and Banta (1996) proposed a model for the evolution of brooding in cheilostomes based on their detailed anatomical study of Recent *Scrupocellaria ferox*. This model incorporated several stages, including the origin of ovicells from proximal spines of the distal zooid which they saw as a means of protecting the embryo from predators after it became too large to be accommodated within the vestibule. Our findings support Santagata and Banta's hypothesis that ovicells evolved as modified spines of the distal zooid but are mute on the issue of whether vestibular brooding was a precursor to ovicellular brooding. However, the wide taxonomic spread of the relatively few cheilostomes with vestibular brooding (e.g. *Smittipora*, *Odontoporella*, *Watersipora* and *Cryptosula*) is better explained by multiple independent switches from ovicellular to vestibular brooding than it is by vestibular brooding being a primitive condition retained in these genera, some of which are relatively advanced, but lost in all other cheilostomes. Again, however, this aspect of the Santagata and Banta model can only be rigorously tested by a broad ranging cladistic analysis. Pending such a study, we suggest a modification of Santagata and Banta's model in which a previously non-brooded larva achieved some selective advantage by being caught up in, and protected by, the mural spines of the distal zooid. Subsequent modification of these spines into a caged structure – the ovicell – would provide better protection. As noted above, there are indications that the ovicell spines in the earliest species of *Stichomicropora* had basal articulations. Basally articulated mural spines have been previously regarded as more primitive than fully calcified spines (see Gordon 2000, p. 20). It is envisaged that articulated mural spines in the early ovicellate cheilostomes, including some species of *Stichomicropora*, evolved fully-calcified bases as a precursor to their lateral fusion with adjoining spines and formation of a rigid ovicell.

ACKNOWLEDGEMENTS

Discussions with Andrew Ostrovsky and Dennis Gordon helped the development of the ideas expressed here. John Everett was instrumental in the success of our field collecting around Wilmington, North Carolina, and we also thank students of the ASU Geology Department for help in the field here and in Alabama. Acknowledgement is made to the Petroleum Research Fund, administered by The American Chemical Society, for partial support of this research.

REFERENCES

Bassler, R.S. 1953. Bryozoa. In R.C. Moore (ed.), *Treatise on invertebrate paleontology. Part G.* Boulder and Lawrence: Geological Society of America and University of Kansas Press.

Brydone, R.M. 1914. Note on new or imperfectly known Chalk Polyzoa. *Geological Magazine* (Decade 6) 1: 97-99.

Brydone, R.M. 1936. *Further notes on new or imperfectly known Chalk Polyzoa.* London: Dulau

Canu, F. & Bassler, R. S. 1920. North American Early Tertiary Bryozoa. *Bulletin of the United States National Museum* 106:1-879.

Canu, F. & Bassler, R. S. 1926. Bryozoa. In B. Wade, The fauna of the Ripley Formation of Coon Creek, Tennessee. *United States Geological Survey Professional Paper* 137: 32-37.

Cheetham, A.H. 1954. A new Early Cretaceous cheilostome bryozoan from Texas. *Journal of Paleontology* 28: 177-184.

Cheetham, A.H. 1975. Taxonomic significance of autozooid size and shape in some early multiserial cheilostomes from the Gulf Coast of the U.S.A. *Documents du Laboratoire de Géologie de la Faculté des Sciences de Lyon, Hors Série* 3(2): 547-564.

Cheetham, A.H & Cook, P.L. 1983. General features of the Class Gymnolaemata. In R.A.Robison (ed.), *Treatise on invertebrate paleontology. Part G. Revised. Volume 1*: 138-207. Boulder and Lawrence: Geological Society of America and University of Kansas Press.

Favorskaya, T. A. 1987. [Maastrichtian bryozoans of eastern Turkmenia]. *Ezheg Vses Paleontol O-va* 30: 82-107. [in Russian]

Gordon, D.P. 2000. Towards a phylogeny of cheilostomes-morphological models of frontal wall/shield evolution. In A. Herrera Cubilla & J.B.C. Jackson (eds) *Proceedings of the 11th International Bryozoology Association Conference*: 17-37. Balboa, Republic of Panama: Smithsonian Tropical Research Institute.

Hagenow, F. v. 1839. Monographie der Rügenschen Kreideversteinerungen. Neues Jahrbuch für Mineralogie 1839: 253-296.

Harmer, S.F. 1902. On the morphology of the Cheilostomata. *Quarterly Journal of Microscopical Science* 46: 263-350.

Hincks, T. 1881. Contributions towards a general history of the marine Polyzoa. VI. Polyzoa from Bass's Straits. *Annals and Magazine of Natural History* (Series 5) 8: 1-14.

Hincks, T. 1882. Contributions towards a general history of the marine Polyzoa. IX. Foreign Cheilostomata (Miscellaneous). *Annals and Magazine of Natural History* (Series 5) 9: 79-90.

Nielsen, C. 1985. Ovicell formation in *Tegella* and four cellularioids (Bryozoa, Cheilostomata). In C. Nielsen & G.P. Larwood (eds), *Bryozoa: Ordovician to Recent*: 213-220. Fredensborg: Olsen & Olsen.

Orbigny A. d'. 1851-4. *Paléontologie Française. Terrains Crétacés. 5. Bryozoaires.* Paris: Masson.

Ostrovsky, A. N. 1998. Comparative studies of ovicell anatomy and reproductive patterns in *Cribrilina annulata* and *Celleporella hyalina* (Bryozoa: Cheilostomatida). *Acta Zoologica* 79: 287-318.

Reuss, A.E. 1846. *Die Versteinerungen der böhmischen Kreideformation. Abt. II.* Stuttgart.

Reuss, A.E. 1872. Die Bryozoen des unteren Quaders. *Palaeontographica* 20: 97-144.

Santagata, S. & Banta, W.C. 1996. Origin of brooding and ovicells in cheilostome bryozoans: interpretive morphology of *Scrupocellaria ferox*. *Invertebrate Biology* 115: 170-180.

Ström, R. 1977. Brooding patterns of bryozoans. In R.M. Woollacott & R.L. Zimmer (eds), *Biology of bryozoans*: 23-55. New York: Academic Press.

Taylor, P.D. 1986. *Charixa* Lang and *Spinicharixa* gen. nov., cheilostome bryozoans from the Lower Cretaceous. *Bulletin of the British Museum (Natural History) (Geology Series)* 40: 197-222.

Taylor, P.D. 1987. Bryozoans. In A.B Smith (ed.), *Fossils of the Chalk*: 30-49. London: Palaeontological Association Field Guides to Fossils.

Taylor, P.D. 1988. Major radiation of cheilostome bryozoans: triggered by the evolution of a new larval type? *Historical Biology* 1: 45-64.

Taylor, P.D. & Larwood, G.P. 1990. Major evolutionary radiations in the Bryozoa. In P.D. Taylor & G.P Larwood (eds), *Major evolutionary radiations*: 209-233. Oxford: Clarendon Press.

Voigt, E. 1924. Beiträge zur Kenntis der Bryozoenfauna der subherzynen Kreidemulde. *Palaeontologischen Zeitschrift* 6(2): 93-247.

Voigt, E. 1949. Cheilostome bryozoen aus der Quadratenkreide Nordwestdeutchlands. *Mitteilungen aus dem Geologischen Staatinstitut in Hamburg* 19: 1-49.

Voigt, E. 1967. Oberkreide-Bryozoen aus den asiatischen Gebieten der UdSSR. *Mitteilungen aus dem Geologischen Staatinstitut in Hamburg* 36: 5-95.

Voigt, E. 1981. Répartition et utilisation stratigraphique des Bryozoaires du Crétacé moyen (Aptien-Coniacien). *Cretaceous Research* 2: 439-462.

Voigt, E. 1985. The Bryozoa of the Cretaceous-Tertiary boundary. In C. Nielsen & G.P. Larwood (eds), *Bryozoa: Ordovician to Recent*: 329-342. Fredensborg: Olsen & Olsen.

Voigt, E. 1989. Beiträg zur Bryozoen-Fauna des sächsischen Cenomaniums. *Abhandlungen des Staatlichen Museums für Mineralogie und Geologie zu Dresden* 36: 8-87.

Bryozoan Studies 2001, Wyse Jackson, Buttler & Spencer Jones (eds)
© 2002 Swets & Zeitlinger, Lisse, ISBN 90 5809 388 3

The effects of increased external Ca^{++} and K^{+} concentrations on the waveform dynamics of bryozoan spermatozeugmata

M. H. Temkin
St. Lawrence University, Canton, New York, United States of America

ABSTRACT: To investigate how *Membranipora membranacea* (Linnaeus 1767) sperm modulate their waveforms, including the production of type III waveforms that are important for sperm spawning and entry into maternal individuals, unencapsulated sperm aggregates or spermatozeugmata were exposed to elevated concentrations of external Ca^{++} and K^{+}. Increasing Ca^{++} levels by 20 and 30 mM above ambient caused spermatozeugmata to increase the duration, but not the frequency of type III waveform events. Increasing ambient levels of Ca^{++} by 40 mM or K^{+} by 10 or 20 mM produced a waveform, type IV, that propagated from tail to head and caused the anterior ends of the spermatozeugmata go through an effective stroke/recovery stroke motion comparable to the movement of a cilium. Type IV waveforms propelled spermatozeugmata forward through the water column and may be the waveform used by spermatozeugmata to pull themselves into intertentacular organs to enter maternal zooids.

1 INTRODUCTION

The gymnolaemate bryozoan *Membranipora membranacea* (Linnaeus 1767) has a fertilization biology that involves the transfer of spermatozeugmata between paternal and maternal zooids (Temkin 1994). Bryozoan spermatozeugmata are unencapsulated aggregates of either 32 or 64 sperm cells arranged parallel to one another so that all of the sperm heads are at one of the aggregate and all of the tails are at the other end (Bonnevie 1907, Franzén 1956, Zimmer & Woollacott 1974). Sperm within an aggregate adhere tightly to one another along most of their lengths, except for the anterior portions of the midpieces and the tips of the tails. Sperm in spermatozeugmata may be attached to each other by mucus (Franzén 1998).

In *M. membranacea*, spawned spermatozeugmata enter maternal bodies and sperm fuse with eggs internally. Functional male zooids of *M. membranacea* release spermatozeugmata tail first through the two distomedial tentacles into the exhalant currents of colonies (Temkin 1994). Spawned spermatozeugmata in the water column are drawn into lophophores by colony feeding currents, and subsequently crawl head first into the intertentacular organ (ITO) of maternal zooids (Temkin 1994). The ITO is a secondary sex organ formed in the gymnolaemate genera *Alcyonidium, Conopeum, Electra, Farella, and Membranipora* by the basal fusion of the two distomedial tentacles (Marcus 1938, Silén 1966). The ITO is considered an extension of the supraneural pore and is also the organ through which eggs are spawned from maternal zooids (Marcus 1938, Silén 1966). Spermatozeugmata inside the maternal zooid disassociate and sperm move to the surface of the ovary, although the sequence of these two events is uncertain. In *M. membranacea*, monospermic sperm-egg fusion occurs during or shortly after ovulation (Temkin 1994).

Although ambient and colony generated water currents are largely responsible for the transfer of spermatozeugmata from paternal to maternal zooids, sperm motility is required for spawning and entry into maternal zooids. Temkin & Bortolami (submitted) studied the motility patterns of *M. membranacea* spermatozeugmata and described three types of waveforms generated in the unbound portions of sperm aggregates based on wave amplitude and the direction of wave propagation. Type I waveforms consist of low amplitude waves that translate along the spermatozeugma from the heads to the tails. Type II waveforms have high amplitude head to tail waves that cause spermatozeugmata to undulate with "snake-like" movements. Type III waveforms have waves that propagate in the reverse direction, moving from the tail end to the head end of spermatozeugmata. The amplitude of waves produced during type III waveforms are intermediate in size between the amplitudes of waves of type I and type II waveforms. During the generation of type III waveforms, spermatozeugmata develop prominent bends just posterior to the heads and at the junction of the bound and unbound portions of the midpieces. These bends produce a strong curvature in the anterior portion of spermatozeugmata so that spermatozeugmata look like a "question mark" in appearance. Waves generated during type II and III waveforms are produced synchronously by all the members of a spermatozeugma, whereas the production of waves of the type I waveform does not appear to be coordinated among members of an aggregate.

Spermatozeugmata modulate the waveforms that they generate depending on their location (Temkin & Bortolami submitted). Type I represent the predominant waveform in the paternal visceral coelom and the water column. Type II occur infrequently in the paternal coelom and water column, but are often produced by spermatozeugmata in lophophores after being drawn in by feeding currents. Type III waveforms are periodically produced in both the paternal visceral coelom and water column. However, spermatozeugmata within the paternal visceral coelom produce type III waveforms more often and for longer periods of time than spermatozeugmata in the water column. Spermatozeugmata produce type III waveforms in the paternal lophophoral coelom that propel spermatozeugmata backwards through the paternal tentacles during spawning. Type III waveforms also are generated within milliseconds of a spermatozeugma contacting an ITO of a maternal lophophore. The production of type III waveforms by spermatozeugmata at the ITO bends the head ends of spermatozeugmata down into the distal pore of the ITO. The modulation of waveforms to type III at contacting the ITO may represent an initial species specific response of sperm to tissues of conspecific maternal individuals (Temkin and Bortolami submitted).

Since spermatozeugmata generate type III waveforms during both spawning from paternal zooids and entering the ITO of the maternal zooids, it is important to understand how spermatozeugmata regulate the generation of type III waveforms. The production of tail to head waves during type III waveforms is an unusual phenomenon and only the sperm of a turbellarian flatworm (Hendelberg 1965), the annelid *Myzostomum cirriferum* (Afzelius 1983), and three species of Tephritid flies (Baccetti et al. 1989) have been reported to change the direction of their wave propagation. Studies of sperm motility in other species indicate that increases in intracellular Ca^{++} and or K^{+} from extracellular sources may regulate the changes in beat and/or direction of wave propagation of flagella (e.g. Gibbons 1980, Baccetti et al. 1989, Brokaw 1991, Cook et al. 1994). Here, to investigate how *M. membranacea* sperm modulate their waveforms, spermatozeugmata were exposed to elevated concentrations of external Ca^{++} and K^{+}.

2 METHODS AND MATERIALS

For this study, spermatozeugmata were obtained from *M. membranacea* colonies collected from waters near the Darling Marine Center (DMC), Walpole, Maine, USA. Video records of spermatozeugmata movements in response to each treatment were made using either a standard video camera (60 frame/s) that was mounted on an Olympus compound microscope or a NAC HSV-200 video camera (200 frames/s) that was mounted on a Zeiss Photo III compound microscope. A simple repeated measures experimental design was used to determine the effects of increased external Ca^{++} levels of 10, 20, 30 mM on the waveform dynamics of spermatozeugmata. For each trial, a recently spawned spermatozeugma was collected in 50 µl of sea water and placed in the well of a depression slide containing 50 µl of 0.2 µm filtered sea water (FSW). Spermatozeugmata were video recorded for at least 36 seconds (36-202 seconds) during the pretreatment period of trials. In control trials, the pretreatment period was followed by the addition of 100 µl of FSW containing only ambient levels of Ca^{++} (i.e. 0 mM additional Ca^{++}) to the well of the depression slide containing the spermatozeugma. In experimental trials, pretreatment periods were followed by the addition of 100 µl of FSW containing twice the final concentration of Ca^{++} to the well of the depression slide. Following the addition (FSW or Ca^{++}-FSW solution), spermatozeugmata were recorded for up to 6 minutes or until they contacted the glass slide. The number of spermatozeugmata used to test the effects of each treatment and the number of type III waveforms measured during pretreatment and treatment periods is shown in Table 1.

A frame-by-frame analysis was made of the videotape records to determine the number and durations of type III waveform events for spermatozeugmata during the pretreatment and treatment periods of each trial. Paired samples t-tests were calculated using Systat 6 to compare type III wave event frequencies and durations between pretreatment and treatment periods for each Ca^{++} concentration. For these paired samples t-tests, the treatment period of control trials was defined as from the addition of FSW up to 3 minutes after the addition and the treatment period of experimental trials was defined as starting from 1 minute after the addition of

Table 1. Number of type III waveform events measured during pretreatment and treatment periods for each Ca^{2+} concentration.

Treatment	N	Number of Type III events		
		Pretreatment	Treatment	Total
0 mM Ca^{++}	9	126	87	213
10 mM Ca^{++}	8	36	126	162
20 mM Ca^{++}	11	111	202	313
30 mM Ca^{++}	11	142	202	344
Total	39	415	617	1032

the Ca solution up to 4 minutes after the addition. By defining the treatment periods in this way, each treatment period represents a 3 minutes window of time that includes the first minute following the addition of FSW in control trials when turbulence should have had its greatest effect on spermatozeugmata and excludes the first minute after addition in experimental trials so that spermatozeugmata may establish a response to the elevated Ca^{++} level. For an individual trial, the type III waveform event durations of pretreatment and treatment periods were taken as the average event duration for each period.

To examine the effects of 40 mM Ca^{++} and elevated K^{+} (10 and 20 mM) two protocols were used. One followed the simple repeated measures design described above. The other protocol was to collect spermatozeugmata in 20 µl of seawater which was added to 180 µl of FSW containing 1.1× the final concentration of either Ca^{++} or K^{+}. For this protocol special well slides were prepared by attaching the sides of a BEEM capsule lid to a glass microscope slide.

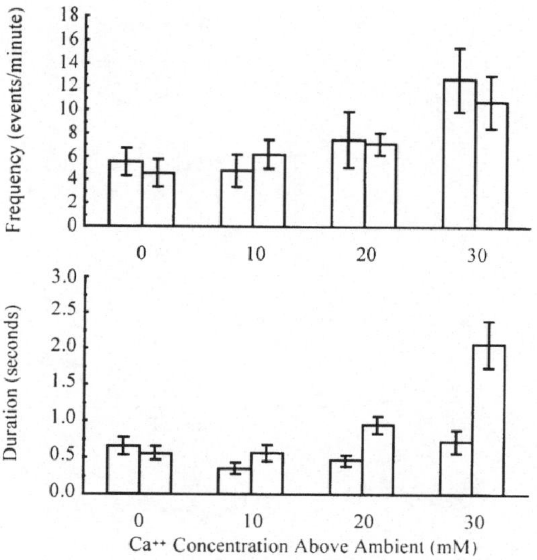

Figure 1. Average type III waveform event frequencies (top) and duration (bottom) for control and elevated Ca^{++} concentration treatments. Bars on the left side are pretreatment values and bars on the right side are treatment values. Error bars = 1 standard error of the mean.

3 RESULTS

Increasing external Ca^{++} and K^{+} concentrations did effect the waveform dynamics of *M. membranacea* spermatozeugmata. Figure 1 shows the effects of FSW control treatments and increased external Ca^{++} concentrations of 10, 20 and 30 mM on the frequency and duration of type III waveform events. A comparison of type III waveform event frequencies and durations during control pretreatment and treatment periods revealed no significant differences (Table 2). Therefore, the addition of solutions to the depression slide alone did not alter the frequency or duration of type III waveform events produced by spermatozeugmata. Paired samples t-tests also detected no significant differences in type III waveform event frequencies between pretreatment and treatment periods for any elevated Ca^{++} concentration (Table 2). However, the duration of type III waveform events did significantly increase when spermatozeugmata were exposed to Ca^{++} concentrations of 20 and 30 mM above ambient (Table 2). Durations of type III waveform events more than doubled for spermatozeugmata in sea water containing an additional 20 mM of Ca^{++}, increasing from an average of 0.46 ± 0.08 seconds during the pretreatment period to an average of 0.95 ± 0.12 seconds during the treatment period, and almost tripled for spermatozeugmata exposed to 30 mM Ca^{++} above ambient, increasing from an average of 0.72 ± 0.15 seconds during the pretreatment period to 2.07 ± 0.33 seconds during the treatment period (Fig. 1).

Exposure of spermatozeugmata to 40 mM Ca^{++} or 10 and 20 mM K^{+} above ambient almost always changed the waveform produced by spermatozeugmata from type I to type IV, a new waveform type (Fig. 2). The waves generated during type IV waveforms propagated in the same direction as waves of type III waveforms, moving along spermatozeugmata from tail to head. However during the generation of type IV waveforms, the anterior portion of a spermatozeugma bent in a large arc so that the tips of the heads nearly touched the midpieces (Fig. 2c-k). As a spermatozeugma unbent, the heads were pulled in close to the spermatozeugma (Fig. 2l-n), until the heads unfolded at the end of the waveform (Fig. 2o-r). The movements of spermatozeugmata during type IV waveforms propelled spermatozeugmata through the water column in a forward direction. An asymmetry in the production of the waveform caused spermatozeugmata to rotate around the long axis of the aggregate. In two instances, 10 mM K^{+} caused spermatozeugmata to generate type

Table 2. Paired samples t-tests comparing type III waveform event frequencies and durations between pretreatment and treatment periods for each Ca^{++} concentration.

Treatment	N	Frequency		N	Duration	
		t	P		t	P
0 mM Ca^{++}	9	1.806	0.109	9	1.418	0.194
10 mM Ca^{++}	8	-1.071	0.320	7	-1.572	0.167
20 mM Ca^{++}	11	0.154	0.881	9	-4.985	0.001
30 mM Ca^{++}	11	1.312	0.219	10	-3.297	0.009

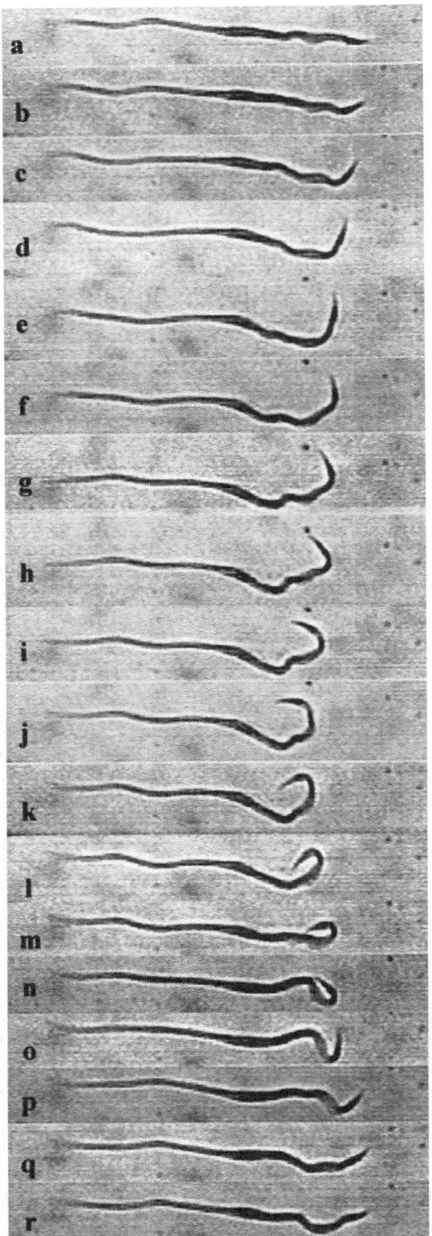

Figure 2. A spermatozeugma generating a type IV waveform. Images are 0.005 seconds apart. The spermatozeugma is about 250 μm long.

II waveforms (Fig. 3) that also propelled spermatozeugmata in a forward direction.

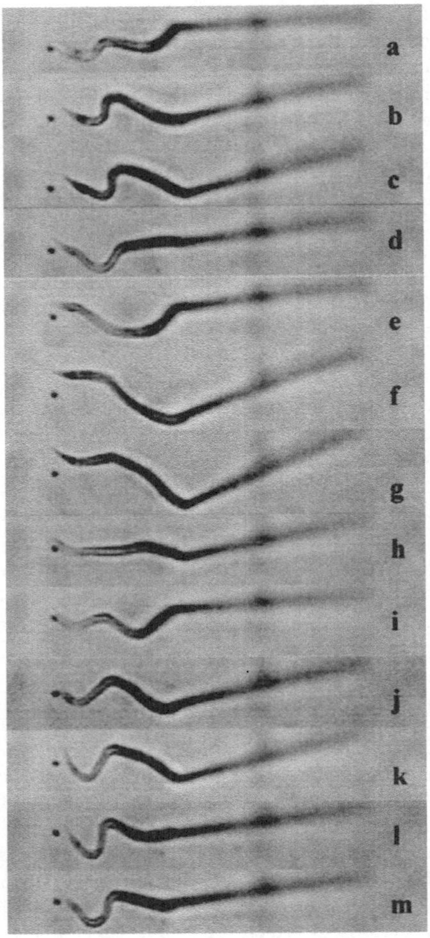

Figure 3. A spermatozeugma generating a type II waveform. Images are 0.005 seconds apart. The spermatozeugma is about 250 μm long.

4 DISCUSSION

Spermatozeugmata of *M. membranacea* generate at least four types of waveforms even though sperm in aggregates are tightly bound to one another long most of their lengths. Three of the four types of waveforms generate enough locomotory force to allow spermatozeugmata to crawl or swim. Type II and type IV waveforms move spermatozeugmata in a forward direction, whereas type III waveforms move spermatozeugmata in a backwards direction. The underlying waves of both type III and IV waveforms propagate from the tail end to the head end of spermatozeugmata. When spermatozeugmata crawl through the paternal tentacles during spawning, tail to head waves push against the walls of the tentacles driving the spermatozeugmata backwards (Temkin & Bortolami submitted). The tail to head waves generated during type IV waveforms cause the anterior portions of spermatozeugmata to undergo movements functionally similar to the effective stroke/recovery stroke of a cilium. The effective stroke portion of the waveform occurs as the head end of a spermatozeugma swings out from the body in a large arc toward the midpieces and the recovery stroke portion of the waveform occurs close to the body of the spermatozeugma as the head end returns to its original position. The rotation around the long axis of sperm aggregates during type IV waveforms allows spermatozeugmata to move linearly forward rather than spinning in circles.

Increases in intracellular Ca^{++} may mediate the generation of tail to head waves during type III and IV waveforms. Spermatozeugmata exposed to elevated external Ca^{++} levels above 20 mM experienced either a significant increase in type III waveform event duration or began to generate a type IV waveform continuously. Spermatozeugmata in the paternal coelom have longer and more frequent type III waveform events than spermatozeugmata in the water column (Temkin & Bortolami submitted). These differences in waveform dynamics suggest that either the coelomic fluid of *M. membranacea* has a higher concentration of Ca^{++} than the ambient seawater or agents within the coelom increase the permeability of sperm cells to Ca^{++}. However, increased Ca^{++} levels did not increase the frequency of type III waveform events. Consequently, increased Ca^{++} levels alone do not completely account for differences in waveform dynamics between spermatozeugmata in the paternal coelom and water column.

For internal fertilization to occur in *M. membranacea*, spermatozeugmata need to recognize and enter ITOs of conspecific maternal individuals. Spermatozeugmata appear to recognize conspecific individuals after the head end contacts the surface of the ITO and spermatozeugmata begin to produce continuous tail to head waves of a type III waveform (Temkin & Bortolami submitted). The production of tail to head waves when spermatozeugmata contact ITOs occurs within milliseconds (Temkin & Bortolami submitted) and may indicate the presence of a receptor on the outer surface of sperm cells that recognizes conspecific maternal individuals and mediates a rise in intracellular Ca^{++}. The production of tail to head waves bends the anterior portion of spermatozeugmata so that they may enter the distal pore of ITOs. However, once the head end of a spermatozeugma has entered an ITO, the waveform should change to either a type II or a type IV so that spermatozeugmata move in forward direction. If entering ITOs causes the sperm cells of an aggregate to depolarize, similar to exposure of elevated K^{+}, then it is likely that spermatozeugmata use type IV waveforms to crawl into the maternal body.

5 ACKNOWLEDGEMENTS

I thank the following people for providing me with laboratory space and facilities: Kevin Eckelbarger and Tim Miller at the Darling Marine Maine, Mary Rice at the Smithsonian Marine Station at Fort Pierce, and Robert Woollacott at Harvard University. Karel Liem generously allowed me to use his high speed video equipment. Lee-Ann Hayek provided help with the statistical analysis. I thank John Thorpe and Marianne DiMarco-Temkin for reading and suggesting changes that with along with comments from the editor improved the manuscript. This paper is dedicated to the memory of John D. Soule.

REFERENCES

Afzelius, B.A. 1983. The spermatozoon of Myzostomum cirriferum. *Journal of Ultrastructure Research* 83:58-68.

Bonnevie K. 1907. Untersuchunger uber Keimzellen. II. Physiologishe polyspermie bei Bryozoen. Jenaische Zeitschrift für Naturwissenschaften. 42, 567-598.

Baccetti, B.H., Gibbons, B. H., & Gibbons, I. R. 1989. Bidirectional swimming in spermatozoa of Tephritid flies. *Journal of Submicroscopic Cytology and Pathology* 21: 619-625.

Brokaw, C.J. 1991. Calcium sensors in sea urchin sperm flagella. *Cell Motility Cytoskeleton* 18:123-130.

Cook, S.P., Brokaw, C. J., Muller, C. H., & Babcock, D. F. 1994. Sperm chemotaxis: egg peptides control cytosolic calcium to regulate flagellar responses. *Developmental Biology* 165:10-19.

Franzén, Å. 1981. Comparative ultrastructural studies of spermatids and spermatozoa in Bryozoa. In G.P. Larwood and C. Nielsen (eds.) Recent and Fossil Bryozoa: 83-92. Fredenborg, Olsen & Olsen.

Franzén, Å. 1956. On spermiogenesis, morphology of the spermatozoan, and biology of fertilizaton among invetebrates . Zoologiska Bidrag från Uppsala 31, 355-482.

Franzén, Å. 1998. Spermiogenesis, sperm structure and spermatozeugmata in the gymnolaematous bryozoan *Electra pilosa* (Bryozoa, Gymnolaemata) *Invertebrate Reproduction and Development* 34: 55-63.

Gibbon, B.H. 1980. Intermittent swimming in live sea urchin sperm. *Journal of Cell Biology* 84: 1-12.

Hendelberg, J. 1965. On different types of spermatozoa in Polycladida, Tubellaria. *Arkiv für Zoologi* 18:267-304.

Marcus, E. 1938. Bryozoarios marinhos brasileiros II. *Boletim da Faculdade de Filosofia, Ciencias e Letras, Universidade de Sao Paulo, Zoologia* 4, 1-196.

Silén, L. 1966. On the fertilization problem in the gymnolaematous Bryozoa. *Ophelia* 3, 113-140.

Temkin, M.H. 1996. Comparative fertilization biology of gymnolaemate bryozoans. *Marine Biology* 127, 329-339.

Temkin, M.H. 1994. Gamete spawning and fertilization in the gymnolaemate bryozoan *Membranipora membranacea*. *Biological Bulletin* 187, 143-155.

Temkin, M.H. & Bortolami, S. submitted. The waveform dynamics of spermatozeugmata during the transfer from paternal to maternal individuals of *Membranipora membranacea*. *Biological Bulletin.*

Zimmer R.L. & Woollacott R.M. 1974. Morphology and biochemical modifications of the spermatozoan mitochondria of *Membranipora* (Bryozoa). *Journal of Cell Biology* 63, 385a.

Bryozoan Studies 2001, Wyse Jackson, Buttler & Spencer Jones (eds)
© 2002 Swets & Zeitlinger, Lisse, ISBN 90 5809 388 3

New bryozoan faunas from the Miocene of Burgenland (Austria)

N.R.Vávra
Institute of Palaeontology, University of Vienna, Geozentrum, Althanstraße 14, A-1090 Wien/Vienna, Austria

ABSTRACT: Taxonomical studies of two bryozoan faunas (Stotzing: 43 taxa [except for the Celleporids] and Forchtenstein: approximately 100 taxa) from the Badenian (Middle Miocene) of Burgenland (Austria) have yielded among others additional results concerning a number of species which are either new for the Austrian Miocene (*Umbonula spinosa,* ? *Triporula*) or had been only insufficiently described until now (e.g. *Schizostomella grinzingensis, Smittistoma polyomma, Kionidella moravica*). Detailed descriptions and data for seven of these taxa are given. Some of these species are of interest also in connection with problems of palaeobiogeography.

1 INTRODUCTION

The Middle Miocene (Badenian) of the Central Paratethys can be regarded as one of the "classical" areas of studies concerning bryozoan faunas from the Tertiary. Since the years of the very first studies (Reuss 1848, 1874), which had been continued by Manzoni (1877, 1878) a number of foreign specialists have contributed from time to time detailed revisions (e.g. Canu & Bassler 1925, David & Pouyet 1974). Nevertheless there are still nowadays good chances for new discoveries; this will be shown in the following for two faunas from the Badenian (= Middle Miocene) of Burgenland. One of them is really completely "new" in the strictest sense of this word, having been collected by a young colleague (Dr H. Temmel, Weiz) in 1997/98 at a recently discovered new bryozoan locality (Stotzing); the second fauna represents material from an ancient, well-known locality (Forchtenau = Forchtenstein) which had never been published except for a few single pictures or remarks. This fauna had been collected mainly by the late C.A. Bobies in 1956. Both faunas are remarkable under at least two different aspects: the material from Stotzing consists of a high number of rather unusual large zoaria; the fauna from Forchtenstein however shows a remarkable high biodiversity with approximately 100 different taxa. Both faunas include also a number of species being usually rather rare but nevertheless of great interest for bryozoologists for various reasons. A complete description of these two faunas being far outside the scope of this publication, this paper will be restricted to documenting some important faunal elements which have until now not been adequately described from other localities.

2 LOCALITIES

2.1 *Forchtenstein* (= *Forchtenau*)

"Forchtenau" - as the locality had been called by Bobies, is part of the municipality "Forchtenstein" now. It is situated approximately 20 km SW from Eisenstadt (capital city of Burgenland). The accurate position of this outcrop can not be reconstructed with certainty any more; Bobies had indicated "Blunzengasse" (now: Antonigasse). At the present time this outcrop is not accessible, having now become a residential district. Generally speaking the area of Forchtenstein is well-known for invertebrate faunas that show a rather high biodiversity; Sieber (1956) summarizing all data available at the time for the Badenian of Mattersburg and Forchtenstein area, mentions Forchtenstein as a fauna with highest diversity of taxa from the whole Austrian Miocene. He gave in his lists 314 species of molluscs, 2 brachiopods, and 8 corals. In respect to the exact biostratigraphical age, he suggested "tieferes (unteres) Torton" (now: Early Badenian), a standpoint which had to be later changed: The Badenian of Forchtenstein being attributed now to the *Spiroplectammina* Zone (Vávra 2000 and references therein). The first few bryozoan taxa from Forchtenau had been mentioned already by Reuss (1874).

The material is kept within the 'Bobies-collection' and is the property of the Museum of Natural History, Vienna.

2.2 *Stotzing*

The small village of Stotzing is situated approximately 8 km NNE from Eisenstadt. In the immediate neighbourhood of this village exists a medium-sized sand pit ("Sandgrube Mayer") where Dr H. Temmel (Weiz) discovered a thin layer of bryozoan-bearing sediment of Badenian age. Closer details in respect to biostratigraphy respectively sedimentology are not yet available. An outstanding feature of this material is the occurrence of rather large and well-preserved zoaria; one disadvantage however, is the rather prevalent recrystallization which is also to be observed in many - if not most - specimens at the "classical" bryozoan locality at Eisenstadt ("*Terebratula* sands"). The diversity of this fauna is distinctly lower than at Forchtenau: until now (Celleporidae excluded) only 43 taxa have been identified.

The material from Stotzing is kept within the bryozoan collection at the Institute of Palaeontology, University of Vienna.

3 TAXONOMY AND SYSTEMATICS

As already mentioned above only very few taxa from these two faunas will be described here: six species of Cheilostomata and one species of Cyclostomata. This is of course only a very small portion of these important two faunas. The selection has been made in such a way to highlight species being completely new for the Austrian Miocene (? *Triporula, Umbonula spinosa*) or still not adequately described (*Tremogasterina areolata, Kionidella moravica, Smittistoma polyomma, Schizostomella grinzingensis, Bobiesipora fasciculata*).

Tremogasterina areolata (Reuss, 1874)
(Pl. 1, Fig. 1)

1874 *Lepralia areolata* m. - Reuss, p. 156, pl. 4, figs 9, 10.
1928 *Tremogasterina areolata* - Canu & Bassler, p. 44.
1974 *Tremogasterina areolata* (Reuss, 1874) - David & Pouyet, p. 112, pl. 6, fig. 7.
1977 *Tremogasterina areolata* (Reuss) - Cook, p. 128-129, fig. 6A.
1997 *Tremogasterina areolata* (Reuss, 1874) - Pouyet, p. 36-37.
2000 *Tremogasterina areolata* - Vávra, p. 397.

Description: Two fragments of a colony of *Metrarabdotos maleckii* (belonging obviously together) are encrusted by a well-preserved zoarium of this extremely rare species (Pl. 1, Fig. 1). The zooecia are very often arranged in a rather irregular pattern, distinctly separated and bordered by a row of areolae. The frontal wall shows the typical foramina of the sort already illustrated by Reuss (1874) and described by Pouyet (1997); the frontal area on which they are situated is however less distinct than in the material described by Pouyet from Poland. The typically shaped aperture shows cardellae and - situated at its proximal border, a rather inconspicous, broad mucro; a few (e.g. four) hollow spines surround the distal part of the aperture. Small but prominent avicularia situated laterally to the apertures can be observed but are far rarer than could be expected according to the descriptions given by Reuss (1874) and Pouyet (1997).

Remarks: This species is extremely rare; the material from Stotzing is the first find within Austria outside its type locality. There exist obviously only a few specimens: two colonies on a coral (*Porites incrustans*) being the type material from Baden, one specimen in the collection of Canu (Baden, Museum of Natural History at Paris), two specimens from Poland, and this new find from Stotzing. A revision of the taxonomy of the genus *Tremogasterina* has been published by Cook (1977), while the possible palaeobiogeographical "background" has been summarized by Vávra (2000).

Occurrence: Austria: Baden, Stotzing; Poland: Olimpow.

Distribution: Austria: Lower Badenian (Baden), Badenian; Poland: Middle Miocene.

Schizostomella grinzingensis David & Pouyet, 1974
(Pl. 1, Figs 2-4)

?1877 *Eschara minax* Manzoni - Manzoni, p. 65, pl. 11, fig. 37.
?1877 *Eschara neglecta* Manzoni - Manzoni, p. 65, pl. 11, fig. 38.
1925 *Schizostoma neglecta* Manzoni 1877. - Canu & Bassler, p. 682, pl. 24, fig. 6-9.
1974 *Schizostomella grinzingensis* nov. sp. - David & Pouyet, p. 203-204, pl. 14, fig. 7-8.
1977 *Schizostomella grinzingensis* David & Pouyet, 1974. - Vávra, p. 149.
2000 *Schizostomella grinzingensis*.- Vávra, pl. 1, 4.

Description: The erect, bifoliate, branching often rather flat zoaria may sometimes also show a circular cross section and even transitions to hollow subcylindrical branches. Mostly however they are of "eschariform" type; due to heavy calcification the general appearance of zoaria may show considerable variability (e.g. Pl. 1, Figs 2, 4). The distinctly separated zooecia are arranged in alternating rows yielding thus a very regular general pattern of the surface. Pores indicate the border of each zooecium, additional pores are scattered on the frontal wall.

The apertures show a well-developed peristome, the thickness depending on the degree of calcification of the zoarium. Distinct pointed small avicularia of a rather elegant shape are very common (Pl. 1, Fig. 2) and are found on each zooecium; their orientation and position may differ considerably. The pores mentioned by David & Pouyet (1974: 204) "correspondant sans aucun doute à un aviculaire" are the result of heavy calcification of the zoarial surface: if the frontal wall is less calcified the avicularia are very obvious and show their differences in respect to position and orientation. They may be situated lateral to the aperture very close to the peristome or they may be on the frontal wall, and may point in different directions. On the basis of the rich material from different localities being at my disposal three additional morphological features could be observed: (1) there exists a dimorphism of zooecia; the rows forming the rim of the flat branches are different in their appearance: no avicularia, frontal wall completely covered with pores (Pl. 1, Fig. 4). (2) Additional, interzooecial avicularia occur: they are extremely rare, very slender and have approximately the length of a zooecium (Pl. 1, Fig. 3). (3) Hollow, subcylindrical branches may be formed (see above); transitions between different zoarial shapes can be observed. The heavy calcification mentioned above is rather common, some branches show the transition between more and less calcified zoaria (e.g. material from Forchtenstein, coll. Bobies). Ovicells could unfortunately not be observed.

Remarks: On the basis of a rather rich material from different Austrian bryozoan localities (four of them being first finds) a rather detailed description of this species can be given; a number of additional morphological features could be observed for the first time. The dimorphism of zooecia as mentioned above is of special interest as dimorphism of just this type can also be observed in *Adeonella polystomella* - both species belong to the family Adeonidae. The necessity to establish a new species as suggested by David & Pouyet (1974) with all its implications shall not be discussed here. Material from Steinebrunn: collected by Dr Rögl, det. B. Schmid; Grund: gift from Mr Suttner, Vienna, Gainfarn: gift from Dr Stojaspal, Vienna. A possible occurrence of this species from the Late Miocene of Morocco has been mentioned by El Hajjaji (1992). This author has also claimed a remarkable similarity with *Schizostomella helvetica* from the Middle Miocene of the Loire Basin (France). In the course of my own studies at the Musée d'Histoire Naturelle at Paris I have moreover realized that there exists a considerable number of species (e.g. *Laminophora miocenica, Meniscopora bifurcata, M. elliptica, M. milleporacea*) being more or less close to *Schizostomella grinzingensis*; in fact a detailed revision of all these forms would be very useful indeed.

Occurrence: Austria: Ehrenhausen, Eisenstadt, Forchtenstein, Gainfarn, Grinzing, Grund, Mailberg (=Locatelliwald), Steinebrunn; Czech Republic: Sedlec (=Porzteich); Morocco: Irhachamène (?); Romania: Lapugy.

Distribution: Austria: Lower and Middle Badenian; Czech Republic and Romania: Badenian; Morocco (?): Late Miocene.

Smittistoma polyomma (Reuss, 1848)
(Pl. 1, Figs 5, 6)

1848 *Eschara polyomma* m. - Reuss, p. 71, pl. 8, fig. 33.
1877 *Eschara polyomma* Reuss - Manzoni, p. 63, pl. 8, fig. 27, pl. 9, fig. 28.
?1925 *Schizostoma polyomma* Reuss, 1847 - Canu & Bassler, p. 682.
1974 *Escharina polyomma* (Reuss, 1848) - David & Pouyet, p. 165-166, pl. 14, fig. 3, 4.
1977 *Escharina polyomma* (Reuss, 1848) - Vávra, p. 122-123.
1997 *Escharina polyomma* (Reuss, 1848) - Pouyet, p. 58, pl. 5, figs 6, 11.

Description: This description is based on material collected by the late C.A. Bobies in 1956/57 at Eisenstadt respectively Forchtenstein. The erect, bifoliate, sometimes dichtotomously branching zoaria show the zooecia arranged in alternating rows. The appearance of the surface depends very much on the degree of calcification of the colony: exact form and shape of the single zooecium is not often discernible with certainty, the borders being indicated mostly by pores only. The aperture - sometimes showing a peristome-like rim, is situated in the distal part of the zooecium and usually shows a well-developed, distinct, short and narrow rimulus. A small avicularium is often present - sometimes situated left, sometimes right of the aperture but generally proximally. The most remarkable discovery in the material being at our disposal are three ovicells on one zoarial fragment from Forchtenstein (Pl. 1, Fig. 6): they are in general appearance and morphological details identical to those illustrated by Canu (1907/08: pl. 22 [= 8], fig. 9) for *Smittistoma mortisaga*, the type species of the genus.

Remarks: The most important observation in respect to this material is the first find of ovicells; David & Pouyet (1974) as well as Pouyet (1997) in her study of material from the Middle Miocene of Poland confirmed that ovicells were unknown. Obviously Canu & Bassler (1925) had been the very first to observe ovicells in this species. In the course of my studies I could not relocate their material in the collections of the Musée d'Histoire Naturelle at Paris - possibly their specimens are kept in the collections of the

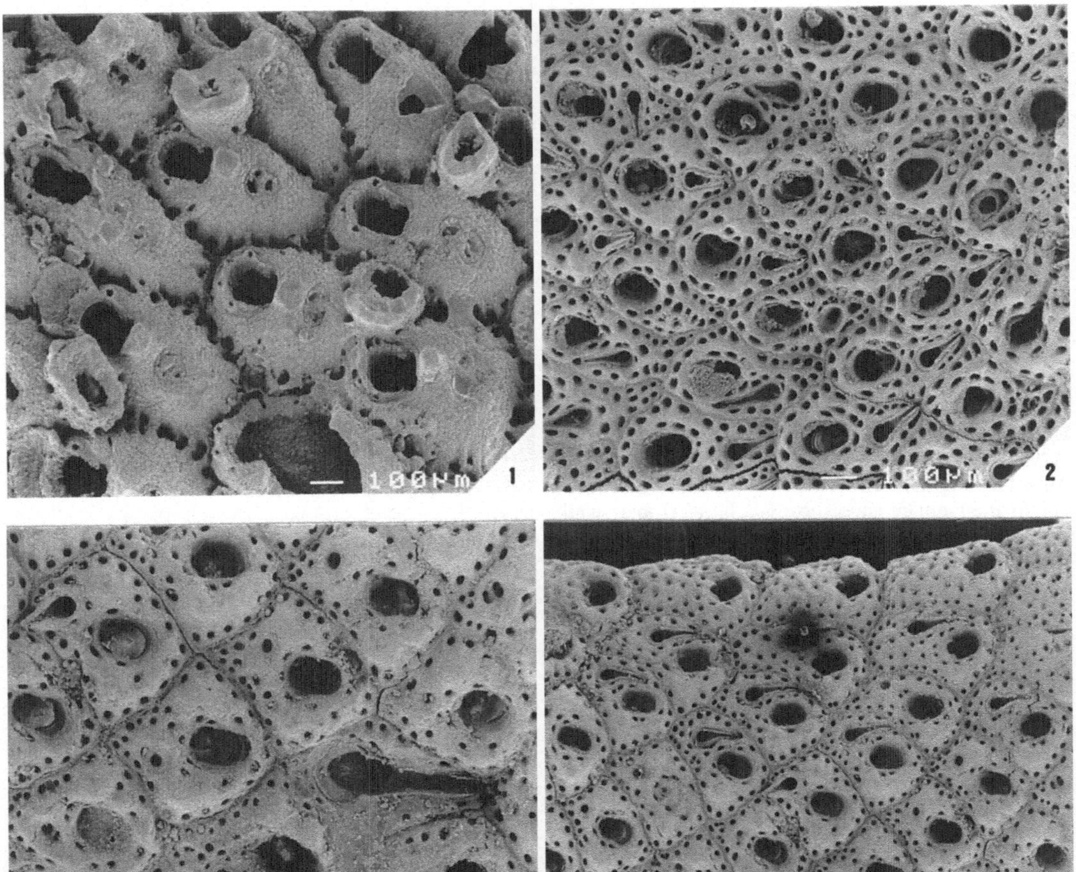
100µm
1
2
3
4
5
6

Plate 1 (opposite). Fig. 1: *Tremogasterina areolata*, Stotzing, Burgenland, Badenian, UWPI; part of zoarium with heavily calcified, broken avicularia (bar: 100 µm).
Figs. 2-4: *Schizostomella grinzingensis*, Forchtenstein, Burgenland, Badenian, NHM; 2: perfectly preserved surface of zoarium, low calcification; 3: heavily calcified surface, large interzooecial avicularium; 4: heavily calcified, zooecia at the border without avicularia (dimorphism); bars: 100 µm.
Figs. 5-6: *Smittistoma polyomma*, Forchtenstein, Burgenland, Badenian, NHM; 5: general view of zoarium with three ovicells; 6: single ovicell; bars: 100 µm
Abbreviations used: UWPI: material kept in the collection of the Institute of Palaeontology, University of Vienna (accession number: 3111). NHM: material kept in the bryozoan collection (coll. Bobies) of the Museum of Natural History, Vienna.

Smithsonian Institution, Washington. Going through the diagnosis of the genus *Smittistoma* as given by Canu (1907/08) one will note some differences in respect to avicularia; realizing however the striking identity of the ovicells for both species, I would regard these differences as being of minor significance.

Occurrence: Austria: Baden, Eisenstadt, Forchtenstein, Prinzendorf, Steinebrunn; Poland: Olimpow; Slovakia.

Distribution: Austria: Badenian; Poland: Middle Miocene; Slovakia: Middle Miocene (Pouyet 1997).

?*Triporula* sp.
(Pl. 2, Fig. 2)

Description: Among material from Forchtenstein (coll. Bobies, collected in 1956, Museum of Natural History, Vienna) there exist nine specimens determined by C.A. Bobies as "*Triporula*" which he has neither published nor described. Most of these zoarial fragments are of rather poor quality; a few however show rather well-preserved zooecia showing a morphology strikingly similar to that as described for *Triporula*. The zoaria have encrusted some substratum but later developed to form multi-layered colonies. The single zooecia are arranged in a rather irregular way - possibly as the result of frontal budding. The frontal wall shows a varying number of rather large, irregular pores showing sometimes the "stellate" appearance as described for the type species (*Escharipora stellata* Smitt, 1873). Well-preserved and less calcified specimens show two avicularia close to the aperture arranged in the way as described for the type species. The third one - being very small, but regarded as characteristic for this genus - could not be detected. Nevertheless the general appearance and morphology of these few zoarial fragments justify at least a tentative attribution of these fossils to the genus *Triporula*.

Remarks: Among material collected by the late C.A. Bobies from Eisenstadt (name of locality: Hartl) in 1957 there is one more specimen determined by him as "*Triporula*"; this determination can not be verified: the existence of a thin and rather deep rimulus, spines surrounding the aperture, and the completely different appearance of avicularia shows that this specimen must be determined as *Gemelliporella ternata* (Reuss, 1848). The wrong earlier determination can be explained by the heavy calcification of this colony during diagenesis which makes determination rather difficult.

Occurrence: Austria: Forchtenstein.

Distribution: Austria: Badenian.

Umbonula spinosa (Procházka, 1894)
(Pl. 2, Fig. 1)

1894 *Eschara spinosa* nov. spec. - Procházka, p. 67, pl. 12, fig. 5 a-g.
1996 *Umbonula spinosa* (Procházka, 1893) - Svacek, p. 11-12, pl. 4, figs 1-4.

Description: The material from Stotzing includes 12 large (up to nearly 2 cm) fragments of flat, bifoliate zoaria. Unfortunately surfaces are heavily calcified and have suffered a rather strong calcification during diagenesis. This makes the study of the morphology difficult. Nevertheless the following characteristics could be observed: a remarkable regular arrangement of zooecia in strictly alternating rows, prominent, spine-like umbos, additional small umbos situated distally to the aperture or at least close to it, and the areolae bordering the single zooecia. Ovicells could not be detected. Some of the better-perserved zoaria showed transitions in morphology to a rather *Umbonula macrocheila*-appearance. In fact both species seem to follow a rather similar pattern in their morphology. Nevertheless *U. spinosa* should for the present be maintained as a separate species. The type material (probably kept in the collection Procházka) was not at my disposal, but I could compare my material at least with topotypes. The most distinct feature of this Czech material was the fact that it was far better preserved; the zoaria were erect but of different cross sections. By additional small umbos the surfaces had sometimes a true chaotically appearance as observed already by Svacek (1996). Except for these details the material showed a rather striking identity with the zoaria from Stotzing. This material had been collected by P. Svacek (Czech Republic) at Kralice (type locality) during our cooperation during a bryozoan project. In the collection of the late C.A. Bobies (property of Museum of Natural History, Vienna) I found zoaria from Eisenstadt (determined by Bobies as "*Umbonula*

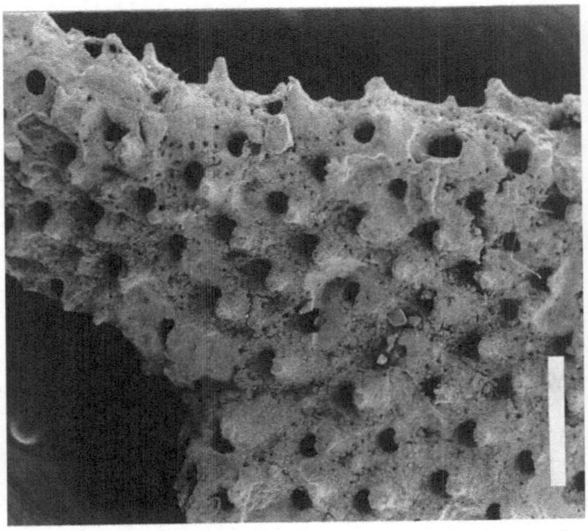
1

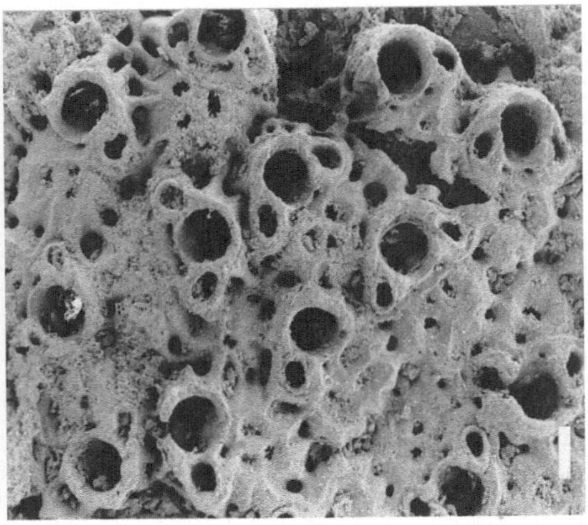
2

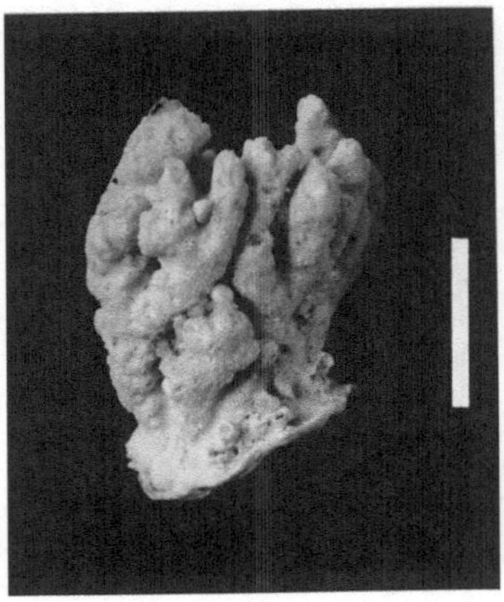
3

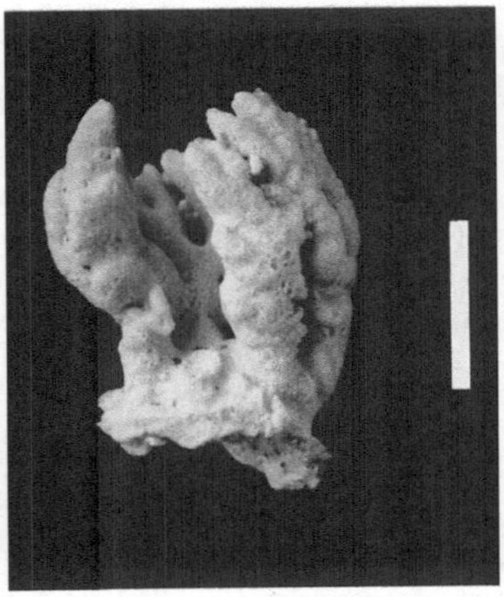
4

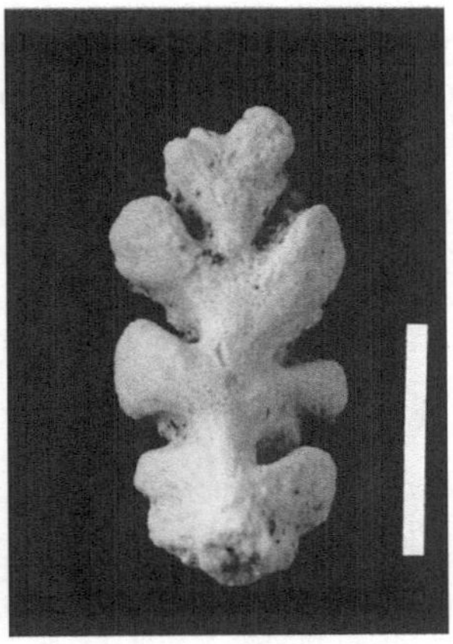
5

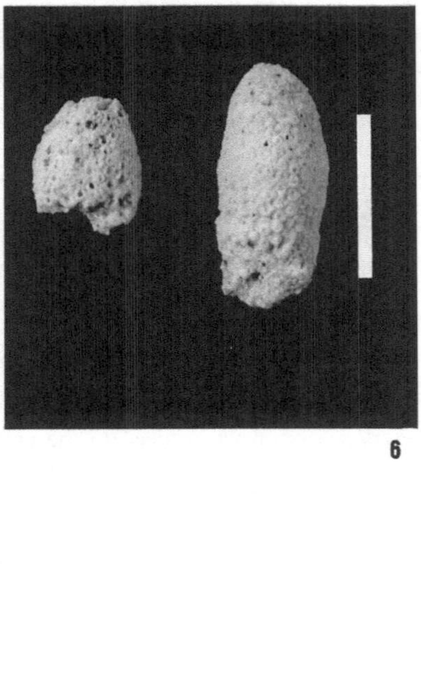
6

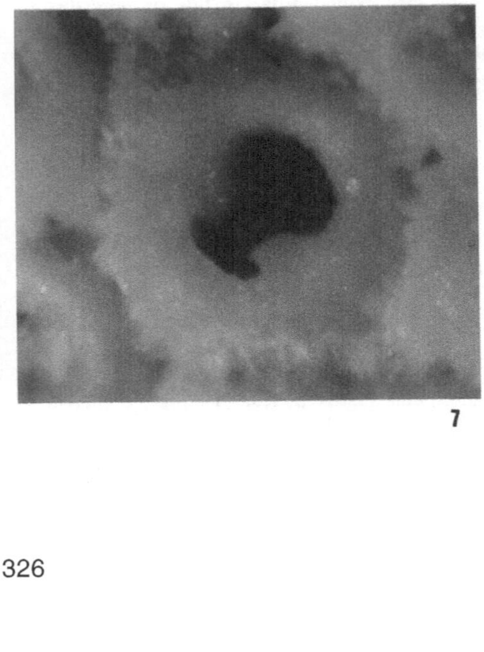
7

Plate 2 (opposite). Fig. 1: *Umbonula spinosa*, Stotzing, Burgenland, Badenian, UWPI; zoarium showing spine-like umbos; bar: 1 mm.
Fig. 2: ?*Triporula* sp., Forchtenstein, Burgenland, Badenian, NHM; bar: 100 μm.
Figs. 3-5: *Bobiesipora fasciculata*, 3, 4: complete zoarium, different views, bars: 5 mm. 5: single branch, bar: 5 mm.
Figs. 6-7: *Kionidella moravica*, Stotzing, Burgenland, Badenian, UWPI; 6: two fragments of zoaria, bar: 5 mm; 7: single zooecium, showing shape of aperture (approximately x 140).
Abbreviations used as in Plate 1.

macrocheila") which can be attributed to *U. spinosa* with certainty. This material shows also the ovicells.

Remarks: This species is new for the Austrian Miocene; in respect to its morphology it is similar to *U. macrocheila* described from at least three other localities (Eisenstadt, Forchtenstein, Oslip) in Burgenland: this species has been revised by Schmid (1989) and quite recently discussed by Pouyet (1997), who found it in Badenian sediments at Olimpow (Poland). Both authors have also discussed the morphology of other species which they regarded as junior synonyms of *U. macrocheila* (e.g. *U. endlicheri* and *U. scarabaeus*). The most outstanding distinctive feature between these two species is the prominent umbo as well as the small additional umbo placed laterally or distally to the aperture as observed in *U. spinosa*. Very important morphological characteristics which both species (*U. macrocheila* and *U. spinosa*) have in common are: (1) the zoarium being erect and bifoliate - "always" for *U. spinosa* and "usually" for *U. macrocheila*, (2) the pattern of arrangement of zooecia, (3) the shape of the aperture, (4) occurrence of a suboral umbo with avicularium, and perhaps even a few more additional features. Future studies should probably include a thorough revision of this group to clarify the situation. One problem in this connection is however the fact that the type material is not accessible at the present time; the collection Procházka could not be located until now (pers. comm. P. Svacek).

Occurrence: Austria: Stotzing, Eisenstadt; Czech Republic: Kralice nad Oslavou (type locality).

Distribution: Lower Badenian (Moravia); Badenian (Austria).

Kionidella moravica Procházka, 1894
(Pl. 2, Figs 6, 7)

1894 *Kionidella Moravica* nov. spec. - Procházka, p. 68-69, pl. 12, fig. 8 a-d.
1977 *Kionidella moravica* Procházka, 1894 - Vávra, p. 160.
1996 *Kionidella moravica* Procházka, 1893 - Svacek, p. 10-11, pl. 5, fig. 1-3.

Description: Two rather large fragments of this extremely rare species could be found among the material from Stotzing. The hollow zoaria of these free-living forms are more or less cone-shaped, one specimen being thicker in the middle part, therefore it is better to describe it as "club-shaped" perhaps. The zooecia are arranged in a very regular pattern, forming alternating rows. Due to rather heavy calcification of the specimens a number of details mentioned in the description of Svacek (1996) can not be confirmed (e.g. areolae bordering the single zooecia). A few well-preserved and less calcified zooecia showed the same aperture as some topotypes from Kralice, an aperture being best described as "keyhole-shaped" (Pl. 2, Fig. 7). The avicularia are situated laterally to single zooecia, details as given by Svacek can not be described because of poor preservation. Ovicells could not be found. Realizing the typical morphology of this species there is no doubt about the identity of the material from Kralice and Stotzing.

Remarks: This species has been described by Procházka (1894) under different names in his paper: *Kionidella Moravicensis,* and *K. moravica.* I follow here the general habit and use "*K. moravica*" - the spelling used by Procházka in his German text. Material of this species from the Austrian Miocene had been found before only at Baden (coll. Prof. E. Voigt, Hamburg). Procházka (1894) had realized that *K. moravica* was very close to two species of the same genus described by Koschinsky (1885) from the Paleogene of Bavaria; he was aware however, also of the morphological characteristics separating these taxa: shape of zooecia and apertures, shape and position of avicularia.

Occurrence: Austria: Baden, Stotzing; Czech Republic: Kralice nad Oslavou (Moravia).

Distribution: Austria: Badenian; Czech Republic: Lower Badenian.

Bobiesipora fasciculata (Reuss, 1848)
(Pl. 2, Figs 3-5)

1848 *Apsendesia fasciculata* m. - Reuss, p. 40, pl. 6, fig. 8.
1977 *Bobiesipora fasciculata* (Reuss, 1848) - Vávra, p. 73.
1978 *Bobiesipora fasciculata* (Reuss, 1848) - Vávra, p. 230-234, pl. 1, fig. 1-8, pl. 2, figs 1-6.
1979 *Bobiesipora fasciculata* (Reuss, 1848) - Vávra, p. 595-596.
1984 *Bobiesipora fasciculata* (Reuss, 1848) - Vávra, p. 228, pl. 1, fig. 8-10.
1989 *Bobiesipora fasciculata* (Reuss, 1848) - Vávra, p. 92, pl. 1, fig. 5.

Description: Three specimens (and one fragment of a possible base) of this remarkable species were found in the material from Stotzing; one of them being a perfectly preserved complete zoarium. Another is a single branch showing the mode of bifurcations. Both specimens contribute therefore considerably to a detailed understanding of the structure of the colony. *Bobiesipora*, having been described in detail already before (Vávra 1978), only some basic information will be given here. A typical zoarium of this genus consists of a circular or oval-shaped base completely covered with kenozooecia with branches developing in radially arranged fascicles similar to those in Lichenoporides. This can be perfectly seen in the one complete colony from Stotzing (Pl. 2, Figs 3, 4), which contains all the morphological details, being usually only available for study in the form of various fragments. Another specimen from Stotzing gives a good idea how a branch may look like after repeated bifurcation has occurred (Pl. 2, Fig. 5). The kenozooecia with their calcareous lamellae sometimes leaving only a tiny pore and nearly closing the openings of the kenozooecia have been described already (Vávra 1978). They could be observed in the material from Stotzing again. The ovicell being situated on the dorsal side in the bifurcation of a branch has been found only once before now (coll. Prof. E. Voigt, Hamburg; Vávra, 1978, Pl. 1, Fig. 7). Among the material from Stotzing no ovicells could be observed.

Remarks: This species was once among the rarest in the Paratethys material; now it can be best described as "being still rare, but rather wide-spread". Unfortunately a reasonable attribution to any of the cyclostomate families is still not possible yet. Perhaps it may be included in the Lichenoporides, but this is at the moment only a suggestion. There exists one genus from the Dano-Montian which is rather similar: *Retecavella.* This genus always shows a typical median lamella, a feature completely unknown in *Bobiesipora.* This relationship has been discussed in some detail (Vávra 1978) on the basis of personal communications with Prof. Voigt; *Retecavella* was described by Voigt a number of years later (Voigt 1987). A species which may be very close to *B. fasciculata* has been described by Procházka (1894) from the Miocene of Kralice (Moravia, Czech Republic) as *Defrancia orbygniana*. One of the specimens from Stotzing encrusted with its base a Horneride, shows in its general appearance a remarkable similarity with the illustration given by Procházka on plate 11, figure 14. As no topotypes from Kralice were at my disposal and the collection of Procházka was not available for study, no accurate answer can be given to this problem.

Occurrence: Austria: Baden, Eisenstadt, Freibühel, Mörbisch, Steinebrunn, Stotzing, Weissenegg; Poland: Korytnica.

Distribution: Austria: Lower and Middle Badenian; Poland: Middle Miocene

4 CONCLUSIONS

On the basis of rich bryozoan finds from the Badenian (Middle Miocene) of Stotzing and Forchtenstein (Burgenland, Austria) a few species have been studied in detail and redescribed. As *Adeonella polystomella* (Forchtenstein and Stotzing) as well as *Adeonellopsis coscinophora* (Forchtenstein) are among this material too, one can regard these studies as a short revision of the Adeonidae in the Austrian Miocene: Four genera of this family (*Adeonella, Adeonellopsis, Schizostomella* and *Smittistoma*) are confirmed in these faunas - an additional one (*Triporula*) remains uncertain however. In addition one species of *Umbonula* (*U. spinosa*), is new for the Austrian Miocene, and two finds of the rare free-living *Kionidella moravica* and a complete, well-preserved zoarium of the cyclostome bryozoan *Bobiesipora fasciculata* are reported.

ACKNOWLEDGEMENTS

The author's bryozoan studies have been supported generously by the "Fonds zur Förderung der wissenschaftlichen Forschung" (Vienna) within the projects P09561-GEO ("Tertiäre und rezente Bryozoenfaunen") and P11442-BIO ("Bryozoenfaunen der Steiermark") - help which is gratefully appreciated. In addition I want to express my sincerest thanks to Dr H. Temmel (Weiz) for the generous donation of the bryozoan material collected at Stotzing and to Mr R. Gold (Institute of Palaeontology, University of Vienna) for his help with photographic work.

REFERENCES

Canu, F. 1907/08. Bryozoaires des terrains des environs de Paris. *Annales de paléontologie* 2: 57-89, 137-160; 3: 61-104.

Canu, F. & Bassler, R.S. 1925. Contributions à l'étude des Bryozoaires d'Autriche et de Hongrie. *Bulletin de la Société Géologique de France 4. Sér.* 24(1924): 672-690.

Canu, F. & Bassler, R.S. 1928. Fossil and recent Bryozoa of the Gulf of Mexico region. *Proceedings of the United States National Museum* 72(14): 1-199.

Cook, P.L. 1977. The genus *Tremogasterina* Canu (Bryozoa, Cheilostomata). *Bulletin of the British Museum (Natural History), Zoology series* 32(5): 103-165.

David, L. & Pouyet, S. 1974. Révision des Bryozoaires cheilostomes miocènes du Bassin de Vienne - Autriche. *Documents des Laboratoires de Géologie Lyon* 60: 83-257.

El Hajjaji, K. 1992. Les Bryozoaires du Miocène Supérieur du Maroc nord-oriental. *Documents des Laboratoires de Géologie Lyon* 123: 1-355.

Koschinsky, C. 1885. Ein Beitrag zur Kenntnis der Bryozoenfauna der älteren Tertiärschichten des südlichen Bayerns. I. Abth. Cheilostomata. *Palaeontographica* 32: 1-73.

Manzoni, A. 1877. I Briozoi fossili del Miocene d'Austria ed Ungheria, II. *Denkschriften der Kaiserlichen Akademie der Wissenschaften, Mathematisch-Naturwissenschaftliche Classe* 37: 49-78.

Manzoni, A. 1878. I Briozoi fossili del Miocene d'Austria ed Ungheria, III. *Denkschriften der Kaiserlichen Akademie der Wissenschaften, Mathematisch-Naturwissenschaftliche Classe* 38: 1-24.

Pouyet, S. 1997. Les Bryozoaires du Badénien (Miocène Moyen) d'Olimpow (Pologne). *Documents des Laboratoires de Géologie Lyon* 145: 1-125.

Procházka, V. J. 1894. Das Miocaen von Kralic nächst Námest in Mähren. *Sitzungsberichte der königlichen boehmischen Gesellschaft der Wissenschaften, Mathematisch-Naturwissenschaftliche Classe* 16(1893): 1-71.

Reuss, A. E. 1848. Die fossilen Polyparien des Wiener Tertiärbeckens. *Naturwissenschaftliche Abhandlungen* 2: 1-109.

Reuss, A. E. 1874. Die fossilen Bryozoen des österreichisch-ungarischen Miocäns. *Denkschriften der Kaiserlichen Akademie der Wissenschaften, Mathematisch-Naturwissenschaftliche Klasse* 33: 141-190.

Schmid, B. 1989. Cheilostome Bryozoen aus dem Badenien (Miozän) von Nußdorf (Wien). *Beiträge zur Paläontologie von Österreich* 15: 1-101.

Sieber, R. 1956. Die Tortonfauna von Mattersburg und Forchetnau (Burgenland). *Verhandlungen der Geologischen Bundesanstalt* 1956(3): 236-249.

Smitt, F.A. 1873. Floridan Bryozoa collected by Count L.F. de Pourtalès, II. *Kongliga Svenska Vetenskapsakademiens handlingar* 11(4): 1-83.

Svacek, P. 1996. *Lower Badenian Bryozoa of the South Moravian part of the Carpathian foredeep.* Brno: Masaryk University [unpublished diploma thesis].

Vávra, N. 1977. *Bryozoa tertiaria.* In H. Zapfe (ed.), *Catalogus Fossilium Austriae*: 5b(3). Vienna: Österreichische Akademie der Wissenschaften.

Vávra, N. 1978. *Bobiesipora* n. g. - eine neue Gattung der Cyclostomata (Bryozoa) aus dem österreichischen Neogen. *Annalen des Naturhistorischen Museums in Wien* 81: 229-235.

Vávra, N. 1979. Bryozoa from the Miocene of Styria (Austria). In G.P. Larwood & M.B. Abbott (eds), *Advances in Bryozoology*: 585-610. London, New York, San Francisco: Academic Press.

Vávra, N. 1984. A littoral bryozoan assemblage from the Korytnica Clays (Middle Miocene; Holy Cross Mountains, Central Poland). *Acta geologica Polonica* 34(3-4): 223-237.

Vávra, N. 1989. Bryozoen aus dem Badenien (Mittelmiozän) von Weissenegg bei Wildon (Steiermark). *Annalen des Naturhistorischen Museums Wien* 90A: 83-102.

Vávra, N. 2000: Biogeographical Aspects of Bryozoan Faunas of the Central Paratethys from the Miocene. In A. Herrera Cubilla & J. B. C. Jackson (eds), *Proceedings of the 11th International Bryozoology Association Conference:* 392-399. Balboa: Smithsonian Tropical Research Institute.

Voigt, E. 1987. Die Bryozoen des klassischen Dano Montiens von Mons (Belgien). *Mémoires pour servir à l'Explication des Cartes Géologiques et Minières de la Belgique* 17(3): 1-161.

Bryozoan Studies 2001, Wyse Jackson, Buttler & Spencer Jones (eds)
© 2002 Swets & Zeitlinger, Lisse, ISBN 90 5809 388 3

Functional morphology of the anastomosing frondose growth form reported in *Heterotrypa frondosa* (d'Orbigny) (Bryozoa: Trepostomata) from the Cincinnatian (Late Ordovician) of Ohio

D.A. Waugh
Department of Geology, Kent State University, Kent OH, 44242, USA

J.M. Erickson
Department of Geology, St. Lawrence University, Canton NY, 13617, USA

ABSTRACT: : It has been rare for paleontologists to reconstruct bryozoan zoaria from the fragmented remains that generally represent these structures in Paleozoic rocks. Without reconstructed specimens, however, there can be little understanding of the functions of the colony, its range of phenotypic variation, or the range of ecological conditions for which a species was adapted. A limestone-to-shale interval in the lower Bellevue Formations has yielded large portions of the skeleton of *Heterotrypa frondosa* (d'Orbigny) that was re-assembled from more than 60 pieces. External currents drove water through the colony whereas ciliary beat of lophophores created food-capturing opportunities in this passive flow situation. Study in a re-circulating flume using a thermistor-based flow meter and dye, allowed analysis of water movements. First, water velocity was slowed in the chambers; secondly, turbulence was created within the chambers. With slowed velocity, zooids could continue to feed at higher ambient current velocities. Turbulence in chambers was beneficial, allowing the interior polypides of the colony to feed with minimal re-filtering; the turbulence immediately mixed fresh and food-depleted water, retaining food-rich waters in the water mix longer before flushing, as a non-chambered system would have done. Chambers and coalesced blades created a robust, internally-buttressed zoarium that was able to withstand normal, shallow-water, wave conditions as well as occasional storm surges.

1 INTRODUCTION

Paleontology of Middle Cincinnatian bryozoans from the type area on the Cincinnati Arch has focused on taxonomy, paleogeography, biostratigraphy and gross morphology (i.e. ramose, frondose, encrusting and massive). Knowledge of colony-wide functional morphology of Cincinnatian bryozoans or their Holocene homeomorphs is limited (McKinney 1977b) (but see Chamberlain & Graus 1975 on branched corals). Study of Ordovician bryozoan paleoecology and use of Paleozoic bryozoans as paleobathymetric indicators are strengthened when the autecology of species is explored. Function of bryozoan colonies (zoaria/skeletons) studied at the larger scale (Cowen & Rider 1972) exhibited by most Ordovician trepostomes requires re-assembly of taphonomically broken colonies in order to study colony-environment interactions.

The specimen we describe was collected at Sharonville (Hamilton County, Ohio) near the contact between the Bellevue and Fairview limestones. The purpose of this paper is to describe the morphology of this unusually complete colony of *Heterotrypa frondosa* (d'Orbigny), to experimentally ascertain the hydrodynamics of the re-assembled and reconstructed colony, and to show the scientific value derived from re-assembly of fragmented colonial growth forms. "Re-assembled" as used here refers to broken bryozoan fragments glued together and "reconstructed" refers to the addition of missing skeletal fragments using clay or other molding compounds.

1.1 *Growth form significance*

Colony morphology and environment are the major variables that alter the hydrodynamic interactions of a bryozoan colony. The interaction of sessile organisms with the water column provides a source of nutrient, a waste removal system and a vector of reproductive dispersal; but water also harbors the destructive forces of breakage and dislodgment (Wainwright & Koehl 1976). Bryozoan growth forms have been shown to correlate with water depth and have been successfully used as paleobathymetric indicators (Hageman 1997, Lagaaij & Gautier 1965, Moissette 2000, Pedley 1976, Schopf 1969, Stach 1936). Smith (1995) provided a review of bryozoans as paleoenvironmental indicators. Description of bryozoan growth forms will increase the

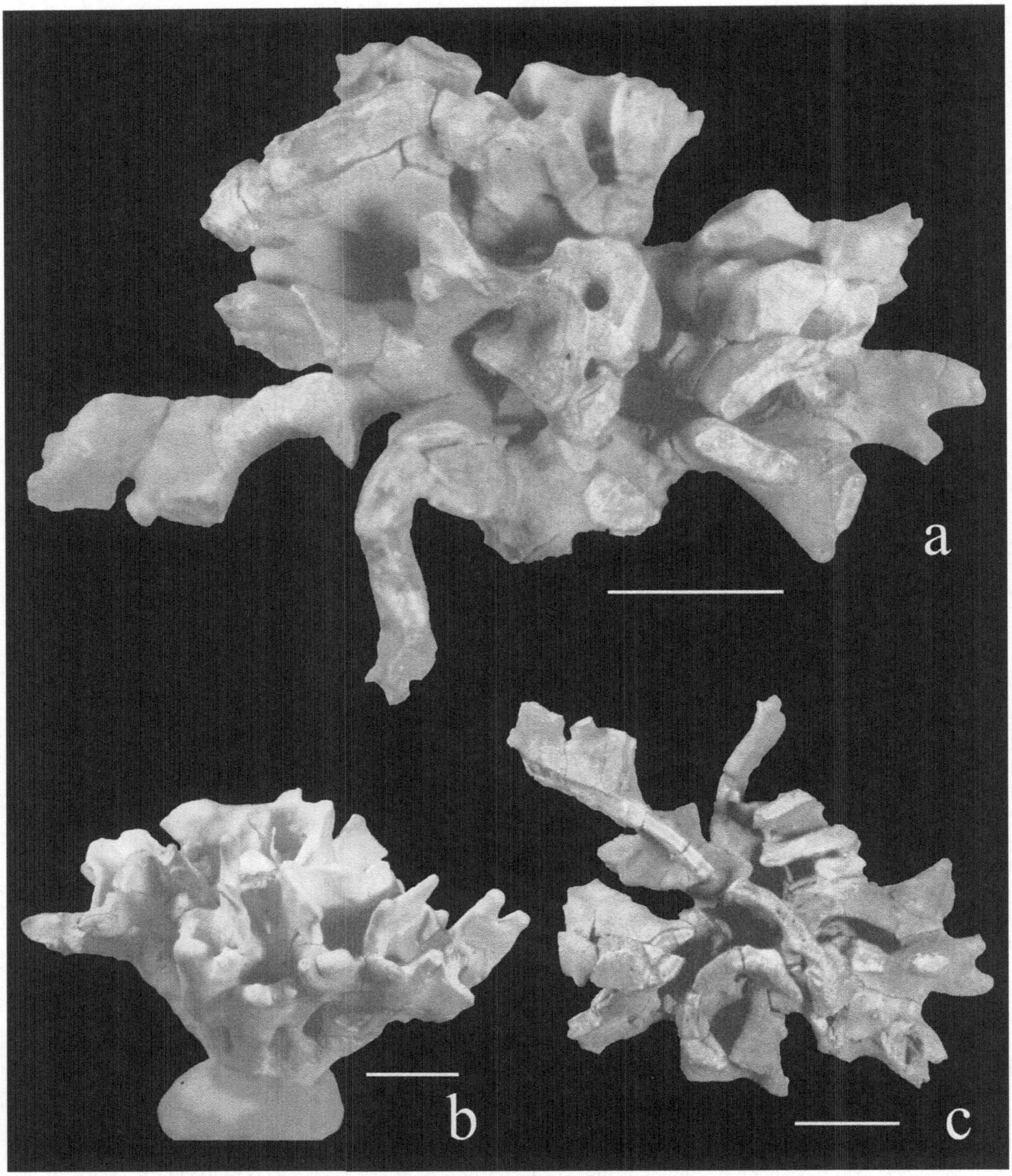

Figure 1. Views of reassembled *H. frondosa* (SLU# FR525). 1a. Oblique view of upper surface; 1c. Bottom view of same colony; 1b. Partially reconstructed cast of *H. frondosa*. Scale bars = 50 mm.

resolution of paleoecologic and sedimentologic studies of Paleozoic rocks.

1.2 *Significance of colony hydrodynamics*

Experimental studies of Paleozoic bryozoan hydrodynamics are few. The only flume study has been that of Stratton and Horowitz (1976) who examined the functional morphology of fenestrate bryozoans. Cheetham and Thomsen (1981) have measured drag and breaking strength of Holocene and fossil cheilostomes. Studies of colony-generated currents in extant bryozoans are more numerous (Best & Thorpe 1983, Best & Thorpe 1986, Cook 1977, Grunbaum 1992, Grunbaum 1995, Mckinney 1990, Okamura 1984, Okamura 1985, Riisgard & Manriquez 1997, Thorpe et al. 1985, Winston 1979) , but many of these focus on smaller-scale interactions than those in the present paper. Coral hydrodynamic studies such as Chamberlain and Graus (1975) are considered more applicable to larger colony forms seen in larger Ordovician bryozoans (Winston 1979).

Flow regime realized by a coral polyp could be radically different from those that the colony experiences (Koehl 1976, Wainwright & Koehl 1976); a similar condition should apply to bryozoans. Many bryozoan colonies are small enough to completely reside in the boundary layer (Lidgard 1981), whereas larger colonies, such as the one studied here, cannot seek refuge from ambient currents at the colony scale, and their growth forms must adapt to, or move from, habitats with unfavorable conditions.

The term "colony-wide current" (Cook 1977, McKinney et al. 1986) has been used in reference to metabolic currents generated by lophophores greater than the sum of an individual autozooid. Especially in taller erect colonies dominated by ambient currents, the colony-wide current is not generally metabolic. Therefore, use of the term may be misleading when discussing the colony-scale ambient currents in larger bryozoans.

H. frondosa morphology, previously known from fragmentary materials can now be studied at a meaningful colonial scale, enabling experimental investigations of colony function. To understand how a colony of *H. frondosa* interacts with flowing water, a plaster replica with some reconstruction was studied in a re-circulating flume.

1.3 *Re-assembly of fragmented material*

Colonial form and function can not always be studied from the fragmentary nature normally characterizing fossil bryozoans, but functional morphology of a limited number of bryozoans has been documented in several well-known works (Blake 1976, Cheetham 1971, Cheetham & Thomsen 1981, Cowen & Rider 1972, McKinney 1977a, b, Stratton & Horowitz 1976, Wyse Jackson et al. 1999).

An assumption often made when working with extinct colonial animals is that the general colony form can be reconstructed from partial or fragmentary material. When a colonial archetype exists the recognition of growth form from such material is possible. This archetype is based when possible on extant forms; in species with no apparent modern analog, archetypes are rare, and unless fully documented may vary between workers. This leads to the use of informal, but useful terms like ramose and frondose (see discussion in Hageman et al. 1998). The re-assembly of fragmented bryozoans is a time consuming and often frustrating process. Increased knowledge in Paleozoic growth forms should make the effort worthwhile.

2 MATERIALS AND METHODS

The material used in this study was collected from a 30-45 cm thick shale bed above the contact between the Maysvillian Fairview and Bellevue formations (perhaps 10-12 metres above the contact) at the Sharonville Industrial Park, Hamilton County, Ohio (OH-HA-0030 in Davis & Cuffey 1998). Approximately 60 fragments were recovered, cleaned and reassembled with cyanoacrylate glue. Epizoans and surface discoloration were often useful markers in the re-assembly process. Breakage surfaces, discordantly re-cemented by limonite were re-broken, cleaned and re-glued. The reassembled portions of the colony were un-glued into manageable pieces and were molded in silicon elastic and later cast in plaster; PVA was added to the plaster to increase the strength of the casts. This replica used in the flume was waterproofed by immersing the cast in polyurethane. Other plaster casts were cut in various orientations to observe internal colony morphology. The specimen is deposited in the St. Lawrence University Geology Faculty Research Collection as SLU # FR525.

Hydrologic interactions were observed in a re-circulating flume, formerly located at St. Lawrence University, Canton NY, now disassembled. Flow directions were mapped with dye, and a thermistor-based flow meter described by Vogel (1981) was used to measure water velocities. Small size and lack of moving parts allow thermistor-based flow meters to function in close proximity to objects with minimal flow disturbance. The plaster cast observed in the flume was conservatively reconstructed with oven-hardening clay replacing holes created by "missing pieces" and by reconstruction of inferred parts not recovered. During the flume study the colony was stabilized on a domal base molded from clay.

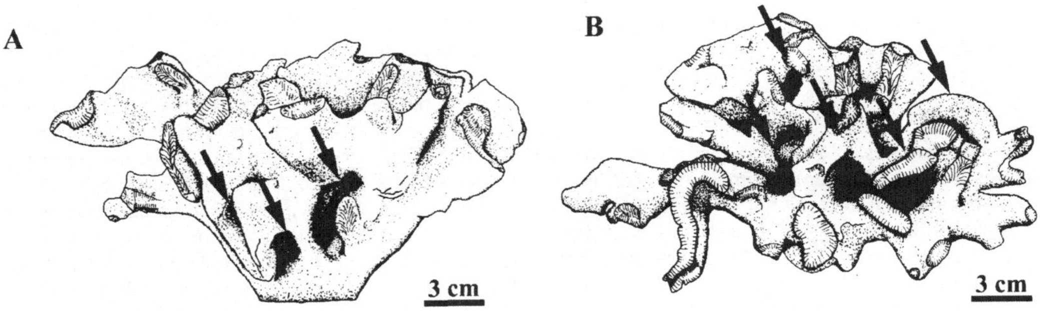

Figure 2. Reassembled *H. frondosa.* 2A. Lateral view of colony, arrows indicate windows. 2B. Oblique view, chambers labeled with arrows.

3 OBSERVATIONS

3.1 *Morphologic observations*

The reassembled colony shown in Figure 1a has a height of 13 cm and maximum width of 26 cm. The base of the colony (Fig. 1b) expands upward and outward radially, at 5 cm above the base the angle of outward expansion increases (Fig. 1c). Ramose protrusions or incipient frond "buds" of not more than a few cm appear sporadically on the interior branch surface. Bilaminate skeletal blades range from 1 to 3 cm in thickness. Thicker portions are situated at locations where colony blades have linked and overgrown one another. The colony shows various stages of intra-colony fusion, where two sheets grew in close proximity and later fused. Lower in the zoarium, blades arise from one main frond, grow apart and then merge in an anastomosing characteristic.

The colony is robust, but maintains an internally open architecture. The bilaminate sheets or blades form a vertically-expanding structure, effectively creating lateral windows that open into the colony interior, which is partitioned into vertical chambers (Figs 1c, 2). Windows are vertically elongate on the lateral perimeter and range from 3 to 7 cm in height and 1 to 2 cm in width. Chamber apertures ranging from 0.5 to 3 cm appear on the (present) upper surface of the colony; they continue uninterrupted for a few centimeters before opening into the central portion of the colony now exposed due to its state of preservation. Most chambers do not have continuous lateral walls; internally chambers become less defined. The preserved portion of the colony has 11 chambers, 5 of which are well formed and intact.

3.2 *Observation of flume study*

Dye injected into the water current upstream of the replica entered the colony in the lower 7 cm through the vertical windows (Waugh, 1999). Some water passed completely through the colony horizontally, a portion of the current was re-directed upward through the chambers to the upper surface; water that approached the colony from above 7 cm was deflected up and over the colony surface. (Fig. 3) This same pattern of water flow was observed in all rotations of the colony with respect to flow direction. Dye injected on the downstream portion of the colony and directly behind sheets normal to flow, was retained for a time before being cleared down stream.

Mainstream velocity in the flume was held at 6.9 cm s^{-1} throughout the experiment. Unlike the mainstream, velocity observed inside colony chambers was highly dynamic. Increased turbulence made precise measurement of velocity difficult, therefore minimum and maximum velocities were recorded over a period of 10 seconds at each chamber site (Fig. 4). Velocity measurements did not vary greatly when the velocity probe was moved to different locations within the same chamber.

4 DISCUSSION

4.1 *Discussion of colony taphonomy*

Colonies entombed in shales and broken under lithostatic pressure are ideal candidates for reassembly because fragments may be easily removed from the matrix and tend not to be widely distributed, most Ordovician bryozoan assemblages are both fragmented and transported, making reassembly impossible. Recognition, collection and re-

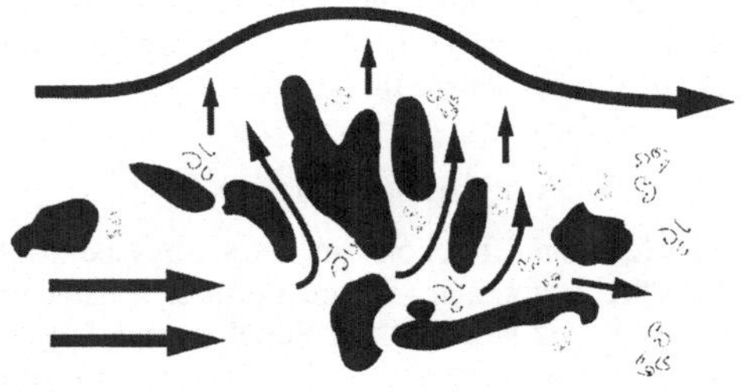

Figure 3. Water flow paths superimposed on a lateral cross-section of colony near the long axis. Turbulent regions are represented, by small squiggles.

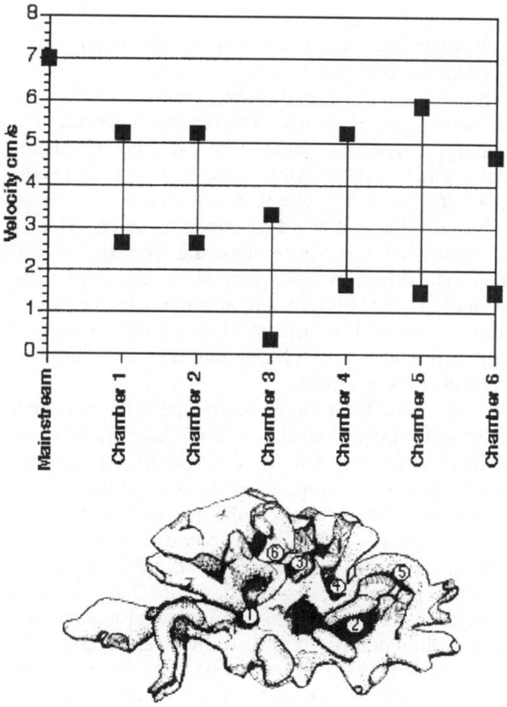

Figure 4. Ranges of velocity measurements taken inside colony reported by chamber. The range of values suggests roughly the same amount of turbulence in each chamber. All chambers had lower velocities than those of the ambient flume flow. Chambers are labeled on line drawing of specimen. Specimen is 26 mm in width.

assembly of undisturbed colony remains are important if subtleties of colony growth forms are to be recognized.

Three types of breaks were recognized on the colony: fresh, exposed (including weathering and boring), and breaks in various stages of re-growth. Raised monticules are rare to absent on the colony surface, but differential weathering of thin walled autozooids may make smooth monticules with thicker walls appear raised. The colony re-assembled in this paper may represent the core of a more extensive colony; breakage surfaces clearly indicate that material from the upper and lower portions of the colony has been lost either from incomplete collection or dispersed during the taphonomic process, or both. The extent of the missing material is unclear (Erickson and Waugh this volume), but fresh breaks represent material lost during exposure on the outcrop or incomplete collection, other abraded or fouled surfaces are indicative of taphonomic loss. Anastomosing-frondose colonies with resistant cores could be transported by rolling considerable distances on the sea floor without complete fragmentation (i.e. "tumble weeds"). Transport of this type could bias "frondose" bryozoan signatures.

The colony surface was moderately infested with the following epizoans and borers: *Cornulites* sp. (encrusting), *Cornulites* sp. (overgrown), *Cuffeyella arachnoidea*, *Trypanites* sp., and unidentified encrusting bryozoans. *Cornulites* sp. overgrown by the colony were contemporaneous. *Trypanites* sp. borings, penetrating the blades normal to the base of its broken surface and localized on restricted lateral surfaces represent exposure during partial burial or a subsequent partial exhumation. The temporal significance of other epizoans is unclear; infestation may represent fouling of dead portions of the colony or exposure before final burial. If trepostome bryozoans were covered with living tissue, even over parts of the colony without actively feeding zooids (Boardman 2001), epizoan fouling should be minimized during life. A cuticle that excludes epizoans and borers is desirable especially if borings substantially weaken colony bases as observed in corals (Chamberlain 1978).

4.2 *Discussion of flume study results*

A passive plaster replica is not a perfect substitute for a living colony with active lophophores. The cuticle covering trepostomate bryozoans would alter skeletal surface conditions, but the greater disadvantage in using dead colonies or plaster replicas is that bryozoans generated their own currents, a condition not easily recreated in non-living material. We agree with Stratton & Horowitz (1976) that metabolic currents generated by the lophophores may be responsible for the final movement of food to the mouth, in growth forms that rely on passive flow. If ambient currents are the controlling factor in the development of *H. frondosa*, then the passive model used in the flume experiments is an acceptable analog.

Although the colony is robust, an open internal architecture maintained water circulation. Water flowed into the lower 7 cm of the base; water higher in the column was deflected by the colony and passed above. Internal flow was turbulent and dampened from that of the mainstream. Turbulence mixes filtered and nutrient rich water (Vogel 1994, p. 192), preventing autozooids from filtering water significantly reduced in food particles and allowing nutrients in mainstream flow to enter the boundary zone inhabited by the zooids (Winston 1979, Vogel 1994, p. 174). A zone of reduced velocity and increased turbulence formed behind the colony, similar to that observed by Chamberlain & Graus (1975) in branched corals. Protruding sheets and other leeward facing surfaces were also observed to have similar turbulent zones, allowing these zooids also to benefit from increased turbulence and decreased velocity.

Schopf et al. (1980), working with comparatively small colonies (less than 15 mm) suggested that lophophore-generated currents are more important than ambient currents. Ordovician bryozoans did not live exclusively in cryptic or sheltered habitats and because of their height had greater reliance on ambient current. In the 0.6 cm above the sediment water interface off the coast of Oregon in 200 metres of water, flow is significantly reduced in the boundary layer and viscous sublayer (Caldwell & Chriss 1979). This zone would have little effect on larger colonies, although smaller growth forms residing in this zone are certainly controlled by it. Flow boundary zones significantly influence bryozoans (Anstey 1987, Lidgard 1981, Winston 1979). Lidgard (1981) demonstrated that species of *Membranipora* encrusting algal fronds are protected from the mainstream flow by a boundary zone. Stenolaemate bryozoans because they possess smaller lophophores may use this zone to further "hide" from mainstream velocity thereby increasing their ecologic range. If recent bryozoans have been shown to use the boundary layer as protection from mainstream flow, then stenoloemates with diminutive lophophores may have been able to feed actively in higher energy regimes.

Winston (1979) correctly recognized that branched corals in which ambient currents control environmental interactions should be used as analogies to branching bryozoans. Winston realized that the comparison was not a perfect analogy because of the large scale of lophophores relative to the branches and the small scale of the bryozoans studied. In larger Ordovician trepostomes, lophophores are small compared to branch size, and colony size is greater, more nearly on par with corals. Dynamic similitude as observed in corals by Chamberlain & Graus (1975) should extend to bryozoans (McKinney & Jackson 1989). This concept of dynamic similitude might explain the environmental association of thick and thin ramose and frondose bryozoans observed by Anstey & Perry (1973) in the Eden Shale.

4.3 *Discussion of* H. frondosa *growth form*

The genus *Heterotrypa* contains species with ramose, frondose, and encrusting morphologies (Boardman & Utgaard 1966). Our subsequent interpretations are limited to the morphology found in *H. frondosa* and other similar growth forms. The morphology suggests growth from a three bladed encrusting base, recognized throughout Late Cincinnatian hardground surfaces (Erickson and Waugh this volume). Encrusting bases such as this seem most suitable for an organism living in higher energy environments. Schopf (1969) observed an increase in the ratio of encrusting to flexible bases at depths below 35 metres in living bryozoans; this may not be true in the Cincinnatian forms. Undoubtedly, exposed bryozoan branches or sheets if broken would continue to grow even in soft substrates, but loss of wave resistance would result unless the colony was re-cemented to the substrate.

Intracolony linkages reduce colony surface area somewhat, subsequently feeding and reproductive capacity, suggesting some other functional advantages. The robust anastomosing frondose morphology of *H. frondosa* made it physically competent in strong currents and acted to dampen water velocity and increasing turbulence allowing zooids to feed when mainstream velocity exceeds lophophore competence. These effects of anastomosing linkages should increase the fitness of a colony in regions with increased water velocity and regions with periodic higher energy events.

To see how the colony could deal with sediment loaded onto it by wave action, sand- and gravel-sized particles were dropped on the top of the colony. Gravity funneled sediment down through the colony and out basal window openings; the maximum particle size passed through the colony was 1.3 cm. Sediment funneling, in conjunction with metabolic particle removal (Cook 1977) would have allowed the colony to recover from periods of rapid sand and gravel sedimentation. If the colony were adapted to handle periodic high energy events, the colony must have been able to manage the concomitant repercussions of such events (i.e. sediment, abrasion etc.) until destruction took place.

If functional interpretations made from this study are to be applied, the first step is recognition of growth forms from fossils that are mostly fragmented. The presence of chambers and anastomosing structure may be the diagnostic morphologies indicating presence of high-energy fauna.

5 CONCLUSIONS

- Colony re-assembly and reconstruction yield important paleontological information.
- Robust colony cores transported along the substrate may bias the "thick frondose" taphonomic signature.
- The scale of larger trepostomate bryozoans makes comparisons to corals the most useful analog in understanding hydrodynamic functional morphology.
- Anastomosing frondose colonies have been experimentally shown to baffle water flow and increase turbulence.
- *H. frondosa* morphology is consistent with life in higher energy flow regimes and periodic storm events.

ACKNOWLEDGEMENTS

We wish to thank Roger Cuffey his for help in bryozoan identification and his continued encouragement in Bryozoology. William Kirchgasser and his students graciously shared their experience in casting complex shapes. Chuck Crawford and the Physics Department at St. Lawrence University advised DAW during construction of the flow meter. Roger Trammel generously allowed collection on his property. David Griffing assisted in the excavation of the material used in this study. Communication and encouragement from Steven Vogel proved very helpful during construction of the flowmeter used in the work.

REFERENCES

Anstey, R.L. 1987. Colony patterning and functional morphology of water flow in Paleozoic Stenolaemate Bryozoans. In Ross, J.R.P. (ed.) *Bryozoa: Present and past*: 1-8. Bellingham: Western Washington University.

Anstey, R.L. & Perry, T.G. 1973. Eden Shale bryozoans: a numerical study (Ordovician, Ohio Valley). Publications of the Museum - Michigan State University Paleontological Series 1(1): 1- 80.

Best, M.A. & Thorpe, J.P. 1983. Effects of particle concentration on clearance rate and feeding current velocity in the marine bryozoan *Flustrellidra hispida*. *Marine Biology* 77: 85-92.

Best, M.A. & Thorpe, J.P. 1986. Effects of food particle concentrations on feeding current velocity in six species of marine Bryozoa. *Marine Biology* 93: 255-262.

Blake, D.B. 1976. Functional morphology and taxonomy of branch dimorphism in the Paleozoic bryozoan genus *Rhabdomeson*. *Lethaia* 9: 169-178.

Boardman, R.S. 2001. The Growth and function of skeletal diaphragms in the colony life of lower Paleozoic Trepostomata (Bryozoan). *Journal of Paleontology* 75(2): 225-240.

Boardman, R.S. & Utgaard, J. 1966. A revision of the Ordovician bryozoan genera *Monticulipora, Peronopora, Heterotrypa,* and *Dekayia*. *Journal of Paleontology* 40(5): 1082-1108.

Caldwell, D.R. & Chriss, T. M. 1979. The viscous sublayer at the Sea Floor. *Science* 205: 1131-11332.

Chamberlain, J.A. 1978. Mechanical properties of coral skeleton: compressive strength and its adaptive significance. *Paleobiology* 4(4): 419-435.

Chamberlain, J.A. & Graus, R.R. 1975. Water flow and hydromechanical adaptations of branched reef corals. *Bulletin of Marine Science* 25(1): 112-125.

Cheetham, A.H. 1971. *Functional morphology and biofacies distribution of Cheilostome Bryozoa in the Danian Stage (Paleocene) of southern Scandinavia.* Washington: Smithsonian Contributions to Paleobiology 6.

Cheetham, A.H. & Thomsen, E. 1981. Functional morphology of arborescent animals: strength and design of cheilostome bryozoan skeletons. *Paleobiology* 7(3): 355-383.

Cook, P.L. 1977. Colony-wide water currents in living Bryozoa. *Cahiers De Biology Marine* 18: 31-47.

Cowen, R. & Rider, J. 1972. Functional analysis of fenestellid bryozoan colonies. *Lethaia* 5: 145-164.

Davis, R.A. & Cuffey, R.J., (eds). 1998. *Sampling the layer cake that isn't: The stratigraphy and paleontology of the type-Cincinnatian.* Columbus: Ohio Department of Natural Resources. Guidebook 13.

Grunbaum, D. 1992. *Local Processes and Global patterns: Biomathematical models of bryozoan feeding currents and density dependent aggregations in Antarctic Krill.* Ithaca: Cornell University.

Grunbaum, D. 1995. A model of feeding currents in encrusting bryozoans shows interference between zooids within a colony. *Journal of Theoretical Biology* 174: 409-425.

Hageman, S.J., Bock, P.E., Bone, Y. & McGowran, B. 1998. Bryozoan growth habits: classification and analysis. *Journal of Paleontology* 72(3): 418-436.

Hageman, S.J., Bone, Y., & McGowran. 1997. Bryozoan colonial growth-forms as paleoenvironmental indicators: evaluation of methodology. *Palaios* 12: 405-419.

Koehl, M.A.R. 1976. Mechanical design in sea anemones. In Mackie, G. O., editor. *Coelenterate Ecology and Behavior*: 23-31, New York: Plenum Press.

Lagaaij, R. & Gautier, Y.V. 1965. Bryozoan assemblages from marine sediments of the Rhône delta, France. *Micropaleontology* 11(1): 39-58.

Lidgard, S. 1981. Water flow, feeding, and colony form in an encrusting cheilostome. In Larwood, G.P. & Nielsen, C., (eds), *Recent and Fossil Bryozoa*: 135-142. Fredensborg: Olsen&Olsen.

McKinney, F.K. 1977a. Functional interpretation of lyre-shaped Bryozoa. *Paleobiology* 3: 90-97.

McKinney, F.K. 1977b. Paraboloid colony bases in Paleozoic stenolaemate bryozoans. *Lethaia* 10: 209-217.

McKinney, F.K. 1990. Feeding and associated colonial morphology in marine bryozoans. *Aquatic Sciences* 2(2): 255-280.

McKinney, F.K. & Jackson, J.B.C. 1989. *Bryozoan Evolution.* Boston: Unwin Hyman.

McKinney, F.K., Listokin, M.R.A. & Phifer, C.D. 1986. Flow and polypide distribution in the cheilostome bryozoan *Bugula* and their inference in *Archimedes*. *Lethaia* 19: 81-93.

Moissette, P. 2000. Changes in bryozoan assemblages and bathymetric variations. Examples from the Messinian of northwest Algeria. *Palaeogeography, Palaeoclimatology, Palaeoecology* 155: 305-326.

Okamura, B. 1984. The effects of ambient flow velocity, colony size, and upstream colonies on the feeding success of Bryozoa. I. *Bugula stolonifera* Ryland, an arborescent species. *Journal of Experimental Marine Ecology* 83: 179-193.

Okamura, B. 1985. The effects of ambient flow velocity, colony size, and upstream colonies on the feeding success of Bryozoa. II. *Conopeum reticulum* (linnaeus), an encrusting species. *Journal of Experimental Marine Ecology* 89: 69-80.

Pedley, H.M. 1976. A palaeoecological study of the Upper Coralline Limestone, *Terebratula-Aphelesia* Bed (Miocene, Malta) based on bryozoan growth-form studies and brachiopod distributions. *Palaeogeography, Palaeoclimatology, Palaeoecology* 20: 209-234.

Riisgard, H.U. & Manriquez, P. 1997. Filter-feeding in fifteen marine ectoprocts (Bryozoa): particle capture and water pumping. *Marine ecology Progress Series* 154: 223-239.

Schopf, T.J.M., Collier, K.O. & Bach, B.O. 1980. Relation of the morphology of stick-like bryozoans at Friday Harbor, Washington, to bottom currents, suspended matter and depth. *Paleobiology* 6(4): 466-476.

Schopf, T.J.M. 1969. Paleoecology of Ectoprocts (Bryozoans). *Journal of Paleontology* 43(2): 234-244.

Smith, A.M. 1995. Palaeonvironmental interpretation using bryozoans: a review. In: Bosence, D. W. J. & Allison,

P.A., (eds). *Marine Palaeoenvironmental Ananlysis from Fossils*: 231-243.

Stach, L.W. 1936. Correlation of zoarial form with habitat. *Journal of Geology* 44: 60-65.

Stratton, J.F. & Horowitz, A.S. 1976. Studies of the flow of water through models of *Polypora*. *Docum. Lab. Geol. Fac. Sci. Lyon H.S. 3* (fasc. 2): 425-438.

Thorpe, J.P., Clarke, D.R.K. & Best, M.A. 1985. Natural variations in the tentacle number in marine bryozoans and the possible effects of intraspecific and interspecific ecological competition for food. In Nielsen, C. & G.P. Larwood (eds), *Bryozoa: Ordovician to Recent*: 319-327. Fredensborg: Olsen & Olsen.

Vogel, S. 1981. *Life in Moving Fluids: the Physical Biology of Flow*. Boston: Willard Grant Press.

Vogel, S. 1994. *Life in Moving Fluids: the Physical Biology of Flow*. Princeton: Princeton university Press.

Wainwright, S.A. & Koehl, M.A.R. 1976. The nature of flow and the reaction of benthic Cnidaria to it. In Mackie, G. O., (ed). *Coelenterate Ecology and Behavior*: 5-21. New York: Plenum Press.

Waugh, D.A. 1999. Description and functional morphology of a uniquely-complete colony of *Heterotrypa frondosa* (Bryozoa) (Upper Ordovician: Maysvillian) from Ohio. Unpublished senior thesis, St. Lawrence University, Canton, NY

Winston, J.E. 1979. Current-related morphology and behavior in some Pacific Coast bryozoans. In Larwood, G.P. & Abbott, M.B (eds), *Advances in bryozoogy*: 247-268. London: Academic Press.

Wyse Jackson, P.N., Taylor, P.D., & Tilsley, J.W 1999. The 'Balladoole Coral' from the Lower Carboniferous of the British Isles, reinterpreted as the unusual cystoporate bryozoan *Meekoporella* Moore & Dudley, 1944. *Proceedings of the Yorkshire Geological Society* 52(3): 257-268.

Bryozoan Studies 2001, Wyse Jackson, Buttler & Spencer Jones (eds)
© 2002 Swets & Zeitlinger, Lisse, ISBN 90 5809 388 3

Freshwater bryozoans: a zoogeographical reassessment

T.S. Wood
Department of Biological Sciences, Wright State University, Dayton, OH 45435 USA

ABSTRACT: Significant changes have occurred since the publication of Bushnell's 1973 paper on the zoogeography of freshwater bryozoans. There have been important new field surveys, the number of recognized species has nearly doubled and the taxonomic foundation is much more solid. Taxonomic revisions and the re-examination of misidentified material have diminished the reported ranges of many species. Such no-longer-widespread species include *Plumatella emarginata, P. repens, P. longigemmis, Stolella indica, Fredericella sultana*, and *Gelatinella toanensis*. An assumption of dispersal by migrating waterfowl explains the narrow, intercontinental ranges of several species. In other instances the dispersal patterns suggest human activities as a major contributing factor. Some species are clearly more vagile than others for reasons currently unknown. Asia, Africa, and South America are still largely unexplored for freshwater bryozoans. Nearly half of all known species are reported from only one or two collections. At a time when extinctions among freshwater animals have reached an unprecedented high rate, endemic bryozoan species may warrant special vigilance and protection. New initiatives are needed to inventory species in large, unexplored regions.

1 INTRODUCTION

1.1 *Biogeography*

With many groups of organisms, biogeography is an inexact science. It seeks to understand spatial patterns in species distribution and biological diversity. It examines, among other things, the roles of biotic, climatic, historical, and other factors in shaping a species' current and past occurrence. In pursuit of this knowledge, those of us asking biogeographic questions assume first that we have accurate distribution data, and second that all the species involved have been correctly identified. However, species inventories are often very uneven, and experienced taxonomists are scarce, making the biogeographic foundation rather unsteady. With full knowledge of these shortcomings, we forge ahead with a third assumption: that the unavoidable uncertainty will not make a significant difference in the long run.

1.2 *Bushnell (1973) paper*

At the Second Conference of the International Bryozoology Association a thoughtful paper was introduced by John Bushnell (1973) on the zoogeography of freshwater bryozoans. In this contribution Bushnell acknowledged the restricted distribution of many phylactolaemate species: holarctic for *Cristatella, Paludicella, Plumatella fungosa*, and *P. fruticosa*; neaarctic for *Pectinatella*, mostly Ethiopian for the gelatinous lophopodids, and so on. Several species were seen as cosmopolitan, including *Plumatella repens, P. emarginata, Hyalinella punctata*, and *Fredericella sultana*. Other species were regarded as highly restricted in their distribution: *Stephanella hina, Gelatinella toanensis, Internectella bulgarica*, and *Hyalinella orbisperma*. Bushnell attributed the differences in species distribution mainly to the nature of their primary disseminules, the dormant statoblasts. Some statoblasts have adherent qualities for dispersal by waterfowl, others were thought to achieve an ideal balance of size and weight for dispersal by wind. Additional factors might include statoblast resistance to desiccation and freezing as well as statoblast productivity and the timing of statoblast release.

In the thirty years since these ideas were published there have been significant additions to the knowledge of phylactolaemate distribution and taxonomy. This paper provides a brief summary of those advances and their biogeographic implications.

2 DEVELOPMENTS SINCE 1973

2.1 *Biological surveys*

Table 1 summarize the major regional surveys for freshwater bryozoans conducted over the past thirty years. All but one of these have been conducted in temperate regions of the world, leaving the freshwater tropics still largely unexplored.

2.2 *Endemic species*

Bushnell noted that among his 39 described phylactolaemate species, 33% were known from only one or two specimens or sites. This proportion of so-called endemic species has now increased to 45%. As expected, most of these occur in Africa, Asia, and South America where the least work has been conducted. Only *Stephanella hina* has come off the "endemic" list, its originally described range in eastern Asia now extended to both coasts of North America (Smith 1989a, Marsh & Wood, this volume). Yet to be confirmed is the likelihood that the "endemic" North American *Plumatella orbisperma* also occurs in both the United Kingdom and in northern Europe.

2.3 *Advances in taxonomy*

The taxonomy of phylactolaemate bryozoans has advanced considerably, mainly through the examination of statoblasts by scanning electron microscopy. The sclerotized outer surface of statoblasts carries a richly detailed surface relief that remains constant through successive generations, even under varying environmental conditions (e.g., Wood 1996, 2001a). These features have become very useful in taxonomic work. As a result, Bushnell's original list of 39 described species has now nearly doubled to 77, including 24 new species and the confirmation of 14 others. Species once considered to be *Fredericella sultana* now include *F. indica*, which itself probably comprises several distinct species (Økland & Økland 2001); *F. australiensis* and *F. browni* also are independently valid species (Wood & Wood 2000). Species once identifiable as *Plumatella repens* are now known also to include *P. nitens, P. nodulosa, P. orbisperma, P. recluse, P. rugosa*, and *P. similirepens* (Wood 2001c). Similarly, *Plumatella reticulata* and *P. mukaii* are now distinguished from P. *emarginata* (Wood 1988, 2001b). There appear to be at least two species currently identified as *P. fungosa* (Wood and Okamura, in prep.).

This clearer diagnosis of phylactolaemate species has also revealed a significant number of misidentifications in the literature. Returning to original specimens we find, for example, that Rogick's (1943) *Stolella indica* in Pennsylvania is actually *Plumatella rugosa*; that Lacourt's (1968) *Plumatella toanensis* does not really occur in either South America or Australia (Wood 1998), and that Annandale's (1911) *Plumatella fruticosa* and *Plumatella punctata* were not found in India after all (Wood, in prep.). In most cases the taxonomic errors are understandable. Species diagnoses were until recently based largely on colony morphology, which is often unreliable.

Table 1. Specific references to collections and surveys summarized in Figure 1.

Asia	
India	Rao 1985, Wiebach 1974a, Annandale 1911
Indonesia	Vorstman 1928a, b, 1930
Japan	Toriumi 1941a
Korea	Toriumi 1941b
Taiwan	Toriumi, 1942
Central America	
Costa Rica	Roush 1998
Europe	
Belarus	Mikaevich, in prep.
Germany (Rhine)	Franz 1992
Ireland	Smyth 1994
Israel	Massard & Geimer 1991a, 1994, Massard et al. 1992
Italy	Viganò 1965
Luxembourg	Geimer & Massard 1986
Sweden	Borg, 1936
Tenerife	Massard & Geimer 1990, 1991b
United Kingdom	Mundy 1980, Wood & Okamura, in prep.
North America	
Illinois	Marsh & Wood, in prep.
Kansas	Ellis, in prep.
Massachusetts	Smith 1989b.
Mexico	Rioja 1940a, b, Bushnell 1968
Michigan	Bushnell 1965a, b, c
Louisiana	Everitt 1973.
Ohio	Rogick 1935, Wood 1989
Ontario/ Quebec	Ricciardi & Reiswig 1994.
Pacific NW	Marsh & Wood (this volume)
Pacific	
Australia	Riek 1946, Wood 1998
New Zealand	Wood et al. 1998
South America	
Argentina	Cazzaniga 1989.
Brazil	Wiebach 1967, 1970a, b, 1974b Marcus 1941, 1942 Bonetto & Cordiviola 1965

2.4 *Revision of distribution data*

With the recognition of new species and the re-examination of old ones has come a revision of distribution data. From a scan of the existing literature, Bushnell (1973) concluded that *Fredericella sultana, Plumatella repens*, and *P. emarginata* were "cosmopolitan" species with a worldwide distribution. Now that these species have been split into several others their ranges have diminished considerably. In addition to the species list above, ranges have also been trimmed for *Plumatella javanica* and *Stolella evelinae* (Wood & Wood 2000). Only *Plumatella casmiana* now approaches cosmopolitan status, although it is not yet reported from South America. From these changes comes the growing realization that freshwater bryozoans are not dispersed as rapidly or as easily as we once thought.

For example, nearly a century after *Paludicella articulata* was reported from the Dunedin city waterworks in New Zealand (Hamilton 1902), Wood et al. (1998) found that it was still there and not to be found elsewhere in the country. *Plumatella nitens* still occurs in a narrow band across northern United States, not venturing far into Canada nor migrating below the 41st north parallel (Wood 1996).

3 EXPLAINING DISJUNCT POPULATIONS

Spotty field work across the globe has also revealed in apparently disjunct populations and other odd distribution patterns. *Plumatella reticulata* is abundant in North America from Ohio to Oregon, extending at least as far south as Panama (Wood, unpublished). However, it has also been documented from a single site in Israel (Massard et al. 1992). *Plumatella bushnelli*, is much less common, known only from similar habitats in North Carolina and New Zealand (Wood 2001a). *Asajirella gelatinosa* is reported in Asia from Korea and Japan to the Indian subcontinent, but it also occurs in Panama (Wood & Okamura 1999).

3.1 *Importance of migratory waterfowl*

Bushnell pointed out the likely importance of migrating waterfowl in distributing bryozoans. The possibility of waterfowl to transport viable statoblasts on their feathers or in their gut was first demonstrated by Brown (1933). Microsatellite analysis, revealing gene flow among populations of *Cristatella mucedo* along a major migratory waterfowl route, has strengthened this view (Freeland et al. 2000a). Among the effective agents of such passive dispersal would be those birds breeding in the circumpolar arctic, and subarctic regions which then migrate along flyways that roughly follow continental coastlines. Such birds tend to be waterfowl that congregate in fresh water habitats: Pacific loon, Yellow-billed loon, Black-bellied plover, Sanderling, and others (Elphick 1995, Scott & Rose 1996).

Passive dispersal of statoblasts along these migratory routes offers the best explanation for the distribution of such bryozoan species as *Plumatella mukaii* and *Stephanella hina*. *Plumatella mukaii*, long mistaken for *P. emarginata*, is known from India, Indonesia, and Japan (Wood 2001b); it has also been recently documented in the Western Hemisphere: in Oregon and Chile (Fig. 1). What appears to be a highly disjunct distribution may, in fact, be a continuous range along the major north-south flyways. This would be confirmed if the species were found in a band along the western regions of both North and South America. Unfortunately, most of these areas are completely unexplored in terms of their phylactolaemate species. *Stephanella hina*, was originally known only from Japan and Korea (Oka 1908, Toriumi 1955), but is now known to occur in North America: in Oregon (Marsh & Wood, this volume), New England (Smith 1989a) and Virginia (Wood, unpubl.). Here again, the widely scattered sites occur along flyways of circumpolar breeding birds, suggesting that this bryozoan may actually have a narrow, continuous range through the coastal states. Other bryozoan species benefitting from this avian dispersal may include *Pectinatella magnifica*, a North American species now appearing in Japan and Korea; and *Fredericella indica*, the major fredericellid in North America which is also represented both in Asia and northern Europe. Yet if birds were the primary dispersing agent, one would expect these species to occur also in western Alaska and eastern Siberia where Asian and American flyways converge. To my knowledge, these regions also remain unexplored.

If *Plumatella mukaii* and *Stephanella hina* can be dispersed in this way, should we not expect to find other Asian species as well along coastal North America? Where are *Plumatella bombayensis, P. vorstmani, P. longigemmis*, and *Hyalinella minuta*? It is possible that the arctic climate functions as a filter for many species. Successful dispersal along this route requires thriving populations to be maintained throughout the nesting range of the migratory birds. Only a few species may be adapted to such varied conditions. For some time I have noted that statoblasts of many tropical species tolerate desiccation well, but do not survive cold storage. Statoblasts that remain viable after long refrigeration include those of *Cristatella mucedo, Plumatella fungosa, P. fruticosa,*

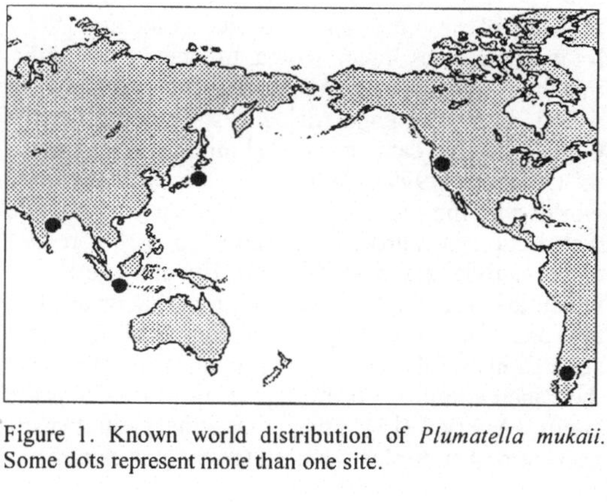

Figure 1. Known world distribution of *Plumatella mukaii*. Some dots represent more than one site.

and *Fredericella sultana*. It therefore comes as no surprise that all of these species have been reported in northern regions of Europe, Asia, and North America. By the same cold storage criterion, we might expect *Fredericella indica* and *Plumatella nitens* to have moved from North America to Asia along avian migration routes as well, but so far no such Asian presence has been reported.

3.2 *Evidence of human influence*

In cases where recent range expansions have been documented, human activity has probably played a major role. *Lophopodella carteri*, described as widespread and abundant in India (Annandale 1911), first appeared in North America around 1930 (Dahlgren 1934) in a shipping canal near Princeton, New Jersey. From there it spread westward across several mountain ranges and major river basins until it now reaches into Oregon (Marsh and Wood, this volume). Masters (1940) cited evidence that the species was being transported with the commercial trade in aquatic plants.

Neck and Fullington (1983) describe how *Pectinatella magnifica*, once known in Texas only from eastern sites, spread across the state as the construction of new reservoirs provided suitable new habitat. The mechanisms for such incremental dispersal cannot be known with certainty, but probably included human activities along with local waterfowl. Dormant statoblasts of any species may be carried from lake to lake on recreational boats and equipment.

Pectinatella magnifica is a common North American species, but within the past three decades it has moved into Japan (Oda 1974) and Korea (Dongsoo Kong, pers. com.). While this is not normally a cold tolerant species, it is conceivable that migratory waterfowl were again responsible. However, it is also very likely that statoblasts were transported with aquatic plants, fish, or other commercial items.

Asajirella gelatinosa is a species whose distribution appears to be very disjunct. It is a difficult species to miss, forming loose, gelatinous colonies more than 20 cm long, and producing the largest statoblasts of any bryozoan species (Lacourt 1968). Its range includes Japan, Korea, Indonesia, and India, but the species also occurs in the Panama Canal as well as in nearby Lago Alajuela, separated from the Canal by a high dam and several miles of river (Wood & Okamura 1999). With no other sites known worldwide, the most likely explanation for this odd distribution is human intervention, possibly related to the activities surrounding the Panama Canal into which the lake drains. Although most labor to build the canal came from the West Indies, over one thousand Chinese laborers were brought in to help build the Panama Railroad from Aspinwall (now Colón) to Gatun (Avery 1915). These immigrants could easily have carried statoblasts with them in their clothing or other articles washed in lake water. If this were the case, one could also expect to find *Asajirella* in California, where over 270,000 Chinese immigrants were registered during 1850-1880 (Chiu, 1960). Unfortunately, there are no published records of phylactolaemate bryozoans anywhere in California. There is also the possibility that shipping traffic through the Canal has introduced statoblasts, just as ships have carried other freshwater species in bilge water to North America from the Ponto-Caspian basin (Ricciardi & Rasmussen, 1999). The salt tolerance of dormant bryozoan statoblasts is likely to be high, and several major Asian ports are situated at the mouth of large rivers. On the other hand, no other Asian bryozoan species are known in Panama. In any case, it is unlikely that *A. gelatinosa* arrived from arctic nesting grounds of migratory waterfowl. Although its cold tolerance has never been tested, evidence so far suggests this is strictly a warm water species.

3.3 *Unexplained distributions*

It is less clear how to explain the widely scattered distribution of the distinctive *Plumatella vaihiriae*. Originally described from a mountain pond in Tahiti (Hastings 1929), the species has since been reported from Argentina (Cazzaniga 1988), Hawaii (Baily-Brock & Hayward 1984), Utah (Rogick, & Brown 1942), Arizona, Wisconsin, South Carolina (Wood & Marsh 1999), and North Carolina (Wood, unpubl.). That all these sites are linked by avian flyways is unquestioned: the east Asian flyway extends across the Pacific through Tahiti and Hawaii to western North America (Miyabayashi & Mundkur, 1999). Rather, it is the apparent scarcity of this species that is so striking. Most of these sites were quite small and isolated: three were wastewater treatment plants, one was a small stream, another was a pond for the culture of freshwater prawns. Such minor sites are often frequented by ducks and other waterfowl, although generally not as stopovers in a long migration route. The only known large population of *Plumatella vaihiriae* is in the highly productive coastal waters of Pamlico Sound, North Carolina. How can we account for this odd distribution in a few very small, isolated sites? Two points are relevant: First, wherever *P. vaihiriae* occurs it dominates, with very rapid growth and an unusually high production of statoblasts. In at least one wastewater plant the bryozoan biomass is regularly harvested and carried away by the truckload (Wood & Marsh 1999). Secondly, what all known sites have in common is a highly eutrophic habitat in which the species thrives. It seems possible that enormous growth and reproductive potential of *P. vaihiriae* in a narrowly defined habitat compensates for the relative scarcity of natural populations. Nevertheless, this apparent scarcity is surely also due to limited recognition in the field. Not a single thorough search has been conducted in the vicinity of any of the known sites.

4 SUMMARY

Advances in phylactolaemate systematics and steady progress in regional surveys have provided valuable new information about the zoogeography of freshwater bryozoans. Since Bushnell=s (1973) paper the number of species worldwide has nearly doubled. Many of those species are known only from highly restricted sites, while a few exhibit wide ranges that parallel seasonal migration routes of circumpolar breeding birds. We need stronger field survey efforts in these areas to more fully understand these patterns. The nesting regions of the migratory waterfowl would be the obvious places to begin. An inventory of freshwater bryozoans along the western coasts of North and South America would also be very useful.

In addition to traditional morphological taxonomy, we are beginning to apply molecular tools to taxonomic and systematic problems among phylactolaemates. A phylogenetic tree at the family level has already been constructed using data from 18S ribosomal DNA (Lore, in prep.). We are next planning to examine genetic differences among species using ITS-1 and perhaps other regions of rDNA. Among other things, this is expected to help unravel certain taxonomic dilemmas among plumatellid and fredericellid bryozoans. At the same time, microsatellite data will continue to be useful for tracking the historical movements of subpopulations (Freeland et al. 2000a).

While some progress is being made in inventorying bryozoans in temperate regions, the phylactolaemate fauna of Asia, Africa, and South America is still practically unknown. The seasonally flooded Amazon forests, for example, have not been touched. Existing data suggest that tropical species tend to have relatively small geographic distribution, yet these are the very areas about which we know the least.

Finally, we are living in a time when documented freshwater extinctions in North America rival those of the forested tropics (Ricciardi and Rasmussen 1999). Similar but unseen species losses may be occurring in freshwater habitats elsewhere. Phylactolaemates comprise the single major group of suspension feeders in fresh water, yet the number of people studying any aspect of these animals could easily meet around a small table. Among other things, we need to be asking why some phylactolaemate species are more easily dispersed than others. Do they have more specialized habitat requirements? Are their statoblasts less tolerant of desiccation, freezing, or long dormancy? Do endemic species warrant special vigilance or protection as freshwater habitats are altered? How are bryozoans being affected by competition with invading zebra mussels and other foreign species? Answers to these questions begin with good field work followed by simple and inexpensive experimental studies. It is not too late to begin.

REFERENCES

Annandale, N. 1911. Freshwater sponges, hydroids, and Polyzoa. *Fauna of British India*. London. 3 (Bryozoans): 161-251.

Avery, R. 1915. *The Greatest Engineering Feat in the World at Panama*. New York: Leslie-Judge Co.

Baily-Brock, J. & Hayward, P. 1984. A freshwater bryozoan, *Hyalinella vaihiriae* Hastings (1929), from Hawaiian prawn ponds. *Pacific Science* 38(3): 199-204.

Bonnetto, A. & Cordiviola de Yuan. 1965. Notas sobre briozoos (Endoprocta y Ectoprocta) del Rio Parana. III. *Fredericella sultana* (Blumenbach) en el Parana Medio. *Physis* 25(70): 255-262.

Borg, F. 1936. Über die Süsswaserbryozoen Schwedens. In *Festskrift tillägnad Prof. Sven Ekman. Zool. Bidr. Uppsala* 20: 479-494.

Brown, C. 1933. A limnological study of certain fresh-water Polyzoa with special reference to their statoblasts. *Transactions of the American Microscopical Society* 52: 271-313.

Bushnell, J. 1965a. On the taxonomy and distribution of freshwater Ectoprocta in Michigan. Part I. *Transactions of the American Microscopical Society* 84: 231-244.

Bushnell, J. 1965b. On the taxonomy and distribution of freshwater Ectoprocta in Michigan. Part II. *Transactions of the American Microscopical Society* 84: 339-358.

Bushnell, J. 1965c. On the taxonomy and distribution of freshwater Ectoprocta in Michigan. Part III. *Transactions of the American Microscopical Society* 84: 529-548.

Bushnell, J. 1968. Aspects of architecture, ecology, and zoogeography of freshwater Ectoprocta. *Atti Della Società Italiana Di Scienze Naturali E Del Museo Civico Di Storia Naturale Di Milano* 108: 129-151.

Bushnell, J. 1973. The freshwater Ectoprocta: a zoogeographical discussion. In G.P Larwood (ed.), *Living and Fossil Bryozoa: Recent Advances in Research*. London: Academic Press.

Cazzaniga, N. 1988. *Hyalinella vaihiriae* (Ectoprocta Phylactolaemata) en la Provincia de San Juan Argentina). *Revista de la Asociacion de Ciencias Naturales del Litoral* 19(2): 205-208

Cazzaniga, N. 1989. Registro des tres especies de Ectoprocta Phylactolaemata en el sur de la Provincia de Buenos Aires (Argentina). *Boletin de la Sociedad de Biologia de Concepción, Chile* 60: 43-49

Chiu, P. 1960. *Chinese Labor in California 1850-1880: An Economic Study*. Ph.D. Thesis. Madison: Univ. Wisconsin.

Cordiviola de Yuan, E. 1977. Notas sobre briozoos del Rio Parana. V. *Plumatella emarginata* Allman (Ectoprocta, Phylactolaemata). *Neotropica* 23(69): 3-6.

Dahlgren, U. 1934. A species and genus of freshwater bryozoan new to North America. *Science* 7 (2057): 510.

Elphick, J. 1995. *The Atlas of Bird Migration*. New York: Random House.

Everitt, B. 1975. Fresh-water Ectoprocta: distribution and ecology of five species in southeastern Louisiana. *Transactions of the American Microscopical Society* 94: 130-134.

Franz, H. 1992. *Der Rhein und seine Beseidlung im Wandel: Schwebstoffzehrende Organismen (Hydrozoa, Kamptozoa und Bryozoa) als Indikatoren fur den ökologischen Zustand eines Gewässers.* Bad Dürkheim, Germany: Pollichia-Buch No. 25,

Freeland, J., Noble, L. & Okamura, B. 2000a. Genetic consequences of the metapopulation biology of a facultatively sexual freshwater invertebrate. *Journal of Evolutionary Biology* 3(2000): 383-395.

Freeland, J., Romualdi, C. & Okamura, B. 2000b. Gene flow and genetic diversity: a comparison of freshwater bryozoan populations in Europe and North America. *Heredity* 85(5): 498-508.

Geimer, G. & Massard, J. 1986. *Les bryozoaires du Grand-Duché de Luxembourg et des régions limitrophes.* Luxembourg: Musée de l'Histoire Naturel Marché-aux-Poissons.

Hamilton, A. 1902. On the occurrence of *Paludicella* in New Zealand. *Transactions and Proceedings of the New Zealand Institute* 12: 301-303.

Hastings, A. 1929. Notes on some little-known phylactolaematous Polyzoa and description of a new species from Tahiti. *Annals of the Magazine of Natural History* 10(3): 300-311.

Lacourt, A. 1968. A monograph of the freshwater Bryozoa - Phylactolaemata. *Zoologische Verhandelingen* 93:1-159.

Marcus, E. 1941. Sôbre Bryozoa do Brasil. *Boletin da Faculdade de Filosofia, Ciencias, e Letras, Universidade de São Paulo, Zoologia* 5: 3-208.

Marcus, E. 1942. Sôbre Bryozoa do Brasil II. *Boletin da Faculdade de Filosofia, Ciencias, e Letras, Universidade de São Paulo, Zoologia* 6: 57-96.

Marsh, T. & Wood, T.S. (this volume).

Massard, J.A. & Geimer, G. 1990. Note on the freshwater Bryozoa (Ectoprocta, Phylactolaemata) of Tenerife. *Vieraea,* 19: 327-338.

Massard, J.A. & Geimer, G. 1991a. Note on the freshwater Bryozoa of Israel (Phylactolaemata). In F. Bigey (ed.) & J.-L. d'Hondt (collab.): Bryozoaires actuels et fossiles: Bryozoa living and fossil. *Bulletin De La Société Des Sciences Naturelles De L'Ouest De La France, Mémoire. HS* 1: 243-253, Nantes.

Massard, J.A. & Geimer, G. 1991b. Additional note on the freshwater Bryozoa of Tenerife (Phylactolaemata). *Bulletin de la Société des Naturalistes Luxembourgeois* 92 (1991): 149-157.

Massard, J.A. & Geimer, G. 1994. Distribution of freshwater and brackish-water Bryozoa (Phylactolaemata, Gymnolaemata) in Israel. In P.J. Hayward, J.S. Ryland & P.D. Taylor: *Biology and Palaeobiology of Bryozoans. Proceedings of the 9th International Bryozoology Conference, School of Biological Sciences, University of Wales, Swansea, 1992*: 117. Fredensborg: Olsen & Olsen.

Massard, J., Geimer, G., Bromley, H. & Dimentman, C. 1992. Additional note on the fresh and brackish water Bryozoa of Israel (Phylactolaemata, Gymnolaemata). *Bulletin de la Société des Naturalistes Luxembourgeois* 93: 199-214.

Masters, C. 1940. Notes on subtropical plants and animals in Ohio. *Ohio Journal of Science* 40: 147-148.

Miyabayashi, Y. & Mundkur, T. 1999. *Atlas of Key Sites for Anatidae in the East Asian Flyway.* Selangor, Malaysia: Wetlands International - Asia Pacific.

Mundy, S. 1980. *A key to the British and European Freshwater Bryozoans.* London: Freshwater Biological Association, Scientific Publication. No. 41.

Neck, R. & Fullington, R. 1983. New records of the freshwater ectoproct *Pectinatella magnifica* in eastern Texas. *Texas Journal of Science* 35(3): 269-271.

Oda, S. 1974. *Pectinatella magnifica* occurring in Lake Shoji, Japan. *Proceedings of the Japanese Society of Systematic Zoology* 10: 31-39.

Oka, A. 1908. Ueber eine neue Gattung von Süsswasserbryozoen von Japan. *Annotationes Zoologicae Japonenses* 6(2): 117-123.

Økland, K. & Økland, J. 2001. Freshwater bryozoans Bryozoa) of Norway II. Distribution and ecology of two species of *Fredericella. Hydrobiology* (in press).

Rao, K. 1985. Studies on freshwater Bryozoa V. Observations on Central Indian materials. In C. Neilsen & G. Larwood (eds.) *Bryozoa: Ordovician to Recent.* Fredensbork, Denmark: Olsen & Olsen.

Ricciardi, A. & Rassmussen, J. 1999. Extinction rates of North American freshwater fauna. *Conservation Biology* 13(5): 1220-1222.

Ricciardi, A & Reiswig, H. 1994. Taxonomy, distribution, and ecology of the freshwater bryozoans (Ectoprocta) of eastern Canada. *Canadian Journal of Zoology* 72: 339-359.

Riek, E. 1946. *Studies on the freshwater bryozoans of Queensland.* M.Sc. Thesis. Braisbane: University of Queensland.

Rioja, E. 1940a. Esponjas, Hidrozoarios y briozoos del Lago de Patzcuaro. *Anales Del Instituto De Biología, Universidad Nacional Autónoma De México* 11(2): 443-448.

Rioja, E. 1940b. Contribucion al conocimiento de los briozoarios del Lago de Xochimilco. *Anales Del Instituto De Biologia de Mexico* 11(2): 585-592.

Rogick, M. 1935. Studies on freshwater Bryozoa. II. The Bryozoa of Lake Erie. *Transactions of the American Microscopical Society* 54(3): 245-263.

Rogick, M. 1943. Studies on freshwater Bryozoa. XIV. The occurrence of *Stolella indica* in North America. *Annals of the New York Academy of Science* 45(4): 163-178.

Rogick, M. & Brown, C. 1942. Studies on freshwater bryozoa. XII. A collection from various sources. *Annals of the New York Academy of Science* 43(3): 123-144.

Roush. S. 1998. *A survey of the phylactolaemate bryozoans (Ectoprocta) of the Guanacaste Conservation Area, Costa Rica.* M.S. Thesis. Dayton, Ohio: Wright State University.

Scott, D. & Rose, P. 1996. *Atlas of Anatidae populations in Africa and Western Eurasia.* Netherlands: Wetlands International.

Smith, D. 1989a. On *Stephanella hina* Oka (Ectoprocta: Phylactolaemata) in North America, with notes on its morphology and systematics. *Journal of the North American Benthological Society* 7: 253-259.

Smith, D. 1989b. *Keys to the freshwater macroinvertebrates of Massachusetts. No. 4: Benthic colonial phyla, including the Cnidaria, Entoprocta, and Ectoprocta (colonial hydroids, moss animals).* Westborough, Massachusetts: Mass. Div. Water Pollution Control.

Smyth, T. 1994. The distribution of freshwater Bryozoa in Ireland. *Bulletin of the Irish Biogeographic Society* 17: 9-21.

Toriumi, M. 1941a. Studies on freshwater Bryozoa of Japan. I. *Science Reports of Tôhoku Imperial Univeristy* 16(2): 193-215.

Toriumi, M. 1941b. Studies on freshwater Bryozoa of Japan. II. Freshwater bryozoa of Tyôsen (Korea). *Science Reports of Tôhoku Imperial Univeristy* 16(4): 413-425.

Toriumi, M. 1942. Studies on freshwater Bryozoa of Japan. IV. Freshwater bryozoa of Taiwan (Formosa). *Science Reports of Tôhoku Imperial Univeristy* 17(2): 207-214.

Toriumi, M. 1955. Taxonomical study on fresh-water Bryozoa XI. *Stephanella hina* Oka. *Science Reports of Tôhoku University.* Ser. 4, 21(2):131-136.

Viganò, A. 1965. Nuovi dati sui Briozoi dell acque interne italiane. *Bollettinodi Zoologia* 32(2): 913-928.

Vorstman, A. 1928a. Some fresh-water Bryozoa of West Java. *Treubia* 10:1-13.

Vorstman, A. 1928b. Freshwater Bryozoa from East Java. *Treubia* 10: 163-165.

Vorstman, A. 1930. Bryozoen aus Java und Sumatra. *Archiv für Hydrobiologie Supplement-Band* 8: 109-110.
Wiebach, F. 1967. Amazonische Moostiere (Bryozoa). *Amazoniana* 1(2): 173-187.
Wiebach, F. 1970a. Amazonische Moostiere (Bryozoa) II. *Amazoniana* 2(3): 353-362.
Wiebach, F. 1970b. Süsswasser-Bryozoen aus Brasilien und Zentralafrika. *Revue de Zoologie et de Botanique Africaines* 81(1-2): 62-81.
Wiebach, F. 1974a. Indische Süsswasser-Bryozoen. *Gewässer und Abwässer* 53/54: 69-84.
Wiebach, F. 1974b. Amazonische Moostiere (Bryozoa) III. *Amazoniana* 5(2): 293-303.
Wood, T.S. 1988. *Plumatella reticulata* sp. nov. in Ohio. *Ohio Journal of Science* 88(3): 101-104.
Wood, T.S. 1989. Ectoproct bryozoans of Ohio. *Ohio Biological Survey Bulletin* 8(2)NS: 1-70.
Wood, T.S. 1996. *Plumatella nitens*, a new species of freshwater bryozoan from North America (Ectoprocta: Phylactolaemata), previously misidentified. *Hydrobiologia* 328: 147-153.
Wood, T.S. 1998. Reappraisal of Australian freshwater bryozoans with two new species of *Plumatella* (Ectoprocta: Phylactolaemata). *Invertebrate Taxonomy* 12: 257-272.
Wood, T. 2001a. Three new species of plumatellid bryozoans (Ectoprocta: Phylactolaemata) defined by statoblast nodules. *Journal of the North American Benthological Society* 20(1): 133-143.
Wood, T. 2001b. *Plumatella mukaii*: a new plumatellid species from Asia and South America (Ectoprocta: Phylactolaemata). *Hydrobiologia* 445: 51-56.
Wood, T. & Marsh, T. 1999. Biofouling of waste water treatment plants by the freshwater bryozoan, *Plumatella vaihiriae* (Hastings, 1929). *Water Research* 33(3): 609-614.
Wood, T. & Okamura, B. 1999. *Asajirella gelatinosa* in Panama: a bryozoan range extension in the Western Hemisphere (Ectoprocta: Phylactolaemata. *Hydrobiologia* 390: 19-23.
Wood, T. & Wood, L. 2000. Statoblast morphology in historical specimens of phylactolaemate bryozoans. In A. Herrera Cubilla and J.B.C. Jackson (eds). *Proceedings of the 11th International Bryozoology Association Conference, 26-31 January 1998, Balboa, Panama.* Balboa, Panama: Smithsonian Tropical Research Institute.
Wood, T., Wood, L., Geimer, G. & Massard, J. 1998. Freshwater bryozoans of New Zealand: a preliminary survey. *New Zealand Journal of Marine and Freshwater Research* 32: 639-648.

Bryozoan Studies 2001, Wyse Jackson, Buttler & Spencer Jones (eds)
© 2002 Swets & Zeitlinger, Lisse, ISBN 90 5809 388 3

The reproductive cycle of *Plumatella casmiana* (Phylactolaemata: Plumatellidae)

Emmy R. Wöss
Department of Limnology, Institute of Ecology and Conservation Biology, University of Vienna, Althanstrasse 14, A-1090 Vienna, Austria

ABSTRACT: Freshwater bryozoans are unique in their reproduction in that they produce a great variety of asexual propagules; among this group, *Plumatella casmiana* has developed the highest degree of complexity. It produces three types of statoblasts, two of them with heavily chitinised valves – the sessoblasts and floatoblasts – while the leptoblasts show a very weak degree of chitinisation. Sessoblasts are fixed to the substrate by an attachment apparatus, whereas floatoblasts and leptoblasts develop a swim-ring that enables them to drift in the water. While sessoblasts and floatoblasts undergo an obligate dormancy, the thin-walled leptoblasts as well as the sexually produced larvae develop further within a few hours. A study focussing on the role of the different kinds of propagules in *P. casmiana* was carried out in the Laxenburg pond in Eastern Austria (48°4' N, 16°22' E) where this species was found to be the dominant bryozoan over a period of several years. From May to September, colonies were collected weekly from artificial substrates that had been exposed in this waterbody for one year. The colonies were narcotised, preserved, stained and examined by light microscopy for their gonadal condition and their production of sessoblasts, floatoblasts and leptoblasts. Results on the investment of *P. casmiana* in its different modes of propagation show that although testes and ovaria are present from late May to late August, the development of fertilized eggs in the embryosacs was rarely detected. Asexual reproduction and in particular the propagation by leptoblasts seems to play the major role in the reproductive cycle throughout the year.

1 INTRODUCTION

In temperate zones the occurrence of freshwater bryozoans is limited to a certain period of the year because freshwater bryozoans generally do not overwinter in these regions (Wöss 1996). To survive unfavorable conditions, a variety of propagation units are formed, all of which are of asexual origin. This is necessary because the sexual propagule, the larva, can only exist for a shorter period (Braem 1890). The asexually produced resting stages, however, such as the statoblasts in the case of the Phylactolaemates and the hibernaculae in the Gymnolaemates, consist of a chitinised shell that protects the germinal mass against desiccation or severe temperatures (Hyman 1959). The statoblasts are unique to Phylactolaemates and are divided in two groups. The first are the floatoblasts which possess a gas-filled annulus that facilitates dispersal and colonisation of new substrates. The second group consists of the sessoblasts and piptoblasts, which both lack the capability to float on the water surface. In the case of the sessoblasts, the part of the shell that is directed to the substrate possesses an attachment apparatus (Mukai 1982). This fixes the resting stage to the substrate and thus promotes local persistence of the bryozoan (Karlson 1994). In addition, within the group of floatoblasts special forms have developed, such as the leptoblasts (Wiebach 1963), which can be distinguished by the low degree of chitinisation of their shell. They are fragile, germinate immediately after release from the colony (Vigano 1968, Wood 1973, 1989) and thus can hardly act as overwintering stages, in contrast to all other types of statoblasts (Wöss 1996). In a number of genera it is quite common that a particular species produces more than one type of statoblast. *Plumatella casmiana*, a widespread species, reaches the highest complexity within the group of Phylactolaemates: sessoblasts, floatoblasts and leptoblasts are present (Rogick 1941, 1943, Toriumi 1955, Mukai et al. 1983). According to Wiebach (1963) a third type of floating propagule is produced. It has been designated an intermediate form between a normal floatoblast and leptoblast.

The Laxenburg pond has a rich and diverse bryozoan fauna and therefore served as a permanent study site within the last decade (Eder & Wöss 2000,

Wöss 1994, 1996, 2000). Experimental set-ups with artificial substrates were installed to study the colonization and development of freshwater bryozoan communities. For the present study, colonies of *P. casmiana* were removed weekly from the panels to examine the reproductive status of the zooids. The goal was to determine the investment in the different kinds of asexual and sexual propagules throughout the year in this species.

2 MATERIAL AND METHODS

In May 1995, 750 artificial substrates (15x15 cm, plexiglas) linked together in vertical arrangements of five substrates each were installed at three rafts in the Laxenburg Pond in Lower Austria (Wöss 2000, Eder & Wöss, 2000). One part of the substrates was monitored in the field throughout the exposure period, which ended in late October 1997. From the other part, in all three years panel arrangements were chosen randomly to be removed weekly to biweekly for more detailed laboratory examination of the bryozoan aufwuchs, e.g, to determine the origin of the colonies. This study especially focuses on the *P. casmiana* colonies collected from substrates in the second year of exposure. The goal was to collect 8 colonies of *P. casmiana* per week and to make whole mounts of a minimum of 10 zooids per colony. *P. casmiana* colonies were narcotized on the plexiglas panels with chloral hydrate and finally 1% neutralized formaldehyde was added.

Table 1: Parameters used to describe the reproductive status of a zooid:

Asexual status	Sexual status
Statoblasts in formation	Testis early stage
Leptoblasts weakly chitinized	Testis late stage
Leptoblasts heavily chitinized	Ovarium
Leptoblasts released	Oecium early stage
Floatoblasts weakly chitinized	Oecium late stage
Floatoblasts heavily chitinized	
Floatoblasts released	
Sessoblasts chitinized	

Leptoblasts (floatoblasts) weakly chitinized: initial chitinization is only present in the most distal link of the statoblast arrangement
Leptoblasts (floatoblasts) heavily chitinized: chitinization is present in more than one statoblast or (in the case of the funiculus or parts of it are demolished) one statoblast with thick cuticula is found
Leptoblasts (floatoblasts) released: at least one statoblast gets detached from the funiculus
Testis early stage: presence of spermatogonia
Testis late stage: presence of spermatocyta or later stages
Oecium early stage: size of oecium is not larger than the ovarium
Oecium late stage: size of oecium is larger than the ovarium

Then, colonies having a minimum size of about 0.25 cm^2 or parts of larger colonies could be removed from the panels and transferred into vials containing 4% neutralized formaldehyde. The samples were stained with borax carmine (Grenacher 1879). Zooid preparation took place after embedding on object slides in the epoxy resin Araldite M. The zooids were dissected from the zooaria and positioned on their sides on the slides; this helped ensure that the ovaria and ooecia (embryosacs) were not masked by the polypid.

The analysis of the reproductive condition of a colony involved recording the presence/absence of the characteristics listed in Table 1. If a character was present, 50 points were assigned; the number assigned to a census is the result of the total number of points for a given character weighted by the number of colonies it was found in.

In whole mounts it is much more difficult to detect an early stage of an ovarium than of a testis; therefore, a two part scale was not used for the ovarium.

3 RESULTS AND DISCUSSION

Plumatella casmiana, although in general not a common bryozoan species in Austria, was clearly the dominant species on natural as well as on artificial substrates from 1991 until 1995 in the Laxenburg pond (Eder & Wöss 2000, Wöss 2000).

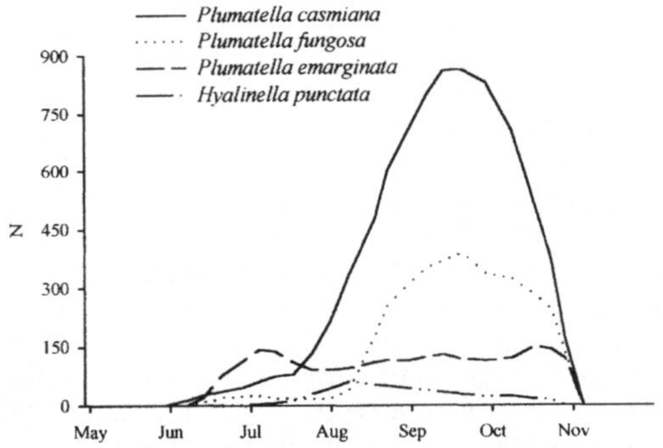

Figure 1: Colony settlement on artificial substrates in the Laxenburg pond: Number of colonies of the four species on 90 panels from 23 censi, May 3(- Nov. 6, 1995

Figure 1 shows the number of colonies of the four most important species on artificial substrates of those weekly monitored in 1995. Forty-nine percent of all of the colonies belonged to *Plumatella casmiana*. In general, increased settlement of colonies was observed from early August on, with *P. casmiana* being twice as abundant as the second most common species, *Plumatella fungosa*.

Initial settlement of the panels in June and July 1995 involved mainly leptoblasts and floatoblasts. In Figure 2 recruitment data refer only to living colonies; additionally, from the beginning of July until mid-August between 4 and 120 empty and opened leptoblast shells were found per sampling date on the substrates with no further remnants of hatched zooids and thus were not counted.

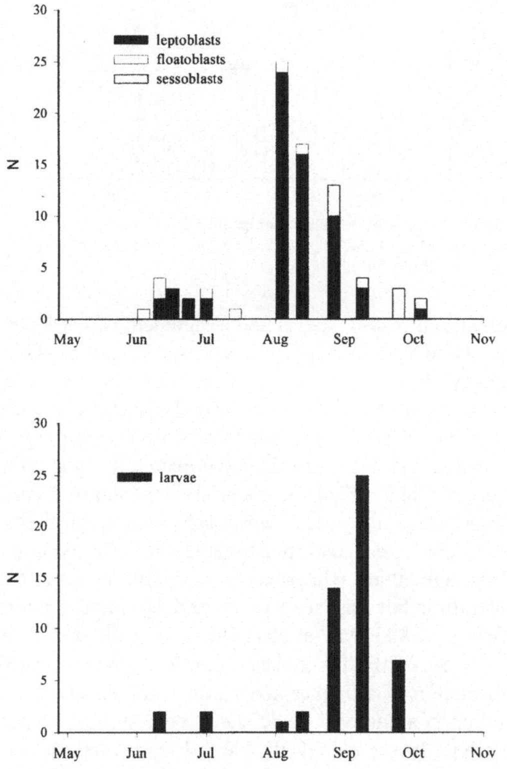

Figure 2: Colony recruitment of *Plumatella casmiana* in 1995: number of colonies of known origin between June 07 and October 20; 10 panels examined per census (14 censi in total)

Vast numbers of colonies hatching simultaneously from leptoblasts are documented in the Laxenburg pond in previous years (Wöss 2000) and were also reported by several authors during early summer, e.g., Mukai et al. (1983) describes numerous settlement of leptoblast-colonies in the end of May and June.

In late August and early September recruitment by larvae became larger in number than from asexual propagules. Colonies hatched from sessoblasts occurred not earlier than August 31 and achieved a proportion of 6.1 % throughout the year (Table 2). As with the floatoblasts they played a minor role, while about half of the settlement originated from leptoblasts, followed by larvae (40.5 %) in the first year.

Table 2: Colony recruitment of *Plumatella casmiana* in 1995: 14 censi with 10 panels at each census

	N	%
Leptoblasts	63	48.0
Floatoblasts	7	5.3
Sessoblasts	8	6.1
Larvae	53	40.5

The panels were heavily covered by sessoblasts in the very beginning of the second year which resulted in a vast simultaneus hatching of colonies in spring from sessile resting stages and led to a coverage of up to 100 % after a few weeks. *P. fungosa* and *P. emarginata* colonies were reported from mid May on in higher numbers, while *P. casmiana* started to germinate with the delay of one week. In contrast to former investigation periods, from 1996 on, *P. casmiana* was found in decreasing numbers on the artificial substrates and only 9.9 % of the colonies could be assigned to that species in the annual record of that year. Although from end of May until the end of June *P. casmiana* colonies still hatched, the beginning of July was marked by a dramatic change in abundance leaving only a few colonies behind (Fig. 3).

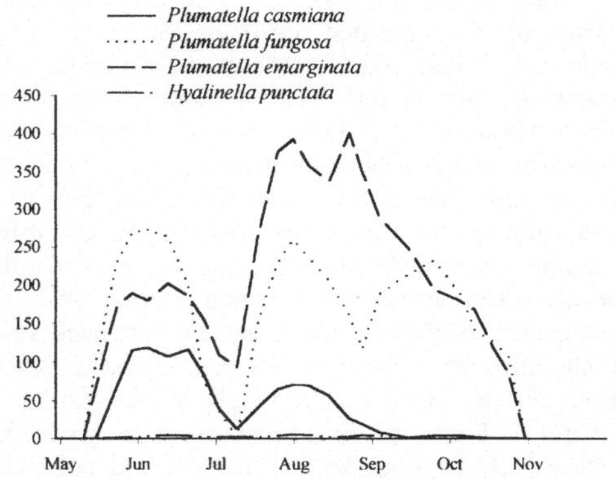

Figure 3: Colony settlement on artificial substrates in the Laxenburg pond: Number of colonies of the four species on 30 panels from 25 censi, May 11- Nov. 1, 1996

A similar dying of colonies on natural as well as artificial substrates was reported by Mukai et al. (1983) from mid-July to the end of August in a pond in Japan. In contrast, in the Laxenburg pond, in mid-August another smaller peak in colony abundance was observed. However, from early September on, *P. casmiana* gradually disappeared from the panels, which was two months earlier than in previous years (Wöss 2000). The comparison of the two years of colony settlement showed that *P. casmiana* proved to be excellent in colonizing new substrates but seemed to be competitively inferior towards other bryozoan species.

The fluctuation and final breakdown of colonies also had consequences on the sampling programme dealing with the examination of the reproductive status of the zooids. In total, in 15 censi 90 colonies were removed from the end of May to the beginning of September (Table 3). However, due to the bad condition of the colonies the last two censi could not be analysed. In total a number of about 3500 zooids were investigated which meant an average of 40 zooids per colonies. As the reproductive status of single zooids within a colony turned out to be quite variable, it was essential to examine a larger number of zooids per colony.

Table 3: Number of *P. casmiana* colonies collected from panels in 1996 for the determination of the reproductive status

Date	N	Date	N	Date	N
May 24	1	Jul.03	2	Aug.08	9
May 30	9	Jul.11	8	Aug.16	2
Jun.05	10	Jul.18	6	Aug.22	5
Jun.20	13	Jul.26	9	Aug.28	1
Jun.26	4	Aug.02	10	Sep.06	1

P. casmiana hatching from sessoblasts was first recorded on May 16 on the panels (Fig. 3). In the following week (May 24), the collection of colonies started and from the first census on formation of statoblasts was observed. Formation and chitinization of sessoblasts was slightly prior to that of the floatoblasts. Mukai et al. (1983), however, suggested that there was no inherent or intrinsic regularity in the order of production of the different types of statoblasts by a colony, as he found sessoblasts produced prior, simultaneously and after the apperance of floatoblasts. In the process of removing colonies from the panels parts of the sessoblasts usually remained stuck to the substrate, therefore the sessile resting stages were not described quantitatively in the results. In general, colonies in the Laxenburg pond proceeded very rapidly to sessoblast formation and continued doing so throughout the season until September with the exception of a decrease in production in midsummer. Wood (1973) described the appearance of sessoblasts of *P.* casmiana at the Sawhill pond as rare and sporadic, but also most frequent during the spring and autumn months.

Chitinization of leptoblasts was observed first one week later (May 30) with 100% of the colonies showing weakly and 44% heavily chitinized shells (Fig. 4). In June the degree of heavy chitinization reached up to 100%. Seventy-three percent of the colonies had leptoblasts that detached from the funiculus to leave the colony through the vestibular pore (Mukai 1982). The tubes then filled with several

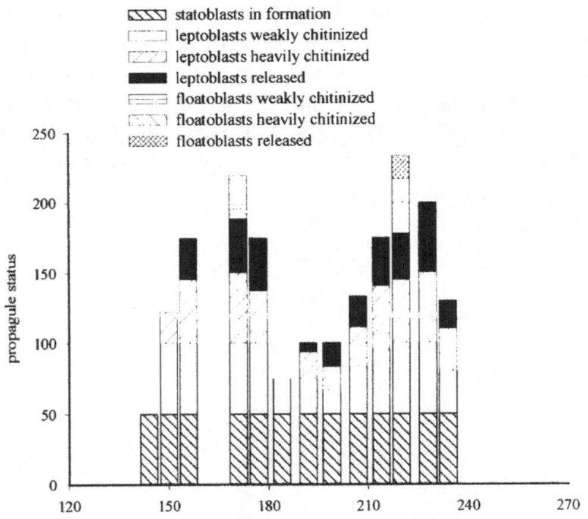

Figure 4: status of the asexual propagules in 88 colonies, collected between May 24 and August 22, 1996 (13 censi); maximum point per character = 50

chitinized resting stages and assignment of the leptoblasts to a certain polypid was no longer possible. Often, shells that were split were found outside adjacent to bryozoan zooaria. Additionally, a smaller number of leptoblasts that failed to be released germinated and decayed within the maternal cystide. On June 20, in 31 % of the material formation of floatoblasts was observed. Leptoblasts and floatoblasts, both chitinized, occurred together in the same colonies, something which has been controversial in the literature but has been confirmed by some authors (Rogick 1943, Mukai et al.1983). The difference, in the case of the floatoblasts, consisted merely in the formation of a thicker and harder cuticula which resulted in a darker capsule and a more spindle-shaped outline. In the ellipitically-shaped leptoblasts the enclosed gerninal mass was discernible through the translucent shell. Except for this census no further floatoblasts could be detected in spring.

In July the remaining colonies still showed a weaker degree of sclerotization of the leptoblast (average value: 50% weak and 32% heavy chitinization, 23% releasing leptoblasts). In August, however, heavily chitinized leptoblasts were even more dominant than in June reaching a degree up to 100%. Additionally, on August 8, in more than half of the colonies floatoblasts were observed a second time in this season, a third of them ready for release.

According to several authors these different types of floating propagules are produced at different times of the year. For Rogick (1943) the "thin-walled type" (leptoblast) represents an early summer form, while the "capsuled type" (floatoblast) is more abundant later in the season, in August especially. Toriumi's

(1955) results demonstrate the occurrence of leptoblasts from May to October with highest abundance in August; floatoblasts, however, are found from May to December and most abundant in October. For Mukai et al. (1983) the formation of thick-walled floatoblasts occur towards the terminal phase of leptoblast production, though occasionally the two types are formed simultaneously in the same colony.

Sexual activity of colonies is described as ripening of the gonads. Except for the first census, testes and ovaria were present in the cystids throughout the year (Fig. 5). This would be in accordance with sexual reproduction taking place in former years (Wöss 1996, 2000), where colonies with embryos, and, as mentioned above, larval recruitment were observed.

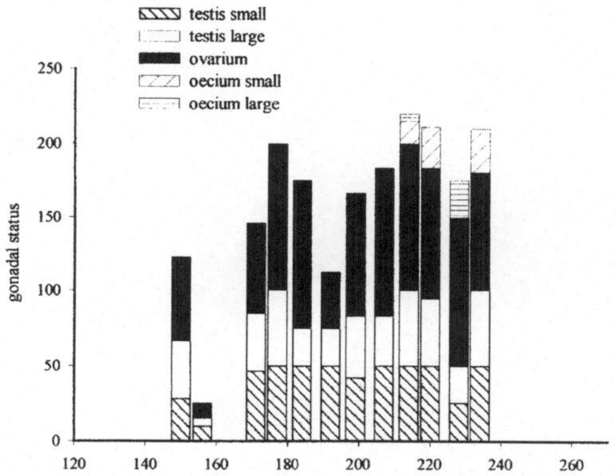

Figure 5: status of the gonads in 88 colonies, collected between May 24 and August 22, 1996 (13 censi); maximum point per character = 50 (testis, oecium) or 100 (ovarium)

However, several authors refer to a strictly limited period of sexual propagation (Kraeplin 1887, Braem 1890, Hyman 1959, Mukai 1982, Mukai et al.1983) with the possibility of total omission (Wesenberg-Lund 1939, Wood, 1973). In detail, Rogick (1943) describes as a rarity three ciliated larvae found in zooecial tubes of colonies of *P. casmiana* in July, as well as Mukai et al. (1983) mentions a small number of larvae in June and early July. In the Laxenburg pond, in 1996 from the end of May until the beginning of June in 20% to 55% of the colonies early stages of testes were detected and a small percentage of mature testes (10%-20%). From mid June on, early stages were observed seven times in 100% of the colonies, three times they ranged between 50% and 92%. Large testes were found most frequently in late June and most of August (100%) and, even in July, never fell below 50 %. In absence of statoblasts, the funiculus became almost completely covered with testes (Mukai 1982). Ovaria were common as well. They were least abundant in early June (11%) and mid July (38%) but found in up to 100% of colonies from late July on. It definitely can be excluded that *P. casmiana* is a protandric hermaphrodite (Braem, 1897), as testes and ovaria were produced simultaneously.

In striking contrast to the frequent gamete production throughout the year, formation of oecia was quite rare and restricted to the last four censi. In August, 33% to 60 % of the colonies showed embryosacs, all but 16% of small size. In general, within a colony the sexual status of the zooids was even more variable than the asexual condition. Well-developed buds showed testis, in contrast to the maternal polypid. On the other hand, disintegrated polypids left intact ovaria and testis behind, while in younger, but mature parts of the colony they could not be observed. Embryogenesis took place in only 10% to 20% of the zooids of the colonies labelled with the character "oecium".

4 CONCLUSIONS

Colonies of *P. casmiana* hatching from overwintering asexual resting stages at the end of May rapidly proceed to gametogenesis and statoblast production. Asexual propagation takes place by three different kinds of statoblasts, by the sessoblasts, by a thin-walled floatoblast, (leptoblast) and by a normal, thick-walled floatoblast. Sessoblasts are observed in a constant but small number throughout the year. Leptoblasts are present throughout the year as well, but are produced in larger quantities throughout late spring and from August until the end of the season. Leptoblasts represent the most important means of propagation. Despite being classified as floating resting stages, leptoblasts are neither buoyant nor long–living. Leptoblasts as well as the sessoblasts thus promote local persistence of the bryozoan. Leptoblasts immediately germinate after their release which lead to vast numbers of colonies simultaneously hatching from asexual "resting stages" particularly in early summer. In the Laxenburg pond, formation of thick-walled floatoblasts were very rare and only observed once in June and once in August. Floatoblasts occur in the same colonies as leptoblasts; the trigger of their formation, such as the end of the life-span of the first generation of colonies or deterioration of the environmental condition, is still speculative. Investment in sexual propagation is observed by production of testes and ovaria from June to August. Formation of larvae is restricted to August, however, larval occurrence appears to be variable in different years, just as colony abundance of the species.

ACKNOWLEDGEMENTS

This work was part of the projects 10806-BIO funded by the Fonds zur Förderung der wissenschaftlichen Forschung (FWF) and ÖNB 7587 funded by the Österreichischen Nationalbank.

REFERENCES

Braem, F. 1890. Untersuchungen über die Bryozoen des süßen Wassers. *Zoologica* 2:1-134.

Braem, F. 1897. Die geschlechtliche Entwicklung von *Plumatella fungosa. Zoologica* 23: 1-96.

Eder, M. & Wöss E.R. 2000. Monitoring of freshwater bryozoans on artificial substrates in the Laxenburg pond (Lower Austria): experimental setup. In A. Herrera Cubilla & J.B.C. Jackson (eds), *Proceedings of the 11th International Braozoology Association Conference*: 219-224. Balboa, Republic of Panama: Smithsonian Tropical Research Institute.

Grenacher, A. 1879. Einige Notizen zur Tinktionstechnik, besonders zur Kernfärbung. *Archiv für Mikroskopische Anatomie* 16: 469-471.

Hyman, L.H. 1959. *The Invertebrates, Vol. 5, Smaller Coelomate Groups.* New York: McGraw-Hill

Karlson. R.H. 1994. Recruitment and catastrophic mortality in *Plumatella emarginata* Allman. In P.J. Hayward, L.F. Ryland & P.D. Taylor (eds), *Biology and palaebiology of bryozoans:* 93-97. Fredensborg: Olsen and Olsen.

Kraeplin, K. 1887. Die Deutschen Süßwasserbryozoen. I. Anatomisch-systematischer Teil, *Abhandlungen des naturwissenschaftlichen Vereins Hamburg* 10: 1-168.

Mukai, H. 1982. Development of Freshwater Bryozoans (Phylactolaemata). In F. W. Harrison & R. R. Cowden (eds), *Developmental Biology of Freshwater Invertebrates:* 535-576. New York: A. R. Liss Inc.

Mukai, H., T. Niwa, M. Tsuchiya & T. Nemoto 1983. The Growth and the Production of Statoblasts in a Freshwater Bryozoan, *Plumatella casmiana. Science Reports of the Faculty of Educaton Gunma Univiversity Vol.* 32: 55-75.

Rogick, M.D. 1941. Studies on fresh-water Bryozoa. X. The occurrence of *Plumatella casmiana* in North America. *Transactions of the American Microscopical Society* 60(2): 211-220.

Rogick, M.D. 1943. Studies on fresh-water Bryozoa. XIII. Additional *Plumatella casmiana* data. *Transactions of the American Microscopical Society* 62(3): 265-270.

Toriumi, M. 1955. Taxonomical study on freshwater Bryozoa. X. *Plumatella casmiana* Oka. *The Science Reports of the Tohoku University*, 4th *ser.* 21(4): 67-77.

Viganò, A. 1968. Note su *Plumatella casmiana* Oka (Bryozoa). *Riv. Idrobiologia* 7: 421-468.

Wiebach, F. 1963. Studien über *Plumatella casmiana* Oka (Bryozoa). *Vie Milieu 14:* 579-596.

Wesenberg-Lund, C. 1939. *Biologie der Süßwassertiere.* Wien: Springer.

Wood, T. S. 1973. Colony development in species of *Plumatella* and *Fredericella* (Ectoprocta). In R. S. Boardman, A. H. Cheetham & W. A. Oliver jr. (eds), *Animal colonies:* 395-432. Stroudsburg, Pennsylvania: Dowden, Hutchinson and Ross Inc.

Wood, T.S. 1989. Ectoproct bryozoans of Ohio. *Ohio Biological Survey Bulletin New Series* 8(2): 1-70.

Wöss, E.R. 1994. Seasonal fluctuations of bryozoan populations in five water bodies with special emphasis of the life cycle of *Plumatella fungosa.* In P. J. Hayward, J. F. Ryland & P. D. Taylor (eds), *Biology and paleobiology of bryozoans*: 219-222. Fredensborg: Olsen and Olsen.

Wöss, E.R. 1996. Life-history variation in freshwater bryozoans. In D.P. Gordon, A.M. Smith & J.A. Grant-Mackie (eds), *Bryozoans in space and time:* 319-399. Wellington: National Institute of Water and Atmospheric Research Ltd.

Wöss, E.R. 2000. Colonization and development of freshwater bryozoan communities on artificial substrates in the Laxenburg pond (Lower Austria). In A. Herrera Cubilla & J.B.C. Jackson (eds), *Proceedings of the 11th International Braozoology Association Cconference*: 431-438. Balboa, Republic of Panama: Smithsonian Tropical Research Institute.

Bryozoan Studies 2001, Wyse Jackson, Buttler & Spencer Jones (eds)
© 2002 Swets & Zeitlinger, Lisse, ISBN 90 5809 388 3

A comparison of the early astogeny and life history of *Celleporella carolinensis* and *Celleporella hyalina*.

P.J. Wright & R.N. Hughes
School of Biological Sciences, University of Wales, Bangor, Gwynedd, LL57 2UW, UK

ABSTRACT: Early astogeny and life history in two species of the cheilostome genus *Celleporella*, whose ranges overlap at Woods Hole, Massachusetts, were compared in a controlled laboratory environment. *C. hyalina* and *C. carolinensis* showed distinctive early budding patterns and differed in certain life-history parameters. *C. carolinensis* became reproductive earlier and at a smaller size than *C. hyalina*, and also behaved differently in reproductive isolation. Both species produced basal and frontal sexual zooids, contradicting earlier reports suggesting that a unilaminar colony is one of the main features that set *C. carolinensis* apart from *C. hyalina*. Furthermore, the astogeny and reproductive behaviour of *C. hyalina* from Woods Hole differ from those of *C. hyalina* from North Wales.

1 INTRODUCTION

Two species of the cheilostome genus *Celleporella* occur in abundance around Cape Cod, Massachusetts. In this region, *C. hyalina* (L.) is a common epibiont of the red alga *Chondrus crispus* whereas *C. carolinensis* (Ryland 1979) is found associated with the brown alga *Sargassum filipendula* (Ryland 1979).

In his 1979 paper reporting the presence of a second species of *Celleporella*, *C. carolinensis*, on the Atlantic coast of North America, Ryland listed several differences between this new species and the already recognised *C. hyalina*. *C. carolinensis* was characterised as having a symmetrical early astogeny, smaller zooids and persistently unilaminar growth.

The present study compares the early astogeny of these two species, together with certain aspects of their life history including timing of first reproductive activity, based on colonies reared in a controlled laboratory environment.

2 MATERIALS AND METHODS

Colonies of both species were collected during July 2000, at Woods Hole, Cape Cod, Massachusetts. *C. hyalina* was found growing on *Chondrus crispus* at Nobska Point and *C. carolinensis* on *Sargassum filipendula* attached to a jetty opposite the Woods Hole Marine Laboratory. Algal fronds bearing colonies, were brought to the Bangor laboratory on 25 July 2000 and placed in 2 litre plastic drinks bottles containing 1 litre of 2 μm-filtered, U.V.-irradiated seawater. 100 ml of the cryptophyte alga *Rhinomonas reticulata* was added daily at a concentration of approximately 700 cells μl^{-1}. This alga has previously proved to be a satisfactory food source, allowing Welsh *C. hyalina* to form compact, circular colonies in the laboratory, showing a pattern of astogeny typical of wild specimens (Hunter & Hughes 1993a). Aeration maintained a constant circulation in each container. An acetate sheet, which had been conditioned in running seawater for 1 week, was wrapped around the interior of each bottle as a substratum for larval settlement. Water changes were carried out 3 times each week, during which the acetate sheets were checked under a binocular microscope for recently formed ancestrulae. Each ancestrula was isolated from the settlement sheet by cutting out a surrounding, 10x10 mm square of acetate, which was subsequently glued with cyanoacrylate onto another piece of acetate of 38x75 mm, the size of a large microscope slide. For colonies that were to be monitored in reproductive isolation, the above procedure was carried out before the first feeding to prevent any uptake and storage of allosperm by the ancestrula (Hughes et al. in prep.). Acetates bearing colonies whose growth was to be monitored in a common-garden environment were placed into a slide rack, which in turn was placed in a 2 litre plastic drinks bottle set up and maintained as described above. Acetates bearing colonies whose growth was to be monitored in isolation were placed in 300 ml glass jars containing 225 ml of 2 μm-filtered, U.V-

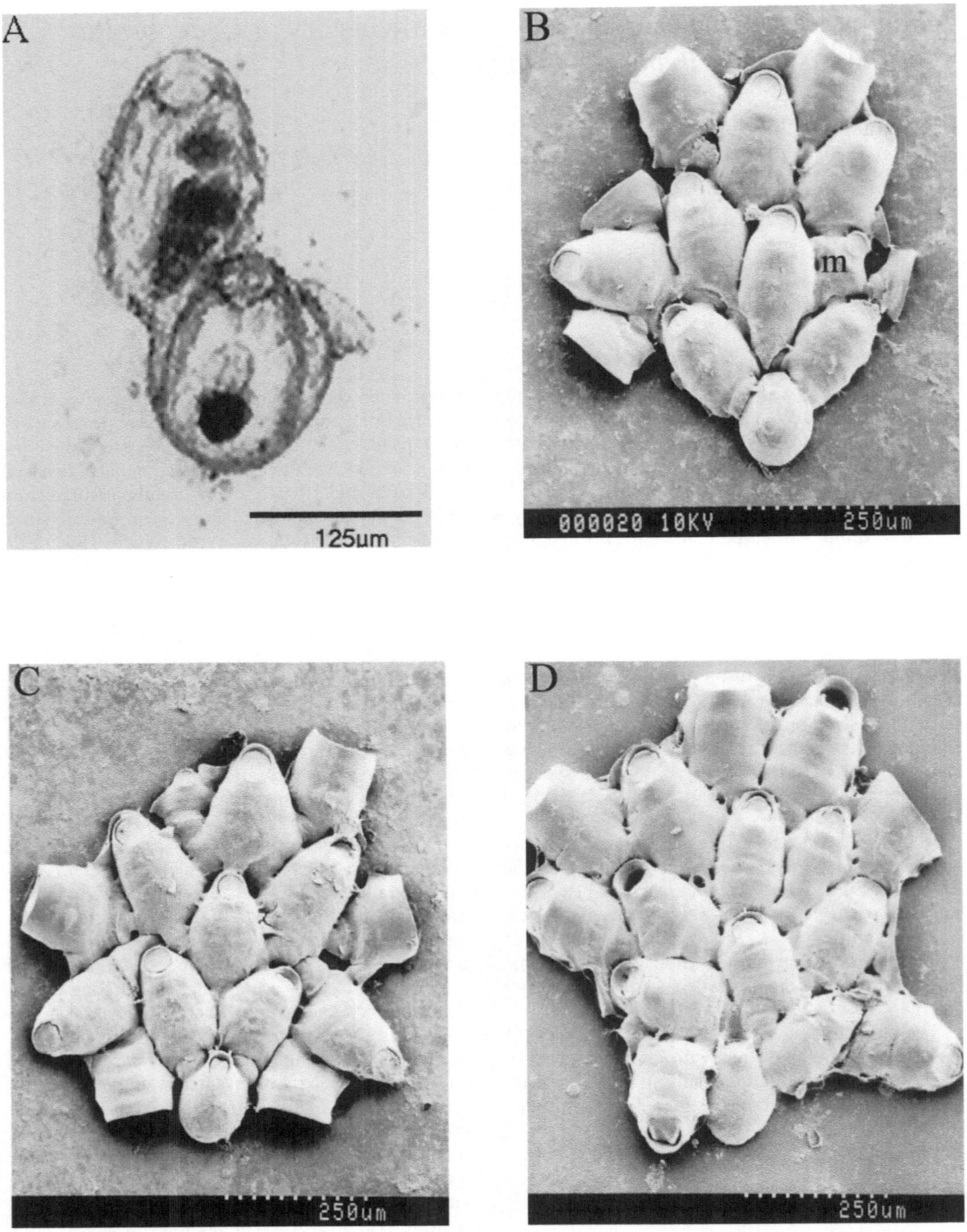

Figure 1. (A) *C. carolinensis*, ancestrula and first fully formed daughter zooid. The second disto-lateral bud is visible on the right of the ancestrula. (B) Young *C. carolinensis* colony with disto-lateral zooids completely separated by third daughter zooid. First male zooid is also shown (m). (C) Young *C. carolinensis* colony showing first two daughter zooids in contact proximally. (D) Young *C. hyalina* colony.

irradiated seawater. These jars were fitted with lids designed to reduce the risk of accidental allosperm transfer (Manríquez et al. in press). The water in these jars was also replaced 3 times a week and 20 ml of the algal diet was added daily.

In order to record the growth of colonies, monitoring was carried out twice weekly. Colonies were placed under a Leica MZ APO binocular microscope mounted with a JVC colour video camera. Images were recorded onto S-VHS video cassette.

For scanning electron microscopy, colonies were soaked in distilled water for two hours before being dried on a hotplate and coated with gold.

3 RESULTS

3.1 *Celleporella carolinensis*

The ancestrula invariably produced two disto-lateral buds simultaneously. However, the growth of one was halted at an early stage while the other continued until it had formed a complete zooid, (Figure 1A). Once growth of the first daughter zooid was completed, the ancestrula stopped feeding and its polypide regressed. During regression of the ancestral polypide the newly formed zooid commenced feeding and growth of the remaining bud recommenced. The third daughter zooid originated between the two disto-lateral zooids and usually commenced growth as the second daughter zooid neared completion. The first two daughter zooids were occasionally seen to be in contact proximally, but more frequently were completely separated by the third (Figure 1B, C). After these initial three zooids had completed their growth one at a time, continual interstitial budding, simultaneously on both sides of the ancestrula, resulted in the formation of a compact circular colony. Zooids in the vicinity of the ancestrula were significantly smaller than zooids produced in later stages of growth (Table 1, $t = 6.65$, $d.f. = 98$, $P < 0.001$).

Production of male and female zooids occurred at a relatively early stage in colonies grown in a common-garden environment. Male production usually began before female production around the end of the first week of growth, when the colony was just 4-6 zooids in size. The first female zooid was usually produced when the colony was 12-15 zooids in size, often being completed early in the third week of growth. It was not uncommon, however, for a female zooid to be produced much earlier than this, often before the first male zooid had been formed. In one colony studied, the third daughter zooid developed into a functional female. Oocytes were usually first observed in the female body 1-2 days before the ovicell had been fully formed and were invariably transferred to the ovicell within 3 days of its completion. Brooding of embryos in this species usually commenced during the third week, but was seen to commence as early as the first week in a few colonies.

Table 1. Autozooid lengths. Zooid measurements (mean ± standard deviation).
Number of measurements in parentheses.

	C. carolinensis	*C. hyalina*
	μm	μm
Close to Ancestrula	282±22 (50)	268±25 (50)
Distant from ancestrula	310±18 (50)	370±23 (50)

Colonies grown in reproductive isolation did not produce female zooids in over 9 weeks of observation. They did produce male zooids, however, but usually not until after the second week of growth.

All colonies remained unilaminar, sexual zooids being formed in the basal layer. However, 6 months after routine observations had terminated, it was discovered that all surviving colonies had produced a large number of frontal female zooids, as well as a smaller number of frontal males and autozooids.

3.2 *Celleporella hyalina*

Invariably one disto-lateral bud was produced that could arise from either the left or right side of the ancestrula. During the growth of this first daughter zooid, another bud formed between it and the ancestrula. Once the first daughter zooid had started feeding the ancestral polypide regressed and the second daughter zooid completed its growth. The third daughter zooid budded from between the first and second daughter zooids and completed its growth when the second daughter zooid had commenced feeding. After these initial three zooids had completed their growth one at a time, continual simultaneous budding from between zooids, in all directions, resulted in the formation of a compact circular colony. Zooids in the vicinity of the ancestrula were significantly smaller than those produced in later stages of growth (Table 1, $t = 20.85$, $d.f. = 98$, $P < 0.001$).

Production of sexual zooids, both in reproductive isolation and in a common-garden environment, occurred later in this species than in *C. carolinensis*. Moreover, no differences were found between the development of colonies reared in either environment. Male zooids were always produced before female zooids and the first of these was usually produced early in the fourth week of growth, when the colony was 60-70 zooids in size. The first female zooid was usually produced when the colony was 100-120 zooids in size and was often completed early in the fifth week of growth. Sexual zooids were produced by frontal budding, which commenced in the newer, peripheral region of the colony. The production of sexual zooids rarely occurred before the colony had attained a compact circular form. Occasionally at an earlier stage in the growth of the colony, a male zooid was produced in the basal layer. As with *C. carolinensis*, oocytes were usually first observed in the female body 1-2 days before the ovicell had been fully formed and were invariably transferred to the ovicell within 3 days of its completion. Brooding in this species usually commenced during week five, in both a common-garden environment and in reproductive isolation.

Occasional female zooids were produced in the basal layer of colonies, in addition to males. Colonies reared in both a common-garden environment and in reproductive isolation were found with female zooids in their basal layers when about 2 months old.

4 DISCUSSION

Within the Hippothoidae, the pattern of early astogeny is thought to be specifically distinct (Ryland & Gordon 1977, Hastings 1979, Gordon & Hastings 1979). Indeed, in the current study *C. carolinensis* and *C. hyalina* displayed a distinctive early budding pattern (Figure 1B, C, D), the only variation being found in *C. carolinensis* with the first and second daughter zooids sometimes being in contact proximally and sometimes being completely separated by the third (Figure 1B, C). No variations in budding pattern were found in *C. hyalina*, although a previous study of colonies collected from North Wales revealed the production of two disto-lateral zooids and one disto-medial zooid in a small number of colonies (Cancino & Hughes 1988). Although the production of two disto-lateral buds occurs in certain other species of *Celleporella* (Ryland & Gordon 1977, Moyano 1986, César-Aldariz et al. 1999), it has not been reported in such cases whether both disto-lateral zooids complete their growth simultaneously or whether, as is the case with *C. carolinensis*, the growth of one is halted until the other has completed its growth and commenced feeding. However, an SEM plate (author unknown) of an early *C. antarctica* colony and of an early *C. galaica* colony (César-Aldariz et al. 1999) each show the ancestrula with one fully formed and one incomplete disto-lateral zooid. This aspect of the production of two disto-lateral buds may be species-specific and therefore a useful taxonomic character.

When grown in a common-garden environment, colonies of *C. carolinensis* become reproductively active much earlier than *C. hyalina*, both in terms of colony age and colony size. *C. carolinensis* was collected on the fronds of *Sargassum filipendula*, whereas *C. hyalina* was collected on the more stable substratum of *Chondrus crispus* stipes. The less stable substratum occupied by colonies of *C. carolinensis* may have caused selection for the ability to reproduce relatively early during astogeny. The ability to produce completely developed female zooids at a very early stage in colonial growth has also been reported for *C. antarctica* (Moyano & Gordon 1980). *C. carolinensis* and *C. hyalina* also employ different strategies when grown in reproductive isolation, with *C. carolinensis* colonies showing reduced levels of male production and a complete lack of female investment. *C. hyalina* on the other hand produced male and female zooids when reared both in reproductive isolation and in a common-garden environment. Moreover, females produced in reproductive isolation brooded embryos and released abundant viable larvae (see also Hughes et al. this volume).

C. hyalina and *C. carolinensis* produced female zooids in both the basal layer and frontally, although numbers of basal females in *C. hyalina* were low and the timing of the onset of frontal budding in *C. carolinensis* is unknown. These findings contradict those of Ryland (1979) who reported *C. carolinensis* as being unilaminar throughout, and of Hayward & Ryland (1979) who reported female zooids as "possibly invariably developing from frontal buds".in *C. hyalina*. The ability of both species to produce basal and frontal sexual zooids may lead to confusion and/or errors being made when trying to identify older specimens of *Celleporella* from the Woods Hole region, in which the pattern of early astogeny may have been obscured by frontal zooids.

The present study has revealed that colonies of *C. hyalina* collected from Woods Hole, Massachusetts, commence production of frontal male and female zooids in peripheral areas of the colony, produce basal females and also, when reared in reproductive isolation, are able to produce copious viable larvae. Colonies of *C. hyalina* from North Wales begin frontal budding of sexual zooids in the centre of the colony (Cancino & Hughes 1988) and when reared in reproductive isolation endure high levels of abortion and produce larvae of poor viability (Hunter & Hughes 1993b, Hughes et al. this volume). We also observed slight morphological differences between colonies from North Wales and Woods Hole. Furthermore, the Woods Hole population has proved to be cladistically distinct within *C. hyalina* sensu lato (A. Gomez, pers. comm.).

REFERENCES

Cancino, J.M. & Hughes, R.N. 1988. The zooidal polymorphism and astogeny of *Celleporella hyalina*. *Journal of Zoology*. 215: 167-181.

César-Aldariz, J., Fernández-Pulpeiro, E. & Reverter-Gil, O. 1999. Anew species of the genus *Celleporella* (Bryozoa: Cheilostomatida) from the European Atlantic coast. *Journal of the Marine Biological Association of the United Kingdom*. 79: 51-55.

Gordon, D.P. & Hastings, A.B. 1979. The interzooidal communications of *Hippothoa sensu lato* (Bryozoa) and their value in classification. *Journal of Natural History* 13: 561-579.

Hastings, A.B. 1979. The genus *Hippothoa* Lamouroux [Polyzoa (Bryozoa) Cheilostomata]. *Journal of Natural History* 13: 535-560.

Hayward, P.J. & Ryland, J.S. 1979. British Ascophoran Bryozoans. In D.M. Kermack & R.S.K. Barnes (eds), *Synopses of the British Fauna*, 14: 252-254. London & New York: Academic Press.

Hughes, R.N., Manríquez, P.H. & Bishop, J.D.D. 2001. Female investment is enhanced by reception of allosperm in the sessile hermaphrodite *Celleporella hyalina* (Bryozoa). (In prep).

Hughes, R.N., Wright, P.J., Manríquez, P.H. & Bishop, J.D.D. 2002. Predominance of obligate outbreeding in the simultaneous hermaphrodite *Celleporella hyalina* sensu lato. *This volume*.

Hunter, E. & Hughes, R.N. 1993a. Effects of diet on life-history parameters of the marine bryozoan, *Celleporella hyalina* (L.). *Journal of Experimental Marine Biology and Ecology* 167(2): 163-177.

Hunter, E. & Hughes, R.N. 1993b. Self-fertilisation in *Celleporella hyalina*. *Marine Biology* 115: 495-500.

Manríquez, P.H., Hughes, R.N. & Bishop, J.D.D. 2001. Age-dependent loss of fertility in water-borne sperm of the bryozoan *Celleporella hyalina* (L.). *Marine Ecology Progress Series*. (In press).

Moyano, G.H.I. 1986. Bryozoa marinos chilenos VI. Cheilostomata Hippothoidae: South eastern Pacific species. *Boletín de la Sociedad de Biología de Concepción, Chile* 57: 89-135.

Moyano, G.H.I. & Gordon, D.P. 1980. New species of Hippothoidae (Bryozoa) from Chile, Antarctica and New Zealand. *Journal of the Royal Society of New Zealand* 10: 75-95.

Ryland, J.S. 1979. *Celleporella carolinensis* sp. nov. (Bryozoa Cheilostomata) from the Atlantic coast of America. In G.P. Larwood & M.B. Abbott (eds), *Advances in Bryozoology*: 611-620. London & New York: Academic Press.

Ryland, J.S. & Gordon, D.P. 1977. Some New Zealand and British species of *Hippothoa* (Bryozoa: Cheilostomata). *Journal of the Royal Society of New Zealand* 7: 17-49.

Bryozoan Studies 2001, Wyse Jackson, Buttler & Spencer Jones (eds)
© 2002 Swets & Zeitlinger, Lisse, ISBN 90 5809 388 3

Palaeoenvironmental interpretation of the Tramore Limestone Formation (Llandeilo, Ordovician) based on bryozoan colony form

Patrick N. Wyse Jackson
Department of Geology, Trinity College, Dublin 2, Ireland

Caroline J. Buttler
Department of Geology, National Museums and Galleries of Wales, Cathays Park, Cardiff CF10 3NP, Wales, UK

Marcus M. Key, Jr.
Department of Geology, Dickinson College, Carlisle, Pennsylvania, 17013-2896, USA

ABSTRACT: The Tramore Limestone Formation (Llandeilo, Ordovician), exposed in southeast Ireland, contains well-preserved colonies of the trepostome bryozoan *Diplotrypa petropolitana*. Two palaeoenvironmental settings - a deep water carbonate basinal facies and a shallow water carbonate shelf facies - are represented. Shape, size and attitude to bedding of *Diplotrypa* colonies were measured in the field. These data were compared with results gathered from experiments carried out to test the behaviour in a wave tank of representative models of bryozoan colonies. Results have demonstrated a clear environmental partitioning of colonies, and suggest that dome-shaped colonies are more dominant in deep water facies than bell-shaped colonies, which occur more abundantly in the shallow water facies. Morphological analysis of bryozoan colonies can be a useful tool for palaeoenvironmental interpretation.

1 INTRODUCTION

Bryozoan colony form may be utilised to determine the environmental and palaeoenvironmental conditions under which bryozoans grew. The classic study of this approach was by Stach (1932). Recent studies by Kelly & Horowitz (1987), Smith (1995), and Hageman *et al.* (1998) have refined and improved our understanding of the influences of environmental parameters on colony morphology, so much so that we can now make reasonable estimations of the bryozoan colony morphological types expected to grow in any environment, and expected to be found preserved in different facies.

Within the Tramore Limestone Formation (Llandeilo, Ordovician), exposed in southeast Ireland, are well-preserved colonies of the trepostome bryozoan *Diplotrypa petropolitana*. In 1999 and 2000 colonies were collected for taxonomic study from several sites along the Waterford coast west of the town of Tramore. On preliminary examination it appeared that these colonies exhibited palaeoenvironmental trends in relation to the size of colonies, to their state of preservation, and in their orientation relative to the substrate. Material was collected from two palaeoenvironmental settings (as indicated by independent sedimentological evidence) - a deep water carbonate basinal facies exposed at Stage Cove in the western part of the study area, and a shallow water carbonate shelf facies exposed in the eastern part of the study area southeast of Tramore at Lady Doneraile's Cove and at Lady Elizabeth's Cove (Fig. 1).

This paper sets out to do several things: (1) It discusses the geological setting from where the bryozoans were collected. (2) It will outline these apparent colony form trends. (3) The morphology of the bryozoan colonies from the two areas will be examined to see what can be discerned from them concerning the nature of the different environments in which they grew. (4) Laboratory-based experiments will be described which attempted to determine the hydrological characteristics of colonies of different morphologies, in order to shed light on the sedimentological and hydrological regimes that might have been present during the Llandeilo in southeast Ireland; (5) The paper will attempt to suggest and explain the controlling influences that caused these morphological trends.

2 GEOLOGICAL SETTING

The Lower Palaeozoic rocks of southern County Waterford consist of shallow and deep water carbonates and shales overlain by volcanic material (Tietzsch-Tyler & Sleeman 1994a, b, Sleeman & McConnell 1995a, b). The nature and age of sedimentation and volcanism have been a matter of debate for some time.

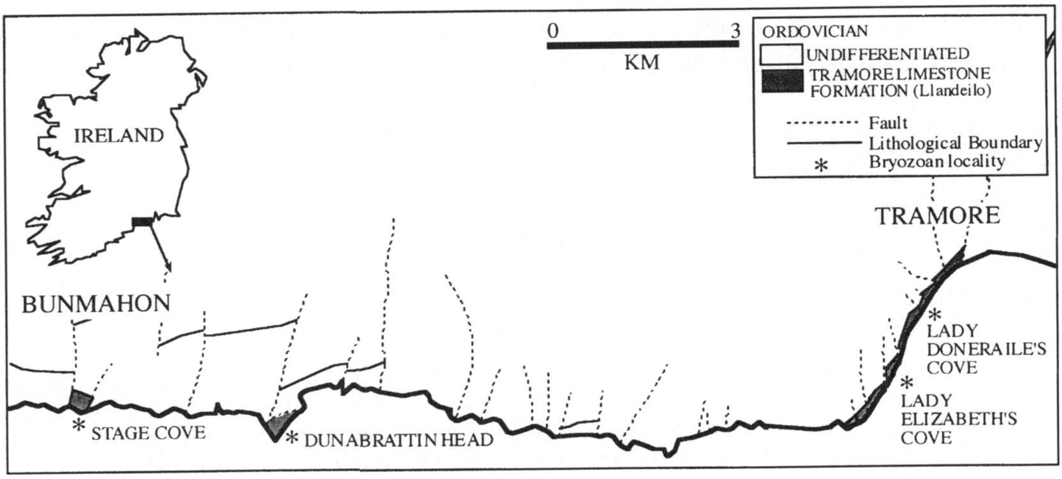

Figure 1. Geological map of south-east Ireland showing the distribution of the Tramore Limestone Formation and the localities mentioned in the text from where bryozoans were collected. Modified from Tietzsch-Tyler & Sleeman 1994a and Sleeman & McConnell 1995a.

Cambrian, Arenig, or Llanvirn sediments are overlain unconformibly by dark shales of the Llanvirn Tramore Shale Formation (to the east) and the Dunabrattin Shale Formation (to the west). These are in turn overlain by the Tramore Limestone Formation, which consists of interbedded shales and carbonates. It is best exposed southwest of the town of Tramore (Ball 1978), Dunabrattin Head (Stillman 1978) and Stage Cove, Knockmahon (Stillman 1978) (Fig. 1), from where specimens of *Diplotrypa petropolitana* have been collected. During the Caradoc volcanism produced a succession of rhyolites, breccias and tuffs (Stillman 1978) which have been assigned to the Campile Formation. It has been postulated that this succession represents a submarine volcanic island arc formed at the margins of the Avalonian continental plate margin above subducting Iapetus Ocean.

The Tramore Limestone Formation crops out in southeast County Waterford, Ireland from Tramore Bay in the east to Ballydowne Bay in the west. Exposures of the Tramore Limestone Formation are excellent along the 30-60 m high coastal sea cliffs but are only accessible by the occasional entrenched gulley, and the inland outcrops are obscured by glacial drift.

The formation consists of grey calcareous mudstones, siltstones, and sandstones with interbedded blue-grey argillaceous nodular limestone beds containing a rich shelly fauna (Murphy 1958, Mitchell *et al.* 1972, Tietzsch-Tyler & Sleeman 1994b, Sleeman & McConnell 1995b, Buttler & Wyse Jackson 1997). The more carbonate-rich beds represent buildups of shelly faunas in shallow water adjacent to the volcanic arc to the southeast (Tietzsch-Tyler & Sleeman 1994a). In response to contemporaneous faults downthrown to the northwest, the Tramore Limestone Formation grades from a shallower shelf facies in the east to a deeper basinal facies in the west. The transition from shelf to basinal facies occurs between Black Rock and Dunabrattin Head (Sleeman & McConnell 1995a). The shelf facies is more fossiliferous and carbonate-rich than the muddier basinal facies (Phillips *et al.* 1976, Carlisle 1979, Boland 1983).

The shelf facies reaches a maximum 65 m thick and comprises calcareous sandstones, siltstones, and mudstones, and limestones which have yielded a rich and diverse fauna. These light to dark grey indurated nodular limestones dip steeply 78° northwest. This shelf facies is best exposed to the east towards Tramore.

The basinal facies, which reaches a maximum thickness of 450 m, is dominated by calcareous shales with more widely spaced, more argillaceous, and less continuous limestones than the shelf facies, and contains immature brachiopods and trilobites. It is well exposed at Boat Strand at Dunabrattin Head and at Stage Cove.

Bryozoans have been reported as most abundant at Tramore from Lady Doneraile's Cove and Lady Elizabeth's Cove to Great Newtown Head, and at Dunabrittin Head, Stage Cove, and Ballydowane Bay (Fig. 1), where they commonly occur in the more carbonate-rich nodular limestones. The shalier beds often contain fragmentary bryozoan, brachiopod, and trilobite remains.

3 FAUNA AND CORRELATION OF THE TRAMORE LIMESTONE FORMATION

The earliest record of fossils from the Tramore area were those noted by Frederick M'Coy in a monograph on Irish Lower Palaeozoic fossils published in 1846 (M'Coy 1846). He illustrated a coral *Favosites petropolitana* var. *lycopodites* Emmons, which is now known as the trepostome bryozoan *Diplotrypa petropolitana*, the subject of the present study.

In the 1860s the officers of the Geological Survey of Ireland collected extensively from limestones (the Tramore Limestone Formation) and black shales in the area. Baily (1865) listed many taxa including 6 graptolite species (from the shales), 22 brachiopods, 25 trilobites, some gastropods, cephalopods, corals, echinoderms, and four bryozoans including *Stenopora fibrosa* var. *lycoperdon* [*Diplotrypa petropolitana*]. *D. petropolitana* has a complex nomenclatural history which requires evaluation by the ICZN. Until our application (Wyse Jackson, Buttler & Key 2001) is considered and the ICNZ publishes its opinion on the nomenclatural status of the taxon, we prefer to leave it in open nomenclature.

At the end of the last century Reed (1899) noted many species and described several trilobite species from the formation. Little work was carried out until Carlisle (1979) who mapped the area as part of a Ph.D. study. Although she collected many brachiopod taxa, these still remain to be fully described (Parkes & Harper 1996). Assessment of the complete fauna has given an unequivocal Llandeilo age for the Tramore Limestone Formation, and has allowed for correlation with the Courtown Limestone Formation 75 km to the east. Initially correlation was on the basis of the common occurrence of the rare trilobite *Eirelithus* (Crimes & Crossley 1968), but subsequently on the basis of additional faunal elements (Mitchell et al. 1972, Carlisle 1979). Twelve brachiopod genera including *Dalmanella*, *Glyptorthis*, *Hesperorthis*, *Howellites*, *Leptestiina*, *Palaeostrophomena*, *Philhedrella*, *Platystrophia*, *Porambonites*, *Salopia* and *Sowerbyella* and four trilobite genera are common to both formations with the stratigraphically restricted brachiopod *Glyptorthis crispa* (M'Coy) and trilobite *Eirelithus* being particularly important for their correlation.

Figure 2. *Diplotrypa petropolitana* colonies form the TLF showing the two end members of colony form. The dome-shaped colony (h/w = 1.0) on the left is from Stage Cove [TCD.49189]; the bell-shaped colony (h/w = 0.55) on the right is from Lady Elizabeth's Cove [TCD.49192]. Scale bars = 1cm.

4 MATERIAL

This paper is based on material collected by the authors during fieldwork in June 1999 and February 2000, and on specimens contained in various museum collections: Trinity College, Dublin [prefix TCD.], Geological Survey of Ireland, Dublin, The Natural History Museum, London, and the National Museum and Galleries of Wales, Cardiff.

5 METHODS

5.1 *Field analysis*

Detailed examination of colonies *in situ* at the two sites (Stage Cove and Lady Elizabeth's Cove) near Tramore was carried out. We measured colony height, basal diameter and noted orientation relative to bedding (i.e. upright or overturned) as well as the spatial distribution of colonies (i.e. colonies clustered into lenses or dispersed). We were also able to access the nature of preservation of colonies at both sites. In addition photographs were taken for later colony shape and size analysis.

5.2 *Laboratory analysis*

Eighty nine specimens of the trepostome *Diplotrypa petropolitana* held in various museums were examined for this study. Colony height and basal diameter were measured to the nearest mm. From these measurements, size (height x basal diameter = mm^2) and shape (Height/basal diameter = dimensionless) were calculated.

Eight replicates of 12 different models both of actual colonies, and of dome-shaped and bell-shaped hypothetical colonies of various sizes were cast in resin. The resin had a density of 1.06 g/cm^3 which was chosen to be similar to the density of living bryozoan colonies. These models were used in wave tank experiments to access the effect of colony shape on settling, and on movement under different wave conditions. This was done in order to understand the features observed in the field.

The wave tank measured 45 cm long x 30 cm wide x 30 cm deep. Water depth: 20 cm. Substrate thickness: 1 cm. Substrate type: very well sorted fine sand.

When doing experimental wave tank work, one must be careful of edge effects (Kaneko 1985). As each model on average only occupied 0.34% of the tank's water volume this effect should have been minimal.

Several experiments were carried out: (1) Settling pattern of colonies. Models were dropped into the water from three initial orientations: right side up, upside down, and sideways. They were simply allowed to fall through the water column and their final attitude in the sediment recorded (upright or overturned). (2) Effect of water agitation on colonies placed on substrate in living position. For each type of colony model, the eight models were placed upright on the sediment surface in the wave tank. They were then subjected to 3 different wave velocities (Low = 2 cm/sec; Medium = 6 cm/sec; High = 8 cm/sec) and in each case their final attitude to the substrate recorded. These velocities were created by dipping a paddle of dimensions 21.5 cm x 13.5 cm to depths of 6, 12, and 18 cm respectively and moving it backwards and forwards across the tank three times.

6 RESULTS

6.1 *Field observations of the nature of bryozoan colonies at each site*

1. Colonies in Stage Cove appeared to be smaller than those at Lady Elizabeth's Cove.
2. Colonies at Stage Cove appeared to be more dome-shaped than bell-shaped.
3. Colonies at Lady Elizabeth's Cove appeared to be more bell-shaped than dome-shaped.
4. At Stage Cove 58% [73] of colonies observed were upright while 42% [53] were overturned.
5. At Lady Elizabeth's Cove 83% [112] of colonies were upright while 17% [23] were overturned.

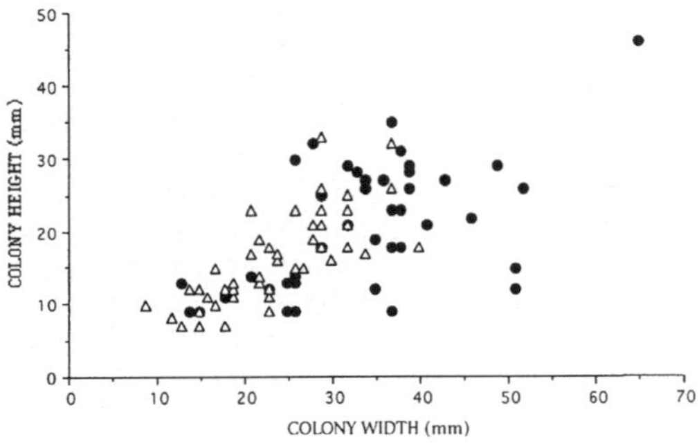

Figure 3. Graph showing colony height plotted against colony basal diameter for colonies from Stage Cove (open triangle) N=47, and from Lady Elizabeth's Cove (solid circle) N=42.

6.2 *Laboratory observations*

6.2.1 *Colony size and shape:*

Colonies of *Diplotrypa petropolitana* broadly form nodular masses that range from 9 to 75 mm in diameter and 5 to 32 mm in height (Fig. 2). The majority of colonies have flat or concave bases. Several colony morphologies have been identified in this study. Conical colonies narrow towards their pointed apices at angles of roughly 45°. Domes are similar to cones but have rounded apices, and may be depressed. Mound-shaped colonies have straight or sub-parallel sides or may show marked increase or decrease in colony width. Some mounds show marked asymmetrical development which suggests that these colonies may have grown on slopes of up to 20°. Nodular and irregular-shaped colonies are present, but rare.

Eighty nine colonies of *Diplotrypa petropolitana* were analyzed for shape, with 53% coming from Stage Cove and 47% from Lady Elizabeth's Cove. These comprised 61% relatively complete hand samples and 39% photographs of colonies in outcrop showing cross-sectional views. The *Diplotrypa petropolitana* colonies from Stage Cove were significantly smaller in both height and width than those from Lady Elizabeth's Cove (Table 1; *t*-tests, $p < 0.01$). A simple shape parameter was calculated from the height/basal diameter ratio that removed the effect of size (Table 1). 93 % of all of the colonies had values less than 1 (Fig. 3). Based on this shape parameter, *Diplotrypa petropolitana* colonies from Stage Cove were more dome shaped (mean height/width ratio = 0.67) whereas those from Lady Elizabeth's Cove were more bell shaped (mean height/width ratio = 0.62), but this difference was insignificant (*t*-test, $p = 0.24$).

In the resin models domes were slightly larger (mean size = 614 mm^2) than bells (mean size ≐ 612 mm^2) (*t*-test, $p > 0.07$). Domes were significantly more cube-shaped (mean height/basal diameter ratio = 0.73), whereas bells were more pancake-shaped (Mean height/basal diameter ratio = 0.48) (*t*-test, $p = 0.028$) (Table 2).

6.2.2 *Settling experiments*

There was no significant correlation ($p > 0.05$) between height, basal diameter or size, and initial orientations in the settling experiments. Most colonies (91%) tended to settle upside down regardless of shape. Bells were less likely to become orientated right side up than domes, but this was barely insignificant (R2 = 0.251, $p = 0.097$). The correlation improves with quadratic regression ($p < 0.05$) indicating that bells were less likely to become orientated right side up than pancakes and domes.

6.2.3 *Wave experiments*

Specimens with right side up final orientation often moved along the substrate but did not flip over during exposure to wave currents. The specimens (particularly the bell-shaped variety) tended to end up with a fringe of sediment covering their lateral margins after settling in medium energy and under high energy were often buried. As the current velocity increased, the percentage of models that overturned increased from 9% (low velocity) to 30% (medium velocity) to 58% (high velocity). There was no correlation ($p > 0.05$) between size, height, or basal diameter and orientation in low or high energy regimes. There was an inverse correlation between shape and orientation in medium and high energy

Table 1. Field observations.

	Stage Cove (Basinal facies)	Lady Elizabeth's Cove (Shelf facies)
	n, mean, sd	n, mean, sd
Colony size*	47, 379 mm^2, 247 mm^2	42, 707 mm^2, 486 mm^2
Colony shape†	47, 0.67, 0.18	42, 0.62, 0.22
Colonies upright	58%	83%
Colonies overturned	42%	17%
Spatial distribution	Clustered (lenses)	Dispersed

n = number of colonies; sd = standard deviation; *size = height x basal diameter; †shape = height / basal diameter.

Table 2. Model size and shapes.

Specimen	Height (mm)	Basal diameter (mm)	Size (mm^2)	Shape (height/basal diameter)
DOMES				
Large (M) - TCD.50340	25	28	700	0.89
Skewed (C) - TCD.49181	27	33	891	0.82
Medium (M) - TCD.50341	25	32	800	0.78
Small (M) - TCD.50342	20	28	580	0.69
Small (C) - TCD.49189	23	23	529	1.00
Flat (C) - TCD.49165	7	29	203	0.24
Averages: x (sd)	21 (7)	29 (3)	614 (238)	0.73 (0.26)
BELLS				
Extra large (M) - TCD.50343	21	45	945	0.47
Large (M) - TCD.50344	21	40	840	0.53
Medium (M) - TCD.50345	21	39	819	0.54
Small (C) - TCD.28155	10	18	180	0.56
Flat (M) - TCD.50346	13	36	468	0.36
Small flat (M) - TCD.50347	12	32	384	0.38
Averages: x (sd)	17 (6)	35 (9)	612 (324)	0.48 (0.10)

Table 3. Results of colony settling experiments using models listed in Table 2.

	Right side up initial Orientation		Upside down initial Orientation		Sideways initial Orientation		Average %	
Final Orientation	Right side up	Upside down	Right side up	Upside down	Right side up	Upside down	Right side up	Upside down
Domes (n=48)	12 [25%]	36 [75%]	2 [4%]	46 [96%]	7 [15%]	41 [85%]	15	85
Bells (n=48)	5 [10%]	43 [90%]	0 [0%]	48 [100%]	0 [0%]	48 [100%]	3	97

Table 4. Results of wave action experiments using models listed in Table 2.

	Low Energy		Medium Energy		High Energy	
	Right side up	Upside down	Right side up	Upside down	Right side up	Upside down
Domes (n=48)	45 [94%]	3 [6%]	31 [65%]	17 [35%]	17 [35%]	31 [65%]
Bells (n=48)	42 [88%]	6 [12%]	36 [75%]	12 [25%]	23 [48%]	25 [52%]

regimes. Put simply, the models that were more dome-shaped were more likely to flip over - but this correlation was insignificant ($p > 0.05$). A higher proportion of bell-shaped colonies remained upright at medium and high wave velocities, but again this correlation was not significant (t-test, $p > 0.05$).

7 DISCUSSION

Colonies were not cemented to the substrate and so were liable to being turned and rolled in water currents. Such mechanisms explain the development of irregular and nodular colony forms, and the occasional occurrence of colonial regeneration at different basal angles. Flat-bottomed colonies (such as those developed at Stage Cove) or concave-bottomed colonies (as seen in more specimens at Lady Elizabeth's Cove) probably sat on the sediment surface and were usually held in a normal orientation by their weight.

The differences in size and shape of *Diplotrypa petropolitana* colonies between Stage Cove and Lady Elizabeth's Cove may be due to a combination of several factors. Perhaps the smaller colonies are found in Stage Cove because they grow more slowly due to their deeper water conditions. Perhaps size was partially controlled by the availability of nutrients which would have been more plentiful in the shallow water environments of Lady Elizabeth's Cove, rather than in the deeper water basins to the west. Colony width may have been controlled by the coarseness of the sediment - those colonies from Lady Elizabeth's Cove grew in coarse shallow water skeletal carbonates, while those at Stage Cove grew in deep water argillaceous carbonates. A further alternative view, but one which is less favoured by us, is that the smaller colonies are found in Stage Cove because they were preferentially transported due to their smaller size to the deeper environment from the shallow Lady Elizabeth's Cove environment where they lived.

We suggest that the slight predominance of bell-shaped colonies in the shallow water facies was an adaptation to the higher energy regimes that would have been present. As has been clearly demonstrated in the settling and wave experiments, these bell-shaped colonies were more stable than the dome-shaped counterparts. The fact that, firstly, in the settling experiments the bulk of the bell-shaped colonies settled upside down, and that secondly this is not seen in the exposures at Lady Elizabeth's Cove, suggests that few in fact were entrained by wave action. The majority of colonies were found in life position on horizontal bedding planes, and not in clusters which would have suggested transportation, except in one exceptional horizon where colonies accumulated at the bottom of foreset crossbeds. The characteristics of the carbonates suggest that they were deposited in shallow water conditions.

The deeper water facies at Stage Cove contained a higher proportion of dome-shaped colonies which we have demonstrated are more susceptible to being overturned than their bell-shaped counterparts. This was reflected in the counts of colony orientation made in the field. Examination of the exposures where they are best seen at Stage Cove revealed that the bulk of specimens were preserved in small lense-like accumulations, where colonies had been moved and sorted by current activity. Although the settling experiment showed that more dome-shaped colonies would settle upright when entrained, we suggest that the dominance of overturned specimens at the localities is easily explained by the fact that most of the specimens are small and dome-shaped and most would have moved in moderate current activity. It is likely that all but the smallest colonies would simply have rolled over rather than be lifted into the water column.

We postulate that the source of the wave currents that agitated the colonies in the basinal facies may have been episodic storms which would have produced the tempstites seen at Stage Cove. Similarly shaped colonies in the mid-Ordovician in Pennsylvania also seem affected primarily by storm processes (Cuffey 1997)

Spjeldnaes (1996) has discussed the application of morphological shape of Lower Ordovician bryozoans from Öland, Sweden in determining sediment - water interface conditions. Concave-based colonies which are commonly developed in *Diplotrypa petropolitana* were thought to indicate a slight winnowing or removal of surface sediment which allowed the basal layer to grow downwards. We are in agreement with his findings, and further suggest that flat-based colonies (such as those in the basinal facies at Stage Cove) grew at a time when sedimentation rate was low.

8 CONCLUSIONS

There is clear morphological partitioning in colonies of *Diplotrypa petropolitana* from the Llandeilo of Tramore, Co. Waterford. Those that grew in shallow water shelf facies were larger, more bell-shaped, and more frequently preserved in life position. By way of contrast those preserved in the basinal facies were smaller, more dome-shaped, and were more susceptible to being overturned by current activity. Colony morphology (shape and size) in *Diplotrypa petropolitana* is controlled by the prevailing environmental and sedimentological characteristics of each site, and provide a useful tool for palaeonenvironmental interpretation.

REFERENCES

Baily, W.H. 1865. Palaeontological Notes. In G.V. Du Noyer, *Explanation to accompany sheets 167, 168, 178, and 179 of the maps, and sheet 13 of the longitudinal sections of the Geological Survey of Ireland*: 18-30. Dublin: Memoirs of the Geological Survey of Ireland.

Ball, D.M. 1978. Lady Doneraile's Cove, Tramore Strand. In C.J. Stillman, South-east County Waterford and south Tipperary: Ordovician volcanic and sedimentary rocks and Silurian turbidites: 53. *Geological Survey of Ireland Guide Series* 2.

Boland, M.A. 1983. *The geology of the Ballyvoyle-Kilmacthomas-Kilfarrasy area, County Waterford with an account of the Lower Palaeozoic geology of County Waterford.* Unpublished Ph.D. dissertation. University of Dublin.

Buttler, C.J. & Wyse Jackson, P.N. 1997. Trepostome bryozoans from the Tramore Limestone (Ordovician), County Waterford, Ireland. *Palaeontology Newsletter* 36: 35.

Carlisle, H. 1979. Ordovician stratigraphy of the Tramore area, County Waterford, with a revised Ordovician correlation of south-east Ireland. In A.L. Harris, C.H. Holland, C.H. & B.L. Leake (eds), *The Caledonides of the British Isles - reviewed*: 545-554. *Geological Society Special Publication* 2.

Crimes, T.P. & Crossley, J.D. 1968. The stratigraphy, sedimentology, ichnology and structure of the Lower Palaeozoic rocks of part of north-eastern Co. Wexford. *Proceedings of the Royal Irish Academy* 67B: 185-215.

Cuffey, R.J. 1997. *Prasopora*-bearing event beds in the Coburn Limestone (Bryozoa; Ordovician; Pennsylvania). In C.E. Brett & G.C. Baird (eds), *Paleontological events; stratigraphic, ecological, and evolutionary implications*: 110-130. New York: Columbia University Press.

Hageman, S.J., Bock, P.E., Bone, Y. & McGowan, B. 1998. Bryozoan growth habits: classification and analysis. *Journal of Paleontology* 72: 418-436.

Kaneko, A. 1985. Formation of beach cusps in a wave tank. *Coastal Engineering* 9: 81-89.

Kelly, S.M. & Horowitz, A.S. 1987. Growth forms and paleoecology of Mississippian bryozoans: critical application of Stach's 1936 model, eastern United States. In J.R.P. Ross (ed.), *Bryozoa: present and past*: 137-144. Bellingham: Western Washington University.

M'Coy, F. 1846. *A synopsis of the Silurian fossils of Ireland.* Dublin: Dublin University Press.

Mitchell, W. I., Carlisle, H., Hiller, N. & Addison, R. 1972. A correlation of the Ordovician rocks of Courtown (Co. Wexford) and Tramore (Co. Waterford). *Proceedings of the Royal Irish Academy* 72B: 83-89.

Murphy, G.J. 1958. Excursion to Tramore Bay. In J.C. Brindley & W.D. Gill (eds.), Summer field meeting in southern Ireland, 1957: 255-257. *Proceedings of the Geologists' Association* 69: 244-261.

Parkes, M.A. & Harper, D.A.T. 1996. Ordovician brachiopod biogeography in the Iapetus suture zone of Ireland: provincial dynamics in a changing ocean. In P. Copper & J. Jin (eds), *Brachiopods: proceedings of the third international brachiopod congress*: 197-202. Rotterdam: Balkema.

Phillips, W.E.A. Stillman, C.J. & Murphy, T. 1976. A Caledonian plate tectonic model. *Journal of the Geological Society of London* 132: 579-609.

Reed, F.R.C. 1899. The Lower Palaeozoic bedded rocks of County Waterford. *Quarterly Journal of the Geological Society of London* 55: 718-772.

Sleeman, A.G. & McConnell, B. 1995a. *Geology of east Cork - Waterford. Sheet 22. Scale 1:100,000.* Dublin: Geological Survey of Ireland.

Sleeman, A.G. & McConnell, B. 1995b. *Geology of east Cork - Waterford. A geological description of east Cork, Waterford and adjoining parts of Tipperary and Limerick to accompany the bedrock geology 1:100,000 scale map series, Sheet 22, east Cork - Waterford.* Dublin: Geological Survey of Ireland.

Smith, A.M. 1995. Palaeoenvironmental interpretation using bryozoans: a review. In D. Bosence & P. Allison (eds), *Marine palaeoenvironmental analysis from fossils*: 231-243. *Geological Society Special Publication* 83.

Spjeldnaes, N. 1996. Bryozoan colonies as indicators of bottom conditions in the Lower Ordovician. In D.P. Gordon, A.M. Smith & J.A. Grant-Mackie (eds), *Bryozoans in Space and Time*: 315-319. Wellington: National Institute of Water & Atmospheric Research Ltd.

Stach, L.W. 1936. Correlation of zoarial form with habit. *Journal of Geology* 44: 60-65.

Stillman, C.J. 1978. South-east County Waterford and south Tipperary: Ordovician volcanic and sedimentary rocks and Silurian turbidites. *Geological Survey of Ireland Guide Series* 2: 41-60.

Tietzsch-Tyler, D. & Sleeman, A.G. 1994a. *Geology of south Wexford. Sheet 23. Scale 1:100,000.* Dublin, Geological Survey of Ireland.

Tietzsch-Tyler, D. & Sleeman, A.G. 1994b. *Geology of south Wexford. A geological description of south Wexford and adjoining parts of Waterford, Kilkenny and Carlow to accompany the bedrock geology 1:100,000 scale map series, Sheet 23, South Wexford.* Dublin, Geological Survey of Ireland.

Wyse Jackson, P.N., Buttler, C.J. & Key, M.M. Jr. 2001. *Dianulites petropolitana* Dybowski, 1877 and *Diplotrypa petropolitana* Nicholson, 1879 (Trepostomata, Bryozoa): proposed conservation of the specific names. *Bulletin of Zoological Nomenclature* 58(3): 215-219.

Bryozoan Studies 2001, Wyse Jackson, Buttler & Spencer Jones (eds)
© 2002 Swets & Zeitlinger, Lisse, ISBN 90 5809 388 3

An Upper Eocene bryozoan fauna from Perwang-1 borehole, Salzburg, Austria

K. Zágoršek
National Museum, Václavské nám. 68, CZ - 115 79 Prague 1, Czech Republic

ABSTRACT: An Upper Eocene bryozoan marl from Perwang-1 borehole (Salzburg, Austria) was studied and found to contain 129 bryozoan species in four sections. Three studied sections are very similar and represent a repetition of the same environment: a cool-water and/or deep-water basin with rapid and large-scaled oscillations of relative sea level. One studied section represents a different environment: a very shallow basin with a hard bottom. The described bryozoan fauna is similar to that from Helmberg-1 borehole from the Austrian Molasse Zone and to that of the Buda marl from Budapest, Hungary.

1 INTRODUCTION

Upper Late Eocene sediments containing Bryozoa are common in the Alpine-Carpathian region. Bryozoans occur in three major types of sediments: bryozoan marl, calcareous sandstone and bioclastic marl with large foraminifera. The highest diversity of the bryozoan associations is known from the bryozoan marls (Zágoršek & Kázmér 2000). When present in these marls, bryozoans form more than 30 weight percent, Zágoršek 1996b. This fauna is characterized by more than a hundred bryozoan species (mostly erect Cheilostomatous) and are often associated with small amounts of echinoid fragments, planktonic foraminifera, small shells of the brachiopod *Argyrotheca*, and rare bivalve fragments (Zágoršek & Kázmér 2000).

Rasser et al. (1999) and Rasser (2000) reported rich associations of bryozoans from the Upper Austrian Molasse zone. He found bryozoan marl in cores of boreholes Helmberg-1 and Perwang-1. The material from Perwang-1 is summarized in this paper.

1.1 *Geological background*

Borehole Perwang-1, situated about 10 km northeast of Salzburg (Austria) (Fig. 1), penetrates sediments belonging to the Molasse Zone

The Austrian Molasse Zone is a part of the Alpine - Carpathian Foreland Basin. The Molasse sedimentation started during the Oligocene, as input of clastic sediments from the uplifted Alpine thrust front. As the studied Late Eocene sediments contain only clastic sediments coming from the Bohemian Massif (Rasser et al. 1999, Rasser 2000) they belong to the Molasse underground or so-called "Premolasse".

Figure 1. Geographical sketch showing the position of the bryozoan locality Perwang-1 (arrow) and Alpine-Carpathian region.

The transgression of sediments was caused by an asymmetric subsidence of the Alpine Foreland Basin during the subduction of the European plate under the Adriatic Plate during the Eocene. The advancing Alpine nappe system subducted the Rhenodanubic Flysch Zone, which was replaced by the Molasse basin during the Oligocene by accumulating debris from the uplifting alpine orogenic system, which covered the area and obscured the older Eocene sediments. Consequently these Eocene sediments of Austrian Molasse Zone are known only from deep wells (Rasser et al 1999, Rasser 2000). Later, during the Miocene, the Molasse Zone was crushed, deformed and over trusted. It is possible now to distinguish autochthonous and allochthonous components of this succession.

Both components were studied in Perwang-1 borehole. The lowest part of the borehole is autochthonous (Section 4 from a depth of 3301 - 3215 m); the upper five thrust sheets (sections one to three) belong to the allochthonous part of the Molasse Zone (Rasser 1999).

1.2 *Samples and methods*

Studied material comes from the Perwang-1 borehole that was drilled by the Rohöl AG Vienna. The samples were collected in Pettenbach (Upper Austria), where the cores are stored.

Rasser et al (1999) attributed the sediments studied here to nanoplankton zones NP 19-20 - upper Eocene Priabonian, after Martini (1971).

The borehole penetrated lower Cretaceous to lower Neogene Molasse sediments, which were divided into five thrusts sheets and an autochthonous part. Bryozoa occur only in the second and the third thrust sheets and in the autochthonous portion Samples were collected from 8 different drill cores, which represent two different thrust sheets and autochthonous part as repetitions of the upper Eocene sedimentary succession. The number of the drill core is indicated by the first number of the sample.

Detailed descriptions of Perwang-1 drill cores are given in Rasser (1999).

The lowest (autochthonous) part of the deep well comprises quartzitic sandstone followed by *Discocyclina* marl and is topped by massive algal limestone, so-called *Lithothamnium* limestones. Only the interval from 3275.5 m to 3256.5 m was sampled. Altogether 6 samples were taken from this part, but only 5 of them contained bryozoans (these form Section 4 described here – see Table 1).

The dominant lithology in the third thrust sheet is marl, party *Discocyclina* marls with thin algal (*Lithothamnium*) limestone intercalations topped by glauconitic sandstone. Two different sections were sampled within the third tectonic slide. The lower one (Section 3 here) produced 5 samples that came from the depth interval 2380.7 m to 2368.8 m. The upper one (Section 2 here) yielded 7 samples from the depth 2321.1 m to 2310.0 m. Only 3 of the samples however contained bryozoans.

The second thrust sheet is formed by algal (*Lithothamnium*) limestone. *Discocyclina* marls and sandstone occur in the lower part of the sheet, marls formed the upper-most pat of the core. Altogether 8 samples were taken from *Discocyclina* marls from depth interval 2084.5 m to 2071.5 m (these make up Section 1 here).

The samples represent partly litified, calcareous and *Discocyclina* marl. There were no bryozoans in 5 of the samples, from a total of 26 examined.

For details about the depth, samples, studied sections, thrust sheets and lithology see Table 1 and Figure 2.

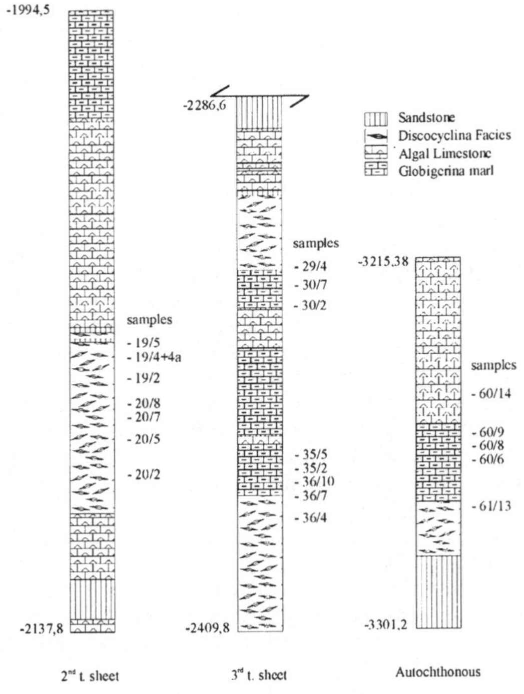

Figure 2. Lithology of the drill cores, thrust sheets and position of the studied sections and samples. Modified from Rasser (1999).

The samples with bryozoans were prepared in different methods. Seven were treated in concentrated acetic acid; the remaining 14 samples were cleaned in Quaternary "O"™. Before taking photographs all the specimens were cleaned in an ultrasonic cleaner with intensity of 80A for 4-8 seconds. For details about these methods see Zágoršek & Vávra (2000).

2 RESULTS

2.1 *Samples*

Average times for acetic acid treatment were 23-24 weeks, but 2 samples were treated more than 36 weeks. The bryozoan colonies were clean; some of the details were however eroded.

The rest of the 14 samples were first macerated in 1% solution of acetic acid for more than one month and than cleaned in Quaternary "O"™ about one day to one week. The bryozoans were very clean without any traces of erosion.

The photos were produced in the Institute of Palaeontology of Vienna University.

The samples from the deeper part of the borehole were more litified and required treatment in high concentrated acid. Acetic acid was used to dissolve some samples from Section 3. The samples from Sections 1 and 2 were disintegrated using only cleaning in Quaternary "O"™. The method used to clean the entire sample is given in Table 1.

The studied and figured material from Perwang-1 borehole has been stored in the Institute of Palaeontology, Vienna University, Austria.

2.2 *Systematic palaeontology*

Together 129 species were identified: 33 cyclostome species and 96 cheilostome species (Table 2).

The taxa indicated as new on Table 2 have been already described from different Upper Eocene basins from the Alpine Carpathian region, but are not yet in print. The identification of some specimens remains uncertain. Further thorough taxonomic study is needed to decide if these are new species. One cyclostome species seems to be new:

Polyascosoecia? sp.
Figure 3, 1 - 3

Diagnosis: The colonies are fixed-walled, erect, rod-like. Zooecia arranged in fascicles, compound from two to four zooecial tubes. There are many large pores between the fascicles in regular position. The frontal part of the colony, between the rows of fascicles, is nonporous and flat. The dorsal side of the colony is formed by parallel kenozooecia closed by diaphragms. Gonozooecium is developed on the dorsal side of the colony with perforated frontal wall. Oeciopore is situated on the distal edge of the gonozooecium

Table 1. Description of the samples studied from Perwang-1 with details of contained fauna.

Tectonic slide	Studied section	depth in m	clean. method	sample core/box	Bryozoa Cheilo	Cyclo	foraminifera	brachiopoda	echinoidea	worm	coral	other fauna
2nd	**1**	2071.5	QO	19/5	***, f	**, f	*	*				
		2072.3	QO	19/4	***	***						
		2072.8	QO	19/4a	**	**		*		***		
		2074.5	QO	19/2	*	*				***		
		2078.5	QO	20/8	***	**						
		2079.5	QO	20/7	*	*		*	*, f			
		2081.5	QO	20/5	*	*	***, b	*				
		2084.5	QO	20/2	**	*	*			***		
3rd	**2**	2310.0	QO	29/4	*	*			*, s		***	molluscs
		2316.1	QO	30/7	***	*	**, b					
		2321.1	QO	30/2	*	*	***,b		*, s	**	*	molluscs
3rd	**3**	2368.8	AC	35/5	*	*	*					
		2371.8	QO	35/2	**, f	*,f		*	*, f			
		2374.7	AC	36/10	*	*	**, p		*, f			
		2377.7	QO	36/7	**, f	*, f						ostracoda
		2380.7	QO	36/4	*	*		*				
auto-chtho-nous	**4**	3256.5	AC	60/14	*	**						
		3261.5	AC	60/9	*	***		*	*, f			
		3262.5	AC	60/8	*	***		*	*, f			
		3264.5	AC	60/6	*, f	**, f						
		3275.5	AC	61/13	*	*	***					

* = present, ** = abundant, *** = dominant. f = fragmented, s = spines, p = plankton, b = benthos, QO = Quaternary "O", AC = acetic acid.

Table 2. List of determined species from Perwang-1 borehole, their distribution in the entire samples and indication of the cleaning procedure of each sample.

depth of sample in the borehole (in m)	3275.5	3264.5	3262.5	3261.5	3256.5	2380.7	2377.7	2374.7	2371.8	2368.8	2321.1	2316.1	2310.0	2084.5	2081.5	2079.5	2078.5	2074.5	2072.8	2072.3	2071.5
taxa / samples (core/core box)	61/13	60/6	60/8	60/9	60/14	36/4	36/7	36/10	35/2	35/5	30/2	30/7	29/4	20/2	20/5	20/7	20/8	19/2	19/4	19/4a	19/5
Adeonella minor (R)	*	*	*	*	*	*	*	*	*	*					*	*	*	*	*	*	*
Adeonella ornatissima (Sto)							*			*						*	*		*		
Adeonellopsis coscinophora (R)									*	*											
Adeonellopsis giampietroi (Zág)	*					*	*	*		*		*	*						*	*	*
Adeonellopsis porina (Ro)	*	*	*	*	*	*	*	*	*	*	*	*	*			*	*	*	*	*	*
Adeonellopsis sp.n. 2										*									*	*	
Alderina subtilimargo (R)						*											*	*	*	*	
Amphiblestrum appendiculata (R)																				*	
Arthropoma rugosa (Re)					*																
Arthropoma sparsipora (R)						*	*											*			
Bactridium hagenowi R	*																				
Batopora multiradiata R	*		*	*	*	*		*		*		*		*	*	*	*				
Berenicea rotula (R)																				*	
Biflustra savartii texturata (R)			*														*				*
Bobiesipora fasciculata (R)										*		*		*					*		
Bryochaperia spinella (Zág)																					*
Calpensia polysticha (R)					*											*					
Cellaria reussi d'Orb																*					
Celleporaria conglomerata (Gold)	*		*	*				*		*						*					
Celleporaria globularis (Br)	*	*	*	*		*		*	*	*		*	*	*			*	*	*	*	*
Crassimarginatella macrostoma (R)			*	*											*	*	*				*
Crisia hoernesii R						*						*						*			
Cyclicopora laticella C & B						*	*														
Cyphonella nodosa Ko						*							*							*	
Diastopora flabellum R																			*		
Diplosolen brendolensis (Wat)																			*		
Disporella coronula (R)																					*
Disporella grignonensis M-E							*									*	*	*	*	*	
Disporella radiata (S-A)																*	*		*		*
Escharella grotriani (Sto)					*		*									*			*	*	*
Escharella tenera (R)		*	*	*	*	*			*	*		*		*				*	*	*	*
Escharoides aliferus (R)														*		*				*	*
Escharoides coccinea (Ab)																				*	
Exidmonea atlantica D., M. & P.	*	*	*	*		*		*				*				*	*		*	*	
Exidmonea giebeli (Sto)			*	*		*	*	*	*	*					*	*	*		*	*	*
Exidmonea hoernesi (Sto)	*	*	*	*	*	*	*	*	*	*		*			*	*	*	*	*	*	*
Exidmonea villaltae (Reg)				*	*	*	*		*								*				
Exochoecia compressa (R)			*	*		*	*									*	*	*	*	*	*
Fedora bidentata (Re)								*						*							
Filisparsa tenella Sto						*	*	*		*									*	*	*
Frondipora sp.												*									
Gaeleopsis subquadrangularis (R)				*		*			*										*	*	*
Gen. nov.1 sp.n.																*					*
Gen. nov.2 sp.n.						*															
Gephyrotes convexa C & B									*												
Gigantopora duplicata (R)							*					*				*	*		*	*	*
Gigantopora lyratostoma (R)						*	*		*							*					*
Gordoniella diporica (Zág)																*					
Herentia hyndmanni (John)							*									*			*		*
Heteropora subreticulata R	*				*					*		*					*				
Hippomenella bragai Zág									*	*	*					*	*				*
Hippomonavella exarata (R)			*	*	*		*		*	*		*				*		*		*	*
Hippomonavella stenosticha (R)	*		*	*	*	*	*	*	*							*	*	*	*	*	*

Table 2. List of determined species from Perwang-1 borehole (continued).

taxa / samples (core/core box)	61/13	60/6	60/8	60/9	60/14	36/4	36/7	36/10	35/2	35/5	30/2	30/7	29/4	20/2	20/5	20/7	20/8	19/2	19/4	19/4a	19/5
Hornera concatenata R	*	*	*	*	*	*	*	*	*	*		*		*		*	*	*	*	*	*
Hornera frondiculata frondiculata M	*	*	*	*	*	*	*	*	*	*					*	*	*	*	*	*	*
Hornera simplicissima B & B															*	*	*				
Hornera subannulata Phil				*															*	*	
Hornera verrucosa R			*					*											*		
Cheilonella prominens R																*	*				
Cheilopora orbifera C & B							*														
Chlidoniopsis sp.n.																					*
Iodictyum labellatum (Zág)										*											
Iodictyum rubeschi (R)						*										*	*		*	*	
Kionidella excelsa Ko						*								*		*	*				
Lacrimula perfecta (Acc)								*													
Lagenipora urceolaris Gold																			*		
Lunulites quadrata (R)											*	*	*	*							
Margaretta cereoides (E & S)																*	*				
Mecynoecia anomala (R)						*	*														*
Mecynoecia geinitzi R																			*		
Mecynoecia proboscidea (M-E)				*	*	*	*	*	*		*	*		*	*	*	*	*	*	*	*
Mecynoecia pulchella (R)	*	*	*	*	*	*	*	*	*	*		*	*	*	*	*	*	*	*	*	*
Meniscopora syringopora (R)							*									*	*	*	*	*	*
Metradolium obliquum C & B								*								*			*	*	*
Metrarabdotos maleckii Cheet					*		*									*	*			*	*
Micropora hexagona (Zág)			*						*									*			
Nematifera susannae Zág			*	*				*													
Oncousoecia biloba (Re)																*	*				
Onychocella subpyriformis (d'Ar)				*		*	*		*			*				*	*		*	*	*
Opaeomorpha michaliki (Zág)							*									*	*			*	
Orbitulipora petiolus Lon	*							*		*									*	*	
Parasmittina saccoi (C)																*			*	*	
Perigastrella granulata Zág						*	*									*	*		*	*	*
Perigastrella oscitans C & B																			*		*
Plagiosmittia denticulifera C & B							*												*	*	*
Pleuronea fenestrata (Busk)																*					
Polyascosoecia coronopus (C & B)	*		*	*	*	*	*	*	*	*		*		*		*	*	*	*	*	
Polyascosoecia? sp.							*				*							*	*	*	*
Porella clavula (C & B)												*				*			*	*	*
Porina coronata (R)			*	*	*	*	*	*	*	*		*				*	*	*	*	*	*
Porina duplicata (R)							*										*		*		
Porina labrosa (R)																			*		
Porina peristomica (Zág)					*				*												
Prenantia phymatopora (R)					*					*						*					
Puellina (Cribrilaria) haueri (R)							*									*	*		*	*	*
Puellina (Cribrilaria) manzonii (R)																*			*		
Puellina (Cribrilaria) radiata (Moll)		*	*	*	*		*		*							*	*	*	*	*	*
Reteporella simplex (Busk)										*					*	*	*		*	*	*
Reteporella subovata (Sto)	*		*		*	*	*	*	*	*						*	*		*	*	*
Reteporella tamaninii A., B & F.					*			*		*											
Reteporella tuberculata (R)									*						*				*	*	
Reusia (Smittina) regularis (R)		*	*		*				*			*		*			*	*	*	*	*
Schizolepralia fissimargo (R)					*												*				*
Schizolepralia scrobiculata (R)					*				*	*		*				*	*		*	*	*
Schizoporella cf. errata (R)																*					
Schizoporella dunkeri (R)						*						*				*	*			*	*
Schizoporella geminipora (R)							*					*									
Schizoporella tetragona (R)				*	*																

Table 2. List of determined species from Perwang-1 borehole (continued).

taxa / samples (core/core box)	61/13	60/6	60/8	60/9	60/14	36/4	36/7	36/10	35/2	35/5	30/2	30/7	29/4	20/2	20/5	20/7	20/8	19/2	19/4	19/4a	19/5
Smittina cervicornis (Pallas)		*			*	*	*	*		*		*			*	*	*	*	*	*	*
Smittoidea angulata (Bronn)																					*
Smittoidea excentrica (R)							*				*					*	*	*	*		*
Steginoporella cucullata (R)					*	*										*	*		*	*	
Steginoporella elegans chattien. P&D																	*				
Steginoporella firma (R)												*									
Steginoporella haidingeri (R)				*		*	*		*							*	*		*	*	*
Stenosipora protecta (Ko)														*		*					
Stenosipora reusii (Sto)	*		*	*	*	*		*		*		*					*				
Stenosipora simplex (Ko)					*																
Stephanollona otophora (Re)																*					
Teichopora cf. clavata Gre					*																
Tervia serrata (R)						*			*						*	*	*	*	*		*
Tubucella mammillaris (M-E)							*			*						*					
Tubucella papillosa (R)	*					*			*			*	*		*	*	*	*	*		*
Tubulipora flabelaris (Fab)			*																		*
Tychinella schreibersi (R)							*					*									*
Umbonula monoceros (R)							*					*		*		*					
Vibracella trapezoidea (R)	*									*				*		*			*		
Vicariopora chelys (Ko)						*										*	*				
Ybselosoecia typica (Man)																	*				
Total number of species 129	20	12	27	28	33	41	44	27	31	33	6	31	7	16	14	66	55	27	61	53	56

2.3 *Comparison with other Upper Eocene localities*

The succession described from Perwang-1 borehole was the only example where tectonic repetition of bryozoan marls intercalated in algal limestone, *Discocyclina* marl and *Globigerina* marls occurs.

The bryozoan association of Perwang-1 borehole was compared with other known bryozoans occurrences in the Eocene of Alpine Carpathians region using the Kojumdgieva similarity coefficient (Braga et al. 1996). The Kojumdgieva similarity coefficient is counted as

$$K = (C / A * 100 + C / B * 100) / 2$$

where A = the number of species from locality a; B = the number of species from locality b; C = the number of species common both to a and b (Braga et al 1996). Table 3 displays the results obtained from the Kojumdgieva similarity coefficient analysis. If the Kojumdgieva similarity coefficient is higher than 50, the faunas are considered very similar.

All the localities compared were very similar to Perwang-1. It can be caused by similar host rock (bryozoan marl), which represents perhaps a similar environmental condition. The most similar association has been described from Helmberg-1 (Zágoršek, in press), where, however, algal limestone terminated the succession. Very similar also are the associations from Buda marls - Hungary (Zágoršek & Kázmér 2001) that exhibit a similar faunal content. The highest number of common species was reported from Hybica (Slovakia). The list of known species from each locality is taken from Zágoršek & Kázmér 2000

Table 3. Kojumdgieva similarity coefficients among the studied localities (Zágoršek & Kázmér 2000, modified). H + S = Hybica and Štrba localities (Slovakia), Hb1 = borehole Helmbreg-1 (Austria), Bm = Buda marls (Hungary), It = Italian localities, Ro = Romanian localities, WZ = Waschberg Zone (Austria) Pe-1 = Perwang-1.

	Hb1	Bm	It	Ro	WZ	Pe-1
H + S	50.9	59.2	64.1	52.4	52.7	59.6
Hb1		64.4	50.1	56.5	57.9	74.2
Bm			60.7	50.2	60.4	69.2
It				62.1	53.2	57.6
Ro					50.3	57.9
WZ						69.4

The most similar locality in faunal content to that of Perwang-1 is the closely situated Helmberg-1 borehole. Similarity indices show that the associations in nearby localities are more similar than in more distant localities even though the sediments are more similar to these distant localities (Zágoršek & Kázmér 2000). This is probably caused by the capability of lecitotrophic bryozoan larvae, which are not able to swim long distances and, as well, cannot survive long-distance planktonic transport (Braga 1987). Any geographic barrier did not limit the dispersal of bryozoan faunas.

All samples from the Perwang-1 borehole were compared one to each other to obtain the similarity coefficient between the studied sections. The similarity coefficients, between samples in the same position within the thrust sheets are very high. For example the similarity coefficient between the lowest sample from the autochthonous Section 4 slide; (61/13) and the lowest sample from the third tectonic slide (36/4) is 55.8 what is very high. All the values are given in the Table 4.

The similarity coefficient between the samples from the Section 2 and other three sections are however very different, which shows the different position of the section in the Upper Eocene succession and perhaps as well the different environmental conditions that were present.

2.4 *Palaeoecology*

Bryozoan marl represents generally a deep and/or cool water association that may have formed by upwelling of cool water into the shallow tropical environment (Zágoršek 1993, 1996b).

According to the similarity coefficients, between samples from the first, third and fourth studied Sections (Table 4) the environmental conditions were probably the same. Because the fauna from these sections are highly similar to others localities with bryozoan marl, one can infer that the environmental conditions were perhaps very similar.

The original paleoenvironment at these localities may have comprised the margin of the continental shelf, or perhaps the continental slope to depths of at least 200 m and perhaps as much as 500 m (Zágoršek 1996a). The appearance of deep-water faunas in successions of continental shelves could be explained by upwelling of cold water in a time of starting of global climate deterioration (Zágoršek 1996a) and/or by the filling the ecological niches opened after the decimation of larger foraminifer faunas on the same shelves (Kázmér 1999).

The fauna from the Section 2 is different and the similarity coefficients with other samples are low (see Table 4). It is the only section, where *Lunulites* occurs. *Lunulites* is known to live in high-energy areas and usually in shallow environments. It is also the only section where free-living bryozoans dominated over erect forms. The bryozoans with free-living forms were prevalent in shallow water and on unconsolidated substrates on the sea bed. Also the occurrences of the other fauna are specific: remains of ahermatypic corals and molluscs occur only in the samples from this section (for details see Table 1).

According to the ecological needs of these organisms, I suppose to regard the original environment of sediment from second studied section as shallow below the wave base, probably on photic zone, with high energy and low sedimentation rate. The bottom was probably formed by unconsolidated sand.

Section 2 is situated on the top of the third thrust sheet and represents a shallow water environment. Section 3 is situated below the second one and represents a deep-water environment (see Fig. 2). It suggests, that the third thrust sheet represents a shallowing upwards sequence (or a tectonically reversed sequence). This inverted order may suggest also the presence of sandstone on the top of the third thrust sheet; on the other thrust sheets the sandstone forms the bottom part (Fig. 2). The study of the ages of these horizons (Rasser 1999) did not indicate the reversed stratigraphic sequence.

Table 4. Similarity coefficient of the samples within the same position in the studied sections

4. slide			3. slide			1. slide			2. slide	
depth	sample	similarity	depth	sample	similarity	depth	sample	similarity	depth	sample
3256.5	60/14	54.5	2368.8	35/5				34.6	2310.0	29/4
					60.1	2071.5	19/5			
3261.5	60/9	64.6	2371.8	35/2	55.9	2072.3	19/4			
					56.2	2072.8	19/4a			
					62.4	2074.5	19/2			
3262.5	60/8	66.7	2374.7	36/10	46.9	2078.5	20/8			
3264.5	60/6	54.8								
		37.5	2377.7	36/7	62.2	2079.5	20/7	42.7	2316.1	30/7
3275.5	61/13	55.8	2380.7	36/4	47.9	2081.5	20/5	13.6	2321.1	30/2
					34.8	2084.5	20/2	26.3		

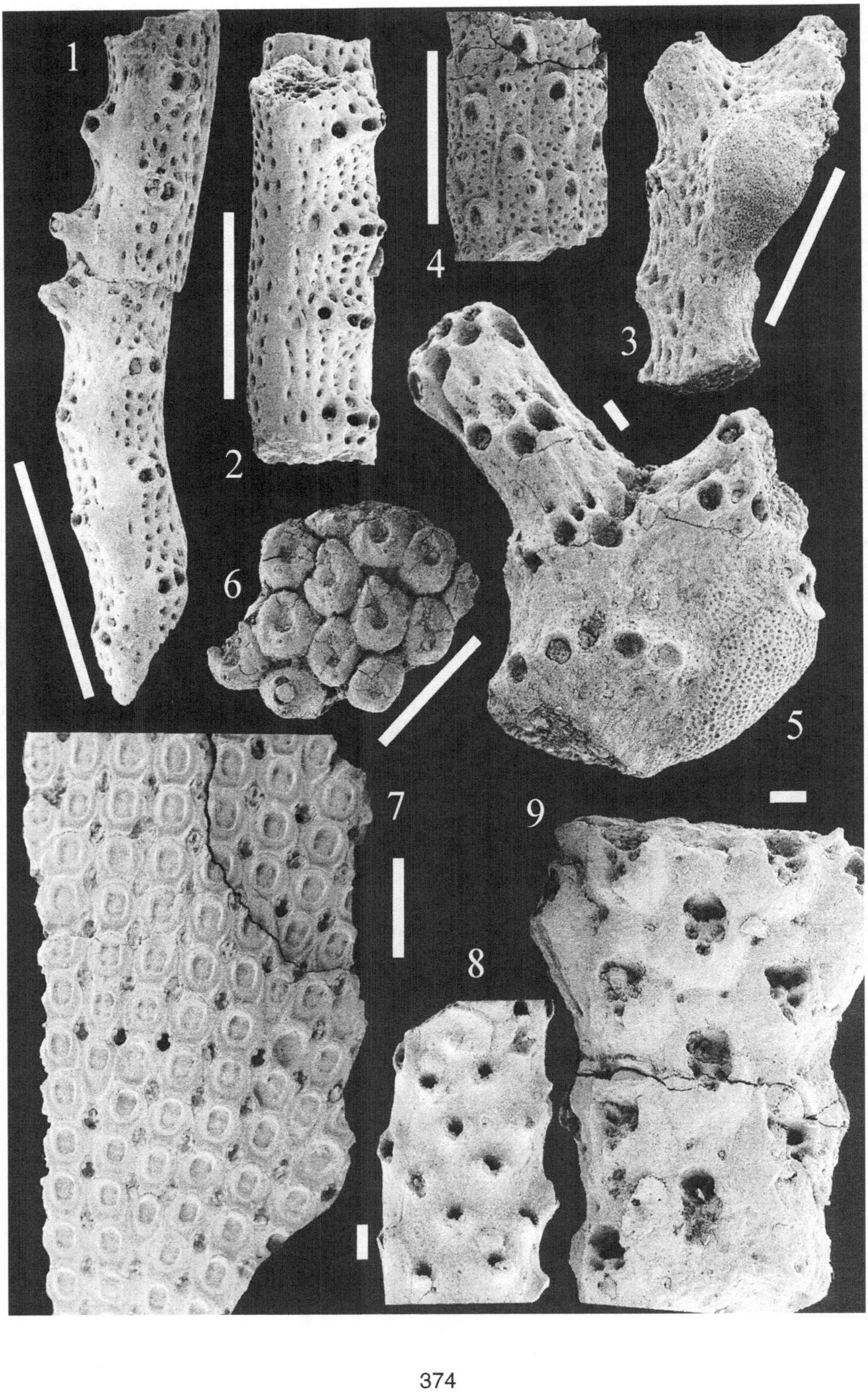
1
2
3
4
5
6
7
8
9

Figure 3. Upper Eocene Bryozoa from Perwang-1 borehole, Austria. 1. *Polyascosoecia*? sp. - frontal view showing nonporous frontal part between rows of fascicles. 2. *Polyascosoecia*? sp. - lateral view showing regularly arranged pores between fascicles. 3. *Polyascosoecia*? sp. - dorsal view showing gonozooecium and parallel kenozooecia closed by diaphragms (so called dactylethrae) 4. *Schizoporella* (?) *scrobiculata* (Reuss) - general view shown position of the avicularia. 5. *Mecynoecia pulchella* (Reuss) - part of the colony with bifurcation and gonozooecium. Scale bar = 100 µm. 6. *Stenosipora protecta* (Koschinsky) - part of the colony shown zooecia with one small avicularium. 7. *Lunulites quadrata* (Reuss) - large fragment of the colony. 8. *Iodictyum rubeschi* (Reuss) - middle part of the colony showing frontal zooecia situated on some autozooecia. Scale bar = 100 µm. 9. *Galeopsis subquadrangularis* (Reuss). - general view showing characteristic development of the proximal part of the zooecial aperture. Scale bar = 100 µm. Unless otherwise indicated, all scale bars = 1 mm.

3 CONCLUSIONS

Aggressive methods (acetic acid treatment) have to be used for preparing and cleaning the samples from deeper part of the borehole than from the shallower one. It suggests that the samples, from the deepest studied section were more litified than the shallower samples from the first, second and third sections.

The similarity coefficients are very high between the samples from Sections 1, 3 and 4 studied here, because of the tectonical repetitions. The paleoenvironmental conditions were probably the margin of the shelf with active upwelling of cold water with muddy bottom.

Section 2 is the only one, where *Lunulites,* remains of ahermatypic corals and molluscs occur. *Lunulites,* corals and molluscs dominated usually in very shallow environment. The similarity coefficients with other samples are also low (see Table 4). This suggests that the original environment of sediment from Section 2 was very shallow with clear water and with sandy bottom.

The most similar locality in terms of bryozoan content is Helmberg-1, which proves the theory that the geographically close localities are more similar than the distant localities

ACKNOWLEDGEMENTS

Many thanks go to Prof. N. Vávra (Institute of Palaeontology, Vienna University, Austria) and Dr Yvonne Bone (Department of Geology and Geophysics, University of Adelaide, Australia) who helped me with kind and very useful comments, and to Dr Michael W. Rasser (Geologische Bundesanstalt, Vienna, Austria) who gave me the material and helped me with fieldwork. The kind assistance of the Rohöl AG Vienna, namely Dr W. Nachtmann and Dr H. Polessny, who supported the sampling and who gave me the possibility to study the core material, was highly appreciated. I am very thankful also to Dr Patrick Wyse Jackson (Department of Geology, Trinity College Dublin, Ireland), who proof read the paper and improved my English. This research was supported by FWF (Fonds zur Förderung der wissenschaftlichen Forschung, Austria) through the Project M517-GEO. I would like to express my sincerely thanks to all of them.

REFERENCES

Braga, G. 1987. Tethyan migration of some Tertiary Bryozoa. - In K.G. McKenzie, (eds), *Shallow Tethys 2. Proceedings of the International Symposium on Shallow Tethys 2 Wagga Wagga*: 79-85. Rotterdam: Balkema.

Braga, G., Zágoršek, K. & Kázmér, M. 1996. Comparison between Venetian and Western Carpathian Late Eocene Bryozoan Faunas. *Supplemento agli Annali dei Musei Civici do Rovereto, Sezione Archeologia, Storia e Scienze Naturali* 11: 259-270.

Kázmér, M. 1999. Climate and carbonate deposits in the Palaeogene Paratethys. In B. Ratanasthien & S.L. Rieb (eds), *Proceedings of the International Symposium on Shallow Tethys 5: Chiang Mai, Thailand, 1-5 February 1999:* 496-497. Chiang Mai: Chiang Mai University.

Martini, E. 1971. Standard Tertiary and Quaternary calcareous nannoplankton zonation. *Proceedings of the 2nd Planktonic Conference*: 739-785. Rome.

Rasser, M. W. 1999. Facies and palecology of Late Eocene red Algal Limestones of the Upper Austrian Molasse Zone. *Dissertation zur Erlangug eines Doktorgrades der Formal- und Naturwissenschaftlichen Fakultat der Universitat Wien.*

Rasser, M., Less, Gy. & Báldi-Beke, M. 1999. Biostratigraphy and facies of the late Eocene in the Western Austrian Molasse zone with special reference to the large Foraminifera. *Abhandlungen der Geologischen Bundesanstalt* 56(2): 679-698.

Rasser, M. 2000. Coralline Red Algal Limestones of the Late Eocene Alpine Foreland Basin in Upper Austria: Component Analysis, facies and paleoecology. *Facies* 42: 59-92.

Zágoršek, K. 1993. Changes in Bryozoa community in the Upper Eocene sequence of Mátyáshegy (Hungary). *Öslénytani Viták (Discussiones Palaeontologicae)* 39: 91-96.

Zágoršek, K. 1996a. Eocénne koilostégne machovky Bryozoa: Coilostega zo Západných Karpát. *Mineralia Slovaca* 28: 523-534.

Zágoršek, K. 1996b. Paleoecology of the Eocene Bryozoan Marl in the Alpine-Carpathian Region. In D.P. Gordon, A.M. Smith & J.A. Grant-Mackie (eds), *Bryozoans in Space and Time*: 413-422. Wellington: National Institute of Water & Atmospheric Research.

Zágoršek, K. in press. Upper Eocene Bryozoa from the Alpine Foreland Basin in Salzburg, Austria (Borehole Helmberg-1). In W.E. Piller & M.W. Rasser (eds), *Paleogene of the Eastern Alps*. Österreichische Akademie der Wisssenschaften, Schriftenreihe der Erdwissenschaftlichen Kommissionen, 14: 505-608.

Zágoršek, K. & Kázmér, M. 2000. Late Eocene Bryozoan Faunas in the Alpine-Carpathian Region - a comparison. *Acta Paleontol. Romaniae* 2: 493-504.

Zágoršek, K. & Vávra, N. 2000. New method for the extraction of Bryozoans from hard rocks from the Eocene of Austria. *Jahrbuch der Geologischen Bundesanstalt* 142(2): 249-258.

Zágoršek, K. & Kázmér, M. 2001. Eocene Bryozoa from Hungary. *Courier Forschungsinstitut Senckenberg* 231: 1-158.

Miscellaneous

Bryozoan Studies 2001, Wyse Jackson, Buttler & Spencer Jones (eds)
© 2002 Swets & Zeitlinger, Lisse, ISBN 90 5809 388 3

Author index to the conference volumes (1-12) of the International Bryozoology Association (1968-2002)

Patrick N. Wyse Jackson
Department of Geology, Trinity College, Dublin 2, Ireland

This index lists all papers and abstracts published in the first twelve conference volumes of the International Bryozoology Association. Items are listed chronologically under first author.

Bibliographic details of conference volumes

1: Annoscia, E. (ed.), 1968. *Proceedings of the First International Conference on Bryozoa.* Atti della Societa di Scienze Naturalia del Museo Civico di Storia Naturale di Milano. Volume 108, Milano. [1st IBA Conference, Milan, Italy, 12th -16th August, 1968]
2: Larwood, G.P. (ed.), 1973. *Living and Fossil Bryozoa: recent advances in research.* Academic Press, London and New York. [2nd IBA Conference, Durham, UK, 6th - 16th September, 1971]
3: Pouyet, S. (ed.), 1975. *Bryozoa 1974. Proceedings of the Third Conference. International Bryozoology Association.* Documents des Laboratoires de Geologie de la Faculte des Sciences de Lyon, HS 3 (fasc. 1, 2). Universite Claude Bernard, Lyon. [3rd IBA Conference, Lyons, France, 2nd - 11th September, 1974]
4: Larwood, G.P. and Abbott, M.B. (eds), 1979. *Advances in Bryozoology.* Academic Press, London. (The Systematics Association, Special Volume No.13). [4th IBA Conference, Woods Hole, Massachusetts, USA, 7th - 17th September, 1977]
5: Larwood, G.P. and Nielsen, C. (eds), 1981. *Recent and Fossil Bryozoa.* Olsen & Olsen, Fredensborg. [5th IBA Conference, Durham, United Kingdom, 1st - 6th September, 1980]
6: Nielsen, C. and Larwood, G.P. (eds), 1985. *Bryozoa: Ordovician to Recent.* Olsen & Olsen, Fredensborg. [6th IBA Conference, Wien, Austria, 18th - 23rd July, 1983]
7: Ross, J.R.P. (ed.), 1987. *Bryozoa: Present and Past.* Western Washington University Press, Bellingham. [7th IBA Conference, Bellingham, Washington, USA, 4th - 9th August, 1986]
8: Bigey, F.P. and J.L. d'Hondt (eds), 1991. *Bryozoaires Actuels et Fossiles.* Bulletin de la Societe Sciences Naturelles de la Ouest de France, Memoire H.S. 1. Nantes. [8th IBA Conference, Paris, France, 17th - 22nd August, 1989]
9: Hayward, P.J., Ryland, J.S. and Taylor, P.D. (eds) 1994. *Biology and Palaeobiology of Bryozoans.* Olsen & Olsen, Fredensborg. [9th IBA Conference, Swansea, Wales, 25th July - 1st August, 1992]
10: Gordon, D.P., Smith, A.M. and Grant-Mackie, J.A. (eds), 1996. *Bryozoans in Space and Time.* NIWA, Wellington. [10th IBA Conference, Wellington, New Zealand, 30th January - 3rd February 1995]
11: Herrera-Cubilla, A. and Jackson, J.B.C. (eds), 2000. *Proceedings of the 11th International Bryozoology Association Conference.* Smithsonian Tropical Research Institute, Panama. [11th IBA Conference, Panama, 26th - 31st January 1998]
12: Wyse Jackson, P.N., Buttler, C.J. and Spencer Jones, M.E. (eds), 2002. *Bryozoans Studies: Dublin 2001.* Balkema, Rotterdam. [12th IBA Conference, Dublin, Ireland, 16th - 21st July, 2001]

Author index

Abbott, M.B. - Seasonal diversity and density in bryozoan populations of Block Island Sound (New York, U.S.A.). **2:** 37-51.
Abbott, M.B. - Relationship of temperature to patterns of sexual reproduction in some Recent encrusting Cheilostomata. **3:** 37-50.
Agrawal, V. and Rao, K.S. - On a report of the elemental composition of some Indian freshwater Ectoprocta. **6:** 343.
Androsova, E.I. - Bryozoa Cyclostomata and Ctenostomata of Antarctic and Subantarctic. **1:** 258-260.
Androsova, E.I. - Bryozoa Cheilostomata (Anasca) of the Antarctic and Subantarctic. **2:** 369-373.
Annoscia, E. - Paleontological and stratigraphical research carried out by AGIP Direzione mineraria

Paleontological Laboratory. **1:** 25-27.
Annoscia, E. - Problems of methodology in studying and describing Bryozoa. **1:** 237-240.
Annoscia, E. - Status of the bryozoological studies and collections in Italy. **1:** 247-257.
Annoscia, E. and Fierro, G. - Bryozoan ecology in relation to sediment texture in the "Golfo dell'Asinara" (Sardinia, Italy). **2:** 53-64.
Annoscia, E. - Status of the bryozoological studies in Italy up to 1974. **3:** 617-625.
Annoscia, E. - Bryozoan studies in Italy in the last decade: 1973 to 1982. **6:** 1-7.
Anstey, R.L and Perry, T.G. - Biometric procedures in taxonomic studies of Paleozoic Bryozoa. **1:** 241-244.
Anstey, R.L. - Colony patterning and functional morphology of water flow in Paleozoic stenolaemate bryozoans. **7:** 1-8.
Aprosi, G. - The rapid and widespread growth of bryozoans in the cooling water circuits of a power plant. **8:** 567.
Ascoli, P. and Cook, P.L. - Conference report. **1:** 13-24.
Astrova, G.G. - Polymorphism and its development in the trepostomatous Bryozoa. **2:** 1-10.
Backus, B.T. - Bryozoan chromosomes: some evolutionary implications. **4:** 1-9.
Bader, B. - Life cycle, growth rate and carbonate production of *Cellaria sinuosa*. **11:** 136-144.
Bader, B. - Bryozoan communities in the Weddell Sea, Antarctica: a first overview. **12:** 1-6.
Balduzzi, A., Barbieri, M. and Gristina, M. - Morphology and life strategies of *Aetea* (Bryozoa Cheilostomata) living on some western Mediterranean *Posidonia oceanica* meadows. **8:** 1-12.
Balson, P.S. - Facies-related distribution of bryozoans of the Coralline Crag (Pliocene) of eastern England. **5:** 1-6.
Balson, P.S. and Taylor, P.D. - Palaeobiology of large cyclostome bryozoans from the Pliocene of England. **5:** 309.
Banta, W.C. - The body wall of the encrusting cheilostome, *Watersipora nigra* (Canu & Bassler) (Bryozoa). Preliminary report. **1:** 93-96.
Banta, W.C. - The significance of areolae in cheilostome Bryozoa. **2:** 209-219.
Banta,W.C. - Origin and early evolution of cheilostome Bryozoa. **3:** 565-582.
Banta, W.C and Wass, R.E. - Preliminary results from a study of the Family Catenicellidae. **4:** 461-465.
Barnes, D.K.A. - Low levels of colonisation in Antarctica: the role of bryozoans in early community development. **10:** 19-28.
Barnes, D.K.A. - Life patterns of encrusting and erect Antarctic Cheilostomatida. **11:** 145-153.
Barnes, D.K.A. and Sanderson, W.G. - Latitudinal patterns in the colonization of marine debris. **11:** 154-160.
Barnes, D.K.A. and De Grave, S. - Modelling multivariate determinants of growth in Antarctic bryozoans. **12:** 7-17.
Barnes, D.K.A. and Maughan, B.C. - Seasonality and inter-annual variability in recruitment patterns of temperate encrusting fauna. **12:** 19-27.
Bartley, J.W and Anstey, R.L. - Growth of monilae in the Permian trepostome *Tabulipora carbonaria*: evidence for periodicity and a new model of stenolaemate wall calcification. **7:** 9-16.
Bayer, M.M and Todd, C.D. - Effect of polypide regression and other parameters on colony growth in the cheilostomate *Electra pilosa* (L.). **10:** 29-38.
Best, B. and Winston, J.E. - Zooid strength in encrusting cheilostomes. **6:** 344.
Best, M.A. and Thorpe, J.P. - Feeding current velocity and rates of particle clearance in marine Bryozoa. **6:** 343.
Best, M.A. and Thorpe, J.P. - Bryozoan faecal pellets: parameters and production rates. **7:** 17-24.
Best, M.A. and Thorpe, J.P. - The uptake of 14C radioisotope tracer from dissolved glycine and glucose by the marine bryozoan *Flustrellidra hispida* (Fabricius). **8:** 13-23.
Best, M.A. and Thorpe, J.P. - An analysis of potential food sources available to intertidal bryozoans in Britain. **9:** 1-7.
Best, M.A. and Thorpe, J.P. - Particle size, clearance rate and feeding efficiency in marine Bryozoa. **9:** 9-14.
Best, M.A. and Thorpe, J.P. - The effect of suspended particulate matter (silt) on the feeding activity of the intertidal ctenostomate bryozoan *Flustrellidra hispida* (Fabricius). **10:** 39-45.
Best, M.A. and Thorpe, J.P. - Use of radioactive labelled food to assess the role of the funicular system in the transport of metabolites in the cheilostome bryozoan *Membranipora membranacea* (L.). **12:** 29-35.
Best, M.A. and Thorpe, J.P. - Use of radioactive labelled silt to show depletion of food supply by upstream colonies of *Flustrellidra hispida* (Fabricius). **12:** 37-40.
Bigey, F.P. - Devonian Bryozoa from the southeastern Armorican Massif, western France. **2:** 375-383.
Bigey, F.P. and Feist, R. - Cryptostomata (Bryozoaires) des "Calcaires a polypiers siliceux" (Devonien Inferieur de la Montagne Noire, France Meridionale). Note preliminaire. **3:** 257-271.
Bigey, F.P. - Microstructure of the main types of Devonian Bryozoa from the Ferques area, Boulonnais, northern France. **4:** 153-177.
Bigey, F.P. - Overgrowths in Palaeozoic Bryozoa: examples from Devonian forms. **5:** 7-17.
Bigey, F.P. - Biogeography of Devonian Bryozoa. **6:** 9-23.
Bigey, F.P. - Paleoecological aspects of Devonian bryozoan assemblages in Brittany and Boulonnais, France. **7:** 25-32.
Bigey, F.P. - Peculiar structures different from typical skeleton Bryozoa: possible relations to the polypide life cycle. **8:** 25-38.

Bigey, F.R. - Bryozoans as components of a Devonian fauna in the Safi area (Morocco). **9:** 15-21.
Billing, I. - Bryozoan growth on brachiopod spines in the Carboniferous of England. **8:** 39- 47.
Bishop, J.D.D. - Survival and reproductive output in relation to substrate type in a bryozoan encrusting disarticulated bivalve shells. **9:** 23-28.
Bizzarini, F. - *Braiesopora voigti* n. gen. n. sp. (cyclostome bryozoan) in the S. Cassiano Formation in the Eastern Alps (Italy). **6:** 25-33.
Bizzarini, F. and Braga, G. - *Corynotrypoides ladina* gen. et sp. nov., a questionable cyclostomatous bryozoan from the Upper Triassic of the Eastern Dolomites (NE Italy). **9:** 29-32.
Blake, D.B. - Acanthopore ultrastructure in the Palaeozoic bryozoan family Rhabdomesidae. **2:** 221-229.
Blake, D.B. - The Order Cryptostomata resurrected. **3:** 211-223.
Blake, D.B. - The Arthrostylidae and articulated growth habits in Paleozoic bryozoans. **4:** 337-344.
Blake, D.B. and Snyder, E.M. - Phenetic and cladistic analyses of the Rhabdomesina (Bryozoa) and similar taxa: a preliminary study. **7:** 33-40.
Boardman, R.S. - Potential use of Paleozoic Bryozoa in subsurface exploration. **1:** 28-30.
Boardman, R.S. - Colony development and convergent evolution of budding pattern in "rhombotrypid" Bryozoa. **1:** 179-184.
Boardman, R.S. - Body walls and attachment organs in some Recent cyclostomes and Palaeozoic trepostomes. **2:** 231-246.
Boardman, R.S. - Taxonomic characters for phylogenetic classifications of cyclostome Bryozoa. **3:** 595-606.
Boardman, R.S. - Origin of the post Triassic stenolaemates. **5:** 309.
Boardman, R.S and McKinney, F.K. - Soft part characters in stenolaemate taxonomy. **6:** 35-44.
Boardman, R.S. - Colony life then and now: Lower Paleozoic trepostomes (500-350 mya) and living cyclostomes: a review. **12:** 41-51.
Bock, P.E. and Cook, P.L. - Occurrence of three phases of growth with taxonomically distinct zooid morphologies. **9:** 33-36.
Bock, P.E and Cook, P.L. - Concatenellidae, a new family of Catenicelloidea (Cheilostomatida) for *Concatenella airensis* (Maplestone) from the Tertiary of Victoria. **10:** 47-53.
Bock, P.E. and Cook, P.L. - Early astogeny of *Adeona* colonies. **11:** 161-167.
Bone, Y. and James, N.P. - Bryozoans from temperate Pleistocene deep-water mounds, Great Australian Bight, Australia. **12:** 53-59.
Braga, G.P. - Bryozoa from the "Marne Rosse" of Spilecco (Lessini Mountains, Verona). **1:** 303-311.
Braga, G.P. - Reuss' collection of cheilostome bryozoans from Venetia stored in natur-historisches Museum Wien: a proposed revision. **8:** 49-59.
Brock, B.J. - South Australian Fouling Bryozoans. **6:** 45-49.
Brood, K. - Stenolaematous ectoprocts from the Danian deposits of Sweden and Denmark (Bryozoa) (Preliminary report). **1:** 300-302.
Brood, K. - Palaeozoic Cyclostomata (a preliminary report). **2:** 247-256.
Brood, K. - Paleoecology of Silurian Bryozoa from Gotland (Sweden). **3:** 401-414.
Brood, K. - Hirnantian (Upper Ordovician) Bryozoa from Baltoscandia. **5:** 19-27.
Brown, G.D. Jr and Daly, E.J. - Evolutionary trends in *Parvohallopora* Singh in the Dillsboro Formation (Upper Ordovician) of southeastern Indiana. **5:** 29-38.
Brown, G.D Jr. - Analysis of evolution of structural characters of *Parvohallopora* Singh from the Dillsboro Formation, Cincinnatian Series, Indiana. **6:** 51-58.
Brown, G.D. Jr and Daly, E.J. - Trepostome bryozoan distributions in the Dillsboro Formation (Cincinnatian Series), Indiana: water depth controls? **7:** 41-47.
Brown, K.M., Schmidt, R. and Bone, Y. - Observations on ecological adaptations of *Lanceopora smeatoni* (MacGillivray), West Island, South Australia. **12:** 61-65.
Buge, E. - Distribution des bryozoaires lunulitiformes ser les cotes du Bresil. **3:** 439-448.
Buge, E. and Reguant, S. - Les bryozoaires Oligocenes de la region Cantabrique (Province de Santander, Espagne). Etude Preliminaire. **3:** 483-495.
Bulman, O.M.B. - Graptolite periderm structure and budding patterns: a resumé. **2:** 11-20.
Bushnell, J.H. - Aspects of architecture, ecology and zoogeography of freshwater Ectoprocta. **1:** 129-151.
Bushnell, J.H. - The freshwater Ectoprocta: a zoogeographical discussion. **2:** 503-521.
Bushnell, J.H and Rao, K.S. - Freshwater Bryozoa: Micro-architecture of statoblasts and some Aufwuchs animal associations. **4:** 75-91.
Buss, L.W. - Mechanisms of competition between *Onychocella alula* (Hastings) and *Antropora tincta* (Hastings) on an easern Pacific rocky shoreline. **5:** 39-49.
Butler, K.L and Cuffey, R.J. - Reduced bryozoan diversity and paleoenvironmental stress in the Saluda Dolomite. **10:** 55-61.
Buttler, C.J. - Possible brooding structure in rhinoporid cystoporate bryozoans. **8:** 61- 70.
Buttler, C.J. - Ordovician (Ashgill) microbryozoans from the Kildare Limestone, Ireland. **9:** 37-40.
Buttler, C.J. and Massa, D. - Late Ordovician bryozoans from carbonate buildups, Tripolitania, Libya. **10:** 63-68.
Cadeé, G.C. - *The Cupuladria canariensis* complex. **4:** 443-459.
Cadeé, G.C. - The shallow soft-bottom faunas of the Java Sea and Banda Sea. **7:** 49-56.

Cadeé, G.C. - Carapaces of the shore crab *Carcinas maenas* as a substrate for encrusting organisms. **8:** 71-79.
Cadman, P.S and Ryland, J.S. - The characters, reproduction, and growth of *Alcyonidium mytili* Dalyell, 1848 (Ctenostomatida). **10:** 69-79.
Cancino, J.M., Orellana, M.C. and Munõz, M. - Consequences on colony size on the metaholic rates of the bryozoans *Membranipora isabelleana* and *Cauloramphus spiniterum.* **8:** 567-568.
Cancino, J.M., Castañeda, B. and Orellana, M.C. - Reproductive strategies in bryozoans: experimental test of the effects of conspecific neighbours. **8:** 81- 88.
Cancino, J.M., Hughes, R.N. and Orellana, M.C. - A comparative study of larval release in bryozoans. **9:** 41-46.
Cancino, J.M. and Gallardo, J.A. - Effects of light on bryozoan larval settlement. **11:** 168-175.
Cancino, J.M., Torres, F. and Moyano G., H.I. - Larval release pattern in Antarctic bryozoans. **12:** 67-72.
Carson, R. - Bryozoans of Northumberland Strait, Gulf of St. Lawrence. **6:** 59-64.
Carthew, R. - The cyclostome bryozoan *Corynotrypa* from the Silurian of Gotland, Sweden. **7:** 57-64.
Chadwick, S.R and Thorpe, J.P. - An investigation of some aspects of bryozoan predation by dorid nudibranchs (Mollusca: Opisthobranchia). **5:** 51-58.
Chaney, H.W. - Evidence of histocompatibility in the cheilostome bryozoan *Thalamoporella californica* Levinsen, 1909. **6:** 344.
Cheetham, A.H. - Evolution of zooecial asymmetry and origin of poricellariid cheilostomes. **1:** 185-194.
Cheetham, A.H. - Study of Cheilostome polymorphism using Principal Components Analysis. **2:** 385-409.
Cheetham, A.H. - Taxonomic significance of autozooid size and shape in some early multiserial cheilostomes from the Gulf Coast of the U.S.A. **3:** 547-564.
Cheetham, A.H. and Thomsen, E. - Morphology and strength of some Tertiary arborescent cheilostome bryozoans. **5:** 310.
Cheetham, A.H. - Patterns of morphologic change in Late Tertiary cheilostome bryozoans from the Dominican Republic: Metrarabdotos revisited. **6:** 65-66.
Cheetham, A.H. and Jackson, J.B.C. - Neogene history of cheilostome Bryozoa in tropical America. **11:** 1-16.
Cheetham, A.H. - Asexual propagation in the cheilostomes *Metrarabdotos* and *Coscinopleura*: effects on genetic variation and larval productivity. **12:** 73-80.
Cheetham, A.H. - Origin and early development of the International Bryozoology Association. **12:** 81.
Cocito, S. and Ferdeghini, F. - Morphological variations in *Pentapora fascialis* (Cheilostomatida, Ascophorina). **11:** 176-181.
Cocito, S., Ferdeghini, F., Pisaroni, S. and Bedulli, D. - Influence of colony morphology on associated biota diversity in four Bryozoa. **12:** 83-88.
Colón-Urban, R.D. - Behavioural effects of morphine on several species of bryozoans. **6:** 67-70.
Colón-Urban, R.D., Urban, C.M and Oppenheim, J.D. - Biologically active substances from bryozoans. **8:** 89-95.
Cook, P.L. - Observations on living Bryozoa. **1:** 155-160.
Cook, P.L. - Settlement and early colony development in some Cheilostomata. **2:** 65-71.
Cook, P.L. - The genus *Tropidozoum* Harmer. **3:** 161-168.
Cook, P.L. - Mode of life of small, rooted 'sand fauna' colonies of Bryozoa. **4:** 269-281.
Cook, P.L. and Chimonides, P.J. - Early astogeny of some rooted cheilostome Bryozoa. **5:** 59-64.
Cook, P.L. and Chimonides, PJ. - Larval settlement and early astogeny of *Parmularia* (Cheilostomata). **6:** 71-78.
Cook, P.L. and Bock, P.E. - The astogeny and morphology of *Rhabdozoum wilsoni* Hincks (Anasca, Buguloidea). **9:** 47-50.
Cook, P.L. and Bock, P.E. - *Characodoma* Maplestone, a senior synonym of *Cleidochasma* Harmer (Cleidochasmatidae). **10:** 81-88.
Cook, P.L. and Bock, P.E. - Occurrence of two colony growth phases in *Dimetopia hirta* MacGillivray from Australia. **11:** 182-188.
Craig, S.F. - Intraspecific fusion in the encrusting bryozoan *Fenestrulina* sp. **9:** 51-54.
Craig, S.F. and Wasson, K. - Self/non-self recognition and fusion in the bryozoan *Hippodiplosia insculpta.* **11:** 189-196.
Cuffey, R.J. - An improved classification, based upon numerical-taxonomic analyses, for the higher taxa of entoproct and ectoproct bryozoans. **2:** 549-564.
Cuffey, R.J. and Foerster, B. - Autoecology of the cyclostome bryozoan *Crisia eburnea* in the modern reefs of Bermuda. **3:** 357-368.
Cuffey, R.J., Gebelein, C.D., Fonda, S.S., Bliefnick, D.M., Kosich, D.F and Soroka, L.G. - A note on living Bahamian bryozoan reefs. **4:** 345-346.
Cuffey, R.J and Drexler, W.W. - Castle Hayne hornerid cyclostomes (Bryozoa, Eocene, North Carolina). **4:** 491-502.
Cuffey, R.J., Stadum, C.J and Cooper, J.D. - Mid-Miocene bryozoan coquinas on the Aliso Viejo Ranch, Orange County, southern California. **5:** 65-72.
Cuffey, R.J. and Sorrentino, A.V. - Globular *Ceriopora* species (Cyclostomida, Bryozoa) from the Virginia-Carolinas Pliocene, and the status of *Atelesopora.* **6:** 79-86.
Cuffey, R.J. and Turner, R. F. - Modern bryozoans on the Kodiak Shelf, southern Alaska. **7:** 65-72.
Cuffey, R.J. and Blake, D.B. - Cladistic analysis of the Phylum Bryozoa. **8:** 97-108.
Cuffey, R.J., Gooden-Seay, J.A and Heckel, P.H. - Incipient bryozoan reef-mounds in the Upper Pennsylvanian of

Kansas (U.S.A.). **8:** 109-120.
Cuffry, R.J. - Cretaceous bryozoan faunas of North America - preliminary generalizations. **9:** 55-56.
Cuffey, R.J. - Identification key for North American Ordovician trepostome families. **12:** 89-92.
Cutler, J.F. - Nature of "acanthopores" and related structures in the Ordovician bryozoan *Constellaria*. **2:** 257-260.
d'Hondt, J.-L. - Etude anatomique et cytologique comparée de quelques larves de bryozoaires cténostomes. **3:** 125-134.
d'Hondt, J.-L. - Bryozoaires cténostomes bathyaux et abyssaux de l'Atlantique Nord. **3:** 311-333.
d'Hondt, J.-L. - Ultrastructural characteristics of the various larval cell categories of the gymnolaematous Bryozoa. **4:** 47-58.
d'Hondt, J.-L. - Neurosecretions in Bryozoa larvae. **5:** 311.
d'Hondt, J.-L. - Digestive enzymatic activities in *Alcyonidium polyoum* (Hassall, 1841), Bryozoa Ctenostomida. **6:** 153-160.
d'Hondt, J.-L. and Goyffon, M. - Comparative electrophoretic study of some *Alcyonidium gelatinosum* (Linné, 1761) (Bryozoa, Ctenostomida) populations from west Europe. **7:** 121-128.
d'Hondt, J.-L. - The Bryozoa of the Lamouroux Collection. **8:** 161-168.
d'Hondt, J.-L.., Goyffon, M. and Queinnec, E. - Contributions of electrophoretic techniques to the knowledge of bryozoan Systematics. **8:** 169-177.
d'Hondt, J.-L. and Goyffon, M. - Experimental transplantations of Bryozoa: utilization of electrophoretic techniques to show environmental effects. **9:** 75-81.
d'Hondt, J.-L. - Discovery of some types of cheilostomatous Bryozoa described from 1812 to 1824 by J.V.F. Lamouroux. **11:** 211-218.
David, L., Mongereau, N., Pouyet, S. and Ritzkowski, S. - Le Bryozoaires du Chattien de la region de Kassel (Hesse - Allemagne). **1:** 327-344.
David, L., Pouyet, S. and Vigne, H. - Documentation automatiqe et bryozoaires. **3:** 627-645.
Debourle, A. - New Bryozoa of south-west Aquitaine (France). **3:** 535-538.
Dewel, R.A., Winston, J.E. and McKinney, F.K. - Deconstructing bryozoans: origin and consequences of a unique body plan. **12:** 93-99.
Di Geronimo, I, Rosso, A. and Sanfilippo, R. - Bryozoans as sedimentary indicators. **8:** 568.
Dick, M.H and Ross, J.R.P. - Intertidal cheilostome bryozoans in rock-pile habitat at Narrow Strait, Kodiak, Alaska. **6:** 87-93.
Dick, M.H. - A proposed mechanism for chimney formation in encrusting bryozoan colonies. **7:** 73-80.
Dick, M.H., Freeland, J.R., Williams, L.P. and Coggeshall-Burr, M. - Use of 16S mitochondrial ribosomal DNA sequences to investigate sister-group relationships among gymnolaemate bryozoans. **11:** 197-210.
Dunaeva, N.N. - On the mode of sexual reproduction of some trepostomatous Bryozoa. **1:** 61-63.
Dunaeva, N.N. - Carboniferous Bryozoa of the Donetz Basin. **2:** 73-77.
Dunaeva, N.N. and Morozova, I.P. - Revision of the Suborder Fenestelloidea. **3:** 225-233.
Dunaeva, N.N. - A review of the Paleozoic bryozoan works of the USSR. **3:** 647-665.
Dyrynda, P.E.J. - Functional allelochemistry in temperate waters: chemical defences of bryozoans. **6:** 95-100.
Dyrynda, P.E.J. - A preliminary study of patterns of polypide generation-degeneration in marine cheilostome Bryozoa. **5:** 73-81.
Dyrynda, P.E.J. - Hydrodynamic gradients and bryozoan distributions within an estuarine basin (Poole Harbour, UK). **9:** 57-63.
Dzik, J. - Possible solitary bryozoan ancestors from the early Palaeozoic and the affinities of the Tentaculita. **8:** 121-131.
Eder, M. and Wöss, E.R. - Videomonitoring of freshwater bryozoans on artificial substrates in Laxenburg pond (lower Austria): experimental setup and preliminary results. **11:** 219-224.
El Safori, Y.A. - Early Eocene bryozoans of the Farafra oasis, Egypt. **11:** 225-231.
El Safori, Y.A. - Paleoenvironmental implications of Middle Eocene bryozoans in Egypt. **11:** 232-237.
Emschermann, P. - Cuticular pores and spines, ion regulatory organs in Entoprocta. **2:** 261-270.
Emschermann, P. - Factors inducing sexual maturation and influencing the sex determination of *Barentsia discreta* Busk (Entoprocta, Barentsiidae). **6:** 101-108.
Engel, B.A. - A preliminary review of Australian Permian bryozoans. **7:** 81-88.
Erickson, J.M. and Waugh, D.M. - Colony morphologies and missed opportunities during the Cincinnatian (Late Ordovician) bryozoan radiation: examples from *Heterotrypa frondosa* and *Monticulipora mammulata*. **12:** 101-107.
Ernst, A. - Systematics and biogeography of the Permian bryozoans in Europe. **12:** 109-112.
Ernst, H. - Biomuration of folliculinids in Upper Cretaceous cheilostome Bryozoa. **6:** 345.
Farmer, J.D. - Morphology and function of zooecial spines in cyclostome Bryozoa: implications for paleobiology. **4:** 219-246.
Ferdeghini, F., Cocito S., Morri, C. and Bianchi C.N. - Living bryozoan buildups: *Schizoporella errata* (Waters, 1848) (Cheilostomatida, Ascophorina) in the northwestern Mediterranean (preliminary observations). **11:** 238-244.
Flor, F.D. - The variation of *Spiropora verticillata* Goldfuss, 1827 from Upper Cretaceous deposits. **1:** 161-164.
Flor, F.D. - Variation and habitat in some erect fossil Bryozoa (preliminary report). **2:** 411-416.

Forester, R.M., Sandberg, P.A. and Anderson, T.F. - Isotopic variability of cheilostome bryozoan skeletons. **2:** 79-94.
Franzén, Å. - Comparative ultrastructural studies of spermatids and spermatozoa in Bryozoa and Entoprocta. **5:** 83-92.
Franzén, Å. - Sperm ultrastructure in the Bryozoa. **7:** 89-96.
Gautier, T.G. - Growth in bryozoans of the Order Fenestrata. **2:** 271-274.
Gautier, Y.V. - Bryozoa as microfossils which may be useful in oil research. **1:** 31-36.
Geraci, S. - Infralittoral Bryozoa settled on artificial substrata in the Ligurian Sea. **3:** 335-346.
Ghiurca, V. - Le biotope récifal á bryozoaires du Miocéne de la Roumanie. **1:** 165-173.
Ghiurca, V. - Types of Miocene Bryozoa described in Roumania. **2:** 417-419.
Ghiurca, V. - Les Bryozoaires Neogenes de la Paratethys. **3:** 497-518.
Ghiurca, V. and Mongereau, N. - La faune de bryozoaires Éocènes d'Europe orientale. **5:** 93-100.
Ghiurca, V. and Ghiurca, C. - La contribution des hommes de science de l'Autriche aux études des bryozoaires de Roumanie. **6:** 109-113.
Gilmour, E.M. and Vogel, D. - Bryozoans of the Toroweap Formation, Southern Nevada, U.S.A. **5:** 310-311.
Gontar, V.I. - On the Cheilostomata (Bryozoa) of the Kurile Islands. **5:** 101-103.
Gordon, D.P. - A fine-structure study of brown bodies in the gymnolaemate *Cryptosula pallasiana* (Moll). **2:** 275-286.
Gordon, D.P. - Ultrastructure of communication pore areas in two bryozoans. **3:** 187-192.
Gordon, D.P. - The deep-sea Bryozoa of the New Zealand region. **7:** 97-104.
Gordon, D.P and Parker, S.A. - The plectriform apparatus - an enigmatic structure in malacostegine Bryozoa. **8:** 133-145.
Gordon, D.P. - A new family of Bifaxarioidea from the deep sea with platyglenid affinities. **9:** 65-68.
Gordon, D.P and Voigt, E. - The kenozooidal origin of the ascophorine hypostegal coelom and associated frontal shield. **10:** 89-107.
Gordon, D.P. - Towards a phylogeny of cheilostomes-morphological models of frontal wall/shields evolution. **11:** 17-37.
Gordon, D.P. - Late Cretaceous-Paleocene "porinids" - mixed frontal shields and evidence of polyphyly. **12:** 113-124.
Gorjunova, R.V. - On the polymorphism of the bryozoan order Rhabdomesonida. **3:** 59-64.
Grant, A. and Hayward, P.J. - Bryozoan benthic assemblages in the English Channel. **6:** 115-124.
Grischenko, A.V. and Mawatari, S.F. *Kubaninella*: a new genus of Adeonidae (Bryozoa: Cheilostomata) from the Western Kamchatka shelf of the Sea of Okhotsk. **12:** 125-130.
Hageman, G.S. - Accessory reproductive structures in *Membranipora serrilamella*: a fine-structural study. **5:** 311.
Hageman, S.J. - Discrete morphotaxa from a Mississippian fenestrate faunule: presence and implications. **8:** 147-150.
Hageman, S.J., Bone, Y., McGowran, B. and James, N.P. - Bryozoan species distributions on the cool-water Lacepede Shelf, southern Australia. **10:** 109-116.
Hageman, S.J., Bayer, M. and Todd, C.D. - Partitioning phenotypic variation: implications for morphometric analyses (Bryozoa). **12:** 131-140.
Håkansson, E. - Mode of growth of the Cupuladriidae (Bryozoa, Cheilostomata). **2:** 287-298.
Håkansson, E. - Population structure of colonial organisms: a palaeoecological study of some Cretaceous Bryozoans. **3:** 385-399.
Håkansson, E. - Breeding, brooding and 'parental care' in Recent *Lunulites*. **5:** 312.
Håkansson, E. and Winston, J.E. - Interstital bryozoans: unexpected life forms in a high energy environment. **6:** 125-134.
Håkansson, E. and Thomsen, E. - Clonal propagation in fossil cheilostomes. **6:** 345-346.
Håkansson, E. and Jürgensen, T. - *Tricornicella tupilaqi*, a marginal bryozoan from the Paleocene of Greenland. **7:** 105-112.
Håkansson, E. and Madsen, L. - Symbiosis - a plausible explanation of gigantism in Permian trepostome bryozoans. **8:** 151-159.
Hamdane, A. and Moissette - Bryozoan distribution in a Messinian coral reef complex of western Algeria. **12:** 141-149.
Harmelin, J.-G. - Morphological variations and ecology of the Recent cyclostome bryozoan "*Idmonea*" *atlantica* from the Mediterranean. **2:** 95-106.
Harmelin, J.-G. - Relations entre la forme zoariale et l'habitat chez les bryozoaires cyclostomes. Consequences taxonomiques. **3:** 369-384.
Harmelin, J.-G. - Evolutionary trends within three Tubuliporina families (Bryozoa, Cyclostomata). **3:** 607-616.
Harmelin, J.-G. - On some stomatoporiform species (Bryozoa Cyclostomata) from the bathyal zone of the northeastern Atlantic Ocean. **4:** 403-421.
Harmelin, J.-G. - Bryozoan-dominated assemblages in Mediterranean cryptic environments. **6:** 135-143.
Harmelin, J.-G. - Relationships between bryozoans and ahermatypic scleractinians. **8:** 568-569.
Harmelin, J.-G., Boury-Esnault, N. and Vacelet, J. - A bryozoan-sponge symbiosis: the association between *Smittina cervicornis* and *Halisarca* cf. *dujardini* in the Mediterranean. **9:** 69-74.

Harmelin, J-G. - Ecology of cave and cavity dwelling bryozoans. **11:** 38-53.
Harmelin, J.-G. and Capo, S. - Effects of sewage on bryozoan diversity in Mediterranean rocky bottoms. **12:** 151-158.
Hayami, T. - The Pliocene cheilostomatous Bryozoa from Shikoku on the Pacific coast of Japan. **5:** 105-112.
Hayward, P.J. - Preliminary observations on settlement and growth in populations of *Alcyonidium hirsutum* (Fleming). **2:** 107-113.
Hayward, P.J. - Observations on the bryozoan epiphytes of *Posidonia oceanica* from the island of Chios (Aegean Sea). **3:** 347-356.
Hayward, P.J. - The geography of *Exochella.* **8:** 569.
Healey, N.D. - Ultrastructure of the skeleton of the cystoporate bryozoans *Ceramophylla, Crassaluna* and *Cystodictya.* **4:** 179-194.
Herrera, A. and Jackson, J.B.C. - Environmental variation in feeding organ morphology of Panamanian cheilostome Bryozoa. **8:** 569-570.
Herrera, A. and Jackson, J.B.C. - Life history variation among "dominant" encrusting cheilostomate Bryozoa. **10:** 117-123.
Herrera, Y.A. - Pliocene Cheilostomata in the Isthmus of the Tehuantepec region (Mexico). **6:** 145-151.
Hillmer, G. - Artificial moulds for studying the internal structure of paleontological objects. **1:** 37-42.
Hillmer, G. - On the variation of gonozooecia of encrusting "*Berenicea*"-forms (Lower Cretaceous) (Bryozoa). **1:** 64-70.
Hillmer, G. - Convergent formation of comparable morphotypes of two cyclostomatous Bryozoans from the Cretaceous and the Holocene. **6:** 346.
Hillmer, G. and Schallreuter, R. - Ordovician bryozoans from erratic boulders of northern Germany and Sweden. **7:** 113-119.
Hillmer, G. - Convergent morphotypes of bryozoans encrusting seagrass from Cretaceous and Recent. **8:** 570.
Hillmer, G., Scholz, J. and Dullo,W.-C. - Two types of bryozoan nodules from the Gulf of Aqaba, Red Sea. **10:** 125-130.
Hinds, R.W. - Intrazooecial structures in some tubuliporinid cyclostome Bryozoa. **2:** 299-306.
Horowitz, A.S., Stratton, J.F and Waters, J.A. - Quantitative clustering applied to some polyporan data from the Middle Devonian of Indiana. **4:** 373-381.
Horowitz, A.S and Pachut, J.F. - Diversity of Cenozoic Bryozoa: a preliminary report. **10:** 133-137.
Horowitz A.S. and Pachut J.F. - The fossil record of bryozoan species diversity. **11:** 245-248.
Hu Zhao-Xun and Spjeldnaes, N. - Early Ordovician bryozoans from China. **8:** 179-185.
Hughes, R.N., Wright, P.J., Manríquez, P.H. and Bishop, J.D.D. - Predominance of obligate outbreeding in the simultaneous hermaphrodite *Celleporella hyalina sensu lato.* **12:** 159-162.
Humphries, E.M. - Seasonal settlement of bryozoans in Rehoboth Bay, Delaware, U.S.A. **2:** 115-128.
Humphries, E.M. - A new approach to resolving the question of speciation of smittinid bryozoans (Bryozoa: Cheilostomata). **3:** 19-35.
Humphries, E.M. - Selected features of growth in *Parasmittina nitida.* **4:** 195-218.
Hunter, E. and Hughes, R.N. - Growth laboratory cultured colonies of *Celleporella hyalina* (L.). **8:** 187-191.
Hunter, E. and Hughes, R.N. - The influence of temperature, food ration and genotype on zooid size in *Celleporella hyalina* (L.). **9:** 83-86.
Hunter, E. and Fusetani, N. - Studies on the effects of larval swimming time on settlement, metamorphosis and post-larval development of *Bugula neritina* (Cheilostomatida). **10:** 139-148.
Hunter, E., Hughes, R.N. and Goldson, A. - Environmental and genetic control of somatic and sexual performance in *Celleporella hyalina* (L.). **10:** 149-156.
Ikezawa, H., Nodasaka, Y. and Mawatari, S.F. - Larval morphology and preancestrula formation of three *Celleporina* species (Bryozoa: Cheilostomatida) from Hokkaido, Japan. **9:** 87-92.
Illies, G. - On the gonozooecium of *Collapora straminea* (Phillips) (Bryozoa Cyclostomata). **1:** 71-74.
Illies, G. - On some cyclostomatous Bryozoa from the Middle Jurassic of southern Germany. **1:** 298-299.
Illies, G. - Different budding patterns in the genus *Stomatopora* (Bryozoa, Cyclostomata). **2:** 307-315.
Illies, G. - On the genus *Stomatoporina* Balavoine 1958 (Bryozoa Cyclostomata). **3:** 51-57.
Illies, G. - The budding pattern of *Stomatopora gracilis* (Milne Edwards) (Bryozoa, Cyclostomata). **4:** 423-430.
Illies, G. - The budding pattern of *Proboscina angustata* (d'Orbigny, 1851) (Bryozoa, Cyclostomata). **5:** 113-120.
Illies, G. - New uniserial *Stomatopora* (Bryozoa Cyclostomata) from the Upper Cretaceous of northern Germany. **7:** 129-136.
Illies, G. - Budding patterns of some stomatoporid cyclostomatous bryozoans from the Jurassic and Cretaceous of Great Britain. **8:** 193-199.
Jackson, J.B.C. and Winston, J.E. - Modular growth and longevity in bryozoans. **5:** 121-126.
Jackson, J.B.C. and Wertheimer, S.P. - Patterns of reproduction in five common species of Jamaican reef-associated bryozoans. **6:** 161-168.
Jackson, J.B.C. and Cheetham, A.H. - Are skeletons enough to tell bryozoan species apart? **8:** 570-571.
Jackson J.B.C. and Herrera Cubilla , A. - Adaptation and constraint as determinants of zooid and ovicell size among encrusting ascophoran cheilostome Bryozoa from opposite sides of the Isthmus of Panama. **11:** 249-258.

Jaroshinskaya, A.M. - Ordovician Bryozoa of the Altai-Sayan region. **2:** 421-428.

Jebram, D. - A cultivation method for saltwater Bryozoa and an example for experimental biology. **1:** 119-128.

Jebram, D. - The importance of different growth directions in the Phylactolaemata and Gymnolaemata for reconstructing the phylogeny of the Bryozoa **2:** 565-576.

Jebram, D. - Effects of different foods on *Conopeum seurati* (Canu) (Bryozoa Cheilostomata) and *Bowerbankia gracilis* Leidy (Bryozoa Ctenostomata). **3:** 97-108.

Jebram, D. - Interrelations of nutrition, food uptake, and growth in bryozoans. **4:** 121-140.

Jebram, D. - Interpretations of the results of feeding experiments concerning the physiology, ecology, and phylogeny of the Bryozoa. **5:** 312.

Jebram, D. - The polyphyletic origin of the "Cheilostomata" (Bryozoa). **8:** 571.

Jones, K.E., Marsh T.G. and Wood T.S. - Surveying for phylactolaemate bryozoans by sieving lentic sites for their statoblasts. **11:** 259-264.

Karklins, O.L. - Zooecial boundary patterns in the Middle Ordovician genera of the families Rhinidictyidae, Stictoporellidae and Ptilodictyidae (Bryozoa). **1:** 77-84.

Karklins, O.L. - Bryozoan biostratigraphy of the Upper Middle Ordovician and Upper Ordovician of Kentucky. **6:** 346-347.

Karlson, R.H. - Recruitment and catastrophic mortality in *Plumatella emarginata* Allman. **9:** 93-96.

Kaufmann, K.W. - The biological role of *Bugula*-type avicularia. **1:** 54-60.

Kelly, S.M and Horowitz, A.S. - Growth forms and palaeoecology of Mississippian bryozoans: critical applications of Stach's 1936 model, eastern United States. **7:** 137-144.

Key, M.M, Jr. - Partitioning of morphologic variation across stability gradients in Upper Ordovician trepostomes. **7:** 144- 152.

Key, M.M Jr. - How to build a ramose trepostome. **8:** 201-207.

Key, M.M., Jr, Lev, S.M. and Lighthart, A. - Colony control over skeletal growth rates in trepostome bryozoans. **9:** 97-100.

Key, M.M, Jr., Jeffries, W.B., Voris, H.K and Yang, C.M. - Epizoic bryozoans and mobile ephemeral host substrata. **10:** 157-165.

Key M.M. Jr., Jeffries W.B., Voris H.K. and Chang M.Y. - Bryozoan fouling pattern on the horseshoe crab *Tachypleus gigas* (Müller) from Singapore. **11:** 265-271.

Key, M.M., Thrane, L. and Collins, J.A. - Functional morphology of maculae in a giant ramose bryozoan from the Permian of Greenland. **12:** 163-170.

Kohring, R. and Hörnig, A. - Freshwater bryozoan remains from the Molteno Formation (Upper Triassic) of South Africa. **12:** 171-174.

Kuklinski, P. - Bryozoa of the high Arctic fjord – a preliminary study. **12:** 175-182.

Kulich, J. - Extracting fossil material of Bryozoa by electrochemical dissolving of limestone. **8:** 209-211.

Labracherie, M and Sigal, J. - Les Bryozoaires cheilostomes des formations Eocene Inferieur du Site 246 (crosiere 25, Deep Dea Drlling Project). **3:** 449-466.

Labracherie, M. - Quelques Bryozoaires cheilostomes de la Falaise de Handia (Biarritz, France). **1:** 312-326.

Labracherie, M. - Functional morphology and habitat of Bryozoa in the Eocene of the northern Aquitaine Basin, France. **2:** 129-138.

Lagaaij, R. - First fossil finds of six genera of Bryozoa Cheilostomata. **1:** 345-360.

Lagaaij, R. - Shallow-water Bryozoa from deep-sea sands of the Principe Channel, Gulf of Guinea. **2:** 139-151.

Lagaaij, R. - *Crisia* and *Crisidia* (Bryozoa Cyclostomata) from the Upper Maastrichtian near Maastricht. **3:** 583-594.

Larsen N. and Håkansson E. - Microfacies mosaics across latest Maastrichtian bryozoan mounds in Denmark. **11:** 272-281.

Larwood, G.P. - Frontal wall structures of Cretaceous and Recent Cribrimorpha and other cheilostoma Bryozoa (Preliminary report). **1:** 97-100.

Larwood, G.P. - New species of *Pyripora* d'Orbigny from the Cretaceous and the Miocene. **2:** 463-473.

Larwood, G.P. - Preliminary report on Early (pre-Cenomanian) cheilostome Bryozoa. **3:** 539-545.

Larwood, G.P. - Colonial integration in Cretaceous cribrimorph Bryozoa. **4:** 503-520.

Larwood, G.P. - Revision of the Cretaceous cribrimorph Bryozoa described by Hennig from the Swedish Chalk. **5:** 127-134.

Larwood, G.P. and Taylor, P.D. - Mesozoic bryozoan evolution: response to increasing predation pressure? **5:** 312-313.

Larwood, G.P. - Form and evolution of Cretaceous myagromorph Bryozoa. **6:** 169-174.

Lidgard, S. - Water flow, feeding, and colony form in an encrusting cheilostome. **5:** 135-142.

Lidgard, S. - Budding process and geometry in encrusting cheilostome bryozoans. **6:** 175-182.

Lidgard, S. and Buckley, G.A. - Toward a morphological species concept in cheilostomates: phenotypic variation in *Adeonellopsis yarraensis* (Waters). **9:** 101-105.

Lidgard, S. - Zooidal skeletal morphogenesis of some Australian and New Zealand *Adeonellopsis* (Cheilostomatida). **10:** 167-177.

López de la Cuadra, C.-M[a]. and García-Gómez, J.-C. - A new species of *Hemicyclopora* (Bryozoa: Cheilostomata) from the southern coast of Spain. **8:** 213-218.

López de la Cuadra, C.-Mª. and García-Gómez, J.-C. - Zoogeographical study of the Cheilostomatida from the Straits of Gibraltar. **9:** 107-112.
Lutaud, G. - The great tentacle sheath nerve as the path of an innervation of the frontal wall structures in the cheilostome *Electra pilosa* (Linné). **2:** 317-326.
Lutaud, G. - The probability of a plexus in the calcified wall of *Crisidia cornuta* (Linné). **4:** 33-45.
Lutaud, G. - The innervation of the external wall in the carnose ctenostome *Alcyonidium polyoum* (Hassall). **5:** 143-150.
Lutaud, G. - The structure of the funiculus in *Electra pilosa* (Linné): formation and evolution of tubular funicular strands. **5:** 313.
Lutaud, G. - Preliminary experiments on interzooidal metabolic transfer in anascan bryozoans. **6:** 183-191.
Lutaud, G. - The construction of cuticle and early stages of calcification in the zooidal bud of malacostegans. **7:** 153-160.
Lutaud, G. - The organization of mesenchymal tissues in a cheilostomate zooecium. **8:** 219- 228.
Mackie, J.A., Keough, M.J., Norman, J.A. and Christidis, L. - Mitochondrial evidence of geographical isolation within *Bugula dentata* Lamouroux. **12:** 199-206.
Madsen, L. - Growth and polypide morphology in some ramose trepostome bryozoans from the permo-Carboniferous of the Arctic. **7:** 169-176.
Madsen, L. - Species differentiation in trepostome bryozoans - a discussion based upon morphological variability in *Tabulipora*. **8:** 229-236.
Madsen, L. - Bryozoans from North Greenland. **8:** 571-572.
Madsen, L. - Exozonal budding in trepostome Bryozoa. **9:** 113-115.
Marcus, E.duB.R. - Ernst Marcus 8.VI.1893 - 30.VI. 1968 **1:** 7-12
Markham, J.P and Ryland, J.S. - Morphometry and function in the lophophore of *Flustrellidra hispida* (Bryozoa). **8:** 237-242.
Marsh, T. and Wood, T.S. - Results of a freshwater bryozoan survey in the Pacific Northwestern United States. **12:** 207-214.
Massard, J.A. and Geimer, G. - Note on the freshwater Bryozoa of Israel (Phylactolaemata). **8:** 243-253.
Massard, J.A. and Geimer, G. - Distribution of freshwater and brackish water Bryozoa (Phylactolaemata, Gymnolaemata) in Israel. **9:** 117.
Massard, J.A. and Geimer, G. - On the occurrence of *Fredericella indica* Annandale, 1909 (Phylactolaemata) in Europe. **10:** 187-192.
Matricardi, G., Montagna, P. and Pisano, E. - Settlement and growth of *Electra posidoniae* Gautier on *Posidonia oceanica* (L.) Delile. **8:** 255-262.
Maturo, F.J.S. - The distributional pattern of the Bryozoa of the east coast of the United States exclusive of New England. **1:** 261-284.
Maturo, F.J.S. - Offspring variation from known maternal stocks of *Parasmittina nitida* (Verrill). **2:** 577-584.
Maturo, F.J.S. and Thorpe, J.P. - The relationships of *Schizomavella linearis* (Hassall) and *S. hastata* (Hincks). **4:** 431-441.
Maturo, F.J. Jr. - Self-fertilization in gymnolaemate Bryozoa. **8:** 572.
Maturo, F.J. Jr. - A new bryozoan structure: an ovipositor. **8:** 572-573.
Mawatari, S. and Mawatari, S.F. - Development and metamorphosis of the cyphonautes of *Membranipora serrilamella* Osburn. **3:** 13-18.
Mawatari, S.F. - Post-larval development of *Pseudocelleporina triplex* (Bryozoa, Cheilostomata). **7:** 177-182.
Mawatari, S.F. - Cheilostomatous bryozoans from Sesoko, Okinawa. **8:** 573
McKinney, F.K. and Cuffey, R.J. - Information services desired by bryozoologists. **2:** 601-606.
McKinney, F.K. - Autozooecial budding patterns in dendroid stenolaemate bryozoans. **3:** 65-76.
McKinney, F.K. - Some paleoenvironments of the coiled fenestrate bryozoan *Archimedes*. **4:** 321-335.
McKinney, F.K and Stedman, T.G. - Constancy of characters within helical portions of *Archimedes*. **5:** 151-157.
McKinney, F.K and Wass, R.E. - The double helix form of branches and its relation to polymorph distribution in *Spiralaria florea* Busk. **5:** 159-167.
McKinney, F.K. and Boardman, R.S. - Zooidal biometry of Stenolaemata. **6:** 193-203.
McKinney, F.K. - Paleobiological interpretation of some skeletal characters of Lower Devonian fenestrate Bryozoa, Prague Basin, Czechoslovakia. **7:** 161-168.
McKinney, F.K. - Colonial feeding currents of *Exidmonea atlantica* (Cyclostomata). **8:** 263-270.
McKinney, F.K. and McKinney, M.J. - Preferences for settlement conditions by larvae of *Schizotheca serratimargo* (Hincks, 1886), an erect bryozoan from protected habitats. **9:** 119-124.
McKinney, F.K., Galloway, G.M. and McKinney, M.J. - Colony shapes and sizes: some life-history attributes of encrusting cyclostomate bryozoans (Eocene, North Carolina). **10:** 179-185.
McKinney, F.K. - Phylloporinids and the phylogeny of the Fenestrida. **11:** 54-65.
McKinney, F.K. - Colony sizes and occurrence patterns among Bryozoa encrusting disarticulated bivalves in the northeastern Adriatic sea. **11:** 282-290.
McKinney, F.K. - Chesterian (Carboniferous) *Septopora* (Order Fenestrida), eastern North America. **12:** 183-189.
McKinney, M.J. and Dewel, R.A. - The ctenostome collar – an enigmatic structure. **12:** 191-197.
Menon, N.R. - Vertical and horizontal distribution of fouling bryozoans in Cochin backwaters, southwest coast of

India. **2:** 153-164.
Moissette, P. - Encrusting bryozoans from two Messinian coral reefs of western Algeria. **6:** 205-212.
Moissette, P. and Pouyet, S. - Bryozoan masses in the Miocene - Pliocene and Holocene of France, North Africa and the Mediterranean. **8:** 271-279.
Moissette, P. - The cheilostomate bryozoan *Batopora rosula* (Reuss, 1848): a paleobathymetric indicator in the Mediterranean Neogene. **10:** 193-198.
Moissette, P. - The use of Neogene bryozoans for a better understanding of the ecology of some recent species. **11:** 291-297.
Moissette, P., Spjeldnaes, N. and Georgiades-Dikeoulia, E. - Highly diverse bryozoan faunas from the Plio-Pleistocene of the Greek island of Rhodes. **12:** 215-220.
Mongereau, N. and Ghiurca,V. - Note preliminaire sur les bryozoaires cyclostomes du Bassin de Transylvanie (Roumanie). **5:** 169-176.
Moore, P.G. - Bryozoa as a community component on the northeast coast of Britain. **2:** 21-36.
Morozova, I.P. - New and little-known structures of Fenestelloidea. **2:** 327-333.
Morris, B. and Prinsep, M. - Metabolites from the marine ctenostomate *Amathia wilsoni*. **10:** 199-205.
Morris, P.A. - Pacific Coast *Celleporina* Gray (1848): Fossil and Recent. **4:** 467-489.
Morris, P.A. - Zoogeographical distribution patterns in some cosmopolitan Bryozoa. **5:** 313.
Morris, P.A. - Bryozoan encrusting patterns on *Cryptopleura ruprechtiana* (Rhodophyta) from Vancouver Island, British Columbia. **8:** 574.
Morris, P.A., von Bitter, P.H., Schenk, P.E. and Wentworth, S.J. - Interactions of bryozoans and microbes in a chemosynthetic hydrothermal vent system: Big Cove Formation (Lower Codroy Group, Lower Carboniferous, Middle Viséan/Arundian), Port au Port Peninsula, Western Newfoundland, Canada. **12:** 221-227.
Morris, P.A. see also Pinter, P.A. and Pinter Morris, P.A.
Moyano G, H.I. - Posición sistematica de los generos *Romancheina*, *Escharoides*, *Cellarinella* y *Systenopora* (Bryozoa, Cheilostomata, Ascophora). **1:** 195-211.
Moyano, G., H.I. - Bryozoa from Antarctic bays: some ecological aspects. **4:** 383-402.
Moyano G., H.I. - Magellanic Bryozoa: some ecological and zoogeographical aspects. **5:** 314.
Moyano G., H.I. - Chilean cribrimorphs. **6:** 347.
Moyano, G., H.I. - Bryozoa from deep-sea water in Chile: Cyclostomata. **8:** 281-290.
Moyano G, H.I. - Bryozoa from Easter Island: a pictorial view. **8:** 574.
Moyano G., H.I. - Bryozoa Microporidae from the south-eastern Pacific: two new species and a review. **9:** 125-132.
Moyano G., H.I. - Holocene bryozoan links between Australia, New Zealand, southern South America, and Antarctica - a preliminary evaluation. **10:** 207-219.
Moyano G., H.I. - Bryozoa from the Magellanic continental slope near Cape Horn: an unexpected collection. **11:** 298-304.
Moyano G., H.I. - Bryozoa from Oceanic south eastern Pacific Islands: diversity and zoogeography. **12:** 229-238.
Mundy, S.P. - Some effects of low concentrations of copper ions on *Cristatella mucedo* (Bryozoa: Phylactolaemata). **5:** 177-184.
Mundy, S.P., Taylor, P.D. and Thorpe, J.P. - A reinterpretation of phylactolaemate phylogeny. **5:** 185-190.
Nakrem, H.A. - Distribution of bryozoans in the Permian succession of Svalbard (preliminary data). **8:** 291-298.
Nakrem, H.A. - Environmental distribution of bryozoans in the Permian of Spitsbergen. **9:** 133-137.
Nakrem, H.A. - A Moscovian (Carboniferous) bryozoan buildup from Svalbard. **12:** 239-245.
Newman, D.J. - Bryostatin: from bryozoan to cancer drug. **10:** 9-17.
Nielsen, C.: Morphology and reproduction of '*Hippodiplosia*' *insculpta* (Hincks). **5:** 314.
Nielsen, C. - Ovicell formation in *Tegella* and four cellularioids (Bryozoa, Cheilostomata). **6:** 213-220.
Nielsen, C. - A revision of the genus *Loxosoma* (Entoprocta) with description of three new species from Phuket, Thailand. **8:** 575.
Nielsen, C. - The phylogenetic position of Ectoprocta and Entoprocta. **11:** 66-73.
Nye, O.B. - Aspects of microstructure in post-Paleozoic Cyclostomata. **1:** 111-114.
O'Dea, A. and Okamura, B. - Cheilostome bryozoans as indicators of seasonality in the Neogene epicontinental seas of western Europe. **11:** 316-320.
Occhipinti Ambrogi, A. and d'Hondt, J.-L. - Distribution of bryozoans in brackish waters of Italy. **5:** 191-198.
Occhipinti Ambrogi, A. - The zonation of bryozoan along salinity gradients in the Venice Lagoon (Northern Adriatic). **6:** 221-231.
Occhipinti Ambrogi, A. and Ambrogi, R. - Short-term changes in a brackish water assemblage of Bryozoa with particular reference to *Conopeum seurati* (Canu). **7:** 183-190.
Occhipinti Ambrogi, A. - The spread of *Tricellaria inopinata* into the lagoon of Venice: an ecological hypothesis. **8:** 299-308.
Occhipinti Ambrogi, A. and d'Hondt, J.-L. - The invasion ecology of *Tricellaria inopinata* into the lagoon of Venice: morphological notes on larva and ancestrula. **9:** 139-144.
Occhipinti Ambrogi, A. - Recent developments in the history of the bryozoans of the lagoon of Venice: biodiversity and environmental stress. **11:** 305-315.

Oda, S. and Nakamura, R.M. - The occurrence of double polypides in freshwater Bryozoa. **2:** 523-528.
Oda, S. - Germination of the statoblasts of *Pectinatella magnifica*, a freshwater bryozoan. **4:** 93-112.
Oda, S. and Mukai, H. - Fine surface structure of the statoblasts of higher phylactolaemate bryozoans. **6:** 233-244.
Oda, S. - Abnormal polypides in *Pectinatella magnifica*, a freshwater bryozoan. **7:** 191-196.
Okamura, B. - Seasonal changes in zooid size and feeding activity in epifaunal colonies of *Electra pilosa*. **7:** 197-203.
Okamura, B. - Variation in local populations of the freshwater bryozoan *Cristatella mucedo*. **9:** 145-149.
Okamura, B. - Metapopulation biology of freshwater bryozoans. **11:** 74-86.
Orellana, M.C. and Cancino, J.M. - Effects of delaying larval settlement on metamorphosis and early colonial growth in *Celleporella hyalina* (Bryozoa: Cheilostomata). **8:** 309-316.
Orellana, M.C., Cancino, J.M. and Hughes, R.N. - Is settlement in lecithotropic bryozoan larvae constrained by energy reserves? **10:** 221-226.
Ostrovsky, A.N. - Brood chambers in cribrimorphs evolved by fusion of costae: further arguments. **12:** 247-255.
Owrid, G.M.A. and Ryland, J.S. - Sexual reproduction in *Alcyonidium hirsutum* (Bryozoa: Ctenostomata). **8:** 317-326.
Owrid, G.M.A. - Environmental effect on the growth and reproduction of *Alcyonidium hirsutum* (Bryozoa: Ctenostomatida). **9:** 151-156.
Pachut, J.F and Horowitz, A.S. - Multivariate discrimination and classification of some North American Mississippian species of *Fistulipora* M'Coy. **7:** 205-212.
Palumbi, S.R. and Jackson, J.B.C. - Intracolonial gradients and zooid senescence in two Bryozoa from Cryptic Reef environments. **5:** 315.
Patzkowsky, M.E. - Inferred water flow patterns in the fossil *Fistulipora* M'Coy (Cystoporata, Bryozoa). **7:** 213-219.
Perry, T.G., Horowitz, A.S. and Anstey, R.L. - Restudy of some Nicholson's types of Palaeozoic bryozoans. **2:** 429-436.
Peterson, N.L. - Neurotoxin found in the freshwater bryozoan *Lophopodella carteri*. **12:** 257-260.
Phifer, C.D. - Polypide and brown body distribution in *Bugula turbinata* (Cheilostomata). **6:** 348.
Pinter, P.A. - The *Hippothoa hyalina* (L.) complex with a new species from the Pacific coast of California. **2:** 437-446.
Pinter, P.A. - see also Pinter-Morris, P.A. and Morris, P.A.
Pinter Morris, P.A. - A comparative study of decalcification of Mollusca shells by various bryozoans. **3:** 109-113.
Pinter Morris, P.A. - see also Morris, P.A. and Pinter, P.A.
Pisano, E. and Balduzzi, A. - Bryozoan colonisation along a infralittoral cliff in the Ligurian Sea (North-western Mediterranean). **6:** 245-256.
Pisano, E. - Cytogenetics in bryozoans. **8:** 327-335.
Pohowsky, R.A. - A Jurassic cheilostome from England. **2:** 447-461.
Pohowsky, R.A. - Boring Bryozoa. **3:** 255-256.
Poluzzi, A. and Sartori,R. - Report on the carbonate mineralogy of Bryozoa. **3:** 193-210.
Poluzzi, A. and Coppa, M.G. - Zoarial strategies to win substratum space in *Calpensia nobilis* (Esper). **8:** 337-360.
Popeko, L.I. and Gorelova, N.G. - On the possibility of *Fenestella* genus objective grouping. **3:** 235-245.
Porter, J.S., Bloor, P., Stokell, B.L. and Ryland, J.S. - Intra- and inter-specific variation in tentacle number in the genus *Alcyonidium* (Bryozoa, Ctenostomatida). **11:** 321-328.
Potter, R. - Bryozoan karyotypes and genome sizes. **4:** 11-32.
Pouyet, S. and David, L. - Revision of the genus *Steginoporella* (Bryozoa Cheilostomata). **4:** 565-584.
Prinsep, M.R and Morris, B.D. - Secondary metabolites from New Zealand marine Bryozoa. **10:** 227-235.
Prud'Homme, J. - Révision de quelques types de la collection d'Orbigny (Crétacé) conservée au Muséum d'Histoire naturelle de Paris. **1:** 212-218.
Rao, K.S. - Studies on freshwater Bryozoa - III. The Bryozoa of the Narmada River system. **2:** 529-537.
Rao, K.S., Agrawal, V., Diwan, A.P. and Shristastava, P. - Studies on freshwater Bryozoa. V. Observations on Central Indian materials. **6:** 257-264.
Rao, K.S., Agrawal, V. and Diwan, A.P. - Studies on water quality monitoring with freshwater Ectoprocta as indicator organisms. **6:** 348-349.
Reed, C.G. - Phylogenetic implications of the morphogenesis of the cystid epithelium during the metamorphosis of gymnolaemate bryozoans. **7:** 221-228.
Reguant, S. and Serra, J. - Bryozoa from the Quaternary platform sediments between Blanes and Palamos (Prov. Gerona, Spain). Preliminary note. **3:** 467-482.
Reguant, S. and Zamarreño, I. - Bryozoan bioherms on the Mediterranean continental shelf (Northeastern Spain). **7:** 229-236.
Reguant, S., Fernandez, J., Rodriguez-Fernandez, J. and Serra-Kiel, J. - Bryozoan biofacies, zoarial forms and sedimentary environments in the Tertiary of Spain. **8:** 361-370.
Reguant, S. and Mayoral, E. - The encrusting Bryozoa on disarticulated bivalve shells (chiefly of *Glycymeris insubrica*) in the Lower Pliocene of Huelva (SW Spain). **9:** 157-160.
Rider, J and Enrico, R. - Structural and functional adaptations of mobile anascan ectoproct colonies (ectoproctaliths). **4:** 297-319.

Ristedt, H. - Skeletal untrastructure and astogenetic development of some cribrimorph Bryozoa. **4:** 141-152.
Ristedt, H. - Morphogenesis of the frontal shield and the ovicells in *Cribrilaria* n. sp. **5:** 315.
Ristedt, H. - Bryozoa as epizoa of *Chlamys nobilis* from the Seychelles. **6:** 349.
Ristedt, H. - Ancestrula and early astogeny of some Anascan Bryozoa: their taxonomic importance and possible phylogenetic implications. **8:** 371-382.
Ristedt, H. - Initial frontal budding in some nodular cheilostomate Bryozoa. **10:** 237-242.
Ross, C.A and Ross, J.R.P. - Biogeographical influences on late Palaeozoic faunal distributions. **5:** 199-212.
Ross, C.A and Ross, J.R.P. - Lower Permian bryozoans, Diamond Range, Nevada. **7:** 237-244.
Ross, J.R.P. - Calcification in the colonial stenolaemate, *Heteropora*. **2:** 475-483.
Ross, J.R.P. - Organisation of calcified tissues in stenolaemate ectoprocts. **3:** 179-186.
Ross, J.R.P. - Ectoproct adaptations and ecological strategies. **4:** 283-293.
Ross, J.R.P. - Late Palaeozoic ectoproct biogeography. **5:** 213-220.
Ross, J.R.P. - Biogeography of Ordovician Ectoproct (bryozoan) faunas. **6:** 265-271.
Ross, J.R.P. - Lower/Middle Ordovician bryozoans from the oil Creek Formation, Arbuckle Mountains, Oklahoma, (U.S.A.). **8:** 383-390.
Ross, J.R.R - Ordovician (Llanvirn) bryozoans from the Amadeus Basin, central Australia. **9:** 161-165.
Ross, J.R.P. and Ross, C.A. - Bryozoan evolution and dispersal and Paleozoic sea-level fluctuations. **10:** 243-258.
Ross, J.R.P., Hamedi, M.A. and Wright, A.J. - Late Ordovician (Caradoc) bryozoans from the Kerman district, East-Central Iran. **11:** 329-336.
Ross, J.R.P. and Ross, C.A. - Bryozoans in Ordovician depositional sequences, Cincinnati Arch region, USA. **12:** 261-274.
Rosso, A. - *Bryobaculum carinatum* gen. et sp. nov.: a new Mediterranean Pliocene deep-water bryozoan. **12:** 275-283.
Roush, S.A. - Distribution pattern and habitat preference of phylactolaemate bryozoans in the Guanacaste conservation area, Costa Rica. **11:** 337-342.
Rubin, J.A. - Settlement, growth and spatial competition in bryozoans in a fouling community. **6:** 349.
Rubin, J.A. and Moate, R.M. - Epiphytic and epilithic cheilostomes from southern Corsica (Corse du Sud). **8:** 575.
Rucker, J.B. - Skeletal mineralogy of cheilostome Bryozoa. **1:** 101-110.
Russ, G. - The relationships of fish grazing to bryozoan and ascidian communities at Portsea Pier, Victoria, Australia (Preliminary report). **4:** 295.
Ryland, J.S. - Terminological problems in Bryozoa. **1:** 225-236.
Ryland, J.S. - The analysis of spatial distribution patterns. **2:** 165-172.
Ryland, J.S. - Parameters of the lophophore in relation to population structure in a bryozoan community. **3:** 123-124.
Ryland, J.S. - *Celleporella carolinensis* sp. nov. (Bryozoa Cheilostomata) from the Atlantic coast of America. **4:** 611-619.
Ryland, J.S. - Colonies, growth and reproduction. **5:** 221-226.
Ryland, J.S. - Gonozooid placement and branching patterns in some species of *Crisia* (Cyclostomatida). **11:** 343-354.
Ryland, J.S. - Life history characters and ecology of some encrusting ctenostomates. **12:** 285-291.
Sakagami, S.. - Study on the Upper Paleozoic Bryozoa of Japan and the Thai-Malayadistricts. **1:** 295-297.
Sanderson, W.G., Thorpe, J.P. and Clarke, A. - A preliminary study of feeding rates in the Antarctic cheilostomate bryozoan *Himantozoum antarcticum*. **9:** 167-171.
Sanderson, W.G. and Thorpe, J.P. - Effects of temperature on the feeding activity of some temperate intertidal Bryozoa. **10:** 271-281.
Sanderson, W.G., Harding, S.P. and Thorpe, J.P. - An investigation of the effects of tidal fluctuations on the availability of seston as food for intertidal Bryozoa. **10:** 259-269.
Sanderson, W.G., Thorpe, J.P. and Clarke, A. - Aspects of structure and function in the feeding of the Antarctic cheilostomate bryozoan, *Himantozoum antarcticum* (Calvet). **11:** 355-364.
Santagata, S. and Zimmer, R.L. - Comparing cell patterns of coronate bryozoan larvae with fluorescent probes. **11:** 365-365.
Schäfer, P. - Significance of soft part morphology in the classification of Recent tubuliporoid cyclostomes. **6:** 273-284.
Schäfer, P. - Colonization patterns of bryozoans on living *Chlamys islandica* (Müller), Troms, Northern Norway. **8:** 576.
Schäfer, P. - Growth strategies of Arctic Bryozoa in the Nordic Seas. **9:** 173-176.
Schager, S.A.N. - Remarks on the genus *Floridina* Jullien, 1881 (Bryozoa). **1:** 219-224.
Schattleitner, M. - Contribution to the taxonomy of some encrusting cyclostome Bryozoa from the Miocene of the Central Paratethys. **8:** 391-397.
Schmid, B.M. - Recent bryozoan species from the Miocene of Nußdorf (Vienna) and their use for ecological interpretation. **8:** 399-408.
Schmidt, R. and Bone, Y. - Eocene Bryozoan Assemblages of the St Vincent Basin, South Australia. **12:** 293-298.
Scholz, J. - Monsoon-related settlement patterns in reef- and harbor bryozoans, Philippines. **8:** 409-416.
Scholz, J. and Krumbein, W.E. - Microbial mats and biofilms associated with bryozoans. **10:** 283-298.

Scholz, J., Sterflinger, K., Junge, C. and Hillmer, G. - A preliminary report on bryostromatolites. **11:** 376-384.
Schopf, T.J.M. - Generalizations regarding the phylum Ectoprocta in the deep-sea (200-6000 m). **1:** 152-154.
Schopf, T.J.M. - Population genetics of ectoprocts: status as of January, 1972. **2:** 585-592.
Schopf, T.J.M. - Bryozoan DNA and evolutionary systematics. **5:** 315.
Schopf, T.J.M., Collier, K.O. and Bach, B.O. - Relation of the morphology of stick-like bryozoans at Friday Harbor, Washington, to bottom currents, suspended matter and depth. **5:** 316.
Schopf, T.J.M. - A genomic library and the genome size for the cheilostome bryozoan *Watersipora*. **6:** 285-292.
Scolaro, R.J. - Paleoecologic interpretation of some Florida Miocene Bryozoa. **1:** 174-177.
Seo, J.E. - On the geographical distribution of cheilostomate Bryozoa in Korean waters. **10:** 299-304.
Simonsen, A.H. - Fenestrate bryozoans of the Drum Limestone, Pennsylvanian (Missourian), of southeastern Kansas (U.S.A). **8:** 417-429.
Smith, A.M. and Nelson, C.S. - Selectivity in sea-floor processes: taphonomy of bryozoans. **9:** 177-180.
Smith, A.M and Nelson, C.S. - Differential abrasion of bryozoan skeletons: taphonomic implications for paleoenvironmental interpretation. **10:** 305-313.
Smith, D.B. - The Magnesian Limestone (Upper Permian) reef complex of north-east England 316
Smith, L.W. - Ultrastructure of the tentacles of *Flustrellidra hispida* (Fabricius). **2:** 335-342.
Snyder, E.M. - Bryozoan succession in the Warsaw Formation (Valmeyeran, Mississippian) of the Mississippi Valley, USA. **7:** 245-252.
Snyder, E.M. - Revised taxonomic approach to acanthocladiid Bryozoa. **8:** 431-445.
Söderqvist, S.T. - Observations on extracellular body wall structures in *Crisia eburnea* L. (Ectoprocta, Crisiidae). **1:** 115-118.
Soule, D.F. - Morphogenesis of giant avicularia and ovicells in some Pacific Smittinidae. **2:** 485-495.
Soule, D.F and Soule, J.D. - Species groups in Watersiporidae. **3:** 299-309.
Soule, D.F., Soule, J.D and Henry, C.A. - The influence of environmental factors on the distribution of estuarine bryozoans, as determined by multivariate analysis. **4:** 347-372.
Soule, D.F and Soule, J.D. - Effects of oceanographic phenomena such as 'El Nino' on the zoogeography and endemism of tropical/subtropical Pacific Watersiporidae. **6:** 293-300.
Soule, D.F., Soule, J.D. and Chaney, H.W. - Evolution, systematics and biogeography of Thalamoporellidae (Cheilostomata, Anasca). **7:** 253-260.
Soule, D.F., Soule, J.D. and Chaney, H.W. - Some little-known genera of Thalamoporellidae: *Thairopora*, *Diploporella*, and new genera *Marsupioporella* and *Thalamotreptos*. **8:** 447-464.
Soule, D.F., Soule, J.D. and Morris, P.A. - Changing concepts in species diversity in the northeastern Pacific. **12:** 299-306.
Soule, J.D. - Histological and histochemical studies on the bryozoan-substrate interface. **2:** 343-347.
Soule, J.D. and Soule, D.F. - *Spathipora*, its anatomy and phylogenetic affinities. **3:** 247-253.
Soule, J.D. and Soule, D.F. - Heavy metals uptake in bryozoans. **5:** 227-233.
Soule, J.D., Soule, D.F and Chaney, H.W. - New tropical Pacific and Indian Ocean Cleidochasmatidae (Cheilostomata: Ascophora). **8:** 465-486.
Southwood, D.A. - Ovicells in some Fenestrata from the Permian of N.E. England. **6:** 301-310.
Spjeldnaes, N. - Silurian bryozoans which grew in the shade. **3:** 415-424.
Spjeldnaes, N. - Symbiosis involving some Ordovician-Silurian bryozoans. **5:** 317.
Spjeldnaes, N. - Bryozoan colonies as indicators of bottom conditions in the Lower Ordovician. **10:** 315-319.
Spjeldnaes, N. - Cryptic bryozoans from west Africa. **11:** 385-391.
Stebbing, A.R.D. - Observations on colony overgrowth and spatial competition. **2:** 173-183.
Stevens, L.M., Gregory, M.R. and Foster, B.A. - Fouling Bryozoa on pelagic and moored plastics from northern New Zealand. **10:** 321-340.
Stratton, J.F. - Ovicells in *Fenestella* from the Speed Member, North Vernon Limestone (Eifelian, Middle Devonian) in southern Indiana, U.S.A. **3:** 169-177.
Stratton, J.F and Horowitz, A.S. - Studies of the flow of water through models of *Polypora*. **3:** 425-438.
Stricker, S.A. - Ultrastructure of the apical organ in a cyphonautes larva. **7:** 261-268.
Taticchi, M.I. - Some observations on the feeding freshwater Bryozoa in nature (Lake Trasimeno, Perugia, Italy). **8:** 576.
Tavener-Smith, R. - Skeletal structure and growth in the Fenestellidae (Bryozoa) (Preliminary Report). **1:** 85-92.
Tavener-Smith, R. - Some aspects of skeletal organization in Bryozoa. **2:** 349-359.
Taylor, P.D. - Functional morphology and evolutionary significance of differing modes of tentacle eversion in marine Bryozoans. **5:** 235-247.
Taylor, P.D., Larwood, G.P and Balson, P.S. - Some British field localities with fossil Bryozoa. **5:** 249-262.
Taylor, P.D. - Polymorphism in melicerititid cyclostomes. **6:** 311-318.
Taylor, P.D. - Skeletal morphology of malacostegan grade cheilostome Bryozoa. **7:** 269-276.
Taylor, P.D. - Observations on symbiotic associations of bryozoans and hermit crabs from the Otago Shelf of New Zealand. **8:** 487-495.
Taylor, P.D. and Badve, R.M. - The mid-Cretaceous bryozoan fauna from the Bagh Beds of central India: composition and evolutionary significance. **9:** 181-186.
Taylor, P.D. and Weedon, M.J. - Skeletal ultrastructure and affinities of eleid (melicerititid) cyclostomate

bryozoans. **10:** 341-350.
Taylor, P.D. and Wilson, M.A. - *Cuffeyella*, a new bryozoan genus from the Late Ordovician of North America, and its bearing on the origin of the post-Paleozoic cyclostomates. **10:** 351-360.
Taylor, P.D. - Cyclostome systematics: phylogeny, suborders and the problem of skeletal organization. **11:** 87-103.
Taylor, P.D. and McKinney, F.K. - Brooding in the Cretaceous bryozoan *Stichomicropora* and the origin of ovicells in cheilostomes. **12:** 307-314.
Temkin, M.H. - The effects of increased external Ca++ and K+ concentrations on the waveform dynamics of bryozoan spermatozeugmata. **12:** 315-320.
Thoelen, M.J. - The Bryozoa fauna from the sands of Deurne, Upper Miocene, at Deurne (Antwerpen, Belgium). **1:** 361-368.
Thorpe, J.P. - Behaviour and colonial activity in *Membranipora membranacea* (L.). **3:** 115-121.
Thorpe, J.P. - A model using deterministic equations to describe some possible parameters affecting growth rate and fecundity in Bryozoa. **4:** 113-120.
Thorpe, J.P and Beardmore, J.A. - Genetic variation in natural populations of marine bryozoans. **5:** 263-272.
Thorpe, J.P., Clarke, D.R.K. and Best, M.A. - Natural variation in tentacle number in marine bryozoans and the possible effects of intraspecific and interspecific ecological competition for food. **6:** 319-327.
Thorpe, J.P and Ryland, J.S. - Some theoretical limitations on the arrangement of zooids in encrusting Bryozoa. **7:** 277-283.
Todd, J.A. - The role of bioimmuration in the exceptional preservation of fossil ctenostomates, including a new Jurassic species of *Buskia*. **9:** 187-192.
Todd, J.A. - The central role of ctenostomes in bryozoan phylogeny. **11:** 104-135.
Turner, R.F. - A new Upper Cretaceous cribrimorph from North America with calcareous opercula. **3:** 273-279.
Turner, R.F. - North American Cretaceous Tricephaloporidae. **4:** 521-540.
Vávra, N.R. - Cyclostomatous Bryozoa from the Austrian Miocene. **3:** 519-533.
Vávra, N.R. - Bryozoa from the Miocene of Styria (Austria). **4:** 585-609.
Vávra, N.R. - Bryozoa from the Eggenburgian (Lower Miocene, Central Paratethys) of Austria. **5:** 273-280.
Vávra, N.R. - Bryozoa from the Early Miocene of the Central Paratethys: biogeographical and biostratigraphical aspects. **7:** 285-292.
Vávra, N.R. - Contributions to the taxonomy and morphology of *Polyascosoecia* (Bryozoa: Cyclostomata) and related forms. **8:** 497-504.
Vávra, N.R. - Bryozoa from the Oligocene of Austria: *Chlidoniopsis* and its possible affinities with Palaeozoic bryozoans. **9:** 193-196.
Vávra, N.R. - Biogeographical aspects of bryozoan faunas of the central Paratethys from the Miocene. **11:** 392-399.
Vávra, N.R. - New bryozoan faunas from the Miocene of Burgenland (Austria). **12:** 321-329.
Viganò, A. and Taticchi-Viganò, M.I. - Some notes on size variability of *Plumatella fruticosa* Allman (Bryozoa) statoblasts collected in some Italian freshwater habitat. **3:** 135-147.
Vinogradov, A.V. - Evolutionary paths in the phylactolaemates. **8:** 577.
Viskova, L.A. - On the morphology and systematics of some Late Cretaceous cyclostomatous Bryozoa. **2:** 497-502.
Voigt, E.- Homeomorphy in cyclostomatous Bryozoa as demonstrated in *Spiropora*. **1:** 43-53.
Voigt, E. - Environmental conditions of bryozoan ecology of the hardground biotope of the Maastrichtian Tuff-Chalk, near Maastricht (Netherlands). **2:** 185-197.
Voigt, E. - Heteromorphy in Cretaceous Bryozoa. **3:** 77-95.
Voigt, E. - The preservation of slightly or non-calcified fossil Bryozoa (Ctenostomata and Cheilostomata) by bioimmuration. **4:** 541-564.
Voigt, E. - Upper Cretaceous Bryozoan-seagrass association in the Maastrichtian of the Netherlands. **5:** 281-298.
Voigt, E. - The Bryozoa of the Cretaceous-Tertiary boundary. **6:** 329-342.
Voigt, E. - Thalassinoid burrows in the Maastrichtian chalk tuff near Maastricht (The Netherlands) as a fossil hardground microcavern biotope of Cretaceous bryozoans. **7:** 293-300.
Voigt, E. - Mono- or polyphyletic evolution of cheilostomatous bryozoan divisions? **8:** 505-522.
Voigt, E. - Ceriopora-like features in idmoneiform cyclostomate bryozoans. **9:** 197-200.
Voigt, E. and Hillmer, G. - Soft-tissue moulds preserved by bioummuration in the new bryozoan genus *Kunradocella* (Cheilostomatida). **10:** 361-366.
Ward, M.A. and Thorpe, J.P. - The distribution and species composition of bryozoan assemblage encrusting horse mussels in the North Irish Sea. **8:** 578.
Wass, R.E. - Frontal wall calcification in a Recent cribrimorph genus from Australia. **2:** 361-367.
Wass, R.E. - Cyst-like bodies in some Permian Bryozoa from Australia. **3:** 155-159.
Wass, R.E. and Yoo, J.J. - Distribution and taxonomy of some recent catenicelliform Bryozoa from Australia. **3:** 281-297.
Wass, R.E., Vail, L.L. and Boardman, R.S. – Early growth of Bryozoa at Lizard Island, Great Barrier Reef. **5:** 299-303.
Wass, R.E. - Early growth stages, ecology and biochemistry of catenicellid Bryozoa. **5:** 317.

Wass, R.E. - Intracolonial variation in the cheilostome genera, *Adeona* and *Adeonellopsis*. **8:** 523-529.
Wass, R.E. - Astogenetic and ontogenetic development in three species of cheilostomate Bryozoa from southwestern and Western Australia. **9:** 201-203.
Waugh, D.A. and Erickson, J.M. - Functional morphology of the anastomosing frondose growth form reported in *Heterotrypa frondosa* (d'Orbigny) (Bryozoa: Trepostomata) from the Cincinnatian (Late Ordovician) of Ohio. **12:** 331-338.
Weedon, M.J. and Taylor, P.D. - Skeletal ultrastructure of primitive cheilostome bryozoans. **11:** 400-411.
Whitehead, J.W., Seed, R. and Hughes, R.N. - Factors controlling spinosity in the epialgal bryozoan *Flustrellidra hispida* (Fabricius). **10:** 367-375.
Whittlesea, P.S. - Effects of sedimentary regimes on relative abundance of bryozoan colony forms: an example from Late Campanian allochthonous chalks of Norfolk, England. **10:** 377-382.
Wiebach, F. - Taxonomical and other remarks on freshwater Bryozoa. **1:** 245-246.
Wiebach, F. - Preliminary notes on a revision of the genus *Hyalinella*. **2:** 539-547.
Wiebach, F. - Specific structures of sessoblasts (Bryozoa Phylactolaemata). **3:** 149-154.
Winston, J.E. - Current-related morphology and behaviour in some Pacific coast bryozoans. **4:** 247-268.
Winston, J.E. and Jackson, J.B.C. - Life histories of cryptic reef bryozoans. **5:** 317-318.
Winston, J.E. - Life history studies of *Disporella* and *Drepanophora* in Jamaica. **6:** 350.
Winston, J.E. - Avicularian behaviour - a progress report. **8:** 531-540.
Winston, J.E. and Hayward, P.J. - Bryozoa of the US Antarctic Research Program: Preliminary report. **9:** 205-210.
Winston, J.E., Hayward, P.J. and Craig, S.F. - Marine bryozoans of the northeast coast of the United States: new and problem species. **11:** 412-420.
Wood, T.S. - Significance of morphological features in bryozoan statoblasts. **4:** 59-73.
Wood, T.S and Marsh, T.G. - The sinking floatoblasts of *Lophopodella carteri* (Bryozoa: Phylactolaemata). **10:** 383-389.
Wood, T.S. and Wood, L.J. - Statoblast morphology in historical specimens of phylactolaemate bryozoans. **11:** 421-430.
Wood, T.S. - Freshwater bryozoans: a zoogeographical reassessment. **12:** 339-345.
Wöss, E.R. - On the taxonomy and faunistics of phylactolaemates in the riverine forests of the Austrian Danube. **8:** 541-549.
Wöss, E.R. - Seasonal fluctuations of bryozoan populations in five water bodies with special emphasis on the life cycle of *Plumatella fungosa* (Pallas). **9:** 211-214.
Wöss, E.R. - Life-history variation in freshwater bryozoans. **10:** 391-399.
Wöss, E.R. - Colonization and development of freshwater bryozoan communities on artificial substrates in the Laxenburg pond (lower Austria). **11:** 431-438.
Wöss, E.R. - The reproductive cycle of *Plumatella casmiana* (Phylactolaemata: Plumatellidae). **12:** 347-351.
Wright, P.J. and Hughes, R.N. - A comparison of the early astogeny and life history of *Celleporela carolinensis* and *Celleporella hyalina*. **12:** 353-357.
Wyer, D.W. and King, P.E. - Relationships between some British littoral and sublittoral bryozoans and pycnogonids. **2:** 199-207.
Wyse Jackson, P.N., Bancroft, A.J. and Somerville, I.D. - Bryozoan zonation in a trepostome-dominated buildup from the Lower Carboniferous of North Wales. **8:** 551-559.
Wyse Jackson, P.N. - Bryozoa from the Lower Carboniferous of Ireland. **8:** 578-579.
Wyse Jackson, P.N. and Bancroft, A.J. - Possible opercular structures in the fenestrate bryozoan *Thamniscus* from the Upper Carboniferous of northern England. **9:** 215-218.
Wyse Jackson, P.N. and Buttler, C.J. - Fossil bryozoan localities in Ireland. **9:** 219-222.
Wyse Jackson, P.N and Spencer Jones, M. - W.H.Harvey's bryozoans from Australia (1854-1856). **10:** 401-412.
Wyse Jackson, P.N., Buttler, C.J. and Key, M.M. - Palaeoenvironmental interpretation of the Tramore Limestone Formation (Llandeilo, Ordovician) based on bryozoan colony form. **12:** 359-365.
Wyse Jackson, P.N. - Author index to the conference volumes (1-12) of the International Bryozoology Association (1968-2002). **12:** 379-394.
Yang, Jing-zhi and Lu, Lin-huang - Geographical and stratigraphical distributions of Permian Bryozoa in China. **5:** 305-307.
Yaroshinskaya, A.M. - The Early Devonian and Eifelian Bryozoa from the Altai. **1:** 285-294.
Zágorsek, K. - Paleoecology of the Eocene bryozoan marl in the Alpine-Carpathian region. **10:** 413-422.
Zágorsek, K. - An Upper Eocene bryozoan fauna from Perwang-1 borehole, Salzburg, Austria. **12:** 367-376.
Ziko, A and Hamza, F. - Bryozoan fauna from a post-Pliocene outcrop north of the Giza Pyramids Plateau, Egypt. **7:** 301-308.
Ziko, A. - Bryozoans and microfacies of the Middle Miocene of Mersa Matruh area, Egypt. **8:** 561-566.
Ziko, A. - Palaeoenvironmental implications of the Miocene bryozoans of Egypt: a preliminary note. **9:** 223-226.
Ziko, A., Hamza, F.H. and El-Safori, Y. - Palaeoecology and palaeobiogeography of the Miocene bryozoans from the western part of the Clysmic area, Egypt. **9:** 227-231.
Zimmer, R.L. - Morphological and developmental affinities of the Lophophorates. **2:** 593-599.
Zimmer, R.L. - Obituary: Christopher Gardiner Reed (1951-1990). **9:** 233-234.

Zimmerman, L.S. and Cuffey, R.J. - Species involved in Permian bryozoan bioherms, West Texas. 7: 309-316.

List of Participants

BADER Dr Beate
Institut für Geowissenschaften der Universität zu Kiel, Olshausenstr. 40, D-24118 Kiel, GERMANY
BARNES, Dr David K.A.
Department of Zoology and Animal Ecology, University College, Lee Maltings, Cork, IRELAND
BIGEY, Dr Françoise P.
Université P. & M. CURIE (PARIS VI), Laboratoire de Micropaléontologie, C.104 4, Place Jussieu, F - 75252 PARIS cédex 05, FRANCE
BOARDMAN, Dr Richard S.
3612 East Forest Lake Drive, Sarasota, Florida 34232, USA
BOCK, Mr Philip
School of Ecology and Environment, Deakin University, 662 Blackburn Road, Clayton, Victoria 3168, AUSTRALIA
BONE, Dr Yvonne
Department of Geology and Geophysics, The University of Adelaide, South Australia, 5005, AUSTRALIA
BRAGA, Professor Giampietro
Dipartimento di Geologia, Paleontologia e Geofisica, Università di Padova, Via Giotto 1, I-35137 Padova, ITALY
BUTTLER, Dr Caroline J.
Department of Geology, National Museums and Galleries of Wales, Cathays Park, Cardiff CF1 3NP, Wales, UK
CADÈE, Dr Gerhard C.
Neth. Inst. Sea Research, PO Box 59, 1790 AB den Burg, Texel. THE NETHERLANDS
CANCINO, Dr Juan M.
Facultad de Ciencias, Universidad Catolica de la Santisima Concepcion, Concepcion, CHILE
CHEETHAM, Dr Alan H.
MRC: NHB121, Smithsonian Institution, Washington DC 20560, USA
COCITO, Dr Silvia
ENEA CRAM - Marine Environment Research Centre, P.O. Box 316, 19100 La Spezia, ITALY
COOK, Ms Patricia L.
18/60 Avendon Boulevard, Glen Waverley, Victoria 3150, AUSTRALIA
CUFFEY, Professor Roger
Department of Geoscience, 412 Deike Building, Pennsylvania State University, University Park, PA 16802, USA
DEWEL, Dr Ruth Ann
Department of Biology, Appalachian State University, Boone, North Carolina 28608, USA
d'HONDT, Dr Jean-Loup
Laboratoire de Biologie des Invertébrés Marins et Malacologie, Museum National d'Histoire Naturelle, 57, rue Cuvier, F-75231 Paris Cedex 05, FRANCE.
DICK, Dr Matthew H.
Department of Biology, Middlebury College, Middlebury VT 05753, USA
EL SAFORI, Dr Yasser A.
Ain Shams University, Faculty of Science, Geology Department, Cairo, EGYPT
ERICKSON, Professor J. Mark
Department of Geology, St. Lawrence University, Canton, NY, 13617, USA
ERNST, Dr Andrej
Institut für Geowissenschaften der Universität zu Kiel, Olshausenstr. 40, D-24118 Kiel, GERMANY
FLORENCE, Wayne K.
Department of Zoology, University of the Western Cape, Private Bag X17, Bellville, 7535, Cape Town, SOUTH AFRICA
GILMOUR, Professor Ernest H.
Department of Geology, Eastern Washington University, Cheney 99004, USA
GLUHAK, Tea
Jurkoviceva 10, 10000 Zagreb, CROATIA

GOMEZ, Dr Africa
Department of Biological Sciences, University of Hull, Hull HU6 7RX, UK
GORDON, Dr Dennis P.
National Institute of Water & Atmospheric Research, P.O. Box 14-901, Kilbirnie, Wellington, NEW ZEALAND
GRISCHENKO, Andrei V.
Division of Biological Sciences, Graduate School of Science, Hokkaido University, Sapporo 060-0810, JAPAN
HAGEMAN, Dr Steven J.
Department of Geology, Appalachian State University, Boone, North Carolina 28608, USA
HÅKANSSON, Dr Eckart
Institute of Geology, Øster Voldgade 10, DK - 1350 Copenhagen, DENMARK
HAMDANE, Miss Assia
10, rue Curie El Biar, Algiers, ALGERIA
HARA, Dr Urszula
Department of Geological Museum, Polish Geological Institute, Warsaw 00-975, POLAND
HARMELIN, Dr Jean-Georges
Centre d'Océanologie de Marseille, Station Marien d'Endoume, CNRS - UMR 65 40: DIMAR, rue Batterie des Lions - F-13007 Marseille, FRANCE
HAYWARD, Dr Peter J.
School of Biological Sciences, University of Wales Swansea, Singleton Park, Swansea SA2 8PP, Wales, UK
HERRERA-CUBILLA, Dr Amalia
Center for Tropical Paleoecology and Archeology, Smithsonian Tropical Research Institute, P.O. Box 2072, Balboa, REPUBLIC OF PANAMA
HILLMER, Professor Gero
Universität Hamburg, Geologische Paläontologisches Institut, Bundesstrasse 55, D-20146 Hamburg, GERMANY
HOERNIG, Anna
Institut fuer Palaeontologie, Freie Universitaet Berlin, Malteserstraße 74-100, Haus D, 12249 Berlin, GERMANY
HUGHES, Professor Roger
Brambell Building, School of Biological Sciences, University of Wales (Bangor), Deniol Road, Bangor, Gwynedd LL57 2UW, UK
ISETO, Tohru
Department of Chemistry, Biology and Marine Science, Faculty of Science, University of the Ryukyus, Dr Hirose's Lab., Nishihara, Okinawa 903-0213, JAPAN
JACKSON, Professor Jeremy B.C.
Geosciences Research Division, Scripps Institution of Oceanography, University of California, San Diego, La Jolla, California 92093-0244 USA
JACOBSEN, Sarah
Institute of Geology, Øster Voldgade 10, DK - 1350 Copenhagen, DENMARK
KASELOWSKY, Juergen
Forschungsinstitut und Naturmuseum Senckenberg, Sektion Marine Evertebraten III (Bryozoologie), Senckenberg-anlage 25, 60325 Frankfurt, GERMANY
KAZMER, Miklos
Department of Palaeontology, Eotvos University, H-1083 Budapest, Ludovika ter 2, HUNGARY
KEY, Dr Marcus M., jr.
Department of Geology, Dickinson College. Carlisle, PA 17013-2896, USA
KOHRING, Rolf
Institut fuer Palaeontologie, Freie Universitaet Berlin, Malteserstraße 74-100, Haus D, 12249 Berlin, GERMANY
KUKLINSKI, Piotr
Institute of Oceanology, Polish Academy of Science, 81-712 SOPOT, ul. Powstanców, Warszawy 55, POLAND
LANE, Jennifer
Department of Geoscience, 412 Deike Building, Pennsylvania State University, University Park, PA 16802, USA
LEWIS, Jane E.
Institute of Marine Biology, National Taiwan Ocean University, 2 PeiNing Road, Keelung 202, TAIWAN

LIU HUNG-CHUN, Betty,
Institute of Marine Biology, National Taiwan Ocean University, 2 PeiNing Road, Keelung 202, TAIWAN
LORE, Michael
251 West Dayton Yellow Springs Road, Apt. # 124, Fairborn, OH 45324, USA
McKINNEY, Frank K.
Department of Geology, Appalachian State University, Boone, North Carolina 28608, USA
McKINNEY, Marjorie J.
Department of Geology, Appalachian State University, Boone, North Carolina 28608, USA
MACKIE, Joshua
University of Melbourne Zoology / Museum of Victoria, GPO Box 666E, Melbourne, Victoria 3010, AUSTRALIA
MARSH, Professor Terrence,
North Central College, P.O. Box 3063, Naperville, IL 60566-7063, USA
MASSARD, Dr Jos
1A, rue des Romains, L-6478 Echternach, LUXEMBOURG
MASSARD-GEIMER, Dr Gaby
1A, rue des Romains, L-6478 Echternach, LUXEMBOURG
MATURO, Dr Frank J., jr.
Zoology Department, University of Florida, P.O. Box 118525, Gainesville, FL 32611-8525, USA
MAWATARI, Professor Shunsuke F. Division of Biological Sciences, Graduate School of Science, Hokkaido University, Sapporo 060-0810, JAPAN
MOISSETTE, Dr Pierre
UFR des Sciences de la Terre/ERS 2042, Universite de Lyon I, 27, Bd du 11 novembre, 69622 Villeurbanne Cedex FRANCE
MORRIS, Dr Penny A,
Department of Natural Sciences, University of Houston-Downtown, One Main Street, Houston, Texas 77002, USA
MOYANO G., Professor Hugo I.
Departamento de Zoología, Universidad de Concepción, Casilla 160-C, Concepción, CHILE
NAKREM, Dr Hans Arne
Paleontologisk Museum, Universitetet i Oslo, Sars Gate 1, N-0562 Oslo 5, NORWAY
NIELSEN, Dr Claus
Zoologisk Museum, Universitetsparken 15, DK-2100, Copenhagen, DENMARK
O'DEA, Dr Aaron
14 Ninetree Hill, Kingsdown, Bristol BS1 3SQ, UK
O'SULLIVAN, Gearoid
"Cuan Baoi", Ballinveltig, Curraheen Road, Bishopstown, Cork, IRELAND
OKAMURA, Dr Beth
School of Animal and Microbial Sciences, The University of Reading, Whiteknights. PO Box 228, Reading RG6 6AJ, UK
OSTROVSKY, Dr Andrew N.
Universität Kiel, Institut für Geowissenschaften, Olshausenstr. 40, 24118, GERMANY
PECK, Professor Lloyd
British Antarctic Survey, High Cross, Madingley Road, Cambridge, UK
PETERS, Lars
Inst. für pharmazeutische Biologie, Uni Bonn, Nußallee 6, 53115 Bonn, GERMANY
PETERSEN, Nancy
North Central College, P.O. Box 3063, Naperville, IL 60566-7063, USA
POPIJAC, Aleksandar
Laboratory for Animal Ecology, Department of Zoology, Faculty of Science, University of Zagreb, Rooseveltov trg 6, 10000 Zagreb, CROATIA
PORTER, Dr Joanne S.,
School of Biological Sciences, University of Wales Swansea, Singleton Park, Swansea SA2 8PP, Wales, UK

POWERS, Susan K.
Department of Geology, Appalachian State University, Boone, North Carolina 28608, USA
PRATT, Marney
Box 90325, Rm 139 BioSci, Biology Dept., Duke University, Durham, NC 27708, USA
RAMALHO, Lais Vieira
Condominio Pontal do Atalaia - Casas Brancas n.30, Praia Grande, Arraial do Cabo, Rio de Janeiro, BRAZIL
REGUANT, Professor Salvador
Dpt. Estratigrafía i Paleontologia, Facultat de Geologia, Universitat de Barcelona, 08028 Barcelona, SPAIN
ROSS, Dr Charles A.
Department of Geology, Western Washington University, Bellingham, WA 98225, USA
ROSS, Professor June R.P.
Department of Biology, Western Washington University, Bellingham, WA 98225, USA
ROSSO, Dr Antonietta
Dipartimento di Scienze Geologiche, Sezione di Oceanologia e Paleoecologia, I-95129- Catania, ITALY
RYLAND, Professor John S.
School of Biological Sciences, University of Wales Swansea, Singleton Park, Swansea SA2 8PP, Wales, UK
SCHÄFER, Dr Priska
Geolog.-Paläont. Institut, Christian Albrechts Universitat zu Kiel, Olshausenstraße 40, D-2300 Kiel 14, GERMANY.
SCHAGER, Sten
Grev Magnigatan 18 114 55, Stockholm, SWEDEN
SCHMIDT, Rolf
Department of Geology and Geophysics, Adelaide University, Adelaide, SA 5001, AUSTRALIA
SCHOLZ, Dr Joachim
Forschungsinstitut und Naturmuseum Senckenberg, Sektion Marine Evertebraten III (Bryozoologie), Senckenberg-anlage 25, 60325 Frankfurt, GERMANY
SMITH, Dr Abigail M.
Department of Marine Science, University of Otago, P.O. Box 56, Dunedin, NEW ZEALAND
SNYDER, Professor Edward
Institute for Environmental Studies, Shepherd College, Shepherdstown, WV 25443, USA
SOULE, Professor Dorothy F.
Allan Hancock Institute of Marine Studies, University of Southern California, Los Angeles, CA 90089-0371, USA
SPENCER JONES, Mary E.
Department of Zoology, The Natural History Museum, Cromwell Road, London, SW7 5BD, UK
STEGER, Kate
Department of Marine Science, University of Otago, P.O. Box 56, Dunedin, NEW ZEALAND
STEINTHÓRSDÓTTIR, Margrét
Institute of Geology, Øster Voldgade 10, DK - 1350 Copenhagen, DENMARK
TAYLOR, Dr Paul D.
Department of Palaeontology, The Natural History Museum, Cromwell Road, London SW7 5BD, UK
TEMKIN, Dr Michael H.
Biology Department, St Lawrence University, Canton, New York 31617, USA
THORPE, Dr John P.
Port Erin Marine Laboratory, University of Liverpool, Port Erin, Isle of Man IM9 6JA, UK
TILBROOK, Kevin J.
Department of Zoology, The Natural History Museum, Cromwell Road, London, SW7 5BD, UK
TOPS, Sylvie
School of Animal and Microbial Sciences, The University of Reading, Whiteknights. PO Box 228, Reading RG6 6AJ, UK
TORRENS, Professor Hugh S.
Lower Mill Cottage, Furnace Lane, Madeley, Crewe CW3 9EU, UK

VÁVRA, Professor Norbert
Institute of Paleontology, University of Vienna, A-1090 Vienna, AUSTRIA
WÄSCHENBACH, Andrea
School of Biological Sciences, University of Wales Swansea, Singleton Park, Swansea SA2 8PP, Wales, UK
WASS, Dr Robin
78 Vermont Street, Wodonga 3690, AUSTRALIA
WAUGH, David
Department of Geology Kent State University Kent, OH 44242-6001, USA
WHITTLESEA, Paul S.
8 Eaton Old Hall, Hurd Road, Eaton, Norwich, Norfolk NR4 7BE, UK
WINSTON, Dr Judith E.
Virginia Museum of Natural History, 1001 Douglas Ave., Martinsville, VA 24112, USA
WOOD, Dr Timothy S.
Department of Biological Sciences, Wright State University, 3640 Colonel Glenn Highway, Dayton, OH 45435, USA.
WÖSS, Dr Emmy
Department of Limnology, Institute of Ecology and Conservation Biology, University of Vienna, Althanstraße 14, A-1090 Wien, AUSTRIA
WRIGHT, Peter
Brambell Building, School of Biological Sciences, University of Wales (Bangor), Deniol Road, Bangor, Gwynedd LL57 2UW, UK
WYSE JACKSON, Dr Patrick N.
Department of Geology, Trinity College, Dublin 2, IRELAND
ZÁGORSEK, Dr Kamil
Rumunská, 250 88 Celákovice, CZECH REPUBLIC

Bryozoan Studies 2001, Wyse Jackson, Buttler & Spencer Jones (eds)
© 2002 Swets & Zeitlinger, Lisse, ISBN 90 5809 388 3

Taxonomic index